CAMBRIDGE LIBRARY COLLECTION

Books of enduring scholarly value

Astronomy

From ancient times, humans have tried to understand the workings of the world around them. The roots of modern physical science go back to the very earliest mechanical devices such as levers and rollers, the mixing of paints and dyes, and the importance of the heavenly bodies in early religious observance and navigation. The physical sciences as we know them today began to emerge as independent academic subjects during the early modern period, in the work of Newton and other 'natural philosophers', and numerous sub-disciplines developed during the centuries that followed. This part of the Cambridge Library Collection is devoted to landmark publications in this area which will be of interest to historians of science concerned with individual scientists, particular discoveries, and advances in scientific method, or with the establishment and development of scientific institutions around the world.

The Scientific Papers of Sir William Herschel

By the time of his death, William Herschel (1738–1822) had built revolutionary telescopes, identified hundreds of binary stars, and published astronomical papers in over forty volumes of the Royal Society's *Philosophical Transactions*. This two-volume collection, which originally appeared in 1912, was the first to gather together his scattered publications. It draws also on a wealth of previously unpublished material, from personal letters to numerous papers presented to the Philosophical Society of Bath. Although Herschel is best known for his discovery of Uranus, this collection highlights the true range of his observations and interests. Focusing on his later work, Volume 2 includes notes on some of the moons of Uranus, studies of solar heat and the atmosphere of Saturn, and some practical experiments investigating the capabilities of contemporary telescopes. It also features an appendix of work compiled by his son, John Herschel, and sister Caroline.

Cambridge University Press has long been a pioneer in the reissuing of out-of-print titles from its own backlist, producing digital reprints of books that are still sought after by scholars and students but could not be reprinted economically using traditional technology. The Cambridge Library Collection extends this activity to a wider range of books which are still of importance to researchers and professionals, either for the source material they contain, or as landmarks in the history of their academic discipline.

Drawing from the world-renowned collections in the Cambridge University Library and other partner libraries, and guided by the advice of experts in each subject area, Cambridge University Press is using state-of-the-art scanning machines in its own Printing House to capture the content of each book selected for inclusion. The files are processed to give a consistently clear, crisp image, and the books finished to the high quality standard for which the Press is recognised around the world. The latest print-on-demand technology ensures that the books will remain available indefinitely, and that orders for single or multiple copies can quickly be supplied.

The Cambridge Library Collection brings back to life books of enduring scholarly value (including out-of-copyright works originally issued by other publishers) across a wide range of disciplines in the humanities and social sciences and in science and technology.

CAMBRIDGE LIBRARY COLLECTION

Books of enduring scholarly value

Astronomy

From ancient times, humans have tried to understand the workings of the world around them. The roots of modern physical science go back to the very earliest mechanical devices such as levers and rollers, the mixing of paints and dyes, and the importance of the heavenly bodies in early religious observance and navigation. The physical sciences as we know them today began to emerge as independent academic subjects during the early modern period, in the work of Newton and other 'natural philosophers', and numerous sub-disciplines developed during the centuries that followed. This part of the Cambridge Library Collection is devoted to landmark publications in this area which will be of interest to historians of science concerned with individual scientists, particular discoveries, and advances in scientific method, or with the establishment and development of scientific institutions around the world.

The Scientific Papers of Sir William Herschel

By the time of his death, William Herschel (1738–1822) had built revolutionary telescopes, identified hundreds of binary stars, and published astronomical papers in over forty volumes of the Royal Society's *Philosophical Transactions*. This two-volume collection, which originally appeared in 1912, was the first to gather together his scattered publications. It draws also on a wealth of previously unpublished material, from personal letters to numerous papers presented to the Philosophical Society of Bath. Although Herschel is best known for his discovery of Uranus, this collection highlights the true range of his observations and interests. Focusing on his later work, Volume 2 includes notes on some of the moons of Uranus, studies of solar heat and the atmosphere of Saturn, and some practical experiments investigating the capabilities of contemporary telescopes. It also features an appendix of work compiled by his son, John Herschel, and sister Caroline.

The Scientific Papers of Sir William Herschel

Including Early Papers Hitherto Unpublished

Volume 2

Edited by
John Louis Emil Dreyer

Cambridge University Press

CAMBRIDGE
UNIVERSITY PRESS

University Printing House, Cambridge, CB2 8BS, United Kingdom

Published in the United States of America by Cambridge University Press, New York

Cambridge University Press is part of the University of Cambridge.
It furthers the University's mission by disseminating knowledge in the pursuit of education, learning and research at the highest international levels of excellence.

www.cambridge.org
Information on this title: www.cambridge.org/9781108064637

This edition first published 1912
This digitally printed version 2013

ISBN 978-1-108-06463-7 Paperback

THE SCIENTIFIC PAPERS

OF

SIR WILLIAM HERSCHEL

KNT. GUELP., LL.D., F.R.S.

Sir William Herschel
From an oil painting by Artaud 1819.
In the possession of Sir W. J. Herschel, Bart.

THE SCIENTIFIC PAPERS

OF

SIR WILLIAM HERSCHEL

KNT. GUELP., LL.D., F.R.S.

INCLUDING EARLY PAPERS HITHERTO UNPUBLISHED

COLLECTED AND EDITED UNDER THE DIRECTION
OF A JOINT COMMITTEE OF THE ROYAL SOCIETY
AND THE ROYAL ASTRONOMICAL SOCIETY

WITH A BIOGRAPHICAL INTRODUCTION COMPILED MAINLY FROM
UNPUBLISHED MATERIAL BY J. L. E. DREYER

VOL. II

LONDON: PUBLISHED BY
THE ROYAL SOCIETY AND THE ROYAL ASTRONOMICAL SOCIETY
AND SOLD BY DULAU & CO., LTD.
37 SOHO SQUARE, LONDON, W.
1912

CONTENTS OF VOL. II

PAPERS PUBLISHED IN THE PHILOSOPHICAL TRANSACTIONS OF THE ROYAL SOCIETY AND ELSEWHERE

APPENDIX

PLATES IN VOL. II

PAPERS

PUBLISHED IN THE

Philosophical Transactions of the Royal Society

AND ELSEWHERE

XL.

On the Discovery of four additional Satellites of the Georgium Sidus. *The retrograde Motion of its old Satellites announced; and the Cause of their Disappearance at certain Distances from the Planet explained.*

[*Phil. Trans.*, 1798, pp. 47–79.]

Read December 14, 1797.

HAVING been lately much engaged in improving my tables for calculating the places of the Georgian satellites, I found it necessary to recompute all my observations of them. In looking over the whole series, from the year of the first discovery of the satellites in 1787 to the present time, I found these observations so extensive, especially with regard to a miscellaneous branch of them, that I resolved to make this latter part the subject of a strict examination.

The observations I allude to relate to the discovery of four additional satellites: to surmises of a large and a small ring, at rectangles to each other: to the light and size of the satellites: and to their disappearance at certain distances from the planet.

In this undertaking, I was much assisted by a set of short and easy theorems I had laid down for calculating all the particulars respecting the motions of satellites; such as, finding the longitude of the satellite from the angle of position, or the position from the longitude; the inclination of the orbit from the angle of position and longitude; the apogee; the greatest elongation; and other particulars. Having moreover calculated tables for reduction; for the position of the point of greatest elongation ; and for the distance of the apogee, or opening of the ellipsis ; and also contrived an expeditious application of the globe for checking computations of this sort, I found many former intricacies vanish.

By the help of these tables and theorems, I could examine the miscellaneous observations relating to additional satellites, on a supposition that their orbits were in the same plane with the two already known, and that the direction of their motion was also the same with that of the latter.

And here I take an opportunity to announce, that the motion of the Georgian satellites is retrograde.

This seems to be a remarkable instance of the great variety that takes place among the movements of the heavenly bodies. Hitherto, all the planets and satellites of the solar system have been found to direct their course according to the order of the signs : even the diurnal or rotatory motions, not only of the primary planets, but also of the sun, and six of their secondaries or satellites, now are known to follow the same direction ; but here we have two considerable celestial bodies completing their revolutions in a retrograde order.

I return to the examination of the miscellaneous observations, the result of which has been of considerable importance, and will be contained in this paper. The existence of four additional satellites of our new planet will be proved. The observations which tend to ascertain the existence of rings not appearing to be satisfactorily supported, it will be proper that surmises of them should either be given up, as ill founded, or at least reserved till superior instruments can be provided, to throw more light upon the subject. A remarkable phænomenon, of the vanishing of the satellites, will be shewn to take place, and its cause animadverted upon.

I shall now, in the first place, relate the observations on which these conclusions must rest for support, and afterwards join some short arguments, to shew that my results are fairly deduced from them.

For the sake of perspicuity, I shall arrange the observations under three different heads ; and begin with those which relate to the discovery of additional satellites.

A great number of observations on supposed satellites, that were afterwards found to be stars, or of which it could not be ascertained whether they were stars or satellites, for want of clear weather, will only be related. For, to enter into the particular manner of recording these supposed satellites, or to give the figures which were delineated to point them out, would take up too much time, and be of no considerable service to our present argument. It ought however to be mentioned, that nearly the same precaution was taken with all the related observations as, it will be found, was used in those that are given in the words of the journals that contain them. The former will be distinguished under the head *Reports*, the latter under that of *Observations*.

Investigation of additional Satellites.

Reports.

Feb. 6, 1782. A very faint star was pointed out as probably a satellite, but Feb. 7 and 8 was found remaining in its former situation.

March 4, 1783. A satellite was suspected, but March 8 was found to be a star.

April 5, 1783. A suspected satellite was delineated, but the 6th it was seen remaining in its former place.

Nov. 19, 1783. A supposed satellite was marked down, but no opportunity could be had to account for it afterwards.

Nov. 16, 1784. Supposed 1st and 2d satellites were pointed out, but not accounted for afterwards.

Many other fruitless endeavours for the discovery of satellites were made; but, finding my instrument, in the NEWTONIAN form, not adequate to the undertaking, the pursuit was partly relinquished. The additional light however which I gained, by introducing the Front-view in my telescope, soon after gave me an opportunity of resuming it with more success.

Jan. 11, 1787. Three supposed satellites were observed: a first, a second, and a third. Jan. 12, the 1st and 2d were gone from the places in which I had marked them, but the 3d was remaining, and therefore was a fixed star.*

Jan. 14. A supposed 3d satellite was delineated, but on the 17th it was found to be a star.

Jan 17. Supposed 3d, 4th, and 5th satellites were marked, but were found remaining in their former places on the 18th.

Jan. 24. Supposed 3d and 4th satellites were noted, but the weather proving bad on the succeeding nights, till February 4, they were lost in uncertainty.

Feb. 4. A 3d satellite was marked, but not being afterwards accounted for remains lost.

Feb. 7. A supposed 3d satellite was proved to be a star the 9th.

Feb. 10. Supposed 3d and 4th satellites have not been afterwards accounted for.

Feb. 13. Supposed 3d, 4th, and 5th satellites proved stars the 16th.

Feb. 16. A 3d satellite proved a star the 17th.

Feb. 19. Supposed 3d and 4th satellites were proved to be stars the same evening, by being left in their places, while the planet was moving on.

Feb. 22. The supposed 3d and 4th of the 19th were seen remaining in their former places; and new 3d, 4th, and 5th satellites were marked; but these were lost through bad weather, which lasted till March 4.

March 5. A supposed 3d satellite proved to be a star the 7th.

* It has already been shewn, in a former paper, that the removed satellites were those two which now are sufficiently known.

March 7. The position of a 3d was taken and a 4th also marked; but March 8 they were both proved to be fixed stars.

October 20. A very small star was seen near the planet, but lost, for want of opportunity to account for it.

March 13, 1789. The positions of 3d and 4th satellites were taken, but the 14th they were found to be stars.

March 16. Supposed 3d and 4th satellites were well laid down, but March 20 were found to be stars.

March 26. The places of supposed 3d and 4th satellites were ascertained, but no opportunity could be had of deciding whether they were stars or satellites.

Dec. 15. A supposed 3d satellite was accurately delineated, but proved to be a star the 16th.

Observations.

Jan. 18, 1790. 6^h 51′.* A supposed 3d satellite is about 2 diameters of the planet following; excessively faint, and only seen by glimpses.

7^h 57′. I cannot perceive the 3d.

Reports.

Jan. 18, 1790. A supposed 4th satellite was described, but was found to be a star the 19th.

Jan. 20. A 3d satellite was perceived, and its angle of position ascertained; but was afterwards lost, for want of opportunity to examine its place again.

Observations.

Feb. 9, 1790. 6^h 28′. There is a supposed 3d satellite, in a line with the planet and the 2d satellite.

6^h 40′. Configuration of the Georgian planet and satellites. See fig. 1.

Clouds prevent further observations.

Feb. 11. The supposed 3d satellite of the 9th of February I believe is wanting; at least I cannot see it, though the weather is very clear, but windy.

FIG. 1.

Feb. 12. The supposed 3d satellite of the 9th is not in the place where I saw it that night.

Reports.

Feb. 11, 1790. Supposed 3d and 4th satellites were laid down, but on the 12th they were both found remaining in their former places.

Feb. 16. A 3d satellite was delineated, but on the 17th it proved to be a star.

* All the times have been corrected so as to be true, sidereal; but are only given here to the nearest minute.

March 5. Supposed 3d and 4th satellites were laid down, but on the 8th were seen remaining in their places.

Feb. 4, 1791. A 3d satellite was marked, but has not been accounted for afterwards.

Feb. 5. Supposed 3d, 4th, and 5th satellites were delineated, but no opportunity could afterwards be found to ascertain their existence.

March 5. Supposed 3d, 4th, and 5th satellites were put down. They could not be seen March 6, but were proved to be small stars the 7th.

Feb. 12, 1792. A third satellite was delineated, but was left behind by the planet the same evening, and also seen in its former place the next night.

Feb. 13. A 3d satellite was put down, but proved to be a star the 14th.

Feb. 20. The position of a 3d satellite was taken, but 4 hours after was found to be left behind by the planet. It was also seen in its former place Feb. 21.

Feb. 26. A 3d satellite, between the planet and 2d, was observed; which, 3^h $37'$ afterwards, was thought to be left behind, but was so faint as hardly to be perceivable. A fourth was also put down. Neither of them have been accounted for afterwards.

March 8, 1793. The position of a supposed 3d satellite was taken, but the next day it was found to be a star.

March 9. A supposed 3d satellite was observed, at 5 or 6 times the distance of the 1st, but was not accounted for afterwards.

March 14. Supposed 3d and 4th satellites were observed, but no opportunity could be had afterwards to see them again.

Observations.

Feb. 25, 1794. With 320, there is a small star *a*, fig. 2, about 15 degrees north preceding the planet; and another *b*, about 30 degrees north preceding: also one *c*, directly preceding. There is a very small fourth star *d*, making a trapezium with the other three; and two more *e f*, preceding this 4th star, are in a line with it.

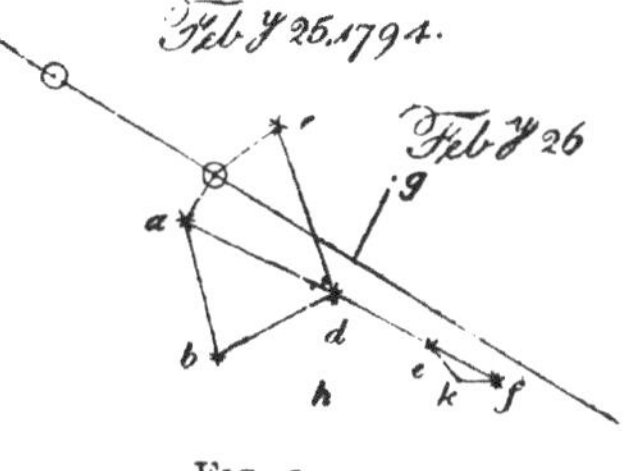

FIG. 2.

Feb. 26. The stars, in figure 2, marked *f e d a*, are in a line. There is a star *g*, at rectangles to *f e d a*: the perpendicular falls upon *d*: it is towards the south. There is also a star *h*, north of *f e d a*; but it is too faint to admit of a determination of its place: I can only see it now and then by imperfect glimpses.

Feb. 28. 6^h $40'$. The stars *f e d a* of the 26th are in their places. *c* is in the place where I have marked it. The star *g* is in the place where I marked it. I see also the very small star *h*.

6^h $50'$. There is a very small star *k*, but not so small as *h*, very near to, and north following *f*, which I did not see on the 26th. It is not quite half way between *f* and *e*, but nearer to *f* than to *e*. It makes an obtuse triangle with *f* and *e*.

9^h 43′. The motion of the planet this evening, since the first observation, is very visible.

10^h 7′. I cannot perceive the star *k*. The weather is not so clear as it was.

10^h 21′. I cannot perceive the star *k* in the place where it was 6^h 50′.

March 4, 1794. Power 320. 6^h 46′. The stars *a b c d e f g* of Feb. 28, fig. 3, are in their places, but I cannot see the small star *k*. The evening is not very clear.

9^h 51′. I cannot see the star *k*.

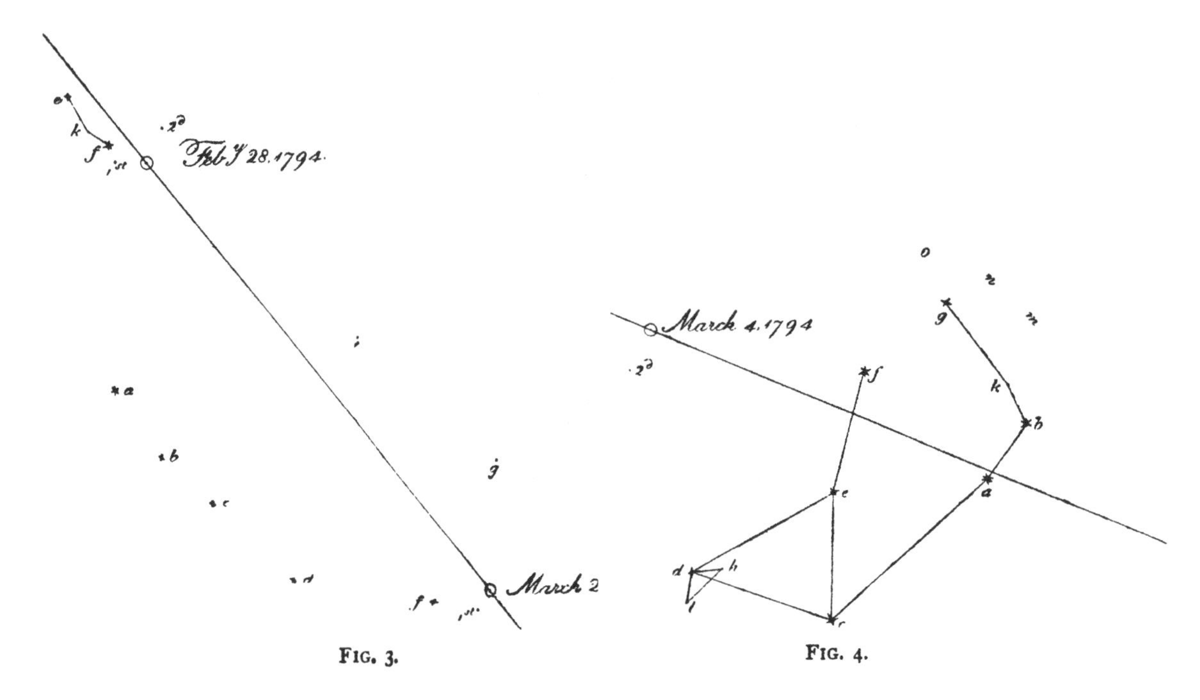

Fig. 3. Fig. 4.

10^h 25′. I suppose *a*, in figure 4, to be the star towards which the planet is moving. *c a b* are in a crooked line. *c e f* are nearly in a line; *f* is a little preceding. *c d e* form a triangle. There is a small star *h*, preceding *d*. There is an exceeding small star *k*, in the line *b k g*, but a little preceding and nearer *b*. *a b c* are large stars. *d e g* are also pretty large. *f* and *h* are small. Power 157. With 320, there is also a very small star *l*, near *d*, forming an isosceles triangle *h d l*, on the preceding side.

March 5. 7^h 39′. Power 320. The stars *a b c d e f g h k l* are in the places where they were marked last night.

9^h 37′. There is a very small star *n*, south of *g*; another *m*, preceding *g*; and a third *o*, south following *g*.

10^h 19′. I suspect a very small star, south following the planet, at one-third of the distance of the 1st satellite; but cannot verify it with 480. With 600, the same suspicion continues.

March 7. 9^h 48. The stars *a b c d e f g h k l* are in their places. *n m o* are in their places. The planet has passed between the stars *e f*, pretty near to *f*.

Reports.

March 21 1794. Power 320. A small star was suspected south of the planet, or about 85° south following. It could not be verified with 480, nor with 600; and was even supposed to have been a deception; but the 22d was found remaining in the place where the planet had left it.

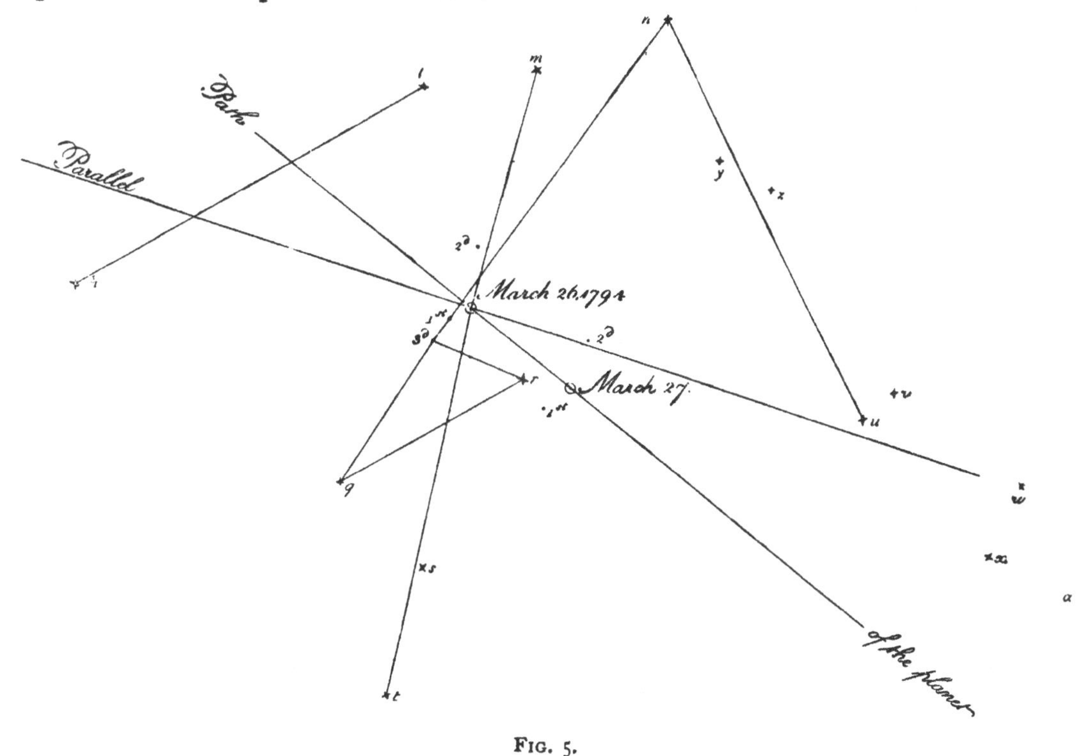

FIG. 5.

Observations.

March 26, 1794. 9^h 35′. With 480, I see the 1st satellite much better than with 320. I suspected, with 320, a 3d satellite, directly north of the planet, a little farther off than the 1st, and this power almost verifies the suspicion. See figure 5.

9^h 44′. With 600, I still suspect the same, but cannot satisfy myself of the reality.

11^h 32′. I see the supposed 3d satellite perfectly well now. It is much smaller than the 1st, and in a line with the planet and the 1st; so that probably it is a fixed star; since it preceded the 1st, when I saw it before, I think more than the quicker motion of the 1st satellite would account for. If it be a fixed star, it makes almost a rectangular triangle with *qr*, the shorter leg being 3d *r*; or it is almost in a line with *q* and *n*.

N. B. The lines in the description are truer than in the figure, as the latter is only intended to point out the stars in question.

March 27. $8^h\ 37'$. Power 320. The same small star, observed last night at $11^h\ 32'$, is gone from the place where I saw it. From its light last night, compared to *r*, which to-night is very near the planet, and scarcely visible, I am certain that it must be bright enough to be perceived immediately, if it were in the place pointed out by my description.

$10^h\ 20'$. The planet is considerably removed from the star *r*.

$11^h\ 41'$ I had many glimpses of small stars or supposed satellites: one of them in a place agreeing with the 3d satellite of last night, (supposing it to have moved with the planet;) that is, a little farther off, and after the 1st. Another preceding the 1st, but nearer. Some others south, at a good distance; but not one of them could I see for any constancy. They were only lucid glimpses.

Reports.

March 27, 1794. A supposed 4th satellite was delineated, but proved to be a star the 28th.

Observations.

March 4, 1796. Configuration of the Georgian planet and fixed stars for $10^h\ 3'$. See fig. 6.

March 5. $9^h\ 50'$. I suspected a very small star between *c* and *h*, which was not there last night. I had a pretty certain glimpse of it. It is in a line from the planet towards *f*: power 320. With 600, I see the satellite better than before; but cannot perceive the suspected small star.

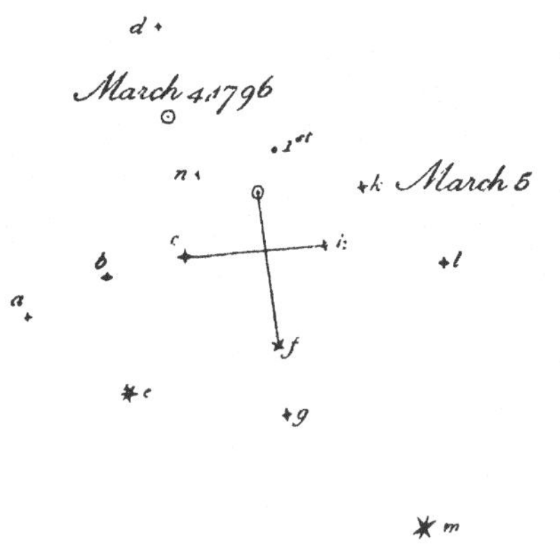

$10^h\ 17'$. The air is remarkably clear at present, but I cannot perceive the suspected star.

March 9. $11^h\ 23'$. As the probability of other satellites is, that they revolve in the same plane with the 1st and 2d, I chiefly look for them in the direction of their orbits, which is now nearly a straight line.

April 5, 1796. There is no star in the line of the transverse, that can be taken for a satellite: the evening is very beautiful, and I examined that line with 300, at a distance; and with 600, within the orbits of the two satellites.

Fig. 6.

March 23, 1797. Three very small stars O P Q, are in the path of the planet; they form an obtuse triangle.

March 25. $11^h\ 4'$. A very bright star S, at almost the distance of the field of view is a little south of the path of the planet. It has a small north preceding star T, which points to two more V W, towards the north. Between the triangle of March 23d and the four last mentioned stars, is a very small star X.

March 28. $10^h\ 52'$. I see the stars S T V W X of March 25th.

11^h 25′. From X towards the triangle O P Q of March 23d, is an exceeding small star Y, about four times the distance of the 2d satellite, and nearly in the line of the greatest elongation. I do not remember to have seen it the 25th.

11^h 41′. The distance of Y from X is about $\frac{1}{4}$ of the distance of X from the triangle. It requires much attention to see it; but I have a very complete view of it, by drawing the planet just out of the field, and the star X almost on the preceding side.

Arguments upon the Reports and Observations.

From the reports of the great number of supposed satellites, compared with the select observations which are given at length, it must be evident that the method of looking for difficult objects, and of marking them down by lines and angles, with every other possible advantage for finding them again, has been completely understood and put in practice. So guarded against deceptions, we cannot but allow, that even a single glimpse of a very small star is a considerable argument in favour of its existence. What I call verifying a suspicion, which is generally done with a higher power than that which caused the suspicion, is obtaining a steadier view of the existence of the object in question; that is, to see it in such a manner as to be able to fix an eye upon it, and to compare it with other surrounding objects; and thus to be able to ascertain its relative situation with those other objects in a satisfactory manner.

An interior Satellite.

The observation of Jan. 18, 1790, says, "a supposed 3d satellite is about two diameters of the planet following." There is not the least doubt expressed about the existence of the satellite, or object in question, which therefore must be looked upon as ascertained. Now, the angle of the greatest elongation of the Georgian satellites, by my new tables, at the time of observation, was 81° 33′ N.F. Therefore, the angle of the apogee was 8° 27′ S.F.; and since, by observation, the satellite was "following," without any mention of degrees being made, we may admit it to have been not far from the parallel; suppose 11 or 12 degrees S.F. In this case, the satellite would be in the apogee about the time of the 2d observation, at 7^h 57′; which says, "I cannot perceive the satellite." But it will be shewn hereafter, when I come to treat of the vanishing of the satellites, that it would become invisible in this situation. Indeed, without the supposition of the satellite's coming to the apogee, it might easily happen that the least change in the clearness of the air, during a time of 1^h 5′ which elapsed between the first and second observation, might render an object invisible, which, as the first observation says, was "excessively faint, and could only be seen by glimpses."

From the observed distance, which is put at "2 diameters of the planet," we may conclude what would be the distance of its greatest elongation. For, 2 diameters from the disk of the planet give $2\frac{1}{2}$ from the centre. Now, the distance of the apogee at this time, by my tables, was ·64, supposing that of the greatest elongation 1; therefore we have the radius of its orbit $\frac{2{\cdot}5 \times 4''{\cdot}12}{{\cdot}64} = 16''{\cdot}1$

This calculation is not intended to determine precisely the distance of the satellite, but only to shew that its orbit is more contracted than that of the 1st, and that consequently it is an interior satellite.

If any doubt should be entertained about the validity of this observation, we have a second, and very striking one, of March 5, 1794; where an interior satellite was suspected south following the planet, at one-third of the distance of the 1st. March 4, when a description was made of the stars, as in figure 4, this satellite was not in the place where it was observed the 5th. And, by an examination of the same stars March 7, it appears, that even the smallest stars *n m o*, of the 5th, were seen in their former places, but not the satellite. The observation therefore must be looked upon as decisive with regard to its existence. If any doubt should arise, on account of the suspicion not being verified with 480, I must remark, that being used to such imperfect glimpses, it has generally turned out, even when I have given up as improbable the existence of a supposed satellite seen in that manner, that it has afterwards nevertheless been discovered that a small star remained in the place where the satellite had been suspected to be situated. An instance of this may be seen in the report of the observations that were made March 21 and 22, 1794. Besides, in the present case, it is additionally mentioned, that the same object was examined with a power of 600, which continued the suspicion.

From the assigned place of this satellite, at $\frac{1}{3}$ of the distance of that of the first, it appears that this observation belongs to the interior satellite of Jan. 18, 1790, which has already been examined. The 1st satellite was this evening at its greatest elongation, one-third of which is about 11″. The apogee distance of a satellite whose greatest distance is 16″·1 would have been 6″·1 on the day of our observation; but, not being come to the apogee, by many degrees, it could not be so near the planet.

For the sake of greater precision, let us admit that the satellite was exactly south following; that is, 45 degrees from the parallel, and 45 from the meridian; then, by calculation, a satellite whose orbit is at 16″·1 from the planet, would, in the situation now admitted, have been 7″·1 from its centre, which might coarsely be rated at $\frac{1}{3}$ of the distance of the first. But the estimation of 11″ is probably more accurate than that in the 1st observation, where 2 diameters are given. And, by calculating from this quantity, we find that the greatest elongation distance of the satellite is 25″·5; now, putting $2\frac{1}{2}$ diameters in the first observation, instead of 2, the distance deduced from it will come out 19″·3; which is certainly an agreement sufficiently near to admit both observations to belong to the same satellite.

March 27, 1794. We find a third observation, which will assist in supporting the two former ones. A glimpse of a satellite is mentioned, which was preceding the 1st, but nearer the planet. The position of the 1st satellite the same evening was, by measuring, found to be 62°·1 N.F. which is still a considerable way from its greatest elongation ; but our new satellite preceded it, and was therefore more advanced in its orbit, or nearer its greatest distance ; and yet the observation says, that it was not so far from the planet as the 1st ; notwithstanding this latter was in a more contracted part of its orbit. It follows therefore that this was also an interior satellite. Now, since we may allow these three observations to belong to the same, we ought not to make a distinction ; but admit, as sufficiently established, the existence of at least one interior satellite of our new planet.

An intermediate Satellite.

March 26, 1794. A satellite was suspected, directly north of the planet. At first it could not be verified, but was seen perfectly well afterwards. It was supposed that probably it might be a star, but this was left undecided. The observation of March 27th however removes all doubt upon the subject ; as it fully affirms that the small star observed the 26th, at 11^h 32′, was gone from the place in which it was the day before. Such strong circumstances are mentioned in confirmation, that we cannot hesitate placing this among the list of existing satellites. It was not the interior satellite of Jan. 18, 1790 ; for both the 1st and 2d known satellites were in full view March 26th ; see figure 5 ; and the observation places this new one in a line drawn from the planet continued through the 1st ; with the remark, that it was a little farther from the planet than the 1st. The 2d was then near its greatest southern elongation, and we may see from the figure, as well as from the above description, that the orbit of this new satellite is situated between the orbits of the other two.

We have a second observation of the same satellite March 27, 1794 ; where, among the glimpses of additional satellites at 11^h 41′, is mentioned " one in a place probably agreeing with the new satellite of March 26th," which, by its motion, must have been carried forward, so as to be where the observation of the 27th says it was, namely, " a little farther off and after the 1st ; " that is, at a little greater distance from the planet than the 1st, and not so far advanced in its orbit as that satellite. This amounts not only to an additional proof, but even announces the recognition of the satellite, and its motion in the course of one day.

An exterior Satellite.

Feb. 9, 1790. A new satellite was seen, in a line with the planet and the 2d satellite. See figure 1. To convince us that this was not a fixed star, we have the observations of two other nights, the 11th and 12th of February, where the removal of it, from the place in which it was Feb. 9, is clearly demonstrated. As it was in a line continued from the planet through the second satellite, its orbit must evidently be of a greater dimension than that of the 2d ; I shall therefore put it down as an exterior satellite.

Most likely this satellite also was seen among the supposed satellites south of the planet, March 27, 1794; where we find mention made of "some others south, at a good distance." In that case, this will make a second observation.

We have a third observation of the same new satellite March 5, 1796; when a very small star was seen, in a place where the evening before there had been none; as appears by the configuration of the 5th of March. See figure 6. At the time of the observation, the planet was come to the longitude of the place where the star was perceived to be; which agrees with the idea of its having been brought to that situation by the planet. It may be objected, that the star could not be verified with a power of 600; but here we have more than a bare suspicion of the satellite, for the observation says, "I had a pretty *certain* glimpse of it;" and this appears also from the assigned place of the star at the intersection of two given lines. For, such a delineation could not have been made, without having perceived it with a considerable degree of steady vision. Its distance, to judge by the description, will agree sufficiently with the two foregoing observations of this new exterior satellite.

The most distant Satellite.

On Feb. 28, 1794, a star was perceived where on the 26th there was none. This star was larger than a very small star which was observed the 26th, not far from the place of the new supposed satellite; and a configuration having been made expressly, by way of ascertaining what stars might afterwards come into a situation where they could be mistaken for satellites, our new star or satellite would not have been omitted, when a smaller one very near it was scrupulously recorded. The motion of the planet, in 3 hours and 3 minutes, is mentioned as very visible. The place of the star, which was a new visitor this evening, was very particularly delineated, at $6^h\ 50'$. From its situation, it is evident the motion of the planet must have carried this star, if it was one of its satellites, towards the large star *f*, figure 3; in the light of which a dim satellite would be lost. This accordingly happened; for at $10^h\ 7'$ and $10^h\ 21'$ it was no longer visible. The direction of the planet's motion is plainly pointed out, by the place of the planet March 2d.

With respect to the orbit of this satellite, it appears, from its situation near the apogee, where it was seen, that its distance was to that of the second satellite, which was then near its greatest elongation, as 8 to 5. And, since the apogee distance, on the day of observation, was only ·37, we have its greatest elongation as $\frac{8}{\cdot 37}$ to 5; that is, as 21·6 to 5, or above 4 to 1. From which we may conclude, that its orbit must lie considerably without the before mentioned exterior satellite of Feb. 9, 1790.

We have a second observation of it March 27, 1794; which, though not very strong, yet adds confirmation to the former. For that evening, which was uncommonly fine, other satellites, south, at a good distance, were perceived. This must relate principally to our present satellite, which may certainly be said to be at a good distance from the planet, and which, by that time, was probably in the southern part of its orbit, and near its greatest elongation.

There is a third observation, March 28, 1797, which probably also belongs to this satellite. For the exceedingly small star Y, which is mentioned as not having been seen the 25th, when the delineation of the stars was made, will agree very well with the two former observations; and, being near the greatest elongation, the distance of this satellite is well pointed out, and agrees remarkably well with the calculation of the first observation of it.

It remains now only to be mentioned, that in such delicate observations as these of the additional satellites, there may possibly arise some doubts with those who are very scrupulous; but, as I have been much in the habit of seeing very small and dim objects, I have not been detained from publishing these observations sooner, on account of the least uncertainty about the existence of these satellites, but merely because I was in hopes of being able soon to give a better account of them, with regard to their periodical revolutions. It did not appear satisfactory to me to announce a satellite, unless I could, at the same time, have pointed out more precisely the place where it might be found by other astronomers. But, as more time is now already elapsed than I had allowed myself for a completion of the theory of these satellites, I thought it better not to defer the communication any longer.

The arrangement of the four new and the two old satellites together will be thus:
First satellite, the interior one of Jan. 18, 1790.
Second satellite, the nearest old one of Jan. 11, 1787.*
Third satellite, the intermediate one of March 26, 1794.
Fourth satellite, the farthest old one of Jan. 11, 1787.†
Fifth satellite, the exterior one of Feb. 9, 1790.
Sixth satellite, the most distant one of Feb. 28, 1794.

Observations and Reports tending to the Discovery of one or more Rings of the Georgian Planet, and the flattening of its polar Regions.‡

Nov. 13, 1782. 7-feet reflector, power 460. "I perceive no flattening of the polar regions."

April 8, 1783. "I surmise a polar flattening."

Feb. 4, 1787. 20-feet reflector, power 300. "Well defined; no appearance of any ring; much daylight."

March 4. "I begin to entertain again a suspicion that the planet is not round. When I see it most distinctly, it appears to have double, opposite points. See figure 7. Perhaps a double ring; that is, two rings, at rectangles to each other."

* [Titania.—Ed.] † [Oberon.—Ed.]
‡ The observations are distinguished from the reports by marks of quotation (" ").

March 5. The Georgian Sidus not being round, the telescope was turned to Jupiter. I viewed that planet with 157, 300, and 480, which showed it perfectly well defined. Returning to the Georgian planet, it was again seen affected with projecting points. Two opposite ones, that were large and blunt, from preceding to following; and two others, that were small and less blunt, from north to south. See figure 7.

March 7. Position of the great ring R, from 70° S.P. to 70° N.F. Small ring *r*, from 20° N.P. to 20° N.F. 600 shewed R and *r*. 800 R and *r*. 1200 R and *r*.

March 8. "R and *r* are probably deceptions."

Nov. 9. "The suspicion of a ring returns often when I adjust the focus by one of the satellites, but yet I think it has no foundation."

Feb. 22, 1789. A ring was suspected.

March 16. 7^h 37′. "I have turned my speculum 90° round. A certain appearance, owing to a defect which it has contracted by exposure to the air since it was

FIG. 7. FIG. 8. FIG. 9. FIG. 10.

made, is gone with it; (see fig. 9 and 10;) but the suspected ring remains in the place where I saw it last."

"7^h 50′. Power 471 shews the same appearance rather stronger. Power 589 still shews the same."

"*Memorandum.* The ring is short, not like that of Saturn. It seems to be as in figure 8; and this may account for the great difficulty of verifying it. It is remarkable that the two *ansæ* seem of a colour a little inclined to red. The blur occasioned by the fault of the speculum is, to-night, as represented in figure 9. The other evening it was as in figure 10; and the ring is likewise as it was the same evening."

March 20. "7^h 53′. When the satellites are best in focus, the suspicion of a ring is the strongest."

Dec. 15. "The planet is not round, and I have not much doubt but that it has a ring."

Feb. 26, 1792. "6^h 34′. My telescope is extremely distinct; and, when I adjust it upon a very minute double star, which is not far from the planet, I see a very faint ray like a ring crossing the planet, over the centre. This appearance is of an equal length on both sides, so that I strongly suspect it to be a ring. There is, however, a possibility of its being an imperfection in the speculum, owing to some slight scratch: I shall take its position, and afterwards turn the speculum on its axis."

"8^h 39′. Position of the supposed ring 55°·6 from N.P. to S.F."

"9^h 56′. I have turned the speculum one quadrant round; but the appearance of the very faint ray continues where it was before, so that the defect is not in the speculum, nor is it in the eye-glass. But still it is now also pretty evident that it arises from some external cause; for it is now in the same situation, with regard to the tube, in which it was 3½ hours ago: whereas, the parallel is differently situated, and the ring, of course, ought to be so too."

March 5, 1792. "I viewed the Georgian planet with a newly polished speculum, of an excellent figure. It shewed the planet very well defined, and without any suspicion of a ring. I viewed it successively with 240, 300, 480, 600, 800, 1200, and 2400; all which powers my speculum bore with great distinctness. I am pretty well convinced that the disk is flattened." The moon was pretty near the planet.

Dec. 4, 1793. "7-feet reflector, power 287. The Georgian planet is not so well defined as, from the extraordinary distinctness of my present 7-feet telescope, it ought to be. There is a suspicion of some apparatus about the planet."

Feb. 26, 1794. "20-feet reflector, power 480. The planet seems to be a little lengthened out, in the direction of the longer axis of the satellites' orbits."

April 21, 1795. "10-feet reflector, power 400. The telescope adjusted to a neighbouring star, so as to make it perfectly round. The disk of the planet seems to be a little elliptical. With 600, also adjusted upon the neighbouring star, the disk still seems elliptical."

Remarks upon the foregoing Observations.

With regard to the phænomena which gave rise to the suspicion of one or more rings, it must be noticed, that few specula or object-glasses are so very perfect as not to be affected with some rays or inequalities, when high powers are used, and the object to be viewed is very minute. It seems, however, from the observations of March 16, 1789, and Feb. 26, 1792, that the cause of deception, in this case, must be looked for elsewhere. It has often happened, that the situation of the eye-glass, being on one side of the tube, which brings the observer close to the mouth of it, has occasioned a visible defect in the view of a very minute object, when proper care has not been taken to keep out of the way; especially when the wind is in such a quarter as to come from the observer across the telescope. The direction of a current of air alone may also affect vision. Without, however, entering further into the discussion of a subject that must be attended with uncertainty, I will only add, that the observation of the 26th seems to be very decisive against the existence of a ring. When the surmises arose at first, I thought it proper to suppose, that a ring might be in such a situation as to render it almost invisible; and that, consequently, observations should not be given up, till a sufficient time had elapsed to obtain a better view of such a supposed ring, by a removal of the planet from its node. This has now sufficiently been obtained in the course of ten years; for, let the node of the ring have been in any situation whatsoever, provided it kept to the same, we must by this time have had a pretty good view of the ring itself. Placing therefore great confidence on the observation of March 5, 1792, supported by my late views of the planet, I venture to affirm, that it has no ring in the least resembling that, or rather those, of Saturn.

The flattening of the poles of the planet seems to be sufficiently ascertained by many observations. The 7-feet, the 10-feet, and the 20-feet instruments, equally confirm it; and the direction pointed out Feb. 26, 1794, seems to be conformable to the analogies that may be drawn from the situation of the equator of Saturn, and of Jupiter.

This being admitted, we may without hesitation conclude, that the Georgian planet also has a rotation upon its axis, of a considerable degree of velocity.

*Reports and Observations relating to the Light and Size of the Georgian Satellites, and to their vanishing at certain Distances from the Planet.**

Jan. 14, 1787. A star was put down, as a supposed very faint satellite; but, on the 17th, the planet being removed, it appeared nearly as bright as two considerable stars that had also been noted.

Jan. 17. "The 1st satellite is the smallest in appearance."

Jan. 24. "The 2d satellite is brighter than the first."

Feb. 9, 1787. "The 1st satellite is larger than the second."

Feb. 10. The planet was supposed to go to a triangle of pretty bright stars. The 11th it was between them, and the stars of the triangle were so dim, that, had they not been seen before, they might have been supposed to be satellites.

Sept. 19, 1787. 4^h 24′. "I can still see the satellites, though daylight is already very strong: they are fainter than the faintest of Saturn's satellites." †

Feb. 22, 1791. "I cannot perceive the 1st satellite, probably owing to its nearness to the planet."

March 2, 1791. "The 1st satellite is hardly to be seen; I have however had several perfect glimpses of it. It seems to be about the most contracted part of its orbit."

March 6. The supposed 3d and 4th satellites of March 5th were imagined to have been gone from their former places; but were seen the 7th, with this memorandum. "I mistook them last night for other stars, they being so large that I did not know them again."

March 9. "The 2d satellite is nearer the planet than the first, and on that account appears smaller."

Dec. 9, 1791. "I do not perceive the 1st satellite."

Feb. 13, 1792. "6^h 16′. The 3d supposed satellite of last night is a considerable star; not much less than *b*."

When the supposed third was pointed out the night before, it is said to be

* [In the following, the first and second satellite are respectively Titania and Oberon.—Ed.]

† Five satellites of Saturn were only known at that time.

smaller than the 1st and 2d satellites. By the figure, it did not exceed the distance of the 2d; and *b* is called a pretty large star.

Feb. 20, 1792. The 2d satellite, being at a great distance, was mistaken for a pretty large star, till about four hours after, when its motion along with the planet was perceived.

Feb. 21, 1792. "7^h 36′. I cannot see the 2d satellite. By calculation, it should be about 8°·6 S.F. and I suspect it to be there, but cannot get the least assurance."

March 15, 1792. "I cannot see the 1st satellite with 300; nor with 480; nor with 600."

March 19. "8^h 35′. I cannot see the 2d satellite with 300. With 480 I see it very well. I see it also with 800; and very well with 1200. With 2400 and 4800 the satellite cannot be seen; but there seems to be a whitish haziness coming on."

March 4, 1794. The 1st satellite could not be seen.

March 7. The 1st was invisible.

March 17. Both 1st and 2d were invisible.

March 21. The 1st was invisible, though looked for with all the powers of the instrument.

March 22. The 2d was hardly visible.

March 23. The 2d was not to be seen.

March 26. The 1st was but just visible.

March 5, 1796. The 2d was invisible.

April 4, 1796. The 1st was invisible.

March 17, 1797. "Power 600. Neither of the satellites are visible to-night; with 300 I cannot see them. The night is very beautiful, and I have a field bar to hide the planet; but, notwithstanding this, I cannot see either of the satellites."

March 21. The 1st satellite was invisible.

March 23. The 2d was invisible. The 1st could not be seen immediately, but, having been informed where exactly to look for it, according to my calculation of its place, it was perceived; and with 600 seen very well.

March 25. Both satellites were invisible.

Remarks on the foregoing Observations.

From the observations of Jan. 14, Feb. 10, March 6, 1787, and Feb. 13, 1792, it appears, that all very small stars, when they come near the planet, lose much of their lustre. Indeed, every observation that has been recorded before, of supposed satellites that have been proved to be stars afterwards, has fully confirmed this circumstance; for they were always found to be considerable stars, and their being mistaken for satellites was owing to their loss of light when near the planet. This would hardly deserve notice, as it is well known that a superior light will obstruct an inferior one; but some circumstances which attend the operation of the affections of light upon the eye, when objects are very faint, are so remarkable, that they must not be passed over in silence.

After having been used to follow up the satellites of Saturn and Jupiter, to the very margin of their planets, so as even to measure the apparent diameter of one of Jupiter's satellites by its entrance on the disk,* I was in hopes that a similar opportunity would soon have offered with the Georgian satellites : not indeed to measure the satellites, but to measure the planet itself, by means of the passage of the satellite over its disk. I expected also to have settled the epochs of the satellites, from their conjunctions and oppositions, with more accuracy than I have yet been able to do from their various positions in other parts of their orbits. A disappointment of obtaining these capital advantages deserves to have its cause investigated ; but, first of all, let us cast a look upon the observations.

The satellites, we may remark, become regularly invisible, when, after their elongation, they arrive to certain distances from the planet. In order to find what these distances are, we will take the first observation of this kind, as an example.

Feb. 22, 1791, the first satellite could not be seen. Now, by my lately constructed tables, its longitude from the apogee, at the time of observation, was 204·5 degrees ; that is, 24·5 degrees from the most contracted part of its orbit, on the side that is turned to us, which, as its opposite is called the apogee, I shall call the perigee. By my tables also for the same day, we have the distance of the apogee from the planet, which is 60 ; supposing the greatest elongation distance to be 1. This being given, we may find an easy method of ascertaining the distance of the satellite, when it is near the apogee or perigee : for it will be sufficiently true for our purpose to use the following analogy. Cosine of the distance of the satellite from the apogee or perigee is to the apogee distance from the planet, as the greatest elongation is to the distance of the satellite from the planet. When the ellipsis is very open, this theorem will only hold good in moderate distances from the apogee or perigee ; but, when it is a good deal flattened, it will not be considerably out in more distant situations : and it will also be sufficiently accurate to take the natural cosine from the tables to two places of decimals only. When this is applied to our present instance, we have 91 for the natural cosine of 24·5 degrees ; and the distance of the satellite from the planet will come out $\frac{\cdot 6 \times 33''}{\cdot 91} = 21''\cdot 8$.

By this method, it appears that the satellite, when it could not be seen, was nearly 22″ from the planet.

We must not however conclude, that this is the given distance at which it will always vanish. For instance, the same satellite, though hardly to be seen, was however not quite invisible March 2, 1791. Its distance from the planet, computed as before, was then only $\frac{\cdot 6 \times 33''}{1} = 19''\cdot 8$.

* See *Phil. Trans.* for 1797, Part II. page 335. [Vol. I. p. 586.]

The clearness of the atmosphere, and other favourable circumstances, must certainly have great influence in observations of very faint objects; therefore, a computation of all the observations where the satellites were not seen, as well as a few others where they were seen, when pretty near the apogee or perigee, will be the surest way of settling the fact. The result of these computations is thus.

First satellite invisible.			Second satellite invisible.		
1791.	Feb. 22	at 21″·8	1792.	Feb. 21	at 23″·3
	Dec. 19	at 16·9	1794.	March 17	at 20·7
1792.	March 15	at 18·4		March 23	at 17·9
1794.	March 4	at 18·5	1796.	March 5	at 9·3
	March 7	at 12·5		March 17	at 6·3
	March 17	at 17·0		March 23	at 6·2
	March 21	at 15·5		March 25	at 8·7
	April 4	at 8·5			
1797.	March 17	at 4·8			
	March 21	at 4·6			
	March 25	at 4·8			
First satellite visible.			**Second satellite visible.**		
1791.	March 2	at 19″·8	1794.	March 22	at 17″·5
1794.	Feb. 26	at 14·1			

Thus, having the observations and calculated distances under our inspection, we find that both the satellites became always invisible when they were near the planet: that the 1st was generally lost when it came within 18″ of the planet, and the 2d at the distance of about 20″. In very uncommon and beautiful nights, the 1st has once been seen at 13″·8, and the 2d at 17″·3; but at no time have they been visible when nearer the planet.

I shall now endeavour to investigate the cause which can render small stars and satellites invisible at so great a distance as 18 or 20″.

A dense atmosphere of the planet would account for the defalcation of light sufficiently, were it not proved that the satellites are equally lost, whether they are in the nearest half of their orbits, or in that which is farthest from us. But, as a satellite cannot be eclipsed by an atmosphere that is behind it, a surmise of this kind cannot be entertained. Let us then turn our view to light itself, and see whether certain affections between bright and very bright objects, contrasted with others that take place between faint and very faint ones, will not explain the phænomena of vanishing satellites.

The light of Jupiter or Saturn, for instance, on account of its brilliancy, is diffused, almost equally, over a space of several minutes all around these planets. Their satellites also, having a great share of brightness, and moving in a sphere that is strongly illuminated, cannot be much affected by their various distances from the planets. The case then is, that they have much light to lose, and comparatively lose but little.

The Georgian planet, on the contrary, is very faint; and the influence of its feeble light cannot extend far, with any degree of equality. This enables us to see the faintest objects, even when they are only a minute or two removed from it. The satellites of this planet are very nearly the dimmest objects that can be seen in the heavens; so that they cannot bear any considerable diminution of their light, by a contrast with a more luminous object, without becoming invisible. If then the sphere of illumination of our new planet be limited to 18 or 20″, we may fully account for the loss of the satellites when they come within its reach; for they have very little light to lose, and lose it pretty suddenly.

This contrast, therefore, between the condition of the Georgian satellites and those of the brighter planets, seems to be sufficient to account for the phænomenon of their becoming invisible.

We may avail ourselves of the observations that relate to the distances at which the satellites vanish, to determine their relative brightness. The 2d satellite appears generally brighter than the 1st; but, as the former is usually lost farther from the planet than the latter, we may admit the 1st satellite to be rather brighter than the 2d. This seems to be confirmed by the observation of March 9, 1791; where the 2d appeared to be smaller than the 1st, though the latter was only 25″ from the planet, while the other was 30″·8.

The first of the new satellites will hardly ever be seen otherwise than about its greatest elongations, but cannot be much inferior in brightness to the other two; and, if any more interior satellites should exist, we shall probably not obtain a sight of them; for the same reason that the inhabitants of the Georgian planet perhaps never can discover the existence of our earth, Venus, and Mercury.

The 2d new or intermediate satellite is considerably smaller than the 1st and 2d old satellites. The two exterior, or 5th and 6th satellites, are the smallest of all, and must chiefly be looked for in their greatest elongations.

Periodical Revolutions of the new Satellites.

It may be some satisfaction to know what time the four additional satellites probably employ in revolving round their planet. Now, as this can only be ascertained with accuracy by many observations, we must of course remain in suspense, till a series of them can be properly instituted. But, in the mean time, we may admit the distance of the interior satellite to be 25″·5, as our calculation of the estimation of March 5, 1794, gives it; and from this we compute that its periodical revolution will be 5 days, 21 hours, 25 minutes.

If we place the intermediate satellite at an equal distance between the two old ones, or at 38″·57, its period will be 10 days, 23 hours, 4 minutes.

By the figure of Feb. 9, 1790, it seems that the nearest exterior satellite is about double the distance of the farthest old one; hence, its periodical time is found to be 38 days, 1 hour, 49 minutes.

The most distant satellite, according to the calculation of the observation of Feb. 28, 1794, is full four times as far from the planet as the old 2d satellite; it will therefore take at least 107 days, 16 hours, 40 minutes, to complete one revolution.

It will hardly be necessary to add, that the accuracy of these periods depends entirely upon the truth of the assumed distances; some considerable difference, therefore, may be expected, when observations shall furnish us with proper *data* for more accurate determinations.*

Slough, near Windsor,
September 1, 1797.

* [Compare paper LXVII., *infra*; also Holden's and Lassell's papers in the *Monthly Notices R.A.S.*, xxxv, pp. 16 and 22.—Ed.]

XLI.

A Fourth Catalogue of the comparative Brightness of the Stars.

[*Phil. Trans.*, 1799, pp. 121–144.]

Read February 21, 1799.

Lustre of the stars in Auriga.

1	f	5	2 , 1 4 , 1	33	δ	4	10 -, 33 , 8 33 ; 10 Camel
2	g	5 . 6	4 , 2 , 1 2 . 4	34	β	2	23 . 34 66 Gemin . 34 34 ; 23
3	ι	4	3 - 37 5 Arietis - 3 - 44 Persei				23 , 34
			3 . 44 Persei	35	π	6	35 . 29 35 , 46
4	ω	5	4 , 2 2 . 4 , 1	36		6	29 -, 36
5		6	5 - 6	37	θ	4	3 - 37 24 Gemin , 37
6		6	5 - 6 6 - 12	38		6 . 7	39 . 38 , 42
7	ε	4	7 - 10 7 -, 10	39		6 . 7	39 . 38
8	ζ	4	10 , 8 33 , 8 10 =, 8 -, 30	40		6	31 . 40 - 28
9		6 . 5	7 Camelop -, 9 11 Camelop . 9 -,	41		6	46 - 41
			12 Camelop	42		6	38 , 42 . 43 47 , 42 . 43
10	η	4	7 - 10 , 8 10 -, 33 7 -, 10 =, 8	43		6	42 . 43
11	μ	5	15 ; 11 - 20 15 -, 11 -, 21	44	κ	4 . 5	44 , 136 Tauri 44 - - 49
12		6	6 - 12	45		6	45 . 46
13	α	1	13 - - 3 Lyræ	46		5	35 , 46 - 41 45 . 46
14		5	16 , 14 . 19	47		6	47 , 42
15	λ	5	15 ; 11 15 -, 11	48		6	49 -, 48 - 53
16		6	16 , 14	49		5 . 6	44 - - 49 -, 48
17		7 . 6	19 . 17 , 18	50		5 . 6	50 - 52 50 . 16 Lyncis
18		8	17 , 18	51		5 . 6	52 - 51
19		6	14 . 19 . 17	52		5	50 - 52 - 51
20	ρ	6	11 - 20 21 . 20	53		6	53 . 28 Gemin 48 - 53 , 54
21	σ	5 . 6	11 -, 21 . 20	54		6	28 Gemin , 54 - 25 Gemin
22		6	26 , 22				53 , 54
23	γ	2	13 Arietis , 23 . 34 34 ; 23	55		5	55 - 58
			23 , 34	56		6	58 , 56 . 57 16 Lyncis , 56
24	φ	5 . 6	25 -, 24	57		6	56 . 57
25	χ	5 . 6	25 -, 26 25 -, 24	58		4 . 5	55 - 58 , 56
26		6	25 -, 26 , 22 27 , 26	59		6	62 . 59 . 60
27	ο	6	27 , 26	60		6	59 . 60
28		7	40 - 28	61		6	61 . 62
29	τ	5	32 , 29 , 31 35 . 29 -, 36	62		6 . 7	61 . 62 . 59
30	ξ	6	8 -, 30 42 Camel . 30 - 31	63		4 . 5	63 . 65
			Camel	64		5	66 ; 64
31	υ	6	29 , 31 . 40	65		5	63 . 65 , 66
32	ν	5	32 , 29	66		5	65 , 66 ; 64

Lustre of the stars in Draco.

1	λ	3 . 4	1 . 5
2		6	5 – – 2 3 ; 2
3		6	3 ; 2
4		6	6 ⁝ 4
5	κ	3	1 . 5 – – 2 11 – 5 , 13
6		6	6 ⁝ 4
7		6	9 , 7
8		6	10 – 8 , 9
9		6	8 , 9 , 7
10	*i*	5	10 – 8
11	*a*	2	11 – 5
12	ι	3	40 Herculis – 12 22 –, 12 27 Herc – 12 , 67 Herculis
13	θ	3	5 , 13
14	η	3	1 Ursæ min – 14 14 – – 22 14 – 23 14 –, 57
15	A	4	15 . 18
16		5	17 – 16
17		5	17 – 16
18	*g*	5	19 , 18 –, 20 15 . 18 21 . 18
19		5	19 , 18
20	*h*	6	18 –, 20
21	μ	5 . 4	21 . 18 30 , 21
22	ζ	2	14 – – 22 –, 12 57 –̦ 22
23	β	2 . 3	23 , 40 Herculis 14 – 23 23 – 27 Herculis
24	ν¹	4	24 . 25
25	ν²	4	24 . 25 – 26
26		6	25 – 26
27	*f*	5	28 – 27 , 34 27 . 42
28	ω	4	31 – 28 – 27
29		6	34 – –, 29
30		6	30 , 21
31	ψ¹	7	31 – 28
32	ξ	3	44 –, 32 – 43
33	γ	2	7 Ursæ min – 33 5 Coronæ –̦ 33 55 Ophiu – 33
34	ψ²	4 . 5	27 , 34 35 . 34 – –, 29 34 – 37
35		6	35 . 34 35 – 40 + 41
36		6	42 . 36
37		6	34 – 37 , 38
38		6	37 , 38
39	*b*	5	45 , 39 . 46
40		5	41 – 40
41		5	41 – 40
42		6	27 . 42 . 36
43	φ	5	32 – 43
44	χ	4	44 –, 32
45	*d*	5	45 , 39
46	*c*	5	39 . 46 ; 47
47	*o*	4	46 ; 47 – – 51 47 – 54
48		6	49 . 48
49		6	51 . 49 . 48
50		4 . 5	52 . 50 –̦ 59 73 , 50 55 , 50 52 – 50
51		5 . 6	47 – – 51 . 49
52	υ	4 . 5	60 – 52 52 . 50 52 – 50
53		5	54 , 53
54		5	47 – 54 , 53
55		6	61 –, 55 55 , 50
56		6	Does not exist.
57	δ	3 . 4	14 –, 57 –̦ 22
58	π	4	63 –, 58 – 61 58 ; 67
59		6	50 –̦ 59
60	τ	4 . 5	60 – 52
61	σ	4 . 5	58 – 61 –, 55 67 , 61
62		6	Does not exist.
63	ε	5 . 6	63 –, 58
64	*e*¹	5 . 6	64 –, 65
65	*e*²	6 . 5	64 –, 65 –, 70 65 , 69
66		6	68 . 66 . 71
67	ρ	5	58 ; 67 . 61
68		6	68 . 66
69		6	65 , 69
70		6	65 –, 70
71		6	66 . 71
72		6	Does not exist.
73		5 . 6	73 , 50 73 – 77 73 , 78
74		6	75 – 74 . 76
75		6	75 – 74
76		5	74 . 76
77		5	73 – 77 78 , 77
78		5	73 , 78 , 77 16 Cephei , 78
79		7	80 – 79
80		6	80 – 79

Lustre of the stars in Lynx.

1		5 . 6	1 , 5
2		4	2 – – 5
3		6	8 – 3 , 10
4		6	4 . 6
5		6	2 – – 5 – 6 1 , 5
6		6 . 7	5 – 6 4 . 6

Lustre of the stars in Lynx.

7		6 . 7	Does not exist.	28		7	36 – 28
8		6 . 7	8 , 41 Camelop 8 – 3	29		5	29 – 56 Camelop 29 , 58 Came-
9		7	11 – 9				lop 29 , 30
10		6 . 7	41 Camelop – 10 3 , 10	30		6	29 , 30
11		6	14 , 11 – 9	31		5	31 –, 27
12		7	15 ; 12 –, 14	32		7	33 ; 32
13		6	13 , 14	33		6	33 ; 32
14		5	12 –, 14 13 , 14 , 11 19 , 14 – 23	34		6	26 , 34 36 , 34
15		5	15 ; 12	35		7	Does not exist.
16		6	50 Aurigæ . 16 , 56 Aurigæ	36		6	36 , 34 37 . 36 – 28
17		7	18 –, 17	37		5 . 6	37 . 36 37 , 42
18		6	18 –, 47 Camelop 18 –, 17	38		6	38 – 40
19		5	24 – 19 , 14	39		4	12 Ursæ , 39 39 –, 10 Leonis
20		6	21 – – 20				min 39 , 10 Ursæ
21		5	22 – 21 – – 20	40		6	38 – 40
22		6	22 – 21	41		4	25 Ursæ . 41 – 12 Ursæ
23		7	14 – 23	42		6	37 , 42 . 45 15 Leonis min
24		5	24 – 19				42 – – 14 Leonis min
25		6	26 –, 25	43		6	43 – 44
26		5	26 –, 25 26 , 34	44		7 . 6	43 – 44
27		5	27 – 50 Camelop 31 –, 27	45		5 . 6	29 Ursæ – – 45 42 . 45

Lustre of the stars in Lyra.

1	κ	5	1 –, 2 6 – 1	11	δ^1	4 . 5	12 –, 11 18 , 11 . 9 16 , 11
2	μ	6	1 –, 2	12	δ^2	4	12 –, 11 12 . 13 12 – 18 12 – 16
3	α	1	16 Bootis – – 3	13	π	6	12 . 13 20 , 13
4	ϵ	5	6 ; 4 . 5	14	γ	3	6 Cygni ÷ 14 , 85 Herculis
5		6	4 . 5	15	λ	6	15 , 9 15 . 17 15 ÷ 19
6	ζ	5	6 – – 7 6 – 1 6 ; 4	16	ρ	6	12 – 16 , 11
7		5	6 – – 7	17		6	15 . 17
8	ν^1	6	9 , 8	18	ι	5	12 – 18 , 11
9	ν^2	6	11 . 9 , 8 15 , 9 , 8	19		6	15 ÷ 19
10	β	3	10 . 14 14 ÷ 10 14 –, 10	20	η	6	21 . 20 , 13 20 , 21
			14 – – – 10 $\overline{6 + 7}$ ∸ 10	21	θ	6	21 . 20 20 , 21

Lustre of the stars in Monoceros.

1		6	3 –, 1 –, 2	10		6	5 –, 10 –, 9 10 – 7
2		6	1 –, 2 , 4	11		5	22 Orionis – 11 – 5 30 ; 11 ÷ 26
3		6	3 –, 1	12		5	13 –, 12 , 14
4		6	2 , 4 . 6	13		4	8 – 13 = ÷ 14 13 –, 12
5		4 . 5	11 – 5 –, 10 5 , 8	14		5 . 6	13 = ÷ 14 12 , 14
6		6 . 7	4 . 6	15		4	8 ÷ 15 , 17
7		6	10 – 7	16		6	17 – 16
8		4	5 , 8 – 13 8 ÷ 15 8 , 18	17		5	15 , 17 – 16
9		5	10 –, 9	18		4	8 , 18

Lustre of the stars in Monoceros.

19		5	22 - 19 , 20	26		4 . 5	11 -; 26 -; 29
20		6	19 , 20 , 25	27		5	29 - 27 28 , 27
21		5	22 -, 21 , 24	28		5	29 , 28 , 27
22		4 . 5	22 -, 21 22 - 19	29		6	29 - 27 26 -; 29 , 28 29 , 31
23		6 . 7	24 , 23	30		6	30 ; 11
24		6	21 , 24 , 23	31		4	29 , 31
25		6	20 , 25				

Lustre of the stars in Perseus.

1		6	4 , 1 - 3 4 - 1	31		5 . 6	29 . 31
2	g	6	3 , 2	32	l	6	32 - 30 32 - 36
3		6 . 7	1 - 3 , 2	33	α	2 . 3	33 - - 26 21 Androm , 33 . 43 Andromedæ
4		6	4 , 1 4 - 1 4 - 9	34		6	35 - 34 34 - 29
5		6	9 - 5 . 7 7 . 5	35	σ	5	37 . 35 - 34
6	h	6	65 Androm , 6 , 63 Androm	36		6 . 7	32 - 36 - 30
7	χ	7 . 6	5 . 7 . 8 8 ; 7 . 5	37	ψ	5	37 , 35
8		7	7 . 8 9 -, 8 ; 7	38	o^1	6	46 , 38 , 59 41 - 38 - 46
9	i	6	4 - 9 -, 8 9 - 5	39	δ	3	45 - - 39 - 25
10		7	Does not exist.	40	o^2	6	52 , 40 . 42 40 , 42
11		7	27 ; 11 - 13	41	ν	4	25 - 41 , 46 41 - 38
12	q	6	28 . 12 . 14	42	n	6	40 . 42 40 , 42 54 ; 42 - 55
13	θ	4	11 - 13 , 18 18 -; 13	43	A	5	43 , 20
14		6	12 . 14 24 , 14	44	ζ	3	3 Aurigæ - 44 3 Aurigæ . 44 44 , 45
15		6	Lost	45	ε	3	44 , 45 - - 39
16	p^1	4	16 , 22 16 - - 20	46	ξ	5	41 , 46 , 38 38 - 46 46 - 58
17	r	5 . 6	17 , 28	47	λ	4	51 . 47 , 53
18	τ	5	13 , 18 18 -; 13	48	c	5	48 , 51
19		6	Does not exist	49		6 . 7	50 , 49
20	p^2	6	16 - - 20 28 - 20 43 , 20	50		6 . 7	52 , 50 , 49
21		4 . 5	28 . 21	51	μ	4	48 , 51 . 47
22	π	4	16 , 22 28	52	f	5	52 , 50 53 . 52 , 40 58 - 52
23	γ	3	23 , 25 23 -, 4 Trianguli	53	d	6	47 , 53 . 52
24	s	6	28 . 24 , 14	54		6	54 ; 42
25	ρ	4	23 , 25 - 41 39 - 25	55		6	42 - 55 ; 56
26	β	2 . 3	26 , 25 26 ; 25 26 - - - 25 6 Arietis , 26 - 23	56		7	55 ; 56
27	κ	5 . 4	27 ; 11	57	m	6	59 . 57
28	ω	5	22 , 28 . 12 17 , 28 . 24 28 . 21 28 - 20	58	e	5	46 - 58 - 52
29		6	34 - 29 . 31	59		6	38 , 59 . 57
30		6	32 - 30 36 - 30				

Lustre of the stars in Sextans.

1		5	1 - 2	4		6	7 , 4 4 - 2
2		5	1 - 2	5		6	3 , 5
3		6	8 -, 3 , 5 6 , 3	6		6	8 - 6 , 3

Lustre of the stars in Sextans.

7		6	7 , 4	25		6	25 . 26
8		6	8 -, 3 8 - 6	26		6	24 -, 26 25 . 26 23 , 26 - 31
9		6	12 , 9 . 13	27		6	27 , 28
10		6	29 Leonis = -, 10 - 11	28		5	29 - 28 . 24 27 , 28 - 32
11		5 . 6	10 - 11	29		5	30 - 29 - 28
12		6	4 - 12 , 9	30		5	32 Hydræ - 30 - 29
13		6	9 . 13	31		6	26 - 31
14		6	14 . 19	32		6	28 - 32
15		4	35 Hydræ , 15 - 32 Hydræ	33		6	33 , 24
16		6	19 , 16	34		6	35 - 34 , 37
17		6	22 , 17 . 18	35		6	35 - 34
18		6	17 . 18 - 20	36		6	37 . 36
19		6	14 . 19 , 16	37		6	55 Leonis -, 37 , 38 34 , 37 . 36
20		6	18 - 20 , 21	38		6	37 , 38
21		6	20 , 21	39		7	41 -, 39 40 , 39
22		6	22 , 17	40		6	41 - 40 , 39
23		5	23 , 26	41		6	41 -, 39 41 - 40
24		6	28 . 24 -, 26 33 , 24				

Lustre of the stars in Taurus.

1	*o*	4	1 - 2	29	u^1	6	29 , 40
2	*ξ*	4	1 - 2 , 35	30	*e*	5	5 - 30 - 4 66 , 30 , 46
3		6	Does not exist	31	u^2	6	40 , 31
4	*s*	6	5 -, 4 -, 6 30 - 4	32		6	53 , 32 . 33
5	*f*	5	38 - 5 -, 4 5 - 30	33		7	32 . 33
6	*t*	6	4 -, 6 , 12	34		7	39 - 34
7		6	7 . 66 Arietis	35	λ	4	2 , 35 . 38 123 -, 35
8		6	Does not exist	36		7	43 ; 36
9		6	Lost	37	A	5	65 , 37 - - 39 37 - - 43
10		4 . 5	38 , 10 , 49	38	*ν*	4	35 . 38 - 5 38 , 10
11		6	21 . 11 , 22	39		6	37 - - 39 51 - 39 43 - 39 - 34
12		6	6 ; 12	40		7	29 , 40 , 31 46 ; 40 , 45
13		6	13 , 14 13 - 14	41		6	42 ; 41 ; 44
14		6	13 , 14 13 - 14	42	*ψ*	5	52 . 42 ; 41
15	*n*	6	Does not exist	43	$ω^1$	6	37 - - 43 - 39 43 ; 36
16	*g*	7	18 . 16 . 21	44	*p*	6	41 ; 44 , 59
17	*b*	5	27 . 17 . 20	45		7	40 , 45
18	*m*	7	28 , 18 . 16	46		7	47 , 46 ; 40 30 , 46 . 93
19	*e*	5	20 , 19 . 23	47		7	49 -, 47 , 46 47 . 60
20	*c*	6	17 . 20 , 19	48		7	58 . 48
21	*k*	6 . 7	16 . 21 . 11	49	*μ*	4	10 , 49 49 -, 47 88 , 49
22	*l*	7	11 , 22 , 26	50	$ω^2$	6	65 - 50 - 56 50 , 67
23	*d*	5	19 . 23 -, 28	51		7	53 , 51 - 39
24	*p*	7	26 , 24	52	*φ*	5	52 . 42
25	*η*	3	27 -, 27	53		7	56 , 53 , 51 53 , 32
26	*s*	7 . 8	22 , 26 , 24	54	*γ*	3	77 -, 54 - - 58 54 -, 61 74 ; 54
27	*f*	6	25 -, 27 . 17	55		7	63 , 55
28	*h*	7 . 8	23 -, 28 , 18	56		7	50 - 56 , 53

Lustre of the stars in Taurus.

57		6.7	58 , 57 . 60
58	*h*	7	54 - - 58 . 48 58 , 57 73 - 58 -, 76 58 , 83
59	χ	5	44 , 59
60		7	57 . 60 47 . 60
61	δ^1	4	54 -̦ 61 , 68
62		7	72 - 62 65 ; 62
63		6	64 -, 63 , 55
64	δ^2	4	68 - 64 -, 63
65	κ	5	65 - 69 65 ; 62 65 - 50 65 , 37
66	γ	5	66 , 30
67	κ^2	5	69 - 67 . 72 50 , 67 ; 72
68	δ^3	6	61 , 68 - 64
69	υ^1	5	65 - 69 - 67 94 . 69
70		7	80 - 70
71		7	71 ; 75
72	υ^2	6	72 - 62 67 . 72 67 ; 72 72 -, 95
73	π	5	73 - 58 86 , 73
74	ϵ	3.4	78 , 74 ; 54
75		7	71 ; 75 , 81
76		7	58 -, 76 83 - 76
77	θ^1	5	77 -̦ 54
78	θ^2	5	78 , 74
79	*b*	5	90 - - 79 , 83
80		7	81 , 80 - 70 80 - 84 80 . 85
81		7	75 ; 81 , 80
82		7	Does not exist
83		7	79 , 83 58 , 83 - 76
84		7	80 - 84
85		7	80 . 85
86	ρ	5	86 , 73
87	α	1	58 Orionis - - 87 19 Orionis = -̦ 87 87 - - 78 Gemin
88	*d*	5	90 , 88 , 49
89		7	91 ; 89
90	c^1	5	90 , 88 90 - - 79 90 - - 93
91	σ^1	6	92 , 91 ; 89
92	σ^2	6	92 , 91
93	c^2	6	46 . 93 90 - - 93
94	τ	5	94 . 69
95		6.7	72 -, 95
96		6	4 Orionis - - 96 97 -, 96
97	*i*	6	4 Orionis , 97 -, 96
98	*k*	6	106 - 98 - 99 103 . 98
99		6	98 - 99 , 107
100		6	Lost
101		6	107 -̦ 101
102	ι	4	102 = -̦ 106 102 , 104
103		6	103 . 98
104	*m*	6	102 , 104 -, 106
105		6	106 , 105 - 107
106	l^1	6	102 = -̦ 106 , 105 106 . 109 104 -, 106 - - 107 106 - 98
107	l^2	6	105 - 107 108 . 107 106 - - 107 99 , 107 -̦ 101
108		7	109 -, 108 109 , 108 . 107
109	*n*	6	114 - 109 -, 108 106 . 109 , 108
110		7	116 - 110 . 113 115 - 110 . 117 120 . 110
111		6.7	111 , 116 111 , 115 119 - 111
112	β	2	112 -, 24 Orionis
113		6	110 . 113
114	*o*	5	114 - 109
115		7.8	111 , 115 - 110 122 . 115
116		6.7	111 , 116 - 110
117		7	110 . 117
118		6	121 - 118
119		7	119 - 111
120		7	120 . 110
121		6	121 - 118
122		7	126 - 122 -, 129 122 , 130 122 . 115
123	ζ	3	123 -̦ 35 13 Gemin -, 123 , 7 Gemin
124		6.7	Does not exist
125		3	132 , 125
126		6	126 - 122
127		6	130 - - 127
128		6	129 , 128
129		6	122 -, 129 , 128
130		6	122 , 130 - - 127
131		6	133 , 131 . 132 135 , 131 , 137
132		4	139 , 132 , 125 131 . 132 . 135
133		6	134 , 133 , 131 134 - 133 , 131
134		6	134 , 133 134 - 133
135		6	132 . 135 - 138 135 , 131
136		5	136 ⁝ 139 44 Aurigæ , 136
137		5	135 - 137 131 , 137
138		6	Does not exist
139		6	1 Gemin - 139 , 132 136 ⁝ 139 , 132
140		6	141 . 140
141		6	141 . 140

Lustre of the stars in Triangulum.

1	*d*	6	3 . 1	8	δ	5	9 -, 8 - 7
2	*a*	4	4 -, 2 -, 9 31 Androm -, 2	9	γ	4	2 -, 9 -, 8
			2 - 99 Piscium	10	*a*	6	6 , 10 ; 12
3	ϵ	6	7 - 3 . 1	11	*d*	7	12 , 11 , 13
4	β	4	6 Arietis - 4 23 Persei -, 4 -, 2	12	*c*	6	10 ; 12 -, 13 12 , 11
			4 =, 31 Andromedæ	13		7	12 -, 13 11 , 13
5		7	7 - 5	14		6	7 ; 14 ; 15
6	ι	6	6 , 7 6 , 10	15		7	14 ; 15
7	η	6	8 - 7 - 3 6 , 7 - 5 7 ; 14	16		7	16 , 30 Arietis 33 Arietis -, 16

Notes to Auriga.

23 Is 112 Tauri.

30 Is 32 Camelopardali.

45 " Oct. 5, 1798. The time of this star, in the observation of FLAMSTEED, Vol. II. page 189, is marked : : but it cannot be much out, as the star seems to be in the place assigned to it by the British catalogue."

61 The R.A. in the *Atlas Cœlestis* requires a correction of $-42'$.

Notes to Draco.

10 Is 87 Ursæ.

12 and 13 Were never observed by FLAMSTEED, but are in LA CAILLE's Catalogue of northern stars.

14 M. DE LA LANDE says the star is not to be found. See Mr. BODE's *Ast. Jahrbuch* for 1795, page 198.

I observed this star in a sweep of the heavens, June 2d, 1788. Its brightness was estimated Sept. 11, 1795; Sept. 24, 1796; Sept. 30, 1796; and Dec. 28, 1798; so that, if M. DE LA LANDE is sure no cloud intervened when he looked for it, we may suspect it to be a changeable star.

15 The British catalogue requires $+30'$ in R.A.

35 The expression " $35 - \overline{40 + 41}$ " in my estimation of brightness, means that, with the naked eye, 35 is a very little brighter than 40 and 41 together, taken as one star. For they are so near each other, that the eye alone cannot distinguish them from a single star. The British catalogue gives them 3′ farther asunder than they ought to be according to FLAMSTEED's observation, page 463. See also Mr. BODE's *Ast. Jahrbuch* for 1785, page 173.

40 The estimation " 40 – 41 " was made with a 7-feet reflector, power 460.

56 Does not exist. FLAMSTEED has no observation of it.* My double star II. 31, called 56 Draconis, is a star situated between 59 and 50, about 1½ degree from 59 towards 50.†

62 Does not exist. FLAMSTEED has no observation of it; but, if an error of two hours be supposed in the calculation of one of the observations of 31 Draconis, it will account for the insertion of this star.

72 Does not exist. There is an observation, page 173, which produced it; but, if we admit an error of 3′ in time in that observation, it will then belong to 71.‡

Notes to Lynx.

7 Does not exist in the place pointed out by the British catalogue; but, in FLAMSTEED's observations, page 286, its time is marked : : and there is probably some considerable error.§

* [It is = 59 Draconis.—ED.]

† When I say 1½ degree from 59 towards 50, it is to be understood that I express myself in degrees of a great circle. I have always used the same method of description in my catalogues of double stars; and, as these objects were pointed out for being viewed with telescopes of great magnifying power, which are generally not fixed, and therefore can give no right ascension, I am rather surprised to find that, in a catalogue published not many years ago, the author has taken my degrees of a great circle for degrees of right ascension. For instance, the double star IV. 82, where, in pointing out its place, I say, "above ¾ degree following the 16 Cephei, in a line parallel to β and α Cassiopeæ," is placed in the zone from 15 to 19° of that author's catalogue, only 2′ 47″ 5 of time following 16 Cephei, when it ought to have been at least 10′ or 11′ following.

I take this opportunity to mention that, in general, the same author's account of my double stars is extremely erroneous.

‡ [72 Draconis is = P. XX. 162, 8·3 mag.—ED.]

§ [Flamsteed's Declination is 1° too great.—ED.]

20, 21, 22 The place of these stars in the heavens does not seem to agree with their situation in the Atlas.

30 Is 58 Camelopardali.

35 FLAMSTEED has no observation of this star; but, as it is marked 7m in the British catalogue, and has a line allotted to it, my Atlas and stars have been numbered so as to take it in; and the numbers I have used with double stars and other objects where the stars in Lynx after the 35th are concerned, must be reckoned accordingly.*

37 " Dec. 4, 1796, This star is nearer to 25 than it is marked in the Atlas." The R.A. should be corrected + 1°.

Notes to Lyra.

10 This is one of our periodical stars discovered by Mr. GOODRICKE; its period is about 6 days 9 hours. See *Phil. Trans.* Vol. LXXVI. page 197. The greatest variation of its light, as far as I have observed, is from " 10 . 14 to $\overline{6+7}$ ⸗ 10." The expression $\overline{6+7}$ is borrowed from algebra, and is always to be understood as has been explained in the note to 35 Draconis.

16 The British catalogue requires a correction of − 9° in P.D.; and this star will then agree with 12 Lyræ HEVELII.

19 The British catalogue requires a correction of + 8° in P.D.

Notes to Perseus.

5 FLAMSTEED has no observation of this star; but there is a star exactly in the place pointed out by the British catalogue.

10 Does not exist. FLAMSTEED never observed it.†

12 " Sept. 5, 1798, This star, which has no time in FLAMSTEED'S observations, is placed a little too forward; or requires about +10′ in R.A."

14 " Sept. 4, 1798, The time of this star is marked doubtful by FLAMSTEED, page 214; but it seems to be in the situation where the British catalogue places it."

15 Is lost. FLAMSTEED observed it Jan. 17, 1693, page 186; but it is not to be seen in the place pointed out by that observation. See BODE'S *Ast. Jahrbuch* for 1794, page 97.‡

19 Does not exist. There is an observation in page 185, which has produced this star, but it belongs to 18; for the star is lettered τ, and a memorandum says, " Post transitum." See also BODE'S *Ast. Jahrbuch* for 1788, page 172.

24 " Sept. 4, 1798, The place of this star in the British catalogue wants a correction of + 56′ in P.D. and − 45′ in R.A."

26 Is a periodical star. It has been noticed in the last century as subject to change, by MONTANARI and MARALDI; but its being periodical was discovered by Mr. GOODRICKE, in 1783, who fixed the time of its change at 2 days 20 hours 48′ 56″. See *Phil. Trans.* Vol. LXXIV. page 287. I have seen it when brightest, " 6 Arietis , 26 − 23 ", and when most diminished, " 26 , 25 ".

38 " Sept. 5, 1798, The British catalogue requires nearly + 2° in R.A., and − 13′ in P.D.; at least there is no other star that can be taken for it."

42 " Sept. 4, 1798, The British catalogue requires a correction of + 13 in P.D."

Notes to Sextans.

1 Is 10 Leonis.

10 Is 25 Leonis.

11 Is 28 Leonis.

12 " March 17, 1797, This star is misplaced in the British catalogue; the P.D. should be corrected + 1°."

28 " March 21, 1797, This star is misplaced in the British catalogue, and requires a correction of + 20′ in R.A., and + 1° in P.D." §

29 " March 21, 1797, The P.D. of this star in the British catalogue requires + 1°."

* [See Baily's note to 1240.—ED.]

† [See Baily's note to 293.—ED.]

‡ [It was observed north of the zenith and not south; = η Persei.—ED.]

§ [See Baily's note to No. 1486.—ED.]

Notes to Taurus.

3 Does not exist. FLAMSTEED never observed it.

8 " Jan. 10, 1796, This star does not exist. FLAMSTEED has no observation of it. There is a star about 9m not far from the place."

9. " Dec. 28, 1798, This star is lost." M. DE LA LANDE says it is not to be found. See Mr. BODE's *Ast. Jahrbuch* for 1795, page 198. FLAMSTEED has two complete observations of it, page 86, and page 506. We can hardly admit what Mr. BODE suggests, that this star, like the rest, has found its way into the British catalogue by some error of writing, or of calculating the observations ; it will therefore be advisable to look for a future re-appearance of it, as it may prove to be a periodical or changeable one.*

15 Does not exist. FLAMSTEED has no observation of it.

34 The estimation " 39 – 34 " belongs to a star very nearly in the place where, according to FLAMSTEED's observation, 34 should be ; but, as we know by calculation that the Georgian planet was about the situation where, the 13 of Dec. 1690, FLAMSTEED observed the supposed 34th, there can be no doubt but that he must have seen it, and taken it for a fixed star. The magnitude, 6m, which he assigned to 34, agrees perfectly well with the lustre of the planet, compared with other stars which the same author has marked 6m ; and, with his telescope, he could not have the most distant suspicion of its being any other object than a fixed star of about the 6th magnitude.

40 " March 4, 1796. The R.A. in the Atlas requires a correction of about + 20′."

55 In the British catalogue, the P.D. requires – 8′.

56 The R.A. in the British catalogue requires – 15′.

82 Does not exist. FLAMSTEED did not observe this star, unless we admit a correction of the British catalogue – 1° 5′ in P.D.

99 FLAMSTEED has no observation of this star ; but, as there is one in the heavens, about a degree more north, the British catalogue requires probably a correction of – 1° in P.D.

100 This star is lost. FLAMSTEED settled its place, page 369, and the observation seems to be a very good one.†

103 FLAMSTEED has no observation of this star. How it came to be inserted in the British catalogue does not appear. I have given it as a double star V, 114, and here also estimated its brightness ; but it must be remembered that my estimations do not strictly ascertain the place of objects. If, therefore, 103 does not exist, my double star, as well as the one here estimated, must be some star not far from the place assigned to 103 in the British catalogue.‡

112 Is 23 Aurigæ.

118 The Atlas should be corrected – 30′ in R.A.

124 Does not exist ; unless we admit a correction of + 1° 4′ in R.A. of the British catalogue.

138 Does not exist ; but, as there is no time in FLAMSTEED's observation of this star, it is probably misplaced in the British catalogue, for there are several considerable stars in the neighbourhood.§

Notes to Triangulum.

1 " Nov. 2, 1798, This star, which has the time and zenith distance in FLAMSTEED's observations doubtful, seems to be nearly in the place where the British catalogue gives it. It should perhaps be a few minutes more north."

Slough, near Windsor,
Jan. 28, 1799.

* [D.M. + 22°·518, 7 mag., does not seem to be variable.—ED.]

† [It is B. 686 ; see Peters' *Memoir*, p. 71.—ED.]

‡ [103 Tauri = P. IV. 295 ; see Baily's note to 655.—ED.]

§ [See Baily's note to 988.—ED.]

XLII.

On the Power of penetrating into Space by Telescopes ; with a comparative Determination of the Extent of that Power in natural Vision, and in Telescopes of various Sizes and Constructions ; illustrated by select Observations.

[*Phil. Trans.*, 1800, pp. 49–85.]

Read November 21, 1799.

It will not be difficult to shew that the power of penetrating into space by telescopes is very different from magnifying power, and that, in the construction of instruments, these two powers ought to be considered separately.

In order to conduct our present inquiry properly, it will be necessary to examine the nature of luminous bodies, and to enter into the method of vision at a distance. Therefore, to prevent the inaccuracy that would unavoidably arise from the use of terms in their common acceptation, I shall have recourse to algebraic symbols, and to such definitions as may be necessary to fix a precise meaning to some expressions which are often used in conversation, without much regard to accuracy.

By luminous bodies I mean, in the following pages, to denote such as throw out light, whatever may be the cause of it : even those that are opaque, when they are in a situation to reflect light, should be understood to be included ; as objects of vision they must throw out light, and become intitled to be called luminous. However, those that shine by their own light may be called self-luminous, when there is an occasion to distinguish them.

The question will arise, whether luminous bodies scatter light in all directions equally ; but, till we are more intimately acquainted with the powers which emit and reflect light, we shall probably remain ignorant on this head. I should remark, that what I mean to say, relates only to the physical points into which we may conceive the surfaces of luminous bodies to be divided ; for, when we take any given luminous body in its whole construction, such as the sun or the moon, the question will assume another form, as will appear hereafter.

That light, flame, and luminous gases are penetrable to the rays of light, we know from experience ;* it follows therefore, that every part of the sun's disk

* In order to put this to a proof, I placed four candles behind a screen, at ¾ of an inch distance from each other, so that their flames might range exactly in a line. The first of the candles was placed at the same distance from the screen, and just opposite a narrow slit, ⅔ of an inch long, and ¼ broad. On the other side of the screen I fixed up a book, at such a distance from the slit that,

cannot appear equally luminous to an observer in a given situation, on account of the unequal depth of its luminous atmosphere in different places.* This regards only bodies that are self-luminous. But the greatest inequalities in the brightness of luminous bodies in general will undoubtedly be owing to their natural texture; which may be extremely various, with regard to their power of throwing out light more or less copiously.

Brightness I ascribe to bodies that throw out light; and those that throw out most are the brightest.

It will now be necessary to establish certain expressions for brightness in different circumstances.

In the first place, let us suppose a luminous surface throwing out light, and let the whole quantity of light thrown out by it be called L.

Now, since every part of this surface throws out light, let us suppose it divided into a number of luminous physical points, denoted by N.

If the copiousness of the emission of light from every physical point of the luminous surface were equal, it might in general be denoted by c; but, as that is most probably never the case, I make C stand for the mean copiousness of light thrown out from all the physical points of a luminous object. This may be found in the following manner. Let c express the copiousness of emitting light, of any number of physical points that agree in this respect; and let the number of these points be n. Let the copiousness of emission of another number of points be c', and their number n'. And if, in the same manner, other degrees of copiousness be called c^2, c^3, &c. and their numbers be denoted by n^2, n^3, &c. then will the sum of every set of points, multiplied by their respective copiousness of emitting light, give us the quantity of light thrown out by the whole luminous body. That is, $L = cn + c'n' + c^2n^2$, &c.; and the mean copiousness of emitting light, of each physical point, will be expressed by $\frac{cn + c'n' + c^2n^2, \&c.}{N} = C$.

It is evident that the mean power, or copiousness of throwing out light, of every physical point in the luminous surface, multiplied by the number of points, must give us the whole power of throwing out light, of the luminous body. That is $CN = L$.

I ought now to answer an objection that may be made to this theory. Light, as has been stated, is transparent; and, since the light of a point behind the surface of a flame will pass through the surface, ought we not to take in its depth, as well as its superficial dimensions? In answer to this, I recur to what has been said

when the first of the candles was lighted, the letters might not be sufficiently illuminated to become legible. Then, lighting successively the second, third, and fourth candles, I found the letters gradually more illuminated, so that at last I could read them with great facility; and, by the arrangement of the screen and candles, the light of the second, third, and fourth, could not reach the book, without penetrating the flames of those that were placed before them.

* See the Paper on the Nature and Construction of the Sun, *Phil. Trans.* 1795, p. 46. [Vol. I. p. 470.]

with regard to the different powers of throwing out light, of the points of a luminous surface. For, as light must be finally emitted through the surface, it is but referring all light arising from the emission of points behind the surface, to the surface itself, and the account of emitted light will be equally true. And this will also explain why it has been stated as probable, that different parts of the same luminous surface may throw out different quantities of light.

Since, therefore, the quantity of light thrown out by any luminous body is truly represented by CN, and that an object is bright in consequence of light thrown out, we may say that brightness is truly defined by CN. If, however, there should at any time be occasion for distinction, the brightness arising from the great value of C may be called the intrinsic brightness ; and that arising from the great value of N, the aggregate brightness ; but the absolute brightness, in all cases, will still be defined by CN.

Hitherto we have only considered luminous objects, and their condition with regard to throwing out light. We proceed now to find an expression for their appearance at any assigned distance ; and here it will be proper to leave out of the account, every part of CN which is not applied for the purpose of vision. L representing the whole quantity of light thrown out by CN, we shall denote that part of it which is used in vision, either by the eye or by the telescope, l. This will render the conclusions that may be drawn hereafter more unexceptionable ; for, the quantity of light l being scattered over a small space in proportion to L, it may reasonably be looked upon as more uniform in its texture ; and no scruples about its inequality will take place. The equation of light, in this present sense, therefore, is $CN = l$.

Now, since we know that the density of light decreases in the ratio of the squares of the distances of the luminous objects, the expression for its quantity at the distance of the observer D, will be $\frac{l}{D^2}$.

In natural vision, the quantity l undergoes a considerable change, by the opening and contracting of the pupil of the eye. If we call the aperture of the iris a, we find that in different persons it differs considerably. Its changes are not easily to be ascertained ; but we shall not be much out in stating its variations to be chiefly between 1 and 2 tenths of an inch. Perhaps this may be supposed underrated ; for the powers of vision, in a room completely darkened, will exert themselves in a very extraordinary manner. In some experiments on light, made at Bath, in the year 1780, I have often remarked, that after staying some time in a room fitted up for these experiments, where on entering I could not perceive any one object, I was no longer at a loss, in half an hour's time, to find every thing I wanted. It is however probable that the opening of the iris is not the only cause of seeing better after remaining long in the dark ; but that the tranquillity of the retina, which is not disturbed by foreign objects of vision, may render it fit to receive impressions such as otherwise would have been too faint to be perceived. This seems to be supported by telescopic vision; for it has often happened to me, in a fine winter's evening, when, at midnight, and in the absence of the moon, I have taken sweeps of the heavens, of four, five, or six hours duration, that the sensibility of the eye, in consequence of the exclusion of light from surrounding objects, by means of a black hood which I wear upon these occasions, has been very great; and it is evident, that the opening of the iris would have been of no service in these cases, on account of the diameter of the optic pencil, which, in the 20 feet telescope, at the time of sweeping, was no more than ·12 inch. The effect of this increased sensibility was such, that if a star of the 3rd magnitude came towards the field of view, I found it necessary to withdraw the eye before its entrance, in order not to injure the delicacy of vision acquired by long continuance in the dark. The transit of large stars, unless where none of the 6th or 7th magnitude could be had, have generally been declined in my sweeps, even with the 20 feet telescope. And I remember, that after a considerable sweep with the 40 feet instrument, the appearance of Sirius announced itself, at a great distance, like the dawn of the morning, and came on by degrees, increasing in brightness, till this brilliant star at last entered the field of view of the telescope, with all the splendour of the rising sun, and forced me to take the eye from that beautiful sight. Such striking effects are a sufficient proof of the great sensibility of the eye, acquired by keeping it from the light.

On taking notice, in the beginning of sweeps, of the time that passed, I found that the eye, coming from the light, required near 20′, before it could be sufficiently reposed to admit a view of very delicate objects in the telescope; and that the observation of a transit of a star of the 2d or 3d magnitude, would disorder the eye again, so as to require nearly the same time for the re-establishment of its tranquillity.

The difficulty of ascertaining the greatest opening of the eye, arises from the impossibility of measuring it at the time of its extreme dilatation, which can only happen when every thing is completely dark; but, if the variation of a is not easily to be ascertained, we have, on the other hand, no difficulty to determine the quantity of light admitted through a telescope, which must depend upon the diameter of the object-glass, or mirror; for, its aperture A may at all times be had by measurement.

It follows, therefore, that the expression $\frac{a^2 l}{D^2}$ will always be accurate for the quantity of light admitted by the eye; and that $\frac{A^2 l}{D^2}$ will be sufficiently so for the telescope. For it must be remembered, that the aperture of the eye is also concerned in viewing with telescopes; and that, consequently, whenever the pencil of light transmitted to the eye by optical instruments exceeds the aperture of the pupil, much light must be lost. In that case, the expression $A^2 l$ will fail; and therefore, in general, if m be the magnifying power, $\frac{A}{m}$ ought not to exceed a.

As I have defined the brightness of an object to the eye of an observer at a distance, to be expressed by $\frac{a^2 l}{D^2}$, it will be necessary to answer some objections that may be made to this theory. Optical writers have proved, that an object is equally bright at all distances. It may, therefore, be maintained against me, that since a wall illuminated by the sun will appear equally bright, at whatsoever distance the observer be placed that views it, the sun also, at the distance of Saturn, or still farther from us, must be as bright as it is in its present situation. Nay, it may be urged, that in a telescope the different distance of stars can be of no account with regard to their brightness, and that we must consequently be able to see stars which are many thousands of times farther than Sirius from us; in short, that a star must be infinitely distant not to be seen any longer.

Now, objections such as these, which seem to be the immediate consequence of what has been demonstrated by mathematicians, and which yet apparently contradict what I assert in this paper, deserve to be thoroughly answered.

It may be remembered, that I have distinguished brightness into three different sorts.* Two of these, which have been discriminated by *intrinsic* and *absolute* brightness, are, in common language, left without distinction. In order to shew that they are so, I might bring a variety of examples from common conversation; but, taking this for granted, it may be shewn that all the objections I have brought against my theory have their foundation in this ambiguity.

The demonstrations of opticians, with regard to what I call *intrinsic* brightness, will not oppose what I affirm of *absolute* brightness; and I shall have nothing farther to do than to shew that what mathematicians have said, must be understood to refer entirely to the intrinsic brightness, or illumination of the picture of objects on the retina of the eye: from which it will clearly follow, that their doctrine and mine are perfectly reconcileable; and that they can be at variance only when the ambiguity of the word brightness is overlooked, and objections, such as I have made, are raised, where the word brightness is used as *absolute*, when we should have kept it to the only meaning it can bear in the mathematicians' theorem.

The first objection I have mentioned is, that the sun, to an observer on Saturn, must be as bright as it is here on earth. Now by this cannot be meant, that an inhabitant standing on the planet Saturn, and looking at the sun, should *absolutely* receive as much light from it as one on earth receives when he sees it; for this would be contrary to the well known decrease of light at various distances. The objection, therefore can only go to assert, that the picture of the sun, on the retina

* See page [33].

of the Saturnian observer, is as *intensely* illuminated as that on the retina of the terrestrial astronomer. To this I perfectly agree. But let those who would go farther, and say that therefore the sun is *absolutely* as bright to the one as to the other, remember that the sun on Saturn appears to be a hundred times less than on the earth; and that consequently, though it may there be *intrinsically* as bright, it must here be *absolutely** an hundred times brighter.

The next objection I have to consider, relates to the fixed stars. What has been shewn in the preceding paragraph, with regard to the sun, is so intirely applicable to the stars, that it will be very easy to place this point also in its proper light. As I have assented to the demonstration of opticians with regard to the brightness of the sun, when seen at the distance of Saturn, provided the meaning of this word be kept to the *intrinsic* illumination of the picture on the retina of an observer, I can have no hesitation to allow that the same will hold good with a star placed at any assignable distance. But I must repeat, that the light we can receive from stars is truly expressed by $\frac{a^2 l}{D^2}$; and that therefore their absolute brightness must vary in the inverse ratio of the squares of their distances. Hence I am authorised to conclude, and observation abundantly confirms it, that stars cannot be seen by the naked eye, when they are more than seven or eight times farther from us than Sirius; and that they become, comparatively speaking, very soon invisible with our best instruments. It will be shewn hereafter, that the visibility of stars depends on the penetrating power of telescopes, which, I must repeat, falls indeed very short of shewing stars that are many thousands of times farther from us than Sirius; much less can we ever hope to see stars that are all but infinitely distant.

If now it be admitted that the expressions we have laid down are such as agree with well-known facts, we may proceed to vision at a distance; and first with respect to the naked eye.

Here the power of penetrating into space is not only confined by nature, but is moreover occasionally limited by the failure in brightness of luminous objects. Let us see whether astronomical observations, assisted by mathematical reasoning, can give us some idea of the general extent of natural vision. Among the reflecting luminous objects, our penetrating powers are sufficiently ascertained. From the moon we may step to Venus, to Mercury, to Mars, to Jupiter, to Saturn, and last of all to the Georgian planet. An object seen by reflected light at a greater distance than this, it has never been allowed us to perceive; and it is indeed much to be admired, that we should see borrowed illumination to the amazing distance of more than 18 hundred millions of miles; especially when that light, in coming from the sun to the planet, has to pass through an equal space, before it can be reflected,

* See the definition of *absolute* brightness, page [33].

whereby it must be so enfeebled as to be above 368 times less intense on that planet than it is with us, and when probably not more than one-third part of that light can be thrown back from its disk.*

The range of natural vision with self-luminous objects, is incomparably more extended, but less accurately to be ascertained. From our brightest luminary, the sun, we pass immediately to very distant objects; for, Sirius, Arcturus, and the rest of the stars of the first magnitude, are probably those that come next; and what their distance may be, it is well known, can only be calculated imperfectly from the doctrine of parallaxes, which places the nearest of them at least 412530 times farther from us than the sun.

In order to take a second step forwards, we must enter into some preliminary considerations, which cannot but be attended with considerable uncertainty. The general supposition, that stars, at least those which seem to be promiscuously scattered, are probably one with another of a certain magnitude, being admitted, it has already been shewn in a former Paper,† that after a certain number of stars of the first magnitude have been arranged about the sun, a farther distant set will come in for the second place. The situation of these may be taken to be, one with another, at about double the distance of the former from us.

By directing our view to them, and thus penetrating one step farther into space, these stars of the second magnitude furnish us with an experiment that shews what phænomena will take place, when we receive the illumination of two very remote objects, equally bright in themselves, whereof one is at double the distance of the other. The expression for the brightness of such objects, at all distances, and with any aperture of the iris, according to our foregoing notation, will be $\frac{a^2 l}{D^2}$; and a method of reducing this to an experimental investigation will be as follows.

Let us admit that α Cygni, β Tauri, and others, are stars of the second magnitude, such as are here to be considered. We know, that in looking at them and the former, the aperture of the iris will probably undergo no change; since the difference in brightness, between Sirius, Arcturus, α Cygni, and β Tauri, does not seem to affect the eye so as to require any alteration in the dimensions of the iris; a, therefore becomes a given quantity, and may be left out. Admitting also, that the latter of these stars are probably at double the distance of the former, we have D^2 in one case four times that of the other; and the two expressions for the brightness of the stars, will be l for those of the first magnitude, and $\frac{1}{4}l$ for those of the second.

The quantities being thus prepared, what I mean to suggest by an experiment is, that since sensations, by their nature, will not admit of being halved or quar-

* According to Mr. BOUGUER, the surface of the moon absorbs about two-thirds of the light it receives from the sun. See *Traité d'Optique*, p. 122.

† *Phil. Trans.* for the year 1796, p. 166, 167, 168. [Vol. I. p. 530.]

tered, we come thus to know by inspection what phænomenon will be produced by the fourth part of the light of a star of the first magnitude. In this sense, I think we must take it for granted, that a certain idea of brightness, attached to the stars which are generally denominated to be of the second magnitude, may be added to our experimental knowledge; for, by this means, we are informed what we are to understand by the expressions $\frac{a^2 l}{\odot^2}$, $\frac{a^2 l}{\overline{\text{Sirius}}|^2}$, $\frac{a^2 l}{\overline{\beta\ \text{Tauri}}|^2}$.* We cannot wonder at the immense difference between the brightness of the sun and that of Sirius; since the two first expressions, when properly resolved, give us a ratio of brightness of more than 170 thousand millions to one; whereas the two latter, as has been shewn, give only a ratio of four to one.

What has been said will carry us, with very little addition, to the end of our unassisted power of vision to penetrate into space. We can have no other guide to lead us a third step than the same beforementioned hypothesis; in consequence of which, however, it must be acknowledged to be sufficiently probable, that the stars of the third magnitude may be placed about three times as far from us as those of the first. It has been seen, by my remarks on the comparative brightness of the stars, that I place no reliance on the classification of them into magnitudes;† but, in the present instance, where the question is not to ascertain the precise brightness of any one star, it is quite sufficient to know that the number of the stars of the first three different magnitudes, or different brightnesses, answers, in a general way, sufficiently well to a supposed equally distant arrangement of a first, second, and third set of stars about the sun. Our third step forwards into space, may therefore very properly be said to fall on the pole-star, on γ Cygni, ϵ Bootis, and all those of the same order.

As the difference between these and the stars of the preceding order is much less striking than that between the stars of the first and second magnitude, we also find that the expressions $\frac{a^2 l}{\overline{\beta\ \text{Tauri}}|^2}$, and $\frac{a^2 l}{\overline{\text{Polaris}}|^2}$, are not in the high ratio of 4 to 1, but only as 9 to 4, or $2\frac{1}{4}$ to 1.

Without tracing the brightness of the stars through any farther steps, I shall only remark, that the diminution of the ratios of brightness of the stars of the 4th, 5th, 6th, and 7th magnitude, seems to answer to their mathematical expressions, as well as, from the first steps we have taken, can possibly be imagined. The calculated ratio, for instance, of the brightness of a star of the 6th magnitude, to that of one of the 7th, is but little more than $1\frac{1}{3}$ to 1; but still we find by experience, that the eye can very conveniently perceive it. At the same time, the faintness of the stars of the 7th magnitude, which require the finest nights, and the best common eyes to be perceived, gives us little room to believe

* The names of the objects ☉, Sirius, β Tauri, are here used to express their distance from us.

† *Phil. Trans.* for the year 1796, p. 168, 169. [Vol. I. p. 531.]

that we can penetrate much farther into space, with objects of no greater brightness than stars.

But, since it may be justly observed, that in the foregoing estimation of the proportional distance of the stars, a considerable uncertainty must remain, we ought to make a proper allowance for it; and, in order to see to what extent this should go, we must make use of the experimental sensations of the ratios of brightness we have now acquired, in going step by step forward: for, numerical ratios of brightness, and sensations of them, as has been noticed before, are very different things. And since, from the foregoing considerations, it may be concluded, that as far as the 6th, 7th, or 8th magnitude, there ought to be a visible general difference between stars of one order and that of the next following, I think, from the faintness of the stars of the 7th magnitude, we are authorized to conclude, that no star, eight, nine, or at most ten times as far from us as Sirius, can possibly be perceived by the natural eye.

The boundaries of vision, however, are not confined to single stars. Where the light of these falls short, the united lustre of sidereal systems will still be perceived. In clear nights, for instance, we may see a whitish patch in the sword-handle of Perseus,* which contains small stars of various sizes, as may be ascertained by a telescope of a moderate power of penetrating into space. We easily see the united lustre of them, though the light of no one of the single stars could have affected the unassisted eye.

Considerably beyond the distance of the former must be the cluster discovered by Mr. MESSIER, in 1764; north following 1 Geminorum. It contains stars much smaller than those of the former cluster; and a telescope should have a considerable penetrating power, to ascertain their brightness properly, such as my common 10-feet reflector. The night should be clear, in order to see it well with the naked eye, and it will then appear in the shape of a small nebula.†

Still farther from us must be the nebula between η and ζ Herculis, discovered by Dr. HALLEY, in 1714. The stars of it are so small that it has been called a Nebula; ‡ and has been regarded as such, till my instruments of high penetrating powers were applied to it. It requires a very clear night, and the absence of the moon, to see it with the natural eye.

Perhaps, among the farthest objects that can make an impression on the eye, when not assisted by telescopes, may be reckoned the nebula in the girdle of Andromeda, discovered by SIMON MARIUS, in 1612. It is however not difficult to perceive it, in a clear night, on account of its great extent.

From the powers of penetrating into space by natural vision, we proceed now to that of telescopes.

* See the catalogue of a second thousand of new nebulæ and clusters of stars, VI. 33, 34, *Phil. Trans.* Vol. LXXIX. page 251. [Vol. I. p. 361.—ED.]

† [M. 35, N.G.C. 2168.]

‡ In the *Connois. d. T.* for 1783, No. 13, it is described as a nebula without stars. [N.G.C. 6205.]

It has been shewn, that brightness, or light, is to the naked eye truly represented by $\frac{a^2 l}{D^2}$; in a telescope, therefore, the light admitted will be expressed by $\frac{A^2 l}{D^2}$. Hence it would follow, that the artificial power of penetrating into space should be to the natural one as A to a. But this proportion must be corrected by the practical deficiency in light reflected by mirrors, or transmitted through glasses; and it will in a great measure depend on the circumstances of the workmanship, materials, and construction of the telescope, how much loss of light there will be sustained.

In order to come to some determination on this subject, I made many experiments with plain mirrors, polished like my large ones, and of the same composition of metal. The method I pursued was that proposed by Mr. BOUGUER, in his *Traité d'Optique*, page 16, fig. 3; but I brought the mirror, during the trial, as close to the line connecting the two objects as possible, in order to render the reflected rays nearly perpendicular.

The result was, that out of 100 thousand incident rays, 67262 were returned; and therefore, if a double reflection takes place, only 45242 will be returned.

Before this light can reach the eye, it will suffer some loss in passing through the eye glass; and the amount of this I ascertained, by taking a highly polished plain glass, of nearly the usual thickness of optical glasses of small focal lengths. Then, by the method of the same author, page 21, fig. 5, I found, that out of 100 thousand incident rays, 94825 were transmitted through the glass. Hence, if two lenses be used, 89918; and, with three lenses, 85265 rays will be transmitted to the eye.

Then, by compounding, we shall have, in a telescope of my construction with one reflection, 63796 rays, out of 100 thousand, come to the eye. In the NEWTONIAN form, with a single eye lens, 42901; and, with a double eye glass 40681 will remain for vision.

There must always remain a considerable uncertainty in the quantities here assigned; as a newly polished mirror, or one in high preservation, will give more light than another that has not those advantages. The quality of metal also will make some difference; but, if it should appear by experiments, that the metals or glasses in use will yield more or less light than here assigned, it is to be understood that the corrections must be made accordingly.

We proceed now to find a proper expression for the power of penetrating into space, that we may be enabled to compare its effects, in different telescopes, with that of the natural eye.

Since then the brightness of luminous objects is inversely as the squares of the distances, it follows, that the penetrating power must be as the square roots of the light received by the eye.

In natural vision, therefore, this power is truly expressed by $\sqrt{a^2 l}$; and, since we have now also obtained a proper correction x, we must apply it to the incident light with telescopes.

In the NEWTONIAN and other constructions where two specula are used, there will also be some loss of light on account of the interposition of the small speculum; therefore, putting b for its diameter, we have $\overline{A^2 - b^2}$ for the real incident light. This being corrected as above, will give the general expression $\sqrt{xl \times \overline{A^2 - b^2}}$ for the same power in telescopes. But here we are to take notice, that in refractors, and in telescopes with one reflection, b will be $= o$, and therefore is to be left out.

Then, if we put natural light $l = 1$, and divide by a, we have the general form $\frac{\sqrt{x \, . \, \overline{A^2 - b^2}}}{a}$ for the penetrating power of all sorts of telescopes, compared to that of the natural eye as a standard, according to any supposed aperture of the iris, and proportion of light returned by reflection, or transmitted by refraction.

In the following investigation we shall suppose $a = 2$ tenths of an inch, as being perhaps nearly the general opening of the iris, in star-light nights, when the eye has been some moderate time in the dark. The value of the corrections for loss of light will stand as has been given before.

We may now proceed to determine the powers of the instruments that have been used in my astronomical observations; but, as this subject will be best explained by a report of the observations themselves, I shall select a series of them for that purpose, and relate them in the order which will be most illustrating.

First, with regard to the eye, it is certain that its power, like all our other faculties, is limited by nature, and is regulated by the permanent brightness of objects; as has been shewn already, when its extent with reflected light was compared to its exertion on self-luminous objects. It is further limited on borrowed light, by the occasional state of illumination; for, when that becomes defective at any time, the power of the eye will then be contracted into a narrower compass; an instance of which is the following.

In the year 1776, when I had erected a telescope of 20 feet focal length, of the NEWTONIAN construction, one of its effects by trial was, that when towards evening, on account of darkness, the natural eye could not penetrate far into space, the telescope possessed that power sufficiently to shew, by the dial of a distant church steeple, what o'clock it was, notwithstanding the naked eye could no longer see the steeple itself. Here I only speak of the penetrating power; for, though it might require magnifying power to see the figures on the dial, it could require none to see the steeple. Now the aperture of the telescope being 12 inches, and the construction of the NEWTONIAN form, its penetrating power, when calculated according to the given formula, will be $\frac{\sqrt{\cdot 429 \times 120^2 - 15^2}}{2} = 38\cdot 99$. *A*, *b*, and *a*, being all expressed in tenths of an inch.*

From the result of this computation it appears, that the circumstance of seeing so well, in the dusk of the evening, may be easily accounted for, by a power of this telescope to penetrate 39 times farther into space than the natural eye could reach, with objects so faintly illuminated.

This observation completely refutes an objection to telescopic vision, that may be drawn from what has also been demonstrated by optical writers ; namely, that no telescope can shew an object brighter than it is to the naked eye. For, in order to reconcile this optical theory with experience, I have only to say, that the objection is intirely founded on the same ambiguity of the word brightness that has before been detected. It is perfectly true, that the *intrinsic* illumination of the picture on the retina, which is made by a telescope, cannot exceed that of natural vision ; but the *absolute* brightness of the magnified picture by which telescopic vision is performed, must exceed that of the picture in natural vision, in the same ratio in which the area of the magnified picture exceeds that of the natural one ; supposing the *intrinsic* brightness of both pictures to be the same. In our present instance, the steeple and clock-dial were rendered visible by the increased absolute brightness of the object, which in natural vision was 15 hundred times inferior to what it was in the telescope. And this establishes beyond a doubt, that telescopic vision is performed by the absolute brightness of objects ; for, in the present case, I find by computation, that the *intrinsic* brightness, so far from being equal in the telescope to that of natural vision, was inferior to it in the ratio of three to seven.

The distinction between magnifying power, and a power of penetrating into space, could not but be felt long ago, though its theory has not been inquired into. This undoubtedly gave rise to the invention of those very useful short telescopes called night-glasses. When the darkness of the evening curtails the natural penetrating power, they come in very seasonably, to the relief of mariners that are on the look out for objects which it is their interest to discover. Night-glasses, such as they are now generally made, will have a power of penetrating six or seven times farther into space than the natural eye. For, by the construction of the double eye-glass, these telescopes will magnify 7 or 8 times ; and the object glass being $2\frac{1}{2}$ inches in diameter, the breadth of the optic pencil will be $3\frac{1}{8}$ or $3\frac{4}{7}$ tenths of an inch. As this cannot enter the eye, on a supposition of an opening of the iris of 2 tenths, we are obliged to increase the value of *a*, in order to make the telescope have its proper effect. Now, whether nature will admit of such an enlargement becomes an object of experiment ; but, at all events, *a* cannot be assumed less than

* I have given the figures, in all the following equations of the calculated penetrating powers, in order to show the constructions of my instruments to those who may wish to be acquainted with them.

$\frac{A}{m}$. Then, if x be taken as has been determined for three refractions, we shall have $\frac{\sqrt{\cdot853 \times 25^2}}{a} = 6{\cdot}46$ or $7{\cdot}39$.

Soon after the discovery of the Georgian planet, a very celebrated observer of the heavens, who has added considerably to our number of telescopic comets and nebulæ, expressed his wish, in a letter to me, to know by what method I had been led to suspect this object not to be a star, like others of the same appearance. I have no doubt but that the instrument through which this astronomer generally looked out for comets, had a penetrating power much more than sufficient to shew the new planet, since even the natural eye will reach it. But here we have an instance of the great difference in the effect of the two sorts of powers of telescopes; for, on account of the smallness of the planet, a different sort of power, namely, that of magnifying, was required; and, about the time of its discovery, I had been remarkably attentive to an improvement of this power, as I happened to be then much in want of it for my very close double stars.*

On examining the nebulæ which had been discovered by many celebrated authors, and comparing my observations with the account of them in the *Connoissance des Temps* for 1783, I found that most of those which I could not resolve into stars with instruments of a small penetrating power, were easily resolved with telescopes of a higher power of this sort; and, that the effect was not owing to the magnifying power I used upon these occasions, will fully appear from the observations; for, when the closeness of the stars was such as to require a considerable degree of magnifying as well as penetrating power, it always appeared plainly, that the instrument which had the highest penetrating power resolved them best, provided it had as much of the other power as was required for the purpose.

Sept. 20, 1783, I viewed the nebula between FLAMSTEED'S 99th and 105th Piscium, discovered by Mr. MECHAIN, in 1780.†

"It is not visible in the finder of my 7-feet telescope; but that of my 20-feet shews it."

Oct. 28, 1784, I viewed the same object with the 7-feet telescope.

"It is extremely faint. With a magnifying power of 120, it seems to be a collection of very small stars: I see many of them."

At the time of these observations, my 7-feet telescope had only a common finder, with an aperture of the object glass of about $\frac{3}{4}$ of an inch in diameter, and a single eye-lens; therefore its penetrating power was $\frac{\sqrt{\cdot899 \times \overline{7{\cdot}5}|^2}}{2} = 3{\cdot}56$. The

* Magnifying powers of 460, 625, 932, 1159, 1504, 2010, 2398, 3168, 4294, 5489, 6450, 6652, were used upon ϵ Bootis, γ Leonis, α Lyræ, &c. See Cat. of double stars, *Phil. Trans.* Vol. LXXII. page 115, and 147; and Vol. LXXV. page 48. [Vol. I. pp. 60, 80, and 172.—ED.]

† [M. 74 = N.G.C. 628.—ED.]

finder of the 20-feet instrument, being achromatic, had an object glass 1 17 inch in diameter; its penetrating power, therefore, was $\frac{\sqrt{\cdot 85 \times \overline{11 \cdot 7}|^2}}{2} = 4 \cdot 50$.

Now, that one of them shewed the nebula and not the other, can only be ascribed to space-penetrating power, as both instruments were equal in magnifying power, and that so low as not to require an achromatic object glass to render the image sufficiently distinct.

The 7-feet reflector evidently reached the stars of the nebula; but its penetrating and magnifying powers are very considerable, as will be shewn presently.

July 30, 1783, I viewed the nebula south preceding FLAMSTEED'S 24 Aquarii, discovered by Mr. MARALDI, in 1746.*

"In the small *sweeper,* † this nebula appears like a telescopic comet."

Oct. 27, 1794. The same nebula with a 7-feet reflector.

"I can see that it is a cluster of stars, many of them being visible."

If we compare the penetrating power of the two instruments, we find that we have in the first $\frac{\sqrt{\cdot 41 \times \overline{42^2 - 12^2}}}{2} = 12 \cdot 84$; and in the latter $\frac{\sqrt{\cdot 43 \times \overline{63^2 - 12^2}}}{2} = 20 \cdot 25$. However, the magnifying power was partly concerned in this instance; for, in the *sweeper* it was not sufficient to separate the stars properly.

March 4, 1783. With a 7-feet reflector, I viewed the nebula near the 5th Serpentis, discovered by Mr. MESSIER, in 1764. ‡

"It has several stars in it; they are however so small that I can but just perceive some, and suspect others."

May 31, 1783. The same nebula with a 10-feet reflector; penetrating power $\frac{1}{2}\sqrt{\cdot 43 \times \overline{89^2 - 16^2}} = 28 \cdot 67$.

"With a magnifying power of 250, it is all resolved into stars: they are very close, and the appearance is beautiful. With 600, perfectly resolved. There is a considerable star not far from the middle; another not far from one side, but out of the cluster; another pretty bright one; and a great number of small ones."

Here we have a case where the penetrating power of 20 fell short, when 29 resolved the nebula completely. This object requires also great magnifying power to shew the stars of it well; but that power had before been tried, in the 7-feet.

* [M. 2 = N.G.C. 7089.]

† The small sweeper is a NEWTONIAN reflector, of 2 feet focal length; and, with an aperture of 4·2 inches, has only a magnifying power of 24, and a field of view 2° 12′. Its distinctness is so perfect, that it will shew letters at a moderate distance, with a magnifying power of 2000; and its movements are so convenient, that the eye remains at rest while the instrument makes a sweep from the horizon to the zenith. A large one of the same construction has an aperture of 9·2 inches, with a focal length of 5 feet 3 inches. It is also charged low enough for the eye to take in the whole optic pencil; and its penetrating power, with a double eye glass, is $\frac{\sqrt{\cdot 41 \times \overline{92^2 - 21^2}}}{2} = 28 \cdot 57$.

‡ [M. 5 = N.G.C. 5904. Compare below, p. 46.—ED.]

as far as 460, without success, and could only give an indication of its being composed of stars; whereas the lower magnifying power of 250, with a greater penetrating power, in the 10-feet instrument, resolved the whole nebula into stars.

May 3, 1783. I viewed the nebula between η and ρ Ophiuchi, discovered by Mr. MESSIER, in 1764.*

"With a 10-feet reflector, and a magnifying power of 250, I see several stars in it, and make no doubt a higher power, and more light, will resolve it all into stars. This seems to be a good nebula for the purpose of establishing the connection between nebulæ and clusters of stars in general."

June 18, 1784. The same nebula viewed with a large NEWTONIAN 20-feet reflector; penetrating power $\frac{\sqrt{\cdot 43 \times 188^2 - 21^2}}{2} = 61\cdot 18$; and a magnifying power of 157. "A very large and very bright cluster of excessively compressed stars. The stars are but just visible, and are of unequal magnitudes: the large stars are red; and the cluster is a miniature of that near FLAMSTEED's 42d Comæ Berenices. R.A. 17^h 6′ 32″; P.D. 108° 18′."

Here, a penetrating power of 29, with a magnifying power of 250, would barely shew a few stars; when, in the other instrument, a power 61 of the first sort, and only 157 of the latter, shewed them completely well.

July 4, 1783. I viewed the nebula between FLAMSTEED's 25 and 26 Sagittarii, discovered by ABRAHAM IHLE, in 1665.†

"With a small 20-feet NEWTONIAN telescope, power 200, it is all resolved into stars, that are very small and close. There must be some hundreds of them. With 350, I see the stars very plainly; but the nebula is too low in this latitude for such a power."

July 12, 1784. I viewed the same nebula with a large 20-feet NEWTONIAN reflector; power 157. "A most beautiful extensive cluster of stars, of various magnitudes, very compressed in the middle, and about 8′ in diameter, besides the scattered ones, which do more than fill the extent of the field of view: ‡ the large stars are red; the small ones are pale red. R.A. 18^h 23′ 39″; P.D. 114° 7′."

The penetrating power of the first instrument was 39, that of the latter 61; but, from the observations, it is plain how much superior the effect of the latter was to that of the former, notwithstanding the magnifying power was so much in favour of the instrument with the small penetrating power.

July 30, 1783. With a small 20-feet NEWTONIAN reflector, I viewed the nebula in the hand of Serpentarius, discovered by Mr. MESSIER, in 1764. §

"With a power of 200, I see it consists of stars. They are better visible with

* [M. 9 = N.G.C. 6333.—ED.]
† [M. 22 = N.G.C. 6656.—ED.]
‡ This field, by the passage of an equatorial star, was 15′ 3″.
§ [M. 14 = N.G.C. 6402.—ED.]

300. With 600, they are too obscure to be distinguished, though the appearance of stars is still preserved. This seems to be one of the most difficult objects to be resolved. With me, there is not a doubt remaining ; but another person, in order to form a judgment, ought previously to go through all the several gradations of nebulæ which I have resolved into stars."

May 25, 1791. I viewed the same nebula with a 20-feet reflector of my construction, having a penetrating power of $\frac{\sqrt{\cdot 64 \times \overline{188}|^2}}{2} = 75\cdot 08$.

" With a magnifying power of 157, it appears extremely bright, round, and easily resolvable. With 300, I can see the stars. It resembles the cluster of stars taken at 16^h 43′ 40″,* which probably would put on the same appearance as this, if it were at a distance half as far again as it is. R.A. 17^h 26′ 19″ ; P.D. 93° 10′."

Here we may compare two observations ; one taken with the penetrating power of 39, the other with 75 ; and, although the former instrument had far the advantage in magnifying power, the latter certainly gave a more complete view of the object.

The 20-feet reflector having been changed from the NEWTONIAN form to my present one, I had a very striking instance of the great advantage of the increased penetrating power, in the discovery of the Georgian satellites. The improvement, by laying aside the small mirror, was from 61 to 75 ; and, whereas the former was not sufficient to reach these faint objects, the latter shewed them perfectly well.

March 14, 1798. I viewed the Georgian planet with a new 25-feet reflector. Its penetrating power is $\frac{\sqrt{\cdot 64 \times \overline{240}|^2}}{2} = 95\cdot 85$; and, having just before also viewed it with my 20-feet instrument, I found, that with an equal magnifying power of 300, the 25-feet telescope had considerably the advantage of the former.

Feb. 24, 1786. I viewed the nebula near FLAMSTEED'S 5th Serpentis, which has been mentioned before, with my 20-feet reflector ; magnifying power 157.

" The most beautiful extremely compressed cluster of small stars ; the greatest part of them gathered together into one brilliant nucleus, evidently consisting of stars, surrounded with many detached gathering stars of the same size and colour. R.A. 15^h 7′ 12″ ; P.D. 87° 8′."

May 27, 1791. I viewed the same object with my 40-feet telescope ; penetrating power $\frac{\sqrt{\cdot 64 \times \overline{480}|^2}}{2} = 191\cdot 69$; magnifying power 370.

" A beautiful cluster of stars. I counted about 200 of them. The middle of it is so compressed that it is impossible to distinguish the stars."

Here it appears, that the superior penetrating power of the 40-feet telescope

* The object referred to is No. 10 of the *Connoissance des Temps* for 1783, called "*Nébuleuse sans étoiles.*" My description of it is, " A very beautiful, and extremely compressed, cluster of stars ; the most compressed part about 3 or 4′ in diameter. R.A. 16^h 46′ 2″ ; P.D. 93° 46′." [N.G.C. 6254.—ED.]

enabled me even to count the stars of this nebula. It is also to be noticed, that the object did not strike me as uncommonly beautiful ; because, with much more than double the penetrating, and also more than double the magnifying power, the stars could not appear so compressed and small as in the 20-feet instrument : this, very naturally, must give it more the resemblance of a coarser cluster of stars, such as I had been in the habit of seeing frequently.

The 40-feet telescope was originally intended to have been of the NEWTONIAN construction ; but, in the year 1787, when I was experimentally assured of the vast importance of a power to penetrate into space, I laid aside the work of the small mirror, which was then in hand, and completed the instrument in its present form.

" Oct. 10, 1791. I saw the 4th satellite and the ring of Saturn, in the 40-feet speculum, without an eye glass."

The magnifying power on that occasion could not exceed 60 or 70 ; but the great penetrating power made full amends for the lowness of the former ; notwithstanding the greatest part of it must have been lost for want of a greater opening of the iris, which could not take in the whole pencil of rays, for this could not be less than 7 or 8 tenths of an inch.

Among other instances of the superior effects of penetration into space, I should mention the discovery of an additional 6th satellite of Saturn, on the 28th of August, 1789 ; and of a 7th, on the 11th of September, in the same year ; which were first pointed out by this instrument. It is true that both satellites are within the reach of the 20-feet telescope ; but it should be remembered, that when an object is once discovered by a superior power, an inferior one will suffice to see it afterwards. I need not add, that neither the 7 nor 10-feet telescopes will reach them ; their powers, 20 and 29, are not sufficient to penetrate to such distant objects, when the brightness of them is not more than that of these satellites. It is also evident, that the failure in these latter instruments arises not from want of magnifying power, as either of them has much more than sufficient for the purpose.

Nov. 5, 1791. I viewed Saturn with the 20 and 40-feet telescopes.

" 20-feet. The 5th satellite of Saturn is very small. The 1st, 2d, 3d, 4th, 5th, and the new 6th satellite, are in their calculated places."

" 40-feet. I see the new 6th satellite much better with this instrument than with the 20-feet. The 5th is also much larger here than in the 20-feet ; in which it was nearly the same size as a small fixed star, but here it is considerably larger than that star."

Here the superior penetrating power of the 40-feet telescope shewed itself on the 6th satellite of Saturn, which is a very faint object ; as it had also a considerable advantage in magnifying power, the disk of the 5th satellite appeared larger than in the 20-feet. But the small star, which may be said to be beyond the reach of magnifying power, could only profit by the superiority of the other power.

Nov. 21, 1791. 40-feet reflector ; power 370.

"The black division upon the ring is as dark as the heavens about Saturn, and of the same colour."

"The shadow of the body of Saturn is visible upon the ring, on the following side; its colour is very different from that of the dark division. The 5th satellite is less than the 3d; it is even less than the 2d."

20-feet reflector; power 300.

"The 3d satellite seems to be smaller than it was the last night but one. The 4th satellite seems to be larger than it was the 19th. This telescope shews the satellites not nearly so well as the 40-feet."

Here, the magnifying power being nearly alike, the superiority of the 40-feet telescope must be ascribed to its penetrating power.

The different nature of the two powers above mentioned being thus evidently established, I must now remark, that, in some respects, they even interfere with each other; a few instances of which I shall give.

August 24, 1783. I viewed the nebula north preceding FLAMSTEED'S 1 Trianguli, discovered by Mr. MESSIER, in 1764.*

"7-feet reflector; power 57. There is a suspicion that the nebula consists of exceedingly small stars. With this low power it has a nebulous appearance; and it vanishes when I put on the higher magnifying powers of 278 and 460."

Oct. 28, 1794. I viewed the same nebula with a 7-feet reflector.

"It is large, but very faint. With 120, it seems to be composed of stars, and I think I see several of them; but it will bear no magnifying power."

In this experiment, magnifying power was evidently injurious to penetrating power. I do not account for this upon the principle that by magnifying we make an object less bright; for, when opticians have also demonstrated that brightness is diminished by magnifying, it must again be understood as relating only to the *intrinsic* brightness of the magnified picture; its absolute brightness, which is the only one that concerns us at present, must always remain the same.† The real explanation of the fact, I take to be, that while the light collected is employed in magnifying the object, it cannot be exerted in giving penetrating power.

* [M. 33 = N.G.C. 598.—ED.]

† This may be proved thus. The mean intrinsic brightness, or rather illumination, of a point of the picture on the retina, will be *all the light that falls on the picture, divided by the number of its points*; or $C = \frac{l}{N}$. Now, since with a greater magnifying power m, the number of points N increases as the squares of the power, the expression for the intrinsic brightness $\frac{l}{N}$, will decrease in the same ratio; and it will consequently be in general $N \propto m^2$, and $\frac{l}{N}$ or $C \propto \frac{1}{m^2}$; that is, by compounding $CN \propto \frac{m^2}{m^2} = l = 1$; or absolute brightness a given quantity. M. BOUGUER has carefully distinguished intrinsic and absolute brightness, when he speaks of the quantity of light reflected from a wall, at different distances. *Traité d'Optique*, page 39 and 40.

June 18, 1799. I viewed the planet Venus with a 10-feet reflector.

" Its light is so vivid that it does not require, nor will it bear, a penetrating power of 29, neither with a low nor with a high magnifying power."

This is not owing to the least imperfection in the mirror, which is truly parabolical, and shews, with all its aperture open, and a magnifying power of 600, the double star γ Leonis in the greatest perfection.

" It shewed Venus, perfectly well defined, with a penetrating power as low as 14, and a magnifying power of 400, or 600 "

Here, penetrating power was injurious to magnifying power ; and that it necessarily must be so, when carried to a high pitch, is evident ; for, by enlarging the aperture of the telescope, we increase the evil that attends magnifying, which is, that we cannot magnify the object without magnifying the medium. Now, since the air is very seldom of so homogeneous a disposition as to admit to be magnified highly, it follows that we must meet with impurities and obstructions, in proportion to its quantity. But the contents of the columns of air through which we look at the heavens by telescopes, being of equal lengths, must be as their bases, that is, as the squares of the apertures of the telescopes ; and this is in a much higher ratio than that of the increase of the power of penetrating into space. From my long experience in these matters, I am led to apprehend, that the highest power of magnifying may possibly not exceed the reach of a 20 or 25-feet telescope ; or may even lie in a less compass than either. However, in beautiful nights, when the outside of our telescopes is dropping with moisture discharged from the atmosphere, there are now and then favourable hours, in which it is hardly possible to put a limit to magnifying power. But such valuable opportunities are extremely scarce ; and, with large instruments, it will always be lost labour to observe at other times.

As I have hinted at the natural limits of magnifying power, I shall venture also to extend my surmises to those of penetrating power. There seems to be room for a considerable increase in this branch of the telescope ; and, as the penetrating power of my 40-feet reflector already goes to 191·69, there can hardly be any doubt but that it might be carried to 500, and probably not much farther. The natural limit seems to be an equation between the faintest star that can be made visible, by any means, and the united brilliancy of star-light. For, as the light of the heavens, in clear nights, is already very considerable in my large telescope, it must in the end be so increased, by enlarging the penetrating power, as to become a balance to the light of all objects that are so remote as not to exceed in brightness the general light of the heavens Now, if P be put for penetrating power, we have $\sqrt{\frac{P^2 a^2}{x}} = A = 10$ feet 5 2 inches for the aperture of a reflector, on my construction, that would have such a power of 500.

But, to return to our subject ; from what has been said before, we may conclude, that objects are viewed in their greatest perfection, when, in penetrating space, the magnifying power is so low as only to be sufficient to shew the object well; and when, in magnifying objects, by way of examining them minutely, the space-penetrating power is no higher than what will suffice for the purpose; for, in the use of either power, the injudicious overcharge of the other will prove hurtful to perfect vision.

It is remarkable that, from very different principles, I have formerly determined the length of the visual ray of my 20-feet telescope upon the stars of the milky way, so as to agree nearly with the calculations that have been given.* The extent of what I then figuratively called my sounding line, and what now appears to answer to the power of penetrating into space, was shewn to be not less than 415, 461, and 497 times the distance of Sirius from the sun. We now have calculated that my telescope, in the NEWTONIAN form, at the time when the paper on the Construction of the Heavens was written, possessed a power of penetration, which exceeded that of natural vision 61 18 times; and, as we have also shewn, that stars at 8, 9, or at most 10 times the distance of Sirius, must become invisible to the eye, we may safely conclude, that no single star, above 489·551, or at most 612 times as far as Sirius, can any longer be seen in this telescope. Now, the greatest length of the former visual ray, 497, agrees nearly with the lowest of these present numbers, 489; and the higher ones are all in favour of the former computation; for that ray, though taken from what was perhaps not far from its greatest extent, might possibly have reached to some distance beyond the apparent bounds of the milky way; but, if there had been any considerable difference in these determinations, we should remember that some of the data by which I have now calculated are only assumed. For instance, if the opening of the iris, when we look at a star of the 7th magnitude, should be only one-tenth of an inch and a half, instead of two, then a, in our formula, will be $= 1 \cdot 5$; which, when resolved, will give a penetrating power of 81·58; and therefore, on this supposition, our telescope would easily have shewn stars 571 times as far from us as Sirius; and only those at 653, 734, or 816 times the same distance, would have been beyond its reach. My reason for fixing upon two-tenths, rather than a lower quantity, was, that I might not run a risk of over-rating the powers of my instruments. I have it however in contemplation, to determine this quantity experimentally, and perceive already, that the difficulties which attend this subject may be overcome.

It now only remains to shew, how far the penetrating power, 192, of my large reflector, will really reach into space. Then, since this number has been calculated to be in proportion to the standard of natural vision, it follows, that if we admit a star of the 7th magnitude to be visible to the unassisted eye, this telescope will shew stars of the one thousand three hundred and forty-second magnitude.

* *Phil. Trans.* Vol. LXXV. p. 247, 248. [Vol. I. pp. 247 248.]

But, as we did not stop at the single stars above mentioned, when the penetration of the natural eye was to be ascertained, so we must now also call the united lustre of sidereal systems to our aid in stretching forwards into space. Suppose therefore, a cluster of 50,000 stars to be at one of those immense distances to which only a 40-feet reflector can reach, and our formula will give us the means of calculating what that may be. For, putting S for the number of stars in the cluster, and D for its distance, we have $\frac{\sqrt{x}\, A^2\, S}{a} = D$;* which, on computation, comes out to be above $11\frac{3}{4}$ millions of millions of millions of miles! A number which exceeds the distance of the nearest fixed star, at least three hundred thousand times.

From the above considerations it follows, that the range for observing, with a telescope such as my 40-feet reflector, is indeed very extensive. We have the inside of a sphere to examine, the radius of which is the immense distance just now assigned to be within the reach of the penetration of our instruments, and of which all the celestial objects visible to the eye, put together, form as it were but the kernel, while all the immensity of its thick shell is reserved for the telescope.

It follows, in the next place, that much time must be required for going through so extensive a range. The method of examining the heavens, by sweeping over space, instead of looking merely at places that are known to contain objects, is the only one that can be useful for discoveries.

In order therefore to calculate how long a time it must take to sweep the heavens, as far as they are within the reach of my 40-feet telescope, charged with a magnifying power of 1000, I have had recourse to my journals, to find how many favourable hours we may annually hope for in this climate. It is to be noticed, that the nights must be very clear; the moon absent; no twilight; no haziness; no violent wind; and no sudden change of temperature; then also, short intervals for filling up broken sweeps will occasion delays; and, under all these circumstances, it appears that a year which will afford 90, or at most 100 hours, is to be called very productive.

In the equator, with my 20-feet telescope, I have swept over zones of two degrees, with a power of 157; but, an allowance of 10 minutes in polar distance must be made, for lapping the sweeps over one another where they join.

As the breadth of the zones may be increased towards the poles, the northern hemisphere may be swept in about 40 zones: to these we must add 19 southern zones; then, 59 zones, which, on account of the sweeps lapping over one another about 5′ of time in right ascension, we must reckon of 25 hours each, will give 1475 hours. And, allowing 100 hours per year, we find that, with the 20-feet telescope, the heavens may be swept in about 14 years and $\frac{3}{4}$.

* $D = 11765475948678678679$ miles.

Now, the time of sweeping with different magnifying powers will be as the squares of the powers; and, putting p and t for the power and time in the 20-feet telescope, and $P = 1000$ for the power in the 40, we shall have $p^2 : t :: P^2 : \frac{t P^2}{p^2} =$ 59840. Then, making the same allowance of 100 hours per year, it appears that it will require not less than 598 years, to look with the 40-feet reflector, charged with the abovementioned power, only one single moment into each part of space; and, even then, so much of the southern hemisphere will remain unexplored, as will take up 213 years more to examine.

Slough, near Windsor,
June 20, 1799.

XLIII.

Investigation of the Powers of the prismatic Colours to heat and illuminate Objects; with Remarks, that prove the different Refrangibility of radiant Heat. To which is added, an Inquiry into the Method of viewing the Sun advantageously, with Telescopes of large Apertures and high magnifying Powers.

[*Phil. Trans.*, 1800, pp. 255–283.]

Read March 27, 1800.

It is sometimes of great use in natural philosophy, to doubt of things that are commonly taken for granted; especially as the means of resolving any doubt, when once it is entertained, are often within our reach. We may therefore say, that any experiment which leads us to investigate the truth of what was before admitted upon trust, may become of great utility to natural knowledge. Thus, for instance, when we see the effect of the condensation of the sun's rays in the focus of a burning lens, it seems to be natural to suppose, that every one of the united rays contributes its proportional share to the intensity of the heat which is produced; and we should probably think it highly absurd, if it were asserted that many of them had but little concern in the combustion, or vitrification, which follows, when an object is put into that focus. It will therefore not be amiss to mention what gave rise to a surmise, that the power of heating and illuminating objects might not be equally distributed among the variously coloured rays.

In a variety of experiments I have occasionally made, relating to the method of viewing the sun, with large telescopes, to the best advantage, I used various combinations of differently-coloured darkening glasses. What appeared remarkable was, that when I used some of them, I felt a sensation of heat, though I had but little light; while others gave me much light, with scarce any sensation of heat. Now, as in these different combinations the sun's image was also differently coloured, it occurred to me, that the prismatic rays might have the power of heating bodies very unequally distributed among them; and, as I judged it right in this respect to entertain a doubt, it appeared equally proper to admit the same with regard to light. If certain colours should be more apt to occasion heat, others might, on the contrary, be more fit for vision, by possessing a superior illuminating power. At all events, it would be proper to recur to experiments for a decision.

Experiments on the heating Power of coloured Rays.

I fixed a piece of pasteboard, AB [fig. 1], in a frame, mounted upon a stand, CD, and moveable upon two centres. In the pasteboard, I cut an opening, *mn*, a little larger than the ball of a thermometer, and of a sufficient length to let the whole extent of one of the prismatic colours pass through. I then placed three thermometers upon small inclined planes, EF: their balls were blacked with

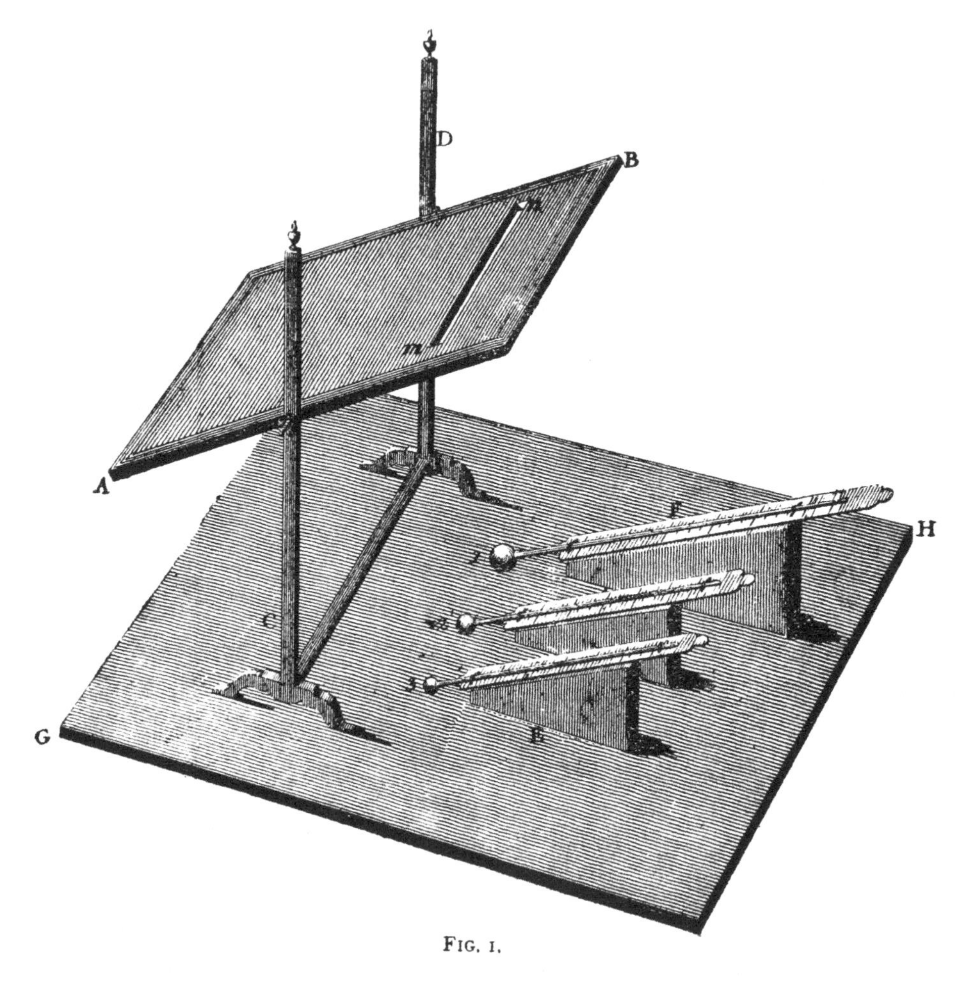

FIG. 1.

japan ink. That of No. 1 was rather too large for great sensibility. No. 2 and 3 were two excellent thermometers, which my highly esteemed friend Dr. WILSON, late Professor of Astronomy at Glasgow, had lent me for the purpose: their balls being very small, made them of exquisite sensibility. The scales of all were properly disengaged from the balls.

I now placed the stand, with the framed pasteboard and the thermometers, upon a small plain board, GH; that I might be at liberty to move the whole apparatus together, without deranging the relative situation of the different parts.

This being done, I set a prism, moveable on its axis, into the upper part of an open window, at right angles to the solar ray, and turned it about till its refracted coloured spectrum became stationary, upon a table placed at a proper distance from the window.

The board containing the apparatus was now put on the table, and set in such a manner as to let the rays of one colour pass through the opening in the pasteboard. The moveable frame was then adjusted to be perpendicular to the rays coming from the prism ; and the inclined planes carrying the three thermometers, with their balls arranged in a line, were set so near the opening, that any one of them might easily be advanced far enough to receive the irradiation of the colour which passed through the opening, while the rest remained close by, under the shade of the pasteboard.

By repeated trials, I found that Dr. WILSON'S No. 2 and mine always agreed in shewing the temperature of the place where I examined them, when the change was not very sudden ; but that mine would require ten minutes to take a change, which the other would shew in five. No. 3 never differed much from No. 2.

1st *Experiment.* Having arranged the three thermometers in the place prepared for the experiment, I waited till they were stationary. Then, advancing No. 1 to the red rays, and leaving the other two close by, in the shade, I marked down what they shewed, at different times.

No. 1 . . .	43½	48	49½	49¾	50
No. 2 . . .	43½	43½	43¼	43¼	43¼
No. 3 . . .	43¼	43½	43¼	43¼	43¼

This, in about 8 or 10 minutes, gave 6¾ degrees, for the rising produced in my thermometer, by the red rays, compared to the two standard thermometers.

2d *Experiment.* As soon as my thermometer was restored to the temperature of the room, which I hastened, by applying it to a large piece of metal that had been kept in the same place, I exposed it again to the red rays, and registered its march, along with No. 2 as a standard, which was as follows.

No. 1 . . .	45	48	51	51	51
No. 2 . . .	45	45	45	44½	44

Hence, in 10 minutes, the red rays made the thermometer rise 7 degrees.

3d *Experiment.* Proceeding in the same manner as before, in the green rays I had,

No. 1 . . .	43	45½	46	46	46
No. 2 . . .	43	43	43	42¾	42¾

Therefore, in ten minutes, the green rays occasioned a rise of 3¼ degrees.

4th *Experiment.* I now exposed my thermometer to the violet rays, and compared it with No. 2.

No. 1 . . .	44	44	44¾	45
No. 2 . . .	44	44	43½	43

Here we have a rising of 2 degrees, in ten minutes, for the violet rays.

5th Experiment. I now exposed Dr. WILSON'S thermometer No. 2 to the red rays, and compared its progress with No. 3.

No. 2 . . .	44	46	46½	46½
No. 3 . . .	44	44	43¾	43¾

Here the thermometer, exposed to red, rose in five minutes 2¾ degrees.

6th Experiment. In red rays again.

No. 2 . . .	44	46	46½	47	47
No. 3 . . .	44	44	43½	43½	43

And here the thermometer, exposed to red, rose in five minutes 4 degrees.

7th Experiment. In green rays.

No. 2 . . .	43½	44½	44½
No. 3 . . .	43½	43½	43

This made the thermometer rise, in the green rays, 1½ degree.

8th Experiment. Again in green rays.

No. 2 . . .	43	44½	44¾
No. 3 . . .	43	42¾	42¾

Here the rising, by the green rays, was 2 degrees.

From these experiments, we are authorised to draw the following results. In the red rays, my thermometer gave 6¾ degrees in the 1st, and 7 degrees in the 2d, for the rising of the quicksilver: a mean of both is 6⅞. In the 3d experiment, we had 3¼ degrees, for the rising occasioned by the green rays; from which we obtain the proportion of 55 to 26, for the power of heating in red to that in green. The 4th experiment gave 2 degrees for the violet rays; and therefore we have the rising of the quicksilver in red to that in violet, as 55 to 16.

A sufficient proof of the accuracy of this determination we have, in the result of the four last experiments. The rising for red rays in the 5th, is 2¾; and in the 6th, 4 degrees: a mean of both is 3⅜. In the 7th experiment, we have 1½, and in the 8th, 2 degrees, for the rising in green: a mean of these is 1¾. Therefore, we have the proportion of the rising in red to that in green, as 27 to 11, or as 55 to 22·4.

We may take a mean of the result of both thermometers, which will be 55 to 24·2, or more than 2¼ to 1, in red to green; and about 3½ to 1, in red to violet.

It appears remarkable, that the most sensible thermometer should give the least alteration, from the exposure to the coloured rays. But since, in these circumstances, there are two causes constantly acting different ways; the one to raise the thermometer, the other to bring it down to the temperature of the room, I suppose, that on account of the smallness of the ball in Dr. WILSON'S No. 2, which is but little more than ⅛ of an inch, the cooling causes must have a stronger effect on the mercury it contains than they can have on mine, the ball of which is half an inch.

More accuracy may hereafter be obtained, by attending to the circumstances of blacking the balls of the thermometers, and their exposure to a more steady and powerful light of the sun, at greater altitudes than it can be had at present; but the experiments which have been related are quite sufficient for my present purpose; which only goes to prove, that the heating power of the prismatic colours is very far from being equally divided, and that the red rays are chiefly eminent in that respect.

Experiments on the illuminating Power of coloured Rays.

In the following examination of the illuminating power of differently-coloured rays, I had two ends in view. The first was, with regard to the illumination itself; and the next, with respect to the aptness of the rays for giving distinct vision; and, though there did not seem to be any particular reason why these two should not go together, I judged it right to attend to both.

The microscope offered itself as the most convenient instrument for this investigation; and I thought it expedient to view only opaque objects, as these would give me an opportunity to use a direct prismatic ray, without running the risk of any bias that might be given to it, in its transmission through the colouring particles of transparent objects.

1st Experiment. I placed an object that had very minute parts, under a double microscope; and, having set a prism in the window, so as to make the coloured image of the sun stationary upon the table where the microscope was placed, I caused the differently-coloured rays to fall successively on the object, by advancing the microscope into their light. The magnifying power was 27 times.

In changing the illumination, by admitting a different colour, it always becomes necessary to re-adjust the instrument. It is well known, that the different refrangibility of the rays will sensibly affect the focal length of object-glasses; but, in compound vision, such as in a microscope, where a very small lens is made to cast a lengthened secondary focus, this difference becomes still more considerable.

By an attentive and repeated inspection, I found that my object was very well seen in red; better in orange, and still better in yellow; full as well in green; but to less advantage in blue; indifferently well in indigo, and with more imperfection in violet.

This trial was made upon one of the microscopic objects which are generally prepared for transparent vision; but, as I used it in the opaque way, I thought that others might be chosen which would answer the purpose better; and, in order to give some variety to my experiments, and to see the effect differently coloured substances might have on the rays of light, I provided the following materials to be viewed. Red paper; green paper; a piece of brass; a nail; a guinea; black paper. Having also found that a higher power might be used, with sufficient convenience for the rays of light to come from the prism to the object, I made the microscope magnify 42 times.

The appearance of the nail in the microscope, is so beautiful, that it deserves to be noticed ; and the more so, as it is accompanied with circumstances that are very favourable for an investigation, such as that which is under our present consideration. I had chosen it on account of its solidity and blackness, as being most likely to give an impartial result, of the modifications arising from an illumination by differently-coloured rays ; but, on viewing it, I was struck with the sight of a bright constellation of thousands of luminous points, scattered over its whole extent, as far as the field of the microscope could take it in. Their light was that of the illuminating colour, but differed considerably in brightness ; some of the points being dim and faint, while others were luminous and brilliant. The brightest of them also admitted of a little variation in their colour, or rather in the intensity of the same colour ; for, in the centre of some of the most brilliant of these lucid appearances, their light had more vivacity and seemed to deviate from the illuminating tint towards whiteness, while on and near the circumference it appeared to take a deeper hue.

An object so well divided by nature, into very minute and differently-arranged points, on which the attention might be fixed, in order to ascertain whether they would be equally distinct in all colours, and whether their number would be increased or diminished by different degrees of illumination, was exactly what I wanted ; nor could I think it less remarkable, that all the other objects I had fixed upon, besides many more which have been examined, such as copper, tin, silver, *&c.*, presented themselves nearly with the same appearance. In the brass, which had been turned in a lathe, the luminous points were arranged in furrows ; and in tin they were remarkably beautiful. The result of the examination of my objects was as follows.

2d Experiment. Red paper.

In the red rays, I view a bright point near an accidental black spot in the paper, which serves me as a mark ; and I notice the space between the point and the spot : it contains several faint points.

In the orange rays, I see better. The bright point, I now perceive, is double.

In the yellow rays, I see the object still better.

In the green rays, full as well as before.

In the blue rays, very well.

In the indigo rays, not quite so well as in the blue.

In the violet rays, very imperfectly.

3d Experiment. Green paper.

Red. I fix my attention on many faint points, in a space between two bright double points.

Orange. I see those faint points better.

Yellow. Still better.

Green. As well as before. I see remarkably well.

Blue. Less bright, but very distinct.

Indigo. Not well.

Violet. Bad.

4th Experiment. A piece of very clean turned brass.

R. I remark several faint luminous points between two bright ones. The colour of the brass makes the red rays appear like orange.

O. I see better, but the orange colour is likewise different from what it ought to be; however, this is not, at present, the object of my investigation.

Y. I see still better.

G. I see full as well as before.

B. I do not see so well now.

I. I cannot see well.

V. Bad.

5th Experiment. A nail.

R. I remark two bright points, and some faint ones.

O. Brighter than before; and more points visible. Very distinct.

Y. Much brighter than before; and more points and lines visible. Very distinct.

G. Full as bright; and as many points visible. Very distinct.

B. Much less bright. Very distinct.

I. Still less bright. Very distinct.

V. Much less bright again. Very distinct.

6th Experiment. I viewed a guinea, at 9 feet 6 inches from the prism; and adjusted the place of the object in the several rays, by the shadow of the guinea. If this be not done, deceptions will take place.

R. Four remarkable points. Very distinct.

O. Better illuminated. Very distinct.

Y. Still better illuminated. Very distinct. The points all over the field of view are coloured; some green; some red; some yellow; and some white, encircled with black about them.

Between yellow and green is the maximum of illumination. Extremely distinct.

G. As well illuminated as the yellow. Very distinct.

B. Much inferior in illumination. Very distinct.

I. Badly illuminated. Distinct.

V. Very badly illuminated I can hardly see the object at all.

7th Experiment. The nail again, at 8 feet from the prism.

R. I attended to two bright points, with faint ones between them. Almost all the points in the field of view are red. Very distinct.

O. I see all the points better: they are red, green, yellow, and whitish, with black about them. Very distinct.

Y. I see better. More bright points, and more faint ones: the points are of various colours. Very distinct.

G. I see as well. The points are mostly green, and brightish-green, inclining to white. Very distinct.

B. Much worse illuminated. Very distinct.

I. Badly illuminated. Very distinct.

V. There is hardly any illumination.

8th Experiment. The nail again, at 9 feet 6 inches from the prism, by way of having the rays better separated.

R. Badly illuminated. The bright points are very distinct.

O. Much better illuminated. The bright points very distinct.

Y. Still better illuminated. All points extremely distinct.

G. As well illuminated, and equally distinct.

B. Badly illuminated. The bright points are distinct; but the others are not so.

I. Very badly illuminated. I do not see distinctly; but I believe it to be for want of light.

V. So badly illuminated that I cannot see the object; or at least but barely perceive that it exists.

9th Experiment. Black paper, at 8 feet from the prism.

R. The object is hardly visible. I can only see a few faint points.

O. I see several bright points, and many faint ones.

Y. Numberless bright and small faint points.

Between yellow and green, is the maximum of illumination.

G. The same as the yellow.

B. Very indifferently illuminated; but not so bad as in the red rays.

I. I cannot see the object.

V. Totally invisible.

From these observations, which agree uncommonly well, with respect to the illuminating power assigned to each colour, we may conclude, that the red-making rays are very far from having it in any eminent degree. The orange possess more of it than the red; and the yellow rays illuminate objects still more perfectly. The maximum of illumination lies in the brightest yellow, or palest green. The green itself is nearly equally bright with the yellow; but, from the full deep green, the illuminating power decreases very sensibly. That of the blue is nearly upon a par with that of the red; the indigo has much less than the blue; and the violet is very deficient.

With regard to the principle of distinctness, there appears to be no deficiency in any one of the colours. In the violet rays, for instance, some of the experiments mention that I saw badly; but this is to be understood only with respect to the number of small objects that could be perceived; for, although I saw fewer of the points, those which remained visible were always as distinct as, in so feeble an illumination, could be expected. It must indeed be evident, that by removing the great obstacle to distinct vision, which is the different refrangibility of the rays of light, a microscope will be capable of a much higher degree of distinctness than it can be under the usual circumstances. A celebrated optical writer has formerly remarked, that a fly, illuminated by red rays, appeared uncommonly distinct, and that all its minute parts might be seen in great perfection; and, from the experiments which have been related, it appears that every other colour is possessed of the same advantage.

I am well aware that the results I have drawn from the foregoing experiments, both with regard to the heating and illuminating powers of differently-coloured rays, must be affected by some little inaccuracies. The prism, under the circumstances in which I have used it, could not effect a complete separation of the colours, on account of the apparent diameter of the sun, and the considerable breadth of the prism itself, through which the rays were transmitted.

Perhaps an arrangement like that in Fig. 16, of the NEWTONIAN experiments, might be employed; if instruments of sufficient sensibility, such as air thermometers, can be procured, that may be affected by the enfeebled illumination of rays that have undergone four transmissions, and eight refractions; and especially when their incipient quantity has been so greatly reduced, in their limited passage through a small hole at the first incidence.

But it appeared most expedient for me, at present, to neglect all further refinements, which may be attempted hereafter at leisure. It may even be presumed that, had there not been some small admixture of the red rays in the other colours, the result would have been still more decisive, with regard to the power of heating vested in the red rays. And it is likewise evident, that at least the red light of the prismatic spectrum, was much less adulterated than any of the other colours; their refractions tending all to throw them from the red. That the same rays which occasion the greatest heat, have not the power of illumination in any strong degree, stands on as good a foundation. For, since here also they have undergone the fairest trial, as being most free from other colours, it is equally proved that they illuminate objects but imperfectly. There is some probability that a ray, purified in the NEWTONIAN manner above quoted, especially in a well darkened room, may remain bright enough to serve the purpose of microscopic illumination, in which case, more precision can easily be obtained.

The greatest cause for a mixture of colours, however, which is, the breadth of the prism, I saw might easily be removed; therefore, on account of the coloured points, which have been mentioned in the 6th and 7th experiments, I was willing to try whether they proceeded from this mixture; and therefore covered the prism in front with a piece of pasteboard, having a slit in it of about $\frac{1}{10}$ of an inch broad.

10*th Experiment*. The nail, at 9 feet 2 inches from the prism.

R. I fix my attention on two shining, red points; they are pretty bright.

O. I see many more points. The object is better illuminated than in the red. The points are surrounded by black; but are orange-coloured.

Y. The points now are yellow, and white surrounded by black. The object is better illuminated than in orange.

The maximum of illumination is in the brightest yellow, or palest green.

G. The points are green and white, as before surrounded by black. Better illuminated than in orange.

B. The illumination is nearly equal to red.

I. Very indifferently illuminated.

V. Very badly illuminated.

The phænomena of the differently-coloured points being now completely resolved, since they were plainly owing to the former admixture of colours, and the illuminating power remaining ascertained as before, I attempted also to repeat the experiments upon the thermometer, with the prism covered in the same manner; but I found the effect of the coloured rays too much enfeebled to give a decisive result.

I might now proceed to my next subject; but it may be pardonable if I digress for a moment, and remark, that the foregoing researches ought to lead us on to others. May not the chemical properties of the prismatic colours be as different as those which relate to light and heat? Adequate methods for an investigation of them may easily be found; and we cannot too minutely enter into an analysis of light, which is the most subtle of all the active principles that are concerned in the mechanism of the operations of nature. A better acquaintance with it may enable us to account for various facts that fall under our daily observation, but which have hitherto remained unexplained. If the power of heating, as we now see, be chiefly lodged in the red-making rays, it accounts for the comfortable warmth that is thrown out from a fire, when it is in the state of a red glow; and for the heat which is given by charcoal, coke, and balls of small-coal mixed up with clay, used in hot-houses; all which, it is well known, throw out red light. It also explains the reason why the yellow, green, blue, and purple flames of burning spirits mixed with salt, occasion so little heat that a hand is not materially injured, when passed through their coruscations. If the chemical properties of colours also, when ascertained, should be such that an acid principle, for instance, which has been ascribed to light in general, on account of its changing the complexion of various substances exposed to it, may reside only in one of the colours, while others may prove to be differently invested, it will follow, that bodies may be variously affected by light, according as they imbibe and retain, or transmit and reflect, the different colours of which it is composed.

Radiant Heat is of different Refrangibility.

I must now remark, that my foregoing experiments ascertain beyond a doubt, that radiant heat, as well as light, whether they be the same or different agents, is not only refrangible, but is also subject to the laws of the dispersion arising from its different refrangibility ; and, as this subject is new, I may be permitted to dwell a few moments upon it. The prism refracts radiant heat, so as to separate that which is less efficacious, from that which is more so. The whole quantity of radiant heat contained in a sun-beam, if this different refrangibility did not exist, must inevitably fall uniformly on a space equal to the area of the prism ; and, if radiant heat were not refrangible at all, it would fall upon an equal space, in the place where the shadow of the prism, when covered, may be seen. But, neither of these events taking place, it is evident that radiant heat is subject to the laws of refraction, and also to those of the different refrangibility of light. May not this lead us to surmise, that radiant heat consists of particles of light of a certain range of momenta, and which range may extend a little farther, on each side of refrangibility, than that of light ? We have shewn, that in a gradual exposure of the thermometer to the rays of the prismatic spectrum, beginning from the violet, we come to the maximum of light, long before we come to that of heat, which lies at the other extreme. By several experiments, which time will not allow me now to report, it appears that the maximum of illumination has little more than half the heat of the full red rays ; and, from other experiments, I likewise conclude, that the full red falls still short of the maximum of heat ; which perhaps lies even a little beyond visible refraction. In this case, radiant heat will at least partly, if not chiefly consist, if I may be permitted the expression, of invisible light ; that is to say, of rays coming from the sun, that have such a momentum as to be unfit for vision. And, admitting, as is highly probable, that the organs of sight are only adapted to receive impressions from particles of a certain momentum, it explains why the maximum of illumination should be in the middle of the refrangible rays ; as those which have greater or less momenta, are likely to become equally unfit for impressions of sight. Whereas, in radiant heat, there may be no such limitation to the momentum of its particles. From the powerful effects of a burning lens, however, we gather the information, that the momentum of terrestrial radiant heat is not likely to exceed that of the sun ; and that, consequently, the refrangibility of *calorific* rays cannot extend much beyond that of *colourific* light. Hence we may also infer, that the invisible heat of red-hot iron, gradually cooled till it ceases to shine, has the momentum of the invisible rays which, in the solar spectrum viewed by day-light, go to the confines of red; and this will afford an easy solution of the reflection of invisible heat by concave mirrors.

Application of the Result of the foregoing Observations, to the Method of viewing the Sun advantageously, with Telescopes of large Apertures and high magnifying Powers.

Some time before the late transit of Mercury over the disk of the sun, I prepared my 7-feet telescope, in order to see it to the best advantage. As I wished to keep the whole aperture of the mirror open, I soon cracked every one of the darkening slips of wedged glasses, which are generally used with achromatic telescopes: none of them could withstand the accumulated heat in the focus of pencils, where these glasses are generally placed. Being thus left without resource, I made use of red glasses; but was by no means satisfied with their performance. My not being better prepared, as it happened, was of no consequence; the weather proving totally unfavourable for viewing the sun at the time of the transit. However, as I was fully aware of the necessity of providing an apparatus for this purpose, since no method that was in use could be applied to my telescopes, I took the first opportunity of beginning my trials.

The instrument I wished to adapt for solar inspection, was a NEWTONIAN reflector, with 9 inches aperture; and my aim was, to use the whole of it open.

I began with a red glass; and, not finding it to stop light enough, took two of them together. These intercepted full as much light as was necessary; but I soon found that the eye could not bear the irritation, from a sensation of heat, which it appeared these glasses did not stop.

I now took two green glasses; but found that they did not intercept light enough. I therefore smoked one of them; and it appeared that, notwithstanding they now still transmitted considerably more light than the red glasses, they remedied the former inconvenience of an irritation arising from heat. Repeating these trials several times, I constantly found the same result; and, the sun in the first case being of a deep red colour, I surmised that the red-making rays, transmitted through red glasses, were more efficacious in raising a sensation of heat, than those which passed through green, and which caused the sun to look greenish. In consequence of this surmise, I undertook the investigations which have been delivered under the two first heads.

As soon as I was convinced that the red light of the sun ought to be intercepted, on account of the heat it occasions, and that it might also be safely set aside, since it was now proved that pale green light excels in illumination, the method which ought to be pursued in the construction of a darkening apparatus was sufficiently pointed out; and nothing remained but to find such materials as would give us the colour of the sun, viewed in a telescope, of a pale green light, sufficiently tempered for the eye to bear its lustre.

To determine what glasses would most effectually stop the red rays, I procured some of all colours, and tried them in the following manner.

I placed a prism in the upper part of a window, and received its coloured spectrum upon a sheet of white paper. Then I intercepted the colours, just before they came to the paper, successively, by the glasses, and found the result as follows.

A deep red glass intercepted all the rays.

A paler red did the same.

From this, we ought not to conclude that red glasses will stop the red rays; but rather, that none of the sun's light, after its dispersion by the prism, remains intense enough to pass through red glasses, in sufficient quantity to be perceptible, when it comes to the paper. By looking through them directly at the sun, or even at day objects, it is sufficiently evident that they transmit chiefly red rays.

An orange glass transmitted nearly all the red, the orange, and the yellow. It intercepted some of the green; much of the blue; and very little of the indigo and violet.

A yellow glass intercepted hardly any light, of any one of the colours.

A dark green glass intercepted nearly all the red, and partly also the orange and yellow. It transmitted the green; intercepted much of the blue; but none of the indigo and violet.

A darker green glass intercepted nearly all the red; much of the orange; and a little of the yellow. It transmitted the green; stopped some of the blue; but transmitted the indigo and violet.

A blue glass intercepted much of the red and orange; some of the yellow; hardly any of the green; none of the blue, indigo, or violet.

A purple glass transmitted some of the red; a very little of the orange and yellow: it also transmitted a little of the green and blue; but more of the indigo and violet.

From these experiments we see, that dark green glasses are most efficacious for intercepting red light, and will therefore answer one of the intended purposes; but since, in viewing the sun, we have also its splendour to contend with, I proceeded to the following additional trials.

White glass, lightly smoked, apparently intercepted an equal share of all the colours; and, when the smoke was laid on thicker, it permitted none of them to pass.

Hard pitch, melted between two white glasses, intercepted much light; and, when put on sufficiently thick, transmitted none.

Many differently-coloured fluids, that were also tried, I found were not sufficiently pure to be used, when dense enough to stop light.

Now, red glasses, and the two last-mentioned resources of smoke, and pitch, any one of which, it has been seen, will stop as much light as may be required, had still a remaining trial to undergo, relating to distinctness; but this I was convinced could only be decided by actual observations of the sun.

As an easy way of smoking glasses uniformly is of some consequence to distinct vision, it may be of service here to give the proper directions, how to proceed in the operation.

With a pair of warm pliers, take hold of the glass, and place it over a candle, at a sufficient distance not to contract smoke. When it is heated, but no more than still to permit a finger to touch the edges of it, bring down the glass, at the side of the flame, as low as the wick will permit, which must not be touched. Then, with a quick vibratory motion, agitate it in the flame from side to side; at the same time advancing and retiring it gently all the while. By this method, you may proceed to lay on smoke to any required darkness. It ought to be viewed from time to time, not only to see whether it be sufficiently dark, but whether any inequality may be perceived; for, if that should happen, it will not be proper to go on.

The smoke of sealing-wax is bad: that of pitch is worse. A wax candle gives a good smoke; but that of a tallow candle is better. As good as any I have hitherto met with, is the smoke of spermaceti oil. In using a lamp, you may also have the advantage of an even flame extended to any length.

Telescopic Experiments.

No. 1. By way of putting my theory to the trial, I used two red glasses, and found that the heat which passed through them could not be suffered a moment; but I was now also convinced that distinctness of vision is capitally injured, by the colouring matter of these glasses.

No. 2. I smoked a white glass, till it stopped light enough to permit the eye to bear the sun. This destroyed all distinctness; and also permitted some heat to come to the eye, by transmitting chiefly red rays.

No. 3. I applied two white glasses, with pitch between them, to the telescope; and found that it made the sun appear of a scarlet colour. They transmitted some heat; and distinctness was greatly injured.

No. 4. I used a very dark green glass, to stop heat; and behind it, or towards the eye, I placed a red glass, to stop light. The first glimpse I had of the sun, was accompanied with a sensation of heat; distinctness also was materially injured.

No. 5. I used a dark green and a pale red; but, the sun not being sufficiently darkened, I smoked the red glass, and, putting a small partition between the two, placed the smoke towards the green glass. This took off the exuberance of light; but did not remedy the inconvenience arising from heat.

No. 6. I used two pale green glasses; smoking that next to the eye, and placing it as in No. 5, so that the smoke might be inclosed between the two. This acted incomparably well; but, in a very short time, the heat which passed the first glass, (though not the second, for I felt no sensation of it in the eye,) disordered the smoke, by drawing it up into little blisters or stars, which let through light; and this composition, therefore, soon became useless.

No 7. I used two dark green glasses, one of them smoked, as in No. 5. These also acted well; but became useless, for the reason assigned in No. 6, though somewhat less smoke had been required than in the former composition. I felt no heat.

No. 8. I used one pale green, with a dark green smoked glass upon it, as in No. 5. It bore an aperture of 4 inches very well, and the smoke was not disordered; but, when all the tube was open, the pale green glass cracked in a few minutes.

No. 9. Placing now a dark green before a smoked green, I saw the sun remarkably well. In this experiment, I had made a difference in the arrangement of the apparatus. The cracking of the glasses, I supposed, might be owing to their receiving heat in the middle, while the outside remained cold; which would occasion a partial dilatation. I therefore cut them into pieces about a quarter of an inch square, and set three of them in a slider, so that I could move them behind the smoked glass, without disturbing it. After looking about three or four minutes through one of them, I moved the slider to the second, and then to the third. This kept the glasses sufficiently cool; but the disturbance of the alterations proved hurtful to vision, which requires repose; and, if perchance I stopped a little longer than the proper time, the glass cracked, with a very disagreeable explosion, that endangered the eye.

No. 10. Two dark green glasses, both smoked, that a thinner coat might be on each; but the smoke still contracted blisters, though less dense than before.

No. 11. To get rid of smoke intirely, I used two dark green glasses, two very dark green, two pale blue, and one pale green glass, together. Distinctness was wanting; nor was light sufficiently intercepted.

No. 12. A dark green and a pale blue glass, smoked. The green glass cracked.

No. 13. A pale blue and a dark green glass, smoked. The blue glass cracked. The eye felt no sensation of heat.

No. 14. Two pale blue glasses, one smoked. The first glass cracked.

It was now sufficiently evident, that no glass which stops heat, and therefore receives it, could be preserved from cracking, when exposed to the focus of pencils. This induced me to try an application of the darkening apparatus to another part of the telescope.

The place where the rays are least condensed, without interfering with the reflections of the mirrors, is immediately close to the small one. I therefore screwed an apparatus to the speculum arm, into which any glass might be placed.

No. 15. A dark green glass close to the small speculum, and smoked pale green in the focus of pencils, as before. I saw remarkably well.

No. 16. The dark green as before; but, that more light might be admitted, a white smoked glass near the eye. Better than No. 15; but the green glass cracked.

No. 17. A very dark green and white smoked glass, as before. Very distinct, but the green glass cracked in about six or seven minutes.

No. 18. A dark blue glass, as in No. 15, and white smoked. This was distinct; and no heat came to the eye. The sun appeared ruddy.

No. 19. A dark blue and a yellow glass, close together, as in No. 15, and a white smoked one, as before. This was not distinct.

No. 20. A purple glass, as in No. 15, with a white smoked one. This gave the sun of a deep orange colour, approaching to scarlet. It was not distinct.

No. 21. An orange glass, as in No. 15, with a white smoked one. The colour of the sun was too red.

No. 22. A white smoked glass, as in No. 15, without any other at the eye. This gave the sun of a beautiful orange colour; but distinctness was totally destroyed.

No. 23. The heat near the small speculum being still too powerful for the glasses, I had a bluish dark green glass made of a proper diameter to be inclosed between the two eye-glasses of a double eye-piece. All glass I knew would stop some heat; and was therefore in hopes that the interposition of this eye-glass would temper the rays, so as in some measure to protect the coloured glass. In the usual place near the eye, I put two white glasses, with a thin coat of pitch between them. These glasses, when looked through by the natural eye, give the sun of a red colour; I therefore entertained no great hopes of their application to the telescope. They darkened the sun not sufficiently; and, when the pitch was thickened, distinctness was wanting.

No. 24. The same glass between the eye-glasses, and a dark green smoked glass at the eye. Very distinct. This arrangement is preferable to that of No. 15; after some considerable time, however, this glass also cracked.

No. 25. I placed a very dark green glass behind the second eye-glass, that it might be sheltered by both glasses, which in my double eye-piece are close together, and of an equal focal length. Here, as the rays are not much concentrated, the coloured glass receives them on a large surface, and stops light and heat, in the proportion of the squares of its diameter now used, to that on which the rays would have fallen, had it been placed in the focus of pencils. And, for the same reason, I now also placed a dark green smoked glass close upon the former, with the smoked side towards the eye, that the smoke might likewise be protected against heat, by a passage of the rays through two surfaces of coloured glass.

This position had moreover the advantage of leaving the telescope, with its mirrors and glasses, completely to perform its operation, before the application of the darkening apparatus; and thus to prevent the injury which must be occasioned, by the interposition of the heterogeneous colouring matter of the glasses and of the smoke.

No. 26. I placed a deep blue glass with a bluish green smoked one upon it, as in No. 25, and found the sun of a whiter colour than with the former composition. There was no disagreeable sensation of heat; though a little warmth might be felt.

No. 27. I used two black glasses, placed as in No. 25. Here there was no occasion for smoke; but the sun appeared of a bright scarlet colour, and an intolerable sensation of heat took place immediately. I rather suspect that these are very deep red glasses, though their outward appearance is black.

In order to have a more sure criterion of heat, I applied Dr. WILSON'S thermometer, No. 2, to the end of the eye-piece, where the eye is generally placed. With No. 25, it rose from 34 to 37 degrees. With No. 26, it rose from 35 to 46; and, with No. 27, it rose, very quickly, from 36 to 95 degrees. I am pretty sure it would have mounted up still higher; but, the scale extending only to 100, I was not willing to run the risk of breaking the thermometer by a longer exposure.

It remains now only to be added, that with No. 25 and 26 I have seen uncommonly well; and that, in a long series of very interesting observations upon the sun, which will soon be communicated, the glasses have met with no accident. However, when the sun is at a considerable altitude, it will be advisable to lessen the aperture a little, in telescopes that have so much light as my 10-feet reflector; or, which will give us more distinctness, to view the sun earlier in the morning, and later in the afternoon; for, the light intercepted by the atmosphere in lower altitudes will reduce its brilliancy much more uniformly than we can soften it, by laying on more smoke upon our darkening glasses. Now, as few instruments in common use are so large as that to which this method of darkening has been adapted, we may hope that it will be of general utility in solar observations.

Slough, near Windsor.
March 8, 1800.

XLIV.

Experiments on the Refrangibility of the invisible Rays of the Sun.

[*Phil. Trans.*, 1800, pp. 284–292.]

Read April 24, 1800.

In that section of my former paper which treats of radiant heat, it was hinted, though from imperfect experiments, that the range of its refrangibility is probably more extensive than that of the prismatic colours; but, having lately had some favourable sunshine, and obtained a sufficient confirmation of the same, it will be proper to add the following experiments to those which have been given.

I provided a small stand, with four short legs, and covered it with white paper [fig. 1]. On this I drew five lines, parallel to one end of the stand, at half an inch distance from each other, but so that the first of the lines might only be ¼ of an inch from the edge. These lines I intersected at right angles with three others; the 2d and 3d whereof were, respectively at 2½ and at 4 inches from the first.

The same thermometers that have before been marked No. 1, 2, and 3, mounted upon their small inclined planes, were then placed so as to have the centres of the shadow of their balls thrown on the intersection of these lines. Now, setting my little stand upon a table, I caused the prismatic spectrum to fall with its extreme colour upon the edge of the paper, so that none might advance beyond the first line. In this arrangement, all the spectrum, except the vanishing last quarter of an inch, which served as a direction, passed down by the edge of the stand, and could not interfere with the experiments. I had also now used the precaution of darkening the window in which the prism was placed, by fixing up a thick dark green curtain, to keep out as much light as convenient.

The thermometers being perfectly settled at the temperature of the room, I placed the stand so that part of the red colour, refracted by the prism, fell on the edge of the paper, before the thermometer No. 1, and about half way or 1¼ inch, towards the second: it consequently did not come before that, or the 3d thermometer, both which were to be my standards. During the experiment, I kept the last termination of visible red carefully to the first line, as a limit assigned to it, by gently moving the stand when required; and found the thermometers, which were all placed on the second line, affected as follows.

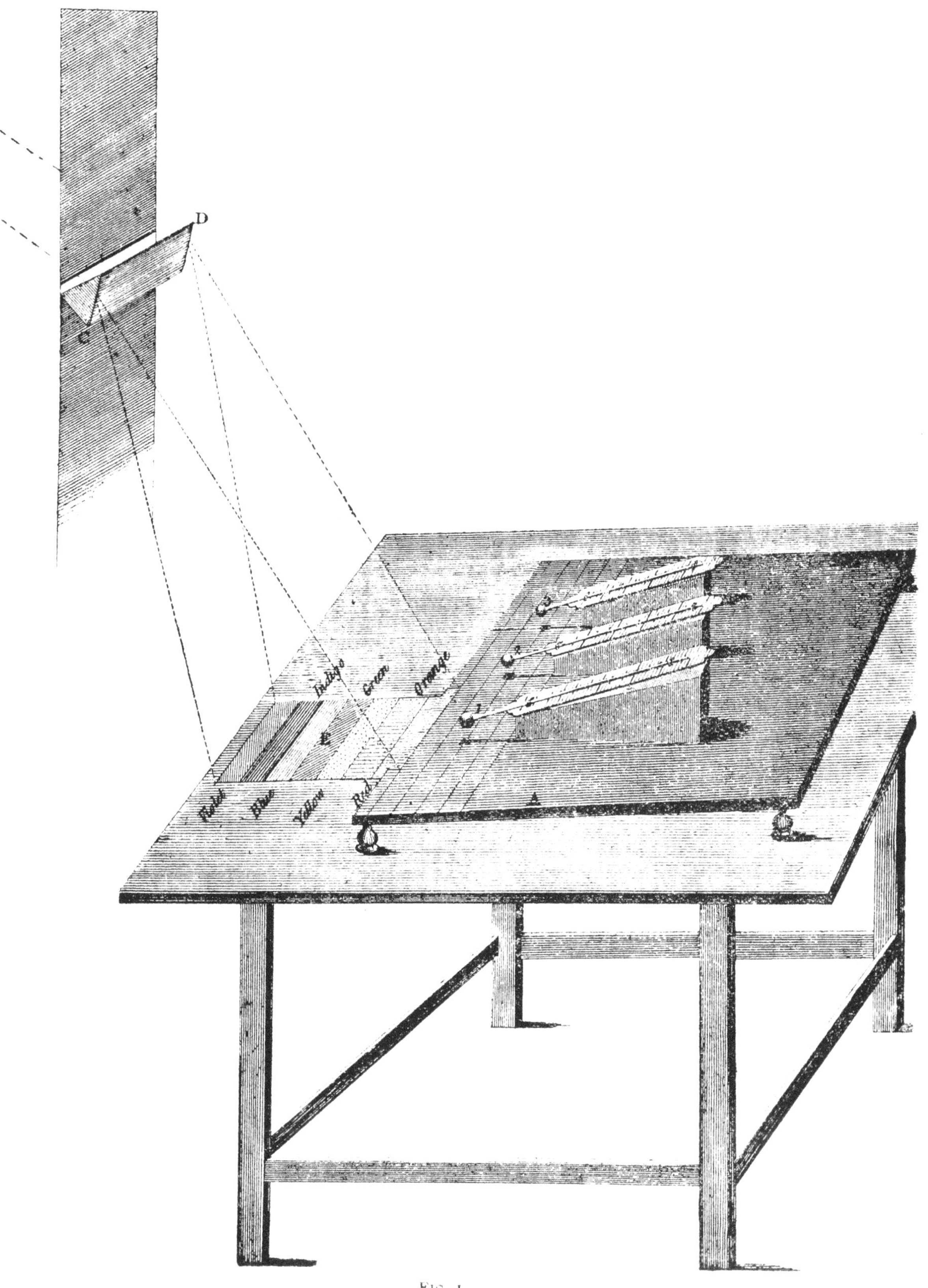

FIG. 1.

No. 1 . . .	45	49	51	$50\frac{1}{4}$
No. 2 . . .	45	45	$44\frac{3}{4}$	$43\frac{3}{4}$
No. 3 . . .	44	44	44	$43\frac{1}{2}$

Here the thermometer No. 1 rose $6\frac{1}{2}$ degrees, in 10 minutes, when its centre was placed $\frac{1}{2}$ inch beyond visible light.

In order to have a confirmation of this fact, I cooled the thermometer No. 1, and placed No. 2 in the room of it: I also put No. 3 in the place of No. 2, and No. 1 in that of No. 3; and, having exposed them as before, arranged on the second line, I had the following result.

No. 2 . . .	44	47	$46\frac{3}{4}$	$46\frac{3}{4}$
No. 3 . . .	44	44	44	44
No. 1 . . .	45	45	45	45

Here the thermometer No. 2 rose $2\frac{3}{4}$ degrees, in 12 minutes; and being, as has been noticed before, much more sensible than No. 1, it came to the temperature of its situation in a short time; but I left it exposed longer, on purpose to be perfectly assured of the result. Its shewing but $2\frac{3}{4}$ degrees advance, when No. 1 shewed $6\frac{1}{2}$, has also been accounted for before.

It being now evident that there was a refraction of rays coming from the sun, which, though not fit for vision, were yet highly invested with a power of occasioning heat, I proceeded to examine its extent as follows.

The thermometers were arranged on the third line, instead of the second; and the stand was, as before, immersed up to the first, in the coloured margin of the vanishing red rays. The result was thus.

No. 1 . . .	46	50	$51\frac{3}{4}$	$52\frac{1}{4}$
No. 2 . . .	46	$46\frac{1}{2}$	$46\frac{3}{4}$	47
No. 3 . . .	$45\frac{3}{4}$	46	$46\frac{1}{4}$	$46\frac{3}{4}$

Here the thermometer No. 1 rose $5\frac{1}{4}$ degrees, in 13 minutes, at 1 inch behind the visible light of the red rays.

I placed now the thermometers on the 4th line, instead of the 3d; and, proceeding as before, I had the following result.

No. 1 . . .	$48\frac{1}{4}$	$51\frac{1}{2}$
No. 2 . . .	$48\frac{1}{4}$	$48\frac{3}{8}$
No. 3 . . .	$47\frac{3}{4}$	$47\frac{7}{8}$

Therefore, the thermometer No. 1 rose $3\frac{1}{8}$ degrees, in 10 minutes, at $1\frac{1}{2}$ inch beyond the visible light of the red rays.

I might now have gone on to the 5th line; but, so fine a day, with regard to clearness of sky and perfect calmness, was not to be expected often, at this time of the year; I therefore hastened to make a trial of the other extreme of the prismatic spectrum. This was attended with some difficulty, as the illumination of the violet rays is so feeble, that a precise termination of it cannot be perceived. However, as well as could be judged, I placed the thermometers one inch beyond the reach of the violet rays, and found the result as follows.

No. 1 . . .	48	48	48	48½	48
No. 2 . . .	48	48	47½	47½	48
No. 3 . . .	47¾	47¾	47	47	47¾

Here the several indications of the thermometers, two of which, No. 1 and 2, were used as variable, while the 3d was kept as the standard, were read off during a time that lasted 12 minutes; but they afford, as may be seen by inspection, no ground for ascribing any of their small changes to other causes than the accidental disturbance which will arise from the motion of the air, in a room where some employment is carried on.

I exposed the thermometer now to the line of the very first perceptible violet light; but so that No. 1 and 2 might again be in the illumination, while No. 3 remained a standard. The result proved as follows.

No. 1 . . .	48	48½	48¾	49
No. 2 . . .	48	48	48½	48½
No. 3 . . .	47¾	47¾	47¾	47¾

Here the thermometer No. 1 rose 1 degree, in 15 minutes; and No. 2 rose ½ degree, in the same time.

From these last experiments, I was now sufficiently persuaded, that no rays which might fall beyond the violet, could have any perceptible power, either of illuminating or of heating; and that both these powers continued together throughout the prismatic spectrum, and ended where the faintest violet vanishes.

A very material point remained still to be determined, which was, the situation of the maximum of the heating power.

As I knew already that it did not lie on the violet side of the red, I began at the full red colour, and exposed my thermometers, arranged on a line, so as to have the ball of No. 1 in the midst of its rays, while the other two remained at the side, unaffected by them.

No. 1 . . .	48½	55½	55½
No. 2 . . .	48½	48½	48½
No. 3 . . .	48	48	48

Here the thermometer No. 1 rose 7 degrees, in 10 minutes, by an exposure to the full red coloured rays.

I drew back the stand, till the centre of the ball of No. 1 was just at the vanishing of the red colour, so that half its ball was within, and half without, the visible rays of the sun.

No. 1 . . .	48½	55½	57
No. 2 . . .	48½	48½	49
No. 3 . . .	48	48	48½

Here the thermometer No. 1 rose 8 degrees, in 10 minutes.

By way of not losing time, in order to connect these last observations the better together, I did not bring back the thermometer No. 1 to the temperature of the room, being already well acquainted with its rate of shewing, compared to that of No. 2, but went on to the next experiment, by withdrawing the stand, till the whole ball of No. 1 was completely out of the sun's visible rays, yet so as to bring the termination of the line of the red colour as near the outside of the ball as could be, without touching it.

No. 1 . . .	57	58½	59	59
No. 2 . . .	49	49¾	50¼	50
No. 3 . . .	48½	49	49¾	49½

Here the thermometer No. 1 rose, in 10 minutes, another degree higher than in its former situation it could be brought up to; and was now 9 degrees above the standard. The ball of this thermometer, as has been noticed, is exactly half an inch in diameter; and its centre therefore was ¼ inch beyond the visible illumination, to which no part of it was exposed.

It would not have been proper to compare these last observations with those taken at an earlier period this morning, in order to obtain a true maximum, as the sun was now more powerful than it had been at that time; for which reason, I caused the line of termination of visible light now to fall again just ½ inch from the centre of the ball; and had the following result.

No. 1 . . .	50½	57¾	58½	58¾
No. 2 . . .	50½	50	50	50
No. 3 . . .	50	49½	49½	49½

And here the thermometer No. 1 rose, in 16 minutes, 8¾ degrees, when its centre was ½ inch out of the visible rays of the sun. Now, as before we had a rising of 9 degrees, and here 8¾, the difference is almost too trifling to suppose, that this latter situation of the thermometer was much beyond the maximum of the heating power; while, at the same time, the experiment sufficiently indicates, that the place inquired after need not be looked for at a greater distance.

It will now be easy to draw the result of these observations into a very narrow compass.

The first four experiments prove, that there are rays coming from the sun, which are less refrangible than any of those that affect the sight. They are invested with a high power of heating bodies, but with none of illuminating objects; and this explains the reason why they have hitherto escaped unnoticed.

My present intention is, not to assign the angle of the least refrangibility belonging to these rays, for which purpose more accurate, repeated, and extended experiments are required. But, at the distance of 52 inches from the prism, there was still a considerable heating power exerted by our invisible rays, one inch and a half beyond the red ones, measured upon their projection on a horizontal plane. I have no doubt but that their efficacy may be traced still somewhat farther.

The 5th and 6th experiments shew, that the power of heating is extended to the utmost limits of the visible violet rays, but not beyond them; and that it is gradually impaired, as the rays grow more refrangible.

The four last experiments prove, that the maximum of the heating power is vested among the invisible rays; and is probably not less than half an inch beyond the last visible ones, when projected in the manner before mentioned. The same experiments also shew, that the sun's invisible rays, in their less refrangible state, and considerably beyond the maximum, still exert a heating power fully equal to that of red-coloured light; and that, consequently, if we may infer the quantity of the efficient from the effect produced, the invisible rays of the sun probably far exceed the visible ones in number.

To conclude, if we call *light*, those rays which illuminate objects, and *radiant heat*, those which heat bodies, it may be inquired, whether light be essentially different from radiant heat? In answer to which I would suggest, that we are not allowed, by the rules of philosophizing, to admit two different causes to explain certain effects, if they may be accounted for by one. A beam of radiant heat, emanating from the sun, consists of rays that are differently refrangible. The range of their extent, when dispersed by a prism, begins at violet-coloured light, where they are most refracted, and have the least efficacy. We have traced these calorific rays throughout the whole extent of the prismatic spectrum; and found their power increasing, while their refrangibility was lessened, as far as to the confines of red-coloured light. But their diminishing refrangibility, and increasing power, did not stop here; for we have pursued them a considerable way beyond the *prismatic spectrum*, into an invisible state, still exerting their increasing energy, with a decrease of refrangibility up to the maximum of their power; and have also traced them to that state where, though still less refracted, their energy, on account, we may suppose, of their now failing density, decreased pretty fast; after which, the invisible *thermometrical spectrum*, if I may so call it, soon vanished.

If this be a true account of solar heat, for the support of which I appeal to my experiments, it remains only for us to admit, that such of the rays of the sun as have the refrangibility of those which are contained in the prismatic spectrum, by the construction of the organs of sight, are admitted, under the appearance of light and colours; and that the rest, being stopped in the coats and humours of the eye, act upon them, as they are known to do upon all the other parts of our body, by occasioning a sensation of heat.

Slough, near Windsor,
March 17, 1800.

XLV.

Experiments on the solar, and on the terrestrial Rays that occasion Heat; with a comparative View of the Laws to which Light and Heat, or rather the Rays which occasion them, are subject, in order to determine whether they are the same, or different.

PART I.

[*Phil. Trans.*, 1800, pp. 293–326.]

Read May 15, 1800.

THE word heat, in its most common acceptation, denotes a certain sensation, which is well known to every person. The cause of this sensation, to avoid ambiguity, ought to have been distinguished by a name different from that which is used to point out its effect. Various authors indeed, who have treated on the subject of heat, have occasionally added certain terms to distinguish their conceptions, such as, latent, absolute, specific, sensible heat; while others have adopted the new expressions of caloric, and the matter of heat. None of these descriptive appellations however would have completely answered my purpose. I might, as in the preceding papers, have used the name radiant heat, which has been introduced by a celebrated author, and which certainly is not very different from the expressions I have now adopted; but, by calling the subject of my researches, the rays that occasion heat, I cannot be misunderstood as meaning that these rays themselves are heat; nor do I in any respect engage myself to shew in what manner they produce heat.

From what has been said it follows, that any objections that may be alleged, from the supposed agency of heat in other circumstances than in its state of radiance, or heat-making rays, cannot be admitted against my experiments. For, notwithstanding I may be inclined to believe that all phænomena in which heat is concerned, such as the expansion of bodies, fluidity, congelation, fermentation, friction, *&c.* as well as heat in its various states of being latent, specific, absolute, or sensible, may be explained on the principle of heat-making rays, and vibrations occasioned by them in the parts of bodies; yet this is not intended, at present, to be any part of what I shall endeavour to establish.

I must also remark, that in using the word rays, I do not mean to oppose, much less to countenance, the opinion of those philosophers who still believe that light itself comes to us from the sun, not by rays, but by the supposed vibrations of an elastic ether, every where diffused throughout space ; I only claim the same privilege for the rays that occasion heat, which they are willing to allow to those that illuminate objects. For, in what manner soever this radiance may be effected, it will be fully proved hereafter, that the evidence, either for rays, or for vibrations which occasion heat, stands on the same foundation on which the radiance of the illuminating principle, light, is built.

In order to enter on our subject with some regularity, it will be necessary to distinguish heat into six different kinds, three whereof are solar, and three terrestrial ; but, as the divisions of terrestrial heat strictly resemble those of solar, it will not be necessary to treat of them separately ; our subject, therefore, may be reduced to the three following general heads.

We shall begin with the heat of luminous bodies in general, such as, in the first place, we have it directly from the sun ; and as, in the second, we may obtain it from terrestrial flames, such as torches, candles, lamps, blue-lights, *&c.*

Our next division comprehends the heat of coloured radiants. This we obtain, in the first place, from the sun, by separating its rays in a prism ; and, in the second, by having recourse to culinary fires, openly exposed.

The third division relates to heat obtained from radiants, where neither light nor colour in the rays can be perceived. This, as I have shewn, is to be had, in the first place, directly from the sun, by means of a prism applied to its rays ; and, in the second, we may have it from fires inclosed in stoves, and from red-hot iron cooled till it can no longer be seen in the dark.

Besides the arrangement in the order of my experiments which would arise from this division, we have another subject to consider. For, since the chief design of this paper is to give a comparative view of the operations that may be performed on the rays that occasion heat, and of those which we already know to have been effected on the rays that occasion light, it will be necessary to take a short review of the latter. I shall merely select such facts as not only are perfectly well known, but especially such as will answer the intention of my comparative view, and arrange them in the following order.

1. Light, both solar and terrestrial, is a sensation occasioned by rays emanating from luminous bodies, which have a power of illuminating objects ; and, according to circumstances, of making them appear of various colours.

2. These rays are subject to the laws of reflection.

3. They are likewise subject to the laws of refraction.

4. They are of different refrangibility.

5. They are liable to be stopped, in certain proportions, when transmitted through diaphanous bodies.

6. They are liable to be scattered on rough surfaces.

7. They have hitherto been supposed to have a power of heating bodies; but this remains to be examined.

The similar propositions relating to heat, which are intended to be proved in this paper, will stand as follows.

1. Heat, both solar and terrestrial, is a sensation occasioned by rays emanating from candent substances, which have a power of heating bodies.

2. These rays are subject to the laws of reflection.

3. They are likewise subject to the laws of refraction.

4. They are of different refrangibility.

5. They are liable to be stopped, in certain proportions, when transmitted through diaphanous bodies.

6. They are liable to be scattered on rough surfaces.

7. They may be supposed, when in a certain state of energy, to have a power of illuminating objects; but this remains to be examined.

Before I can go on, I have to mention, that the number of experiments which will be required to make good all these points, exceeds the usual length of my papers; on which account, I shall divide the present one into two parts. Proceeding therefore now to an investigation of the three first heads that have been proposed, I reserve the three next, and a discussion which will be brought on by the seventh article, for the second part.

1st Experiment. Reflection of the Heat of the Sun.

I exposed the thermometer which in a former paper has been denoted by No. 3, to the eye-end of a ten-feet NEWTONIAN telescope, which carried a *Camera-eye-piece,** but no eye-glass. When, by proper adjustment, the focus came to the ball of the thermometer, it rose from 52 degrees to 110; so that rays which came from the sun, underwent three regular reflections; one, on a concave mirror, and the other two, on two plain ones. Now these rays, whether they were those of light or not, for that our experiment cannot ascertain, had a power of occasioning heat, which was manifested in raising the thermometer 58 degrees.

2d Experiment. Reflection of the Heat of a Candle.

At the distance of 29 inches from a candle, I planted a small steel-mirror, of $3\frac{4}{10}$ inches diameter, and about $2\frac{3}{4}$ inches focal length [see fig. 1]. In the secondary focus of it, I placed the ball of the thermometer which in my paper has been marked No. 2; and very near it, but out of the reach of reflection, the thermometer No. 3. Having covered the mirror till both were come to the temperature of their stations, I began as follows.

	0′	1′	2′	3′	4′	5′
No. 2, in the focus . .	54	55	56	57	$57\frac{1}{4}$	$57\frac{1}{4}$
No. 3, standard . .	54	54	54	54	54	54

* See *Phil. Trans.* Vol. LXXII. p. 176. [Vol. I. p. 98.]

Here, in five minutes, the thermometer No. 2 received $3\frac{1}{4}$ degrees of heat from the candle, by reflected rays. I now covered the mirror, but left all the rest of the apparatus untouched.

	0′	1′	$1\frac{1}{2}$′	6′
No. 2, in the focus . . .	$57\frac{1}{4}$	$55\frac{1}{2}$	55	54
No. 3, standard . . .	54	54	54	54

Here, in six minutes, the thermometer lost the $3\frac{1}{4}$ degrees of heat again, which it had gained before. I uncovered the mirror once more.

	0′	$1\frac{1}{2}$′	$3\frac{1}{2}$′	5′
No. 2, in the focus . . .	54	56	57	$57\frac{1}{4}$
No. 3, standard . . .	54	54	54	54

And, in five minutes, the $3\frac{1}{4}$ degrees of heat were regained. In consequence of which, we are assured that certain rays came from the candle, subject to the laws of reflection, which, though they might not be the rays of light, for that our experiment does not determine, were evidently invested with a power of heating the thermometer placed in the focus of the mirror.

3d Experiment. Reflection of the Heat that accompanies the Solar prismatic Colours.

In the spectrum of the sun, given by a prism, I placed my small steel mirror, with a thermometer in its focus [see fig. 2]. It was covered by a piece of pasteboard, which, through a proper opening, admitted all the visible colours to fall on its polished surface, but excluded every other ray of heat that might be, either on the violet or on the red side, beyond the spectrum. Then, placing the apparatus so as to have the thermometer in the red rays, but keeping the mirror covered up till the thermometer became settled, I found it stationary at 58°. Uncovering the mirror, I had as follows.

	0′	2′
No. 2	58	93

Here, in two minutes, the thermometer rose 35 degrees, by reflected heat. I covered the mirror again, and, in a few minutes, the thermometer, exposed to the direct prismatic red, came down to 58° again. And thus the prismatic colours, if they are not themselves the heat-making rays, are at least accompanied by such as have a power of occasioning heat, and are liable to be regularly reflected.

4th Experiment. Reflection of the Heat of a red-hot Poker.

I placed the small steel mirror at 12 inches from a red-hot poker, set with its heated end upwards, in a perpendicular position, and so elevated as to throw its rays on the mirror [see fig. 1]. The thermometer No. 2 was placed in its secondary focus, and had a small pasteboard screen, to guard its ball from the direct heat of the poker.

		0′	1½′
	No. 2	54½	93
I covered the mirror.		3′	
	No. 2	65	

Here, in 1½ minute, the thermometer rose 38½ degrees, by reflected rays; and, when the mirror was covered up, it fell in the next 1½ minute, 28 degrees. On which account, we cannot but allow, that certain rays, whether it be those that shine or not, issue from an ignited poker, which are subject to the regular laws of reflection, and have a power of heating bodies.

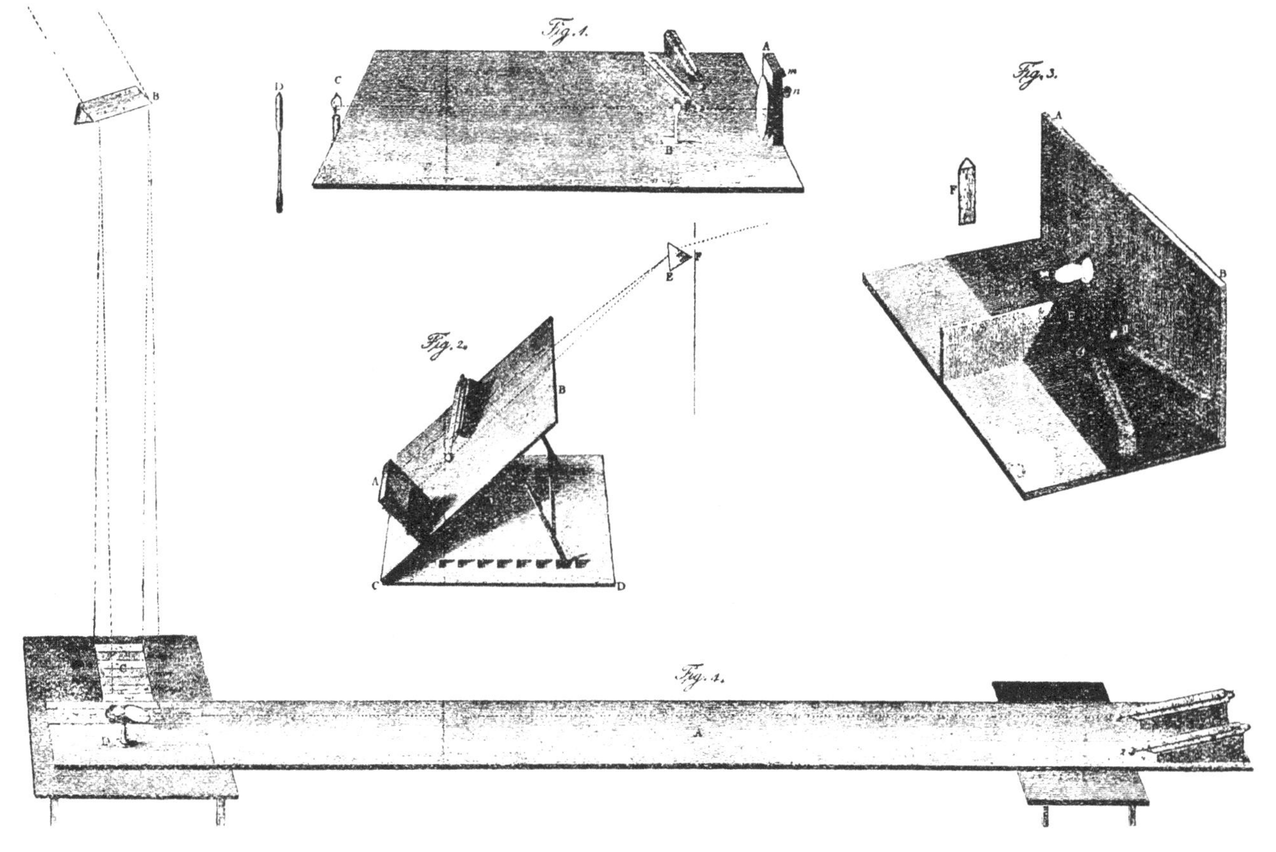

5th Experiment. Reflection of the Heat of a Coal Fire by a plain Mirror.

I placed a small speculum, such as I use with my 7-feet reflectors, upon a stand, and so as to make an angle of 45 degrees with the front of it [see fig. 3]. This was afterwards to face the fire in my parlour chimney, and would make the same angle with the bars of the grate. At a distance of 3½ inches from the speculum, on the reflecting side of it, was placed the thermometer No. 1; and close by it, but out of the reach of the reflected rays, the thermometer No. 4. The whole was guarded in front, against the influence of the fire, by an oaken board 1½ inch thick, which had a circular opening of 1¼ inch diameter, opposite the situation of the plain mirror, in order to permit the fire to shine upon it. The thermometers were divided from the mirror by a wooden partition, which also had an opening in it, that the reflected rays might come from the mirror to No. 1, while No. 4 remained screened from their influence. On exposing this apparatus to the fire, I had the following result.

	0′	1′	2′	3′	4′	5′
No. 1	60	62	64	66	66	67
No. 4	60	60	60	60	60	60½

Here, in five minutes, the heat reflected from the plain mirror raised the thermometer No. 1, 7 degrees; while the change in the temperature of the screened place, indicated by No. 4, amounted only to half a degree: which shews, that an open fire sends out rays that are subject to the laws of reflection, and occasion heat.

6th Experiment. Reflection of Fire-heat by a Prism.

Every thing remaining arranged as in the 5th experiment, I removed the small plain mirror, and placed in its stead a prism, which had one of its angles of 90 degrees, and the other two of 45° each [see fig. 3]. It was put so as to have one of the sides facing the fire, while the other was turned towards the thermometer: the hypotenuse, consequently, made an angle of 45 degrees with the bars of the grate. The apparatus, after having been cooled some time, was exposed to the fire, and the following result was taken.

	0′	1′	2′	4′	5′	8′	10′	11′
No. 1	62½	63	64	64½	65	65¾	66½	67
No. 4	62½	62¾	63	63	63¼	63½	63¾	64¼

Here, in eleven minutes, the rays reflected by the prism raised the thermometer 4½ degrees; but, the temperature of the place having undergone an alteration of 1¾ degrees, we can only place 2¾ to the account of reflection. The apparatus becoming now very hot, it would not have been fair to have continued the experiment for a longer time; but the effect already produced was fully sufficient to shew, that even a prism, which stops a great many heat-making rays, still reflects enough of them to prove, that an open fire not only sends them out, but that they are subject to every law of reflection.

7th Experiment. Reflection of invisible Solar Heat.

On a board of about 4 feet 6 inches long, I placed at one end, a small plain mirror, and at the other, two thermometers [see fig. 4]. The distance of No. 1, from the face of the mirror, was 3 feet 9¾ inches; and No. 2 was put at the side of it, facing the same way, but out of the reach of the rays that were to be reflected by the mirror. The colours of the prism were thrown on a sheet of paper having parallel lines drawn upon it, at half an inch from each other. The mirror was stationed upon the paper; and was adjusted in such a manner as to present its polished surface, in an angle of 45 degrees, to the incident coloured rays, by which means, they would be reflected towards the ball of the thermometer No. 1. In this arrangement, the whole apparatus might be withdrawn from the colours to any required distance, by attending to the last visible red colour, as it shewed itself on the lines of the paper. When the thermometers were properly settled to the temperature of their situation, during which time the mirror had been covered, the apparatus was drawn gently away from the colours, so far as to cause the mirror, which was now open, to receive only the invisible rays of heat which lie beyond the confines of red. The result was as follows.

	0′	—	—	7′	10′
No. 1 . . .	56	57	59	60	60
No. 2 . . .	56	56	56	56	56

Here, in ten minutes, the thermometer No. 1 received four degrees of heat, reflected to it, in the strictest optical manner, by the plain mirror of a NEWTONIAN telescope. The great regularity with which these invisible rays obeyed the law of reflection, was such, that Dr. WILSON's sensible thermometer No. 2, which had been chosen on purpose for a standard, and was within an inch of the other thermometer, remained all the time without the least indication of any change of temperature that might have arisen from straggling rays, had there been any such. I now took away the mirror, but left every thing else in the situation it was. The effect of this was thus.

	0′	5′	8′	10′
No. 1 . . .	60	58	57	56
No. 2 . . .	56	56	56	56

Here, in ten minutes, the thermometer No. 1 lost again the 4 degrees it had acquired, while No. 2 still remained unaltered; and this becomes therefore a most decisive experiment, in proof of the existence of invisible rays, of their being subject to the laws of reflection, and of their power of occasioning heat.

8th Experiment. Reflection, and Condensation, of the invisible solar Rays.

I made an apparatus for placing the small steel mirror at any required angle [see fig. 2]; and, having exposed it to the prismatic spectrum, so as to receive it perpendicularly, I caused the colours to fall on one half of the mirror, which, being covered by a semicircular piece of pasteboard, would stop all visible rays, so that none of them could reach the polished surface. On the pasteboard were drawn several lines, parallel to the diameter, and at the distance of one-tenth of an inch from each other; that, by withdrawing the apparatus, I might have it at option to remove the last visible red to any required distance from the reflecting surface. In the focus of the mirror was placed the thermometer No. 2. I covered now also the other half of the mirror, till the thermometer had assumed the temperature of its situation. Then, withdrawing the apparatus out of the visible spectrum, till the last tinge of red was one-tenth of an inch removed from the edge of the pasteboard, and the whole of the coloured image thus thrown on the semicircular cover, I opened the other half of the mirror, for the admission of invisible rays. The result was as follows.

	0′	1′
No. 2, in the focus of invisible heat . . .	61	80

Here, in one minute, the thermometer rose 19 degrees. I covered the mirror.

	2′	3′	4′
No. 2, ...	72	67	64

Here, in three minutes, the thermometer fell 16 degrees. I opened the mirror again.

	5′	6′
No. 2, in the focus of invisible heat . . .	83	88

Here, in two minutes, the thermometer rose 24 degrees. I covered the mirror once more.

	7′
No. 2, ...	69

And, in one minute, the thermometer fell 19 degrees. Now, by this alternate rising and falling of the thermometer, three points are clearly ascertained. The first is, that there are invisible rays of the sun. The second, that these rays are not only reflexible, in the manner which has been proved in the foregoing experiment, but that, by the strict laws of reflection, they are capable of being condensed. And, in the third place, that by condensation, their heating power is proportionally increased; for, under the circumstances of the experiment, we find that it extended so far as to be able to raise the thermometer, in two minutes, no less than 24 degrees.

9th Experiment. Reflection of invisible culinary Heat.

I planted my little steel mirror upon a small board [see fig. 5]; and at a proper distance opposite to it I erected a slip of deal, ½ inch thick, and 1 inch broad, in a horizontal direction, so as to be of an equal height, in the middle of its thickness, with the centre of the mirror. Against the side, facing the mirror, were fixed the two thermometers No. 2 and No. 3, with their balls within half an inch of each other, and the scales turned the opposite way. A little of the wood was cut out of the slip, to make room for the balls to be freely exposed. That of No. 2 was in the axis of the mirror; and the ball of No. 3 was screened from the reflected rays, by a small piece of pasteboard tied to the scale. The small ivory scales of the thermometers, with the slip of wood at their back, which however was feather-edged towards the stove, intercepted some heat; but it will be seen presently that there was enough to spare. When my stove was of a good heat, I brought the apparatus to a place ready prepared for it.

	0′	1′
No. 2, in the focus . . .	52	91
No. 3, screened . . .	52	53

Here we find that, in one minute, the invisible culinary heat raised the thermometer No. 2, 39 degrees; while No. 3, from change of temperature, rose only one, notwithstanding its exposure to the stove was in every respect equal to that of No. 3, except so far as relates to the rays returned by the mirror; and therefore, the radiant nature of these invisible rays, their power of heating bodies, and their being subject to the laws of reflection, are equally established by this experiment.

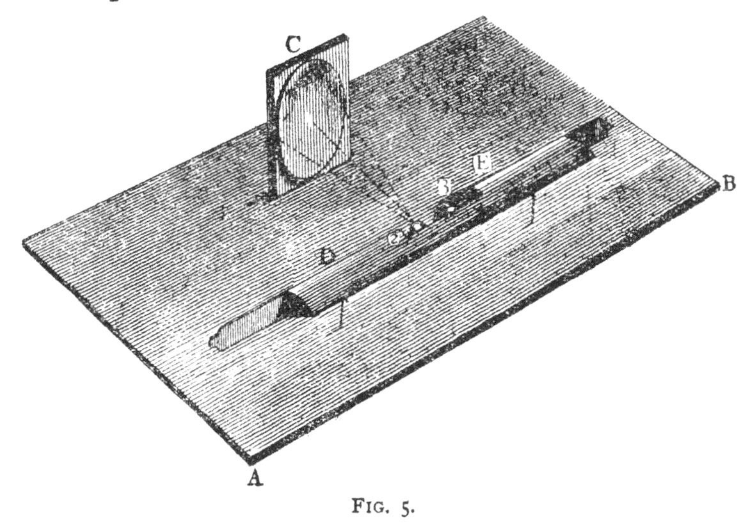

FIG. 5.

10th Experiment. Reflection of the invisible Rays of Heat of a Poker, cooled from being red-hot till it could no longer be seen in a dark Place.

The great abundance of heat in my last experiment would not allow of its being carried on without injury to the thermometer, the scale of which is not extensive; I therefore placed a poker, when of a proper black heat, at 12 inches from the steel mirror [see fig. 1], and received the effect of its condensed rays upon the thermometer No. 2, placed in the focus. Then, alternately covering and uncovering the mirror, one minute at a time, the effect was as follows.

	The mirror covered.	Open.	Covered.	Open.	Covered.	Open.	Covered.
	0′	1′	2′	3′	4′	5′	6′
No. 2 . . .	61	68	61	64	59	61½	58

Here, in six minutes, we have a repeated result of alternate elevations and depressions of the thermometer, all of which confirm the reflexibility, the radiant nature, and the heating power, of the invisible rays that came from the poker.

From these experiments it is now sufficiently evident, that in every supposed case of solar and terrestrial heat, we have traced out rays that are subject to the regular laws of reflection, and are invested with a power of heating bodies; and this independent of light. For though, in four cases out of six, we had illuminating as well as heating rays, it is to be noticed that our proof goes only to the power of occasioning heat, which has been strictly ascertained by the thermometer. If it should be said, that the power of illuminating objects, of these same rays, is as strictly proved by the same experiments, I must remark, that from the cases of invisible rays brought forward in the four last experiments, it is evident that the conclusion, that rays must have illuminating power, because they have a power of occasioning heat, is erroneous; and, as this must be admitted, we have a right to ask for some proof of the assertion, that rays which occasion heat can ever become visible. But, as we shall have an opportunity to say more of this hereafter, I proceed now to investigate the refraction of heat-making rays.

11th Experiment. Refraction of solar Heat.

With a new ten-feet NEWTONIAN telescope, the mirror of which is 24 inches in diameter of polished surface, I received the rays of the sun; and, making them pass through a day-piece with four lenses, I caused them to fall on the ball of the thermometer No. 3, placed in their focus. Those who are acquainted with the lines in which the principal rays and pencils move through a set of glasses, will easily conceive how artfully, in our present instance, heat was sent from one place to another. Heat crossing heat, through many intersecting courses, without jostling together, and each parcel arriving at last safely to its destined place. As soon as the rays were brought to the thermometer, it rose almost instantly from 60 degrees to 130; and, being afraid of cracking the glasses, I turned away the telescope. Here the rays, which occasioned no less than 70 degrees of heat, had undergone eight regular successive refractions; so that their being subject to its laws cannot be doubted.

12th Experiment. Refraction of the Heat of a Candle.

I placed a lens of about 1 4 inch focus, and 1·1 diameter, mounted upon a small support, at a distance of 2 8 inches from a candle [see fig. 6]; and the thermometer No. 2, behind the lens, at an equal distance of about 2·8 inches; but which ought to be very carefully adjusted to the secondary focus of the candle. Not far from the lens, towards the candle, was a pasteboard screen, with an aperture of nearly the same size as the lens. The support of the lens had an eccentric pivot, on which it might be turned away from its place, and returned to the same situation again, at pleasure. This arrangement being made, the thermometer was for a few moments exposed to the rays of the candle, till it had assumed the temperature of its situation. Then the lens was turned on its pivot, so as to intercept the direct rays, which passed through the opening in the pasteboard screen, and to refract them to the focus, in which the thermometer was situated.

	0′	1′	2′	3′
No. 2 . . .	$53\frac{7}{8}$	$55\frac{1}{2}$	$55\frac{3}{4}$	56

Here, in three minutes, the thermometer received $2\frac{1}{8}$ degrees of heat, by the refraction of the lens. The lens was now turned away.

	0′	1′	2′	3′
No. 2 . . .	56	$54\frac{5}{8}$	$54\frac{1}{8}$	$53\frac{7}{8}$

Here, in three minutes, the thermometer lost $2\frac{1}{8}$ degrees of heat. The lens was now returned to its situation.

	0′	1′	2′	3′
No. 2 . . .	$53\frac{7}{8}$	$54\frac{3}{4}$	$55\frac{3}{8}$	56

And, in three minutes, the thermometer regained the $2\frac{1}{8}$ degrees of heat. A greater effect may be obtained by a different arrangement of the distances. Thus,

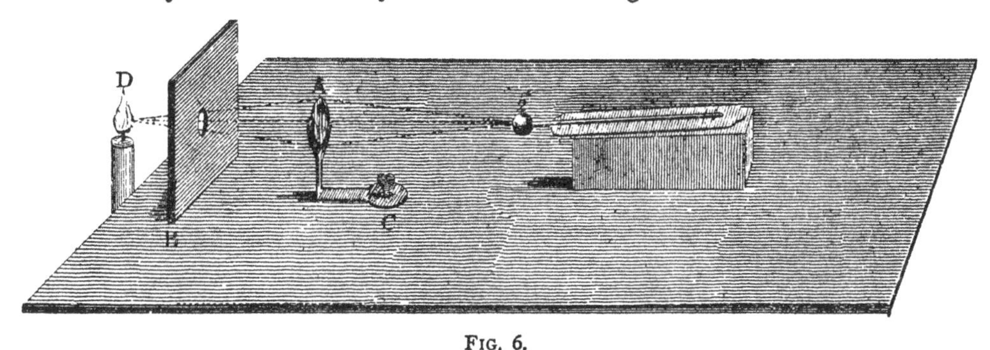

Fig. 6.

if the lens be placed at $3\frac{1}{2}$ inches from a wax-candle, and the thermometer situated, as before, in the secondary focus, we shall be able to draw from 5 to 8 degrees of heat, according to the burning of the candle, and the accuracy of the adjustment of the thermometer to the focus. The experiment we have related shews evidently, that rays invested with a power of heating bodies issue from a candle, and are subject to laws of refraction, nearly the same with those respecting light.

13th Experiment. Refraction of the Heat that accompanies the coloured part of the prismatic Spectrum.

I covered a burning lens of Mr. DOLLOND'S, which is nearly 9 inches in diameter, and very highly polished, with a piece of pasteboard, in which there was an opening of a sufficient size to admit all the coloured part of the prismatic spectrum [see fig. 7]. In the focus of the glass was placed the thermometer No. 3; and, when every thing was arranged properly, I covered the lens for five minutes, that the thermometer might assume the temperature of its situation. The result was as follows.

	The lens covered. 0′	Open. 1′
No. 3 . . .	64	176

Here, in one minute, the thermometer received 112 degrees of heat, which came with the coloured part of the solar spectrum, and were refracted to a focus ; so that, if the coloured rays themselves are not of a heat-making nature, they are at least

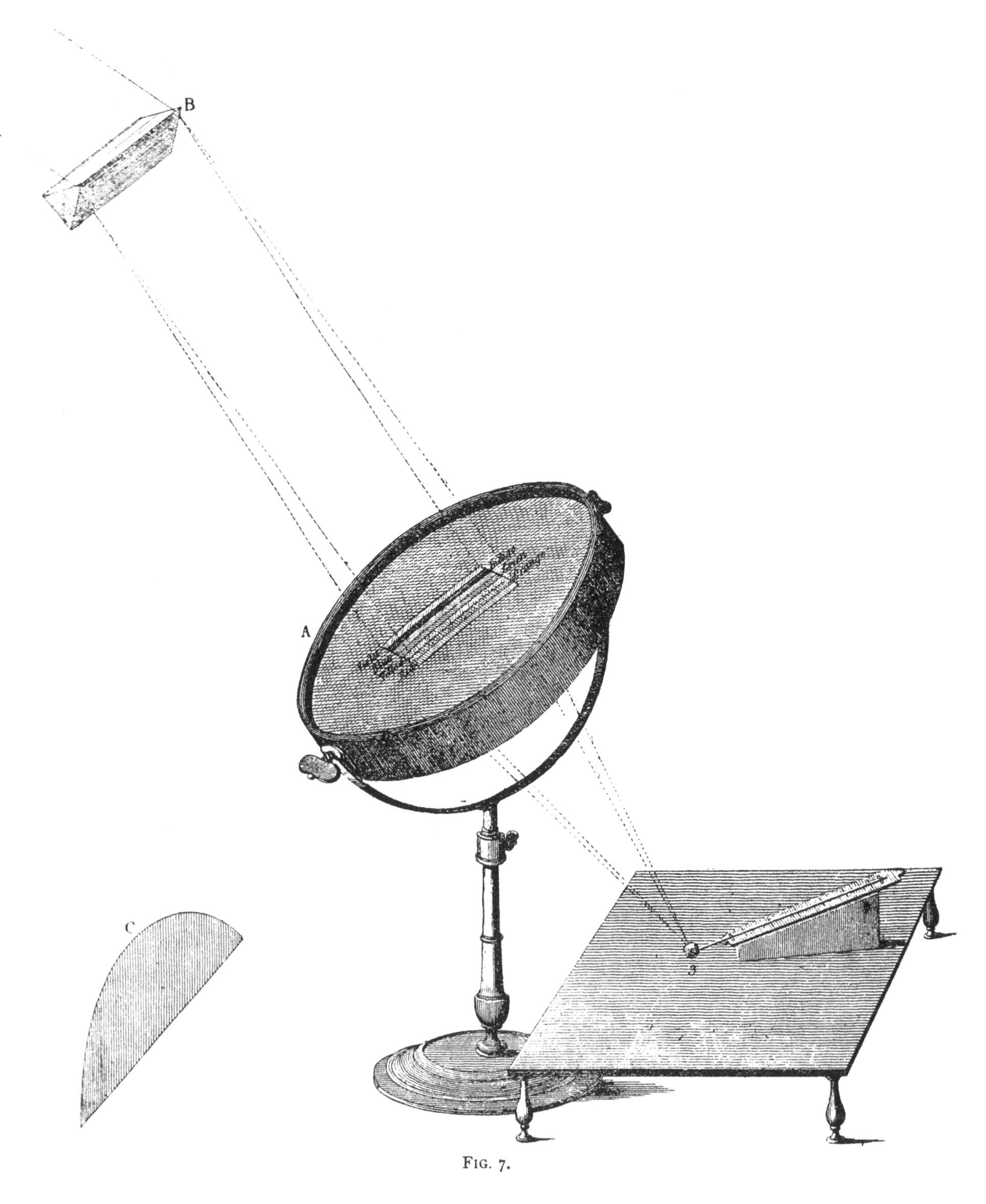

FIG. 7.

accompanied with rays that have a power of heating bodies, and are subject to certain laws of refraction, which cannot differ much from those affecting light.

14th Experiment. Refraction of the Heat of a Chimney-Fire.

I placed Mr. DOLLOND's lens before the clear fire of a large grate [see fig. 8]. Its distance from the bars of the grate was three feet; and, in the secondary focus of it was placed the thermometer No. 1. No. 4 was stationed, by way of standard, at 2½ inches from the former, and at an equal distance from the fire. Before the thermometers was a slip of mahogany, which had three holes in it, $\frac{8}{10}$ of an inch in diameter each. Behind the centre of the 1st hole, ⅜ of an inch from the back, was placed the

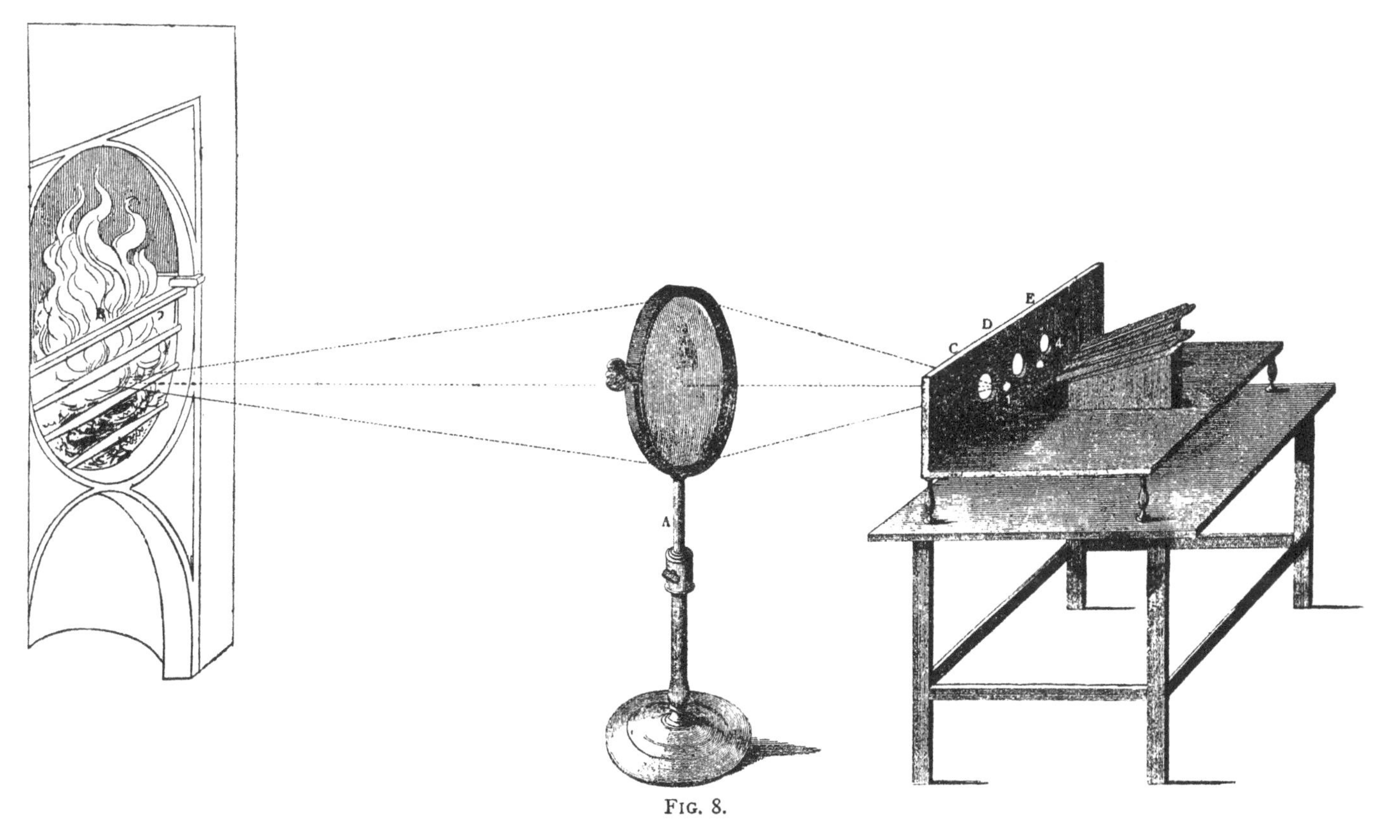

Fig. 8.

thermometer No. 1; and, between the 2d and 3d hole, guarded from the direct rays of the fire by the partition, at the same distance from the back, was put No. 4. Things being thus arranged, the situation of the apparatus which carried the thermometers, and that where the lens was fixed, were marked. Then the thermometers, having been taken away to be cooled, were restored to their places again, and their progress marked as follows.

	0	1½′	3′	5′	7′	9′
No. 1, burning lens	58	65	68	70	71¼	71½
No. 4, screened	58	60	61	61½	61¾	61¾

Here, in nine minutes, the rays coming from the fire, through the burning glass, gave 9¾ degrees of heat more to the thermometer No. 1, than No. 4, from change of temperature, had received behind the screen. Now, to determine whether this was owing merely to a transmission of heat through the glass, or to a condensation of the rays, by the refraction of the burning lens, I took away the lens, as soon as the last observation of the thermometers was written down, and continued to take down their progress as follows.

	9½′	11′	12′	—	14½′
No. 1 . . .	71½	70½	70¼	69½	69¼
No. 4 . . .	61¾	61¾	61¾	61¾	61¾

Here the direct rays of the fire, we see, could not keep up the thermometer No. 1; which lost 2¼ degrees of heat, notwithstanding the lens intercepted no longer any of them. I now restored the burning glass, and continued.

	15′	16′	17′	20′	25′
No. 1 . . .	69¼	69½	70	70¾	71
No. 4 . . .	61¾	61¾	61¾	61¾	61¾

Here again, the lens acted as a condenser of heat, and gave 1¾ degrees of it to the thermometer No. 1. I now once more took away the lens, and continued the experiment.

	25½′	31′
No. 1 . . .	71	68
No. 4 . . .	61¾	61¾

This again confirms the same, by a loss of 3 degrees of heat. I restored the lens once more, and had as follows.

	31½′	35′
No. 1, burning lens . . .	68	69½
No. 4, screened . . .	61¾	61¾

And here the thermometer received 1½ degree of heat again; so that, in the course of 35 minutes, the thermometer No. 1 was alternately raised and depressed five times, by rays which came from the chimney fire, and were subject to laws of refraction, not sensibly different from those which affect light.

15th Experiment. Refraction of the Heat of red-hot Iron.

I caused a lump of iron to be forged into a cylinder of 2½ inches diameter, and 2½ inches long [see fig. 9]. This, being made red-hot, was stuck upon an iron handle fixed on a stand, so as to present one of its circular faces to a lens placed at 2·8 inches distance; its focus being 1·4 inch, and diameter 1·1. Before the lens, at some distance, was placed a screen of wood, with a hole of an inch diameter in it, by way of limiting the object, that its image in the focus might not be larger than necessary. The screen also served to keep the heat from the thermometers. No. 2 was situated in the secondary focus of the lens; and No. 3 was placed within $\frac{3}{10}$ of an inch of it, and at the same distance from the lens as No. 2. By this arrangement, both thermometers were equally within the reach of transmitted heat; or, if there was any difference, it could only be in favour of No. 3, as being behind a part of the lens which, on account of its thinness, would stop less heat than the middle. Now, as the experiment gives a result which differs from what would have arisen from the situation of the thermometers, on a supposition of transmitted heat, we can only ascribe it to a condensation of it by the refraction of the lens; and, in this case, the thermometer No. 3, by its situation, must have been partly within the reach of the heat-image formed in the focus. During the experiment, the thermometers were

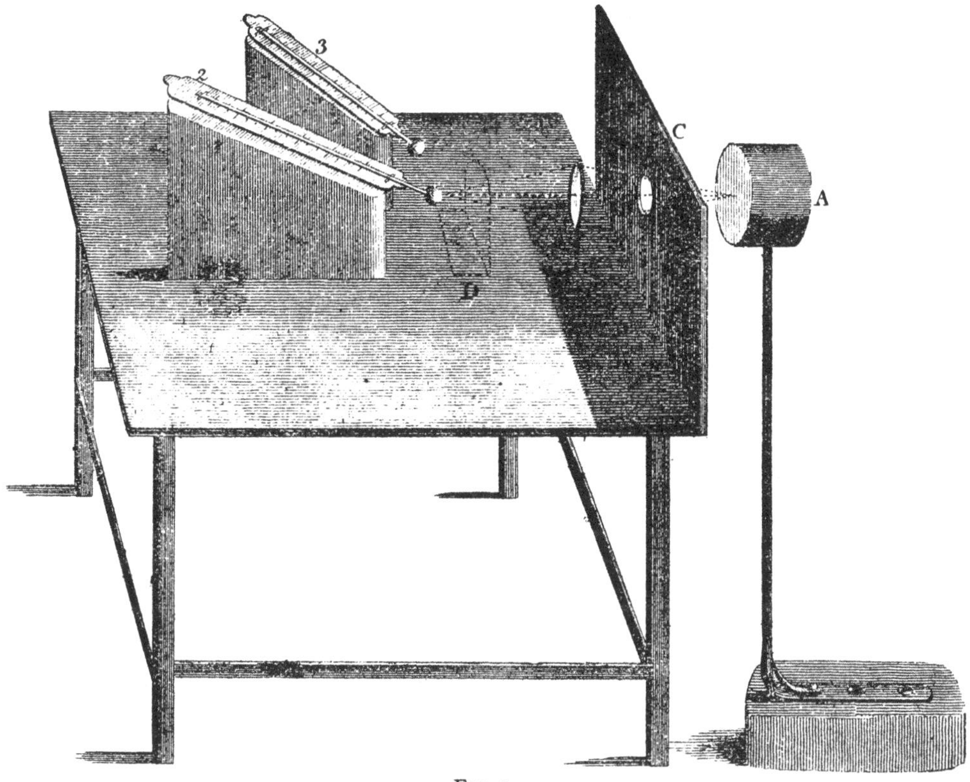

FIG. 9.

alternately screened two minutes from the effects of the lens, and exposed to it for the same length of time; and the result was as follows.

	Screened.	Open.	Screened.	Open.	Screened.	Open.
	0′	2′	4′	6′	8′	10′
No. 2, in the focus	56	62	59	61	$58\frac{1}{4}$	$59\frac{1}{2}$
No. 3, near the focus	56	60	58	59	$57\frac{3}{4}$	$58\frac{1}{4}$

Here, in the first and second minutes, No. 2 gained two degrees of heat more than No. 3. In the third and fourth, it lost one more than No. 3. In the fifth and sixth, it gained one more. In the seventh and eighth, it lost $1\frac{1}{2}$ more; and in the ninth and tenth, it gained $\frac{3}{4}$ more than the other thermometer. This plainly indicates its being acted upon by refracted heat. Lest there should remain a doubt upon the subject, I now removed the lens, and, putting a plain glass in the room of it, I repeated the experiment, with all the rest of the apparatus in its former situation.

	Screened. 0′	Open. 2′	Screened. 4′	Open. 6′	Screened. 8′	Open. 10′
No. 2, in the focus . . .	57¼	62¼	60½	61	60	60¾
No. 3, near the focus . . .	56¾	61¾	60	60½	59½	60¼

Here we find, that both thermometers received heat and parted with it always in equal quantities, which confirms the experiment that has been given. And thus it is evident, that there are rays issuing from red-hot iron, which are subject to laws of refraction, nearly equal to those which affect light; and that these rays are invested with a power of causing heat in bodies.

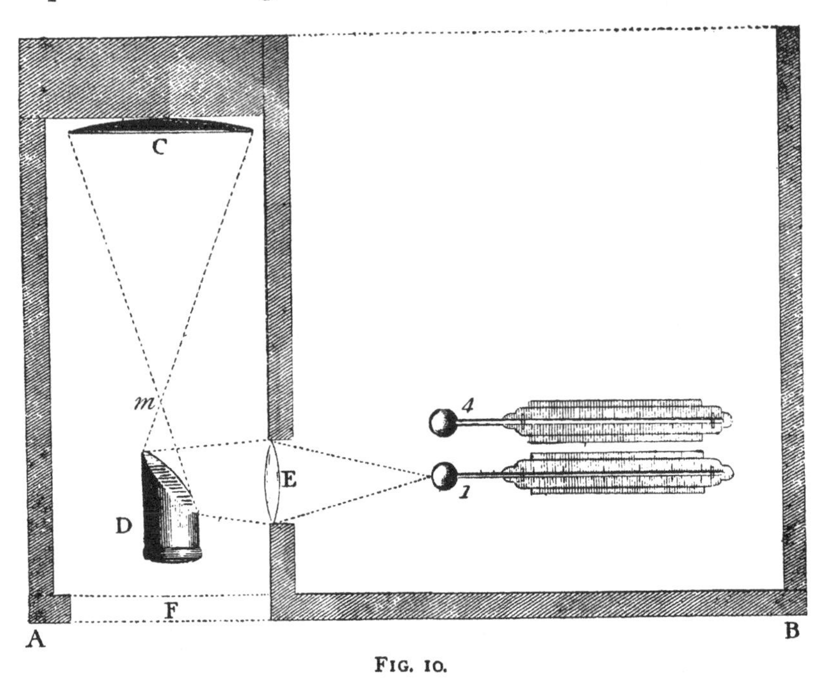

FIG. 10.

16th Experiment. Refraction of Fire-heat, by an Instrument resembling a Telescope.

It occurred to me, that I might use a concave mirror, to condense the heat of the fire in the grate of my chimney, and, reflecting it sideways by a plain mirror, I might afterwards bring it to a secondary focus by a double convex lens; and that, by this construction, I should have an instrument much like a NEWTONIAN telescope [see fig. 10]. The thermometer would figuratively become the observer of heat, by being applied to the place where, in the real telescope of the same construction, the eye is situated to receive light. Having put together the different parts, in such a way as I supposed would answer the end, I tried the effect by a candle, in order to ascertain the proper distance of the object-mirror from the bars of the chimney-grate. The front of the apparatus was guarded by an iron plate, with a thick lining of wood; and the two thermometers which I used, were parted from the mirrors and lens by a partition, which screened them from the heat that was to be admitted through a proper opening in the front plate, to come at the object-mirror. In the partition was likewise an opening, of a sufficient diameter to permit the rays to come from the eye-glass to their focus, on the ball of the thermometer No. 1; while No. 4 was placed by the side of it, at less than half an inch distance. In the experiment, the object-mirror was alternately covered by a piece of pasteboard, and opened again. The thermometers were read off every minute; but, to shorten my account, I only give the last minute of every change.

	The mirror covered.	The mirror open.	Covered.	Open.	Covered.	Open.	Covered.
	0′	8′	16′	21′	27′	37′	47′
No. 1, in the focus .	77½	84	86½	89¾	89¾	91½	84
No. 4, near the focus .	77½	76	79½	81	82¼	83½	77

Here, in the first eight minutes, the thermometer exposed to the effects of the fire-instrument, gained 8 degrees of heat more than the other. In the next 8 minutes, the mirror being covered, it gained 1 degree less than the other. The mirror being now opened again, it gained, in 5 minutes, 1¾ degrees more than the other. When covered 6 minutes, it gained 1¼ degree less than No. 4. In the next 10 minutes, when open, it gained ½ degree more; and, in the last 10 minutes, when the fire began to fail, and the mirror was again covered, it lost 1 degree more than the other thermometer. All which can only be accounted for by the heat which came to the thermometer through the fire-instrument; and, as this experiment confirms what has been said before of the refraction of culinary heat, so it also adds to what has already been proved of its reflection. For, in this fire-instrument, the rays which occasion heat could undergo no less than two reflections and two refractions.

17th Experiment. Refraction of the invisible Rays of solar Heat.

I covered one half of Mr. DOLLOND's burning lens with pasteboard, and threw the prismatic spectrum upon that cover [see fig. 7]; then, keeping the last visible red colour one-tenth of an inch from the margin of the pasteboard, I let the invisible rays beyond the spectrum fall on the lens. In the focus of the red rays, or a very little beyond it, I had placed the ball of the thermometer No. 1; and, as near to it as convenient, the small one No. 2. Now, that the invisible solar rays which occasion heat were accurately refracted to a focus, may be seen by the following account of the thermometers.

	0′	1′
No. 1. in the focus	57	102
No. 2, near the focus . . .	57	57

Here, in one minute, these rays gave 45 degrees of heat to the thermometer No. 1, which received them in the focus, while the other, No. 2, suffered no change.

It is remarkable, that notwithstanding I kept the red colour of the spectrum $\frac{1}{10}$ of an inch upon the pasteboard, a little of that colour might still be seen on the ball of the thermometer. This occasioned a surmise, that possibly the invisible rays of the sun might become visible, if they were properly condensed; I therefore put this to the trial, as follows.

18th Experiment. Trial to render the invisible Rays of the Sun visible by Condensation.

Leaving the arrangement of my apparatus as in the last experiment, I withdrew the lens, till the last visible red colour was two-tenths of an inch from the margin of the semicircular pasteboard cover; then, taking the thermometers, I had as follows.

	0′	1′
No. 2 . . .	57	78
No. 3 . . .	57	57

Here, there was no longer the least tinge of any colour, or vestige of light, to be seen on the ball of the thermometer; so that, in one minute, it received 21 degrees of heat, from rays that neither were visible before, nor could be rendered so by condensation.

To account for the colour which may be seen in the focus, when the last visible red colour is less than two-tenths of an inch from the margin of the pasteboard which intercepts the prismatic spectrum, we may suppose, that the imperfect refraction of a burning lens, which from its great diameter cannot bring rays to a geometrical focus, will bring some scattered ones to it, which ought not to come there. We may also admit, that the termination of a prismatic spectrum cannot be accurately ascertained, by looking at it in a room not sufficiently dark to make very faint tinges of colour visible. And, to this must be added, that the incipient red rays must actually be scattered over a considerable space, near the confines of the spectrum, on account of the breadth of the prism, the whole of which cannot bring its rays of any one colour properly together; nor can it separate the invisible rays intirely from the visible ones. For, as the red rays will be but faintly scattered in the beginning of the visible spectrum, so, on the other hand, will the invisible rays, separated by the parts of the prism that come next in succession, be mixed with the former red ones. Sir ISAAC NEWTON has taken notice of some imperfect tinges or haziness, on each side of the prismatic spectrum, and mentions that he did not take them into his measures.*

19th Experiment. Refraction of invisible culinary Heat.

There are some difficulties in this experiment; but they arise not so much from the nature of this kind of heat, as from our method of obtaining it in a detached state. A red-hot lump of iron, when cooled so far as to be no longer visible, has but a feeble stock of heat remaining, and loses it very fast. A contrivance to renew and keep this heat might certainly be made, and I should indeed have attempted to carry some method or other of this kind into execution, had not the following trials appeared to me sufficiently conclusive to render it unnecessary. Admitting, as has been proved in the 15th experiment, that the alternate rising and falling of a

* NEWTON'S *Optics*, page 23, line 11.

thermometer placed in the focus of a lens, when the ball of it is successively exposed to, or screened from, its effects, is owing to the refraction of the lens, and cannot be ascribed to a mere alternate transmission and stoppage of heat, I proceeded as follows [see fig. 9]. My lens, 1·4 focus, and 1·1 diameter, being placed 2·8 inches from the face of the heated cylinder of iron, the thermometer No. 2, in its focus, was alternately guarded by a small pasteboard screen put before it, and exposed to the effects of condensed heat by removing it.

	Screened.	Open.	Screened.	Open.
	0′	2′	4′	6′
No. 2. . .	55 Very red-hot.	63½ Red-hot.	58 Still red-hot.	60½ Still red.

	Screened.	Open.	Screened.	Open.	Screened.	Open.	Screened.	Open.	Screened.	Open.	Screened.
	8′	10′	12′	14′	16′	18′	20′	22′	24′	26′	28′
No. 2.	57½	59¼	57½	58½	57½	58¼	57¼	58	57½	58	57½
	A little red.	Doubtful.	Not visible in my room darkened.								

Now, the beginning of this experiment being exactly like that of the 15th, with the thermometer No. 3 left out, the arguments that have before proved the refraction of heat in one state, will now hold good for the whole. For here we have a regular alternate rising and falling of the thermometer, from a bright red heat of the cylinder, down to its weakest state of black heat ; where the effect of the rays, condensed by the lens, exceeded but half a degree the loss of those that were stopped by it.

20th Experiment. Confirmation of the 19th.

In order to have some additional proof, besides the uniform and uninterrupted operation of the lens in the foregoing experiment, I repeated the same, with an assistant thermometer, No. 3, placed first of all at ¾ of an inch from No. 2, and more towards the lens, but so as to be out of the converging pencil of its rays, and also to allow room for the little screen between the two thermometers, that No. 3 might not be covered by it.

	Screened.	Open.	Screened.	Open.	Screened.	Open.	Screened.	Open.	Screened.
	0′	1′	2′	3′	4′	5′	6′	7′	8′
No. 2, in the focus .	62½	63¾	62⅞	64	63¾	64⅞	64½	64¾	64½
No. 3, advanced sideways ; always open .	63	64	64	64½	64½	64½	64½	64	64

Here No. 3, being out of the reach of refraction, gradually acquired its maximum of heat, in consequence of an uniform exposure to the influence of the hot cylinder ; after which, it began to decline. No. 2, on the contrary, came to its maximum by alternate great elevations, and small depressions ; and afterwards lost its heat by great depressions, and small elevations. After the first eight minutes, I changed the place of the assistant thermometer, by putting it into a still more decisive situation ; for it was now placed by the side of that in the focus, so as to participate of the alternate screening, and also to receive a small share of one side of the invisible heat-image, which, though unseen, we know must be formed in the focus of the lens. Here, if our reasoning be right, the assistant thermometer should be affected by alternate risings and fallings; but they should not be so considerable as those of the lens.

	Both open. 8′	Both open. $9\frac{1}{2}$′
No. 2, in the focus . .	$64\frac{1}{2}$	$63\frac{3}{4}$
No. 3, in the edge of it . .	64	$63\frac{3}{4}$

	Open. 11′	Screened. $12\frac{1}{2}$′	Open. 14′	Screened. 16′	Open. 18′
No. 2, in the focus	$64\frac{1}{4}$	63	$63\frac{3}{4}$	$62\frac{3}{4}$	$63\frac{3}{4}$
No. 3, in the edge of it	64	$63\frac{1}{4}$	$63\frac{1}{2}$	63	$63\frac{3}{4}$

Here the changes of the thermometer No. 2 were $-\frac{3}{4}+\frac{1}{2}-1\frac{1}{4}+\frac{3}{4}-1+1$; and those of No. 3 were $-\frac{1}{4}+\frac{1}{4}-\frac{3}{4}+\frac{1}{4}-\frac{1}{2}+\frac{3}{4}$. All which so clearly confirm the effect of the refraction of the lens, that it must now be evident that there are rays issuing from hot iron, which, though in a state of total invisibility, have a power of occasioning heat, and obey certain laws of refraction, very nearly the same with those that affect light.

As we have now traced the rays which occasion heat, both solar and terrestrial, through all the varieties that were mentioned in the beginning of this paper, and have shewn that, in every state, they are subject to the laws of reflection and of refraction, it will be easy to perceive that I have made good a proof of the three first of my propositions. For, the same experiments which have convinced us that, according to our second and third articles, heat is both reflexible and refrangible, establish also its radiant nature, and thus equally prove the first of them.

END OF THE FIRST PART.

Slough, near Windsor.
April 26, 1800.

EXPLANATION OF THE FIGURES.

SEE FIGS. 1–10.*

Fig. 1.

Shews the arrangement of the apparatus used in the 2d experiment.
A, is the small mirror with its adjusting screws *m*, *n*.
No. 2, is the thermometer in the focus of the mirror.
No. 3, The assistant thermometer.
B, A small screen for the thermometer No. 2.
C, The candle.
D, The poker which, in the 4th and 10th experiments, is to be placed in the situation of the candle; the rest of the apparatus being brought nearer to it.

* [Plates XII. to XVI. in the *Phil. Trans.*]

Fig. 2.

Shews the apparatus used in the 3d and 8th experiments.
A, The mirror.
No. 2, The thermometer.
BCD, A desk adjustable to different altitudes.
E, The prism receiving the sun's rays through an opening in the window shutter F.

Fig. 3.

AB, is the front of the apparatus, which, in the 5th experiment, is exposed to the fire of the chimney.
C, is the opening in the front plate AB, for the admission of heat.
D, is the small mirror which reflects the rays of heat.
E, is the hole through which the heat passes to the thermometers.
No. 1 and No. 4, are the thermometers.
F, is a prism, which, in the 6th experiment, is to be placed in the room of the mirror D.

Fig. 4.

A, is the board that holds the apparatus used in the 7th experiment.
B, The prism.
C, The spectrum, thrown partly on the paper with parallel lines, and partly on one of the small tables which support the board.
D, The mirror which reflects the rays of heat sideways.
No. 1, The thermometer which receives the reflected rays.
No. 2, The standard thermometer.

Fig. 5.

AB, is the front which, in the 9th experiment, is put close to the flat side of a heated iron stove.
C, is the mirror.
D, The feather-edged slip of deal, on two pins.
No. 2, The thermometer which receives the rays condensed in the focus of the mirror.
No. 3, The standard thermometer.
E, A small screen tied to No. 3, to guard it from reflected heat.

Fig. 6.

A, The lens in the apparatus used for the 12th experiment.
No. 2, The thermometer placed in its focus.
B, The screen with an aperture for admitting the rays of heat.
C, The eccentric pivot for turning away the lens.
D, The candle.

Fig. 7.

A, The burning lens, covered; with the prismatic spectrum thrown upon an opening, left for it, in the pasteboard cover of the 13th experiment.
No. 3, The thermometer placed in its focus.
B, The prism.
C, Semicircular cover, used in the 17th and 18th experiments, instead of the one with a square hole.

Fig. 8.

A, The burning lens of the 14th experiment.
B, The fire in the chimney.
No. 1, The thermometer in the focus of the lens.
No. 4, The standard thermometer.
C, The hole through which the rays of heat pass to No. 1.
D and E, Two holes, between which the ball of the thermometer No. 4 is screened from the direct rays of the fire; while free access is given to the heat which may affect the temperature of the place.

Fig. 9.

A, The iron cylinder, stuck upon its handle, as it is used in the 15th and 19th experiments.
B, The lens.
C, The screen with an opening in it.
No. 2, The thermometer in the focus of the lens.
No. 3, The standard thermometer.
D, The little moveable pasteboard screen.

Fig. 10.

AB, The front, plated with iron, that it may bear to be exposed close to the bars of a chimney fire.
C, The concave mirror.
D, The plain mirror.
E, The lens.
No. 1, The thermometer in the focus of the lens.
No. 4, The standard thermometer.
F, A circular opening in the front plate AB, for admitting the rays of heat to fall on the concave mirror C.
m, The first focus of the rays, from which they go on diverging, to the small mirror, and to the lens; which brings them to a second focus, on the ball of the thermometer No. 1.

Part II.

Read November 6, 1800.

[*Phil. Trans.*, 1800, pp. 437–538.]

In the first part of this Paper it has been shewn, that heat derived immediately from the sun, or from candent terrestrial substances, is occasioned by rays emanating from them; and that such heat-making rays are subject to the laws of reflection, and of refraction. The similarity between light and heat, in these points, is so great, that it did not appear necessary to notice some small difference between them, relating to the refraction of rays to a certain focus, which will be mentioned hereafter. But the next three articles of this Paper will require, that while we shew the similarity between light and heat, we should at the same time point out some striking and substantial differences, which will occur in our experiments on the rays which occasion them, and on which hereafter we may proceed to argue, when the question reserved for the conclusion of this Paper, whether light and heat be occasioned by the same or by different rays, comes to be discussed.

Article iv.—*Different Refrangibility of the Rays of Heat.*

We might have included this article in the first part of this Paper, as a corollary of the former three; since rays that have been separated by the prism, and have still remained subject to the laws of reflection and refraction, as has been shewn, could not be otherwise than of different refrangibility; but we have something to say on this subject, which will be found much more circumstantial and conclusive than what might have been drawn as a consequence from our former experiments.

However, to begin with what has already been shewn, we find that two degrees of heat were obtained from that part of the spectrum which contains the violet rays, while the full red colour, on the opposite side, gave no less than seven degrees; * and these facts ascertain the different refrangibility of the rays which occasion heat, as clearly as that of light is ascertained by the dispersion and variety of the colours. For, whether the rays which occasion heat be the same with those which occasion the colours, which is a case that our foregoing experiments have not ascertained, the arguments for their different refrangibility rests on the same foundation,

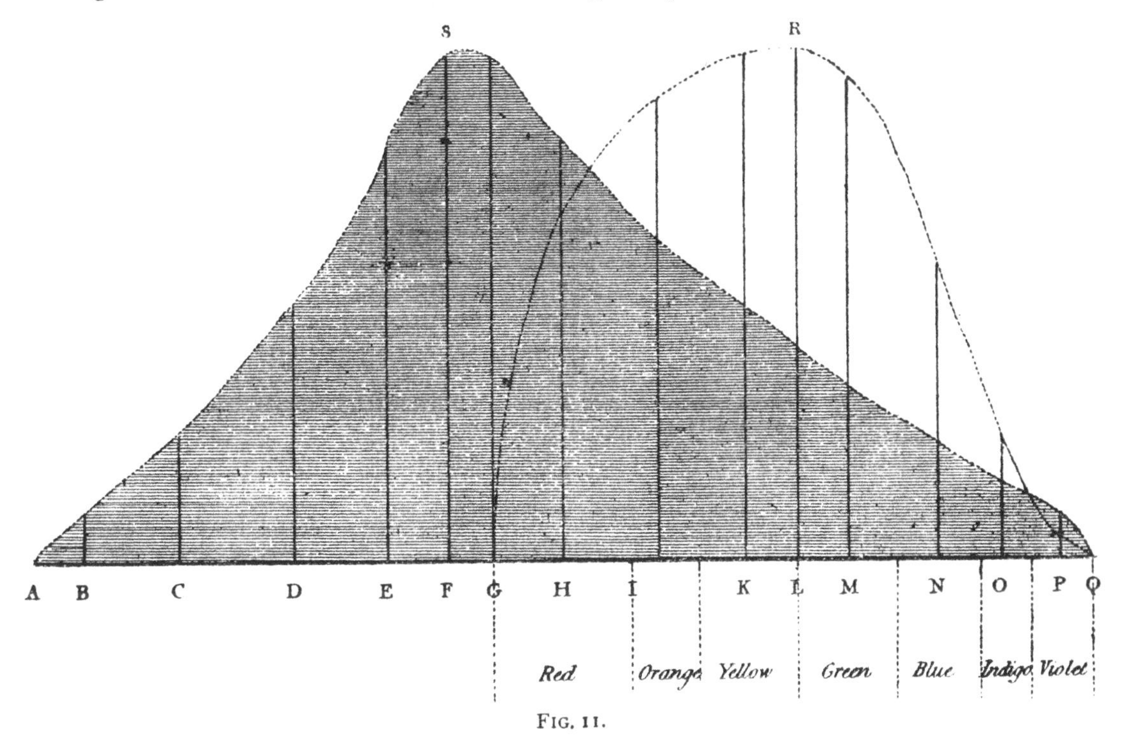

FIG. 11.

namely, their being dispersed by the prism; and that of the rays of light being admitted, the different refrangibility of the rays of heat follows of course. So far then, a great resemblance again takes place.

I must now point out a very material difference, which is, that the rays of heat are of a much more extensive refrangibility than those of light. In order to make this appear, I shall delineate a spectrum of light, by assuming a line of a certain length; and, dividing it into seven parts, according to the dimensions assigned to the seven colours by Sir ISAAC NEWTON, in the fourth figure of the second part of his *Optics*, I shall represent the illuminating power of which each colour is possessed, by an ordinate drawn to that line. And here, as the absolute length of the ordinates is arbitrary, provided they be proportional to each other, I shall assume the length of that which is to express the maximum, equal to $\frac{27}{33}$ of the whole line.

Thus, let GQ [fig. 11] represent the line that contains the arrangement of the

* See 2d and 4th experiments, pages 258 and 259. [Above, p. 55.]

colours, from the red to the violet. Then, erecting on the confines of the yellow and green the line LR = $\frac{27}{33}$ of GQ, it will represent the power of illumination of the rays in that place. For, by experiments already delivered, we have shewn that the maximum of illumination is in the brightest yellow or palest green rays.* From the same experiments we collect, that the illuminations of yellow and green are equal to each other, and not much inferior to the maximum; this gives us the ordinates K and M. Then, by the rest of the same experiments, we obtain also the ordinates H, I, N, O, P, with sufficient accuracy for the purpose here intended. All these being applied to the middle of the spaces which belong to their respective colours, we have the figure GRQG, representing what may be called the spectrum of illumination.

We are now, in the same manner, to find a figure to express the heating power of the refracted prismatic rays, or what may be called the spectrum of heat. In order to determine the length of our base, I examined the extent of the invisible rays, and found, that at a distance of two inches beyond visible red, my thermometer, in a few minutes, acquired 1¼ degree of heat. The extent of the coloured spectrum at that time, or the line which answers to GQ in my figure, measured 2·997 inches. If two inches had been the whole of the extent of the invisible part, it might be stated to be in proportion to the visible one as 2 to 3; but we are to make some allowance for a small space required beyond the last ordinate, that the curve of the heating power drawn through it may reach the base; and indeed, at 2¼ inches beyond visible red, I could still find ½ degree of heat. It appears therefore sufficiently safe, to admit the base of the spectrum of heat AQ, to be to that of the spectrum of light GQ, as 5¼ to 3; or, conforming to the NEWTONIAN figure before mentioned, the base of which is 3·3 inches, as 57¾ to 33. Now, if we assume for the maximum of heat, an ordinate of an equal length with that which was fixed upon for the maximum of light, it will give us a method of comparing the two spectra together. Accordingly, I have drawn the several ordinates B, C, D, E, F, G, H, I, K, L, M, N, O, P, of such lengths as, from experiments made on purpose, it appeared they should be, in order to express the heat indicated by the thermometer, when placed on the base, at the several stations pointed out by the letters.

A mere inspection of the two figures, which have been drawn as lying upon one another, will enable us now to see how very differently the prism disperses the heat-making rays, and those which occasion illumination, over the areas ASQA, and GRQG, of our two spectra! These rays neither agree in their mean refrangibility, nor in the situation of their maxima. At R, where we have most light, there is but little heat; and at S, where we have most heat, we find no light at all!

* See page 262. [Above, p. 57.]

21st Experiment. The Sines of Refraction of the heat-making Rays, are in a constant Ratio to the Sines of Incidence.

I used a prism with a refracting angle of 61 degrees; and, placing the thermometer No. 4 half an inch, and No. 1 one inch, beyond the last visible red colour, I kept No. 2 by the side of the spectrum, as a standard for temperature.

	0′	2′	5′	8′
At ½ inch, No. 4 . . .	64	67	69	69½
At 1 inch, No. 1 . . .	64	66	67	67½
Standard, No. 2 . . .	63½	63½	63½	63½

Here, in eight minutes, the thermometer at half an inch from visible colour, rose 5½ degrees; and, at one inch from the same, the other thermometer rose 3½; while the temperature, as appears by No. 2, remained without change.

I now took a prism with an angle of forty-five degrees, and, placing the thermometers as before, I had as follows:

	0′	2′	5′	8′
At ½ inch, No. 4 . . .	55	59	61	62
At 1 inch, No. 1 . . .	55	57	58	58¾
Standard, No. 2 . . .	55	54¾	55	55

Here we likewise had, in 10 minutes, a rise of 7 degrees in the thermometer No. 4, and of 3¾ in No. 1; while No. 2 remained stationary.

I tried now all the three angles of a prism of whitish glass: they were of 63, 62, and 55 degrees; and I found invisible rays of heat to accompany all the visible spectra given by these angles.

I tried a prism of crown glass, having an angle of 30 degrees; and found invisible heat rays as before.

I tried a prism of flint glass, with so small an angle as 19 degrees, and again found invisible heat rays.

I made a hollow prism, by cementing together three slips of glass of an equal length, but unequal breadth, so as to give me different refracting angles: they were of 51°, 62° 30′, and 66° 30′. Then, filling it with water, and receiving the spectrum, when exposed to the sun, as usual, on the table, I placed the thermometer No. 1 at 45 inch behind the visible red colour, and No. 5 in the situation of the standard. The refracting angle of the prism was 62° 30′; and, in five minutes, the thermometer received 1⅝ degrees of heat from the invisible rays. On trying the other angles, I likewise found invisible heat rays, in their usual situation beyond the red colour.

Now, setting aside a minute inquiry into the degrees of heat occasioned by these invisible rays, I shall here only consider them as an additional part, annexed to the different quantities of heat which are found to go along with the visible spectrum; in the same manner as if, in the spectrum of light, another colour had been added beyond the red. Then, as from the foregoing experiments it appears, that a change of the refracting medium, and of the angle by which the refractions were made, occasioned no alteration in the relative situation of the additional part AG, with respect to GQ; and, as the part GQ is already known to follow the law of refraction we have mentioned, it is equally evident, that the additional heat of AG must follow the same law. We do not enter into the dispersive power of different mediums with respect to heat, since that would lead us farther than the present state of our investigation could authorise us to go; the following experiment however will shew that, as with light so with heat, such dispersive power must be admitted.

22d Experiment. Correction of the different Refrangibility of Heat, by contrary Refraction in different Mediums.

I took three prisms; one of crown glass, having an angle of 25 degrees; another of flint glass, with an angle of 24; and a third of crown glass, with an angle of 10 degrees. These being put together, as they are placed when experiments of achromatic refractions are to be made, I found that they gave a spectrum nearly without colour. The composition seemed to be rather a little over adjusted; there being a very faint tinge of red on the most refracted side, and of violet on the least refracted margin. I examined both extremes by two thermometers; keeping No. 3 as a standard, while No. 2 was applied for the discovery of invisible rays; but I found no heat on either side. After this, I placed No. 2 in the middle of the colourless illumination; and in a little time it rose two degrees, while No. 3 still remained unaltered at some small distance from the spectrum. This quantity was full as much as I could expect, considering the heat that must have been intercepted by three prisms. Thus then it appears, that the different refrangibility of heat, as well as that of light, admits of prismatic correction. And we may add, that this experiment also tends to the establishment of the contents of the preceding one; for the refrangibility of heat rays could not be thus corrected, were the sines of refraction not in a constant ratio to those of incidence.

23d Experiment. In Burning-glasses, the Focus of the Rays of Heat is different from the Focus of the Rays of Light.

I placed my burning lens, with its aperture reduced to three inches, in order to lessen the aberration arising from the spherical figure, in the united rays of the sun; and, being now apprised of the different refrangibility of the rays of heat, and knowing also that the least refrangible of them are the most efficacious, I examined the focus of light, by throwing hair-powder, with a puff, into the air. This pointed out the mean focus of the illuminating rays, situated in that part of the pencil which opticians have shewn to be the smallest space into which they can be collected. That this may be called the focus of light, our experiments, which have proved the maximum of illumination to be situated between the yellow and green, and therefore among the mean refrangible rays of light, have fully established. The mean focus being thus pointed out by the reflection of light on the floating particles of powder, I held a stick of sealing wax 1″·6, or four beats of my chronometer, in the contracted pencil, half an inch nearer to the lens than the focus. In this time, no impression was made upon the wax. I applied it now half an inch farther from the lens than that focus ; and, in 8-tenths of a second, or two beats of the same chronometer, it was considerably scorched. Exposing the sealing wax also to the focus of light, the effect was equally strong in the same time ; from which we may safely conclude, notwithstanding the little accuracy that can be expected, for want of a more proper apparatus, from so coarse an experiment, that the focus of heat, in this case, was certainly farther removed from the lens than the focus of light, and probably not less than $\frac{1}{4}$ of an inch ; the heat, at half an inch beyond the focus of light, being still equal to that in the focus.

ARTICLE V.—*Transmission of heat-making Rays.*

We enter now on the subject of the transmission of heat through diaphanous bodies. Our experiments have hitherto been conducted by the prism, the lens, and the mirror ; these may indeed be looked upon as our principal tools, and, as such, will stand foremost in all our operations ; but the scantiness of this stock cannot allow us to bring our work to perfection. Nor is it merely the want of tools, but rather the natural imperfection of those we have, that hinders our rapid progress. The prism which we use for separating the combined rays of the sun, refracts, reflects, transmits, and scatters them at the same time ; and the laws by which it acts, in every one of these operations, ought to be investigated. Even the cause of the most obvious of its effects, the separation of the colours of light, is not well understood ; for, in two prisms of different glass, when the angles are such as to give the same mean refraction, the dispersive power is known to differ. Their transmissions have been still less ascertained ; and I need not add, that the internal and external reflexions, and the scattering of rays on every one of the surfaces, are all of such a nature as must throw some obscurity on every result of experiments made with prisms. A lens partakes of all the inconveniences of the prism ; to which its own defects of spherical aberrations must be added. And a mirror, besides its natural incapacity of separating the rays of light from the different sorts of heat, scatters them very profusely. But, if we have been scantily provided with materials to act upon rays, it has partly been our own fault : every diaphanous body may become a new tool, in the hands of a diligent inquirer.

My apparatus for transmitting the rays of the sun is of the following construction [fig. 12]. In a box, 12 inches long, and 8 inches broad, are fixed two thermometers. The sides of the box are $2\frac{1}{4}$ inches deep. That part of the box where the balls of the thermometers are, is covered by a board, in which are two holes of $\frac{3}{4}$ inch diameter, one over each of the balls of the thermometers; and the bottom of the box, under the cover, is cut away, so as to leave these balls freely exposed. There is a partition between the two thermometers, in that part of the box which is covered, to prevent

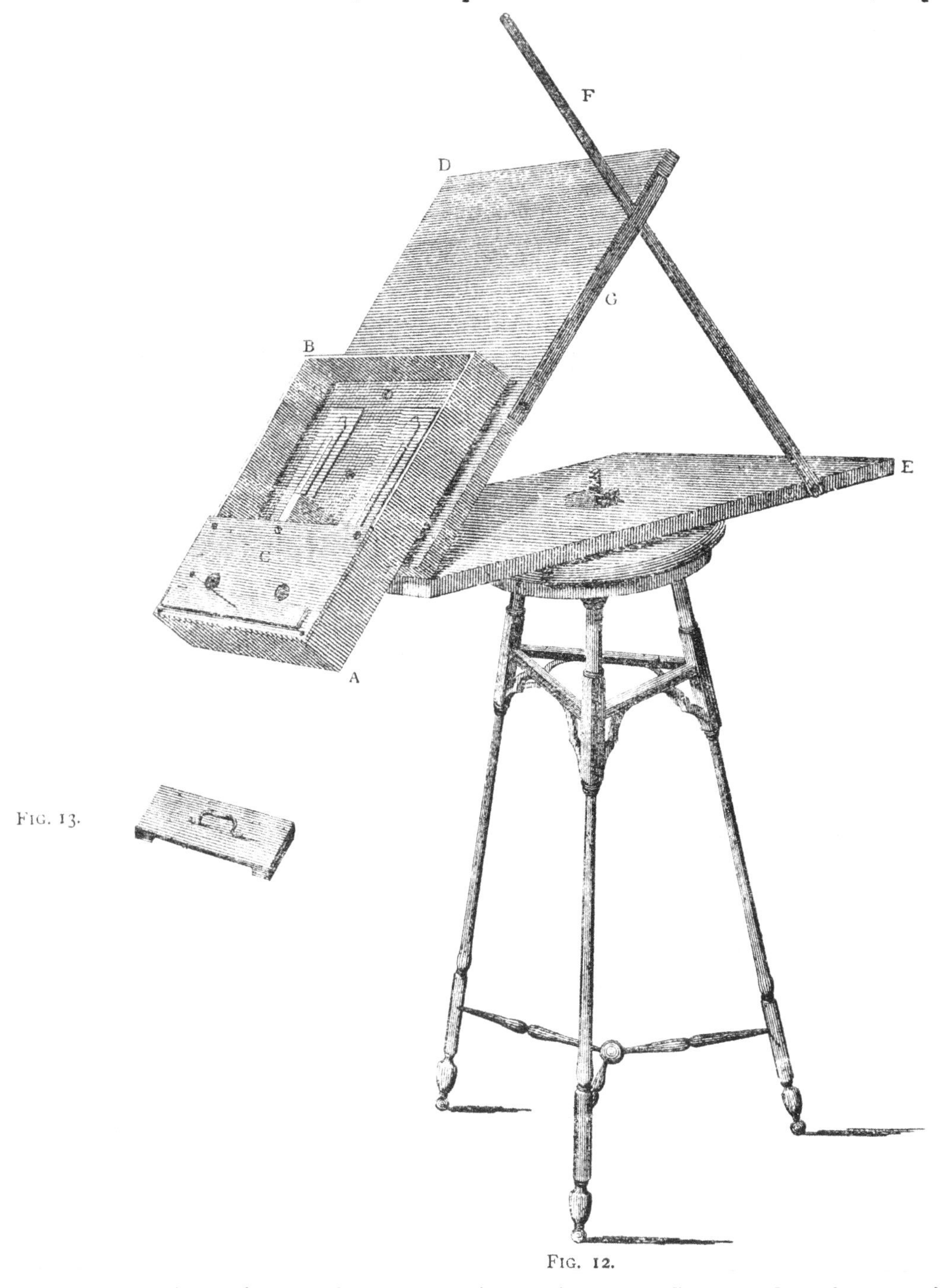

Fig. 13.

Fig. 12.

the communication of secondary scatterings of heat. Just under the opening of the transmitting holes, on the outside of the cover, is fixed a slip of wood, on which may rest any glass or other object, of which the transmitting capacity is to be ascertained. A thin wooden cover is provided [fig. 13], that it may be laid over the transmitting holes, occasionally, to exclude the rays of the sun; and, on the middle of the slip of wood, under the holes, a pin is to be stuck perpendicularly, that its shadow may point out the situation of the box with respect to the sun. The box, thus prepared, is to be fastened upon two short boards, joined together by a pair of hinges. A long slip of mahogany is screwed to the lowest of these boards, and lies in the hollow part of a long spring, fastened against the side of the upper one. The pressure of the spring must be sufficiently strong to keep the boards at any angle; and the slip of mahogany long enough to permit an elevation of about 85 degrees.

In order to see whether all be properly adjusted, expose the apparatus to the sun, and lift up the board which carries the box, till the directing pin throws the shadow of its head on the place where the point is fastened. Then hold a sheet of paper under the box, and, if the thermometers have been properly placed, the shadow of their balls will be in the centre of the rays passing through the transmitting holes to the paper.

A screen of a considerable size [fig. 14], with a parallelogrammic opening should be placed at a good distance, to keep the sun's rays from every part of the apparatus, except that which is under the cover; and no more sun should be admitted into the room, than what will be completely received on the screen, interposed between the window and the apparatus.

FIG. 14.

As one of the thermometers is to indicate a certain quantity of heat coming to it by the direct ray, while the other is to show how much of it is stopped by the glass laid over the transmitting hole, it becomes of the utmost consequence to have two thermometers of equal sensibility.* The difficulty of getting such is much greater than can be imagined: a perfect equality in the size and thickness of the balls is, however, the most essential circumstance. When two are procured, they should be tried in quick and in slow exposures. These terms may be explained by referring to fire heat; for here the thermometers may be exposed so as to acquire, for instance, 30 degrees of heat in a very short time; which may then be called a quick exposure: or they may be placed so as to make it require a good while to raise them so many degrees: on which account the exposure may be called slow. It is true, that we have it not in our power to render the sun's rays more or less efficacious, and therefore cannot have a quick or slow exposure at our command; but a great difference will be found in the heat of a rising, or of a meridian sun: not to mention a variety of other causes, that influence the transmission of heat through the atmosphere. Now, when thermometers are tried in various exposures, they should traverse their scales together with constant equality; otherwise no dependance can be placed on the results drawn from experiments made with them, in cases where only a few minutes can be allowed for the action of the cause whose influence we are to investigate.

The balls must not be blacked; for, as we have already to encounter the transmitting capacity of the glass of which these balls are made, it will not be safe to add to this the transmitting disposition of one or more coats of blacking, which can never be brought to an equality, and are always liable to change, especially in very quick exposures.

Transmission of Solar Heat through colourless Substances.

24th Experiment.

I laid a piece of clear transparent glass, with a bluish-white cast, upon one of the holes of the transmitting machine: the faces of this glass are parallel, and highly polished. Then, putting the cover over both holes, I placed the machine in the situation where the experiment was to be made, and let it remain there a sufficient time, that the thermometers might assume a settled temperature. For this purpose, an assistant thermometer, which should always remain in the nearest

* The theory of the sensibility of thermometers, as far as it depends on the size of the balls, may be considered thus. Let D, d, S, s, T, t be the *diameters*, the points on which the *sun* acts, and the points on which the *temperature* acts, of a large and a small thermometer having spherical balls; and let $x : y$ be the intensity of the action of the sun, to the intensity of the action of the temperature, on equal points of the surface of both thermometers. Then we have $s : S :: d^2 : D^2$, and $t : T :: 4d^2 : 4D^2$. The action of the sun therefore will be expressed by $d^2 x$, $D^2 x$; and that of the temperature by $4d^2 y$, $4D^2 y$; and the united action of both by $\overline{x-4y} \times d^2$, $\overline{x-4y} \times D^2$; which are to each other, as $d^2 : D^2$. Now, the total effect being as the squares of the diameters, while $x : y$ remain in their incipient ratio, and the contents of the thermometers being as the cubes, the sensible effect produced on the particles of mercury, must be as $\frac{d^2}{d^3} : \frac{D^2}{D^3} :: \frac{1}{d} : \frac{1}{D}$; that is, inversely as the diameters. The small thermometer therefore will set off with a sensibility greater than that of the large one, in the same ratio.

convenient place to the apparatus, will be of use, to point out the time when the experiment may be begun; for this ought not to be done, till the thermometers to be used agree with the standard. In order not to lose time after an experiment, the apparatus may be taken into a cool room, or current of air, till the thermometers it contains are rather lower than the standard; after which, being brought to the required situation, they will soon be fit for action.

All these precautions having been taken, I began the experiment by first writing down the degrees of the thermometers; then, opening the cover at the time that a clock or watch showing seconds came to a full minute, I continued to write down the state of the thermometers for not less than five minutes. The result was as follows.

	0′	1′	2′	3′	4′	5′
No. 5, Sun . .	67	$68\frac{3}{4}$	$70\frac{1}{8}$	$71\frac{3}{8}$	$72\frac{3}{8}$	73
No. 1, Bluish-white glass	67	$68\frac{1}{8}$	$69\frac{1}{8}$	70	$70\frac{7}{8}$	$71\frac{1}{2}$...6 : $4\frac{1}{2}$ = ·750

Here the sun communicated, in 5 minutes, 6 degrees of heat to the thermometer No. 5, which was openly exposed to its action; while, in the same time, No. 1 received only $4\frac{1}{2}$ degrees by rays transmitted through the bluish-white glass: then, as 6 : $4\frac{1}{2}$: : 1 : ·750. This shews plainly, that only $\frac{3}{4}$ of the incident heat were transmitted, and therefore that $\frac{1}{4}$ of it was intercepted by the glass.

I shall here, as well as in the following experiments, point out the difference between heat and light, in order, as has been mentioned before, to lead to an elucidation of our last discussion. To this effect, therefore, I have ascertained, with all the accuracy the subject will admit of, the quantity of light transmitted through such glasses as I have used; but, as it would here interrupt the order of our subject, I have joined, at the end of this Paper, a table, with a short account of the method that has been used in making it, wherein the quantity of light transmitted is set down; and to this table I shall now refer.

To render this comparative view more clear, we may suppose always 1000 rays of heat to come from the object: then, 750 being transmitted, it follows, that the bluish-white glass used in our experiment stops 250 of them; and, by the table at the end of this Paper, it stops 86 rays of light; the number of them coming from the object also being put equal to 1000.

It should be remarked, that when I compare the interception of solar heat with that of the light of a candle, it must not be understood that I take terrestrial to be the same as solar light; but, not having at present an opportunity of providing a similar table for the latter, I am obliged to use the former, on a supposition, that the quantity stopped by glasses may not be very different.

25th Experiment.

I took a piece of flint glass, about $2\frac{1}{2}$ tenths of an inch thick, and fastened it over one of the holes of the transmitting apparatus.

	0′	1′	2′	3′	4′	5′
No. 5, Sun . .	$69\frac{3}{4}$	$71\frac{1}{4}$	$72\frac{5}{8}$	$74\frac{1}{8}$	$74\frac{7}{8}$	$75\frac{1}{4}$
No. 1, Flint glass .	$69\frac{3}{4}$	71	$72\frac{1}{8}$	$73\frac{7}{8}$	74	$74\frac{3}{4}$...$5\frac{1}{2}$: 5 = ·909

Here the heat-making rays gave, in 5 minutes, $5\frac{1}{2}$ degrees to the thermometer No. 5; and, by transmission through the flint glass, 5 degrees to No. 1. Then, proceeding as before, we have $\frac{5}{5\frac{1}{2}} = \cdot 909$; which shews that 91 rays of heat were stopped. In the table before referred to, we find that this glass stops 34 rays of light.

Before I proceed, it will be necessary to adopt a method of reducing the detail of my experiments into a narrower compass. It will be sufficient to say, that they have all been made on the same plan as the two which have been given. The observations were always continued for at least five minutes; and, by examining the ratios of the numbers given by the thermometers in all that time, it may be seen that, setting aside little irregularities, there is a greater stoppage at first than towards the end; but, as it would not be safe to take a shorter exposure than five minutes, on account of the small quantity of heat transmitted by some glasses, I have fixed upon that interval as sufficiently accurate for giving a true comparative view. The experiments therefore may now stand abridged as follows.

26th Experiment.

I took a piece of highly polished crown glass, of a greenish colour, and, cutting it into several parts, examined the transmitting power of one of them, reserving the other pieces for some other experiments that will be mentioned hereafter.

	Sun.	Greenish crown glass.
0′	$66\frac{1}{4}$	$66\frac{1}{4}$
5	73	$71\frac{1}{4}$ . . . $6\frac{3}{4}$: 5 = ·741

This glass therefore stops 259 rays of heat, and 203 of light.

27th Experiment.

I cut likewise a piece of coach glass into several parts, and tried one of them, reserving also the other pieces for future experiments.

	Sun.	Coach glass.
0′	$68\frac{7}{8}$	$68\frac{7}{8}$
5	$75\frac{7}{8}$	$74\frac{3}{8}$ . . . 7 : $5\frac{1}{2}$ = ·786

It stops 214 rays of heat, and 168 of light.

28th Experiment.

I examined a piece of Iceland crystal, of nearly two-tenths of an inch in thickness.

	Sun.	Iceland crystal.
0′	67	67
5	$72\frac{5}{8}$	$71\frac{1}{4}$ . . . $5\frac{5}{8}$: $4\frac{1}{4}$ = ·756

It stops 244 rays of heat, and 150 of light.

29th Experiment.

	Sun.	Talc.
0′	$67\frac{1}{2}$	$67\frac{1}{2}$
5	72	$71\frac{3}{8}$ . . . $4\frac{1}{2}$: $3\frac{7}{8}$ = ·861

It stops 139 rays of heat, and 90 of light.

30th Experiment.

	Sun.	An easily calcinable talc.
0′	50	50
5	$54\frac{3}{4}$	$53\frac{7}{8}$ . . . $4\frac{3}{4}$: $3\frac{7}{8}$ = ·816

It stops 184 rays of heat, and 288 of light.

Transmission of solar Heat through Glasses of the prismatic Colours.

31st Experiment.

	Sun.	Very dark red glass.
0′	73	73
5	$79\frac{1}{4}$	$74\frac{1}{4}$ . . . $6\frac{1}{4}$: $1\frac{1}{4}$ = ·200

This glass stops 800 rays of heat, and 9999, out of ten thousand, rays of light; which amounts nearly to a total separation of light from heat.

32d Experiment.

	Sun.	Dark-red glass.
0′	$68\frac{3}{8}$	$68\frac{3}{8}$
5	$72\frac{1}{2}$	$70 \ldots 4\frac{1}{8} : 1\frac{5}{8} = \cdot 394$

This red glass stops only 606 rays of heat, and above 4999, out of five thousand, rays of light.

33d Experiment.

	Sun.	Orange glass.
0′	$67\frac{3}{4}$	$67\frac{3}{4}$
5	$74\frac{3}{8}$	$70\frac{3}{8} \ldots 6\frac{5}{8} : 2\frac{5}{8} = \cdot 396$

This orange-coloured glass stops 604 rays of heat, which is nearly as much as is stopped by the last red one; but it stops only 779 rays of light.

34th Experiment.

	Sun.	Yellow glass.
	$70\frac{1}{2}$	$70\frac{1}{2}$
	$74\frac{1}{4}$	$73 \ldots 3\frac{3}{4} : 2\frac{1}{2} = \cdot 667$

It stops 333 rays of heat, and 319 of light.

35th Experiment.

	Sun.	Pale-green glass.
0′	$70\frac{1}{2}$	$70\frac{1}{2}$
5	$74\frac{1}{4}$	$71\frac{7}{8} \ldots 3\frac{3}{4} : 1\frac{3}{8} = \cdot 367$

It stops 633 rays of heat, and only 535 of light.

36th Experiment.

	Sun.	Dark-green glass.
0′	$67\frac{1}{2}$	$67\frac{1}{2}$
5	$74\frac{1}{8}$	$68\frac{1}{2} \ldots 6\frac{5}{8} : 1 = \cdot 151$

This glass stops 849 rays of heat, and 949 of light. This accounts for its great use as a darkening glass for telescopes.

37th Experiment.

	Sun.	Bluish-green glass.
0′	$69\frac{3}{8}$	$69\frac{3}{8}$
5	$76\frac{3}{8}$	$71 \ldots 7 : 1\frac{5}{8} = \cdot 232$

It stops 768 rays of heat, and 769 of light.

38th Experiment.

	Sun.	Pale-blue glass.
0′	$70\frac{3}{4}$	$70\frac{3}{4}$
5	$76\frac{3}{4}$	$71\frac{7}{8} \ldots 6 : 1\frac{1}{8} = \cdot 188$

The pale blue glass stops not less than 812 rays of heat, and only 684 of light.

39th Experiment.

	Sun.	Dark-blue glass.
0′	71	71
5	$76\frac{7}{8}$	$74\frac{3}{4} \ldots 5\frac{7}{8} : 3\frac{3}{4} = \cdot 638$

The dark-blue glass stops only 362 rays of heat, and 801 of light.

40th Experiment.

	Sun.	Indigo glass.
0′	$61\frac{3}{4}$	$61\frac{3}{4}$
5	$67\frac{7}{8}$	$64 \ldots 6\frac{1}{8} : 2\frac{1}{4} = \cdot 367$

This glass stops 633 rays of heat, and 9997, out of ten thousand, rays of light.

41st Experiment.

	Sun.	Pale-indigo glass.
0′	62	62
5	$67\frac{7}{8}$	$64\frac{3}{4} \ldots 5\frac{7}{8} : 2\frac{3}{4} = \cdot 468$

It stops 532 rays of heat, and 978 of light.

42d Experiment.

	Sun.	Purple glass.
0′	$61\frac{3}{4}$	$61\frac{3}{4}$
5	$67\frac{3}{4}$	$64\frac{1}{4} \ldots 6 : 2\frac{1}{2} = \cdot 417$

It stops 583 rays of heat, and 993 of light.

43d Experiment.

	Sun.	Violet glass.
	$62\frac{1}{4}$	$62\frac{1}{4}$
	$68\frac{1}{8}$	$65\frac{1}{4} \ldots 5\frac{7}{8} : 3 = \cdot 511$

It stops 489 rays of heat, and 955 of light.

Transmission of Solar Heat through Liquids.

I took a small tube, $1\frac{1}{2}$ inch in diameter [fig. 15], and fixed a stop with a hole $\frac{3}{4}$ inch wide at each end, on which a glass might be fastened, so as

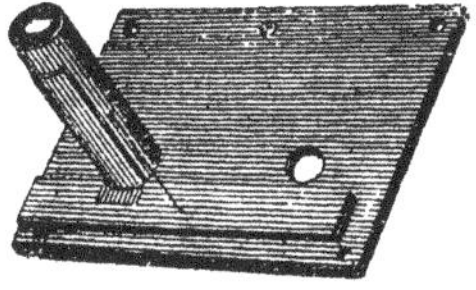

FIG. 15.

to confine liquids. The inner distance, or depth of the liquid, when confined, is three inches. Placing now the empty tube, with its two end glasses fixed, upon the transmitting apparatus, I had as follows:

44th Experiment.

	Sun.	Empty tube, and two glasses.
0′	53	53
5	59	$55\frac{3}{4} \ldots 6 : 2\frac{3}{4} = \cdot 458$

These glasses, with the intermediate air, stop 542 rays of heat, and 204 of light.

45th Experiment.

I filled the tube with well-water, and placed it on the transmitting apparatus.

	Sun.	Well-water.
0′	$52\frac{1}{4}$	$52\frac{1}{8}$
5	$58\frac{3}{4}$	$55 \ldots 6\frac{1}{2} : 2\frac{7}{8} = \cdot442$

Here two glasses, with water between them, stopped 558 rays of heat. The same glasses, and water, stop only 211 rays of light If we were to deduct the effect of the empty machine, there would remain, for the water to stop, only 16 rays of heat, and 7 of light; but it cannot be safe to make this conclusion, as we are not sufficiently acquainted with the action of surfaces between the different mediums on the rays of heat and light; I shall therefore only notice the effect of the compound.

46th Experiment.

I filled now the tube with sea-water, taken from the head of the pier at Ramsgate, at high tide.

Sun.	Sea-water.
$54\frac{1}{2}$	$54\frac{1}{4}$
60	$56 \ldots 5\frac{1}{2} : 1\frac{3}{4} = \cdot318$

The compound stops 682 rays of heat, and 288 of light.

47th Experiment.

	Sun.	Spirit of wine.
0′	$51\frac{5}{8}$	$51\frac{5}{8}$
5	$57\frac{3}{4}$	$54 \ldots 6\frac{1}{8} : 2\frac{3}{8} = \cdot388$

The compound stops 612 rays of heat, and 224 of light.

48th Experiment.

	Sun.	Gin.
0′	52	52
5	$57\frac{3}{4}$	$53\frac{1}{2} \ldots 5\frac{3}{4} : 1\frac{1}{2} = \cdot261$

This compound stops 739 rays of heat, and 626 of light.

49th Experiment.

	Sun.	Brandy.
0′	56	56
5	$60\frac{1}{4}$	$56\frac{7}{8} \ldots 4\frac{1}{4} = \cdot206$

This stops 794 rays of heat, and 996 rays of light.

Other liquids have also been tried; but the experiments having been attended with circumstances that demand a further investigation, they cannot now be given.

Transmission of Solar Heat through scattering Substances.

50th Experiment.

I rubbed one of the pieces of crown glass, mentioned in the 26th experiment, on fine emery laid on a plain brass tool, to make the surface of it rough, which, it is well known, will occasion the transmitted light to be scattered in all directions. Supposing that it would have the same effect on heat, I tried the transmitting capacity of the glass, by exposing it with the rough side towards the sun, over one of the transmitting holes of the apparatus.

	Sun.	Crown glass; one side rubbed on emery.
0′	67	67
5	74	$70\frac{3}{4} \ldots 7 : 3\frac{3}{4} = \cdot536$

The glass so prepared stops 464 scattered rays of heat, and 854 of light. Now, as the same glass, in its polished state, transmitted 259 rays of heat, and 203 of light, the alteration produced in the texture of its surface acts very differently upon these two principles; occasioning an additional stoppage of only 205 rays of heat, but of 651 rays of light.

51st Experiment.

One of the pieces of coach glass, mentioned in the 27th experiment, was prepared in the same manner.

	Sun.	Coach glass; one side rubbed on emery, the rough side exposed.
0′	$66\frac{1}{2}$	$66\frac{1}{2}$
5	$73\frac{1}{2}$	$69\frac{1}{2} \ldots 7 : 3 = \cdot429$

It stops 571 scattered rays of heat, and 885 of light; so that the fine scratches on its surface, made by the operation of emery, have again acted very differently upon the rays of heat, and of light, occasioning an additional stoppage of 375 of the former, but of no less than 717 of the latter.

52d Experiment.

I took another of the pieces of crown glass, mentioned in the 26th experiment, and rubbed both sides on emery.

	Sun.	Crown glass; both sides rubbed on emery.
0′	$69\frac{1}{2}$	$69\frac{1}{2}$
5	$75\frac{1}{2}$	$71\frac{1}{2} \ldots 6 : 2 = \cdot333$

The glass thus prepared, stops 667 scattered rays of heat, and 932 of light.

53*d Experiment.*

Another piece of coach glass, one of those that were mentioned in the 27th experiment, was prepared in the same manner.

	Sun.	Coach glass; both sides rubbed on emery.
0′	$69\frac{5}{8}$	$69\frac{5}{8}$
5	$75\frac{3}{4}$	$71\frac{1}{4} \ldots 6\frac{1}{8} : 1\frac{5}{8} = \cdot 265$

It stops 735 scattered rays of heat, and 946 of light.

54*th Experiment.*

I placed now the coach glass, one side of which had been rubbed on emery, upon the transmitting hole, and over it the crown glass prepared in the same manner, both with the rough side towards the sun; but two slips of card were placed between the glasses, to keep them from touching each other.

	Sun.	Crown glass. Coach glass. } One side of each rubbed on emery.
0′	67	67
5	$73\frac{5}{8}$	$69 \ldots 6\frac{5}{8} : 2 = \cdot 302$

These glasses stop 698 scattered rays of heat, and 969 of light.

55*th Experiment.*

I placed now the coach glass, with both sides rubbed on emery, on the transmitting hole, and over it the crown glass prepared in the same manner, with two slips of card between them, to prevent a contact.

	Sun.	Coach glass. Crown glass. } Both sides of each rubbed on emery.
0′	$69\frac{5}{8}$	$69\frac{5}{8}$
5	$75\frac{7}{8}$	$70\frac{7}{8} \ldots 6\frac{1}{4} : 1\frac{1}{4} = \cdot 200$

These glasses stop 800 scattered rays of heat, and 979 of light.

56*th Experiment.*

I used now all the four glasses; placing them as follows, and putting slips of card between them, to prevent a contact.

	Sun.	Crown glass; the rough side to the sun. Coach glass; ditto. Crown glass; rough on both sides. Coach glass; ditto.
0′	$57\frac{3}{8}$	$57\frac{1}{2}$
5	$62\frac{1}{2}$	$58\frac{1}{4} \ldots 5\frac{1}{8} : \frac{3}{4} = \cdot 146$

These four glasses stop no more than 854 scattered rays of heat, and 995 of light.

57*th Experiment.*

I used now a piece of glass of an olive colour, burnt into the glass, in the manner that glasses are prepared for church windows, which transmits only scattered light.

	Sun.	Olive-coloured glass.
0′	69	69
5	$76\frac{3}{4}$	$70\frac{1}{4} \ldots 7\frac{3}{4} : 1\frac{1}{4} = \cdot 161$

This glass stops 839 scattered rays of heat, and 984 of light.

58*th Experiment.*

	Sun.	Calcined talc.
0′	$51\frac{3}{8}$	$51\frac{3}{8}$
5	$55\frac{1}{8}$	$51\frac{7}{8} \ldots 3\frac{3}{4} : \frac{1}{2} = \cdot 133$

This substance stops 867 scattered rays of heat, and so much light that the sun cannot be perceived through it.*

59*th Experiment.*

	Sun.	White paper.
0′	63	63
5	68	$63\frac{3}{4} \ldots 5 : \frac{3}{4} = \cdot 150.$

This substance stops 850 scattered rays of heat, and 994 of light.

60*th Experiment.*

	Sun.	Linen.
0′	63	63
5	69	$63\frac{1}{2} \ldots 6 : \frac{1}{2} = \cdot 0833$

White linen stops 916 scattered rays of heat, and 952 of light.

61*st Experiment.*

	Sun.	White persian.
0′	70	70
5	$76\frac{1}{4}$	$71\frac{1}{2} \ldots 6\frac{1}{4} : 1\frac{1}{2} = \cdot 240$

This thin silk stops 760 scattered rays of heat, and 916 of light.

62*d Experiment.*

Sun.	Black muslin.
$64\frac{3}{4}$	$64\frac{3}{4}$
70	$66\frac{1}{4} \ldots 5\frac{1}{4} : 1\frac{1}{2} = \cdot 286$

This substance stops 714 scattered rays of heat, and 737 of light.

* See the 175th Experiment.

Transmission of terrestrial Flame-heat through various Substances.

My apparatus for the purpose of transmitting flame-heat is as follows [fig. 16]. A box 22 inches long, $5\frac{1}{2}$ broad, and $1\frac{3}{4}$ deep, has a hole in the centre $1\frac{1}{10}$ inch in diameter, through which a wax candle, thick enough entirely to fill it, is to be put at the bottom; the box being properly elevated for the purpose. There must be two lateral holes in the bottom, 2 inches long, and $1\frac{1}{4}$ broad, one on each side of the candle, to supply it with a current of air, as otherwise it will not give a steady flame, which is absolutely necessary. At the distance of $1\frac{3}{10}$ inch from the candle, on each side, are two screens, 12 inches square, with a hole in each, $\frac{3}{4}$ inch in diameter, through which the heat of the candle passes to the two thermometers, which are to be placed in opposite directions, one on each side of the table. Care must be taken to place them exactly at the same distance from the centre of the flame, as otherwise they will not receive equal quantities of heat. The scales, and their supports also, must be so kept out of the way of heat coming from the candle, that they may not scatter it back on the balls, but suffer all that is not intercepted by them to pass freely forwards in the box, and downwards, through

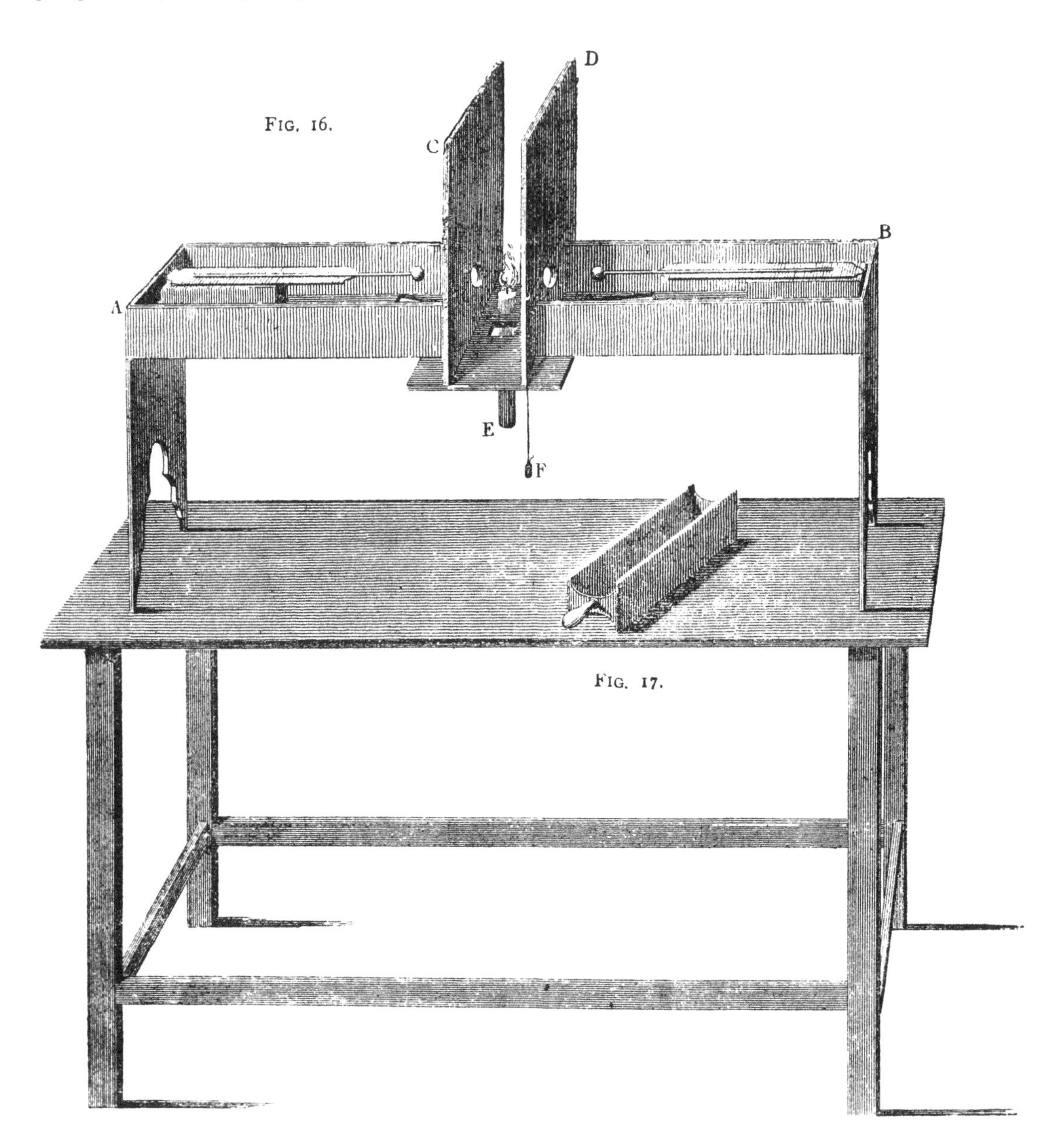

Fig. 16.

Fig. 17.

openings cut in the bottom. Before the transmitting holes, between the two wooden screens, must be two covers of the same material, close to the openings [see fig. 17]; and it will be necessary to join these covers at the side, by a common handle, that they may be removed together, without disturbing any part of the apparatus, when the experiment is to begin.

The glasses are to be put before the thermometer, close to the transmitting hole, by placing them on a small support below, while the upper part is held close to the screen by a light plummet, suspended by a thread which is fastened on one side, and passes over the glass, to a hook on the other side.

In making experiments, many attentions are necessary, such as, keeping the candle exactly to a certain height, that the brightest part of the flame may be just in the centre of the two transmitting holes: that the wick may be always straight, and not, by bending, approach nearer to one thermometer than to the other: that the wax-cup of the candle be kept clean, and never suffered to run over, &c.

Before, and now and then between, the observations also, the thermometers must be tried a few degrees, that it may be seen whether they act equally; and the candle, during the time they cool down to the temperature, must be put out by an extinguisher, large enough to rest on the bottom of the box, without touching any part of the wax. Many other precautions I need not mention, as they will soon be discovered by any one who may repeat such experiments.

63d Experiment.

	Candle.	Bluish-white glass.
0′	$59\frac{3}{8}$	$59\frac{1}{2}$
5	$62\frac{3}{8}$	$60\frac{5}{8}$ . . . $3 : 1\frac{1}{8} = \cdot 375$

From this experiment we find, that while the rays of the candle gave 3 degrees of heat to the thermometer openly exposed to their action, the other thermometer, which received the same rays through the medium of the interposed glass, rose only $1\frac{1}{8}$ degrees. Hence we calculate, that this glass stops 625 rays of flame-heat, out of every thousand that fall on it. It stops only 86 rays of candle-light; but this, having been referred to before, will not in future be repeated.

64th Experiment.

	Candle.	Flint glass.
0′	$59\frac{7}{8}$	$59\frac{5}{8}$
5	$62\frac{5}{8}$	$60\frac{3}{4}$ . . . $2\frac{3}{4} : 1\frac{1}{8} = \cdot 409$

It stops 591 rays of flame-heat, and light as before.

65th Experiment.

	Candle.	Crown glass.
0′	$59\frac{7}{8}$	$59\frac{7}{8}$
5	$62\frac{5}{8}$	$60\frac{7}{8}$ . . . $2\frac{3}{4} : 1 = \cdot 364$

It stops 636 rays of flame-heat.

66th Experiment.

	Candle.	Coach glass.
0′	60	$60\frac{3}{8}$
5	63	62 . . . $3 : 1\frac{5}{8} = \cdot 542$

It stops 458 rays of flame-heat.

67th Experiment.

	Candle.	Iceland crystal.
0′	$58\frac{3}{4}$	$58\frac{3}{4}$
5	$62\frac{1}{8}$	$60\frac{3}{8}$ . . . $3\frac{3}{8} : 1\frac{5}{8} = \cdot 484$

It stops 516 rays of flame-heat.

68th Experiment.

	Candle.	Calcinable talc.
0′	$58\frac{7}{8}$	$58\frac{7}{8}$
5	$61\frac{7}{8}$	$60\frac{3}{4}$ . . . $3 : 1\frac{7}{8} = \cdot 625$

This substance stops only 375 rays of flame-heat.

69th Experiment.

	Candle.	Very dark red glass.
0′	$60\frac{3}{8}$	$60\frac{3}{8}$
5	$63\frac{1}{8}$	$61\frac{3}{8}$ . . . $2\frac{3}{4} : 1 = \cdot 364$

This glass stops 636 rays of flame-heat.

70th Experiment.

	Candle.	Dark red glass.
0′	$60\frac{3}{4}$	$60\frac{3}{4}$
5	$63\frac{1}{8}$	$61\frac{7}{8}$ . . . $2\frac{3}{8} : 1\frac{1}{8} = \cdot 474$

It stops 526 rays of flame-heat.

71st Experiment.

	Candle.	Orange glass.
0′	$60\frac{1}{4}$	$60\frac{1}{4}$
5	$63\frac{3}{8}$	$61\frac{5}{8}$ . . . $3\frac{1}{8} : 1\frac{3}{8} = \cdot 440$

It stops 560 rays of flame-heat.

72d Experiment.

	Candle.	Yellow glass.
0′	$60\frac{5}{8}$	$60\frac{5}{8}$
5	$63\frac{5}{8}$	$61\frac{7}{8}$ . . . $3 : 1\frac{1}{4} = \cdot 417$

It stops 583 rays of flame-heat.

73*d* Experiment.

	Candle.	Pale-green glass.
o′	$60\frac{7}{8}$	$60\frac{7}{8}$
5	$63\frac{7}{8}$	$62\frac{3}{8} \ldots 3 : 1\frac{1}{2} = \cdot500$

It stops 500 rays of flame-heat.

74*th* Experiment.

	Candle.	Dark-green glass.
o′	$61\frac{1}{8}$	$61\frac{1}{8}$
5	64	$61\frac{7}{8} \ldots 2\frac{7}{8} : \frac{3}{4} = \cdot261$

It stops 739 rays of flame-heat.

75*th* Experiment.

	Candle.	Bluish-green glass.
o′	$61\frac{1}{8}$	$61\frac{1}{8}$
5	64	$62\frac{1}{8} \ldots 2\frac{7}{8} : 1 = \cdot348.$

It stops 652 rays of flame-heat.

76*th* Experiment.

	Candle.	Pale-blue glass.
o′	$61\frac{1}{2}$	$61\frac{1}{2}$
5	$64\frac{3}{8}$	$62\frac{5}{8} \ldots 2\frac{7}{8} : 1\frac{1}{8} = \cdot391$

It stops 609 rays of flame-heat.

77*th* Experiment.

	Candle.	Dark-blue glass.
o′	$61\frac{7}{8}$	$61\frac{3}{4}$
5	$64\frac{1}{2}$	$62\frac{3}{4} \ldots 2\frac{5}{8} : 1 = \cdot381$

It stops 619 rays of flame-heat.

78*th* Experiment.

	Candle.	Indigo glass.
o′	$61\frac{7}{8}$	$61\frac{7}{8}$
5	$65\frac{3}{8}$	$63 \ldots 3\frac{1}{2} : 1\frac{1}{8} = \cdot321$

It stops 679 rays of flame-heat.

79*th* Experiment.

	Candle.	Pale indigo glass.
o′	$62\frac{1}{8}$	$62\frac{1}{8}$
5	$64\frac{3}{4}$	$63\frac{1}{4} \ldots 2\frac{5}{8} : 1\frac{1}{8} = \cdot429$

It stops 571 rays of flame-heat.

80*th* Experiment.

	Candle.	Purple glass.
o′	$61\frac{7}{8}$	$61\frac{7}{8}$
5	65	$63\frac{3}{8} \ldots 3\frac{1}{8} : 1\frac{1}{2} = \cdot480$

It stops 520 rays of flame-heat.

81*st* Experiment.

	Candle.	Violet glass.
o′	$59\frac{7}{8}$	$59\frac{7}{8}$
5	$63\frac{3}{8}$	$61\frac{5}{8} \ldots 3\frac{1}{2} : 1\frac{3}{4} = \cdot500$

It stops 500 rays of flame-heat.

82*d* Experiment.

	Candle.	Crown glass; one side rubbed on emery; the rough side exposed.
o′	60	60
5	$63\frac{3}{8}$	$60\frac{7}{8} \ldots 3\frac{3}{8} : \frac{7}{8} = \cdot259$

This glass, so prepared, stops 741 scattered rays of flame-heat.

83*d* Experiment.

	Candle.	Coach glass; one side rubbed on emery; the rough side exposed.
o′	$59\frac{5}{8}$	$59\frac{5}{8}$
5	$63\frac{3}{8}$	$60\frac{7}{8} \ldots 3\frac{3}{4} : 1\frac{1}{4} = \cdot333$

It stops 667 scattered rays of flame-heat.

84*th* Experiment.

	Candle.	Crown glass; both sides rubbed on emery.
o′	$59\frac{3}{4}$	$59\frac{3}{4}$
5	63	$61 \ldots 3\frac{1}{4} : 1\frac{1}{4} = \cdot385$

It stops 615 scattered rays of flame-heat.

85*th* Experiment.

	Candle.	Coach glass; both sides rubbed on emery.
o′	$59\frac{7}{8}$	$59\frac{3}{4}$
5	63	$60\frac{3}{4} \ldots 3\frac{1}{8} : 1 = \cdot320$

It stops 680 scattered rays of flame-heat.

86*th* Experiment.

	Candle.	Crown glass. / Coach glass. } One side of each rubbed on emery.
o′	$55\frac{7}{8}$	$55\frac{7}{8}$
5	59	$56\frac{3}{4} \ldots 3\frac{1}{8} : \frac{7}{8} = \cdot280$

These glasses stop 720 scattered rays of flame-heat.

87*th* Experiment.

Candle.	Crown glass. / Coach glass. } Both sides of each rubbed on emery.
$55\frac{7}{8}$	$55\frac{7}{8}$
$59\frac{1}{4}$	$57 \ldots 3\frac{3}{8} : 1\frac{1}{8} = \cdot333$

These glasses stop 667 rays of flame-heat.

88th Experiment.

	Candle.	Crown glass; the rough side to the candle. Coach glass; ditto. Crown glass; rough on both sides. Coach glass; ditto.
0′	$56\frac{3}{4}$	$56\frac{7}{8}$
5	$59\frac{5}{8}$	$57\frac{1}{4} \ldots 2\frac{7}{8} : \frac{3}{8} = \cdot 130$

These four glasses stop 870 scattered rays of flame-heat.

89th Experiment.

	Candle.	Olive-colour, burnt in glass.
0′	60	60
5	63	$60\frac{5}{8} \ldots 3 : \frac{5}{8} = \cdot 208$

This glass stops 792 scattered rays of flame-heat.

90th Experiment.

	Candle.	White paper.
0′	$57\frac{3}{8}$	$57\frac{1}{4}$
5	$60\frac{3}{8}$	$57\frac{7}{8} \ldots 3 : \frac{5}{8} = \cdot 208$

This substance stops 792 scattered rays of flame-heat.

91st Experiment.

	Candle.	Linen.
0′	$57\frac{3}{8}$	$57\frac{3}{8}$
5	61	$58\frac{1}{2} \ldots 3\frac{5}{8} : 1\frac{1}{8} = \cdot 310$

It stops 690 scattered rays of flame-heat.

92d Experiment.

	Candle.	White persian.
0′	$57\frac{1}{8}$	57
5	$60\frac{1}{2}$	$58\frac{3}{8} \ldots 3\frac{3}{8} : 1\frac{3}{8} = \cdot 407$

It stops 593 scattered rays of flame-heat.

93d Experiment.

	Candle.	Black muslin.
0′	$57\frac{3}{4}$	$57\frac{3}{4}$
5	$60\frac{5}{8}$	$59 \ldots 2\frac{7}{8} : 1\frac{1}{4} = \cdot 435$

It stops 565 scattered rays of flame-heat.

Transmission of the solar Heat which is of an equal Refrangibility with red prismatic Rays.

The apparatus which I have used for transmitting prismatic rays, is of the same construction as that which has already been described under the head of direct solar transmissions [figs. 12 and 13]; but here the holes in the top of the box are only two inches from each other, and no more than $\frac{3}{8}$ths in diameter [fig. 18]. On the face of the box are drawn two parallel lines, also $\frac{3}{8}$ths of an inch distant from each other, and inclosing the transmitting holes: they serve as a direction whereby to keep any required colour to fall equally on both holes. The distance at which the box is to be placed from the prism, must be such as will allow the

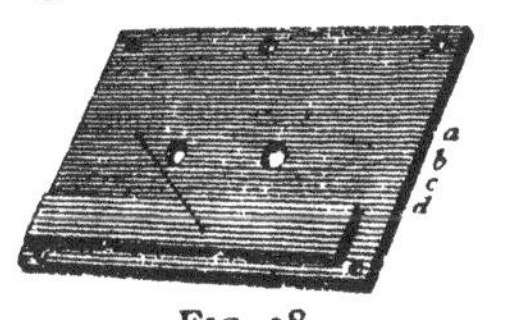

FIG. 18.

rays to diverge sufficiently for the required colour to fill the transmitting holes; and the balls of the thermometers placed under them ought to be less than these holes, that the projected rays may pass around them, and shew their proper adjustment. The diameters of mine, used for this purpose, are $2\frac{1}{4}$ tenths of an inch.

94th Experiment.

I placed my apparatus at five feet from the prism, and so as to cause the red-making rays to fall between the parallel lines, in order to find what heat-making rays would come to the thermometer along with them.

	Red rays. Therm. A.	Bluish white glass. Therm. B.
0′	$75\frac{5}{8}$	$75\frac{3}{8}$
5	$77\frac{5}{8}$	$76\frac{5}{8} \ldots 2 : 1\frac{1}{4} = \cdot 625$

From this experiment it appears, that when a thousand red-making rays fall on each transmitting hole, 375 of them, if they also be the heat-making rays, are stopped by the bluish-white glass which covers one of these holes; or, what requires no other proof than the experiment itself, that 375 rays of heat, of the same refrangibility with the red rays, are intercepted by this glass.

95th Experiment.

Red rays.	Flint glass.
$75\frac{3}{4}$	$75\frac{3}{8}$
$77\frac{1}{2}$	$76\frac{7}{8} \ldots 1\frac{3}{4} : 1\frac{1}{2} = \cdot 857$

This glass stops only 143 rays of heat, which are of the same refrangibility with the red rays.

96th Experiment.

	Red rays.	Crown glass.
0′	$75\frac{7}{8}$	$75\frac{5}{8}$
5	78	$77\frac{1}{8} \ldots 2\frac{1}{8} : 1\frac{1}{2} = \cdot 706$

This glass stops 294 rays of the same sort of heat.

97th Experiment.

	Red rays.	Coach glass.
0′	$54\frac{3}{8}$	$53\frac{3}{4}$
5	$55\frac{5}{8}$	$54\frac{3}{4} \ldots 1\frac{1}{4} : 1 = \cdot 800$

It stops 200 rays of the same sort of heat.

98*th Experiment.*

	Red rays.	Iceland crystal.
0′	$76\frac{1}{8}$	$75\frac{3}{4}$
5	78	$77\frac{1}{4} \ldots 1\frac{7}{8} : 1\frac{1}{2} = \cdot 800$

This substance stops 200 rays of the same sort of heat.

99*th Experiment.*

	Red rays.	Calcinable talc.
0′	$51\frac{3}{4}$	$51\frac{1}{4}$
5	$53\frac{5}{8}$	$52\frac{7}{8} \ldots 1\frac{7}{8} : 1\frac{5}{8} = \cdot 867$

It stops 133 rays of the same sort of heat.

100*th Experiment.*

	Red rays.	Dark-red glass.
0′	$76\frac{1}{2}$	$76\frac{5}{8}$
5	$78\frac{1}{8}$	$77\frac{1}{8} \ldots 1\frac{5}{8} : \frac{1}{2} = \cdot 308$

This glass stops 692 rays of the same sort of heat.

101*st Experiment.*

	Red rays.	Orange glass.
0′	75	$74\frac{3}{4}$
5	77	$75\frac{3}{4} \ldots 2 : 1 = \cdot 500$

It stops 500 rays of the same sort of heat.

102*d Experiment.*

	Red rays.	Yellow glass.
0′	$75\frac{3}{8}$	75
5	$76\frac{7}{8}$	$75\frac{7}{8} \ldots 1\frac{1}{2} : \frac{7}{8} = \cdot 583$

It stops 417 rays of the same sort of heat.

103*d Experiment.*

	Red rays.	Pale-green glass.
0′	$74\frac{1}{4}$	$74\frac{1}{8}$
5	$76\frac{3}{8}$	$75 \ldots 2\frac{1}{8} : \frac{7}{8} = \cdot 412$

It stops 588 rays of the same sort of heat.

104*th Experiment.*

	Red rays.	Dark-green glass.
0′	$68\frac{3}{4}$	$68\frac{7}{8}$
5	$70\frac{1}{2}$	$69\frac{1}{4} \ldots 1\frac{3}{4} : \frac{3}{8} = \cdot 214$

It stops 786 rays of the same sort of heat.

105*th Experiment.*

	Red rays.	Bluish-green glass.
0′	69	$68\frac{7}{8}$
5	$70\frac{5}{8}$	$69\frac{3}{4} \ldots 1\frac{5}{8} : \frac{7}{8} = \cdot 538$

It stops 462 rays of the same sort of heat.

106*th Experiment.*

	Red rays.	Pale-blue glass.
0′	$69\frac{5}{8}$	$69\frac{1}{2}$
5	$70\frac{7}{8}$	$69\frac{7}{8} \ldots 1\frac{1}{4} : \frac{3}{8} = \cdot 300$

It stops 700 rays of the same sort of heat.

107*th Experiment.*

	Red rays.	Dark-blue glass.
0′	67	$67\frac{1}{4}$
5	$68\frac{3}{4}$	$68\frac{7}{8} \ldots 1\frac{3}{4} : 1\frac{5}{8} = \cdot 929$

This glass stops only 71 rays of the same sort of heat.

108*th Experiment.*

	Red rays.	Indigo glass.
0′	$68\frac{1}{2}$	$68\frac{1}{4}$
5	$70\frac{1}{2}$	$69\frac{1}{2} \ldots 2 : 1\frac{1}{4} = \cdot 633$

It stops 367 rays of the same sort of heat.

109*th Experiment.*

	Red rays.	Pale-indigo glass.
0′	69	$68\frac{5}{8}$
5	71	$70 \ldots 2 : 1\frac{3}{8} = \cdot 687$

It stops 313 rays of the same sort of heat.

110*th Experiment.*

	Red rays.	Purple glass.
0′	$56\frac{1}{4}$	56
5	$58\frac{1}{2}$	$57\frac{1}{4} \ldots 2\frac{1}{4} : 1\frac{1}{4} = \cdot 556$

It stops 444 rays of the same sort of heat.

111*th Experiment.*

	Red rays.	Violet glass.
0′	$57\frac{3}{8}$	57
5	$59\frac{1}{4}$	$58\frac{1}{8} \ldots 1\frac{7}{8} : 1\frac{1}{8} = \cdot 600$

It stops 400 rays of the same sort of heat.

112*th Experiment.*

	Red rays.	Crown glass; one side rubbed on emery, rough side exposed.
0′	$49\frac{3}{4}$	49
5	52	$50\frac{3}{8} \ldots 2\frac{1}{4} : 1\frac{3}{8} = \cdot 611$

This glass, so prepared, stops 389 scattered rays of the same sort of heat.

113*th Experiment.*

	Red rays.	Coach glass; one side rubbed on emery, rough side exposed.
0′	$53\frac{3}{8}$	$52\frac{7}{8}$
5	$55\frac{1}{8}$	$53\frac{3}{4} \ldots 1\frac{3}{4} : \frac{7}{8} = \cdot 500$

It stops 500 scattered rays of the same sort of heat.

114*th Experiment.*

	Red rays.	Crown glass; both sides rubbed on emery.
0′	$50\frac{5}{8}$	$49\frac{7}{8}$
5	$52\frac{3}{4}$	$51 \ldots 2\frac{1}{8} : 1\frac{1}{8} = \cdot 529$

It stops 471 scattered rays of the same sort of heat.

115th Experiment.

	Red rays.	Coach glass ; both sides rubbed on emery.
0′	$54\frac{1}{8}$	$53\frac{3}{4}$
5	$55\frac{5}{8}$	$54 \ldots 1\frac{1}{2} : \frac{1}{4} = \cdot 167$

It stops 833 scattered rays of the same sort of heat.

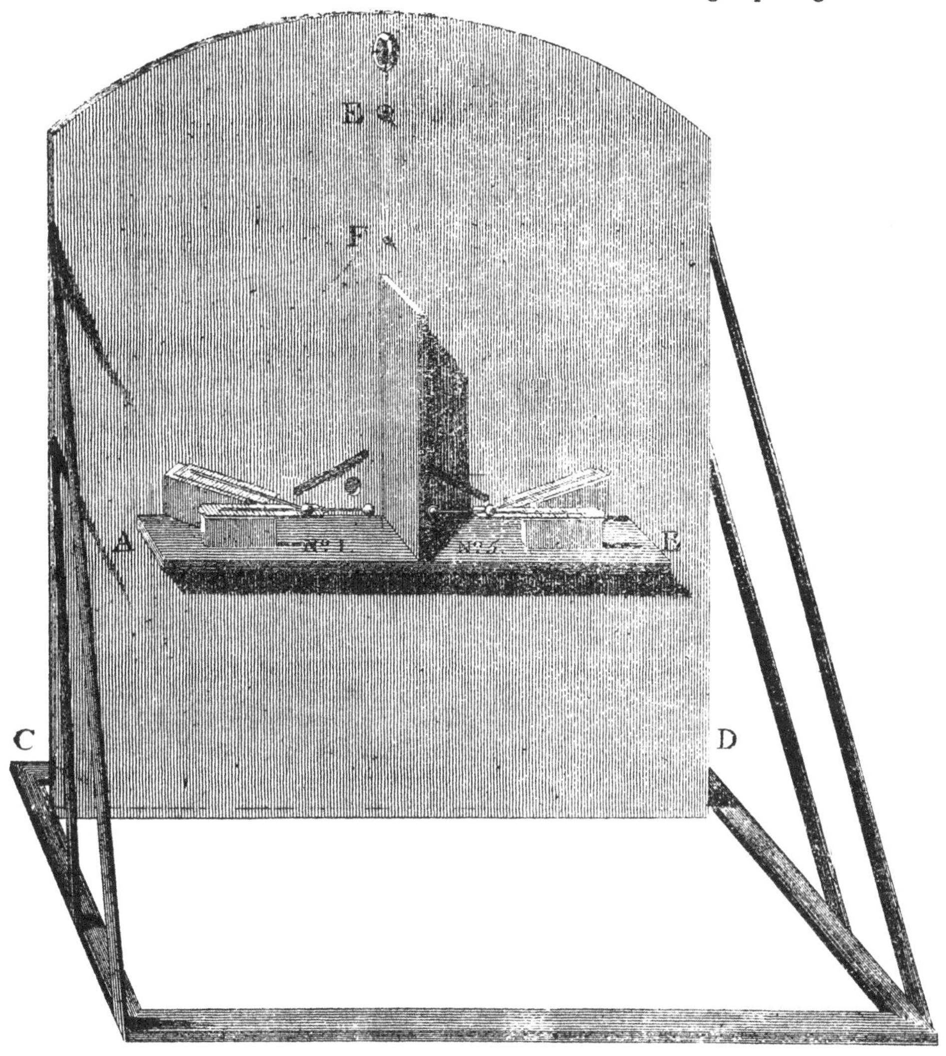

Fig. 19.

116th Experiment.

	Red rays.	Calcined talc.
0′	$51\frac{1}{8}$	$50\frac{1}{2}$
5	$53\frac{1}{2}$	$51\frac{1}{8} \ldots 2\frac{3}{8} : \frac{5}{8} = \cdot 263$

This substance stops 737 scattered rays of the same sort of heat.

Transmission of Fire-Heat through various Substances.

When the same fire is to give an equal heat to two thermometers, at some short distance from each other, it becomes highly necessary that there should be a place of considerable dimensions in its centre, where it may burn with an equal glow, and without flame or smoke. To obtain this, I used a grate 19 inches broad, and $8\frac{3}{4}$ high, having only three bars, which divide the fire into three large openings. In the centre of the middle one of these, when the grate is well filled with large coals or coke, we may, with proper management, keep up the required equality of radiance.

The apparatus I have used is of the following construction [fig. 19]. A screen of wood, 3 feet 6 inches high, and 3 feet broad, lined towards the fire with plates of iron, has two holes, $\frac{3}{4}$ of an inch in diameter, and at the distance of $2\frac{1}{2}$ inches from each other, one on each side of the middle of the screen, and of a height that will answer to the centre of the fire. $2\frac{1}{4}$ inches under the centre of the holes is a shelf, about 22 inches long and 4 broad, on which are placed two thermometers, in opposite directions, fixed on proper stands, to bring the balls, quite disengaged from the scales, directly 2 inches behind the transmitting holes. A small thin wooden partition is run up between the thermometers, to prevent the heat transmitted through one hole from coming to the thermometer belonging to the other.

The screen is fixed upon a light frame, which fits exactly into the opening of the front of the marble chimney-piece; and the ends of the frame are of a length which, when the screen is placed before the fire, will just bring the transmitting holes to be $6\frac{1}{2}$ inches from the front bars of the grate.

A large wooden cover, also plated with iron, shuts up the transmitting holes on the side next to the fire; but may be drawn up by a string on the outside, so as to open them when required.

Two assistant thermometers are placed on proper stands, to bring their balls to the same distance from the screen as those which receive the heat of the fire; but removed sideways as far as necessary, to put them out of the reach of any rays that pass obliquely through the transmitting holes. They are to indicate any change of temperature that may take place during the time of the experiment: for, notwithstanding the largeness of the screen, some heat will find its way round and over it; and this acting as a general cause, its effect must be allowed for.

117th Experiment.

Having tried the apparatus sufficiently to find that the thermometers exposed to the transmitting holes would generally receive 20 or more degrees of heat, without differing more than sometimes $\frac{1}{8}$ or at most $\frac{1}{4}$ of a degree, I now placed the bluish-white glass of the 24th experiment upon a support prepared for the purpose, so as closely to cover one of the transmitting holes. A small spring, moveable on its centre, is always turned against the upper part of the transmitting glasses, to keep them in their situation.

	Fire.	Bluish-white glass.
0′	66	66
5	86	71 . . . 20 : 5 = ·250

This glass stops 750 rays of fire-heat. By looking through it, at the same place in the fire, after the screen was removed, in order to cool the apparatus for the next experiment, I found that this glass can hardly be said to stop any of the light of the fire.

118th Experiment.

	Fire.	Flint glass.
0′	67	67
5	87	72 . . . 20 : 5 = ·250

It stops 750 rays of fire-heat.

119th Experiment.

	Fire.	Crown glass.
0′	67	67
5	$86\frac{3}{4}$	$72\frac{1}{2}$ . . . $19\frac{3}{4}$: $5\frac{1}{2}$ = ·278

It stops 722 rays of fire-heat.

120th Experiment.

	Fire.	Coach glass.
0′	$67\frac{1}{2}$	$67\frac{1}{2}$
5	$86\frac{3}{4}$	73 . . . $19\frac{1}{4}$: $5\frac{1}{2}$ = ·286

It stops 714 rays of fire-heat.

121st Experiment.

	Fire.	Iceland crystal.
0′	68	68
5	$90\frac{1}{2}$	$73\frac{1}{2}$ . . . $22\frac{1}{2}$: $5\frac{1}{2}$ = ·244

This substance stops 756 rays of fire-heat.

122d Experiment.

I took now the piece of talc used in the 30th experiment, and, placing it over the transmitting hole, I had the following result. But, as the unexpected event of a calcination, which took place, was attended with circumstances that ought to be noticed, I shall, instead of the usual abridgment of the experiments, give this at full length.

	Fire. Therm. D.	Talc. Therm. C.
0′	65	65
1	72	67 . . . 7 : 2 = ·289
2	77	$68\frac{3}{4}$ . . . 12 : $3\frac{3}{8}$ = ·281
3	$80\frac{1}{2}$	$69\frac{1}{2}$ . . . $15\frac{1}{2}$: $4\frac{1}{2}$ = ·290
4	83	70 . . . 18 : 5 = ·278
5	85	$70\frac{3}{4}$ . . . 20 : $5\frac{3}{4}$ = ·287

This substance stops 713 rays of fire-heat.

I am now to point out the singularity of this experiment; which consists, as we may see by the above register of it, in the apparently regular continuance of its power of transmitting heat, while its capacity of transmitting light was totally destroyed. For, when I placed this piece of talc over the hole in the screen, it was extremely transparent, as this substance is generally known to be; and yet, when the experiment was over, it appeared of a beautiful white colour; and its power of transmitting light was so totally destroyed, that even the sun in the meridian could not be perceived through it. Now, had the power of transmitting heat through this substance been really uniform during all the five minutes, it would have been quite a new phænomenon ; as all my experiments are attended with a regular increase of it ; but since, by calcination, the talc lost much of its transmitting power, we may easily account for this unexpected regularity.

123d Experiment.

Fire.	Very dark red glass.
66	66
$89\frac{1}{4}$	$75 \ldots 23\frac{1}{4} : 9 = \cdot387$

This glass stops 613 rays of fire-heat.

124th Experiment.

	Fire.	Dark-red glass.
0′	67	67
5	$92\frac{3}{4}$	$78 \ldots 25\frac{3}{4} : 11 = \cdot427$

This glass, which stops 999·8 rays of candle-light, stops only 573 rays of fire-heat ; whereas my piece of thick flint glass, which stops no more than 91 rays of that light, stops no less than 750 of fire-heat. It does not appear, by looking through these glasses, that there is a difference in their disposition to transmit candle-light or fire-light.

125th Experiment.

	Fire.	Orange glass.
0′	66	66
5	80	$71 \ldots 14 : 5 = \cdot357$

It stops 643 rays of fire-heat.

126th Experiment.

	Fire.	Yellow glass.
0′	$61\frac{1}{4}$	$61\frac{1}{4}$
5	83	$68\frac{7}{8} \ldots 21 : 7\frac{6}{8}$. cor. $-1\frac{1}{8}°$. $20\frac{5}{8} : 6\frac{1}{2} = \cdot315$

This experiment being made early in the morning, before the temperature of the room was come to its usual height, the assistant thermometers shewed a gradual rising of $1\frac{1}{8}$ degree in the 5 minutes : they are in general very steady. The glass stops 685 rays of fire-heat.

127th Experiment.

	Fire.	Pale-green glass.
0′	$65\frac{3}{4}$	$65\frac{3}{4}$
5	85	$71\frac{3}{4} \ldots 19\frac{1}{4} : 6 = \cdot312$

It stops 688 rays of fire-heat.

128th Experiment.

	Fire.	Dark-green glass.
0′	68	68
5	$88\frac{3}{4}$	$73\frac{3}{8} \ldots 20\frac{3}{8} : 5\frac{3}{4}$ cor. $-\frac{1}{4}° = \cdot255$

It stops 745 rays of fire-heat.

129th Experiment.

	Fire.	Bluish green glass.
0′	$68\frac{1}{2}$	$68\frac{1}{2}$
5	87	$74\frac{1}{8} \ldots 18\frac{1}{2} : 5\frac{5}{8} = \cdot304$

It stops 696 rays of fire-heat.

130th Experiment.

	Fire.	Pale-blue glass.
0′	$68\frac{1}{2}$	68
5	$86\frac{1}{4}$	$73\frac{3}{4} \ldots 17\frac{3}{4} : 5\frac{3}{4} = \cdot324$

It stops 676 rays of fire-heat.

131st Experiment.

	Fire.	Dark-blue glass.
0′	67	67
5	$84\frac{7}{8}$	$73 \ldots 17\frac{7}{8} : 6$. cor. $-1° = \cdot296$

It stops 704 rays of fire-heat.

132d Experiment.

	Fire.	Indigo glass.
0′	$69\frac{1}{4}$	$69\frac{1}{4}$
5	$85\frac{3}{8}$	$73\frac{3}{4} \ldots 16\frac{1}{8} : 4\frac{1}{2} = \cdot279$

It stops 721 rays of fire-heat.

133d Experiment.

	Fire.	Pale indigo glass.
0′	$67\frac{1}{2}$	$67\frac{1}{2}$
5	$85\frac{7}{8}$	$74 \ldots 18\frac{3}{4} : 6\frac{1}{2}$. cor. $-\frac{1}{4}° = \cdot345$

It stops 655 rays of fire-heat.

134th Experiment.

	Fire.	Purple glass.
0′	69	69
5	83	$73\frac{1}{2} \ldots 14 : 4\frac{1}{2} = \cdot321$

It stops 679 rays of fire-heat.

135th Experiment.

	Fire.	Violet glass.
0′	$66\frac{1}{2}$	$66\frac{1}{2}$
5	$86\frac{1}{2}$	$74\frac{1}{4} \ldots 20 : 7\frac{3}{4} = \cdot385$

It stops 615 rays of fire-heat.

136*th Experiment.*

	Fire.	Crown glass; one side rubbed on emery.
0′	$67\frac{5}{8}$	$67\frac{5}{8}$
5	$89\frac{3}{4}$	$73\frac{3}{4} \ldots 22\frac{1}{8} : 6\frac{1}{8} = \cdot 277$

This glass, so prepared, stops 723 scattered rays of fire-heat.

137*th Experiment.*

	Fire.	Coach glass; one side rubbed on emery.
0′	68	$67\frac{1}{2}$
5	$87\frac{1}{8}$	$72\frac{1}{8} \ldots 19\frac{1}{8} : 4\frac{5}{8} = \cdot 242$

It stops 758 scattered rays of fire-heat.

138*th Experiment.*

	Fire.	Crown glass; both sides rubbed on emery.
0′	$68\frac{1}{2}$	68
5	$92\frac{3}{8}$	$73 \ldots 23\frac{7}{8} : 5 = \cdot 209$

It stops 791 scattered rays of fire-heat

139*th Experiment.*

	Fire.	Coach glass; both sides rubbed on emery.
0′	67	67
5	88	$70\frac{1}{2} \ldots 21 : 3\frac{1}{2}$. cor. $-\frac{1}{2}^{\circ} = \cdot 146$

It stops 854 scattered rays of fire-heat.

140*th Experiment.*

	Fire.	Crown glass. Coach glass. — One side of each rubbed on emery.
0′	66	66
5	86	$69\frac{7}{8} \ldots 20 : 3\frac{7}{8}$. cor. $-1^{\circ} = \cdot 151$

These glasses stop 849 scattered rays of fire-heat.

141*st Experiment.*

	Fire.	Crown glass. Coach glass. — Both sides of each rubbed on emery.
0′	$66\frac{3}{4}$	$66\frac{3}{4}$
5	$83\frac{3}{4}$	$68\frac{1}{2} \ldots 17 : 1\frac{3}{4} = \cdot 103$

These glasses stop 897 scattered rays of fire-heat.

142*d Experiment.*

	Fire.	The four glasses of the two preceding experiments put together.
0′	66	66
5	80	$67\frac{3}{8} \ldots 14 : 1\frac{3}{8} = \cdot 098$

These four glasses stop 902 scattered rays of fire-heat.

143*d Experiment.*

	Fire.	Olive colour, burnt into glass.
0′	$63\frac{3}{4}$	$63\frac{3}{4}$
5	$85\frac{3}{4}$	$67\frac{1}{2} \ldots 22 : 3\frac{3}{4}$. cor. $-\frac{1}{2}^{\circ} = \cdot 151$

This glass stops 849 scattered rays of fire-heat.

144*th Experiment.*

	Fire.	Paper.
0′	$66\frac{1}{2}$	$66\frac{1}{2}$
5	$83\frac{1}{2}$	$68 \ldots 17 : 1\frac{1}{2} = \cdot 0882$

This substance stops 912 scattered rays of fire-heat; it was turned a little yellow by the exposure.

145*th Experiment.*

	Fire.	Linen.
0′	$63\frac{3}{4}$	$63\frac{3}{4}$
5′	$84\frac{3}{4}$	$67 \ldots 21 : 3\frac{1}{4}$. cor. $-1\frac{1}{2}^{\circ} = \cdot 0897$

This substance stops 910 scattered rays of fire-heat.

146*th Experiment.*

	Fire.	White persian.
0′	$65\frac{3}{4}$	$65\frac{3}{4}$
5	$81\frac{1}{8}$	$68\frac{3}{8} \ldots 15\frac{3}{8} : 2\frac{5}{8} = \cdot 171$

This substance stops 829 scattered rays of fire-heat.

147*th Experiment.*

	Fire.	Black muslin.
0′	66	66
5	$82\frac{1}{2}$	$70\frac{1}{2} \ldots 16\frac{1}{2} : 4\frac{1}{2}$. cor. $+\frac{1}{2}^{\circ} = \cdot 294$

This substance stops 706 scattered rays of fire-heat.

Transmission of the invisible Rays of Solar Heat.

The same apparatus which I have used for the transmission of coloured prismatic rays [see fig. 18], will also do for the invisible part of the heat spectrum: it is only required to add two or three more parallel lines, one-tenth of an inch from each other, below the two which inclose the transmitting holes, in order to use them for directing the invisible rays of heat, by the position of the visible rays of light, to fall on the place required for coming to the thermometers.

148*th Experiment.*

	Invisible rays.	Bluish-white glass.
0′	48	47
5	$49\frac{3}{4}$	$48\frac{5}{8} \ldots 1\frac{3}{4} : 1\frac{5}{8} = \cdot 929$

This glass stops only 71 invisible rays of heat.

149*th Experiment.*

	Invisible rays.	Flint glass.
0′	$50\frac{3}{4}$	$49\frac{7}{8}$
5	52	$51\frac{1}{8} \ldots 1\frac{1}{4} : 1\frac{1}{4} = 1 \cdot 000$

This glass stops no invisible rays of heat.

150*th Experiment.*

	Invisible rays.	Crown glass.
0′	$50\frac{1}{2}$	$49\frac{3}{4}$
5	$51\frac{7}{8}$	$50\frac{7}{8} \ldots 1\frac{3}{8} : 1\frac{1}{8} = \cdot 818$

It stops 182 invisible rays of heat.

151*st Experiment.*

	Invisible rays.	Coach glass.
	$54\frac{1}{2}$	$53\frac{7}{8}$
	$55\frac{3}{8}$	$54\frac{5}{8} \ldots \frac{7}{8} : \frac{3}{4} = \cdot 857$

It stops 143 invisible rays of heat.

152*d Experiment.*

	Invisible rays.	Calcinable talc.
0′	$51\frac{3}{8}$	$50\frac{3}{4}$
5	$52\frac{7}{8}$	$51\frac{7}{8} \ldots 1\frac{1}{2} : 1\frac{1}{8} = \cdot 750$

This substance stops 250 invisible rays of heat.

153*d Experiment.*

	Invisible rays.	Dark-red glass.
0′	$47\frac{5}{8}$	$46\frac{3}{4}$
5	$48\frac{5}{8}$	$47\frac{3}{4} \ldots 1 : 1 = 1\cdot 000$

This glass stops no invisible rays of heat. This accounts for the strong sensation of heat felt by the eye, in looking at the sun through a telescope, when red darkening glasses are used.

154*th Experiment.*

	Invisible rays.	Orange glass.
0′	$51\frac{5}{8}$	51
5	53	$52 \ldots 1\frac{3}{8} : 1 = \cdot 727$

It stops 273 invisible rays of heat.

155*th Experiment.*

	Invisible rays.	Yellow glass.
0′	$51\frac{3}{4}$	51
5	53	$52 \ldots 1\frac{1}{4} : 1 = \cdot 800$

It stops 200 invisible rays of heat.

156*th Experiment.*

	Invisible rays.	Pale-green glass.
0′	$51\frac{7}{8}$	$51\frac{1}{4}$
5	$52\frac{7}{8}$	$51\frac{7}{8} \ldots 1 : \frac{5}{8} = \cdot 625$

It stops 375 invisible rays of heat.

157*th Experiment.*

	Invisible rays.	Dark-green glass.
0′	$51\frac{7}{8}$	$51\frac{1}{2}$
5	$52\frac{7}{8}$	$52 \ldots 1 : \frac{1}{2} = \cdot 500$

It stops 500 invisible rays of heat.

158*th Experiment.*

	Invisible rays.	Bluish-green glass.
0′	53	$52\frac{1}{4}$
5	$54\frac{1}{4}$	$52\frac{1}{2} \ldots 1\frac{1}{4} : \frac{1}{4} = \cdot 200$

It stops 800 invisible rays of heat.

159*th Experiment.*

	Invisible rays.	Pale-blue glass.
0′	$51\frac{7}{8}$	$51\frac{1}{4}$
5	$53\frac{3}{8}$	$51\frac{7}{8} \ldots 1\frac{1}{2} : \frac{5}{8} = \cdot 417$

It stops 583 invisible rays of heat.

160*th Experiment.*

	Invisible rays.	Dark-blue glass.
0′	$52\frac{7}{8}$	$51\frac{5}{8}$
5	$52\frac{7}{8}$	$52\frac{1}{4} \ldots \frac{3}{4} : \frac{5}{8} = \cdot 833$

It stops 167 invisible rays of heat.

161*st Experiment.*

	Invisible rays.	Indigo glass.
0′	$52\frac{7}{8}$	$52\frac{1}{4}$
5	$54\frac{3}{8}$	$53 \ldots 1\frac{1}{2} : \frac{3}{4} = \cdot 500$

It stops 500 invisible rays of heat.

162*d Experiment.*

	Invisible rays.	Pale-indigo glass.
0′	$52\frac{3}{4}$	$52\frac{1}{8}$
5	$53\frac{3}{4}$	$52\frac{7}{8} \ldots 1 : \frac{3}{4} = \cdot 750$

It stops 250 invisible rays of heat.

163*d Experiment.*

	Invisible rays.	Purple glass.
0′	$51\frac{1}{4}$	$50\frac{3}{8}$
5	$52\frac{5}{8}$	$51\frac{3}{8} \ldots 1\frac{3}{8} : 1 = \cdot 727$

It stops 273 invisible rays of heat.

164*th Experiment.*

	Invisible rays.	Violet glass.
0′	$53\frac{1}{4}$	$52\frac{3}{8}$
5	$54\frac{1}{4}$	$53\frac{1}{8} \ldots 1 : \frac{3}{4} = \cdot 750$

It stops 250 invisible rays of heat.

165*th Experiment.*

	Invisible rays.	Crown glass; one side rubbed on emery, rough side exposed.
0′	$49\frac{1}{2}$	$48\frac{3}{4}$
5	$50\frac{3}{4}$	$49\frac{1}{4} \ldots 1\frac{1}{4} : \frac{1}{2} = \cdot 400$

This glass, so prepared, stops 600 scattered invisible rays of heat.

166th Experiment.

	Invisible rays.	Coach glass; one side rubbed on emery, rough side exposed.
0′	54	$53\frac{3}{8}$
5	$55\frac{1}{4}$	$54 \ldots 1\frac{1}{4} : \frac{5}{8} = \cdot 500$

It stops 500 scattered invisible rays of heat.

167th Experiment.

	Invisible rays.	Crown glass; both sides rubbed on emery.
0′	50	$49\frac{1}{8}$
5	$51\frac{1}{4}$	$49\frac{5}{8} \ldots 1\frac{1}{4} : \frac{1}{2} = \cdot 400$

It stops 600 scattered invisible rays of heat.

168th Experiment.

	Invisible rays.	Coach glass; both sides rubbed on emery.
0′	$54\frac{3}{4}$	$54\frac{1}{8}$
5	$55\frac{5}{8}$	$54\frac{3}{8} \ldots \frac{7}{8} : \frac{1}{4} = \cdot 286$

It stops 714 scattered invisible rays of heat.

169th Experiment.

	Invisible rays.	Calcined talc.
0′	$51\frac{7}{8}$	$50\frac{7}{8}$
5	53	$51 \ldots 1\frac{1}{8} : \frac{1}{8} = \cdot 111$

This substance stops 889 scattered invisible rays of heat.

Transmission of invisible terrestrial Heat.

This is perhaps the most extensive and most interesting of all the articles we have to investigate. Dark heat is with us the most common of all; and its passage from one body into another, is what it highly concerns us to trace out. The slightest change of temperature denotes the motion of invisible heat; and if we could be fully informed about the method of its transmission, much light would be thrown on what now still remains a mysterious subject. It must be remembered, that in the following experiments, I only mean to point out the transmission of such dark heat as I have before proved to consist of rays, without inquiring whether there be any other than such existing.

My apparatus for these experiments is as follows [fig. 20]. A box 12 inches long, $5\frac{1}{2}$ broad, and 3 deep, has a partition throughout its whole length, which divides it into two parts. At one end of each division is a hole $\frac{3}{4}$ inch in diameter; and each division contains a thermometer, with its ball exposed to the hole, and at one inch distance from the outside of the box. Four inches of the box, next to the holes, are covered; the rest is open. In the front of it is a narrow slip of wood, on which may rest any glass to be tried; and it is held close to the wood at the top, by a small spring applied against it. Two screws are planted upon the front, one on each side, which may be drawn out or screwed in, by way of accurately adjusting the distance of the thermometer from the line of action.

In order to procure invisible terrestrial heat, I have tried many different ways, but a stove is the most commodious of them. Iron is a substance that transmits invisible heat very readily; while, at the same time, it will most effectually intercept every visible ray of the fire by which it is heated, provided that be not carried to any great excess. I therefore made use of an iron stove [fig. 21], having four flat sides, and being constructed so as to exclude all appearance of light. I had it placed close to a wall, that the pipe which conveys away smoke might not scatter heat into the room.

The thermometer box, when experiments are to be made, is to be put into an arrangement of twelve bricks, placed on a stand, with casters [fig. 22]: these bricks, when the stand is rolled close to the stove, which must not be done till an experiment is to begin, form an inclosure, just fitting round the sides, bottom, and covered part of the top of the thermometer box, and completely guard it against the heat of the stove. The box is then shoved into the brick opening, close to the iron side of the stove, where the two front screws, coming into contact with the iron plate, give the thermometers their proper distance; which, in the following experiments, has been such as to bring the most advanced part of the balls to one inch and four-tenths from the hot iron.

It will be necessary to remark, that on calculating the transmissions for the fifth minute, I found that it would not be doing justice to the stopping power of the glasses, to take so long a time; for, notwithstanding the use of brick-work, and the precaution I had taken, of having two sets of it, that one might be cooling while the other was employed, and though neither of them was ever very hot, yet I found that so much heat came to the box, that when it was taken out of the bricks, in order to be cooled, the thermometers continued still to rise, at an average about two, degrees higher than they were. I have therefore now taken the third minute, as a much safer way to come at the truth.

170th Experiment.

	Stove.	Bluish-white glass.
0′	56	$55\frac{3}{4}$
3	$59\frac{3}{4}$	$56\frac{7}{8} \ldots 3\frac{3}{4} : 1\frac{1}{8} = \cdot 300$

This glass stops 700 invisible rays of heat.

171st Experiment.

	Stove.	Flint glass.
0′	$53\frac{3}{4}$	$53\frac{1}{2}$
3	$55\frac{5}{8}$	$54\frac{3}{8} \ldots 1\frac{7}{8} : \frac{7}{8} = \cdot 467$

It stops 533 invisible rays of heat.

174th Experiment.

	Stove.	Iceland crystal.
0′	47	$46\frac{1}{2}$
3	$54\frac{3}{4}$	$48\frac{5}{8} \ldots 7\frac{3}{4} : 2\frac{1}{8} = \cdot 274$

This substance stops 726 invisible rays of heat.

Fig. 21. Fig. 22.

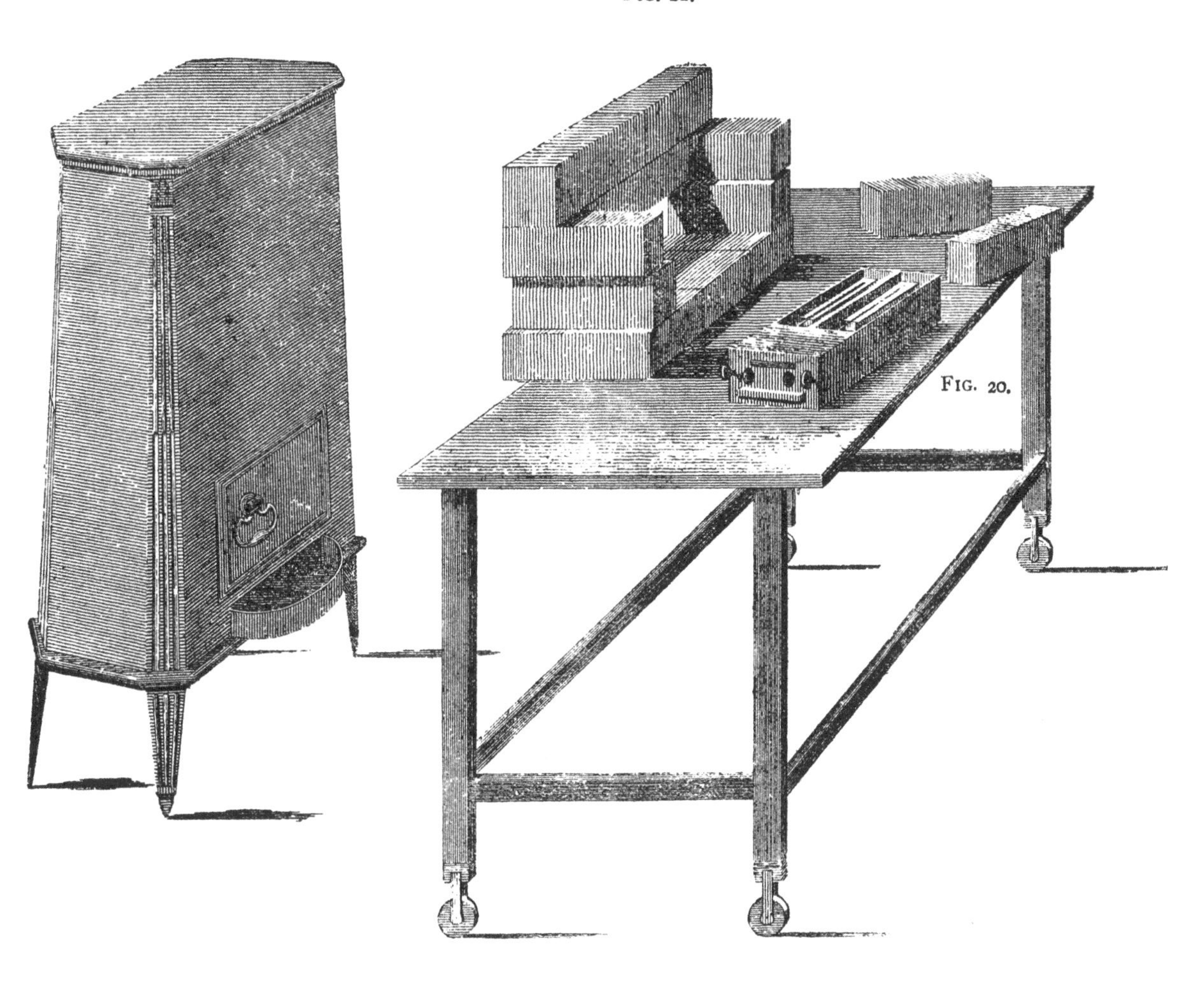

Fig. 20.

172d Experiment.

	Stove.	Crown glass.
0′	$50\frac{1}{2}$	$50\frac{1}{2}$
3	$53\frac{3}{8}$	$51\frac{1}{8} \ldots 2\frac{7}{8} : \frac{5}{8} = \cdot 217$

It stops 783 invisible rays of heat.

173d Experiment.

	Stove.	Coach glass.
0′	$50\frac{1}{2}$	$50\frac{1}{2}$
3	$52\frac{1}{2}$	$51\frac{1}{4} \ldots 2 : \frac{3}{4} = \cdot 375$

It stops 625 invisible rays of heat.

175th Experiment.

	Stove.	Calcinable talc.
0′	51	$51\frac{5}{8}$
3	$57\frac{1}{2}$	$54\frac{1}{4} \ldots 6\frac{1}{2} : 2\frac{5}{8} = \cdot 404$

At the end of five minutes, when the box was taken out of the bricks, the talc was perfectly turned into a scattering substance: as such, it stops 596 scattered invisible rays of heat. The sun cannot be seen through it; but this I find is chiefly owing to its scattering disposition. It stops however 997 scattered rays of light.

176*th Experiment.*

	Stove.	Dark red glass.
0′	58	58
3	$64\frac{3}{4}$	$60\frac{1}{2} \ldots 6\frac{3}{4} : 2\frac{1}{2} = \cdot370$

This glass stops 630 invisible rays of heat.

177*th Experiment.*

	Stove.	Orange glass.
0′	$55\frac{1}{2}$	$55\frac{1}{4}$
3	$60\frac{3}{4}$	$57\frac{3}{4} \ldots 5\frac{1}{4} : 2\frac{1}{2} = \cdot476$

It stops 524 invisible rays of heat.

178*th Experiment.*

	Stove.	Yellow glass.
0′	$57\frac{3}{4}$	$57\frac{1}{4}$
3	$61\frac{3}{4}$	$59\frac{3}{8} \ldots 4 : 2\frac{1}{8} = \cdot531$

It stops 469 invisible rays of heat.

179*th Experiment.*

	Stove.	Pale-green glass.
0′	$51\frac{1}{2}$	$51\frac{1}{2}$
3	$56\frac{1}{4}$	$53\frac{1}{4} \ldots 4\frac{3}{4} : 1\frac{3}{4} = \cdot368$

It stops 632 invisible rays of heat.

180*th Experiment.*

	Stove.	Dark-green glass.
0′	50	$49\frac{1}{2}$
3	$53\frac{3}{4}$	$50\frac{5}{8} \ldots 3\frac{3}{4} : 1\frac{1}{8} = \cdot300$

It stops 700 invisible rays of heat.

181*st Experiment.*

	Stove.	Bluish-green glass.
0′	51	51
3	$55\frac{1}{2}$	$53 \ldots 4\frac{1}{2} : 2 = \cdot444$

It stops 556 invisible rays of heat.

182*d Experiment.*

	Stove.	Pale-blue glass.
	$53\frac{3}{4}$	$53\frac{5}{8}$
	$57\frac{5}{8}$	$55\frac{3}{8} \ldots 3\frac{7}{8} : 1\frac{3}{4} = \cdot452$

It stops 548 invisible rays of heat.

183*d Experiment.*

	Stove.	Dark-blue glass.
0′	$53\frac{1}{2}$	53
5	$55\frac{7}{8}$	$53\frac{7}{8} \ldots 2\frac{3}{8} : \frac{7}{8} = \cdot368$

It stops 632 invisible rays of heat.

184*th Experiment.*

	Stove.	Indigo glass.
0′	$54\frac{1}{8}$	54
3	$59\frac{5}{8}$	$55\frac{7}{8} \ldots 5\frac{1}{2} : 1\frac{7}{8} = \cdot341$

It stops 659 invisible rays of heat.

185*th Experiment.*

	Stove.	Pale indigo glass.
0′	$53\frac{1}{2}$	$53\frac{1}{2}$
3	$59\frac{3}{4}$	$55\frac{3}{8} \ldots 6\frac{1}{4} : 1\frac{7}{8} = \cdot300$

It stops 700 invisible rays of heat.

186*th Experiment.*

	Stove.	Purple glass.
0′	$51\frac{3}{4}$	$51\frac{1}{8}$
3	$56\frac{3}{8}$	$52\frac{3}{8} \ldots 4\frac{5}{8} : 1\frac{1}{4} = \cdot270$

It stops 730 invisible rays of heat.

187*th Experiment.*

	Stove.	Violet glass.
0′	51	$51\frac{1}{2}$
3	$55\frac{3}{4}$	$53 \ldots 4\frac{3}{4} : 1\frac{1}{2} = \cdot316$

It stops 684 invisible rays of heat.

188*th Experiment.*

	Stove.	Crown glass; one side rubbed on emery.
0′	$49\frac{1}{4}$	$49\frac{1}{4}$
3	$54\frac{1}{4}$	$50\frac{3}{8} \ldots 5 : 1\frac{1}{8} = \cdot225$

This glass, so prepared, stops 775 invisible rays of scattered heat.

189*th Experiment.*

	Stove.	Coach glass; one side rubbed on emery.
0′	50	50
3	$57\frac{1}{4}$	$51\frac{7}{8} \ldots 7\frac{1}{4} : 1\frac{7}{8} = \cdot259$

It stops 741 invisible rays of scattered heat.

190*th Experiment.*

	Stove.	Crown glass; both sides rubbed on emery.
0′	52	52
3	58	$53 \ldots 6 : 1 = \cdot167$

It stops 833 invisible rays of scattered heat.

191*st Experiment.*

	Stove.	Coach glass; both sides rubbed on emery.
0′	52	52
3	$55\frac{1}{4}$	$52\frac{3}{4} \ldots 3\frac{1}{4} : \frac{3}{4} = \cdot231$

It stops 769 invisible rays of scattered heat.

192*d Experiment.*

	Stove.	Olive colour, burnt in glass.
0′	$51\frac{1}{2}$	$51\frac{1}{2}$
3	57	$53\frac{1}{2} \ldots 5\frac{1}{2} : 2 = \cdot364$

It stops 636 invisible rays of scattered heat.

193*d Experiment.*

	Stove.	White paper.	
0′	52	52	
3	$57\frac{3}{8}$	$54\frac{1}{2}$	. . . $5\frac{3}{8} : 2\frac{1}{2} = \cdot 465$

This substance stops only 535 invisible rays of scattered heat.

194*th Experiment.*

	Stove.	Linen.	
0′	$53\frac{3}{8}$	$53\frac{3}{8}$	
3	$57\frac{3}{4}$	$55\frac{3}{4}$	. . . $4\frac{3}{8} : 2\frac{3}{8} = \cdot 543$

It stops 457 invisible rays of scattered heat.

ARTICLE VI.—*Scattering of Solar Heat.*

We are now come to a branch of our inquiry which, from its novelty, would deserve a fuller investigation than we can at present enter into. The scattering of heat, is a reflection of it on the rough surfaces of bodies: it is therefore a principle of general influence, since all bodies, even the most polished, are sufficiently rough to scatter heat in all directions. In order, therefore, to compare the effect of rough surfaces on heat with their effect on light, I have made a number of experiments, from which the following are selected, for the purpose of our intended comparative view.

The apparatus I have used for scattering solar heat, is like that which served for transmissions [fig. 12]; but here the holes through which the sun's rays enter [fig. 23], are very exactly $1\frac{1}{2}$ inch in diameter each; and are chamfered away

FIG. 23.

on the under side, that no re-scattering may take place in the thickness of the covering board: the distance of the centre of the holes is 4 inches. A little more than an inch below, and under the centre of the holes, are the balls of the small thermometers A and B, well shaded from the direct rays of the sun, by small slips of wood, of the shape of the ball, and of that part of the stem which is exposed.

Under each thermometer is a small tablet [fig 24] on which the objects intended for scattering the sun's rays are to be placed. The tablets are contrived so as to bring the objects perpendicularly under the openings, and under the centre of the balls of the thermometers, at the distance of exactly one inch from them. Every thing being thus alike on both sides of the box, it is evident, from the equality of the holes, that an equal number of solar rays will fall on each object, and will by them be scattered back on the thermometers, at equal angles, and equal distances.

FIG. 24.

The first five experiments that follow, were made with an apparatus somewhat different from the one here described; and, though the result of them may not be so accurate as if they had been made with the present one, I must give them as they are, since time will not allow of a repetition.

195*th Experiment.*

	Sun.	Message card scattering.	
0′	64	64	
5	$69\frac{3}{4}$	$66\frac{3}{8}$	. . . $5\frac{3}{4} : 2\frac{3}{8} = \cdot 413$

Here an object of a white colour, 3·6 inches long, and 2·6 broad, scattered, in 5 minutes, 413 rays of heat back upon one thermometer, while the other one received a thousand, directly from the sun. Now, in order the better to compare the proportion of light and heat scattered by different objects, we shall put these 413 rays equal to 1000; or, which is nearly the same, multiply them by 2·421. Then, since the message card also scatters 1000 rays of light, as will be found in a table at the end of the transmission table, our present object may be made a standard for a comparison with the four following ones.

196*th Experiment.*

	Sun.	Pink-coloured paper scattering.	
0′	64	64	
5	70	$66\frac{5}{8}$	. . . $6 : 2\frac{5}{8} = \cdot 438$

Here a piece of pink-coloured paper, of the same dimensions with the card of the last experiment, and placed in the same situation, scattered, as we find by the same mode of multiplication, 1060 rays of heat; and, by our table, it scatters 513 of light.

197*th Experiment.*

	Sun.	Pale-green paper scattering.	
0′	$64\frac{1}{8}$	$64\frac{1}{8}$	
5	$69\frac{7}{8}$	$66\frac{1}{4}$	. . . $5\frac{3}{4} : 2\frac{1}{8} = \cdot 370$

This piece of paper scatters 896 rays of heat, and 549 of light.

198*th Experiment.*

	Sun.	Dark-green paper scattering.
0′	$64\frac{5}{8}$	$65\frac{1}{4}$
5	$69\frac{1}{2}$	$67\frac{3}{4} \ldots 4\frac{7}{8} : 2\frac{1}{2} = \cdot 513$

This paper scatters 1242 rays of heat, and only 308 of light.

199*th Experiment.*

Sun.	Black paper scattering.
$65\frac{1}{2}$	66
$70\frac{3}{8}$	$68 \ldots 4\frac{7}{8} : 2 = \cdot 410$

This paper scatters 993 rays of heat, and 420 of light.

From these experiments it seems to be evident, that in scattering heat, the colour of the object is out of the question; or, at least, that it is no otherwise concerned than as far as it may influence the texture of the surface of bodies. For here we find that pale-green, which is brighter, or scatters more light, than dark-green, yet scatters less heat. Even black, so generally known to scatter but little light, scatters much heat. But, in order to put this surmise to a fairer trial, I made the following experiments with my new machine.

200*th Experiment.*

I covered one of the tablets with white paper, and the other with black. The quantity of sunshine admitted through the two openings, of $1\frac{1}{2}$ inch in diameter each, being equal, I found the heat scattered on both thermometers to be as follows.

	White paper.	Black paper scattering.
0′	$71\frac{3}{4}$	72
5	$75\frac{5}{8}$	$75 \ldots 3\frac{7}{8} : 3 = \cdot 774$

I turned now the tablets, and had,

	Black paper.	White paper scattering.
0′	$73\frac{1}{4}$	$72\frac{3}{4}$
5	$75\frac{5}{8}$	$75\frac{7}{8} \ldots 2\frac{3}{8} : 3\frac{1}{8} = \cdot 760$

These results, agreeing sufficiently well together, shew that if we make white paper our standard, and suppose it to scatter 1000 rays of heat, and 1000 of light, then will black paper scatter 767 rays of heat, and 420 of light.

201*st Experiment.*

	White paper.	Black muslin scattering.
0′	$73\frac{3}{4}$	$73\frac{3}{4}$
5	$77\frac{3}{4}$	$77 \ldots 4 : 3\frac{1}{4} = \cdot 813$

This scatters 813 rays of heat; and, when it is suspended so that the rays which pass through it may not be reflected, it scatters only 64 rays of light.

202*d Experiment.*

As my intention at present was to find a black substance that should scatter more heat than a white one, I thought it would be the readiest way to examine the white and black objects separately, that of all the white ones I might afterwards take that which scattered least, and compare it with the black one which scattered most.

	White paper.	White linen scattering.
0′	$74\frac{7}{8}$	75
5	79	$79\frac{1}{8} \ldots 4\frac{1}{8} : 4\frac{1}{8} = 1 \cdot 000$

These objects scatter heat equally, and very nearly also light; for our table gives for linen 1008.

203*d Experiment.*

	White paper.	White cotton scattering.
0′	$74\frac{1}{2}$	$74\frac{5}{8}$
5	$78\frac{3}{8}$	$78\frac{1}{2} \ldots 3\frac{7}{8} : 3\frac{7}{8} = 1 \cdot 000$

These objects scatter heat equally. White cotton scatters 1054 rays of light.

204*th Experiment.*

	White paper.	White muslin scattering.
0′	$73\frac{3}{8}$	$73\frac{3}{8}$
5	$77\frac{3}{8}$	$76\frac{7}{8} \ldots 4 : 3\frac{1}{2} = \cdot 875$

White muslin scatters 875 rays of heat, and 827 of light.

205*th Experiment.*

	White paper.	White persian scattering.
0′	$74\frac{1}{2}$	$74\frac{5}{8}$
5	$77\frac{7}{8}$	$78\frac{1}{4} \ldots 3\frac{3}{8} : 3\frac{5}{8} = 1 \cdot 074$

White persian scatters 1074 rays of heat; and, when suspended like the black muslin in the 201st experiment, it scatters 671 rays of light.

206*th Experiment.*

	White paper.	White knit worsted; rough side outwards.
0′	51	$51\frac{3}{4}$
5	$52\frac{5}{8}$	$53\frac{3}{4} \ldots 1\frac{5}{8} : 2 = 1 \cdot 231$

White worsted scatters 1231 rays of heat, and 620 of light.

207*th Experiment.*

White paper.	White chamois leather; the smooth side exposed.
$74\frac{5}{8}$	$74\frac{5}{8}$
$78\frac{3}{8}$	$79 \ldots 3\frac{3}{4} : 4\frac{3}{8} = 1 \cdot 167$

White chamois leather scatters 1167 rays of heat, and 1228 of light.

208*th Experiment.*

Black paper.	Black velvet scattering.
$75\frac{3}{4}$	$75\frac{7}{8}$
$79\frac{1}{4}$	$80 \ldots 3\frac{1}{2} : 4\frac{1}{8} = 1{\cdot}179$

Making now black paper the standard, and supposing it to scatter 1000 rays of heat, and the same of light, then black velvet scatters 1179 rays of heat, and only 17 of light. This last number we obtain, by dividing the tabular number 7, for black velvet, by ·42, which is the proportion of black paper to white.

209*th Experiment.*

	Black paper.	Black muslin scattering.
0′	$75\frac{1}{8}$	$75\frac{1}{4}$
5	$78\frac{3}{8}$	$79\frac{1}{8} \ldots 3\frac{1}{4} : 3\frac{7}{8} = 1{\cdot}192$

Black muslin scatters 1192 rays of heat, and 43 of light.

210*th Experiment.*

	Black paper.	Black satin scattering.
0	$76\frac{1}{4}$	$76\frac{1}{4}$
5	79	$80\frac{1}{8} \ldots 2\frac{3}{4} : 3\frac{7}{8} = 1{\cdot}409$

Black satin scatters 1409 rays of heat, and 243 of light.

211*th Experiment.*

Having now ascertained, that of all the white and black substances I had tried, white muslin scatters the least, and black satin the most heat, I placed the former on one tablet, while the latter was put on the other.

	White muslin.	Black satin scattering.
0′	$76\frac{3}{8}$	$76\frac{3}{8}$
5	80	$80\frac{1}{4} \ldots 3\frac{5}{8} : 3\frac{7}{8} = 1{\cdot}069$

Here the black object scattered more heat than the white one; but, in order to try again the equality of the tablets and apparatus, I placed the objects under the opposite thermometers, and had as follows.

	Black satin.	White muslin scattering.
0′	78	78
5	$80\frac{5}{8}$	$80\frac{1}{2} \ldots 2\frac{5}{8} : 2\frac{1}{2} = 1{\cdot}050$

So that, notwithstanding some little difference in the apparatus, or other unavoidable circumstances, the black object gave again the greatest scattering of heat; and consequently, as no colour can be more opposite than black and white, colour can have no concern in the laws that relate to the scattering of heat.

212*th Experiment.*

I wished now to try some experiments of the scattering power of metals, and had some plates of iron, brass, and copper, two inches square, set flat, and smooth-filed, by round strokes.

	Iron.	Copper scattering.
0′	74	$73\frac{3}{4}$
5	$78\frac{1}{2}$	$77\frac{7}{8} \ldots 4\frac{1}{2} : 4\frac{1}{8} = {\cdot}917$

213*th Experiment.*

	Tin foil.	Gold-leaf paper scattering.
0′	74	74
5	$77\frac{3}{4}$	$79\frac{5}{8} \ldots 3\frac{3}{4} : 5\frac{5}{8} = 1{\cdot}500$

But the tin foil was considerably tarnished.

214*th Experiment.*

Finding the form of the last experiments inconvenient, for want of a standard, I had recourse again to white paper.

	White paper.	Tin foil scattering.
0′	$50\frac{1}{8}$	52
5	$53\frac{3}{8}$	$54\frac{7}{8} \ldots 3\frac{1}{4} : 2\frac{7}{8} = {\cdot}885$

This substance scatters 885 rays of heat, and 8483 of light.

215*th Experiment.*

	White paper.	Iron.
0′	$51\frac{5}{8}$	$53\frac{1}{2}$
5	$54\frac{5}{8}$	$55\frac{3}{4} \ldots 3 : 2\frac{1}{4} = {\cdot}750$

Some time having elapsed between the former observation and the present one, this plate of iron was not now so bright as before, and seems to have suffered more than brass or copper from having been laid by: it scatters now only 750 rays of heat, and 10014 of light.

216*th Experiment.*

	White paper.	Brass.
0′	50	$51\frac{1}{4}$
5	$53\frac{1}{8}$	$55\frac{3}{8} \ldots 3\frac{1}{8} : 4\frac{1}{8} = 1{\cdot}320$

It scatters 1320 rays of heat, and no less than 43858 of light.

217*th Experiment.*

White paper.	Copper.
$49\frac{7}{8}$	$51\frac{7}{8}$
53	$55\frac{7}{8} \ldots 3\frac{1}{8} : 4 = 1{\cdot}280$

It scatters 1280 rays of heat, and 13128 of light.

218th Experiment.

White paper.	Gold-leaf paper.
$54\frac{3}{4}$	$55\frac{3}{8}$
$56\frac{1}{2}$	$56 \ldots 1\frac{3}{4} : \frac{5}{8} = \cdot 357$

I changed the tablets to see what difference there might be.

Gold paper.	White paper.
$55\frac{3}{8}$	$55\frac{7}{8}$
$56\frac{1}{8}$	$57\frac{3}{8} \ldots \frac{3}{4} : 1\frac{1}{2} = \cdot 500$

A mean between the two gives ·429. Gold paper, therefore, scatters only 429 rays of heat, and no less than 124371 rays of light.

219th Experiment.

	Black velvet.	Gold paper scattering.
0′	52	$51\frac{7}{8}$
5	$53\frac{1}{8}$	$52\frac{1}{2} \ldots 1\frac{1}{8} : \frac{5}{8} = \cdot 556$

I turned the tablets, in order to ascertain the difference.

	Gold paper.	Black velvet.
0′	51	$51\frac{3}{4}$
5	$51\frac{3}{4}$	$53 \ldots \frac{3}{4} : 1\frac{1}{4} = \cdot 600$

From a mean of both it appears, that when black velvet scatters 1000 rays of heat, and only 7 rays of light, gold paper, on the contrary, scatters no more than 578 rays of heat, but 124371 of light.

ART. VII.—*Whether Light and Heat be occasioned by the same, or by different Rays.*

Before we enter into a discussion of this question, it appears to me that we are authorised, by the experiments which have been delivered in this Paper, to make certain conclusions, that will entirely alter the form of our inquiry. Thus, from the 18th experiment it appears, that 21 degrees of solar heat were given in one minute to a thermometer, by rays which had no power of illuminating objects, and which could not be rendered visible, notwithstanding they were brought together in the focus of a burning lens. The same has also been proved of terrestrial heat, in the 9th experiment ; where, in one minute, 39 degrees of it were given to a thermometer, by rays totally invisible, even when condensed by a concave mirror. Hence it is established, by incontrovertible facts, that there are rays of heat, both solar and terrestrial, not endowed with a power of rendering objects visible.

It has also been proved, by the whole tenour of our prismatic experiments, that this invisible heat is continued, from the beginning of the least refrangible rays towards the most refrangible ones, in a series of uninterrupted gradation, from a gentle beginning to a certain maximum ; and that it afterwards declines, as uniformly, to a vanishing state. These phenomena have been ascertained by an instrument, which, figuratively speaking, we may call blind, and which, therefore, could give us no information about light ; yet, by its faithful report, the thermometer, which is the instrument alluded to, can leave no doubt about the existence of the different degrees of heat in the prismatic spectrum.

This consideration, as has been observed, must alter the form of our proposed inquiry ; for the question being thus at least partly decided, since it is ascertained that we have rays of heat which give no light, it can only become a subject of inquiry, whether some of these heat-making rays may not have a power of rendering objects visible, superadded to their now already established power of heating bodies.

This being the case, it is evident that the *onus probandi* ought to lie with those who are willing to establish such an hypothesis; for it does not appear that nature is in the habit of using one and the same mechanism with any two of our senses; witness the vibrations of air that make sound; the effluvia that occasion smells; the particles that produce taste; the resistance or repulsive powers that affect the touch: all these are evidently suited to their respective organs of sense. Are we then here, on the contrary, to suppose that the same mechanism should be the cause of such different sensations, as the delicate perceptions of vision, and the very grossest of all affections, which are common to the coarsest parts of our bodies, when exposed to heat?

But, let us see what light may now be obtained from the several articles that have been discussed in this Paper. It has been shewn, that the effect of heat and of illumination may be represented by the two united spectra, which we have given.* Now, when these are compared, it appears that those who would have the rays of heat also to do the office of light, must be obliged to maintain the following arbitrary and revolting positions; namely, that a set of rays conveying heat, should all at once, in a certain part of the spectrum, begin to give a small degree of light; that this newly acquired power of illumination should increase, while the power of heating is on the decline; that when the illuminating principle is come to a maximum, it should, in its turn, also decline very rapidly, and vanish at the same time with the power of heating. How can effects that are so opposite be ascribed to the same cause? first of all, heat without light; next to this, decreasing heat, but increasing light; then again, decreasing heat and decreasing light. What modification can we suppose to be superadded to the heat-making power, that will produce such inconsistent results?

We must not omit to mention another difference between light and heat, which may be gathered from the same article of the refrangibility of heat-making rays It is, that though light and heat are both refrangible, the ratio of the sines of incidence and refraction of the mean rays is not the same in both. Heat is evidently less refrangible than light; whether we take a mean refrangible ray of each, or, which I believe to be the better way of proceeding, whether we take the maximum of heat and light separately. This appears, not only from the view we have taken of the two spectra already mentioned, but more evidently from the 23d experiment, by which we find, that heat cannot be collected by a lens, to the same focus where light is gathered together.

Our fifth article, in which an account has been given of the proportions of heat and light stopped by glasses and other substances, will afford us now an ample field for pointing out a striking difference between these two principles. From the 24th to the 30th experiment, we have the quantities intercepted by colourless substances as follows.

* [See page 99, and fig. 11.]

TABLE I.

Bluish-white glass stops	250 rays of heat, and	86 of light.	
White flint glass ,,	91 ,, ,,	34 ,,	
Greenish crown glass	259 ,, ,,	203 ,,	
Coach glass ,,	214 ,, ,,	168 ,,	
Iceland crystal ,,	244 ,, ,,	150 ,,	
Talc ,,	139 ,, ,,	90 ,,	
Calcinable talc ,,	184 ,, ,,	288 ,,	

Now, by casting an eye on the above table, it will be seen immediately, that no kind of regularity takes place among the proportions of rays of one sort and of another, which are stopped in their passage. Heat and light seem to be entirely unconnected. The bluish-white and flint glasses, for instance, stop nearly three times as much heat as light ; whereas, the greenish crown glass stops only about one-fourth more of the former than of the latter ; but, as coloured glasses take in a much greater range, I will now also give a tabular result of the experiments that have been given relating to them.

TABLE II.

Very dark red glass stops	800 rays of heat, and	$999\frac{9}{10}$ of light.			
Dark-red ,,	606 ,, ,,	$999\frac{8}{10}$,,			
Orange ,,	604 ,, ,,	779 ,,			
Yellow ,,	333 ,, ,,	819 ,,			
Pale-green ,,	633 ,, ,,	535 ,,			
Dark-green ,,	849 ,, ,,	949 ,,			
Bluish-green ,,	768 ,, ,,	769 ,,			
Pale-blue ,,	812 ,, ,,	684 ,,			
Dark-blue ,,	362 ,, ,,	801 ,,			
Indigo ,,	633 ,, ,,	$999\frac{7}{10}$,,			
Pale-indigo ,,	532 ,, ,,	978 ,,			
Purple ,,	583 ,, ,,	993 ,,			
Violet ,,	489 ,, ,,	955 ,,			

From this table, I shall also point out a few of the most remarkable results. A yellow glass, for instance, stops only 333 rays of heat, but stops 819 of light: on the contrary, a pale blue stops 812 rays of heat, and but 684 of light. Again, a dark blue glass stops only 362 rays of heat, but intercepts 801 of light ; and a dark red glass stops no more than 606 rays of heat, and yet intercepts nearly all the light ; scarcely one ray out of 5000 being able to make its way through it.

Before I proceed to a more critical examination of these results, it will be necessary to add also a table of the same kind, collected from the experiments with liquids.

TABLE III.

Empty tube and 2 glasses stop	542 rays of heat, and	204 of light.
Spring water ,,	558 ,, ,,	211 ,,
Sea water ,,	682 ,, ,,	288 ,,
Spirit of wine ,,	612 ,, ,,	224 ,,
Gin ,,	739 ,, ,,	626 ,,
Brandy ,,	794 ,, ,,	996 ,,

To which may be joined, a table containing the stoppages occasioned by scattering substances.

TABLE IV.

Rough crown glass stops	464 rays of heat, and			854 of light.	
Rough coach glass	,,	571	,, ,,	879	,,
The 1st doubly rough	,,	667	,, ,,	932	,,
The 2d doubly rough	,,	735	,, ,,	946	,,
The 2 first together	,,	698	,, ,,	969	,,
The 2 next together	,,	800	,, ,,	979	,,
The 4 first together	,,	854	,, ,,	995	,,
Olive colour, burnt in	,,	839	,, ,,	984	,,
Calcined talc	,,	867	,, ,,	996	,,
White paper	,,	850	,, ,,	994	,,
White linen	,,	916	,, ,,	952	,,
White persian	,,	760	,, ,,	916	,,
Black muslin	,,	714	,, ,,	737	,,

We shall now enter more particularly into the subject of these four tables, that we may, if possible, find a criterion by which to judge whether heat and light can be occasioned by the same rays or not. Now this I think will be obtained, if we can make it appear that stopping one sort of rays does not necessarily bring on a stoppage of the other sort; for, if it can be shewn that heat and light are in this respect independent of each other, it will follow that they must be occasioned by different rays; and I shall make all possible objections to the arguments I mean to draw from these tables, in order to shew that no hypothesis will evade the force of our conclusions.

It has been noticed, that bluish-white and flint glasses stop nearly three times as much heat as light; whereas, crown glass stops only about one-fourth more of the former than of the latter. Now, in answer to this, it may be alleged, " that the ingredients of which the former glasses are made, dispose them probably to stop the invisible rays of heat, and that consequently a great interception of it may take place, without bringing on a necessity of stopping much light; and that, on the other hand, the different texture of crown glass may stop one sort of heat as well as the other, so that nearly an equality in this respect may be produced."

When a hypothesis is made in order to explain any phænomenon of nature, we ought to examine how it will agree with other facts; and, in this case, we are already furnished with experiments, which are decidedly against the supposition that has been brought forward. For, the 148th and 149th experiments shew that the bluish-white and flint glasses transmit all, or nearly all, the invisible rays of solar heat; whereas crown glass, by the 150th experiment, stops a considerable number of them. But, to assist the objecting argument, let it be alleged, as has been proved by the 94th experiment, that our bluish-white glass stops a considerable portion of the heat that goes with the red rays; then, if the 86 rays of light which this glass stops, are supposed to be all of that sort, the heat which will be stopped in consequence, will, according to the experiment we have mentioned, amount to 86 multiplied by 375, that is, 32 rays of heat; but, since 250 have been stopped there will remain 218 to be accounted for.

In this calculation, a manifest concession has been made, which ought to be explained. When I mention 86 red-coloured or red-making rays, I mean so many of them as will make up 86-thousandths of the whole effect of light; for the quantity of heat and light transmitted, or stopped, in all the experiments that have been given, has been reduced to what proportion it bears to unity; and, having afterwards represented the joint effect of every ray of heat and light by 1000, each mean ray of heat must be the thousandth part of that effect; but, a mean ray of light, although it be likewise the thousandth part of the whole effect of light, will not be so of heat, because the whole effect of the latter is partly owing to rays that have been proved to be invisible. On this account, the 86 mean rays of red light, stopped by our bluish-white glass, cannot even amount to a stoppage of 32 rays of heat, which we have allowed.

As I have made the concession on one hand, I must explain an advantage that may be claimed on the other; which is, that mean rays and promiscuous ones have already, in a former Paper, been proved to differ considerably, and that it remains therefore unknown how many red-making rays we may suppose to be stopped, in order to make up 86 mean rays of light. In answer to this, however, I must observe, that the number of promiscuous rays of light and of heat must always be inversely as their power of occasioning those sensations; so that if, for instance, a red ray is supposed to be twice as heating as a green one, there will only go half the number of them to make up a certain effect of heat; and, on the other hand, if a green ray should have a double power of illuminating, there will be no more than half the number of them necessary to occasion a certain effect of light. But, by my former experiments,* a red ray, though much inferior to a green one, is probably fully equal in illumination to a mean ray of all the colours united together. Now, as red rays have also been proved to be accompanied by the greatest heat, and as our bluish-white glass stops hardly any invisible heat rays, we have certainly gone the full length of fair concessions, by allowing all the light stopped by this glass to be of that sort; and thus it seems to be evident, that the heat which lies under the colours, if I may use this expression, may be stopped, without stopping the colours themselves.

It will not be necessary to lay much stress on this single experiment; our second table affords us sufficient ground on which to rest more forcible arguments. A dark-red glass, for instance, was found to stop 606 rays of heat, and 999·8 of light This, even at the very first view, seems to amount to a total separation of the two principles; but let us discuss the phænomenon with precision.

As only one ray in 5000 can make its way through this glass, it is evident,

* See 10th experiment [above, p. 85].

that if the rays of light be also those of heat, there can hardly come any heat through it but what must be occasioned by rays that are invisible. It will therefore become a question to be examined, how many of this sort we can admit, if we proceed on a supposition that heat consists of light, as far as that will go. Now this, we find, has already been ascertained, in a great measure, by our 13th, 17th, and 18th experiments. In the 13th, one hundred and twenty degrees of heat were given to a thermometer, in one minute, by the rays which accompany the coloured part of the spectrum. In the 17th experiment, on the contrary, we find only 45 degrees of heat communicated to the same thermometer, in the same time, by the invisible rays of the same spectrum. If we would be more scrupulous, the 18th experiment limits the heat from rays totally invisible even to 21 degrees; but, in order to make every possible allowance, let the proportion be the most favourable one of 120 to 45, which, reduced to mean rays of heat, will give 727 of them visible, and 273 invisible, to make up our thousand.

To return to the experiment: if the total number of rays of heat ascribed to light should accordingly be rated at 727, it is evident, from the stoppage of light of this glass, that 726 rays of heat at least must also be intercepted; and, in consequence of the 153d experiment, which shews that our glass opposes no obstruction to any of the invisible rays, we shall require no more. But, by our present experiment, this glass stops only 606 rays of heat; so that 120 of them will remain unaccounted for. Now, the moment we give up the hypothesis that heat is occasioned by the rays of light, the difficulty becomes fully resolved by our 100th experiment, which shews that full three-tenths of the rays that have the refrangibility of the red are actually transmitted.

In order, however, to make a second attempt to overcome this difficulty, without giving up the hypothesis, it may be supposed, "that perhaps the lens which has been used in the 13th, 17th, and 18th experiments might stop a greater number of invisible than visible rays, and that its report therefore ought not to be depended upon." Now, although it does not appear from the 148th experiment that such a supposition can have much foundation, yet, since those experiments were not made with a view to ascertain the proportion of heat contained in each part of the prismatic spectrum, we cannot lay so much stress upon them as the accuracy which is required in this case renders necessary. Let it therefore, contrary to our 100th experiment, be admitted, in order to explain the phænomenon of the red glass, which stops so much light and so little heat, that all the heat which it intercepts consists entirely of the rays which are visible, and that every one of the invisible rays of heat is transmitted. Then will 999·8 intercepted rays of light be equal to 606 rays of heat; and the remaining 394 will be the number of rays we are now to place to the account of the invisible heat which is transmitted.

Having thus also got rid of this difficulty, we are next to examine how other facts, collected in the same table, will agree with our new concession. A violet-coloured glass, for instance, stops 955 rays of light; these, at the rate of 999·8, or say 1000, to 606, must occasion a deficiency of 579 rays of heat. But, by our table, this glass stops only 489 of them; and there will thus be 90 rays of heat left unaccounted for. To enhance the difficulty, this glass, by our 164th experiment, stops also $\frac{1}{4}$ of the supposed 394 invisible rays, which will amount to an additional sum of 98. And our 111th experiment shews, that actually a great number of these rays, that otherwise cannot be accounted for, come from the store of heat, the rays of which are of the refrangibility of red light.

A dark-blue glass stops 801 rays of light; these, if light and heat were occasioned by the same rays, would produce a stoppage of 485 rays of heat; but we find that our glass stops no more than 362, so that 123 rays cannot be accounted for by this hypothesis. To this we should add 66 invisible rays, (that is, $394 \times \cdot 167$,) which, according to our 160th experiment, this glass also intercepts. But the 107th experiment, if we reject the hypothesis, immediately explains the difficulty; for here we plainly see, that only 71 rays of heat of the refrangibility of red light are stopped, whatever may be the stoppage of that light itself.

A yellow glass stops 819 rays of light: these will occasion a stoppage of 496 rays of heat; but this glass intercepts only 333, and therefore 163 rays of heat must also remain unaccounted for. And, turning to the 155th experiment, we find that 79 rays, or $\frac{1}{5}$ of the 394 allowed to be invisible ones, are also to be added to that number.

If in the results of our second table we have had an excess of heat, which the last hypothesis would not account for, we shall, on the contrary, meet with a considerable deficiency, when we come to consider those of the third table.

For instance, our tube filled with well-water, including the glasses at the end, intercepted 211 rays of light. These, at the rate of 606 to the thousand, would produce only a stoppage of 128 rays of heat; but here we find no less than 558 of them intercepted. To evade the pressure of these consequences, it may be said, "that as before every invisible ray was supposed to have been transmitted through glasses, so they may now be all intercepted by liquids." And, granting this also to be possible, though by no means probable, for the great extent of these researches has not allowed sufficient time for many experiments to be made that have been planned for execution; yet, even then, 128 visible and 394 invisible rays to be intercepted, will only make up 522; so that a deficiency of 36 must still remain.

In sea-water, the balance will stand thus: 288 rays of light give 175 rays of heat; these and 394 invisible rays make up 569; but the rays actually intercepted were 682, which argues a deficiency of no less than 113 rays.

But if I have for a moment admitted the entire stoppage of the invisible rays of heat in liquids, the same indulgence cannot be granted for the empty tube, as we know it does neither take place in glasses, nor in air. Therefore we must calculate thus: this compound of glass and air stops 204 rays of light; these can amount only to 124 rays of heat; but it is found to stop 542 of them, so that 418 remain to be accounted for. Now, we certainly can not suppose more than 100 of them to owe their deficiency to the store of invisible heat; so that 318 will still remain unaccounted for.

And thus, from the second table, we have given instances where the assumed hypothesis of visible and invisible heat, in certain proportions, would require a greater stoppage than our experiments will admit; and now, on the contrary, it appears, that interceptions calculated according to the same hypothesis, should be less than the results in the third table give them. From which we conclude, that every other proportion fixed upon, would always be erroneous, either in excess or in defect.

Equal contradictions may be shewn to attend all endeavours to account for the results contained in our fourth table, by admitting any visible heat at all, let the quantity be what it will. To make the proof of this general, let 1000 be the total heat, and assume x for that part of it which we would suppose to be occasioned by visible rays; then will $1000-x$ be the remainder, which must be ascribed to rays that cannot be seen. Now, by our table, we find that crown glass, of which one side has been rubbed on emery, stops 854 rays of light. These alone, if not a particle of invisible heat were stopped, would be equal to $\cdot 854\,x$ visible rays of heat, that must be intercepted by the glass. When the other side of this glass has also been rubbed on emery, it will stop 932 rays of light, which will give $\cdot 932\,x$, for the quantity of heat to be intercepted, on the same supposition, that all invisible rays of heat are transmitted. But, by our fourth table, we have the actual stoppage of heat of these glasses; which will therefore give us the two following equations; $\cdot 854\,x=464$, and $\cdot 932\,x=667$. Then, taking the first from the last, and reducing the remaining equation, we obtain $x=\frac{203}{078}$, for the visible part of the total heat. But $\frac{203}{\cdot 078}$, or 2602, being only a part, comes out greater than 1000, which is the whole; and this being absurd, it follows that visible rays of heat cannot be admitted, in any proportion whatsoever. This will equally hold good with any additional stoppage of invisible heat, provided it be equal in both glasses; and, of this equality, the 165th and 167th experiments can leave us no room to doubt.

But it is high time that we should now take into consideration a more direct proof, which may be drawn from our prismatic experiments. The results of them are here brought into a table, as follows.

TABLE V.

Stoppage of prismatic Heat of the Refrangibility of the Red Rays, and of the Invisible Rays.

	Red rays.	Invisible rays.
Bluish-white glass stops . .	375	71
Flint glass ,, . . .	143	000
Crown glass ,, . . .	294	182
Coach glass ,, . . .	200	143
Iceland crystal ,, . . .	200	——
Calcinable talc ,, . .	133	250
Dark-red glass ,, . . .	692	000
Orange ,, . . .	500	273
Yellow ,, . . .	417	200
Pale-green ,, . . .	588	375
Dark-green ,, . . .	786	500
Bluish-green ,, . . .	462	800
Pale-blue ,, . . .	700	750
Dark-blue ,, . . .	71	167
Indigo ,, . . .	367	222
Pale-indigo ,, . . .	313	250
Purple ,, . . .	444	273
Violet ,, . . .	400	250
Crown glass, one side rough, stops	389	600
Coach glass ,, ,,	500	500
Crown glass, both sides rough, stops	471	600
Coach glass ,, ,,	833	714
Calcined talc ,, ,,	737	889

As a necessary introduction to the decisive experiment I am going to analyse, I must remark, that it has been shewn in a former Paper, that the prism separates invisible heat from the coloured spectrum, by throwing that which is less refrangible than light to one side. But it has also been proved, that heat of the same variety in refrangibility as the different colours, is likewise contained in every part of the coloured spectrum. The question which we are discussing at present, may therefore at once be reduced to this single point. Is the heat which has the refrangibility of the red rays occasioned by the light of these rays ? For, should that be the case, as there will then be only one set of rays, one fate only can attend them, in being either transmitted or stopped, according to the power of the glass applied to them. We are now to appeal to our prismatic experiment upon the subject, which is to decide the question. First, with regard to light, I must anticipate a series of highly interesting observations I have made, but which, though they certainly claim, cannot find room in this Paper. These have given me the means of acting separately upon either of the extremes, or on the middle of the prismatic spectrum ; and by them I am assured that red glass does not stop red rays. Indeed the appearance of objects seen through such coloured glasses, till I can give those observations, will be a sufficient proof to every one that they transmit red light in abundance. Next, with regard to the rays of heat, the case is just the reverse ; for, by our preceding table, the red glass stops no less than 692, out of a thousand, of such rays as are of the refrangibility of red light. The incipient stoppage, moreover, or that in two minutes, of which something will be said hereafter, amounts even to 750 rays.

Now, if it should be suspected, " that on account of the great breadth of prism, some invisible heat may be thrown upon the spot where the red colour falls," I do not only agree to it, but am certain it cannot be otherwise : but this again, will give additional weight to our present argument ; for, by the 153d experiment, as our last table shews, it has already been ascertained, that all such heat will be transmitted through a red glass ; so that, were it not for some of this admixture, the stoppage might be still greater.

Here then we have a direct and simple proof, in the case of the red glass, that the rays of light are transmitted, while those of heat are stopped, and that thus they have nothing in common but a certain equal degree of refrangibility, which, by the power of the glass, must occasion them to be thrown together into the place which is pointed out to us by the visibility of the rays of light.

The manifest use of the union of these rays, arising from their equal refrangibility, will be explained at a future opportunity, when I may perhaps throw out several hints that have already occurred to me, where the contents of this Paper may be applied to the useful purposes of life.

There still remains a general argument, that heat and light are occasioned by different rays, which ought not to be omitted. This, on account of the contracted state in which the experiments have been given, cannot appear from my Paper ; but, by an inspection of them at full length, it is proved, that the stoppage of solar heat, setting aside little irregularities, to which all observations are liable, has constantly been greater in the first, second, or third minute, than in the fourth or fifth ; or, more accurately, nearer the beginning of the five minutes, than about the end of them. Now this does not happen in the transmission of light, which, as far as we know, is instantaneous ; at least a failure in the brightness of an object, when first we look at it through a glass, amounting to one, two, or even three minutes, could not possibly have escaped our observation. This seems to suggest to us, that the law by which heat is transmitted, is different from that which directs the passage of light ; and, in that case, it must become an irrefragable argument of the difference of the rays which occasion them.

The surmise of a difference in the law of the transmission of heat and of light, is considerably supported by an argument drawn from circumstances of a very different nature. In the scattered transmissions arising from rough surfaces, we find, that when crown glass, for instance, has one of its sides rubbed on emery, it will stop 205 rays of heat more than while that surface remained polished ; but the effect of the roughness produced by emery scratches, is far more considerable on the rays of light ; the additional stoppage of them amounting to no less than 651

A confirmation of the same effect we have in coach glass ; which, having also one side rubbed on emery, stops only 357 rays of heat more than it did before, while there is an additional stoppage of rays of light, amounting to no less than 717. Now, since the interior construction of these glasses, before and after having been rubbed on emery, remains the same, these remarkable effects can only be ascribed to the roughness of their surfaces. Hence, we may conclude, that as the same cause, when it acts upon the rays of heat and light, produces effects so very different, it can only be accounted for by admitting the rays themselves to be of a different nature, and therefore subject to a different law in being scattered. It has already been shewn, that the rays of heat are, upon an average, less *refrangible* than those of light ; and now it appears that they are also, if I may introduce a convenient term, less *scatterable*.

We ought now also to take a short review of the phænomena attending the transmission of terrestrial heat. The results of the experiments which have been given, are drawn into one view in the following table.

TABLE VI.

	Rays of flame.	Fire.	Invisible heat.
Bluish-white glass stops . . .	625	750	700
Flint glass ,, ...	591	750	533
Crown glass ,, ...	636	722	783
Coach glass ,, ...	458	714	625
Iceland crystal ,, ...	516	756	726
Talc ,, ...	375	713	615
Very dark red glass ,, ...	636	613	—
Dark-red ,, ...	526	573	630
Orange ,, ...	560	643	524
Yellow ,, ...	583	685	531
Pale-green ,, ...	500	688	632
Dark-green ,, ...	739	745	700
Bluish-green ,, ...	652	696	556
Pale-blue ,, ...	609	676	548
Dark-blue ,, ...	619	704	632
Indigo ,, ...	679	721	659
Pale-indigo ,, ...	571	655	700
Purple ,, ...	520	679	730
Violet ,, ...	500	615	684
Crown glass, one side rough, stops .	741	723	775
Coach glass, ,, ,, ,,	667	758	741
Crown glass, both sides rough ,,	615	791	833
Coach glass, ,, ,, ,,	680	854	769
The two last but two, together ,,	720	849	—
The two last together ,,	667	897	—
The four last together ,,	870	902	—
Olive-colour, burnt in glass ,,	792	849	636
White paper ,, ,,	792	912	535
White linen ,, ,,	690	910	457
White persian ,, ,,	593	829	—
Black muslin ,, ,,	565	706	—

Let us now examine what information we may draw from the facts which are recorded in this table. The first that must occur is, that a candle which emits light, is also a copious source of invisible heat. If this should seem to require a proof, I give it as follows.

That the candle emits heat along with light, the thermometer has ascertained; and, that a considerable share of this at least must be invisible, follows from comparing together the quantity of light and heat which are stopped by different glasses. The bluish-white one, for instance, stops 86 rays of light, and 625 of heat. Hence, if only visible rays of heat came from the candle, a glass stopping more light, as for instance the dark-red glass, which stops 999·8, ought to stop all heat whatsoever; but the fact is, that it even stops one hundred rays less than the former.

This instance alone shews plainly, that the existence of invisible terrestrial heat in the flame of a candle is proved; while, on the contrary, heat derived from rays that are visible, remains yet to be established, by those who would maintain that there are any such. But, for the sake of argument, let us endeavour to explain how visible rays of heat may be reconciled with the contents of our 6th table.

"Now although we must allow," it may be said, "that there is a certain quantity of candle-heat which cannot be seen, we are however at liberty to assign any ratio that this may bear to its visible heat-rays. Let us therefore begin with the bluish-white glass, and make the most favourable supposition we can, in order to explain its phænomena. Visible or invisible, it stops 625 rays of heat, and also 86 of light. Now, as in the last column of the table we have likewise the proportional quantity of invisible heat it intercepts, which is 700 out of a thousand, we may surmise that the 914 rays of light, together with the 300 of the invisible rays which are transmitted, make up the 375 rays of heat which pass through the glass. Hence, by algebra, we have the number of invisible heat-rays 878, and the number of the visible ones 122. Then, to try how this will answer, if 1000 rays of light give 122 of heat, 86 will give 10; and, if out of a thousand invisible rays 700 be stopped, 878 will give 615 to be intercepted. The sum of these will be 625, which is exactly the number pointed out by our table." Now this being a fair solution of one instance, let us see how it will agree with some others.

Before I proceed, however, I cannot help remarking, that the supporters of visible heat-rays must feel themselves already considerably confined, as our present argument will not allow them more than 122 of such rays out of a thousand.

Now, if the assumption that terrestrial heat is owing to a mixture of visible and invisible rays, in the proportion of 122 of the former to 878 of the latter, be well founded, it ought to explain every other phænomenon collected in our table.

The purple-coloured glass stops 993 rays of light, which, according to our present hypothesis, should stop 121 rays of heat: it also stops 730 invisible rays, which will give 641 rays of intercepted heat; therefore this glass should stop 762 rays of heat, out of every thousand that come from a candle; but, from our table, we find that it stops no more than 520, so that 242 rays cannot be accounted for.

The glass with an olive colour burnt into it, stops 984 rays of light, or 120 of heat, and 637 invisible rays, or 559 of heat. The sum is 679 which that glass should stop; but it stops actually 792; so that, as in the foregoing instance we had a deficiency of 242 rays, we now have an excess of 113; which plainly shews, that no hypothesis of any other proportion between the visible and invisible rays of heat can answer to both cases; and that consequently, not only the present, but every other assumption of this kind, must be given up as erroneous.

I shall not enlarge on these arguments, as I take them to be sufficiently clear to decide the question we have had under consideration. I also forbear going into an examination of what our sixth article, which treats of scattered heat, might afford, in addition to the former arguments. It may just be remarked, that the 211th experiment points out a black object, which scatters more heat than a white one; while the case, as to light, is well known to be the reverse. The 219th experiment also shews, that the scattering of heat of gold paper is considerably inferior to that of black velvet; whereas a contrary difference, of a very great extent, is pointed out between these two substances; for black velvet scatters only 7 rays of light, while the scattering of gold paper amounts to more than 124000. I am well aware that this difference will perhaps admit of a solution on other principles than those which relate merely to the laws of scattering, and confess that many experiments are still wanting to complete this article, which cannot now be given; but, as this Paper is already of an unusual length, I ought rather to apologise for having given so much, than for not giving more.

Table of the transmission of terrestrial scattered Light through various Substances; with a short Account of the Method by which the Results contained in this Table have been obtained.

The transmissions here delivered are called terrestrial and scattered, to distinguish them from others, which are direct and solar; and, in the use I have made of them in the foregoing Paper, it has been supposed that light-making rays, whether direct and solar, or scattered and terrestrial, are transmitted in the same manner; or that the difference, if there be any, may not be considerable enough to affect my arguments materially. In this I have only followed the example of an eminent optical writer, who does not so much as hint at a possibility that there may be a difference. Before I describe my apparatus, I ought to mention that it is intirely founded on the principles of the author now alluded to,* and that no other difficulty occurs in the execution of his plan, than how to guard properly against the scatterings of the lamp: for the light which this will throw on every

* See *Traité d'Optique*, page 16, Fig. 5; *Ouvrage posthume de* M. BOUGUER.

object, must not be permitted to come to the vanes ; since these scatterings cannot remain equal on both vanes, when one of them is moveable. In the following construction, the greatest difficulties have been removed ; and a desirable consistency in the results of the experiments, when often repeated, has now been obtained.

A board about fourteen feet long, and six inches broad [fig. 25], has two slips of deal, an inch square, fastened upon the two sides : these make a groove, for two

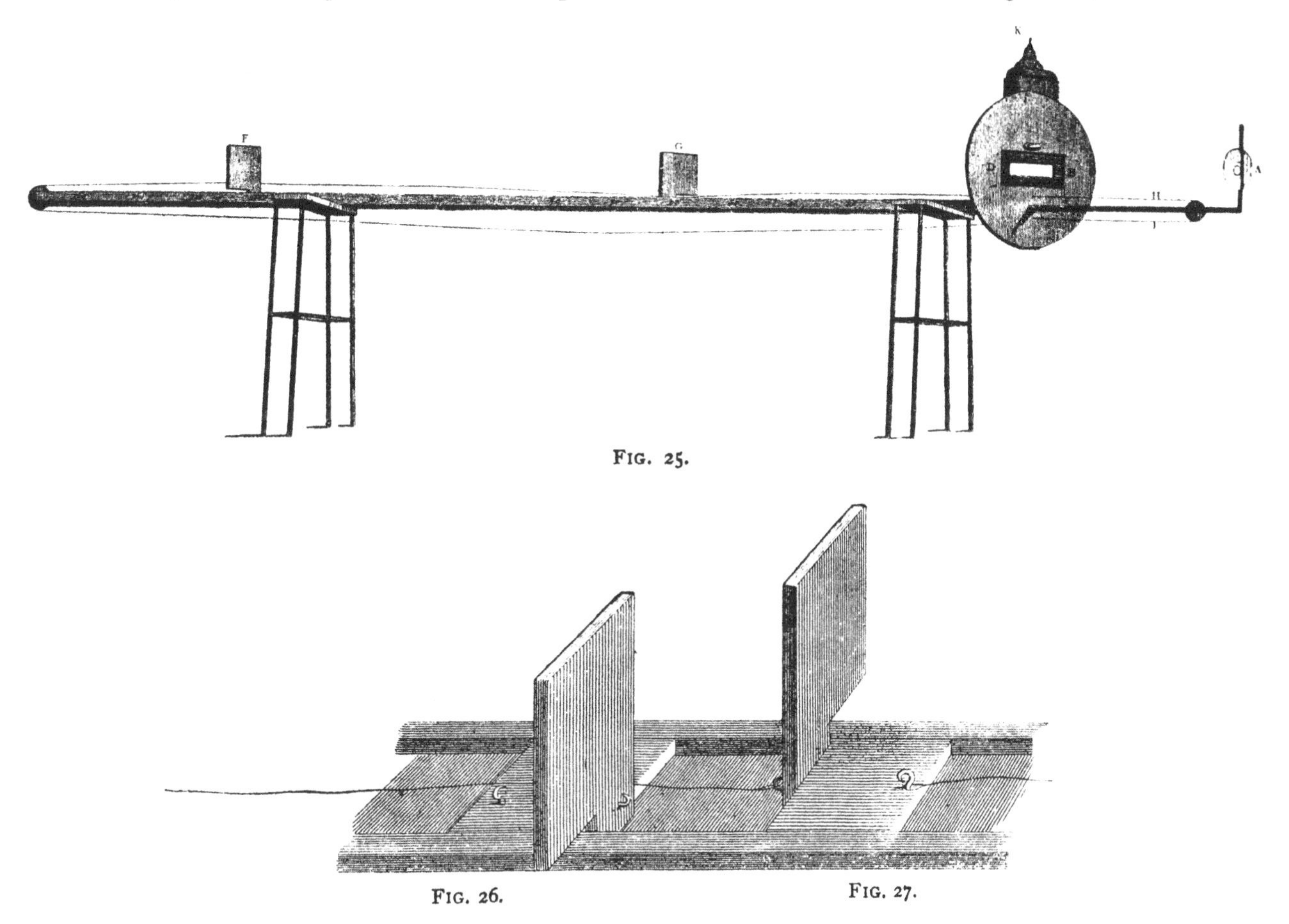

FIG. 25.

FIG. 26. FIG. 27.

short pieces to slide in, backwards and forwards. The two sliding pieces [figs. 26 and 27] carry each a small board or vane ; one towards the right, the other towards the left ; but so as to meet in the middle, and apparently to make but one when placed side by side. The vanes are covered with a piece of fair white paper, which is to reflect, or rather to scatter light in every direction. To one end of the board is fixed a circular piece of wood, with an opening in it, which is afterwards to be shut up by a small moveable piece [fig. 28], intended for placing the transmitting objects upon. This moveable piece contains two holes, at the distance of $1\frac{1}{2}$ inch from centre to centre, and $\frac{3}{4}$ inch in diameter each. Against the circular wooden screen, and close over the opening in it, is placed a lantern containing a lamp [fig. 29]. Its construction is such as to admit a current of air to feed the flame from below, by means of a false bottom, and to let it out by a covered roof; and the whole of the light, by the usual contrivance of dark lanterns, is thus kept within, so as to leave the room in perfect darkness. In the front, that is towards the vanes, the lantern has a sliding door of tin-plate, in which there is a parallelogrammic hole, covered with a spout five inches long, of the same shape. Two or three such doors, with different spouts and openings, will be required to be put in, according to the experiments to be made; but the first will do for most of them.

A narrow arm is fastened to the long board, which advances about three feet beyond the screen, and carries a circular piece of pasteboard, that has an adjustable hole in the centre, through which the observer is to look when the experiments are to be made. At the farther end of the long board is a pulley, over which a

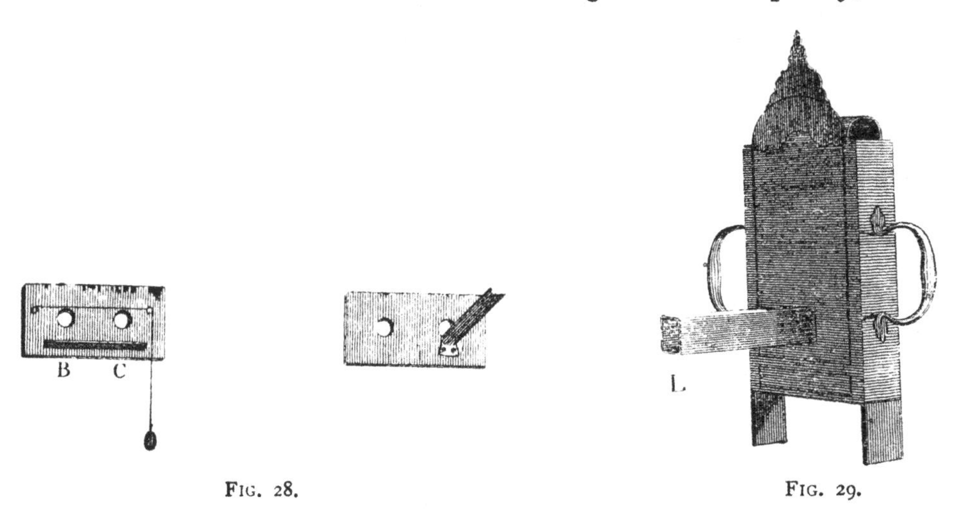

FIG. 28. FIG. 29.

string, fastened to the back of the slider that carries one of the vanes, is made to pass. This string returns under the bottom of the long board, towards the other end, where, close to the observer, another pulley is fixed; and, after going also over this pulley, it returns at the top of the board, to the front of the same vane, to which the other end of it is fastened at the back. By pulling the string either way, the observer may bring forward the moveable vane, or draw it back, at pleasure

At the side of the long board is a scale of tens of inches, numbered from the place of the flame of the lamp, 0, 10, 20, 30, 40, and so on to 160. A pair of compasses being applied from the last ten towards the vane, ascertains its distance from the flame, to as great an accuracy as may be required.

When the transmitting power of a glass is to be tried, it must be placed over one of the holes of the small moveable piece, which then is fastened with a button, upon the opening left for it in the circular wooden screen. Then, looking through the hole of the pasteboard at the two vanes, and bringing that which is seen through the glass near enough to give an image equally bright with that which is seen through the open hole, the observation will be completed. Having measured the odd inches by a pair of compasses, or immediately by a scale, we deduce, as usual, the transmitting power, by taking double the logarithm of the distance of the farthest vane from the lamp, from double the logarithm of the distance of the nearest vane. The remaining logarithm is that of the transmitting power, as compared to the light coming directly to the eye from the other vane.

I have now only to remark, that the use of this instrument requires some practice, especially when coloured glasses are to be examined; it will, however, be found, that the difference of the colour of the two objects, when their light is brought to an equality, may be overcome by a little abstraction, which is required for the purpose; for, by attending only to brightness, it has often happened to me, that both objects appeared at last of the same colour; which proved to be some mean between the two appearances considered separately.

Some glasses stop so much light, that it will be advisable to take them by the assistance of an intermediate one. Thus, instead of comparing the open vane directly to a red glass, I settle first the ratio of the violet one to that vane; then, taking the ratio of the red to the violet, and compounding these two ratios, the result will be more accurate. The reason for this will be easily comprehended, when the construction of the apparatus is considered. For a red glass, immediately compared to the open vane, would require its object to be brought extremely near the lamp, while the other must remain at a very great distance. This would occasion a considerable difference in the angles, both of incidence and of reflection, between the rays falling on one vane, and on the other. But, by dividing the observation into two operations, we avoid the errors that might be occasioned by the former arrangement.

In the following table, the first column contains the names of the different substances through which light has been transmitted. The second column shews the transmission of light, expressed in decimal fractions; or the proportion which it bears to the whole incident light considered as unity. An arithmetical complement to this fraction, or what it wants to unity, will therefore give us the proportion of light which is stopped by each of the substances contained in the first column; and that quantity multiplied by 1000 is placed in the third column.

TABLE VII.

Substances without colour.	Transmission.	Stoppage.
Bluish-white glass	·914	86
Flint glass	·966	34
Crown glass	·797	203
Coach glass	·832	168
Iceland crystal	·850	150
Talc	·910	90
Easily calcinable talc	·712	288

Glasses of the prismatic colours.	Transmission.	Stoppage.
Very dark red glass	·0001335	999·9
Dark-red glass	·000188	999·8
Orange glass	·221	779
Yellow glass	·681	319
Pale-green glass	·465	535
Dark-green glass	·0511	949
Bluish-green glass	·231	769
Pale-blue glass	·316	684
Dark-blue glass	·199	801
Indigo glass	·000281	999·7
Pale-indigo glass	·0218	978
Purple glass	·00675	993
Violet glass	·0452	955
Liquids.		
Empty tube and two glasses	·796	204
Well-water and "	·789	211
Sea-water	·712	288
Spirit of wine	·776	224
Gin	·374	626
Brandy	·00381	996
Scattering Transmissions.		
Crown glass, one side rubbed on emery .	·146	854
Coach glass, " " .	·115	885
Crown glass, both sides rubbed on emery .	·0685	932
Coach glass, " " .	·0542	946
The two first, together	·03158	969
The two next, together	·0208	979
The four first, together	·00456	995
Olive colour, burnt in glass	·0160	984
Calcined talc	·00345	997
White paper	·00556	994
Linen	·0483	952
White Persian	·0841	916
Black muslin	·263	737

Table of the proportional terrestrial Light scattered by various Substances.

The same apparatus which has been used to gain the results of the preceding table, has also been employed for the following one, with no other difference than that while the vane with the white paper remained on one side, the other vane was successively covered by the objects whose power of scattering light was to be ascertained, and both vanes were viewed directly through the two open holes in the screen ; the eye being stationed in the same place as before.

It will be found, that this table contains the scattering of more objects than have been referred to in the preceding paper ; but, as I made these experiments in a certain order, I thought it would be acceptable to give the table at full length.

The first column gives the names of the objects ; and the second contains the number of rays of light scattered by them, when compared to a standard of white paper, which is supposed to scatter one thousand.

TABLE VIII.

White paper scatters	1000	rays of light.
Message card ,, ...	1000	,,
White linen ,, ...	1008	,,
White cotton ,, ...	1054	,,
White chamois leather, smooth side scatters . .	1228	,,
White worsted ,, . .	620	,,
White Persian, suspended ,, .	671	,,
White Persian, on whitish-brown paper ,, . .	719	,,
White Persian, on white Persian ,, . .	818	,,
White muslin ,, . .	827	,,
Red paper ,, . .	158	,,
Deep pink-coloured paper ,, . .	513	,,
Pale pink-coloured-paper ,, . .	621	,,
Orange paper ,, . .	619	,,
Yellow paper ,, . .	824	,,
Pale-green paper ,, . .	549	,,
Dark-green paper ,, . .	308	,,
Pale-blue paper ,, . .	665	,,
Dark-blue paper ,, . .	149	,,
Indigo paper, with a strong gloss ,, . .	144	,,
Dark-violet paper ,, . .	75	,,
Brown paper ,, . .	101	,,
Black paper, with a strong gloss ,, . .	420	,,
Black satin ,, . .	102	,,
Black muslin, suspended ,, . .	64	,,
Black muslin, upon black muslin ,, . .	18	,,
Black worsted ,, . .	16	,,
Black velvet ,, . .	7	,,
Tin-foil ,, . .	8483	,,
Iron ,, . .	10014	,,
Copper ,, . .	13128	,,
Brass ,, . .	43858	,,
Gold-leaf paper ,, . .	124371	,,

I cannot help remarking, that in making these last experiments, I found that black paper could not be distinguished from white ; and that, on bringing it a little nearer to the light than it should be to make them perfectly equal, any of my friends who happened to be present, would mistake the black for the white.

EXPLANATION OF THE FIGURES [Plates XX. to XXVI. in the *Phil. Trans.*].

Fig. 11, represents the spectrum of heat A, S, Q, A ; and of light G, R, Q, G. If a prism be placed in a window, so as to throw the colours of light upon a table, and fig. 11 be laid under the colours, so that they may respectively fall upon the places where their names are inserted, then may these colours be made to fit into their proper spaces, by lowering or raising the prism, at pleasure. When the colours occupy their proper situations, the line A Q will express the space over which the prism, by their different refrangibility, scatters the rays of heat ; and the ordinates to A Q will nearly express the proportional elevations, which a set of equi-changeable thermometers would experience, when placed in the different situations of these ordinates.

Fig. 12. A, B, is the box which holds the two thermometers, No. 1 and No. 5. C is the board which contains the transmitting holes, the slip of wood for supporting the glasses, and the perpendicular pin for adjusting the angle. D, E, are the boards joined together by hinges. F is a slip of mahogany screwed to E. G is the spring to confine the slip F; which will keep the board D up to any angle less than 90 degrees.

Fig. 13, is the cover for shutting the transmitting holes.

Fig. 14, is the screen, which may be elevated, by the usual contrivance of springs at the back, to any required height, so as to permit the rays of the sun to pass through the opening in the middle, and to fall upon the transmitting holes of the box A, B, in Fig. 12.

Fig. 15, is a second upper part of the box A, B, in fig. 12. The first upper part being screwed off, this is to be put on instead of it, when experiments with liquids are to be made. It contains, as before, the two transmitting holes, the slip of wood, and the pin; and it has moreover a small bracket fastened under one of the holes, on which the tube containing the liquid to be tried may be laid.

Fig. 16. A, B, is the box which holds the thermometers. C, D, are the screens, with the transmitting holes in them, opposite the flame of E, the candle. F, is a small weight, stretching a string across the glass, or other object, placed upright against the transmitting hole of the screen D. It may be carried round the screen C, if required, and hold a glass against the hole in it.

Fig. 17, is the double cover: in putting it on, it must be passed over the candle downwards, against the two transmitting holes.

Fig. 18, is a third upper part to the box A, B, of fig. 12. It contains two small holes, for transmitting prismatic rays to the two thermometers A and B, which must now be put into the box, instead of No. 1 and No. 5. The parallel lines *a*, *b*, inclosing the holes, will direct any coloured rays to the thermometers; and, by drawing the red rays down to the lower parallels *c* or *d*, invisible rays may be brought to enter the transmitting holes.

Fig. 19, represents a large screen, with the shelf A, B, carrying four thermometers, No. 1 and No. 5 opposite the transmitting holes, and the other two as represented. When the fire is properly prepared, lift the screen, by taking hold a little above A and B, and set it close to the fire, so that C, D, may touch the bottom of the grate; then take hold of the ring E, and pull the string far enough out to hang that ring on the hook F: this will open the transmission holes, when the experiment is to begin.

Fig. 20, is the box containing the thermometers. The bricks are piled up, as represented in fig. 22. The stove, fig. 21, being prepared, bring the stand, and brick-work, fig. 22, close to it; and set also the two spare bricks, which lie on the stand, upon the front of the stove, that no heat may pass from the top of it to the brick inclosure; then put the box, fig. 20, into the brick-work, close up to the stove, and begin the experiment. The ash-hole should also be covered with a brick.

Fig. 23, is a fourth upper part to the box A, B, fig. 12. It contains two large holes, for admitting the rays of the sun to fall upon the objects on fig. 24.

Fig. 24, represents two tablets, *a*, *b*, united; they may be covered with any objects that are to be examined; for instance, *a* with white paper, and *b* with black velvet. These tablets, by a proper contrivance, are brought under the holes of fig. 23, where a button fastens them at the required distance.

Fig. 25, represents the Photometer. The hole at A, is for the observer to look through, that he may have a fixed station. The vanes F and G are moveable. By pulling the string at H, G will be brought nearer the lamp placed at K; and, by drawing the same string at I, it will be removed towards the vane F; which latter may be fixed at any distance most convenient for the experiment.

Fig. 26 and 27, shew the mechanism of the adjustable vane 26, and moveable one 27. There are, however, hooks on fig. 26, which will occasionally receive the strings from the hooks on fig. 27, when a motion of the left vane, instead of the right, is required.

Fig. 28, contains two limiting holes B, C; over one of which, C, a glass may be laid. This piece is to be buttoned on the rabbet of the screen, at D E, fig. 25. When liquids are to be tried, the second piece of fig. 28, which contains a bracket for supporting the transmitting tube, is to be fastened on D E, fig. 25, instead of the former plate.

Fig. 29, gives a view of the lamp and its sliding door, with the spout L, which, when the lamp is placed at K, fig. 25, conveys the light to the vanes F and G, without permitting it to be scattered on the long board.

XLVI.

Observations tending to investigate the Nature of the Sun, in order to find the Causes or Symptoms of its variable Emission of Light and Heat; with Remarks on the Use that may possibly be drawn from Solar Observations.

[*Phil. Trans.*, 1801, pp. 265–318.]

Read April 16, 1801.

ON a former occasion I have shewn, that we have great reason to look upon the sun as a most magnificent habitable globe; and, from the observations which will be related in this Paper, it will now be seen, that all the arguments we have used before are not only confirmed, but that we are encouraged to go a considerable step farther, in the investigation of the physical and planetary construction of the sun. The influence of this eminent body, on the globe we inhabit, is so great, and so widely diffused, that it becomes almost a duty for us to study the operations which are carried on upon the solar surface. Since light and heat are so essential to our well-being, it must certainly be right for us to look into the source from whence they are derived, in order to see whether some material advantage may not be drawn from a thorough acquaintance with the causes from which they originate.

A similar motive engaged the Egyptians formerly to study and watch the motions of the Nile; and to construct instruments for measuring its rise with accuracy. They knew very well, that it was not in their power to add a single inch to the flowing waters of that wonderful river; and so, in the case of the sun's influence, we are likewise fully aware, that we shall never be able to occasion the least alteration in the operations which are carried on in the solar atmosphere. But, if the Egyptians could avail themselves of the indications of a good Nilometer, what should hinder us from drawing as profitable consequences from solar observations? We are not only in possession of photometers and thermometers, by which we can measure from time to time the light and heat actually received from the sun, but have more especially telescopes, that may lead us to a discovery of the causes which dispose the sun to emit more or less copiously the rays which occasion either of them. And, if we should even fail in this respect, we may at least succeed in becoming acquainted with certain symptoms or indications, from which some judgment might be formed of the temperature of the seasons we are likely to have.

Perhaps our confidence in solar observations made with this view, might not exceed that which we now place on the indications of a good barometer, with regard to rain or fair weather; but, even then, a probability of a hot summer, or its contrary, would always be of greater consequence than the expectation of a few fair or rainy days.

It will be easily perceived, that in order to obtain such an intimate knowledge of the sun as that which is required for the purpose here pointed out, a true information must be first procured of all the phenomena that usually appear on its surface. I have therefore attended to many circumstances, that have either not been noticed at all before, or have not been examined with any particular view of information. The improvement also in the solar apparatus of my ten-feet telescope, by which I can take away as much light and heat as required, has given me additional facility of making a great number of particular observations; and, as they have been all directed to an investigation of certain points, I shall give them here, in the order which the arrangement of my subject will require.

It will be necessary, before I can enter into a detail of the observations, to give notice that, from an improved knowledge of the physical construction of the sun, I have found it convenient to lay aside the old names of *spots, nuclei, penumbræ, faculæ*, and *luculi*, which can only be looked upon as figurative expressions that may lead to error. Nor were these few terms sufficient to describe the more minute appearances on the sun, which I have to point out.

The expressions I have used are *openings, shallows, ridges, nodules, corrugations, indentations*, and *pores*. It will not be amiss to give a short explanation of these terms.

Openings are those places where, by the accidental removal of the luminous clouds of the sun, its own solid body may be seen; and this not being lucid, the openings through which we see it may, by a common telescope, be mistaken for mere black spots, or their nuclei.

Shallows are extensive and level depressions of the luminous solar clouds, generally surrounding the openings to a considerable distance. As they are less luminous than the rest of the sun, they seem to have some distant, though very imperfect, resemblance to penumbræ; which might occasion their having been called so formerly.

Ridges are bright elevations of luminous matter, extended in rows of an irregular arrangement.

Nodules are also bright elevations of luminous matter, but confined to a small space. These nodules, and ridges, on account of their being brighter than the general surface of the sun, and also differing a little from it in colour, have been called faculæ, and luculi.

Corrugations, I call that very particular and remarkable unevenness, ruggedness, or asperity, which is peculiar to the luminous solar clouds, and extends all over the surface of the globe of the sun. As the depressed parts of the corrugations are less luminous than the elevated ones, the disk of the sun has an appearance which may be called mottled.

Indentations are the depressed or low parts of the corrugations; they also extend over the whole surface of the luminous solar clouds.

Pores are very small holes or openings, about the middle of the indentations.

Any other terms I may hereafter use, will be sufficiently explained by the observations in which they occur.

I shall now enter into an examination of all the phenomena that may be observed in viewing the sun through a good telescope, beginning with those that are most common; a critical investigation of which will lead us gradually to such as are more intricate.

It will be seen that I have brought my observations under a number of short heads, or propositions, such as my subject requires. The advantage of this method is, that the tendency of every observation will be immediately understood, while it is read; whereas, had I arranged these observations in the order in which they were made, the mixture of the various points to be ascertained by them must have brought on a considerable obscurity; and, in drawing conclusions from them afterwards, a repetition of the observations which were to support them would have been unavoidable.

I must take notice of what will perhaps be censured in many of the observations; they may be said to be accompanied with surmises, suppositions, or hypotheses which should have been kept separate. In defence of this seeming impropriety, I must say, that the observations are of such a nature, that I found it impossible, at the very time of seeing the new objects that presented themselves to my view, to refrain from ideas that would obtrude themselves. It may even be said, that since observations are made with no other view than to draw such conclusions from them as may instruct us in the nature of the things we see, there cannot be a more proper time for entertaining surmises than when the object itself is in view.

Now, since the suggestions that have been inserted were always such as arose at the moment of the observations, they are so blended with them, that they would lose much of their value as arguments, if they were given separately.

In order not to lengthen this Paper unnecessarily, I have given but a few observations under each head; especially with those propositions which may be looked upon as already sufficiently established by the observations of other astronomers. The whole tenor of the observations I have given, though divided under such numerous heads, is indeed such as must produce a mutual support; so that, frequently one or two particular observations were thought sufficient to establish my point, when I might have added many more, from my journals, in support of it.

OF OPENINGS.

Openings are Places where the luminous Clouds of the Sun are removed.

That those appearances which have been called spots in the sun, are real openings in the luminous clouds of the solar atmosphere, may be concluded from the following observations, where the sides or thicknesses of the borders which limit the openings are distinctly described.

Jan. 4, 1801. There is a large opening much past the centre of the sun, with a shallow about it (see fig. 1 and 2). On the preceding side I see the thickness of the shallow from its surface downwards. On the following side, I also see the edge of the shallow near the opening; but it is sharp, and its thickness cannot be seen. I see also the side of the elevation surrounding the shallow, going curvedly down to the surface of the shallow, on its preceding side. A large collection of openings, of very different sizes, are near the following limb. They all manifest the same kind of optical appearance, but on the side which is contrary to that in the opening before mentioned; for here the thicknesses or depths of the shallows, and of the slopes going down from the upper surface to the shallows, are only visible on the following side, but not on the preceding one.*

Large Openings have generally Shallows about them.

Jan. 24, 1801. The two largest openings of January 19, are completely surrounded by shallows.

Many Openings are without Shallows.

Jan. 22, 1800. There are two openings which have not the least shallow about them. The corrugations go equally between and around them.

Feb. 7, 1800. There are two considerable openings not far from each other: they have no shallows about them.

Small Openings are generally without Shallows.

Dec. 2, 1800. There are a great number of large and small openings: the large ones have shallows about them; the small ones are mostly without.

Openings have generally Ridges and Nodules about them.

Dec. 20, 1794. There are two openings near the preceding margin of the sun; they have elevated extensive luminous ridges about them.

Dec. 23, 1799. A pretty considerable opening, on the following side of the disk, is surrounded by many ridges.

* For a geometrical proof of the depression of the nucleus of a *spot*, as an opening was formerly called, see a most valuable paper of Observations on the Solar Spots, by the late ALEXANDER WILSON, M.D. Professor of Practical Astronomy in the University of Glasgow. *Phil. Trans.* Vol. LXIV. Part I.

Feb. 7, 1800. Following two considerable openings, which are not far from each other, are several irregularly dispersed ridges, more bright than the rest of the sun.

Jan. 4, 1801. Many clustering openings are lately come into the disk; a crowd of ridges and nodules surround, and are interspersed among them.

Openings have a Tendency to run into each other.

Dec. 25, 1799. The large opening of December 23, and the small one near it, are now nearly run into each other.

Jan. 4 1801. The two largest openings of Jan. 2, are nearly joined into one.

Jan. 6, 1801. The largest of the preceding openings, of a set observed Jan. 4, has drawn together all the small ones, and is increased in dimensions.

Jan. 29, 1801. 2^h 10′. A longish opening, observed at 12^h, is increased by the addition of two projections. With more attention, however, I perceive that these projections are united to each other, but separated from the longish opening, by a narrow luminous bridge, or compressed row of luminous clouds.

Jan. 30, 1801. The two united projections of yesterday are now joined to the longish opening.

New Openings break out near other Openings.

Dec. 23, 1799. There is a small opening near the large one observed yesterday, which then was not visible.

Jan. 21, 1800. The preceding of two openings observed before, has now two other small ones near it.

Probable Cause of Openings.

Jan. 18, 1801. Between two clusters of openings, near each other, there are some, as I suspect, incipient openings: they resemble coarse pores of indentations.

Jan. 19, 1801. The incipient openings, between the clusters of yesterday's observation, are completely turned into considerable openings. It seems as if an elastic, but not luminous gas, had come up through the pores or incipient openings, and spread itself on the luminous clouds, forcing them out of the way, and widening its passage.

Feb. 18, 1801. 7^h 44′. The south preceding one of three large connected openings (see fig. 3), has a narrow branch coming from its shallow.

9^h 55′. The opening is broken out at three places (see fig. 4); and the shallow has three projections just opposite. It is plain that the breakings-out and the projections must have the same cause; which probably acts first at the opening, and widens it, then goes forward, and occasions the corresponding projections in the shallows. The shallow is very large on that side where the breakings-out are situated; and, on the contrary, very narrow on the opposite, or, as it may be called, the quiescent side.

10^h 12′ The broadest of the little sprouting shallows is opposite the broadest of the breakings-out, or encroachments of the opening on the general shallow.

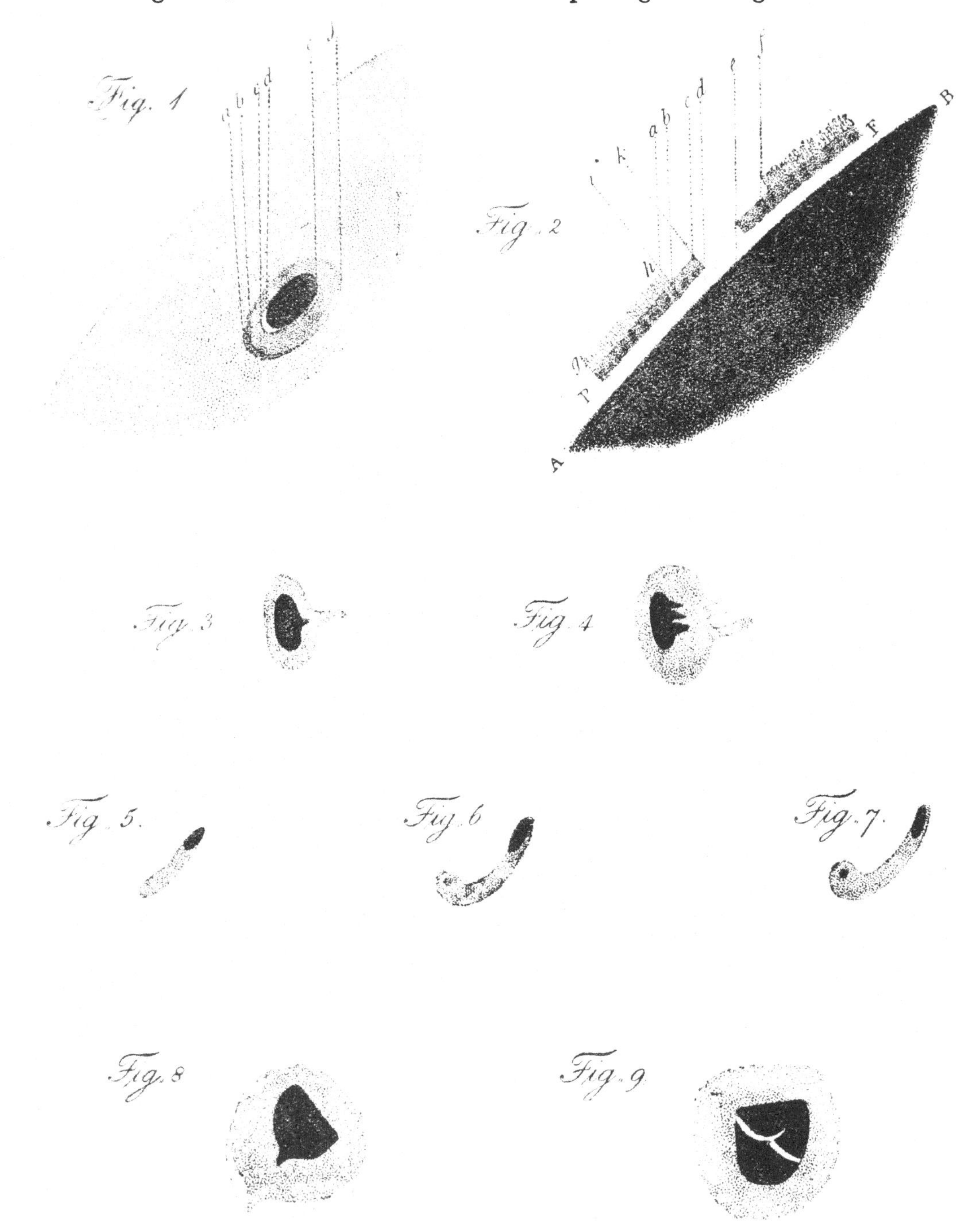

Direction and Operation of the disturbing Cause.

Jan. 24, 1801. A small oblong opening, near a preceding large one, has on its north side a very long shallow. This made me surmise, that the elastic fluid coming out of it might have a strong direction from below, towards that side. Examining therefore this opening all round, I found that the shallow extended only to one side, leaving the other parts full of luminous matter close to the margin of the opening. And, on examining the large opening, I found that the shallow about it was also larger, in the same direction as the shallow about the small one. Eight other small openings, forming together a cluster with the former two, have every one also their incipient shallows on the same side; that is, towards the north-following; and none at all on the other parts of their margins.

3^h 15'. The shallow about the small opening has changed its direction. It was like fig. 5, and is now as in fig. 6. In the farthest end of it, I expect an opening to be coming on.

4^h 30. There is already the appearance of an incipient opening (see fig. 7).

Jan. 6, 1801. The shallows about the large opening and the small one near it, are much altered in situation and dimensions. All the smaller openings near them have undergone great changes; some being gone, and new ones come in different places, while others have altered their appearance. The bias in the direction of the shallows is also considerably changed.

Jan. 30, 1801. 9^h 20'. There is a cluster of small openings, which I expect to see united into one, by the breaking down, driving away, or dissolution, of the intermediate communications of luminous clouds. From having seen it yesterday, I know the largest of the openings to be a generating one. It has an increasing shallow; and the next largest opening of the cluster has an incipient one.

10^h 40'. The incipient shallow is now pretty large.

11^h 2'. There is now also an incipient shallow about the most south of the small openings.

11^h 6'. The shallow of the largest opening is increased.

Maxima of Openings.

Jan. 29, 1801. 11^h 0'. The shallow of the largest of many openings that are visible, is most extended towards the north-following side (see fig. 8). It affects a circular form more than the opening, but is not concentric with it. It has a small lip on the north, which I suppose denotes the direction of the gas coming out of the opening. A similar lip is visible in the opening itself, as if the gas, in coming out, pressed against the luminous clouds which limit the opening, and belong to the flat.

2^h 10'. The lip or projection of the shallow about the opening is filled up at the sides; they being now as broad as the lip's projection. The filling up is marked with points in the figure.

Jan. 30, 1801. The large opening observed yesterday is no longer increased; but seems to be nearly at its maximum.

Feb. 4, 1801. 1^h 10'. The shallow of the large opening is much more round than the opening, though not concentric with it. Hence, its figure being no longer disturbed, I guess that the opening is near its maximum.

There is some Difference in the Colour of Openings.

March 1, 1800. There are two large openings, which seem to be partially covered or rather to have a thin, semi-transparent, luminous veil of clouds still hovering over them; this gives them a fainter black colour than openings generally have.

Openings divide when they are decaying.

Dec. 26, 1799. An opening observed the 25th is reduced; and is divided in the middle by a lucid line (see fig. 9).

Dec. 27, 1799. The luminous bridge or passage across the opening is pretty broad, and has a branch about the middle. This branch of light has the appearance of a luminous cloud, irregularly breaking out from the passage, towards the southern half of the opening.

Dec. 28, 1799. The opening is so completely vanished, that I cannot find the least mark of its former existence. From the appearance of the branch yesterday, in the luminous division resembling a bridge thrown over a cave, I surmise that this branch, as well as the sides of what I call the bridge, have extended themselves, and as it were drawn a lucid curtain over the opaque surface of the sun.

Decaying Openings sometimes increase again.

Feb. 8, 1801. The great opening, which has been gradually diminishing since Feb. 4, is now enlarged again.

When Openings are divided, they grow less, and vanish.

Dec. 29, 1799. The south preceding of two openings observed yesterday is divided into three small ones.

Dec. 31, 1799. The divided opening of the 29th exists no longer.

Feb. 9, 1800. Of two considerable openings observed Feb. 7, the preceding one is divided into several smaller parts.

Feb. 10, 1800. The divided opening observed yesterday is in a vanishing state.

Feb. 11, 1800. The openings are all covered in, and no trace of them can be found.

Decayed Openings sometimes become large Indentations.

Feb. 9, 1800. The preceding of two openings observed the 7th of February is now about half filled up; and that half contains two indentations, with black pores, or rather remaining small openings. They are nearly of the size of the general corrugations of the solar surface at present.

Decaying Openings turn sometimes into Pores.

Feb. 10, 1800. $1^h\ 0'$. Between the half of the opening and that part which was nearly covered in yesterday, is a set of indentations, with pores rather larger than they are in general.

When Openings are vanished, they leave Disturbance behind.

Dec. 28, 1799. 12^h 10′. There is a place among the corrugations, where they are coarser now than they were an hour ago.

1^h. Two considerable openings are broken out, in the place where the corrugations were coarse. They are both so completely dark, and free from thin luminous clouds, that it appears very plainly they were only hidden behind the slight covering of luminous obstructions; one of them is about the place where the opening of yesterday was situated.

Jan. 24, 1800. There is a pretty large place, which contained the openings observed the 22d, the luminous clouds of which are in a state of disturbance: it includes five or six places, where the pores of indentations assume the shapes of incipient openings, and after some time lose them again, more or less.

Feb. 11, 1800. There is a place which I suppose to be that where the now vanished openings were yesterday: it seems to be rather disturbed in the arrangement of its corrugations. One of the indentations is probably an evanescent opening, as it still shews a considerable pore.

Apparent View into the Openings, under luminous Bridges and Shallows.

Dec. 27, 1799. A luminous passage across a large opening has all the appearance of a bridge going over a hollow space; and I have no doubt but that it is at a considerable distance from the opaque surface of the sun.

Jan. 4, 1801. A ridge which separates two openings much past the centre, shews its thickness on the following side: it has a shallow, the preceding side of which I see going down to the opening; and it appears to me, that there is a considerable distance between the lowest part of that shallow and the dark surface of the sun under it. Nor can I help believing that I see aslant under it, towards the preceding side; though I find it difficult to account for such vision, on the principles of solar perspective.

Jan. 6, 1801. The same opening is now further advanced towards the limb of the sun. It appears again to me, that on the preceding side I see far under the shallow. I suspect the part towards P to be of a deeper dark colour than that towards F (see fig. 1 and 2); but the difference, if there be any, is not marked enough to be decisive.

Depth of the Openings, indicated by their Darkness.

Jan. 15, 1801. 11^h 10′. One of the openings, which is near the preceding limb of the sun, is remarkably black. The tint of openings may perhaps assist us to infer their depth; which must be greater in a direction towards us, when the opening is near the circumference, than when it is near the centre, if the distance from the shallows to the opaque surface should be considerable. I have compared the darkness of two openings near the centre, with the appearance of that near the preceding limb. It is decidedly in favour of the blackness of the latter: this is however larger than the former two; which may occasion deceptions. The opening near the limb is certainly darker than the two which are near the centre.

Feb. 5, 1801. An opening very near the preceding margin is of a deep black colour; certainly more so than another which is not far off, but is more towards the centre of the sun.

Feb. 8, 1801. The large opening near the margin, is darker than two other openings which are about the centre of the solar disk: the difference, however, is hardly sufficient to be decisive. The two openings are also much smaller; which may occasion a deception.

Distance between the Shallows and solar Surface, indicated by the free Motion of low Clouds.

Jan. 25, 1801. 9^h 22′. A large opening, which I have been observing since the 19th, is now much advanced towards the limb (see fig. 10). I can see into it; and, on the preceding side, as it appears to me, a good way under the lowest regions of the clouds of which the shallow consists. The upper margin of the shallow is very sharply determined; but the clouds of the lower part of it, on the contrary, are more dispersed; some of them hanging a good way down, towards the surface of the sun's body (see fig. 11).

10^h 20′. The preceding side of the shallow of the large opening, is now more abruptly terminated at the bottom of its thickness; the hanging or projecting clouds being removed towards the following side (see fig. 12).

OF SHALLOWS.

Shallows are depressed below the general Surface of the Sun; and are Places where the luminous solar Clouds of the upper Regions are removed.

That those appearances which have been called penumbræ are real depressions, or shallows, may already be concluded, from what has been related with regard to the slopes from the upper surface of the sun, down to the top of these shallows; and will follow still more evidently, from an observation of one of them on the very limb of the sun.

Dec. 3, 1800. There is a considerable opening just come into the disk, which is followed by another that is actually in the limb of the sun. The uniformity of the circular termination of the limb suffers a small deviation; it being somewhat depressed, owing to the shallow about the opening. I do not yet perceive the opening itself with certainty; but suppose it will appear to be one, when it advances more into the disk.

The Thickness of the Shallows is visible.

Jan. 6, 1801. There is a large opening much advanced beyond the centre, with a shallow round it. On the preceding side of the shallow, its descent down towards the opening is visible; but, on the following side, I see abruptly into the opening.

Sometimes there are Shallows without Openings in them.

Feb. 7, 1801. There is a pretty large shallow inclosed by the ridges which follow some preceding openings.

Feb. 12, 1800. A place where yesterday I saw five or more nodules, at present contains low ridges inclosing some shallows.

Incipient Shallows come from the Openings, or branch out from Shallows already formed, and go forwards.

Jan. 18, 1801. In a cluster of openings, there is an incipient shallow, coming from one of them.

Jan. 19, 1801. The incipient shallow is increased, and has now spread all round the opening.

Jan. 24, 1801. A large opening sends from its shallow already formed, a narrow projection, towards the end of a neighbouring shallow belonging to a smaller opening, as if they were going to meet.

Probable Cause of Shallows.

Jan. 25, 1801. 9^h $20'$. Two branches A B (see fig. 13), of a shallow coming from an opening C, are going towards the south. It seems as if they were destined to meet the incipient shallow of a south-following opening D.

9^h $50'$. The shallow B is now very nearly united to the narrow part of the shallow surrounding the opening D. The shallow A seems to advance, in a direction towards the farthest south-following opening E.

10^h $20'$. The shallow B is now completely run into the shallow about D (see fig. 14); and the shallow A is grown broader towards F.

11^h $30'$. The shallow B is so completely joined to the shallow about D, that it appears as if it had not come from the opening C. The shallow A now ends in a sharp point (see fig. 15).

12^h $50'$. The shape of the shallow A is again much changed; it is no longer pointed, but very broad at the end (see fig. 16); and there is a new branch breaking out at G. These changes seem all to denote, that the shallows are occasioned by something coming out of the openings, which, by its propelling motion, drives away the luminous clouds from the place where it meets with the least resistance; or which, by its nature, dissolves them as it comes up to them. If it be an elastic gas, its levity must be such as to make it ascend through the inferior region of the solar clouds, and diffuse itself among the superior luminous matter.

$1^{h} 10'$. The new branch G increases; and the openings C, D, E, are enlarged. A new branch is also breaking out from the shallow about E. It is marked H in

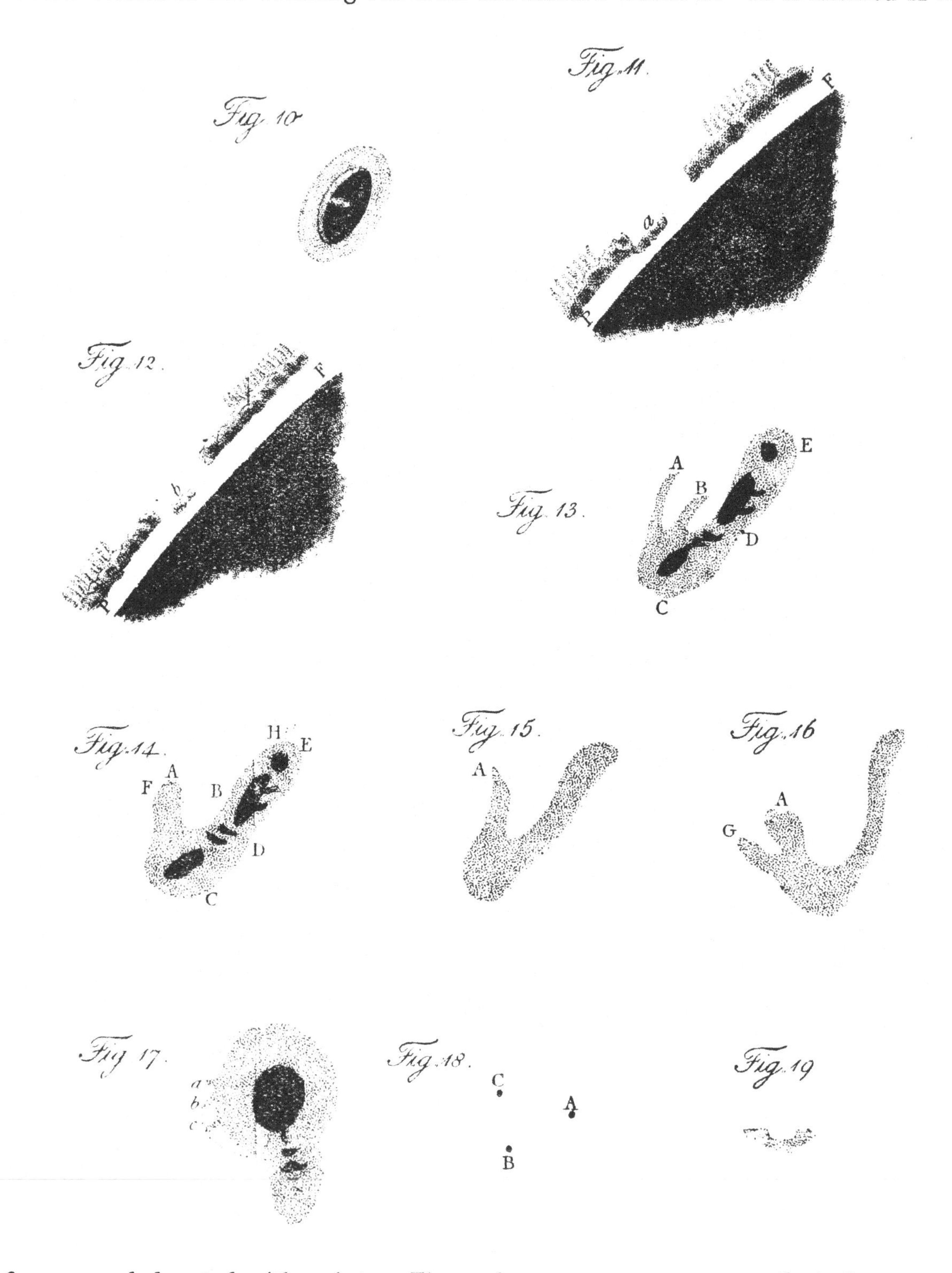

fig. 14, and denoted with points. These changes seem to prove, that the same gas which diffuses itself over the shallows has forced open the passages at first, and is now widening them. Hence, the increase of the openings is an additional circumstance which points out the cause of the shallows.

1^h 20′. From the shallow of a very large preceding opening, which is in an increasing state, are lately projected three small branches *a*, *b*, *c* (see fig. 17).

2^h 30′. The vacancies between the three small projecting shallows are now filled up by the same cause that occasioned them, so as to have given them the shape of an uniform but broader shallow, on the side where the branches come out, as denoted by points.

Shallows have no Corrugations, but are tufted.

Feb. 4, 1801. The great shallow about a large opening has no corrugations.

Feb. 18, 1801. The lower clouds of the shallow of a large opening, though not corrugated, are not smooth, but tufted. They are so closely connected in their tufts, that it makes them appear as if, in every vacancy, there were clouds under clouds, that prevent our looking far into them.

Decay of Shallows.

Jan. 30, 1801. The borders of the shallow belonging to the large opening observed Jan. 29, seem to be remarkably high; so that, if the opening were near the limb, they would probably appear like ridges. The shallow has again a lip, nearly in the place where there was one yesterday. But it seems as if the lip, which is visible now, had a cause contrary to what produced it yesterday. For the luminous clouds all round the shallow seem no longer to be kept off by an issuing elastic fluid, but are probably now breaking in upon the shallow, except in the place of the lip, where some energy, like that exerted yesterday, may still remain in action; in that case, the shallow, as well as the opening, is past its maximum.

OF RIDGES.

Ridges are Elevations above the general Surface of the luminous Clouds of the Sun.

Dec. 27, 1799. On the south-following side of the sun's disk, close to the margin, are some bright ridges. They are all in a direction parallel, or nearly parallel, to the margin; and have the appearance of elevations.

Jan. 29, 1801. Two sets of openings, near the north-following limb, have wide-spreading ridges about them: four other sets, being farther advanced into the disk, do not shew any. This denotes them to be thin elevations, which can only be seen near the circumference, by a side view.

Feb. 8, 1801. Many ridges and nodules are now to be seen about the large opening, which yesterday had none. I suppose they are become visible by its advancement towards the margin, to which it draws near.

Length of a Ridge.

Dec. 27, 1799. I measured one of the longest ridges in view. It extended over an angular space of 2′ 45″·9, which is nearly 75000 miles.

Ridges generally accompany Openings.

Feb. 5, 1801. Three sets of openings near the preceding limb, and two near the following one, are surrounded by luminous ridges.

Ridges are also often in Places where there are no Openings.

Dec. 22, 1799. On the following side of the sun are luminous ridges; but not within 50 or 60 degrees of an opening.

Jan. 4, 1801. Towards the north, near the limb, is a collection of ridges without openings.

Feb. 5. Two of the sets of openings of yesterday are gone; but have left extensive ridges behind.

Ridges disperse very soon.

Dec. 28, 1799. The appearance of the ridges I saw yesterday is much changed: they are less luminous and extensive than they were. The range is much broken; and they appear more in detached irregular elevations.

Dec. 29, 1799. The ridges are so much reduced to the resemblance of the rest of the sun, that had I not known where to look for them, I should hardly have been able to trace any vestige of them.

Feb. 9, 1800. The ridges which followed some openings Feb. 7, exist no longer.

Different Causes of Ridges hinted at.

Jan. 4, 1801. A crowd of ridges and nodules surround, and are interspersed among, a cluster of openings. A ridge which crosses one of the openings like a bar or bridge, is sharp on the following side, but shews thickness on the preceding. It seems probable, that the openings permit a transparent elastic fluid to come out, which disturbs the luminous matter on the top, so as to occasion ridges and nodules. There are not less than 17 openings in the cluster.

Jan. 6, 1801. Following a set of openings lately come into the disk, are many luminous and broad, that is to say, high ridges, without any openings among them; so that the cause which produces them acts probably below the shining matter. Their own levity also may occasion them to go into the higher regions.

Jan. 30, 1801. Near the following margin is an extended plane, full of bright ridges and nodules, with a great number of openings lately broken out, and still breaking out among them. This leads us to suppose that some elastic gas, acting below the luminous clouds, lifts them up, or increases them; and at last forces itself a passage through them, by throwing them aside.

OF NODULES.

Nodules are small, but highly elevated, luminous Places.

Jan. 24, 1800. On the south, near the limb of the sun, is a nodule ; and on the south-following side is another, with two smaller ones near it. They are round, or roundish, bright elevations, of the same nature as the ridges.

Feb. 19, 1800. There are two small bright nodules, on the preceding limb of the sun. Why they should only be seen near the sun's margin, can only be explained by admitting their elevation.

Nodules may be Ridges foreshortened.

Dec. 27, 1799. Mixed with many ridges, nearly parallel to the margin of the sun, are here and there small thicknesses or knotty places. I take them to be ridges in a more central direction, which gives them the shape of nodules.

The most north of three nodules, in one of the ridges, seems to be highest in elevation. If it should last till to-morrow, it will then appear whether that nodule is really more elevated than the rest, or whether it is the foreshortening of a ridge extended in the direction of a radius.

OF CORRUGATIONS.

Corrugations consist of Elevations and Depressions.

Dec. 23, 1799. The corrugations have a mottled appearance. I see the figure of the dark and bright places. Many of the dark places are not round, but a little extended in different directions, and appear to be lower than the bright places. This, if admitted, will explain why the corrugations towards the margin of the sun, cannot so readily be seen as about the middle of the disk.

Jan. 4, 1800. The day being very favourable, I saw the sun uncommonly well. The corrugated surface presented its elevations and depressions, with as much distinctness as the rough surface of the moon.

Corrugations extend all over the Surface of the Sun.

Dec. 23, 1799. The corrugations extend all over the sun. They are less distinct all around towards the limb, than at the centre.

Jan. 22, 1800. I followed the corrugations from the centre to the circumference ; and could trace them every where to within, I suppose, two minutes of the margin.

Jan. 24, 1800. The corrugations are equally spread over the whole surface of the sun. I viewed them distinctly in every part of it ; and traced them with much attention to within, I suppose, half a minute of the margin.

Jan. 4, 1801. The corrugations are extended all over the disk of the sun, and go to the polar regions, as well as to the equatorial parts.

Dispersed Ridges or Nodules make Corrugations.

Nov. 17, 1800. The surface of the sun appears richly filled with very small broken or dispersed ridges, which produce the corrugated appearance.

Feb. 18, 1801. The high parts of the corrugations contain numberless separations, like small nodules, which leave room for the indentations to be seen between them.

Corrugations change their Shape and Situation; they increase, diminish, divide, and vanish quickly.

Dec. 27, 1800. $1^h\ 0'$. There is a pretty large corrugation near a small opening, which serves me as a direction to find the place. Its indentation is about four diameters of the opening from it; and 10 or 16 degrees north-preceding.

$1^h\ 5'$. I have seen the corrugation again; and find its indentation larger than it was, and farther from the opening.

$1^h\ 10'$. It is vanished; and several other such very minute changes have taken place.

$1^h\ 12'$. Within a diameter of the opening, and a little north-following it, is an oval indentation, nearly as large as the small opening.

$1^h\ 15'$. Its shape is altered; and it is divided into a corrugation, with two indentations.

$1^h\ 35'$. Both are entirely gone.

Jan. 18, 1801. Between two clusters of openings, that are near each other, there are some incipient openings, which resemble coarse corrugations, and establish a step between small openings and pores of indentations. I shewed them to my friend Dr. WILSON, who happened to be upon a visit to me, at Slough. He saw the same phenomena; and judged of their being a link in the chain of appearances, as I did.

We drew a small sketch of the place of the phenomena, merely to serve us to communicate our observations to each other (see fig. 18).

$1^h\ 19'$. A, is a small opening, without a shallow, which we had fixed upon, by way of enabling us to find again the minute objects which were to be examined. B, is the indentation, or dark place, of a corrugation I pointed out to Dr. WILSON. C, is a dark place of a corrugation he pointed out to me.

$1^h\ 24'$. We both found the dark part of the corrugation B gone; and C had either changed its place or was vanished.

$1^h\ 34'$. C, was certainly gone.

Dr. WILSON pointed out another round pore, which we had not perceived before, at some distance; also a largish indentation near the opening, which guided our research. Shortly after, we found the indentation gone; and the pore was further removed from the opening.

OF INDENTATIONS.

The dark Places of Corrugations are Indentations.

Dec. 27, 1799. That the low places of the corrugations are not much depressed, is evident from their visibility pretty near the margin of the sun.

Jan. 27, 1800. The corrugations in many places are so coarse, that their indentations resemble small shallows. The indentations go down at the sides, like circular arches, presenting their concavities to us; but the bottom of them is nearly flat (see fig. 19).

Indentations are without Openings.

Jan. 15, 1801. The low places of corrugations do not contain punctures; but seem to be irregularly shaped places, of less luminous matter than the borders which inclose them.

In some Places the Indentations contain small Openings.

Dec. 27, 1799. On examining some of the largest corrugations with a high magnifying power, I see plainly, that the less bright parts, or indentations, are small openings; and that those dark places, which were the coarsest, shew the opaque surface of the sun best: some of them are as black as the large openings.

The Elevations and Indentations of Corrugations are of different Figures.

Feb. 18, 1800. Among all the corrugations, I could hardly perceive any that were round: they were of all shapes; chiefly lengthened.

Indentations change to Openings.

Feb. 10, 1800. Three corrugations, observed an hour ago, are now so enlarged, that their indentations are passed over from their former state, to small openings.

Indentations are of the same Nature as Shallows.

Jan. 30, 1801. The depressed parts, or indentations, of corrugations, are of the colour of shallows; and are probably of the same depth below their elevations, as shallows are below the general surface of the sun.

Indentations are low Places, which often contain very small Openings.

Jan. 2, 1801. That indentations are small hollow places, and that the pores in them are little openings, may be concluded from a set of real openings of different sizes, of which I see no less than 13. Four of them are visible openings; five of them are less than the smallest openings, and larger than the indentations of corrugations; the remaining four may already be called large pores. We cannot expect to see into these pores, as we do into holes, their diameter being too small.

Indentations are of different Sizes.

Jan. 31, 1800. The indentations are very uniform, but not round. It seems they admit of every possible shape.

Indentations are extended all over the Sun.

Dec. 20, 1794. I can follow the indentations, from the centre up to the margin of the sun ; but it requires great attention, as, on account of the sphericity of the disk, they become gradually less conspicuous, the nearer we go to the circumference. I saw them equally well at the north pole of the sun.

Dec. 22, 1799. The whole disk of the sun is strongly indented.

With low magnifying Powers, Indentations will appear like Points.

Feb. 4, 1800. I tried a magnifying power of only 45 times, and the sun then appeared punctulated, instead of indented. The points, or rather darker coloured places in the punctulations, were of different figures ; few of them being round.

OF PORES.

The low Places of Indentations are Pores.

Dec. 20, 1794. The lowest parts of the indented appearances, are almost dark and depressed enough to be called very minute openings.

Pores increase sometimes, and become Openings.

Feb. 10, 1800. Two indentations, observed an hour ago, are increased, and contain large pores, as if they were going to be converted into visible openings, like the indentations of three neighbouring corrugations mentioned under a former article.

Jan. 22, 1800. Between some large and small openings were two pores, that grew darker while I looked at them, and may now be taken for very small openings. This seems to trace the openings to their origin, and perfectly connects the pores with them.

Pores vanish quickly.

Dec. 27, 1800. A small pore, that was north-following a very small opening, which served me as a direction for finding it again, 1^h 30′ ago, is no longer to be seen.

12^h 30′. There were two pores north-preceding the same opening. When I returned to the telescope, in order to describe their situation exactly, they were vanished.

OF THE REGIONS OF SOLAR CLOUDS.

It must be sufficiently evident, from what we have shewn of the nature of openings, shallows, ridges, nodules, corrugations, indentations, and pores, that these phenomena could not appear, if the shining matter of the sun were a liquid ;

since, by the laws of hydrostatics, the openings, shallows, indentations, and pores, would instantly be filled up; nor could ridges and nodules preserve their elevation for a single moment. Whereas, many openings have been known to last for a whole revolution of the sun; and extensive elevations have remained supported for several days. Much less can it be an elastic fluid of an atmospheric nature: this would be still more ready to fill up the low places, and to expand itself to a level at the top. It remains, therefore, only for us to admit this shining matter to exist in the manner of empyreal, luminous, or phosphoric clouds, residing in the higher regions of the solar atmosphere. The following observations will explain and support this idea more at large.

Changes in the Solar Clouds happen continually.

Feb. 19, 1800. In order to find whether the solar clouds were subject to very quick changes, I fixed my attention on several places; but, on looking off, even for a moment, the spots I had marked for the purpose could not be found again.*

There are two different Regions of Solar Clouds.

Feb. 19, 1800. It is not possible to see the sun more distinctly than I do at present. The corrugations are evidently caused by a double stratum of clouds; the lower whereof, or that which is next to the sun, consists of clouds less bright than those which compose the upper stratum. The lower clouds are also more closely connected; while the upper ones are chiefly detached from each other, and permit us to see every where through them.

Feb. 5, 1801. An opening near the preceding limb has no shallow about it. I can see the thickness of the preceding partition, from the top of the luminous clouds down to the vacancy; and perceive that the lower part of the descent is of a less bright nature than the upper one: it is of the colour of an incipient shallow.

The inferior Clouds are opaque, and probably not unlike those of our Planet.

Feb. 5, 1801. The shallows about three considerable openings, on the following side of the sun, are of the same colour with that of a large opening on the preceding side.

Feb. 18, 1801. The tufts of the shallows, or lower clouds, are all of the same colour.

The shallows about the three largest openings are all of the same colour, which is that of all the shallows I have ever seen.

Feb. 4, 1801. The colour of a very small shallow about a little opening, is as faint as that of a large shallow of a very large opening now in view; and, as far as I can remember, all the shallows I have seen have been nearly of the same colour.

* See what has been said of the quick changes among the corrugations and indentations, under the former articles.

Hence we have reason to conclude, that there are two different regions of clouds, the lowest whereof is never affected in colour by the cause which acts upon the upper one, when shallows are generated. If so, these clouds are probably of a very different nature; for, were they not different, they would not be differently affected by the same cause. Perhaps this lower region is a set of dense opaque planetary clouds, like those upon our globe. In that case, their light is only the uniform reflection of the surrounding superior self-luminous region. If this be admitted, it will at once account for the sameness of the colour of the shallows, and of their tufts; and for many other phenomena.

Quantity of Light reflected from the inferior Planetary Clouds.

Feb. 7, 1801. I made an artificial contrivance, for the use of my photometer, to represent a portion of the bright surface of the sun, and of an opening with a shallow surrounding it. The opening was represented by a small patch of black velvet, resembling nearly, in shape, the large opening in the sun which is now visible. This was fastened on the farthest vane of my photometer, covered with white paper, and arranged so as to be in the line of the centre. In the nearest vane, covered with the same paper, and equally prepared so as to move in the centre of the photometer, I cut a hole large enough to shew the black velvet on the farthest vane, with a small margin of its paper about it, which was intended to represent the shallow about the opening. The illumination of the nearest vane was to represent the brightness of the sun. The two vanes were arranged so as to be one behind the other, in a straight line; and a single hole was made in the middle of the moveable piece No. 4, of the photometer described in my last paper.*

I now viewed the opening and shallow in the sun, and immediately after went to the photometer, to examine the artificial phenomenon. By withdrawing the farthest vane, I diminished its illumination, till I found the small visible rim about the velvet less luminous than the paper of the first vane, in the same proportion as I judged the shallow to be less bright than the rest of the sun. Then, going alternately many times to the telescope and to the photometer, and making such little alterations in the apparatus as I thought necessary, I obtained at last a result, which shewed that the rim of paper representing the shallow reflected 469 parts of the incident light.

Hence, if the superior self-luminous clouds of the sun throw the same quantity of light on the inferior region of opaque solar clouds as they send to us, it follows, that those inferior clouds of which the shallows are composed, reflect about 469 rays out of every thousand they receive.

With regard to the solar surface which we see in the openings, I also found that black worsted, which, by my lately published tables, reflects 16 rays out of a thousand, was not dark enough, compared to the blackness of the opening, but

* See *Phil. Trans.* for 1800, p. 528 [above, p. 141].

that black velvet seemed to be nearly of the same intensity; so that, probably, when the luminous surface of the sun sends us 1000, and the flats 469 rays, the solid surface seen in the openings reflects no more than about seven.

Indentations are planetary Clouds, reflecting Light through the open Parts of the Corrugations.

Jan. 15, 1801. The corrugations do not contain pores, but irregularly shaped depressed places, of a darker or rather less luminous matter than the borders of the corrugations; probably owing to the same cause that makes the shallows appear less bright than the general surface.

Feb. 7, 1801. The corrugations go up to the borders of the shallow of an opening observed since Feb. 4. Indeed the high parts of the corrugations themselves consist of the upper self-luminous clouds; while their indentations are the reflection of light from the lower regions of the opaque ones.

The opaque inferior Clouds probably suffer but little of the Light of the self-luminous superior Clouds to come to the Body of the Sun.

Feb. 5, 1801. The shallow of a large opening, though already contracted, is still sufficiently broad not to permit a single direct ray of the superior self-luminous clouds to enter this opening; and the blackness of the opening shews, that but little light can penetrate through the inferior region of planetary clouds of which the shallow consists.

Motion of the inferior Clouds.

Feb. 6, 1801. The great opening of Feb. 4, is much diminished: it is now divided by a branch of the inferior clouds of the shallow, with a few superior ones upon the following half of the division or bridge. The shallow on the other side of the bridge is plainly still free from self-luminous clouds.

11^h 53′. On the preceding side, more of the planetary clouds are advancing to draw themselves over the opening; they are very faint.

Feb. 7, 1801. Another passage is now thrown over the great opening, consisting evidently of the lower clouds. Perhaps a few clouds of the upper regions may be drawn upon it; but, at both sides, the shallow continues to be visible from the bridge to its margin, or confinement by the surrounding self-luminous clouds.

10^h 40. A third passage is beginning to come on from the following side, also consisting of lower clouds. It seems that the curtain will be closely drawn over the opening, before many of the self-luminous clouds can advance.

Motion of the superior Clouds.

Feb. 5, 1801. The large opening of Feb. 4, is in a diminishing state. Its shallow is contracted; and, though it has no corrugations, it seems as if a few self-luminous clouds of the upper regions were here and there scattered over it.

I see their superior brightness, and their elevation above the shallow in the places where they are.

1^h 42′. More of the clouds of the upper regions have scattered themselves over the shallow ; and one of them faintly passes over part of the opening, almost across it. This will probably form a division.

1^h 50′. There is an opposite cloud advancing to meet the former one ; probably that part of the opening where they are, opposes less resistance than the rest.

1^h 53′. The shallow contracts very fast ; the sides of the upper regions of clouds press on ; the borders become irregular, and jagged, according to the advancement of the clouds ; but a coarsely circular form of the shallow in general is still preserved.

Eminent Use of the planetary Clouds.

It has been shewn that the openings, compared to the luminous surface of the sun, reflect less than 16 rays out of a thousand, and probably not more than seven. To account for this extraordinary darkness, it must be remembered, that according to the observations which have been given, hardly any but transmitted rays can ever come to the body of the sun. The shallows about large openings are generally of such a size, as hardly to permit any direct illumination from the superior clouds to pass over them into the openings ; and the great height and closeness of the sides of small ones, though not often guarded by shallows, must also have nearly the same effect. By this it appears, that the planetary clouds are indeed a most effectual curtain, to keep the brightness of the superior regions from the body of the sun.

Another advantage arising from the planetary clouds of the sun, is of no less importance to the whole solar system. We have shewn that corrugations are every where dispersed over the sun ; and that their indentations may be called shallows in miniature. From this we may conclude, that the immense curtain of the planetary solar clouds is every where closely drawn ; and, as our photometrical experiments have proved that these clouds reflect no less than 469 rays out of a thousand, it is evident that they must add a most capital support to the splendour of the sun, by throwing back so great a share of the brightness coming to them from the illumination of the whole superior regions.

OF THE SOLAR ATMOSPHERE.

The Sun has a planetary Atmosphere.

Our observations on the double regions of clouds in the sun, are certainly a sufficient proof of the existence of a solar atmosphere. The clouds of the lower regions of the sun bear such a resemblance to our own, that they can only, like ours, be upheld by a thin elastic medium, in which, like ours in air, they may freely move about in all directions.

The Sun's planetary Atmosphere extends to a great Height.

If we have concluded, from the appearance of the clouds of the lower regions, that they were supported by an atmosphere, the same will hold good with regard to the self-luminous clouds of the upper regions. For, though probably they do not swim or float in the planetary atmosphere of the sun, like the lower ones, it is evident, from observation, that they arrange themselves regularly at certain given altitudes ; which can only be ascribed to the specific gravity of the gases to which they owe their existence. Besides, as the solar atmosphere is elastic, it can be no otherwise confined than by its gravitation to the sun, in the same manner as the air, by its own weight, is kept down to the earth ; and the solar atmosphere must therefore expand itself considerably above the highest ridges and nodules.

The planetary Atmosphere of the Sun is of a great Density.

This may be deduced directly from the known quantity of matter in the sun. Sir ISAAC NEWTON has proved, that the gravitation of bodies on the surface of the sun, is 27 times stronger than it is with us. Hence, the compression of the elastic gases of which the solar atmosphere consists, if similar to our own, must be greater than that of ours, in proportion to the superior force by which they are compressed, namely, their own powerful gravitation towards the sun.

The Solar Atmosphere, like ours, is subject to Agitations, such as with us are occasioned by Winds.

A proof of this may be drawn from the observations which have been given. In several instances, we have seen the planetary clouds move over the openings ; which could not have happened, unless the atmosphere in which they floated had been considerably agitated. In many other instances, we have shewn that a strong bias existed in the direction of the cause which generates the shallows. This indeed is so evident, that I have hardly ever seen a single shallow which had not some excentricity ; the smallest segments of these shallows being always turned towards what I have called the quiescent side of the openings. To this may be added, that the continual luminous decompositions in the superior regions, and the consequent necessary regeneration of the atmospheric gases that serve to carry them on, and which probably are produced below the inferior cloudy regions, near the surface of the sun's body, must unavoidably be attended with great agitations, such as with us might even be called hurricanes.

There is some clear Atmospheric Space, between the solid Body of the Sun and the lowest Region of the Clouds.

From what has just now been said of the agitations which appear to take place in the solar atmosphere, it follows, that those biases which have been shewn to affect the direction of a number of shallows at the same time, must have arisen from a motion of the atmospheric gases under the clouds; and, that there is a considerable vacant space between these clouds and the solid body of the sun, is also to be inferred, from the free motion of clouds considerably lower than usual, which were seen to pass over an opening, and which cannot be supposed to have rolled over the ground in contact with it. Without, however, entering into any particular examination of the amount of the distance from the sun to the first cloudy regions, which, were I to guess by some pretty obvious circumstances, would not be less than some hundreds of miles, we may take it for granted, that the altitude of the clouds will every where be determined by their own density, and by that of the atmosphere in which they are suspended.

The Sun's planetary Atmosphere is transparent.

It will be easily shewn, that the gases of the solar atmosphere are transparent; for we have already given observations that prove our being able to see the reflected light of the corrugations from their indentations; and of the self-luminous regions in general, from the shallows which they surround and illuminate. To this may be added, that we also see clearly down through the space which leads through the openings, as fully appears from our being able to see the thicknesses of the borders which inclose them. We have likewise given an instance of seeing the limb of the sun broken by a vacancy proceeding from a large shallow; though undoubtedly that shallow must have been covered with the solar atmospheric gases.

THEORETICAL EXPLANATION OF THE SOLAR PHENOMENA.

We have admitted, in order to explain the generation of shallows, that a transparent elastic gas comes up through the openings, by forcing itself a passage through the planetary clouds. Our observations seemed naturally to lead to this supposition, or rather to prove it; for, in tracing the shallows to their origin, it has been shewn, that they always begin from the openings, and go forwards. We have also seen, that in one case, a particular bias given to incipient shallows, lengthened a number of them out in one certain direction, which evidently denoted a propelling force acting the same way in them all. I am, however, well prepared to distinguish between facts observed, and the consequences that in reasoning upon them we may draw from them; and it will be easy to separate them, if that should hereafter be required.

If however, it be now allowed, that the cause we have assigned may be the true one, it will then appear, that the operations which are carried on in the atmosphere of the sun are very simple and uniform.

Generation of Pores.

By the nature and construction of the sun, an elastic gas, which may be called empyreal, is constantly formed. This ascends every where, by a specific gravity less than that of the general solar atmospheric gas contained in the lower regions. When it goes up in moderate quantities, it makes itself small passages among the lower regions of clouds: these we have frequently observed, and have called them pores. We have shewn that they are liable to continual and quick changes, which must be a natural consequence of their fleeting generation.

Formation of Corrugations.

When this empyreal gas has reached the higher regions of the sun's atmosphere, it mixes with other gases, which, from their specific gravity have their residence there, and occasions decompositions which produce the appearance of corrugations. It has been shewn, that the elevated parts of the corrugations are small self-luminous nodules, or broken ridges; and I have used the name of self-luminous clouds, as a general expression for all phenomena of the sun, in what shape soever they may appear, that shine by their own light. These terms do not exactly convey the idea affixed to them; but those of meteors, coruscations, inflammations, luminous wisps, or others, which I might have selected, would have been liable to still greater objections. It is true, that when speaking of clouds, we generally conceive something too gross, and even too permanent, to permit us to apply that expression properly to luminous decompositions, which cannot float or swim in air, as we are used to see our planetary clouds do. But it should be remembered that, on account of the great compression arising from the force of gravity, all the elastic solar gases must be much condensed; and that, consequently phenomena in the sun's atmosphere, which in ours would be mere transitory coruscations, such as those of the aurora borealis, will be so compressed as to become much more efficacious and permanent.

Cause of Indentations.

The great light occasioned by the brilliant superior regions, must scatter itself on the tops of the inferior planetary clouds, and, on account of their great density, bring on a very vivid reflection. Between the interstices of the elevated parts of the corrugations, or self-luminous clouds, which, according to the observations that have been given, are not closely connected, the light reflected from the lower clouds will be plainly visible, and, being considerably less intense than the direct illumination from the upper regions, will occasion that faint appearance which we have called indentations.

Cause of the mottled Appearance of the Sun.

This mixture of the light reflected from the indentations and that which is emitted directly from the higher parts of the corrugations, unless very attentively examined by a superior telescope, will only have the resemblance of a mottled surface.

Formation of small Openings, Ridges, and Nodules.

When a quantity of empyreal gas, more than what produces only pores in ascending, is formed, it will make itself small openings ; or, meeting perhaps with some resistance in passing upwards, it may exert its action in the production of ridges and nodules.

Production of large Openings and Shallows.

Lastly, if still further an uncommon quantity of this gas should be formed, it will burst through the planetary regions of clouds, and thus will produce great openings ; then, spreading itself above them, it will occasion large shallows, and, mixing afterwards gradually with other superior gases, it will promote the increase, and assist in the maintenance, of the general luminous phenomena.

If this account of the solar appearances should be well founded, we shall have no difficulty in ascertaining the actual state of the sun, with regard to its energy in giving light and heat to our globe ; and nothing will now remain, but to decide the question which will naturally occur, whether there be actually any considerable difference in the quantity of light and heat emitted from the sun at different times. But, since experience has already convinced us, that our seasons are sometimes very severe, and at other times very mild, it remains only to be considered, whether we should ascribe this difference immediately to a more or less copious emission of the solar beams. Now, as we have lately had seasons of deficiency, that seem to indicate a want of the vivifying principles of light and heat, and as, from the appearance of last summer, and the present mild winter, there seems to be a change that may be in our favour, it will be proper to have recourse to solar observations, in order to compare the phenomena which indicate the state of the sun, with the seasons of these remarkable times. The following two sets, which are selected from my journals, I believe will assist us materially in this inquiry.

SIGNS OF SCARCITY OF LUMINOUS MATTER IN THE SUN.

Visible Deficiency of empyreal Clouds.

July 5, 1795. 1^h 6′. The appearance of the sun is very different from what I have ever seen it before. There is not a single opening in the whole disk ; there are no ridges or nodules ; there are no corrugations.

A perfect Calm in the upper Regions of solar Clouds.

Dec. 9, 1798. 12^h 33′. The sun has no openings of any kind ; nor can I perceive any places that look disturbed, like those where openings have lately been.

Want of Openings, Ridges, and Nodules.

Sept. 18, 1795. There is no opening in the sun. I viewed it with powers from 90 to 460.

April 1, 1798. 11^h 49′. I examined the sun with a power of 230; but could find no openings.

Nov. 27, 1799. The sun is without openings. I cannot however perceive any indication that, by the mere look, would denote a deficiency of light.

Dec. 31, 1799. There are no openings in the disk of the sun.

Jan. 3, 1800. There is no opening visible any where.

Jan. 27, 1800. There is no opening. There are no ridges; nor is there a single nodule any where.

Jan. 30, 1800. There are neither openings, ridges, nor nodules, in the sun.

Jan. 31, 1800. There are neither openings, nor ridges, in the sun.

Feb. 4, 1800. There are neither openings, ridges, nor nodules, any where.

Feb. 11, 1800. There is not an opening, ridge, or nodule, any where in the sun.

Feb. 18, 1800. There are no openings, no ridges, or nodules.

Dec. 22, 1799. In one part of the sun are some vivid ridges; but I cannot find any of them in other parts.

Dec. 27, 1799. Near the following margin are some bright ridges; but there is not a single one to be seen in any other part of the sun's disk.

Many Indentations without, and others with, changeable Pores.

Jan. 3, 1800. The indentations contain fewer black points than last week.

Jan. 4, 1800. The corrugations are punctured with blackish indentations. The sun is more affected in this manner than it was yesterday.

Jan. 27, 1800. The sun is every where coarsely indented, but not punctured; there being no black points in the indentations.

SIGNS OF ABUNDANCE OF LUMINOUS MATTER IN THE SUN.

Visible Increase of empyreal Clouds.

Feb. 12, 1800. The indentations, in many parts, are changed to small shallows of corrugations. There seems to have been a gradual increase of the luminous clouds for some time past. The reason why I am not positively assured of this increase is, that my present method of viewing the sun is so much better than formerly, that, by seeing things to greater advantage, there may be some deception in the seeming change of appearances.

March 5, 1800. I can now entertain no doubt that the luminous clouds are more copious than they were some time ago. The corrugations seem all to be better filled. Hardly any of the indentations have pores.

Many Openings, Ridges, and Nodules.

March 5, 1800. A range of openings has a very fine appearance; there are 55 of them. The most south and largest has a considerable shallow about it. Two, just north of it, have shallows on the northern side, but not towards the south, where the borders of the openings seem to be full as elevated as the highest luminous clouds. Near the south-following margin are extensive ridges, studded here and there with nodules.

Nov. 17, 1800. The sun is beautifully ornamented with openings, shallows, ridges, and nodules.

Dec. 2, 1800. The sun is every where richly covered by luminous clouds. Ridges and nodules are also to be seen in many places.

Dec. 27. A large opening is lately come into the disk; several other small ones are visible; and there are, near the preceding and following limbs, many extensive ridges. The luminous clouds are very plentifully and richly scattered all over.

Jan. 15, 1800. There are three collections of openings in different parts of the disk of the sun, and many ridges and nodules. The small indentations, as I have formerly called them, are so coarse, and of such irregular shapes, that they can be called so no longer. Corrugations, therefore, are that variety and unevenness of the whole surface of the sun, when it appears richly furnished with luminous clouds.

I am now much inclined to believe, that openings with great shallows, ridges, nodules, and corrugations, instead of small indentations, may lead us to expect a copious emission of heat, and therefore mild seasons. And that, on the contrary, pores, small indentations, and a poor appearance of the luminous clouds, the absence of ridges and nodules, and of large openings and shallows, will denote a spare emission of heat, and may induce us to expect severe seasons. A constant observation of the sun with this view, and a proper information respecting the general mildness or severity of the seasons, in all parts of the world, may bring this theory to perfection, or refute it, if it be not well founded.

Jan. 24, 1801. The surface of the sun is every where richly decked out with luminous clouds. An additional opening is lately come into view, attended by many spreading ridges.

Jan. 29, 1801. If openings be a sign of richness in the illuminating and heating disposition of the sun, there are enough of them: considerable ones are scattered in six different regions, taking up a broad zone.

Feb. 4, 1801. Between flying clouds, I counted 31 openings in the sun.

March 2, 1801. There are six different sets of openings in the sun. One of them consists of ten; another of two; the rest are single.

Coarse and luminous Corrugations.

Jan. 4, 1801. The corrugations are very coarse; and the luminous clouds seem to be very rich.

Jan. 3, 1801. The elevations of the corrugations are all very luminous, like so many nodules.

Feb. 18, 1801. The corrugations are every where very luminous.

March 2. The general surface of the sun is so rich, that the indentations are a good deal covered by self-luminous clouds.

From these two last sets of observations, one of which establishes the scarcity of the luminous clouds, while the other shews their great abundance, I think we may reasonably conclude, that there must be a manifest difference in the emission of light and heat from the sun. It appears to me, if I may be permitted the metaphor, that our sun has for some time past been labouring under an indisposition, from which it is now in a fair way of recovering. An application of the foregoing method, however, even if we were perfectly assured of its being well founded, will still remain attended with considerable difficulties.

We see how, in that simple instrument the barometer, our expectations of rain or fair weather, are only to be had by a consideration of many circumstances, besides its actual elevation at the moment of inspection.

The tides also present us with the most complicated varieties in their greatest elevation, as well as in the time when they happen on the coasts of different parts of this globe. The simplicity of their cause, the solar and lunar attractions, we might have expected, would have precluded every extraordinary and seemingly discordant result.

In a much higher degree, may the influence of more or less light and heat from the sun, be liable to produce a great variety in the severity or mildness of the seasons of different climates, and under different local circumstances; yet, when many things which are already known to affect the temperature of different countries, and others which future attention may still discover, come to be properly combined with the results we propose to draw from solar observations, we may possibly find this subject less intricate than we might apprehend on a first view of it.

If, for instance, we should have a warm summer in this country, when phenomena observed in the sun indicate the expectation of it, I should by no means consider it as an unsurmountable objection, if it were shewn that in another country the weather had not been so favourable. And, if it were generally found that our prognostication from solar observations held good in any one given place, I should be ready to say that, with proper modifications, they would equally succeed in every other situation.

Before we can generalize the influence of a certain cause, we ought to confine our experiment to one permanent situation, where local circumstances may be supposed to act nearly alike at all times, which will remove a number of difficulties.

To recur to our instance of the tides, if we were to examine the phenomena which they offer to our inspection in any one given place, such as the mouth of the Thames, we should soon be convinced of their agreement with the motion of the sun and moon. A little reflection would easily reconcile us to every deviation from regularity, by taking into account the direction and violence of winds, the situation of the coast, and other circumstances. Nor should we doubt the truth of the theory of the tides, though high water at Bristol, Liverpool, or Hull, should have been very deficient, at a time when, in the place of our experiments, it had happened to be uncommonly abundant.

Now, with regard to the effects of the influence of the sun, we know already, that in the same latitudes the seasons differ widely in temperature; that it is not hottest at noon, or coldest at midnight; that the shortest day is neither attended with the severest frosts, nor the longest day with the most oppressing heats; that large forests, lakes, morasses, and swamps, affect the temperature one way; and rocky, sandy, gravelly, and barren situations, in a contrary manner; that the seasons of islands are considerably different from those of large continents, and so forth.

But it will now be necessary to examine the accounts we already have of the appearance and disappearance of the solar spots, and to compare them with the temperature of the respective times, as far as history will furnish us with records.

The first thing which appears from astronomical observations is, that the periods of the disappearance of spots on the sun are of much shorter duration than those of their appearance; so that, if the symptoms which have been pointed out, as denoting the state of the sun with regard to light and heat, should be well founded, we ought rather to look upon the absence of spots as a sign of deficiency, than on their presence as one of abundance; and this would justify my expression, of the recovery of the sun from an indisposition, as being a return to its usual splendour.

In going back to early observations, we cannot expect to meet with a record of such minute phenomena as we have attended to. The method of viewing spots on the sun, by throwing their picture, in a dark room, on a sheet of white paper, is not capable of delicacy; nor were the direct views of former astronomers so distinct as, in the present improved state of the telescope, we can have them; a very imperfect account of solar spots may therefore be expected, considering our present inquiry, which would require complete observations of every spot, great or small, that has been on the sun during such periods as will be examined.

With regard to the contemporary severity and mildness of the seasons, it will hardly be necessary to remark, that nothing decisive can be obtained. But, if we are deficient here, an indirect source of information is opened to us, by applying to the influence of the sun-beams on the vegetation of wheat in this country. I do not mean to say, that this is a real criterion of the quantity of light and heat emanated from the sun ; much less will the price of this article completely represent the scarcity or abundance of the absolute produce of the country. For the price of commodities will certainly be regulated by the demand for them ; and this we know is liable to be affected by many fortuitous circumstances. However, although an argument drawn from a well ascertained price of wheat, may not apply directly to our present purpose, yet, admitting the sun to be the ultimate fountain of fertility, this subject may deserve a short investigation, especially as, for want of proper thermometrical observations, no other method is left for our choice.

Our historical account of the disappearance of the spots in the sun, contains five very irregular and very unequal periods.* The first takes in a series of 21 years, from 1650 to 1670, both included. But it is so imperfectly recorded, that it is hardly safe to draw any conclusions from it ; for we have only a few observations of one or two spots that were seen in all that time, and those were only observed for a short continuance. However, on examining the table of the prices of the quarter of nine bushels of the best or highest priced wheat at Windsor, marked in Dr. ADAM SMITH'S valuable Inquiry into the nature and causes of the wealth of nations,† we find that wheat, during the time of the 21 years above mentioned, bore a very high price; the average of the quarter being £2. 10*s*. $5\frac{19}{21}$*d*. This period is much too long to suppose that we might safely compare it with a preceding or following one of equal duration. Besides, no particulars having been given of the time preceding, except that spots in the sun, a good while before, began to grow very scarce, there might even be fewer of them than from the year 1650 to 1670. Of the 21 years immediately following, we know that they certainly comprehend two short periods, in which there were no spots on the sun ; of these, more will be said hereafter ; but, including even them, we have the average price of wheat, from 1671 to 1691, only £2. 4*s*. $4\frac{2}{3}$*d*. the quarter. The difference, which is a little more than as 9 to 8, is therefore still a proof of a temporary scarcity.

Our next period is much better ascertained. It begins in December 1676, which year therefore we should not take in, and goes to April 1684; in all which time, FLAMSTEED, who was then observing, saw no spot in the sun. The average price of wheat, during these 8 years, was £2. 7*s*. 7*d*. the quarter. We cannot justly compare this price with that of the preceding 8 years, as some of the former years of scarcity would come into that period ; but the 8 years immediately following, that is, from 1685 to 1691, both included, give an average price of no more than £1. 17*s*. $1\frac{3}{4}$*d*. The difference, which is as full 5 to 4, is well deserving our notice.

A third but very short period, is from the year 1686 to 1688, in which time CASSINI could find no spot in the sun. If both years be included, we have the average price of wheat, for those three years, £1. 15*s*. $0\frac{2}{3}$*d*. the quarter. We ought not to compare this price with that of the three preceding years, as two of them belong

* See *Astronomie* par M. DE LA LANDE, § 3235.

† See Book I. Chap. XI.

to the preceding period of scarcity ; but the three following years give the average price for the quarter of wheat £1. 12*s*. 10⅔*d*. or, as nearly 11 to 10.

The fourth period on record, is from the year 1695 to 1700, in which time no spot could be found in the sun. This makes a period of 5 years ; for, in 1700 the spots were seen again. The average price of wheat, in these years, was £3. 3*s*. 3⅕*d*. the quarter. The 5 preceding years, from 1690 to 1694, give £2. 9*s*. 4⅘*d*. and the 5 following years, from 1700 to 1704, give £1. 17*s*. 11⅕*d*. These differences are both very considerable ; the last is not less than 5 to 3.

The fifth period extends from 1710 to 1713 ; but here there was one spot seen in 1710, none in 1711 and 1712, and again one spot only in 1713. The account of the average price of wheat, for these four years, is £2. 17*s*. 4*d*. the quarter. The preceding four years, from 1706 to 1709, give the price £2. 3*s*. 7½*d*. and the following years, from 1714 to 1717, it was £2. 6*s*. 9*d*. When the astronomical account of the sun for this period, which has been stated above, is considered, these two differences will be found very considerable ; the first of them being nearly as 4 to 3.

The result of this review of the foregoing five periods is, that, from the price of wheat, it seems probable that some temporary scarcity or defect of vegetation has generally taken place, when the sun has been without those appearances which we surmise to be symptoms of a copious emission of light and heat. In order, however, to make this an argument in favour of our hypothesis, even if the reality of a defective vegetation of grain were sufficiently established by its enhanced price, it would still be necessary to shew that a deficiency of the solar beams had been the occasion of it. Now, those who are acquainted with agriculture may remark, that wheat is well known to grow in climates much colder than ours ; and that a proper distribution of rain and dry weather, with many other circumstances which it will not be necessary to mention, are probably of much greater consequence than the absolute quantity of light and heat derived from the sun. To this I shall only suggest, by way of answer, that those very circumstances of proper alternations of rain, dry weather, winds, or whatever else may contribute to favour vegetation in this climate, may possibly depend on a certain quantity of sun-beams, transmitted to us at proper times ; but, this being a point which can only be ascertained by future observations, I forbear entering farther into a discussion of it.

It will be thought remarkable, that no later periods of the disappearance of the solar spots can be found. The reason however is obvious. The perfection of instruments, and the increased number of observers, have produced an account of solar spots, which, from their smallness, or their short appearance, would probably have been overlooked in former times. If we should in future only reckon the years of the total absence of solar spots, even that remarkable period of scarcity which has fallen under my own observation, in which nevertheless I have now and then seen a few spots of short duration, and of no great magnitude, could not be admitted.

For this reason, we ought now to distinguish our solar observations, by reducing them to short periods of symptoms for or against a copious emission of the solar beams, in which, all the phenomena we have pointed out should be noticed. The most striking of them are certainly the number, magnitude, and duration of the openings. The increase and decrease of the luminous appearance of the corrugations is perhaps full as essential; but, as it is probable that their brilliancy may be a consequence of the abundance of the former phenomena, an attention to the latter, which is subject to great difficulties, and requires the very best of telescopes, may not be so necessary.

What remains to be added is but short. In the first of my two series of observations, I have pointed out a deficiency in what appears to be the symptomatic disposition of the sun for emitting light and heat: it has lasted from the year 1795 to 1800.* That we have had a considerable deficiency in the vegetation of grain, will hardly require any proof. The second series, or rather the commencement of it, for I hope it will last long, has pointed out a favourable return of the rich appearance of the sun. This, if I may venture to judge, will probably occasion a return of such seasons as, in the end, will be attended by all their usual fertility.

The subject, however, being so new, it will be proper to conclude, by adding, that this prediction ought not to be relied on by any one, with more confidence than the arguments which have been brought forwards in this Paper may appear to deserve.

EXPLANATION OF THE 1ST, 2D, 11TH AND 12TH FIGURES.

[See pp. 152 and 158.]

Fig. 1, represents an opening in the luminous solar clouds, with its surrounding shallow, in a situation much past the centre, towards the preceding limb of the sun. The lines marked with the letters *a, b, c, d, e, f*, answer to those which are marked with the same letters in Fig. 2.

Fig. 2, is a section of the same opening. The lines *a, b, c, d, e, f*, are supposed to be drawn from the eye of the observer. *a* and *b* point out the elevation of the corrugations *g, h*, on the preceding side P, above the surface of the shallow *i, k*. *c* and *d* shew the thickness of the shallow; and the line *d* goes through the opening, down into the clear atmospheric space P F, till it meets with the opaque surface of the sun A B. On the following side F, the thickness of the shallow and elevation of the corrugations cannot be seen; since the line *e* goes abruptly into the opening; and *f* goes, as abruptly, from the top of the corrugations, down to the shallow.

Fig. 11, shews a section of the corrugations, shallow, and opening, of Fig. 10, in the same manner as Fig. 2 represents those of Fig. 1. There is a hanging cloud *a*, in Fig. 11, over the preceding part of the opening; and the same cloud is represented at *b*, in Fig. 12, to which place it was seen to move from its former situation, in 58 minutes of time.

The rest of the figures are sufficiently explained in the places of the text which refer to them.

---

* This period should properly have been divided into two small ones; but, for want of intermediate solar observations, I have joined the visible deficiencies in the illuminating and heating powers of the sun, from the year 1795 to 1796, and again from 1798 to 1800, into one.

[The above paper was translated in the *Astronomisches Jahrbuch* for 1805 and 1806. At the end of it, the translator, Ideler, added a footnote, in which he remarked that the price of wheat in a single country could not supply a correct measure of the general fertility of the earth. In reply, Herschel wrote the following letter to Bode, which was printed in translation in the *Jahrbuch* for 1807, p. 242.]

I have received the almanac you have sent me for the year 1806, and am much obliged to you for the communication. I have read the remark which the translator has added to my paper. But he seems to have intirely mistaken the meaning when I referred to the price of wheat *in England*, as some indication of the emission of the rays of heat. It was never intended to say, that a copious emission of the rays of heat would produce a general success in the vegetation of this article. On the contrary, in climates which are full hot enough for wheat, an emission a little more copious than usual might occasion a considerable sterility. It is therefore evident that the success of the vegetation of wheat ought only to be taken in one place, and that an extensive commerce of that article with the whole world is out of the question. Nor has it been affirmed that fruitfulness is *immediately* connected with the copious emission of the sun's rays. On the contrary, I have called the sun the *ultimate* fountain of fertility ; for it is well known and admitted that the solar rays pass through several modifications which give wind, rain, electrical phenomena, etc., all which affect in a different manner the fruitfulness of different vegetables. The remark of the translator therefore seems to resemble an objection that might be brought to the theory of the moon which occasions the tides. For instance, an observer shews that in the place where he lives it is high water six hours after the meridian passage of the moon, and that the tide rises 50 feet. An objector says this cannot be owing to the moon, because in the place where he lives high water is 2½ hours after the meridian passage, and amounts only to three feet. And yet if the movement of the waters in any one place will agree with the phenomena of the motions of the sun and moon, the argument of their being the cause of this movement will be conclusive. It is of little consequence whether you take wheat, rice, the sugar-cane or the bread-fruit, for the vegetable which is to serve the purpose of the experiment, or whether it is the abundance or scarcity which happens when the sun emits its rays most copiously. If the experiment be well ascertained it will be equally conclusive. Wheat being the only vegetable of which an authentic account was to be had in England, was chosen for the purpose, and all the imperfections to which the subject is liable were pointed out in the paper which treats of it. It should also be remembered that the whole theory of the symptomatic disposition of the sun is only proposed as an experiment to be made.

Slough, near Windsor,
May 31, 1804.

XLVII.

Additional Observations tending to investigate the Symptoms of the variable Emission of the Light and Heat of the Sun; with Trials to set aside darkening Glasses, by transmitting the Solar Rays through Liquids; and a few Remarks to remove Objections that might be made against some of the Arguments contained in the former Paper.

[*Phil. Trans.*, 1801, pp. 354–362.]

Read May 14, 1801.

HAVING brought up the solar observations, relating to the symptoms of a copious emission of the light and heat of the sun, to the 2d of March, I give them continued in this Paper to the 3d of May. It will be seen, that my expectations of the continuance of the symptoms which I supposed favourable to such emissions, have hitherto been sufficiently verified; and, by comparing the phænomena I have reported, with the corresponding mildness of the season, my arguments will receive a considerable support.

I have given the following observations without delay, as containing an outline of the method we ought to pursue, in order to establish the principles which have been pointed out in my former Paper. But we need not in future be at a loss how to come at the truth of the current temperature of this climate, as the thermometrical observations, which are now regularly published in the *Philosophical Transactions*, can furnish us with a proper standard, with which the solar phænomena may be compared. This leads me to remark, that, although I have, in my first Paper, sufficiently noticed the want of a proper criterion for ascertaining the temperature of the early periods where the sun has been recorded to have been without spots, and have also referred to future observations for shewing whether a due distribution of dry and wet weather, with other circumstances which are known to favour the vegetation of corn, do or do not require a certain regular emission of the solar beams, yet, I might still have added, that the actual object we have in view, is perfectly independent of the result of any observations that may hereafter be made, on the favourable or defective vegetation of grain in this or in any other climate. For, if the thermometer, which will be our future criterion, should establish the symptoms we have assigned, of a defective or copious emission of the solar rays, or even help us to fix on different ones, as more likely to point out the end we have in view, we may leave it entirely to others, to determine the use to which a fore-knowledge of the probable temperature of an approaching summer, or winter, or perhaps of both, may be applied; but still it may be hoped that some advantage may be derived, even in agricultural economy, from an improved knowledge of the nature of the sun, and of the causes, or symptoms, of its emitting light and heat more or less copiously.

Before I proceed, I must hint to those who may be willing to attend to this subject, that I have a strong suspicion that one half of our sun is less favourable to a copious emission of rays than the other; and that its variable lustre may possibly appear to other solar systems, as irregular periodical stars are seen by us; but, whether this arises from some permanent construction of the solar surface, or is merely an accidental circumstance, must be left to future investigation: it should, however, be carefully attended to.

OBSERVATIONS OF THE SUN.

March 4, 1801. I viewed the sun with a skeleton eye-piece, into the vacancy of which may be placed a moveable trough, shut up at the ends with well-polished plain glasses, so that the sun's rays may be made to pass through any liquid contained in the trough, before they come to the eye-glass (see fig. 1 and 2).

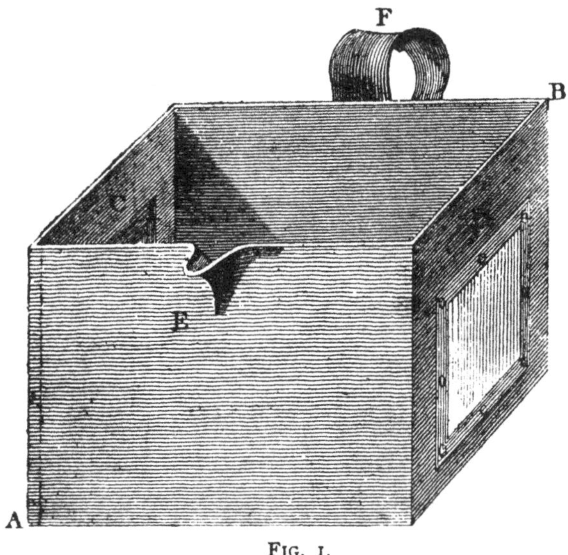

FIG. 1.

Through spirit of wine, I saw the sun very distinctly. There are 10 openings without shallows; and a pretty considerable one with a shallow. The opening is nearly round; and the shallow is concentric with it, and also round. The want of shallows about the small openings, and the roundness of that about the largest, indicate that the elastic empyreal gas which passes through them, is without side-bias in its motion.*

March 8. I viewed the sun through water. It keeps the heat off so well, that we may look for any length of time, without the least inconvenience. There are a few openings, many ridges and nodules.

March 9. The ridges near the preceding limb are more extensive than I have ever seen them; there is a broad zone of them.

† See [page 157 of] my last Paper, "*Probable Cause of Shallows*"; and [page 169] "*The solar Atmosphere, like ours,*" &c.

March 12. There is a cluster of 20 small openings; none of them have any shallows.

March 13. There are 31 openings in the cluster of yesterday: they are contained in a double row, nearly parallel to the sun's equatorial motion; the largest of them has now a shallow of a considerable size, on its north-following side. The number of small openings near each other, indicates a perpendicular ascent of the empyreal gas that breaks through the atmospheric clouds; and their want of shallows shews the same thing.

March 15. The set of openings which began to enter on the 8th, consists now of 29. There are 3 other small openings in different parts of the sun.

March 16. There is an opening lately entered. The cluster of yesterday has undergone considerable changes.

March 18. The opening of the 16th consists now of 8 different ones; none of them have any shallows. The whole space about the cluster of the 8th, is surrounded with luminous ridges in many directions. The corrugations all over the sun are

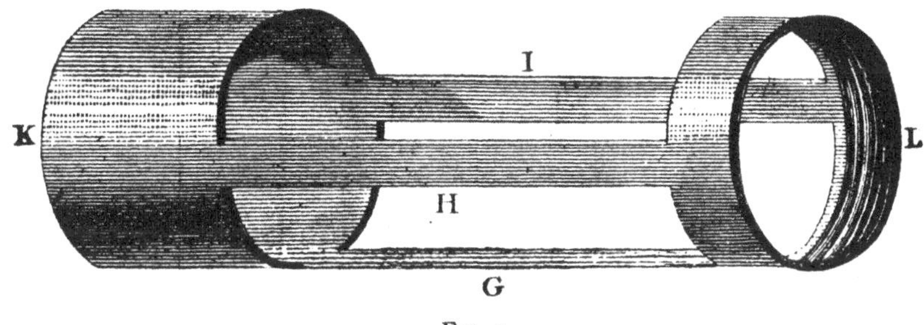

FIG. 2.

beautiful, and coarse; resembling small nodules joined together like irregular honeycomb. In a multitude of places, the corrugations are quite detached, like luminous wisps, or slender tufts, standing upright.

March 19. Another set of ridges has entered the disk; it contains one opening. The corrugations are rich, and may be called luminous wisps, being much disjoined, except at their bottom; they are so rich, that they partake of the yellowish colour of the ridges. The northern ridges extend a good way into the disk, like a zone.

March 21. There are five sets, containing 29 openings, none of which have any shallows. At equal distances from the limb, the corrugations are equally coarse all over the disk of the sun.

March 22. An additional opening, with surrounding ridges, has lately entered the north-following limb. I counted 21 openings.

March 31. An opening very near the preceding limb is surrounded by a shallow, which is bordered by a luminous ridge all round it. The opening itself is also bordered by an elevated edge, which is nearly as high as the general surface of the corrugations; but not so high as that which borders the shallow, and stands above the general surface.

April 1, 11^h $30'$. I saw the opening of yesterday go out of the limb: it was the only one left.

2^h $0'$. The sun is now without any openings; but the corrugations are very luminous and rich.

April 2. A considerable opening has entered the disk, accompanied with ridges. From its present situation, I conclude it must have entered not long after my last observation yesterday. The sun is very rich in luminous corrugations, interspersed with bright nodules towards the south pole.

April 4. There are 4 considerable openings, and many ridges, as well as nodules, on the south and north preceding and following limbs. The north-preceding ridges extend into the sun, till I can no longer distinguish them; and begin again at the north-following side, as far as they generally can be seen from the limb; so that there is probably a whole zone across the disk. Where I lose them, they are generally converted into tufted, rich, coarse corrugations, such as the sun is now every where covered with.

April 6. There are many ridges and rich corrugations; but I can perceive no opening. The air is not clear enough to discover very small ones.

April 8. A cluster of 7 small openings is visible; and many ridges.

April 10. Five sets contain 32 openings. The sun is full of rich tufted corrugations.

April 17. Two sets of openings contain 20 of them.

April 19. I count 45 openings. The corrugations are extremely rich. The whole solar surface seems to be studded with nodules. There are probably two belts of ridges across the sun's disk; for, on the preceding side, as well as on the following, I see two ends of belts of ridges very plainly, extending over all the space where these phænomena can be seen.

April 20. The whole surface of the sun is rich: the corrugations are tufted. I count more than 50 openings; many of them have considerable shallows about them.

April 23, 6^h. There are above 60 openings in the sun. The last set is much towards the sun's north pole; very rich in ridges, and disturbed neighbouring surface.

April 24. I count above 50 openings. The corrugations seem to be closer than they were yesterday.

April 26. I viewed the sun through Port wine, and without smoke on the darkening glasses; but distinctness was much injured.

April 27. I count 39 openings. Many ridges and rich corrugations.

April 29. Six different sets contain 24 openings. There are many sets of ridges and rich corrugations.

4^h. I viewed the sun through a mixture of ink diluted with water, and filtered through paper. It gave an image of the sun as white as snow; and I saw objects very distinctly, without darkening glasses. As one of the largest openings had a considerable shallow, I found, in viewing it through this mixture, that the difference between what I suppose to be the light reflected from opaque, and the direct light of empyreal clouds, is now more striking than I ever had observed it before. The ridges, through this composition, appear whiter than the rest of the sun. The tops of the corrugations are whiter than their indentations, instead of approaching to a yellowish cast, as they do in my former way of seeing through green smoked glasses. The corrugations are very small and contracted to-day. Suspecting that this new way of seeing might represent objects less than they appear, when I view them through an eye-piece that gives them in the manner I have been used to see them, I put on again the former composition ; but found the corrugations as small and close then as they appeared before. I count 36 openings. When the ink mixture is more diluted, the sun's image will become tinged with purple. A solution of green vitriol, with a sufficient number of drops of the tincture of galls to stop as much light as is required, gives a dark blue colour to the sun ; and, by dilution with water, a light blue. It is considerably distinct. With this composition, the corrugations look whiter at the top than in their indentations. The tincture of galls, with as many drops of the solution of green vitriol as will turn it sufficiently black to stop light, makes the sun look of a deep red colour ; and, by dilution, the red will be paler. This composition is not so distinct as the former.

May 2. 5^{h} 20′. There are 36 openings, contained in six sets.

As I have remarked, March 19th, April 4th, and April 19th, that ridges are generally placed in equatorial zones, so I now may add, that the different sets of openings have also been generally arranged in the same directions.

May 3. 11^{h} 56′. Ink mixture. There are 37 openings, arranged in two zones. Four sets in the southern zone contain 27, and three sets in the northern have 10 openings. Through this mixture, I can observe the sun in the meridian, for any length of time, without danger to the eye or to the glasses, with a mirror of nine inches in diameter, and with the eye-pieces open, as they are used for night observations.

Slough, May 4, 1801.

EXPLANATION OF THE FIGURES.

A B, Fig. 1, is a square trough, closed at the two opposite ends C D, by well polished plain glasses. It will hold any liquid through which the sun's rays are to be transmitted. E is a small spout, and F a handle, so that any portion of the liquid may conveniently be poured out, when the rest is to be diluted.

The trough is made to fit into the open part of the skeleton eye-tube, Fig. 2, resting on the bottom G, and being held in its proper situation by the sides H and I. The end K, at the time of observation, is put into a short tube fixed to the NEWTONIAN telescope, and may be turned about, so as always to have the open part H I horizontal.

When the eye-piece Fig. 3, is screwed, by its end M, into the skeleton tube at L, Fig. 2, and the trough Fig. 1, with any liquid to be tried, is placed in the open part G H I, the sun's rays will come from the small mirror of the telescope to K, and, passing through the plain glasses C D, inclosing the liquid, will enter the eye-piece M, and, after the necessary refractions, come to the eye at N.

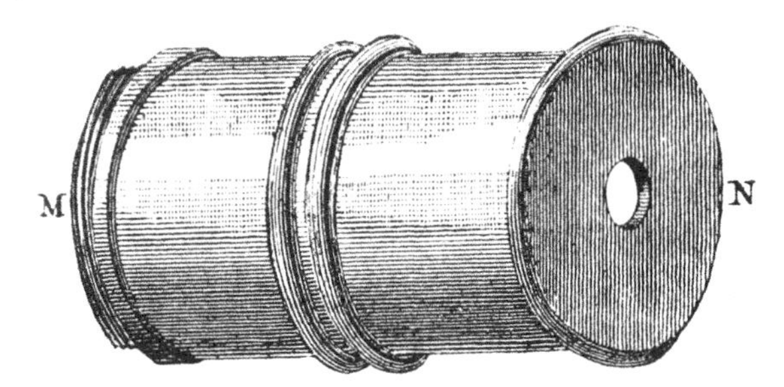

FIG. 3.

Any other, single or double, eye-pieces, of different magnifying powers, may be screwed into L, instead of the piece Fig. 3; and the liquid may easily be tempered so as to intercept a proper quantity of light to suit every eye-glass which is in use, and thus to render the inspection of the sun perfectly convenient.

XLVIII.

Observations on the two lately discovered celestial Bodies.

[*Phil. Trans.*, 1802, pp. 213–232.]

Read May 6, 1802.

In my early account of the moving star discovered by Mr. PIAZZI,* I have already shewn that it is of a remarkably small size, deviating much from that of all the primary planets.†

It was not my intention to rest satisfied with an estimation of the diameter of this curious object, obtained by comparing it with the GEORGIAN planet, and, having now been very successful in the application of the lucid disk micrometer, I shall relate the result of my investigations.

But the very interesting discovery of Dr. OLBERS having introduced another moving star to our knowledge, I have extended my researches to the magnitude, and physical construction, of that also. Its very particular nature, which, from the observations I shall relate, appears to be rather cometary than planetary, will possibly throw also considerable light upon the circumstances belonging to the other celestial body; and, by that means, enable us to form some judgment of the nature of both the two last-discovered phenomena.

As the measures I have taken will oblige me to give a result which must appear extraordinary, it will be highly necessary to be particular in the circumstances of these measures, and to mention the condition and powers of the telescopes that were used to obtain them.

Magnitude of the new Stars.

April 1, 1802. Having placed a lucid disk at a considerable distance from the eye, but so that I might view it with perfect distinctness, I threw the image of Mr. PIAZZI's star, seen in a 7-feet reflector, very near it, in order to have the projected picture of the star and the lucid disk side by side, that I might ascertain their comparative magnitudes. I soon perceived that the length of my garden would not allow me to remove the disk-micrometer, which must be placed at right angles to the telescope, far enough to make it appear no larger than the star; and, not having disks of a less diameter prepared, I placed the smallest I had, as far from me

* [Read February 18, 1802. See Vol. I., unpublished papers.]

† By comparing its apparent disk with that of the GEORGIAN planet, it was estimated, that the real diameter of this new star could not amount to $\frac{3}{8}$ths of that of our moon.

as the situation of the star would allow. Then, bringing its image again by the side of the disk, and viewing, at the same time, with one eye the magnified star, while the other eye saw the lucid disk, I perceived that Ceres, which is the name the discoverer has given to the star, was hardly more than one third of the diameter of the disk, and certainly less than one half of it.

This being repeated, and always appearing the same, we shall not under-rate the size of the star, by admitting its diameter to have been 45 hundredths of the lucid disk.

The power of the telescope, very precisely ascertained, by terrestrial geometrical measures properly reduced to the focus of the mirror on the stars, was 370·42. The distance of the lucid disk from the eye, was 2131 inches; and its diameter 3·4 inches. Hence we compute, that the disk was seen under an angle of 5′ 29″·09; and Ceres, when magnified 370 times, appearing, as we have shewn, 45 hundredths of that magnitude, its real diameter could not exceed 0″·40. Had this diameter amounted to as much as was formerly estimated, the power of 370 would have made it appear of 6′ 10″, which is more than the whole lucid disk.

This extraordinary result raised in me a suspicion, that the power 370 of a 7-feet telescope, and its aperture of 6·3 inches, might not be sufficient to shew the planet's feeble light properly. I therefore adapted my 10-feet instrument to observations with lucid disks; which require a different arrangement of the head of the telescope and finder: I also made some small transparencies, to represent the object I intended to measure.

April 21. The night being pretty clear, though perhaps not quite so proper for delicate vision as I could have wished, I directed my 10-feet reflector, with a magnifying power of 516·54, also ascertained by geometrical terrestrial measures reduced to the focus of the instrument on celestial objects, to Mr. PIAZZI's star, and compared it with a lucid disk, placed at 1486 inches from the eye, and of 1·4 inch in diameter. I varied the distance of the lucid disk many times; and fixed at last on the above-mentioned one, as the best I could find. There was, however, a haziness about the star, which resembled a faint coma; and this, it may be supposed, must render the measure less satisfactory than it would otherwise have been.

From these data we compute, that the disk appeared to the natural eye under an angle of 3′ 14″·33; while Ceres, when magnified 516½ times, was seen by the other eye of an equal magnitude; and that consequently its real diameter, by measurement, was only 0″·38.

April 22. 11^{h} 38′, sidereal time. I used now a more perfect small mirror; the former one having been injured by long continued solar observations. This gave me the apparent diameters of the stars uncommonly well defined; to which, perhaps, the very favourable and undisturbed clearness of the atmosphere might contribute considerably.

With a magnifying power of 881·51, properly ascertained, like those which have been mentioned before, I viewed Dr. OLBERS's star, and compared it with a lucid disk of 1·4 inch in diameter, placed at 1514 inches from the eye, measured, like the rest of the distances, with long deal rods. The star appeared to me so ill defined, that, ascribing it to the eye-glass, I thought it not adviseable to compare the object, as it then appeared, with a well defined lucid disk. Exchanging the glass for that which gives the telescope a magnifying power of 516½, I found Pallas, as the discoverer wishes to have it called, better defined; and saw, when brought together, that it was considerably less in diameter than the lucid disk.

In order to produce an equality, I removed the disk to 1942 inches; and still found Pallas considerably less than the disk.

Before I changed the distance again, I wished to ascertain whether Ceres or Pallas would appear under the largest angle, especially as the air was now more pure than last night. On comparing the diameter of Ceres with that of the lucid disk, I found it certainly less than the disk. By proper attention, and continued examination, for at least an hour, I judged it to be nearly ¾ of the lucid disk.

Then, if we calculate as before, it appears by this observation, in which there is great reason to place confidence, that the angle under which this star appeared, was only 0″·22. For, a lucid disk of 1·4 inch diameter, at the distance of 1942 inches, would be seen under an angle of 2′ 28″·7; three quarters of which are 1′ 51″·52. This quantity, divided by the power 516·54, gives 0″·2159, or, as we have given it abridged, 0″·22.

13^h 7′. I removed the micrometer to the greatest convenient distance, namely, 2136 inches, and compared Dr. OLBERS's star, which, on account of its great altitude, I saw now in high perfection, with the lucid disk. It was, even at this distance, less than the diameter of the disk, in the proportion of 2 to 3.

When, by long continued attention, the appearance of Pallas was reduced to its smallest size, I judged it to bear no greater proportion to the diameter of the lucid disk of the micrometer, than as 1 to 2.

In consequence of these measures, it appears that the diameter of Pallas, according to the first of them, is 0″·17; and, according to the last, where the greatest possible distinctness was obtained, only 0″·13.

If it should appear almost incredible that these curious objects could give so small an image, had they been so much magnified as has been reported, I can say, that curiosity led me to throw the picture of Jupiter, given by the same telescope and magnifying power, on a wall at the distance of 1318 inches, of which it covered a space that measured 12 feet 11 inches. I do not mention this as a measure of Jupiter, for the wall was not perfectly at right angles to the telescope, on which account the projected image would be a little larger than it should have been, nor was I very attentive to other necessary minute circumstances, which would be required for an accurate measure; but we see at once, from the size of this picture, that the power of the telescope exerted itself to the full of what has been stated.

As we generally can judge best of comparative magnitudes, when the measures are, as it were, brought home to us; it will not be amiss to reduce them to miles. This, however, cannot be done with great precision, till we are more perfectly acquainted with the elements of the orbits of these stars. But, for our present purpose, it will be sufficiently accurate, if we admit their mean distances from the sun, as the most recent information at present states them; for Ceres 2·6024; and for Pallas 2·8. The geocentric longitudes and north latitudes, at the time of observation, were, for Ceres, about ♍ 20° 4′, 15° 20′; and for Pallas, ♍ 23° 40′, 17° 30′. With these data, I have calculated the distances of the stars from the earth at the time of observation, partly by the usual method, and, where the elements were wanting, by a graphical process, which is sufficiently accurate for our purpose. My computed distances were 1·634 for Ceres, and 1·8333 for Pallas; and, by them we find, that the diameter of Ceres, at the mean distance of the earth from the sun, would subtend an angle of 0″·35127; and that, consequently, its real diameter is 161·6 miles.

It also follows, that Pallas would be seen, at the same distance from the sun, under an angle of 0″·3199; and that its real diameter, if the largest measure be taken, is 147 miles; but, if we take the most distinct observation, which gives the smallest measure, the angle under which it would be seen from the sun, will be only 0″·2399; and its diameter, no more than $110\frac{1}{3}$ miles.

Of Satellites.

After what has just now been shewn, with regard to the size of these new stars, there can be no great reason to expect that they should have any satellites. The little quantity of matter they contain, would hardly be adequate to the retention of a secondary body; but, as I have made many observations with a view to ascertain this point, it will not be amiss to relate them.

Feb. 25. 20-feet reflector. There is no small star near Ceres, that could be supposed to be a satellite.

Feb. 28. There is no small star within 3 or 4 minutes of Ceres, that might be taken for a satellite.

March 4. 9^h 45′, sidereal time. A very small star, south-preceding Ceres, may be a satellite. See Fig. 1, where C is Ceres, S the supposed satellite, *a b c d e f*, are delineation stars, *c* and *d* are very small. S makes nearly a right angle with them; *e* is larger than either *c* or *d*. There is an extremely faint star *f*, between *e* and *d*.

14^h 16′. Ceres has left the supposed satellite behind.

March 5. There are two very small stars, which may be satellites; see Fig. 2, where they are marked, 1st S, 2d S. The rest, as before, are delineation stars.

March 6. The two supposed satellites of last night remain in their situation, Ceres having left them far behind.

$10^h\ 16'$. There is a very small star, like a satellite, about 75° south-following Ceres. See Fig. 3. It is in a line from C to *b* of last night.

$11^h\ 20'$. Ceres has advanced in its orbit; but has left the supposed satellite behind.

March 30. $9^h\ 35'$. A supposed 1st satellite is directly following Ceres: it is extremely faint. A 2d supposed satellite is north-following. See Fig. 4. The supposed satellites are so small, that, with a 20-feet telescope, they require a power of 300 to be seen; and the planet should be hidden behind a thick wire, placed a little out of the middle of the field of view, which must be left open to look for the supposed satellites.

$12^h\ 17'$. Ceres has changed its place, and left both the supposed satellites behind.

March 31. 9^h 20′. There is a very small star, on the north-preceding side of Ceres, which may be a satellite.

11^h 50′. Ceres has moved forwards in its path; but the supposed satellite remains in its former situation. The nearest star is 20″ of time from Ceres; so that, within a circle of 40″ of time, there certainly is no satellite that can be seen with the space-penetrating power of this instrument.

It is evident, that when the motion of a celestial body is so considerable, we need never be long in doubt whether a small star be a satellite belonging to it, since a few hours must decide it.

May 1. 12^h 51′. I viewed Pallas with the 20-feet reflector, power 300; there was no star within 3′, that could be taken for a satellite.

Of the Colour of the new Stars.

Feb. 13. The colour of Ceres is ruddy, but not very deep.

April 21. Ceres is much more ruddy than Pallas.

April 22. Pallas is of a dusky whitish colour.

Of the Appearances of the new Stars, with regard to a Disk.

Feb. 7. Ceres, with a magnifying power of 516½, shews an ill defined planetary disk, hardly to be distinguished from the surrounding haziness.

Feb. 13. Ceres has a visible disk.

April 22. In viewing Pallas, I cannot, with the utmost attention, and under the most favourable present circumstances, perceive any sharp termination which might denote a disk; it is rather what I would call a nucleus.

April 28. In the finder, Pallas is less than Ceres. It is also rather less than when I first saw it.

Of the Appearances of the new Stars, with regard to an Atmosphere or Coma.

April 21. I viewed Ceres for nearly an hour together. There was a haziness about it, resembling a faint coma, which was, however, easily to be distinguished from the body.

April 22. I see the disk of Ceres better defined, and smaller, than I did last night. There does not seem to be any coma; and I am inclined to ascribe the appearance of last night to a deception, as I now and then, with long attention, saw it without; at which times, it was always best defined, and smallest.

April 28. Ceres is surrounded with a strong haziness. Power 550.

With 516½, which is a better glass, the breadth of the coma beyond the disk may amount to the extent of a diameter of the disk, which is not very sharply defined. Were the whole coma and star taken together, they would be at least three times as large as my measure of the star. The coma is very dense near the nucleus; but loses itself pretty abruptly on the outside, though a gradual diminution is still very perceptible.

April 30. Ceres has a visible, but very small coma about it. This cannot be seen with low powers; as the whole of it together is not large enough, unless much magnified, to make up a visible quantity.

May 1. The diameter of the coma of Ceres is about 5 times as large as the disk, or extends nearly 2 diameters beyond it.

13^h 19′. 20-feet reflector; power 477. The disk of Ceres is much better defined than that of Pallas. The coma about it is considerable, but not quite so extended as that of Pallas.

May 2. 13^h 20′. Ceres is better defined than I have generally seen it. Its disk is strongly marked; and, when I see it best, the haziness about it hardly exceeds that of the stars of an equal size.

Memorandum. This may be owing to a particular disposition of the atmosphere, which shews all the stars without twinkling, but not quite so bright as they appear at other times. Jupiter likewise has an extremely faint scattered light about it, which extends to nearly 4 or 5 degrees in diameter.

April 22. Pallas, with a power of 881½, appears to be very ill defined. The glass is not in fault; for, in the day time, I can read with it the smallest letters on a message card, fixed up at a great distance.

13^h 17′. The appearance of Pallas is cometary; the disk, if it has any, being ill defined. When I see it to the best advantage, it appears like a much compressed, extremely small, but ill defined, planetary nebula.

April 28. Pallas is very ill defined: no determined disk can be seen. The coma about it, or rather the coma itself, for no star appears within it, would certainly measure, at first sight, 4 or 5 times as much as it will do after it has been properly kept in view, in order to distinguish between the haziness which surrounds it, and that part which may be called the body.

May 1. Pallas has a very ill defined appearance; but the whole coma is compressed into a very small compass.

13^h 5′. 20-feet reflector; power 477. I see Pallas well, and perceive a very small disk, with a coma of some extent about it, the whole diameter of which may amount to 6 or 7 times that of the disk alone.

May 2. 13^h 0′. 10-feet reflector. A star of exactly the same size, in the finder with Pallas, viewed with 516½, has a different appearance. In the centre of it is a round lucid point, which is not visible in Pallas. The evening is uncommonly calm and beautiful. I see Pallas better defined than I have seen it before. The coma is contracted into a very narrow compass; so that perhaps it is little more than the common aberration of light of every small star. See the memorandum to the observation of Ceres, May 2.

On the Nature of the new Stars.

From the account which we have now before us, a very important question will arise, which is, What are these new stars, are they planets, or are they comets? And, before we can enter into a proper examination of the subject, it will be necessary to lay down some definition of the meaning we have hitherto affixed to the term planet. This cannot be difficult, since we have seven patterns to adjust our definition by. I should, for instance, say of planets,

1. They are celestial bodies, of a certain very considerable size.

2. They move in not very excentric ellipses round the sun.

3. The planes of their orbits do not deviate many degrees from the plane of the earth's orbit.

4. Their motion is direct.

5. They may have satellites, or rings.

6. They have an atmosphere of considerable extent, which however bears hardly any sensible proportion to their diameters.

7. Their orbits are at certain considerable distances from each other.

Now, if we may judge of these new stars by our first criterion, which is their size, we certainly cannot class them in the list of planets: for, to conclude from the measures I have taken, Mercury, which is the smallest, if divided, would make up more than 31 thousand such bodies as that of Pallas, in bulk.

In the second article, their motion, they agree perhaps sufficiently well.

The third, which relates to the situation of their orbits, seems again to point out a considerable difference. The geocentric latitude of Pallas, at present, is not less than between 17 and 18 degrees; and that of Ceres between 15 and 16; whereas, that of the planets does not amount to one half of that quantity. If bodies of this kind were to be admitted into the order of planets, we should be obliged to give up the zodiac; for, by extending it to them, should a few more of these stars be discovered, still farther and farther deviating from the path of the earth, which is not unlikely, we might soon be obliged to convert the whole firmament into zodiac; that is to say, we should have none left.

In the fourth article, which points out the direction of the motion, these stars agree with the planets.

With regard to the fifth, concerning satellites, it may not be easy to prove a negative; though even that, as far as it can be done, has been shewn. But the retention of a satellite in its orbit, it is well known, requires a proper mass of matter in the central body, which it is evident these stars do not contain.

The sixth article seems to exclude these stars from the condition of planets. The small comas which they shew, give them so far the resemblance of comets, that in this respect we should be rather inclined to rank them in that order, did other circumstances permit us to assent to this idea.

In the seventh article, they are again unlike planets; for it appears, that their orbits are too near each other to agree with the general harmony that takes place among the rest; perhaps one of them might be brought in, to fill up a seeming vacancy between Mars and Jupiter. There is a certain regularity in the arrangement of planetary orbits, which has been pointed out by a very intelligent astronomer, so long ago as the year 1772; but this, by the admission of the two new stars into the order of planets, would be completely overturned; whereas, if they are of a different species, it may still remain established.

As we have now sufficiently shewn that our new stars cannot be called planets, we proceed to compare them also with the other proposed species of celestial bodies, namely, comets. The criteria by which we have hitherto distinguished these from planets, may be enumerated as follows.

1. They are celestial bodies, generally of a very small size, though how far this may be limited, is yet unknown.

2. They move in very excentric ellipses, or apparently parabolic arches, round the sun.

3. The planes of their motion admit of the greatest variety in their situation.

4. The direction of their motion also is totally undetermined.

5. They have atmospheres of very great extent, which shew themselves in various forms of tails, coma, haziness, &c.

On casting our eye over these distinguishing marks, it appears, that in the first point, relating to size, our new stars agree sufficiently well; for the magnitude of comets is not only small, but very unlimited. Mr. PIGOTT's comet, for instance, of the year 1781, seemed to have some kind of nucleus; though its magnitude was so ill defined, that I probably over-rated it much, when, November 22, I guessed it might amount to 3 or 4″ in diameter. But, even this, considering its nearness to the earth, proves it to have been very small.

That of the year 1783, also discovered by Mr. PIGOTT, I saw to more advantage, in the meridian, with a 20-feet reflector. It had a small nucleus, which, November 29, was coarsely estimated to be of perhaps 3″ diameter. In all my other pretty numerous observations of comets, it is expressly remarked, that they had none that could be seen. Besides, what I have called a nucleus, would still be far from what I now should have measured as a disk; to constitute which, a more determined outline is required.

In the second article, their motions differ much from that of comets; for, so far as we have at present an account of the orbits of these new stars, they move in ellipses which are not very excentric.

Nor are the situations of the planes of their orbits so much unlike those of the planets, that we should think it necessary to bring them under the third article of comets, which leaves them quite unlimited.

In the fourth article, relating to the direction of their motion, these stars agree with planets, rather than with comets.

The fifth article, which refers to the atmosphere of comets, seems to point out these stars as belonging to that class ; it will, however, on a more particular examination, appear that the difference is far too considerable to allow us to call them comets.

The following account of the size of the comas of the smallest comets I have observed, will shew that they are beyond comparison larger than those of our new stars.

Nov. 22, 1781. Mr. PIGOTT'S comet had a coma of 5 or 6′ in diameter.

Nov. 29, 1783. Another of Mr. PIGOTT'S comets had a coma of 8′ in diameter.

Dec. 22, 1788. My sister's comet had a coma of 5 or 6′ in diameter.

Jan. 9, 1790. Another of her comets was surrounded by haziness of 5 or 6′ in diameter.

Jan. 18, 1790. Mr. MECHAIN'S comet had a coma of 5 or 6′ in diameter.

Nov. 7, 1795. My sister's comet had a coma of 5 or 6′ in diameter.

Sept. 8, 1799. Mr. STEPHEN LEE'S comet had a coma of not less than 10′ in diameter, and also a small tail of 15′ in length.

From these observations, which give us the dimensions of the comas of the smallest comets that have been observed with good instruments, we conclude, that the comas of these new stars, which at most amount only to a few times the diameter of the bodies to which they belong bear no resemblance to the comas of comets, which, even when smallest, exceed theirs above a hundred times. Not to mention the extensive atmospheres, and astonishing length of the tails, of some comets that have been observed, to which these new stars have nothing in the least similar.

Since, therefore, neither the appellation of planets, nor that of comets, can with any propriety of language be given to these two stars, we ought to distinguish them by a new name, denoting a species of celestial bodies hitherto unknown to us, but which the interesting discoveries of Mr. PIAZZI and Dr. OLBERS have brought to light.

With this intention, therefore, I have endeavoured to find out a leading feature in the character of these new stars ; and, as planets are distinguished from the fixed stars by their visible change of situation in the zodiac, and comets by their remarkable comas, so the quality in which these objects differ considerably from the two former species is that they resemble small stars so much as hardly to be distinguished from them, even by very good telescopes. It is owing to this very circumstance, that they have been so long concealed from our view. From this, their asteroidical appearance, if I may use that expression, therefore, I shall take my name and call them *Asteroids* ; reserving to myself, however, the liberty of changing that name, if another, more expressive of their nature, should occur. These bodies will hold a middle rank, between the two species that were known before; so that planets, asteroids, and comets, will in future comprehend all the primary celestial bodies that either remain with, or only occasionally visit, our solar system.

I shall now give a definition of our new astronomical term, which ought to be considerably extensive, that it may not only take in the asteroid Ceres, as well as the asteroid Pallas, but that any other asteroid which may hereafter be discovered, let its motion or situation be whatever it may, shall also be fully delineated by it. This will stand as follows.

Asteroids are celestial bodies, which move in orbits either of little or of considerable excentricity round the sun, the plane of which may be inclined to the ecliptic in any angle whatsoever. Their motion may be direct, or retrograde; and they may or may not have considerable atmospheres, very small comas, disks or nuclei.

As I have given a definition which is sufficiently extensive to take in future discoveries, it may be proper to state the reasons we have for expecting that additional asteroids may probably be soon found out. From the appearance of Ceres and Pallas it is evident, that the discovery of asteroids requires a particular method of examining the heavens, which hitherto astronomers have not been in the habit of using. I have already made five reviews of the zodiac, without detecting any of these concealed objects. Had they been less resembling the small stars of the heavens, I must have discovered them. But the method which will now be put in practice, will completely obviate all difficulty arising from the asteroidical appearance of these objects; as their motion, and not their appearance, will in future be the mark to which the attention of observers will be directed.

A laudable zeal has induced a set of gentlemen on the Continent, to form an association for the examination of the zodiac. I hope they will extend their attention, by degrees, to every part of the heavens; and that the honourable distinction which is justly due to the successful investigators of nature, will induce many to join in the meritorious pursuit. As the new method of observing the zodiac has already produced such interesting discoveries, we have reason to believe that a number of asteroids may remain concealed; for, how improbable it would be, that if there were but two, they should have been so near together as almost to force themselves to our notice. But a more extended consideration adds to the probability that many of them may soon be discovered. It is well known that the comas and tails of comets gradually increase in their approach to the sun, and contract again when they retire into the distant regions of space. Hence we have reason to expect, that when comets have been a considerable time in retirement, their comas may subside, if not intirely, at least sufficiently to make them assume the resemblance of stars; that is, to become asteroids, in which state we have a good chance to detect them. It is true that comets soon grow so faint, in retiring from their perihelia, that we lose sight of them; but, if their comas, which are generally of great extent, should be compressed into a space so small as the diameters of our two asteroids, we can hardly entertain a doubt but that they would again become visible with good telescopes. Now, should we see a comet in its aphelion, under the conditions here pointed out, and that there are many which may be in such situations, we have the greatest inducements to believe, it would be a favourable circumstance to lead us to a more perfect knowledge of the nature of comets and their orbits; for instance, the comet of the year 1770, which Mr. LEXELL has shewn to have moved in an elliptical orbit, such as would make the time of its periodical return only about $5\frac{1}{2}$ years: if this should still remain in our system, which is however doubtful, we ought to look for it under the form of an asteroid.

If these considerations should be admitted, it might be objected, that asteroids were only comets in disguise; but, if we were to allow that comets, asteroids, and even planets, might possibly be the same sort of celestial bodies under different circumstances, the necessary distinction arising from such difference, would fully authorise us to call them by different names.

It is to be hoped that time will soon throw a greater light upon this subject; for which reason, it would be premature to add any other remarks, though many extensive views relating to the solar system might certainly be hinted at.

Additional Observations relating to the Appearances of the Asteroids Ceres and Pallas.

May 4, 12^h 40′. 10-feet reflector; power $516\frac{1}{2}$. I compared Ceres with two fixed stars, which, in the finder, appeared to be of very nearly the same magnitude with the asteroid, and found that its coma exceeds their aberration but in a very small degree.

12^h 50′. 20-feet reflector; power 477. I viewed Ceres, in order to compare its appearance with regard to haziness, aberration, atmosphere, or coma, whatever we may call it, to the same phenomena of the fixed stars; and found that the coma of the asteroid did not much exceed that of the stars.

I also found, that even the fixed stars differ considerably in this respect among themselves. The smaller they are, the larger in proportion will the attendant haziness shew itself. A star that is scarcely perceptible, becomes a small nebulosity.

10-feet reflector. 13^h 10′. I compared the appearance of Pallas with two equal fixed stars; and found that the coma of this asteroid but very little exceeds the aberration of the stars.

14^h 5′. 20-feet reflector. I viewed Pallas; and, with a magnifying power of 477, its disk was visible. The coma of this asteroid is a little stronger than that which fixed stars of the same size generally have.

XLIX.

Catalogue of 500 new Nebulæ, nebulous Stars, planetary Nebulæ, and Clusters of Stars; with Remarks on the Construction of the Heavens.

[*Phil. Trans.*, 1802, pp. 477–528.]

Read July 1, 1802.

SINCE the publication of my former two catalogues of nebulæ, I have, in the continuation of my telescopic sweeps, met with a number of objects that will enrich our natural history, as it may be called, of the heavens. A catalogue of them will be found at the end of this paper, containing 500 new nebulæ, nebulous stars, planetary nebulæ, and clusters of stars. These objects have been arranged in eight classes, in conformity with the former catalogues, of which the present one is therefore a regular continuation. This renders it unnecessary to give any further explanation, either of the contents of its columns, or the abbreviations which have been used in the description of the objects.

It has hitherto been the chief employment of the physical astronomer, to search for new celestial objects, whatsoever might be their nature or condition; but our stock of materials is now so increased, that we should begin to arrange them more scientifically. The classification adopted in my catalogues, is little more than an arrangement of the objects for the convenience of the observer, and may be compared to the disposition of the books in a library, where the different sizes of the volumes is often more considered than their contents. But here, in dividing the different parts of which the sidereal heavens are composed into proper classes, I shall have to examine the nature of the various celestial objects that have been hitherto discovered, in order to arrange them in a manner most conformable to their construction. This will bring on some extensive considerations, which would be too long for the compass of a single paper; I shall therefore now only give an enumeration of the species that offer themselves already to our view, and leave a particular examination of the separate divisions, for some early future occasions.

In proceeding from the most simple to the more complex arrangements, several methods, taken from the known laws of gravitation, will be suggested, by which the various systems under consideration may be maintained; but here also we shall confine ourselves to a general review of the subject, as observation must furnish us first with the necessary data, to establish the application of any one of these methods on a proper foundation.

ENUMERATION OF THE PARTS THAT ENTER INTO THE CONSTRUCTION OF THE HEAVENS.

I. *Of insulated Stars.*

In beginning our proposed enumeration, it might be expected that the solar system would stand foremost in the list; whereas, by treating of insulated stars, we seem, as it were, to overlook one of the great component parts of the universe. It will, however, soon appear that this very system, magnificent as it is, can only rank as a single individual belonging to the species which we are going to consider.

By calling a star insulated, I do not mean to denote its being totally unconnected with all other stars or systems; for no one, by the laws of gravitation, can be intirely free from the influence of other celestial bodies. But, when stars are situated at such immense distances from each other as our sun, Arcturus, Capella, Lyra, Sirius Canopus, Markab, Bellatrix, Menkar, Shedir, Algorah, Propus, and numberless others probably are, we may then look upon them as sufficiently out of the reach of mutual attractions, to deserve the name of insulated stars.

In order not to take this assertion for granted, without some examination, let us admit, as is highly probable, that the whole orbit of the earth's annual motion does not subtend more than an angle of one second of a degree, when seen from Sirius. In consequence of this, it appears by computation, that our sun and Sirius, if we suppose their masses to be equal, would not fall together in less than 33 millions of years, even though they were not impeded by many contrary attractions of other neighbouring insulated stars; and that, consequently with the assistance of the opposite energies exerted by such surrounding stars, these two bodies may remain for millions of ages, in a state almost equal to undisturbed rest. A star thus situated may certainly deserve to be called insulated, since it does not immediately enter into connection with any neighbouring star; and it is therefore highly probable, that our sun is one of a great number that are in similar circumstances. To this may be added, that the stars we consider as insulated are also surrounded by a magnificent collection of innumerable stars, called the milky-way, which must occasion a very powerful balance of opposite attractions, to hold the intermediate stars in a state of rest. For, though our sun, and all the stars we see, may truly be said to be in the plane of the milky-way, yet I am now convinced, by a long inspection and continued examination of it, that the milky-way itself consists of stars very differently scattered from those which are immediately about us. But of this, more will be said on another occasion.

From the detached situation of insulated stars, it appears that they are capable of being the centres of extensive planetary systems. Of this we have a convincing proof in our sun, which, according to our classification, is one of these stars. Now, as we enjoy the advantage of being able to view the solar system in all its parts, by means of our telescopes, and are therefore sufficiently acquainted with it, there will be no occasion to enter into a detail of its construction.

The question will now arise, whether every insulated star be a sun like ours, attended with planets, satellites, and numerous comets? And here, as nothing appears against the supposition, we may from analogy admit the probability of it. But, were we to extend this argument to other sidereal constructions, or, still farther, to every star of the heavens, as has been done frequently, I should not only hesitate, but even think that, from what will be said of stars which enter into complicated sidereal systems, the contrary is far more likely to be the case; and that, probably, we can only look for solar systems among insulated stars.

II. *Of Binary sidereal Systems, or double Stars.*

The next part in the construction of the heavens, that offers itself to our consideration, is the union of two stars, that are formed together into one system, by the laws of attraction.

If a certain star should be situated at any, perhaps immense, distance behind another, and but very little deviating from the line in which we see the first, we should then have the appearance of a double star. But these stars, being totally unconnected, would not form a binary system. If, on the contrary, two stars should really be situated very near each other, and at the same time so far insulated as not to be materially affected by the attractions of neighbouring stars, they will then compose a separate system, and remain united by the bond of their own mutual gravitation towards each other. This should be called a real double star; and any two stars that are thus mutually connected, form the binary sidereal system which we are now to consider.

It is easy to prove, from the doctrine of gravitation, that two stars may be so connected together as to perform circles, or similar ellipses, round their common centre of gravity. In this case, they will always move in directions opposite and parallel to each other; and their system, if not destroyed by some foreign cause, will remain permanent.

Figure 1 (p. 205) represents two equal stars *a* and *b*, moving in one common circular orbit round the centre *o*, but in the opposite directions of *a t* and *b t*. In Fig. 2 we have a similar connection of the two stars *a b*; but, as they are of different magnitudes, or contain unequal quantities of matter, they will move in circular orbits of different dimensions round their common centre of gravity *o*. Fig. 3 represents equal, and Fig. 4 unequal stars, moving in similar elliptical orbits round a common centre; and, in all these cases, the directions of the tangents *t t*, in the places *a b*, where the stars are, will be opposite and parallel, as will be more fully explained hereafter.

These four orbits, simple as they are, open an extensive field for reflection, and, I may add, for calculation. They shew, even before we come to more complicated combinations, where the same will be confirmed, that there is an essential difference between the construction of solar and sidereal systems. In each solar system, we have a very ponderous attractive centre, by which all the planets, satellites, and comets are governed, and kept in their orbits. Sidereal systems take a greater scope: the stars of which they are composed move round an empty centre, to which they are nevertheless as firmly bound as the planets to their massy one. It is however not necessary here to enlarge on distinctions which will hereafter be strongly supported by facts, when clusters of stars come to be considered. I shall only add, that in the subordinate bodies of the solar system itself, we have already instances, in miniature, as it may be called, of the principle whereby the laws of attraction are applicable to the solution of the most complicated phenomena of the heavens, by means of revolutions round empty centres. For, although both the earth and its moon are retained in their orbits by the sun, yet their mutual subordinate system is such, that they perform secondary monthly revolutions round a centre without a body placed in it. The same indeed, though under very narrow limits, may be said of the sun and each planet itself.

That no insulated stars, of nearly an equal size and distance, can appear double to us, may be proved thus. Let Arcturus and Lyra be the stars: these, by the rule of insulation, which we must now suppose can only take place when their distance from each other is not less than that of Sirius from us, if very accurately placed, would be seen under an angle of 60 degrees from each other. They really are at about 59°. Now, in order to make these stars appear to us near enough to come under the denomination of a double star of the first class, we should remove the earth from them at least 41253 times farther than Sirius is from us. But the space-penetrating power of a 7-feet reflector, by which my observations on double stars have been made, cannot intitle us to see stars at such an immense distance; for, even the 40-feet telescope, as has been shewn,* can only reach stars of the 1342d magnitude. It follows, therefore, that these stars could not remain visible in a 7-feet reflector, if they were so far removed as to make their angular distance less than about 24¼ minutes; nor could even the 40-feet telescope, under the same circumstances of removal, shew them, unless they were to be seen at least 2½ minutes asunder. Moreover, this calculation is made on a supposition that the stars of which a double star is composed, might be as small as any that can possibly be perceived; but if, on the contrary, they should still appear of a considerable size, it will then be so much the more evident that such stars cannot have any great real distance, and that, consequently, insulated stars cannot appear double, if they are

* See *Phil. Trans.* for 1800, Part I. page 83 [above, p. 50].

situated at equal distances from us. If, however, their arrangement should be such as has been mentioned before, then, one of them being far behind the other, an apparent double star may certainly be produced; but here the appearance of proximity would be deceptive; and the object so circumstanced could not be classed in the list of binary systems. However, as we must grant, that in particular situations stars apparently double may be composed of such as are insulated, it cannot be improper to consult calculation, in order to see whether it be likely that the 700 double stars I have given in two catalogues, as well as many more I have since collected, should be of that kind. Such an inquiry, though not very material to our present purpose, will hereafter be of use to us, when we come to consider more complicated systems. For, if it can be shown that the odds are very much against the casual production of double stars, the same argument will be still more forcible, when applied to treble, quadruple, or multiple compositions.

Let us take ζ Aquarii, for an instance of computation. This star is admitted, by FLAMSTEED, DE LA CAILLE, BRADLEY, and MAYER, to be of the 4th magnitude. The two stars that compose it being equal in brightness, each of them may be supposed to shine with half the light of the whole lustre. This, according to our way of reckoning magnitudes,* would make them $4m \times \sqrt{2} = 5\frac{2}{3}m$; that is, of between the 6th and 5th magnitude each. Now, the light we receive from a star being as the square of its diameter directly, and as the square of its distance inversely, if one of the stars of ζ Aquarii be farther off than the stars of between the 6th and 5th magnitude are from us, it must be so much larger in diameter, in order to give us an equal quantity of light. Let it be at the distance of the stars of the 7th magnitude; then its diameter will be to the diameter of the star which is nearest to us as 7 to $5\frac{2}{3}$, and its bulk as 1·885 to 1; which is almost double that of the nearest star. Then, putting the number of stars we call of between the 6th and 5th magnitude at 450, we shall have 686 of the 7th magnitude to combine with them, so that they may make up a double star of the first class, that is to say, that the two stars may not be more than 5″ asunder. The surface of the globe contains 34036131547 circular spaces, each of 5″ in diameter; so that each of the 686 stars will have 49615357 of these circles in which it might be placed; but, of all that number, a single one would only be the proper situation in which it could make up a double star with one of the 450 given stars. But these odds, which are above $75\frac{1}{2}$ millions to one against the composition of ζ Aquarii, are extremely increased by our foregoing calculation of the required size of the star, which must contain nearly double the mass allotted to other stars of the 7th magnitude; of which, therefore, none but this one can be proper for making up the required double star. If the stars of the 8th and 9th magnitudes, of which there will be 896 and 1134, should be taken in, by way of increasing the chance in favour of the supposed composition of our

* The expressions 2m, 3m, 4m, &c. stand for stars at the distance of 2, 3, 4, &c. times that of Sirius, supposed unity.

double star, the advantage intended to be obtained by the addition of numbers, will be completely counteracted by the requisite uncommon bulk of the star which is to serve the purpose; for, one of the 8th magnitude ought to be more than $2\frac{3}{4}$ times bigger than the rest; and, if the composition were made by a star of the 9th magnitude, no less than four times the bulk of the other star which is to enter the composition of the double star would answer the purpose of its required brightness. Hence therefore it is evident, that casual situations will not account for the multiplied phenomena of double stars, and that consequently their existence must be owing to the influence of some general law of nature; now, as the mutual gravitation of bodies towards each other is quite sufficient to account for the union of two stars, we are authorised to ascribe such combinations to that principle.

It will not be necessary to insist any further on arguments drawn from calculation, as I shall soon communicate a series of observations made on double stars, whereby it will be seen, that *many of them have actually changed their situation with regard to each other, in a progressive course, denoting a periodical revolution round each other; and that the motion of some of them is direct, while that of others is retrograde.* Should these observations be found sufficiently conclusive, we may already have their periodical times near enough to calculate, within a certain degree of approximation, the parallax and mutual distance of the stars which compose these systems, by measuring their orbits, which subtend a visible angle.

Before we leave the subject of binary systems, I should remark, that it evidently appears, that our sun does not enter into a combination with any other star, so as to form one of these systems with it. This could not take place without our immediately perceiving it; and, though we may have good reason to believe that our system is not perfectly at rest, yet the causes of its proper motion are more probably to be ascribed to some perturbations arising from the proper motion of neighbouring stars or systems, than to be placed to the account of a periodical revolution round some imaginary distant centre.

III. *Of more complicated sidereal Systems, or treble, quadruple, quintuple, and multiple Stars.*

Those who have admitted our arguments for the existence of real double stars, will easily advance a step farther, and allow that three stars may be connected in one mutual system of reciprocal attraction. And, as we have from theory pointed out, in figures 1, 2, 3, and 4, how two stars may be maintained in a binary system, we shall here shew that three stars may likewise be preserved in a permanent connection, by revolving in proper orbits about a common centre of motion.

In all cases where stars are supposed to move round an empty centre, in equal periodical times, it may be proved that an imaginary attractive force may be supposed to be lodged in that centre, which increases in a direct ratio of the distances. For since, in different circles, by the law of centripetal forces, the squares of the periodical times are as the radii divided by the central attractive forces, it follows, that when these periodical times are equal, the forces will be as the radii. Hence we conclude, that in any system of bodies, where the attractive forces of all the rest upon any one of them, when reduced to a direction as coming from the empty centre, can be shewn to be in a direct ratio of the distance of that body from the centre, the system may revolve together without perturbation, and remain permanently connected without a central body.

Hence may be proved, as has been mentioned before, that two stars will move round a hypothetical centre of attraction. For, let it be supposed that the empty centre *o*, in Fig. 1 and 3, is possessed of an attractive force, increasing in the direct ratio of the distances *oa* : *ob*. Then, since here *ao* and *bo* are equal, the hypothetical

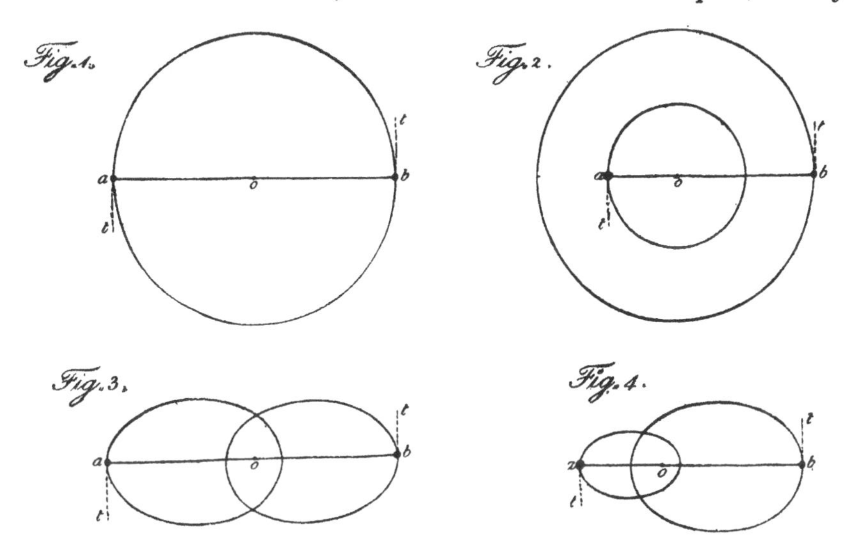

attractions will be equal, and the bodies will revolve in equal times. That this agrees with the general law of attraction, is proved thus. The real attraction of *b* upon *a* is $\frac{b}{ab^2}$; and that of *a* upon *b* is $\frac{a}{ab^2}$; and, since $b = a$, it will be $\frac{b}{ab^2} : \frac{a}{ab^2} :: ao : bo$; which was required.

In Figures 2 and 4, when the stars *a* and *b* are unequal, and their distances from *o* also unequal, let $oa = n$, and $ob = m$; and let the mass of matter in $a = m$, and in $b = n$. Then the attraction of *b* on $a = \frac{b}{ab^2}$, will be to the attraction of *a* on $b = \frac{a}{ab^2}$, as $n : m$; which is again directly as *ao* : *bo*.

I proceed now to explain a combination of three bodies, moving round a centre of hypothetical attraction. Fig. 5 contains a single orbit, wherein three equal bodies *a b c*, placed at equal distances, may revolve permanently. For, the real attraction of *b* on *a* will be expressed by $\frac{a}{ab^2}$; but this, reduced to the direction *ao*. will be only $\frac{b \cdot by}{ab^3}$; for, the attraction in the direction *ba* is to that in the direction *by*, parallel to *ao*, as $\frac{b}{ab^2}$ to $\frac{b \cdot by}{ab^3}$. The attraction also of *c* on *a* is equal to that of *b* on *a*; therefore the whole attraction on *a*, in a direction towards *o*, will be expressed by $\frac{2b \cdot by}{ab^3}$ In the same manner we prove, that the attraction of *a* and *c* on *b*, in the direction *bo*, is $\frac{2a \cdot by}{ab^3}$; and that of *a* and *b* on *c*, in the direction *co*, is $\frac{2c \cdot by}{ab^3}$. Hence, *a b* and *c* being equal, the attractions in the directions *ao*, *bo* and *co* will also be equal; and, consequently, in the direct ratio of these distances. Or rather, the hypothetical

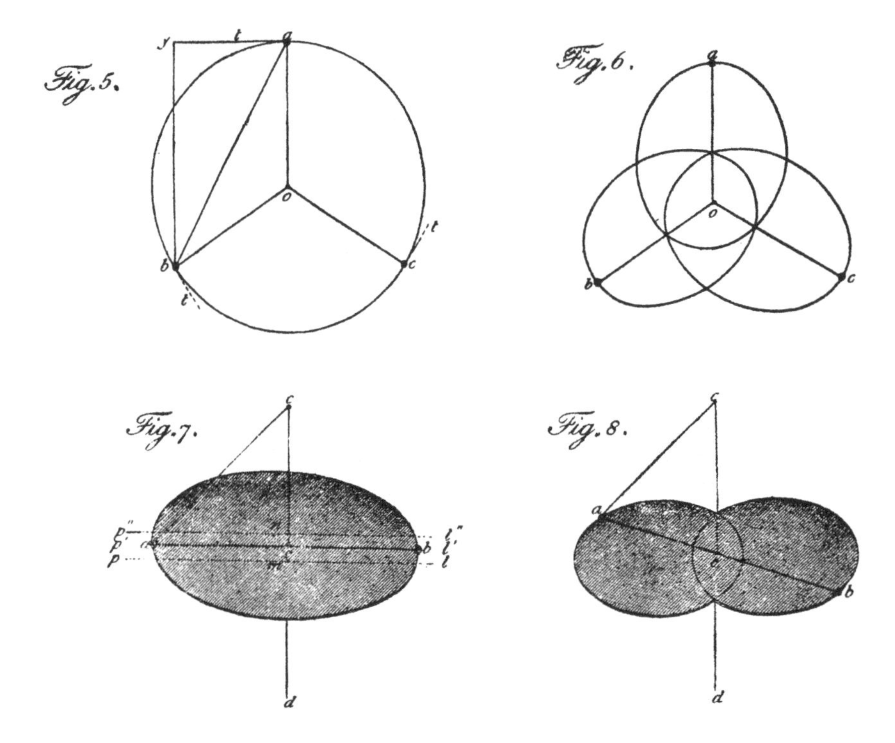

attractions being equal, it proves that, in order to revolve permanently, *a b* and *c* must be equal to each other.

Instead of moving in one circular orbit, the three stars may revolve in three equal ellipses, round their common centre of gravity, as in Fig. 6. And here we should remark, that this centre of gravity will be situated in the common focus *o*, of the three ellipses; and that the absolute attraction towards that focus, will vary in the inverse ratio of the squares of the distances of any one of the stars from that centre, while the relative attractions remain in the direct ratio of their several distances from the same centre. This will be more fully explained, when we come to consider the motion of four stars.

A very singular straight-lined orbit, if so it may be called, may also exist in the following manner. If *a* and *b*, Fig. 7, are two large equal stars, which are connected together by their mutual gravitation towards each other, and have such projectile motions as would cause them to move in a circular orbit about their common centre of gravity, then may a third small star *c*, situated in a line drawn through *o*, and at rectangles to the plane described by the stars *a b*, fall freely from rest, with a gradually acquired motion to *o*; then, passing through the plane of the orbit of the two stars, it will proceed, but with a gradually retarded motion, to a second point of rest *d*; and, in this manner, the star *c* may continue to oscillate between *c* and *d*, in a straight line, passing from *c*, through the centre *o*, to *d*, and back again to *c*.

In order to see the possibility and permanency of this connection the better, let *o* be the centre of gravity of the three bodies, when the oscillating body is at *c*; then, supposing the bodies *a* and *b* to be at that moment in the plane *pl*, and admitting *m* to represent a body equal in mass to the two bodies *a b*, *o* will be the common centre of gravity of *m* and *c*. Then, by the force of attraction, the body *c* and the fictitious body *m* will meet in *o*; that is to say, the plane *pl*, of the bodies *a b*, will now be at $p'l'$. The fictitious body *m* may then be conceived to move on till it comes to *n*, while the body *c* goes to *d*; or, which is the same, the plane of the bodies *a b* will now be in the position $p''l''$, as much beyond the centre of gravity *o*, as it was on the opposite side *m*. By this time, both the fictitious body *m*, now at *n*, and the real body *c*, now at *d*, have lost their motion in opposite directions, and begin to approach to their common centre of gravity *o*, in which they will meet a second time. It is evident that the orbit of the two large stars will suffer considerable perturbations, not only in its plane, but also in its curvature, which will not remain strictly circular; the construction of the system, however, is such as to contain a sufficient compensation for every disturbing force, and will consequently be in its nature permanent.

In order to add an oscillating star, it is not necessary that the two large stars should be so situated as to move in a circular orbit, without the oscillating star. In Fig. 8, the stars *a* and *b* may have such projectile forces given them as would cause them to describe equal ellipses, of any degree of excentricity. If now the small star *c* be added, the perturbations will undoubtedly affect not only the plane of the orbits of the stars, but also their figures, which will become irregular moveable ovals. The extent also of the oscillations of the star *c* will be affected; and will sometimes exceed the limits *c d*, and sometimes fall short of them. All these varieties may easily be deduced from what has been already said, when Fig. 7 was considered. It is however very evident, that this system also must be permanent; since not only the centre of gravity *o* will always be at rest, but *ao*, whatever may be the perturbations arising from the situation of *c*, will still remain equal to *bo*.

It should be remarked, that the vibratory motion of the star *c* will differ much from a cometary orbit, even though the latter should be compressed into an evanescent ellipsis. For, while the former extends itself over the diameter of a globe in which it may be supposed to be inscribed, the hypothetical attractive force being supposed to be placed in its centre, the cometary orbit will only describe a radius of the same globe, on account of its requiring a solid attractive centre.

After what has been said, it will hardly be necessary to add, that with the assistance of any proper one of the combinations pointed out in the four last figures, the appearance of every treble star may be completely explained; especially when the different inclinations of the orbits of the stars, to the line of sight, are taken into consideration.

If we admit of treble stars, we can have no reason to oppose more complicated connections; and, in order to form an idea how the laws of gravitation may easily support such systems, I have joined some additional delineations. A very short explanation of them will be sufficient.

Fig. 9 represents four stars, *a b c* and *d*, arranged in a line; *a* being equal to *b*, and *c* equal to *d*. Then, if $ao = bo$, and $co = do$, the centre of gravity will be in *o*; and, with a proper adjustment of projectile forces, the four stars will revolve in two circular orbits round their common centre. By calculating in the manner already pointed out, it will be found, that when, for instance, $ao = 1$, $co = 3$, and $c = d = 1$, then the mass of matter in $a = b$, will be required to be equal to 1·3492.

It is not necessary that the projectile force of the four stars should be such as will occasion them to revolve in circles. The system will be equally permanent when they describe similar ellipses about the common centre of gravity, which will also be the common focus of the four ellipses. In Fig. 10, the stars *a b c d*, revolving in ellipses that are similar, will always describe, at the same time, equal angles in each ellipsis about the centre of hypothetical attraction; and, when they are removed from *a b c d* to *a′ b′ c′ d′*, they will still be situated in a straight line, and at the same proportionate distances from each other as before. By this it appears, as we have already observed, that the absolute hypothetical force in the situation *a′ b′ c′ d′*, compared to what it was when the stars were at *a b c d*, is inversely as the squares of the distances; but that its comparative exertion on the stars, in their present situation, is still in a direct ratio of their distances from the centre *o*, just as it was when they were at *a b c d*; or, to express the same perhaps more clearly, the force exerted on *a′*, is to that which was exerted on *a* as $\frac{1}{\overline{a'o}|^2} : \frac{1}{\overline{ao}|^2}$. But the force exerted on *a* is to that exerted on *c*, in our present instance, as $ao = 1$ to $co = 3$; and still remains in the same ratio when the stars are at *a′* and *c′*; for the exertion will here be likewise as $a'o = 1$ to $c'o = 3$.

Fig. 11 represents four stars in one circular orbit; and its calculation is so simple, that, after what has been said of Fig. 5, I need only remark that the stars may be of any size, provided their masses of matter are equal to each other.

It is also evident, that the projectile motion of four equal stars is not confined to that particular adjustment which will make them revolve in a circle. It will be sufficient, in order to produce a permanent system, if the stars *a b c d*, in Fig. 12, are impressed with such projectile forces as will make them describe equal ellipses round the common centre *o*. And, as the same method of calculation which has

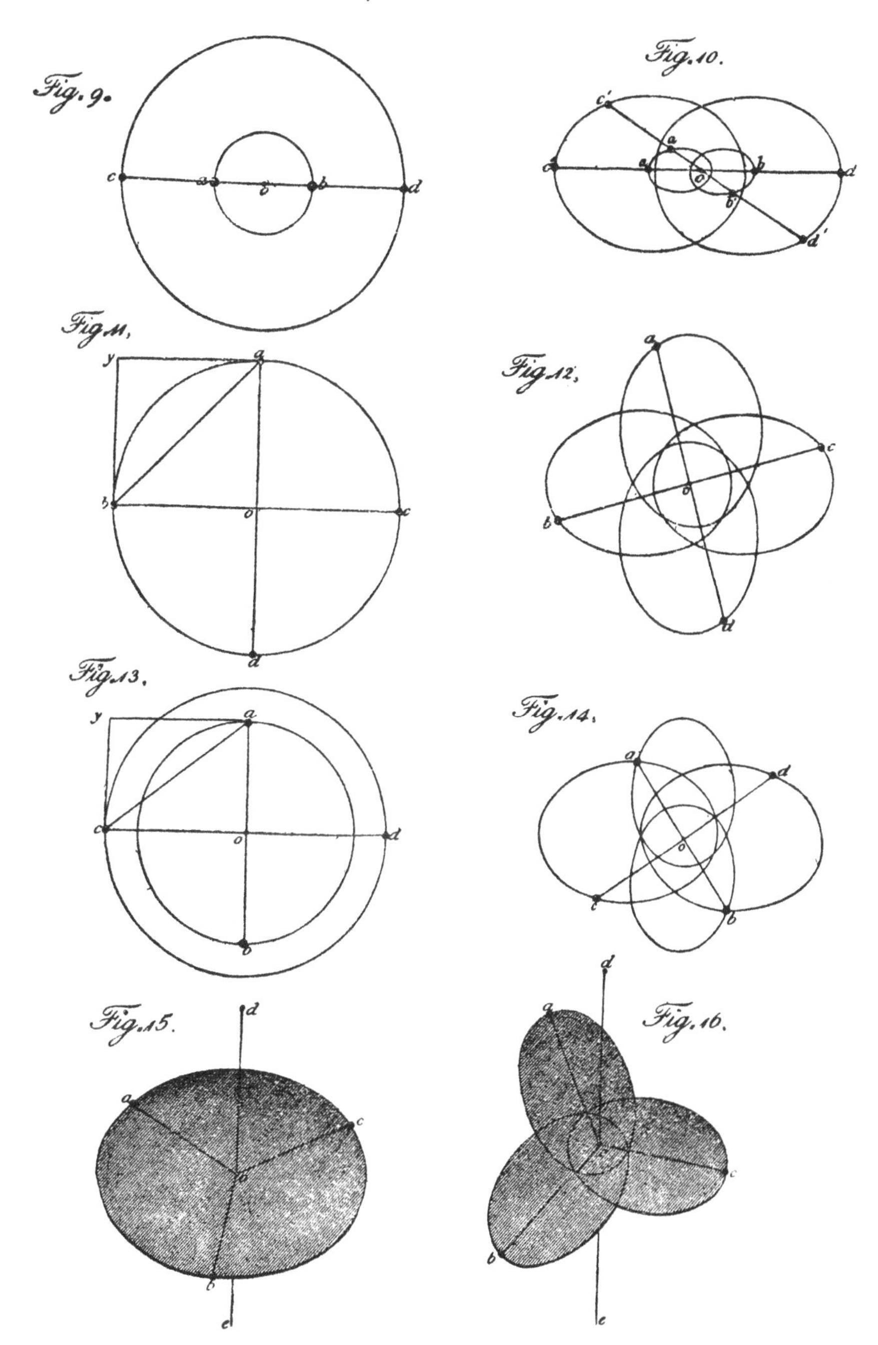

been explained with Figs. 6 and 10 may here be used, it will not be necessary to enter into particulars.

Fig. 13 represents four stars, placed so that, with properly adjusted projectile forces, they may revolve in equal times, and in two different circles, round their common centre of gravity *o*. If $ao = bo = 4$, $co = do = 5$, and $c = d = 1$, then will the mass of matter in $a = b$, required for the purpose, be 1·5136. This arrangement, remarkable as it may appear, cannot be made in all situations; for instance, if the distance $ao = bo$ were assumed equal to 1, that of $co = do$ being 2, it would be impossible to find such quantities of matter in *a* and *b* as would unite the four stars into one system.

As we have shewn how the arrangements in Fig. 10 may be derived from that of Fig. 9, so it will equally appear, that four stars may revolve in different but similar ellipses round their common centre, as in Fig. 14. For here the four stars, when placed at *a b c d*, are exactly in the situation represented in Fig. 13; but, on account of different projectile forces, they revolve, not as before in concentric circles, but in similar elliptical orbits.

Fig. 15 represents three stars, *a b c*, in the situation of Fig. 5, to which a small oscillating star, *d*, is added. The addition of such a star to Fig. 1, has been sufficiently explained in Fig. 7; and, what has been remarked there, may easily be applied to our present figure. As the fictitious body *m*, in Fig. 7, was made to represent the stars *a* and *b*, it will now stand for the three stars *a b* and *c*. If we suppose these stars to be of an equal magnitude in both figures, the centre of gravity *o*, of the three stars, will not be so far from *m* and *n* as in Fig. 7; and the perturbations will be proportionally lessened.

Fig. 16 gives the situation of three stars, *a b c*, moving in equal elliptical orbits about their common focus *o*, while the star *d* performs oscillations between *d* and *e*. What has been said in explaining Fig. 8, will be sufficient to shew, that the present arrangement is equally to be admitted among the constructions of sidereal systems that may be permanent.

We have before remarked, that any appearance of treble stars might be explained, by admitting the combinations pointed out in Figs. 5, 6, 7, and 8; and it must be equally obvious, that quadruple systems, under what shape soever they may shew themselves, whether in straight lines, squares, trapezia, or any other seemingly the most irregular configurations, will readily find a solution from one or other of the arrangements of the eight last figures.

More numerous combinations of stars may still take place, by admitting simple and regular perturbations; for then all sorts of erratic orbits of multiple flexures may have a permanent existence. But, as it would lead me too far, to apply calculation to them, I forbear entering upon the subject at present.

Before I proceed, it will be proper to remark, that it may possibly occur to many who are not much acquainted with the arrangement of the numberless stars of the heavens, that what has been said may all be mere useless surmise; and that, possibly, there may not be the least occasion for any such speculations upon the subject. To this, however, it may be answered, that such combinations as I have mentioned, are not the inventions of fancy: they have an actual existence; and, were it necessary, I could point them out by thousands. There is not a single night when, in passing over the zones of the heavens by sweeping, I do not meet with numerous collections of double, treble, quadruple, quintuple, and multiple stars, apparently insulated from other groups, and probably joined in some small sidereal system of their own. I do not imagine that I have pointed out the actual manner in which they are held together; but it will always be a desirable step towards information, if the possibility of such unions, in many different ways, can be laid before us; and, very probably, those who have more leisure to consider the different combinations of central forces, than a practical astronomer can have, may easily enlarge on what has been laid down in the foregoing paragraphs.

IV. *Of clustering Stars, and the Milky-way.*

From quadruple, quintuple, and multiple stars, we are naturally led to a consideration of the vast collections of small stars that are profusely scattered over the milky-way. On a very slight examination, it will appear that this immense starry aggregation is by no means uniform. The stars of which it is composed are very unequally scattered, and shew evident marks of clustering together into many separate allotments. By referring to some one of these clustering collections in the heavens, what will be said of them will be much better understood, than if we were to treat of them merely in a general way. Let us take the space between β and γ Cygni for an example, in which the stars are clustering with a kind of division between them, so that we may suppose them to be clustering towards two different regions. By a computation, founded on observations which ascertain the number of stars in different fields of view, it appears that our space between β and γ, taking an average breadth of about five degrees of it, contains more than 331 thousand stars; and, admitting them to be clustering two different ways, we have 165 thousand for each clustering collection. Now, as a more particular account of the milky-way will be the subject of a separate paper, I shall only observe, that the above mentioned milky appearances deserve the name of clustering collections, as they are certainly brighter about the middle, and fainter near their undefined borders. For, in my sweeps of the heavens, it has been fully ascertained, that the brightness of the milky-way arises only from stars; and that their compression increases in proportion to the brightness of the milky-way.

We may indeed partly ascribe the increase, both of brightness and of apparent compression, to a greater depth of the space which contains these stars; but this will equally tend to shew their clustering condition: for, since the increase of brightness is gradual, the space containing the clustering stars must tend to a spherical form, if the gradual increase of brightness is to be explained by the situation of the stars.

V. *Of Groups of Stars.*

From clustering stars there is but a short transition to groups of stars; they are, however, sufficiently distinct to deserve a separate notice. A group is a collection of closely, and almost equally compressed stars, of any figure or outline;

it contains no particular condensation that might point out the seat of an hypothetical central force; and is sufficiently separated from neighbouring stars to shew that it makes a peculiar system of its own. It must be remembered, that its being a separate system does not exclude it from the action or influence of other systems. We are to understand this with the same reserve that has been pointed out, when we explained what we called insulated stars.

The construction of groups of stars is perhaps, of all the objects in the heavens, the most difficult to explain; much less can we now enter into a detail of the numerous observations I have already made upon this object. I therefore proceed in my enumeration.

VI. *Of Clusters of Stars.*

These are certainly the most magnificent objects that can be seen in the heavens. They are totally different from mere groups of stars, in their beautiful and artificial arrangement: their form is generally round; and the compression of the stars shews a gradual, and pretty sudden accumulation towards the centre, where, aided by the depth of the cluster, which we can have no doubt is of a globular form, the condensation is such, that the stars are sufficiently compressed to produce a mottled lustre, nearly amounting to the semblance of a nucleus. A centre of attraction is so strongly indicated, by all the circumstances of the appearance of the cluster, that we cannot doubt a single moment of its existence, either in a state of real solidity, or in that of an empty centre, possessed of an hypothetical force, arising from the joint exertion of the numerous stars that enter into the composition of the cluster.

The number of observations I have to give relating to this article, in which my telescopes, especially those of high space-penetrating power, have been of the greatest service, of course can find no room in this enumeration.

VII. *Of Nebulæ.*

These curious objects, which, on account of their great distance, can only be seen by instruments of great space-penetrating power, are perhaps all to be resolved into the three last mentioned species. Clustering collections of stars, for instance, may easily be supposed sufficiently removed to present us with the appearance of a nebula of any shape, which, like the real object of which it is the miniature, will seem to be gradually brighter in the middle. Groups of stars also may, by distance, assume the semblance of nebulous patches; and real clusters of stars, for the same reason, when their composition is beyond the reach of our most powerful instruments to resolve them, will appear like round nebulæ that are gradually much brighter in the middle. On this occasion I must remark, that with instruments of high space-penetrating powers, such as my 40-feet telescope, nebulæ are the objects that may be perceived at the greatest distance. Clustering collections of stars, much less than those we have mentioned before, may easily contain 50000 of them ; and, as that number has been chosen for an instance of calculating the distance at which one of the most remote objects might be still visible,* I shall take notice of an evident consequence attending the result of the computation ; which is, that a telescope with a power of penetrating into space, like my 40-feet one, has also, as it may be called, a power of penetrating into time past. To explain this, we must consider that, from the known velocity of light, it may be proved, that when we look at Sirius, the rays which enter the eye cannot have been less than 6 years and 4½ months coming from that star to the observer. Hence it follows, that when we see an object of the calculated distance at which one of these very remote nebulæ may still be perceived, the rays of light which convey its image to the eye, must have been more than nineteen hundred and ten thousand, that is, almost two millions of years on their way ; and that, consequently, so many years ago, this object must already have had an existence in the sidereal heavens, in order to send out those rays by which we now perceive it.

VIII. *Of Stars with Burs, or Stellar Nebulæ.*

Situated as we are, at an immense distance from the remote parts of the heavens, it is not in the power of telescopes to resolve many phenomena we can but just perceive, which, could we have a nearer view of them, might probably shew themselves as objects that have long been known to us. A stellar nebula, perhaps, may be a real cluster of stars, the whole light of which is gathered so nearly into one point, as to leave but just enough of the light of the cluster visible to produce the appearance of burs. This, however, admits of a doubt.

IX. *Of milky Nebulosity.*

The phenomenon of milky nebulosity is certainly of a most interesting nature : it is probably of two different kinds ; one of them being deceptive, namely, such as arises from widely extended regions of closely connected clustering stars, contiguous to each other, like the collections that construct our milky-way. The other, on the contrary, being real, and possibly at no very great distance from us. The changes I have observed in the great milky nebulosity of Orion, 23 years ago, and which have also been noticed by other astronomers, cannot permit us to look upon this phenomenon as arising from immensely distant regions of fixed stars. Even HUYGENS, the discoverer of it, was already of opinion that, in viewing it, we saw, as it were, through an opening into a region of light.† Much more would he be convinced now, when changes in its shape and lustre have been seen, that its light is not, like that of the milky-way, composed of stars. To attempt even a guess at what this light may be, would be presumptuous. If it should be surmised, for instance, that this nebulosity is of the nature of the zodiacal light, we should then

* See *Phil. Trans.* for 1800, page 83 [above, p. 51].

† See *Systema Saturnium*, page 8 and 9.

be obliged to admit the existence of an effect without its cause. An idea of its phosphorical condition, is not more philosophical, unless we could shew from what source of phosphorical matter, such immeasurable tracts of luminous phenomena could draw their existence, and permanency; for, though minute changes have been observed, yet a general resemblance, allowing for the difference of telescopes, is still to be perceived in the great nebulosity of Orion, even since the time of its first discovery.

X. *Of nebulous Stars.*

The nature of these remarkable objects is enveloped in much obscurity. It will probably require ages of observations, before we can be enabled to form a proper estimate of their condition. That stars should have visible atmospheres, of such an extent as those of which I have given the situation in this and my former catalogues, is truly surprising, unless we attribute to such atmospheres, the quality of self-luminous milky nebulosity. We can have no reason to doubt of the starry nature of the central point; for, in no respect whatever does its appearance differ from that of a star of an equal magnitude; but, when the great distance of such stars is taken into consideration, the real extent of the surrounding nebulosity is truly wonderful. A very curious one of this kind will be found in the 4th class, No. 69, of the annexed catalogue.

XI. *Planetary Nebulæ.*

This seems to be a species of bodies that demands a particular attention. To investigate the planetary nature of these nebulæ, is not an easy undertaking. If we admit them to contain a great mass of matter, such as that of which our sun is composed, and that they are, like the sun, surrounded by dense luminous clouds, it appears evidently that the intrinsic brightness of these clouds must be far inferior to those of the sun. A part of the sun's disk, equal to a circle of 15″ in diameter, would far exceed the greatest lustre of the full moon; whereas, the light of a planetary nebula, of an equal size, is hardly equal to that of a star of the 8th or 9th magnitude. If, on the other hand, we should suppose them to be groups, or clusters of stars, at a distance sufficiently great to reduce them to so small an apparent diameter, we shall be at a loss to account for their uniform light, if clusters; or for their circular forms, if mere groups of stars.

Perhaps they may be rather allied to nebulous stars. For, should the planetary nebulæ with lucid centres, of which the next article will give an account, be an intermediate step between planetary nebulæ and nebulous stars, the appearances of these different species, when all the individuals of them are fully examined, might throw a considerable light upon the subject.

XII. *Of planetary Nebulæ with Centres.*

In my second catalogue of nebulæ, a single instance of a planetary nebula with a bright central point was mentioned; and, in the annexed one, No. 73 of the 4th class, is another of very nearly the same diameter, which has also a lucid, though not quite so regular a centre. From several particularities observed in their construction, it would seem as if they were related to nebulous stars. If we might suppose that a gradual condensation of the nebulosity about a nebulous star could take place, this would be one of them, in a very advanced state of compression, A further discussion of this point, however, must be reserved to a future opportunity.

Catalogue of 500 additional new Nebulæ, and Clusters of Stars.

First Class. Bright Nebulæ.

I.	1788.	Stars.		M. S.		D. M.	Ob.	Description.	N.G.C.
216	Dec. 3	22 Ursæ	*p*	13 52	*s*	3 4	2	*v*B. *p*L. *i*F. *r*. *mb*M. Towards the *sf*, within the nebulosity, is a *v*S. *st*.	2787
217	27	54 Persei	*f*	9 25	*n*	0 46	2	*c*B. *c*L. *mb*M. Stands nearly in the center of a trapezium.	1579
218	31	63 Aurigæ	*f*	26 43	*s*	0 20	1	*c*B. R. *vgmb*M. about 3′ *d*.	2419
219	1789 Mar. 23	55 Ursæ	*f*	5 33	*n*	0 36	1	*v*B. *c*L. *i*F. *vgmb*M.	3665
220	Apr. 12	64 (γ) Ursæ	*p*	43 59	*s*	0 20	2	*c*B. *m*E. 70° *np sf*. 3 or 4′ *l*, 2′ *b*.	3549
221	—	— —	*p*	21 41	*s*	0 37	2	*c*B. R. *vgmb*M, 4 or 5′ *d*.	3718
222	—	— —	*p*	20 20	*s*	0 35	2	*c*B. *i*E. near mer. *gb*M. 2′ *l*.	3729
223	—	— —	*f*	6 4	*s*	2 45	2	*v*B. *m*E. *np sf*. BN. 5′ *l*. 1½′ *b*.	4026
224	—	1 Canum	*p*	9 19	*s*	3 10	2	*c*B. *p*L. *m*E. SN.	4085
225	—	— —	*p*	8 31	*s*	0 46	2	*v*B. *p*L. B*r*N. just *f* a *cst*.	4102
226	14	64 (γ) Ursæ	*p*	33 32	*s*	0 34	1	*c*B. R. SB*r*N and *v*F chev. 4′ *d*.	3631
227	—	— —	*p*	15 28	*n*	2 37	2	*c*B. *c*L. *i*F. *r*. *vgb*M. 3′ *l*. 2′ *b*.	3780
228	—	— —	*p*	5 20	*n*	2 24	2	*v*B. *v*B*i*N. and F. bran. 1½′ *l*. ¾′ *b*.	3898
229	—	— —	*f*	3 46	*n*	1 47	1	The 2d of 2. *v*B. R. *vgb*M. See II. 791.	3998
230	—	83 Ursæ	*f*	20 24	*n*	0 27	2	*c*B. S. E. *sp nf*. *c*BN. and F bran.	5422
231	—	— —	*f*	24 34	*n*	0 10	2	*c*B. *p*S. *i*R.	5473
232	—	— —	*f*	27 7	*n*	0 16	1	The 2d of 2. *c*B. S. R. *vgmb*M. See III. 791.	5485
233	17	44 Ursæ	*f*	1 14	*s*	0 16	2	*c*B. E. 30° *sp nf*. *r*. *mb*M. 3′ *l*. 1½′ *b*.	3448
234	—	74 Ursæ	*f*	1 31	*s*	0 28	2	*c*B. S. *l*E. Just *p* a *p*L *st*.	4500
235	—	12 (ι) Draconis	*p*	66 52	*s*	2 3	2	*c*B. *i*F *vgmb*M. 7′ *l*, 5′ *b*.	5585
236	—	— —	*p*	59 56	*s*	2 13	3	*v*B. S. *i*R. B*ir*N. *vgmb*M.	5631
237	—	— —	*p*	54 10	*s*	0 52	1	B. *i* oval. *vgmb*M.	5678
238	24	69 Ursæ Hev.	*f*	27 55	*s*	0 32	2	*c*B. *p*L. *i*R *vgmb*M.	5376
239	—	— —	*f*	28 10	*s*	0 17	3	*c*B. *p*L. E. *mb*M.	5379
240	—	— —	*f*	28 34	*s*	0 17	2	*c*B. *p*L. E. SBN.	5389

I.	1790.	Stars.		M. S.		D. M.	Ob.	Description.	N.G.C.
241	Feb. 17	19 (ξ) Hyd. Crat.	*p*	14 43	*s*	0 57	1	*c*B. E. 70° *np sf*. *vgb*M 7′ *l*, 4′ *b*. within a parallelogram.	3621
242	Mar. 17	15 (*f*) Ursæ	*p*	15 40	*s*	0 21	1	*v*B. LB*r*N. with *v*F chev.	2681
243	—	77 (ε) Ursæ	*f*	1 47	*n*	2 25	1	*c*B. S. R. *gb*M.	4814
244	18	39 Ursæ	*f*	36 44	*n*	0 40	2	*c*B. R. *vgmb*M. 1½′ *d*.	3619
245	—	— —	*f*	39 27	*n*	1 58	3	*v*B. *c*L. R. *vgb*M.	3642
246	—	66 Ursæ	*p*	29 19	*n*	0 20	2	*c*B. *p*L. E.	3683*
247	—	— —	*p*	28 13	*n*	2 0	2	*v*B. *p*L. *l*E. near par. *mb*M.	3690
248	—	— —	*p*	7 5	*n*	2 52	2	*c*B. *p*L. *i*F.	3894
249	19	17 Ursæ	*p*	9 0	*n*	3 43	2	*c*B. E. near par. *er*. *b*M. 4′ *l*, 2′ *b*. I suppose, with a higher power and longer attention, the stars would become visible.	2742
250	—	— —	*p*	4 47	*n*	3 17	1	*v*B. *c*L. *l*E. LBNM.	2768
251	—	76 Ursæ	*p*	50 48	*s*	2 3	1	*v*B. perfectly R. BN and F chev. *vgb*M. 1½′ *d*.	3945
252	—	— —	*p*	41 11	*s*	0 34	1	*v*B. *c*L. R.	4041
253	—	— —	*p*	41 46	*s*	0 51	1	*v*B. *v*L. E.	4036
254	—	— —	*p*	1 47	*s*	1 8	1	*e*B. E. par. 5′ *l*. all over equally B. except just on the edges.	4605
255	—	69 Ursæ Hev.	*f*	19 26	*n*	1 1	1	*v*B. BENM. 3′ *l*. ½′ b.	5308
256	—	— —	*f*	21 33	*n*	0 13	1	*v*B. *p*L. *i*F. suddenly *mb*M.	5322
257	Oct. 9	12 Eridani	*f*	16 58	*s*	1 58	1	*c*B. *i*R *vgmb*M. 1½′ *d*.	1344
258	Dec. 28	47 (λ) Persei	*p*	3 41	*n*	1 0	1	*v*B. *i*F. *r*. *b*M. 5′ *l*. 4′ *b*. A *p*L star in it towards the *f* side, but unconnected.	1491
	1791.								
259	Mar. 7	17 Hydræ Crat	*f*	18 31	*n*	0 27	1	*c*B. *p*L. *l*E. *gb*M. The brightness takes up a large space of it.	3923
260	Apr. 2	23 (*h*) Ursæ	*p*	1 49	*s*	0 34	1	*v*B. *v*S. *i*R. *mb*M.	2880
	1793								
261	Feb. 4	38 of the *Connois.*	*f*	3 7	*s*	1 35	1	*v*B. *i*R. *vgb*M. 5′ *d*. Seems to have 1 or 2 stars in the middle, or an *i*N; the chev. diminishes *vg*.	1931
262	Apr. 6	1 (λ) Draconis	*p*	2 6	*s*	2 41	1	*c*B. *v*S. *i*F. N. with *v*F chev.	3682*
263	—	4 Draconis	*p*	22 48	*s*	0 23	1	*c*B. *l*E. *b*M.	4128
264	7	— —	*p*	14 18	*n*	1 36	1	*c*B. S. *b*M.	4250*
265	8	37 Ursæ	*p*	16 16	*n*	1 5	1	*c*B. S. *i*R. *vgmb*M.	3182
266	—	— —	*p*	13 35	*s*	0 11	1	*c*B. *p*L. *i*F. *gb*M.	3206
267	—	39 Ursæ	*f*	11 21	*s*	0 10	1	*c*B. *p*L. *i*R. 1¼′ *d*. The greatest part of it almost equally B.	3445
268	—	— —	*f*	12 46	*s*	0 4	1	*v*B. *v*S. R. Stellar.	3458
269	—	— —	*f*	18 1	*n*	0 29	1	*c*B. R. 1′ *d*. just *n* of a S*st*.	3488
270	—	— —	*f*	35 36	*n*	1 42	2	*v*B. *c*L. E. par. SN. E par.	3610
271	—	— —	*f*	35 54	*n*	0 55	1	*v*B. *c*L. E. *mb*M.	3613*

I.	1796.	Stars.		M. S.		D. M.	Ob.	Description.	N.G.C.
272	Mar. 4	Georgian planet	*p*	0 53	*n*	0 6	2	*c*B. S. *i*R. BN. *mb*M. This nebula was seen at 9^h 27′, sidereal time; the telescope being out of the meridian, the estimations may be a little faulty.	3332*
	1797								
273	Nov. 22	A double star	*f*	5 45	*s*	0 39	3	*v*B. *v*L. E. near par. The determining star follows 5 Draconis Hevelii 13′ 54″ in time, and is 0° 23′ more south.	4589
274	—	— —	*f*	10 13	*s*	0 24	3	*c*B. *v*S. *i*R. *b*M.	4648
275	Dec. 10	5 Dracon. Hev.	*f*	1 32	*n*	0 12	2	*c*B. S. R.	4291
276	—	— —	*f*	2 45	*n*	0 12	2	*c*B. *c*L. *i*F. *l*E. *mb*M.	4319
277	—	— —	*f*	6 20	*n*	0 20	2	*v*B. *c*L. *l*E. *mb*M.	4386
278	12	— —	*p*	11 5	*s*	0 15	1	*c*B. *c*L. *i*R. *mb*M.	4133*
279	—	— —	*p*	10 28	*n*	1 38	2	*c*B. *c*L. *l*E. *b*M.	4127*
280	—	16 (ζ) Ursæ min.	*f*	51 33	*n*	0 3	3	*v*B. *c*L. *l*E. *lb*M. The greatest brightness confined to a small point.	6217
	1798								
281	Dec. 9	τ Apps. Sculps. L. C. 95	*p*	1 47	*n*	0 27	1	*c*B. E. *np sf*. NM. 6′ *l*. 1½′ *b*.	613
	1801								
282	Apr. 2	Star 6·7 m. [B. 1446]	*p*	55 17:	*s*	1 10	1	*c*B. *p*L. *i*F.	2977*
283	—	— —	*p*	15 42:	*s*	1 31	1	*c*B. *c*L. *er*.	3183*
284	—	208 (N) Camelop. of BODE'S Cat.	*p*	85 18:	*s*	0 23	1	*c*B. *v*S. *i*F.	3329*
285	Nov. 8	24 (*d*) Ursæ	*f*	13 14	*s*	1 53	1	*v*B. *v*L. E. *np sf*. 6′ *l*. 2′ *b*.	2976
286	—	— —	*f*	30 0	*s*	1 8	1	*v*B. *c*L. R. *vgmb*M. On the north-following side there is a F ray interrupting the roundness.	3077
287	Dec. 7	1 (λ) Draconis	*p*	4 37	*n*	1 13	1	*c*B. *m*E. *np sf*. *mb*M. 3′ *l*, 1′ *b*.	3735
	1802								
288	Sept. 26	184 Camelopar. of BODE'S Cat.	*p*	11 58	*s*	2 34	1	*v*B. *c*L. *l*E. suddenly *mb*M.	2655

Second Class. Faint Nebulæ.

II.	1789.	Stars.		M. S.		D. M.	Ob.	Description.	N.G.C.
769	Feb. 22	81 (*g*) Geminor.	*p*	37 58	*n*	0 4	1	*p*B. *p*L. *i*R. *er*. *b*M.	2339
770	—	62 Ursæ	*p*	13 44	*s*	2 15	1	*p*B. *p*L. R. *lb*M.	3687
771	Mar. 20	26 (χ) Virginis	*p*	7 0	*n*	0 26	2	*p*B. *c*L. *i*F. *er*. *mb*M. 4 or 5′ *d*.	4504
772	—	— —	*f*	3 9	*n*	0 57	2	F. S. E.	4626

II.	1789.	Stars.		M. S.		D. M.	Ob.	Description.	N.G.C.
773	Mar. 20	26 (χ) Virginis	*f*	3 5	*n*	1 1	2	F. S. E. *b*M.	4628
774	—	— —	*f*	6 27	*n*	0 55	2	*p*B. S. *i*R. *mb*M.	4671
775	23	55 Ursæ	*f*	3 31	*s*	0 25	1	*p*B. *c*L. *l*E. *vgmb*M.	3652
776	—	26 (χ) Virginis	*p*	8 19	*s*	0 4	1	F. *v*L. *er*.	4487
777	—	— —	*f*	17 15	*n*	1 9	1	F. S. R. *b*M.	4813
778	—	— —	*f*	21 12	*n*	1 54	1	F. S. *sf*. a double star.	4888
779	—	— —	*f*	22 44	*n*	0 14	1	F. S.	4925
780	26	46 (γ) Hydræ	*f*	1 22	*s*	1 14	1	F. R. *r*. *vglb*M. 4′ d.	5085
781	Apr. 12	1 Canum	*p*	10 55	*s*	0 53	2	A *p*S. *st*. involved in nebulosity of no great extent; the *st*. does not seem to belong to it.	4068*
782	14	64 (γ) Ursæ	*p*	31 7	*n*	0 7	1	*p*B. S. R. *vgmb*M. just *f* a S*st*.	3656
783	—	— —	*p*	18 40	*n*	0 50	1	*p*B. *p*L. *b*M.	3738
784	—	— —	*p*	17 41	*n*	0 37	1	*p*B. *c*L. *l*E. 3′ *l*.	3756
785	—	— —	*p*	7 3	*n*	2 18	1	*p*B. S. *l*E.	3888
786	—	— —	*p*	3 31	*n*	1 39	1	F. E.	3913
787	—	— —	*p*	3 2	*n*	1 27	1	Two nebulæ; the 1st *p*B. S.	3916
788				3 7		1 24		The 2d *p*B. S.	3921
789	—	— —	*f*	1 35	*n*	1 38	1	Two nebulæ; the 1st *p*B. E.	3972
790								The 2d F. S.	3977
791	—	— —	*f*	3 24	*n*	1 48	1	The 1st of 2. *p*B. S. E. See I. 229.	3990
792	—	1 Canum	*p*	3 12	*n*	2 47	1	F. S. R. *b*M.	4172
793	—	— —	*p*	0 57	*n*	2 36	2	F. *p*L. *i*F. *b*M.	4198
794	—	77 (ε) Ursæ	*p*	11 32	*s*	0 49	2	F. S.	4644*
795	—	— —	*p*	8 25	*s*	1 13	2	*p*B. *v*S. *mb*M.	4675*
796	—	— —	*p*	7 20	*s*	1 25	2	*p*B. *c*S. *l*E. B*r*N.	4686*
797	—	81 Ursæ	*p*	3 33	*s*	2 18	2	*p*F. *p*S. R. *vgb*M.	5201*
798	—	83 Ursæ	*f*	0 49	*n*	1 1	1	*p*B. E. 1½′ *l*, ½′ *b*.	5278
799	—	— —	*f*	21 27	*n*	1 7	2	*p*B. *c*L. E.	5443
800	—	— —	*f*	25 7	*n*	1 2	1	*p*B. S.	5475
801	—	— —	*f*	27 27	*n*	0 23	1	F. *c*L.	5486*
802	17	71 Ursæ	*p*	15 20	*n*	1 33	1	F. S. E.	4149
803	—	— —	*p*	13 57	*n*	0 59	2	F. S. R.	4161*
804	—	— —	*p*	5 43	*s*	0 3	1	*p*B. *p*L. *i*F.	4271
805	—	— —	*p*	4 41	*n*	1 20	1	The 2d of 2. *p*B. *p*L. *mb*M. See III. 798.	4290
806	—	— —	*p*	2 13	*n*	1 42	1	*p*B	4335
807	—	12 (ι) Draconis	*p*	55 48	*n*	0 42	1	*p*B. E. *mer*. 1½′ *l*, ¾′ *b*.	5667
808	24	Neb. II., 756	*p*	24 16	*n*	0 41	1	*p*B. S. *i*F. *er*. mixed with some *p*L. stars, which may perhaps belong to it.	5687
809	—	— —	*p*	15 5	*s*	0 26	1	F. S. E.	5751
810	—	21 (μ) Draconis	*p*	46 31	*n*	3 23	1	*p*F. *p*S. *l*E.	6125*
811	—	— —	*p*	44 9	*n*	0 50	1	*p*B. *i*R. *vgvlb*M.	6143
812	—	— —	*f*	10 4	*n*	2 55	1	F. S. R. *vglb*M.	6338
813	26	5 Canum	*p*	10 53	*s*	0 50	1	*p*B. S. *l*E.	4187
814	—	7 Canum	*f*	20 24	*n*	1 20	1	F. S. *vsmb*M.	4732

II.	1789.	Stars.		M. S.		D. M.	Ob.	Description.	N.G.C.
815	Apr. 26	82 Ursæ	*p*	31 48	*s*	0 52	1	F. *v*S. Stellar.	4987
816	—	— —	*p*	26 52	*s*	1 36	1	F. S. *i*R. *vgmb*M.	5040
817	—	— —	*p*	3 42	*s*	1 40	1	*p*B. S. R. *vgb*M.	5250
818	—	12 Drac. Hev.	*p*	40 16	*n*	0 33	1	*p*F. *c*S. R. *vgb*M.	5881*
819	1790 Mar. 8	13 (λ) Hyd. Crat.	*p*	11 58	*n*	0 31	1	*p*F. *p*L. *i*F. *b*M.	3571
820	10	65 Aurigæ	*f*	7 22	*n*	0 1	1	*p*B. S. Stellar.	2387
821	—	70 Geminorum	*p*	1 43	*n*	0 12	1	*p*B. *c*S. *r*. *p* a *cst*.	2415
822	17	27 Lyncis	*p*	25 42	*n*	0 41	1	*p*F. R. *r*. *vgb*M.	2426
823	—	15 (*f*) Ursæ	*p*	12 10	*s*	0 18	1	*p*B. S. R. *mb*M.	2693
824	—	26 Ursæ	*f*	139 17	*s*	0 1	1	*p*B. *m*E. 6′ *l*, 2′ *b*.	3917*
825	—	— —	*f*	139 40	*s*	1 44	1	*p*B. S. *i*F. *b*M.	3922*
826	—	77 (ε) Ursæ	*f*	28 0	*n*	1 42	1	F. S. E.	5109
827	—	— —	*f*	69 19	*n*	3 27	1	*p*B. S. *i*F. *mb*M.	5430*
828	18	17 Ursæ	*p*	6 25	*s*	2 57	1	*p*B. S. *vgmb*M.	2756
829	—	66 Ursæ	*p*	31 14	*n*	1 9	2	F. E. *np sf*. *er*. 1½′ *l*.	3669
830	—	— —	*p*	15 23	*s*	0 20	1	*p*B. E.	3804
831	—	— —	*p*	11 44	*n*	1 22	1	*p*B. *v*S. *l*E.	3838
832	—	— —	*p*	6 53	*n*	2 52	2	*p*B. *p*L. R. The nebulosity of this runs into that of I. 248.	3895
833	—	— —	*p*	1 1	*n*	1 46	1	F. S.	3958
834	19	17 Ursæ	*p*	11 34	*n*	3 10	1	*p*F. *p*S. *i*F *er*.	2726
835	—	29 (*υ*) Ursæ	*f*	5 11	*n*	0 15	2	F. S. E. near par.	3043
836	—	76 Ursæ	*p*	70 41	*s*	0 53	1	F. S. R. *r*. almost of equal light throughout.	3725
837	—	— —	*p*	66 54	*s*	1 0	1	*p*B. *l*E.	3762
838	—	— —	*p*	66 15	*s*	3 9	1	*p*B. S.	3770
839	—	— —	*p*	63 0	*s*	2 28	1	*p*B. *c*S. R. *mb*M.	3796
840	—	— —	*p*	47 30	*s*	2 16	1	F. S. *b*M.	3978
841	—	69 Ursæ Hev.	*f*	4 24	*n*	2 46	2	The 1st of 2. *p*B. S. *i*F.	5216
842	—	— —	*f*	4 35	*n*	2 50	2	The 2d of 2. *p*B. *p*L. *i*F.	5218
843	—	— —	*f*	26 40	*n*	0 42	1	F. S.	5370
844	—	— —	*f*	27 43	*s*	0 29	1	*p*B. *c*L.	3795
845	20	50 (*a*) Ursæ	*f*	22 41	*n*	1 44	3	*p*B. *p*L. *i*R. *b*M.	3668
846	—	76 Ursæ	*p*	23 9	*n*	3 13	1	*p*B. *m*E. *sp nf*. BN. 5′ *l*, ½′ *b*.	4256
847	—	— —	*p*	19 1	*n*	3 8	1	*p*B. S. *l*E.	4332
848	—	— —	*p*	14 21	*n*	2 8	1	F. *i*F. *b*M. Stellar.	4441
849	—	— —	*p*	9 7	*n*	1 15	1	*p*B. *v*S. *l*E. SN.	4521
850	—	— —	*p*	7 16	*n*	0 48	1	*p*B. *p*L. *i*R. *r*. *vgb*M.	4545
851	Oct. 9	72 Pegasi	*f*	18 3	*s*	0 6	2	*p*F. *p*L. *i*R. *lb*M. *sp*. a *v*S*st*.	7773
852	—	σ Fornacis L. C. 285	*p*	4 15	*s*	0 34	1	F. *p*L. *i*R. *gb*M.	1425
853	Nov. 26	29 (π) Androm.	*p*	25 48	*s*	0 24	1	F. S. E. near mer.	29
854	Dec. 25	44 Piscium	*f*	3 49	*n*	0 56	1	*p*B. *v*S. R. *vgmb*M. pretty well defined on the margin.	128
855	—	—	*f*	4 44	*n*	0 10	2	*p*B. *c*L. *i*R. *r*. *vgb*M. *sp*. *v*S*st*.	132

II.	1790.	Stars.		M. S.		D. M.	Ob.	Description.	N.G.C.
856	Dec. 25	44 Piscium	*f*	13 52	*n*	1 8	1	F. S. *vgb*M.	194
857	—	— —	*f*	13 52	*n*	0 53	1	F. S. *vgb*M.	198
858	—	— —	*f*	14 10	*n*	0 58	1	*p*B. S. *vgb*M.	200
859	—	98 (μ) Piscium	*f*	20 28	*n*	0 1	1	*p*B. S. E. near par. *sp*. a S*st*.	693
860	28	MAYER's Zod. Cat. No. 18	*p*	5 48	*n*	0 39	1	*p*F. *v*S. *vgb*M.	196
861	—	57 Aurigæ	*f*	17 30	*n*	1 54	1	*p*B. *p*L. *i*F. *gb*M.	2320
862	—	— —	*f*	23 5	*n*	1 29	1	F. *p*L.	2332?*
863	29	63 (δ) Piscium	*p*	0 39	*n*	0 44	1	*p*L. *l*E. *r*. *gb*M.	257
864	1791 Mar. 7	17 Hyd. Crat.	*f*	16 46	*s*	0 1	1	*p*B. S. R. *vgmb*M. almost resembling a N.	3904
865 } 866 }	—	— —	*f*	34 2	*s*	0 31	1	Two nebulæ, both F. S. R. *b*M. and nearly in the same par.	4105 4106
867	April 2	73 Ursæ	*p*	14 8	*s*	1 12	1	*p*B. *v*S. Stellar.	4194
868 } 869 }	3	14 (τ) Ursæ	*f*	11 8	*n*	0 47	1	Two nebulæ, the 1st F. S. *i*F, the 2d F. *p*L. E. The place is that of the second, the other precedes it about 30^s and is nearly in the same parallel.	2814* 2820
870	—	35 Ursæ	*f*	2 50	*s*	0 36	1	F. S. *i*R. Almost of equal light throughout.	3259
871	—	— —	*f*	3 37	*s*	0 52	1	F. *v*S. *mb*M.	3266
872	—	— —	*f*	21 30	*n*	0 11	1	F. *c*L. *i*R.	3394*
873	May 6	13 (γ) Ursæ min.	*f*	37 53	*s*	1 17	1	F. R. *b*M. 1′ *d*.	6048
874	24	37 (ξ) Bootis	*f*	34 48	*s*	1 12	1	*p*B. *p*L. *i*R. *vgmb*M.	5928
875	30	25 Herculis	*f*	3 10	*n*	2 12	1	*p*B. S. *l*E. *vgmb*M.	6166
876	1792 Apr. 20	22 (*f*) Bootis	*p*	15 58	*n*	0 26	1	*p*B. *v*S.	5492
877	—	— —	*p*	13 27	*n*	1 21	1	*p*B. *p*L. *i*F.	5513
878	Sept. 16	3 Cephei Hev.	*p*	29 15	*s*	0 25	1	*p*B. *i*F. *b*M. contains 2 stars.	6824*
879	1793 Apr. 6	1 (λ) Draconis	*p*	9 49	*s*	2 5	1	*p*B. S. R. *b*M.	3622*
880	—	— —	*p*	7 44	*n*	0 6	2	F. S. *l*E. *sp nf*. but near mer. *gb*M.	3654
881	7	4 Draconis	*p*	45 43	*n*	0 12	1	F. *m*E. *np sf*. but near par. about 1½′ *l*.	3879*
882	8	37 Ursæ	*p*	10 40	*n*	1 3	1	*p*B. *p*L. *l*E. *b*M.	3225
883	—	— —	*p*	8 36	*n*	0 8	1	F. S. R. *b*M.	3238
884	—	39 Ursæ	*f*	22 42	*s*	0 37	1	F. S. R. *b*M.	3517
885	—	— —	*f*	37 41	*n*	0 42	1	F. S. *l*E. *np sf*.	3625
886	—	— —	*f*	44 5	*s*	0 2	1	*p*B. *i*F.	3674
887	9	42 Ursæ	*f*	2 41	*n*	1 56	1	F. *p*L. *i*F. *b*M.	3435
888	—	— —	*f*	7 21	*n*	0 11	1	F. S. R. *b*M.	3470
889	May 12	19 Bootis Hev.	*p*	26 45	*n*	0 20	1	*p*B. *p*L. R. just foll. a S*st*.	5374
890	—	— —	*p*	13 20	*n*	0 33	1	*p*B. *p*L. *i*R.	5491
891	—	— —	*f*	6 44	*n*	0 8	1	*p*B. *p*L. *l*E. BM.	5652

II.	1793.	Stars.		M. S.		D. M.	Ob.	Description.	N.G.C.
892	May 12	19 Bootis Hev.	*f*	7 44	*n*	0 24	1	F. S. E. near mer.	5661
893	—	— —	*f*	9 37	*s*	0 22	1	*p*B. S. *i*F.	5674
894	—	— —	*f*	10 46	*s*	0 31	1	F. S.	5679
895	13	93 (τ) Virginis	*p*	21 54	*s*	0 40	1	F. S. *i*R.	5257
896	—	— —	*p*	21 49	*s*	0 40	1	F. S. *i*R.	5258
897	Sept. 6	53 Aquarii	*p*	16 29	*n*	0 7	1	*p*B. *l*E. *r*. 1½′ *l*. 1¼′ *b*.	7218
898	1794 Mar. 22	Georgian planet	*f*	¾°	*n*	½	1	By coarse estimation. F. 3′ north of a *p*L. red *st*. This nebula was seen at 8^h 49′, sidereal time, the telescope being out of the meridian.	3107
899	1797 Dec. 20	4 (*b*) Ursæ min.	*p*	26 13	*s*	0 40	1	F. S. E. near mer. 1′ *l*.	5323
900	1798 Dec. 10	18 (ε) Eridani	*p*	20 53	*s*	1 5	1	F. E. *sp nf*. near par. 3′ *l*, 1′ *b*.	1247
901	1799 June 29	93 Herculis	*p*	27 30	*s*	0 11	1	F. S. *i*F. *er*. 2′ *l*.	6389
902	—	— —	*f*	7 47	*n*	0 49	1	F. *p*L. R. *vgb*M. 3½′ *d*.	6555
903	1801 Apr. 2	Star 6·7 m. [B. 1446]	*p*	41 21:	*s*	0 6	1	F. *p*L. *r*.	.. *
904	—	— —	*f*	29 49:	*s*	0 25	1	F. *p*L. *lb*M.	.. *
905	—	— —	*f*	61 5:	*s*	0 49	1	*p*B. *p*L.	.. *
906	Nov. 28	11 (*a*) Draconis	*f*	86 13	*n*	0 8	1	F. S. *l*E. *sp nf*. *vglb*M.	5949
907	1802 June 26	2 (μ) Lyræ	*f*	5 21	*n*	0 18	1	F. S. *i*F.	6646
908	Sept. 30	Ursæ 24 Bode	*f*	6 59	*n*	0 9	1	*p*B. *p*L. *er*. I believe I see some of the stars. *i*F.	2650*
909	—	27 Ursæ	*f*	20 3	*s*	0 5	1	F. *p*L. R.	3066*
910	1791 Mar. 24	73 Ursæ	*f*	15 44	*s*	0 48	1	F. S.	4646*

Third Class. Very faint Nebulæ.

III.	1788.	Stars.		M. S.		D. M.	Ob.	Description.	N.G.C.
748	Dec. 3	43 Camelop.	*f*	35 5	*n*	0 29	1	*v*F. *v*S. has a *v*F. branch *nf*.	2366
749	—	22 Ursæ	*p*	12 45	*s*	0 24	1	*c*F. *v*S.	2810
750	31	63 Aurigæ	*f*	48 58	*n*	0 43	1	*v*F. S. R. *lb*M.	2493
751	—	38 Lyncis	*f*	25 15	*s*	0 30	2	*e*F. S.	2965*
752	1789 Feb. 22	16 (ζ) Cancri	*p*	4 19	*n*	0 8	1	*e*F. *l*E. *s* of a *v*S*st*.	2530
753	—	33 (η) Cancri	*p*	8 11	*s*	0 4	1	*v*F. S. R. *vlb*M.	2582*
754	24	6 Corvi	*p*	17 33	*s*	1 43	1	*e*F. *v*S. R.	4087
755	Mar. 20	26 (χ) Virginis	*p*	13 3	*n*	0 20	1	Two nebulæ, both *v*F. *v*S.	4403
756								E. within 1½′ of each other.	4404

III.	1789.	Stars.		M. S.		D. M.	Ob.	Description.	N.G.C.
757	Mar. 20	26 (χ) Virginis	*p*	5 25	*n*	0 38	2	2 *v*S. stars involved in *v*F. nebulosity of no great extent.	4520
758 }	23	— —	*f*	20 55	*n*	1 53	1	Two nebulæ, both *v*F. *v*S.	4878
759 }									4879
760	—	— —	*f*	23 47	*s*	0 9	1	*c*F. *v*S. R.	4928
761	—	— —	*f*	24 55	*n*	0 18	1	*v*F. S.	4942
762	—	102 (υ′) Virginis	*p*	11 30	*n*	0 36	1	*v*F. *v*S.	5478
763	—	105 (φ) Virginis	*p*	1 1	*s*	0 1	1	*e*F. S.	5618
764	26	9 (β) Corvi	*p*	4 55	*n*	0 15	1	*c*F. *p*S. R. Stellar.	4462
765	—	45 (ψ) Hydræ	*p*	1 35	*s*	0 53	1	*v*F. *p*L. *i*F.	4970
766	—	— —	*f*	0 39	*s*	0 16	1	*v*F. *v*S.	4993
767	Apr. 12	64 (γ) Ursæ	*p*	78 24	*s*	3 45	1	*v*F. *p*S. *i*E.	3298
768	—	— —	*p*	30 48	*s*	0 49	2	*v*F. *v*S. Stellar.	3657
769	—	— —	*p*	1 40	*s*	1 44	1	*c*F. S.	3931*
770	14	— —	*p*	39 32	*n*	2 2	1	*v*F. *v*S. Stellar.	3594
771	—	— —	*p*	19 37	*n*	1 8	1	*e*F. S. *i*E. On account of the brightness of 179 Ursæ maj. of BODE's Cat. which was in the field of view with it, I had nearly overlooked it.	3733
772	—	— —	*p*	19 2	*n*	1 16	1	*v*F. Stellar.	3737
773	—	— —	*p*	14 0	*n*	2 32	1	*c*F. *p*S. *l*E. just *f* a *v*S*st*.	3804*
774	—	— —	*p*	10 37	*s*	0 58	2	*v*F. S.	3824*
775	—	— —	*p*	10 17	*s*	1 1	1	*v*F. *v*S.	3829
776	—	— —	*p*	9 33	*n*	2 12	1	*e*F. *p*L. *l*E, time inaccurate. Left doubtful.	3850*
777	—	1 Canum	*p*	1 54	*s*	0 33	1	*e*F. S. Stellar.	4181
778	—	77 (ε) Ursæ	*p*	9 10	*s*	1 4	2	*c*F. S. *l*E. *i*F.	4669*
779	—	— —	*f*	11 36	*n*	0 20	2	*v*F. S.	4964*
780	—	— —	*f*	12 37	*s*	0 19	1	*c*F. S.	4977*
781 }	—	— —	*f*	12 44	*s*	2 20	1	Two nebulæ. Both *v*F. S. Place is that of 2nd, the other is 3′ or 4′ *sp*.	4973*
782 }									4974*
783	—	— —	*f*	12 33	*s*	2 28	1	*v*F. S. E.	4967*
784	—	81 Ursæ	*p*	7 6	*n*	0 9	1	*c*F. S. *i*R.	5164
785	—	83 Ursæ	*f*	4 34	*n*	0 37	1	2 *e*F. *st*. with nebulosity.	5294
786	—	— —	*f*	14 3	*s*	0 22	1	*v*F. *v*S. Stellar.	5368
787	—	— —	*f*	22 27	*s*	0 28	1	*v*F. *v*S.	5447
788	—	— —	*f*	23 47	*s*	0 24	1	*v*F. *v*S.	5461
789	—	— —	*f*	23 54	*s*	0 22	1	*v*F. *v*S.	5462
790	—	— —	*f*	25 23	*s*	0 17	1	*v*F. *p*L.	5477
791	—							The 1st of 2. *v*F. S. 3′ or 4′ dist. from I. 232.	5484*
792	17	44 Ursæ	*p*	2 11	*n*	0 50	1	*v*F. S. E. 20° *sp nf*. *er*.	3398
793	—	48 (β) Ursæ	*f*	1 25	*s*	0 10	1	*v*F. *v*S. Stellar. The brightness of β Ursæ is so considerable, that it requires much attention to perceive this nebula.	3499

III.	1789.	Stars.		M. S.		D. M.	Ob.	Description.	N.G.C.
794	Apr. 17	71 Ursæ	*p*	22 30	*n*	1 8	1	*c*F. S. ver. 300.	4054*
795	—	— —	*p*	16 8	*n*	2 5	2	*v*F. S. *i*F. *r*.	4141
796	—	— —	*p*	11 23	*n*	2 52	1	*e*F.	4195
797	—	— —	*p*	10 56	*n*	3 11	2	*e*F. S.	4199*
798	—	— —	*p*	5 4	*n*	1 20	1	The 1st of 2. *c*F. *l*E. *i*F. See II. 805.	4284
799	—	— —	*p*	1 12	*n*	1 36	1	*v*F. *v*S.	4358
800 }	—	— —	*p*	1 9	*n*	1 37	1	Two, both *c*F. *c*S. R.	4362*
801 }									4364*
802	—	74 Ursæ	*f*	4 54	*n*	0 30	2	The 1st of 2. *v*F. S. *l*E. See III. 807.	4547
803	—	69 Ursæ Hev.	*f*	9 33	*s*	2 53	2	Suspected *e*F. *v*S.	5255*
804	—	— —	*f*	46 59	*s*	2 18	2	*e*F. S. E. *r*.	5526
805	—	— —	*f*	48 9	*s*	0 1	3	*e*F. *v*S. R. Stellar.	5540
806	—	12 (*ι*) Draconis	*p*	34 20	*n*	0 8	1	*v*F. *v*S. *l*E.	5777
807	24	74 Ursæ	*f*	5 26	*n*	0 34	1	The 2d of 2. *e*F. S. E. differently from III. 802.	4549
808	—	69 Ursæ Hev.	*p*	7 35	*s*	2 19	1	*c*F. S. E.	5109*
809	—	— —	*f*	27 7	*s*	1 25	1	*v*F. *v*S.	5372
810	—	— —	*f*	30 44	*s*	0 13	1	*c*F. *v*S. R.	5402
811	—	Neb. II. 756	*f*	0 32	*n*	0 2	1	*v*F. S. E.	5821
812	—	21 (*μ*) Draconis	*p*	55 20	*n*	3 18	1	*v*F. *v*S. *l*E.	6088*
813	—	— —	*p*	36 1	*n*	1 14	1	*v*F. *v*S. *i*R.	6182
814	26	5 Canum	*p*	15 0	*n*	0 32	1	*v*F. S. *er*.	4142
815	—	7 Canum	*f*	18 48	*s*	0 22	1	S. Stellar.	4707
816	—	— —	*f*	25 11	*n*	1 33	1	*e*F. S. *l*E.	4801
817	—	— —	*f*	26 43	*n*	0 45	1	*c*F. S. *i*F.	4834
818	—	— —	*f*	33 4	*s*	1 7	1	*c*F. S. R. *vglb*M.	4932
819	—	82 Ursæ	*p*	32 15	*s*	2 12	1	*v*F.	4998
820	—	— —	*p*	29 17	*s*	2 48	1	2*v*S stars at less than 1′ *d*. with *v*F. nebulosity between them.	5009
821	—	— —	*p*	12 59	*s*	0 7	1	*c*F. Stellar.	5163*
822	—	— —	*p*	6 23	*s*	1 25	1	*c*F. *p*S. *i*R. *lb*M.	5225
823	—	— —	*p*	5 5	*s*	1 18	1	*c*F. *p*L. R. *vlb*M.	5238
824	1790 Mar. 8	7 (*a*) Hyd. Crat.	*f*	7 26	*s*	1 9	1	*v*F. *v*S. *i*R. *glb*M.	3528*
825	10	39 Lyncis	*p*	12 53	*s*	1 31	1	*v*F. S. R. *b*M. *s* of a S*st*.	2746
826	—	— —	*p*	5 55	*s*	1 56	1	*v*F. S. *r*.	2780
827	—	— —	*f*	2 11	*s*	1 29	1	*e*F. *v*S. *sf* a *v*S*st*.	2840
828	—	Hyd. L. C. 1039	*p*	2 1	*s*	1 11	2	*e*F. *p*S. R. *vgb*M. Stellar. just *p* a *v*S*st*.	3885
829	17	27 Lyncis	*p*	23 49	*n*	1 30	1	*e*F. *v*S. R. *b*M.	2431
830	—	— —	*p*	10 40	*n*	1 19	1	*c*F. *p*S. *b*M.	2474
831	—	15 (*f*) Ursæ	*p*	12 8	*n*	0 23	1	*v*F. *v*S.	2692
832	—	— —	*f*	9 39	*n*	0 57	1	*v*F. S. *l*E.	2800
833	—	26 Ursæ	*f*	134 3	*s*	1 43	1	*v*F. *v*S.	3870
834	—	74 Ursæ	*f*	2 4	*s*	1 56	1	*e*F. S. *i*F.	4511
835	—	77 Ursæ	*f*	82 37	*n*	1 52	1	*e*F. S. E. but nearly R.	5526*

III.	1790.	Stars.		M. S.		D. M.	Ob.	Description.	N.G.C,
836	Mar. 18	17 Ursæ	*p*	79 17	*s*	0 33	1	*v*F. *v*S. may be a patch of stars.	2469
837	—	— —	*p*	75 32	*s*	0 40	1	*e*F. *v*S.	2488
838	—	— —	*p*	75 10	*s*	0 15	1	*e*F. *v*S.	2497
839	—	— —	*p*	72 22	*s*	3 40	1	*e*F. *v*S.	2505*
840	—	— —	*p*	63 56	*s*	1 28	1	*c*F. *c*S.	2534*
841	—	— —	*p*	16 9	*s*	1 9	1	*v*F. S.	2710
842	—	43 Ursæ	*p*	5 8	*s*	0 39	1	*v*F. *v*S. R.	3353*
843	—	66 Ursæ	*p*	19 23	*n*	1 52	1	*v*F. Stellar. *np* a S*st.*	3757
844	—	— —	*p*	16 1	*n*	2 2	1	*v*F. S. *m*E.	3795
845	—	69 (δ) Ursæ	*p*	4 55	*n*	1 17	1	*v*F. S. E. in the par.	4154*
846	19	20 Ursæ	*f*	7 53	*s*	2 23	1	*c*F. S. *m*E. very narrow.	2870
847	—	76 Ursæ	*p*	67 53	*s*	2 50	1	*e*F. *v*S. *i*F.	3740
848	—	69 Ursæ Hev.	*p*	19 5	*n*	2 13	1	*v*F. *v*S.	5007*
849	—	— —	*f*	23 53	*s*	0 8	1	*v*F. *v*S.	5342
850	20	76 Ursæ	*p*	26 56	*n*	3 17	1	*v*F. *p*S.	4210
851	—	— —	*p*	25 25	*n*	0 43	1	*e*F. S. *i*F.	4238
852	—	— —	*p*	16 38	*n*	2 12	1	*v*F. Stellar, *nf* a S triangle of B*st.*	4391
853	Apr. 1	30 (φ) Ursæ	*f*	8 55	*n*	1 35	1	*v*F. S. *vglb*M.	3073
854	Oct. 9	72 Pegasi	*f*	15 8	*s*	0 23	2	2 *v*S close *st.* with nebulosity between.	7760
855 } 856 }	—	— —	*f*	27 15	*n*	0 3::	1	Two nebulæ, both *e*F. Stellar. dist. 1′ from 30° *sp* to *nf.*	7805 7806
857	—	σ Fornacis L. C. 285	*p*	12 30	*s*	1 54	1	*v*F. S. *i*F. *lb*M.	1366
858	10	6 Pegasi	*p*	24 40	*n*	0 43	1	*e*F. *p*L. *i*R. *vlb*M. requires great attention to be seen.	7046
859	—	— —	*p*	7 56	*n*	0 17	1	*c*F. *v*S. *i*R. *mb*M. near a *v*S*st.*	7081
860	Nov. 2	72 Pegasi	*p*	5 19	*n*	1 7	1	*v*F. S. *lb*M.	7680
861	—	— —	*f*	37 50	*s*	0 17	1	*e*F. S.	39
862	8	1 Lacertæ Hev.	*p*	3 17	*n*	1 19	1	*e*F. *p*L. *i*R. *r.*	7223
863	—	— —	*f*	3 9	*n*	0 48	1	*v*F. *v*S. *mb*M.	7248
864	—	— —	*f*	4 37	*n*	0 50	1	*v*F. S. *m*E. 75° *np sf.* *b*M.	7250
865	13	26 Aurigæ	*p*	1 9	*n*	1 31	1	*v*F. *v*S. R. *b*M.	1985
866	26	29 (π) Androm.	*p*	27 37	*s*	0 20	1	*v*F. *v*S. The *np* corner of a square.	13
867	Dec. 6	MAYER'S Zod. Cat. 20	*p*	49 19	*s*	1 39	1	*e*F. *p*S. *i*R. *lb*M.	7797
868	—	— —	*p*	39 35	*s*	0 42	1	*e*F. *p*S. *i*F.	12
869	25	44 Piscium	*f*	3 25	*n*	0 55	1	*v*F. *v*S. *b*M. *p.* and in the field with II. 854. *nf.* 2. S*st.*	125
870	—	— —	*f*	12 48	*n*	0 49	1	*v*F. S. *i*R. *vgb*M.	182
871	28	MAYER'S Zod. Cat. 18	*p*	8 1	*n*	1 44	1	*v*F. S. R. *vgb*M.	173
872	—	— —	*p*	5 52	*n*	0 41	1	*v*F. *v*S. *b*M.	192
873	—	— —	*p*	5 32	*n*	0 39	1	*e*F. *c*L. In the field with the foregoing, and with II. 860	201
874	—	57 Aurigæ	*f*	17 56	*n*	1 50	1	*v*F. *v*S. *l*E.	2322

III.	1790.	Stars.		M. S.		D. M.	Ob.	Description.	N.G.C.
875	Dec. 28	57 Aurigæ	*f*	21 42	*s*	0 7	1	*v*F. *v*S.	2329
876	29	51 Piscium	*f*	5 44	*n*	1 43	1	*v*F. *p*L. *i*R. *sf* a S*st* which is partly involved in the nebulosity.	180
	1791								
877	Feb. 23	26 Hydræ	*p*	73 56	*n*	0 22	1	*v*F. *i*R. *r*. 2′*d*. almost of equal light throughout.	2525
878	Apr. 2	14 (τ) Ursæ	*f*	9 14	*n*	0 38	2	*v*F. *c*L. R. *mb*M. near 5′ *d*.	2805
879	—	73 Ursæ	*p*	2 39	*s*	1 12	1	*c*F. S. *i*F.	4384
880	—	— —	*f*	8 13	*s*	1 26	1	*e*F. S.	4566
881	3	35 Ursæ	*f*	21 51	*n*	0 13	1	*v*F. S.	3392
882	May 6	9 Ursæ min.	*p*	34 52	*s*	2 0	1	*v*F. *p*L. R. *b*M.	5671*
883	—	13 (γ) Ursæ min.	*f*	42 41	*s*	1 36	1	*e*F. *v*S. ver. 300.	6071*
884	—	— —	*f*	44 51	*s*	2 22	1	*v*F. *v*S. with 300 *c*L.	6079*
885	24	37 (ξ) Bootis	*p*	3 44	*s*	0 35	1	*e*F. *v*S. E. near par.	5760
886 }	26	7 Serpentis	*p*	15 32	*n*	0 20	1	Two nebulæ, both *e*F. *v*S. the *p* is the most *n*. dist. 1½′.	5851
887 }									5852
888	27	19 (ξ) Coronæ	*p*	6 41	*n*	1 7	1	*e*F. *v*S. R. with 300 *p*L.	6103
889	28	17 (σ) Coronæ	*p*	2 1	*s*	0 52	1	*v*F. S. R. *vglb*M.	6089
890	—	20 (ν′) Coronæ	*f*	8 9	*n*	1 20	1	*v*F. *p*L. *l*E. *lb*M.	6177
891	30	25 Herculis	*p*	3 41	*n*	0 37	1	*e*F. *v*S. R. *lb*M.	6129
892	—	— —	*p*	2 5	*s*	0 9	1	*e*F. S. *b*M.	6142
893	—	44 (η) Herculis	*p*	6 26	*n*	0 8	1	*e*F. *v*S. *i*F. ver. 300.	6195
	1792								
894	Apr. 20	22 (*f*) Bootis	*f*	12 29	*n*	1 15	1	*v*F. *v*S.	5702
895	—	— —	*f*	12 55	*n*	0 47	1	*v*F. *v*S.	5710
896	—	— —	*f*	16 45	*s*	0 25	1	*e*F. S. *vlb*M.	5737
	1793								
897 }	Feb. 4	34 (θ) Gemin.	*p*	1 33	*s*	0 31	1	Two nebulæ. The most *n*. and *p*. *e*F. S. The other *e*F. *v*S. dist. 4′.	2290
898 }									2289
899	—	— —	*f*	15 18	*n*	1 17	1	*v*F. S. nearly R. *b*M.	2333
900 }	—	— —	*f*	36 21	*n*	0 9	1	Two nebulæ just preceding III. 703. Both *e*F.	2385
901 }									2388
902	Mar. 8	18 Navis	*f*	10 36	*n*	0 32	1	*v*F. *l*E. *r*. *b*M.	2578
903	Apr. 6	4 Draconis	*p*	30 43	*n*	0 10	1	*e*F. S. *i*F. *vlb*M.	4034
904	—	— —	*p*	23 25	*n*	0 24	1	*e*F. *v*S. E. mer.	4120
905	7	— —	*p*	37 3	*n*	0 8	1	*e*F. *v*S. ver. 300.	3961
906	—	6 Draconis	*f*	12 31	*n*	1 8	1	*v*F. E. 2′ *l*, ½′ *b*.	4693
907	—	— —	*f*	16 26	*n*	1 35	1	*v*F. E. *np sf*. 1½′ *l*, ½′ *b*.	4749*
908	—	— —	*f*	23 36	*n*	0 10	1	*e*F. *v*S. *i*R. *vlb*M.	4857
909	—	— —	*f*	39 10	*n*	0 35	1	*v*F. *v*S. R.	5034*
910	8	37 Ursæ	*p*	15 47	*n*	0 19	1	*v*F. *p*L. *i*F. *r*. some of the stars visible.	3188
911	—	— —	*p*	11 47	*s*	0 5	1	*v*F. *c*L. *i*F.	3220
912	—	— —	*f*	0 59	*n*	1 27	1	*e*F. *v*S. ver. 300.	3284*
913	—	39 Ursæ	*f*	8 14	*n*	1 14	1	*v*F. *v*S.	3408
914	—	— —	*f*	10 29	*s*	0 2	1	*v*F. S. *l*E.	3440
915	—	— —	*f*	25 35	*n*	0 3	1	*v*F. S.	3530

III.	1793.	Stars.		M. S.		D. M.	Ob.	Description.	N.G.C.
916	Apr. 9	42 Ursæ	*p*	48 48	*n*	0 39	1	*e*F. *v*S. Stellar near a S*st.*	3102
917	—	— —	*p*	15 19	s	0 44	1	Two nebulæ.	3286
918				15 10		0 47		Both *v*F. *p*S. R. *lb*M.	3288
919	—	— —	*p*	0 1	*n*	2 2	1	*v*F. *v*S. near a *v*S*st.*	3407
920	—	— —	*f*	19 23	*n*	2 1	1	*e*F. *v*S. E. near mer.	3543
921	—	— —	*f*	24 11	*n*	1 22	1	*e*F. *p*L. E.	3589
922	—	— —	*f*	35 14	*n*	1 11	1	*v*F. *v*S. 2*v*S. stars in it.	3671
923	May 5	Hydr. L. C. 1179	*p*	1 25	*n*	0 5	1	*v*F. *v*S. R. *lb*M.	5328
924	—	6 Hydræ conti	*f*	11 2	s	1 27	1	*e*F. S. *r.* ver. 300.	5592
925	12	64 Virginis	*f*	1 18	*n*	1 10	1	*c*F. S.	5118
926	—	— —	*f*	13 5	*n*	1 17	1	*v*F. S. *sp* a *c*B*st.*	5224
927	—	19 Bootis Hev.	*p*	0 20	*n*	0 44	1	*v*F. S.	5599
928	13	93 (τ) Virgin.	*p*	26 17	s	0 5	1	*v*F. S.	5227
929	—	— —	*p*	9 25	*n*	0 35	1	*v*F. S. E. mer.	5331
930	Sept. 6	53 Aquarii	*p*	27 19	*n*	0 18	1	*e*F. ver. 300.	7165
931	—	— —	*p*	12 23	s	0 19	1	*e*F. S. *i*R.	7230
932	—	— —	*p*	8 50	*n*	1 11	1	*e*F. S. *l*E. *s* of a S*st.* to which it seems almost to be attached, but is free from it. The star is the 1st of 3, making a S triangle.	7246
933	—	— —	*p*	6 7	*n*	0 58	1	*v*F. S. R. *b*M.	7251
934	1794 Apr. 1	Georgian planet	*p*	0 16	s	0 2	1	*v*F. This nebula was seen at $9^h\ 45'$, sidereal time, the telescope being out of the meridian.	3080
935	19	12 (δ) Hydræ crateris	*f*	15 11	*n*	0 40	1	*e*F. S. *b*M.	3734
936	Oct. 15	5 (α) Cephei	*f*	7 54	*n*	0 16	1	*v*F. *er.*	7076
937	1797 Nov. 22	Neb. I. 274	*f*	25 3	*n*	0 53	1	*v*F. S. *i*R. *b*M.	4954*
938	Dec. 10	A double st*	*p*	9 5	*n*	0 10	1	*e*F. *p*L. *i*F. *See I. 273.	4363
939	—	— —	*f*	4 0	s	0 35	1	*e*F. S.	4572
940	12	5 Dracon. Hev.	*p*	32 24	s	0 49	1	*v*F. S. R. *b*M.	3890*
941	—	— —	*p*	8 21	*n*	0 57	1	*v*F. *p*S. 2 S *nf* stars make a triangle with it.	4159
942	—	— —	*f*	4 16	*n*	0 59	1	*e*F. E. near mer. ver. 300.	4331
943	—	5 (α) Ursæ mi.	*f*	46 2	s	0 28	1	Two nebulæ.	5909
944								Both *v*F. *v*S. *r.* dist. 1½′ par.	5912
945	—	35 Draconis	*p*	47 10	s	1 17	1	*v*F. S. E. *n* of a S*st.*	6324
946	20	4 (*b*) Ursæ mi.	*p*	29 31	*n*	1 57	1	*v*F. *v*S. R.	5295*
947	—	— —	*p*	14 39	*n*	0 42	1	*v*F. *c*L. *i*F. *vlb*M. *s* of a *p*B. *st.*	5452
948	—	— —	*f*	2 20	*n*	1 3	1	*e*F. *v*S. E. near mer.	5547
949	—	— —	*f*	14 44	*n*	2 29	1	*e*F. S. *l*E. near par.	5640*
950	—	— —	*f*	24 18	*n*	1 13	1	*v*F. S. *r.* It is preceded by a S. patch of *st.* which appears almost like this nebula, but more resolved.	5712

III.	1797.	Stars.		M. S.		D. M.	Ob.	Description.	N.G.C.
951	Dec. 20	4 Cephei of BODE's Cat.	*p*	21 18	*s*	1 25	1	*e*F. S. better with 320.	6331
	1798								
952 } 953 }	Dec. 9	2 (π^{2}) Orionis	*p*	10 20	*s*	1 34	1	Two nebulæ within 1′ of each other; mer. Both *v*F. *v*S.	1633 1634
954	10	8 Ceti	*f*	17 5	*s*	1 15	1	*e*F. S.	163*
955	—	21 Ceti	*p*	3 46	*n*	0 4	1	*c*F. *v*S. *i*R.	270
956	—	18 (ε) Eridani	*p*	15 21	*s*	0 53	1	*v*F. *v*S. 2 or 3′ *n* of 2 S*st*.	1284
	1799								
957 } 958 }	June 29	93 Herculis	*p*	3 59	*n*	1 37	1	Two; both *v*F. *v*S; place that of the *f* one, *p* one about 4′ more *s* and 5 or 6ˢ *p*.	6500 6501
959	Dec. 19	16 Eridani	*f*	6 37	*n*	0 26	1	The 2d of 2 *v*F. *v*S. 1½′ *sf* I. 60. I.C.	324*
960	—	19 Eridani	*f*	1 19	*n*	1 13	1	*v*F. *v*S. ver. 300.	1362
961	—	— —	*f*	2 43	*n*	0 46	1	*v*F. *v*S.	1377
962	—	— —	*f*	20 51	*n*	1 15	1	*v*F. *v*S. *sp*. 2*p*B*st*.	1482
	1801								
963	Apr. 2	Star 6·7 m. [B. 1446]	*p*	59 37:	*n*	0 17	1	*e*F. S. *i*F.	..*
964	—	— —	*p*	21 56	*s*	1 32	1	*c*F. S. Stellar. ver. 300. just *p*. a S*st*.	3144*
965	—	— —	*p*	19 24	*s*	1 23	1	*v*F. *v*S.	3155*
966	—	— 8 m [P. IX. 112]	*f*	31 14:	*s*	0 11	1	*v*F. *v*S.	..*
967 } 968 }	—	— 6·7 m [B. 1446]	*f*	25 48:	*s*	0 19	1	Two nebulæ. The 1st *v*F. S. The 2d *nf* the 1st *e*F. *v*S.	3465* ..*
969	—	— —	*f*	60 27	*s*	1 6	1	*e*F. S.	..*
970	—	208 (N) Camel. of BODE's Cat.	*p*	24 19:	*n*	0 28	1	*v*F. *p*L. *r*.	..*
971	—	Star 6·7 m. [B. 1446]	*f*	77 24	*s*	0 58	1	*e*F. *v*S. R.	3890*
972	Nov. 28	50 (α) Ursæ	*p*	4 54	*s*	0 10	1	*v*F. *v*S. R. *b*M.	3471
973	Dec. 6	16 (ζ) Ursæ mi.	*f*	14 15	*n*	1 8	1	*v*F. S. *l*E. mer. *r*.	6068
	1802								
974 } 975 }	Jan. 1	22 (ε) Ursæ mi.	*p*	10 49	*n*	0 37	1	Two nebulæ; the preceding *c*F. S. *b*M. the foll. *v*F. *v*S. it follows the 1st a few seconds, and is about 3′ more north. The place is that of the first.	6251 6252
976	May 21	2 (η) Coronæ	*p*	26 50	*n*	0 2	1	*e*F. S. *i*F.	5789
977	Sept. 26	186 P. Camelo. of BODE's Cat.	*f*	9 49	*s*	1 33	1	*e*F. *v*S. 300 confir.	2908*
978	—	— —	*f*	33 19	*s*	0 58	1	*e*F. *p*L. *l*E. *vlb*M. just *n* of 2 S*st* that are nearly in the parallel.	3057*

III.	1802.	Stars.		M. S.		D. M.	Ob.	Description.	N.G.C.
979 980 981	Sept. 26	191 Camelop. Bo.	*p*	7 44	*s*	0 38	1	Three, the place is that of the last. Two last *v*F*v*S, *p* one stellar; all in a line, about 1′ dist. from each other, *p* one most *n*, about 2′ more than the last.	3210* 3212* 3215*
982 983	30	24 Ursæ Bode	*f*	3 19	*n*	2 39	1	Two, the place is that of the last, the other about 42^s *p*. 6′ *n*; *p* one stellar, S∗1′ *f*; *f* one *v*F. S.	2629* 2641*
984	1784 Nov. 17	86 Pegasi	*p*	3 14	*s*	0 24	1	Suspected, 240 shewed 2 S*st*	7810*
985	1791 Mar. 24	73 Ursæ	*f*	20 11	*s*	1 17	1	*e*F. *p*S.	4695*

Fourth Class. Planetary Nebulæ,

Stars with Burs, with milky Chevelures, with short Rays, remarkable Shapes, &c.

IV.	1789.	Stars.		M. S.		D. M.	Ob.	Description.	N.G.C,
59	Mar. 23	55 Ursæ	*f*	4 51	*n*	0 23	1	*c*B. S. R. BN. The N is considerably well defined, and the chevelure *v*F.	3658
60	Apr. 12	36 Ursæ	*f*	8 37	*s*	2 28	2	*v*B. R. Planetary, but very ill defined. The indistinctness on the edges is sufficiently extensive to make this a step between planetary neb. and those which are described *vsmb*M.	3310
61	—	64 (γ) Ursæ	*f*	3 56	*s*	0 19	2	*c*B. B*r*N with *v*FE branches about 30° *np sf*. 7 or 8′ *l*, 4 or 5′ *b*.	3992
62	14	— —	*f*	2 27	*n*	1 25	1	*c*B. quite R. A large place in the middle is nearly of an equal brightness. Towards the margin it is less bright.	3982
63	24	69 Ursæ Hev.	*f*	1 24	*s*	1 33	1	*c*B. *c*L. *i*R. *er*. *vgmb*M. 4′ diam. I suppose, with a higher power, I might have seen the stars.	5204
64	1790 Mar. 4	6 Navis	*p*	7 41	*s*	1 2	2	A beautiful planetary nebula, of a considerable degree of brightness; not very well defined, about 12 or 15″ diam.	2440

IV.	1790.	Stars.		M. S.		D. M.	Ob.	Description.	N.G.C.
65	Mar. 5	28 Monocerotis	*p*	51 49	*n*	0 26	1	A pretty considerable star, 9 or 10m. visibly affected with *v*F. nebulosity, of very little extent all around. A power of 300 shewed the same, but gave a little more extent to the nebulosity. The 22d Moncerotis was quite free from nebulosity.	2346*
66	18	17 Ursæ	*p*	16 29	*s*	3 6	1	A small star with a *p*B. fan-shaped nebula. The star is on the *p* side of the diverging chevelure, and seems to be connected with it.	2701
67	—	66 Ursæ	*p*	0 39	*n*	1 55	1	*p*B. *p*L. R. The greatest part of it equally B, then fading away *p* suddenly; between 2 and 3′ diam.	3963
68	19	44 Lyncis	*p*	4 15	*n*	1 44	1	*v*B. S. exactly R. BNM. and *v*F. chev. *vg*. joining to the N. In a lower situation the chev. might not be visible, and this neb. would then appear like an ill defined planetary one.	2950
69	Nov. 13	{ 26 Aurigæ or 31 Hevelii	*p* *f*	88 24 24 59	*s* *s*	0 11 1 26	1	A most singular phenomenon; A *st* 8m. with a faint luminous atmosphere of a circular form, about 3′ in diam. The star is perfectly in the centre, and the atmosphere is so diluted, faint, and equal throughout, that there can be no surmise of its consisting of stars, nor can there be a doubt of the evident connection between the atmosphere and the star. Another star, not much less in brightness, and in the same field with the above, was perfectly free from any such appearance.	1514
70	1791 May 6	6 Draconis	*f*	50 27	*n*	0 27	2	*c*B. R. almost equally B throughout, resembling a very ill defined planetary neb. about ½′ diam.	5144

IV.	1791.	Stars.		M. S.		D. M.	Ob.	Description.	N.G.C
71	May 24	37 (ξ) Bootis	*f*	16 5	*s*	0 44	1	A star 7·6m. enveloped in extensive milky nebulosity. Another star 7m. is perfectly free from such appearance.	5856
72	1792 Sept. 15	34 Cygni	*p*	5 10	*n*	0 23	1	A double star of the 8th magnitude, with a faint south-preceding milky ray joining to it, 8′ *l*, and 1½′ broad.	6888
73	1793 Sept. 6	16 (*c*′) Cygni	*f*	2 51	*s*	0 1	1	A bright point, a little extended, like two points close to one another ; as bright as a star of the 8·9 magnitude, surrounded by a very bright milky nebulosity suddenly terminated, having the appearance of a planetary nebula with a lucid centre; the border however is not very well defined. It is perfectly round, and I suppose about half a minute in diam. It is of a middle species, between the planetary nebulæ and nebulous stars, and is a beautiful phenomenon.	6826
74	1794 Oct. 18	7 Cephei	*p*	24 57	*n*	1 22	1	A star 7m. very much affected with nebulosity, which more than fills the field. It seems to extend to at least a degree all around; smaller stars, such as 9 or 10m. of which there are many, are perfectly free from this appearance. A star 7·8m. is perfectly free from this appearance.	7023
75	—	7 Cephei	*f*	14 40	*s*	0 46	2	Three stars about 9m. involved in nebulosity. The whole takes up a space of about 1½′ diam. other stars of the same size are free from nebulosity.	7129
76	1798 Sept. 9	3 (η) Cephei	*p*	10 31	*s*	1 36	1	*c*F. *v*L. *i*F. a sort of BNM. The nebulosity 6 or 7′. The N seems to consist of stars, the nebulosity is of the milky kind. It is a pretty object.	6946

IV.	1798.	Stars.		M. S.		D. M.	Ob.	Description.	N.G.C.
77	Dec. 19	16 Eridani	*f*	4 56	*n*	0 14	1	A star about 9 or 10m. with a nebulous ray to the south-preceding side. The ray is about 1½′ long. The star may not be connected with it.	1325
78	Nov. 8	8 Ursæ min. of BODE's Cat.	*p*	25 0	*n*	0 12	1	*c*B. R. about 1½′ diam. Somewhat approaching to a planetary nebula, with a strong hazy border.	4750

Fifth Class. Very large Nebulæ.

V.	1789.	Stars.		M. S.		D. M.	Ob.	Description.	N.G.C.
45	Apr. 12	64 (γ) Ursæ	*f*	0 9	*s*	1 23	2	*c*B. *i*F. E. mer. LBN. with F. branches 7 or 8′ *l*, 5 or 6′ *b*.	3953
46	17	48 (β) Ursæ	*f*	10 4	*s*	0 41	2	*v*B. *m*E. *r*. 10′ *l*, 2′ *b*. There is an unconnected pretty bright star in the middle.	3556
47	1790 Apr. 1	30 (φ) Ursæ	*f*	10 9	*n*	1 39	1	*v*B. *m*E. *np sf*. *vgmb*M. 8′ *l*, 2′ *b*.	3079
48	Oct. 9	ι Fornacis L. C. 182	*f*	8 7	*s*	0 2	1	*v*B. E. 75° *np sf*. 8′ long. A very bright nucleus, confined to a small part, or about 1′ diam.	1097
49	Dec. 28	41 Persei Hev.	*f*	22 0	*n*	0 15	1	6 or 7 small stars, with faint nebulosity between them, of considerable extent, and of an irregular figure.	1624
50	1793 Mar. 4	ε Pixidis Na. L. C. 831	*f*	35 26	*s*	0 43	1	*v*F. *v*S. *l*E. 15° *sp nf*. *lb*M. 8′ *l*, 5 or 6′ *b*.	2997
51	Apr. 6	4 Draconis	*p*	14 48	*n*	0 20	2	*v*F. *m*E. 70° *np sf*. About 25′ *l*, and losing itself imperceptibly, about 6 or 7′ broad.	4236
52	Nov. 28	50 (α) Ursæ	*p*	17 49	*n*	1 30	1	*c*B. E. mer. *vgb*M. About 5′ *l*. and 3′ broad; the nebulosity seems to be of the milky kind; it loses itself imperceptibly all around. The whole breadth of the sweep seems to be affected with very faint nebulosity.	3359

Sixth Class. Very compressed and rich Clusters of Stars.

Additional Abbreviations. cl. *Cluster*, com. *compressed*, sc. *scattered*, co. *coarsely*.

VI.	1790.	Stars.		M. S.		D. M.	Ob.	Description.	N.G.C.
36	Mar. 4	6 Navis	*p*	8 45	*s*	1 55	2	A *v*. com. cl. of S, and some L*st*. E near mer. The most compressed part is about 8′ *l*, and 2′ *l*. with many scattered to a considerable distance.	2432
37	1791 Feb. 23	26 Hydræ	*p*	79 30	*n*	1 0	1	A *v*. com. and very rich cl. of stars. The stars are of 2 sizes, some considerably L. and the rest next to invisible. The com. part 5 or 6′ in diam.	2506
38	Aug. 25	50 (γ) Aquilæ	*p*	14 50	*s*	1 18	1	*c*B. S. *i*F. *er*. Some of the *st*. are visible.	6804
39	1793 Mar. 3	ζ Pixidis Naut. L. C. 777	*p*	20 39	*s*	0 19	2	A cl. of L*st*. considerably rich *i*R. above 15′ diam.	2571
40	May 12	53 (ν) Serpentis	*p*	48 17	*n*	0 2	1	A very beautiful *e* com. cl. of *st*. extremely rich, 5 or 6′ in diam. gradually more compressed towards the centre.	6171
41	1797 Dec. 12	35 Draconis	*p*	22 6	*s*	1 7	1	R. *r*. about 3′ diam. *vgb*M. I suppose it to be a cluster of stars extremely compressed. 300 confirms the supposition, and shews a few of the stars ; it must be immensely rich.	6412
42	1798 Sept. 9	3 (η) Cephei	*p*	13 26	*s*	1 6	1	A beautiful compressed cl. of S*st*. extr. rich, of an *i*F. The preceding part of it is round, and branching out on the following side, both towards the *n*. and towards the *s*. 8 or 9′ in diam.	6939

Seventh Class. Pretty much compressed Clusters of large or small Stars.

VII.	1788.	Stars.		M. S.		D. M.	Ob.	Description.	N.G.C.
56	Dec. 16	11 (β) Cassiop.	*p*	9 57	*n*	2 6	1	A *p*. com. cl. of S*st*. of several sizes, cons. rich. E. near par. 5 or 6′ *l*.	7790
57	31	40 Aurigæ	*f*	8 28	*n*	1 25	1	A compressed cl. of *v*S. stars *i*F. 6′ diam. consid. rich.	2192

VII.	1790.	Stars.		M. S.		D. M.	Ob.	Description.	N.G.C.
58	Mar. 4	6 Navis	*f*	5 18	*s*	0 29	1	A *p*. com. and rich cl. of S stars *i*R. 7 or 8′ diam.	2479
59	Sept. 11	18 (δ) Cygni	*f*	18 38	*s*	1 4	1	A *v*. rich cl. of L*st*. considerably compressed, above 15′ diam. by the size of the *st*. it is situated in the milky-way, towards us.	6866
60	Dec. 28	47 (λ) Persei	*f*	3 30	*s*	0 50	1	A L. cl. of *c*L. *st*. *p*. com. and very rich. *i*R. 7′ diam.	1513
61	—	41 Persei Hev.	*p*	3 8	*n*	0 56	1	A beautiful cl. of L*st*. *v* rich, and considerably com. about 15′ diam.	1528
62	1791 Aug. 21	19 Aquilæ	*p*	0 26	*s*	1 24	1	A S. *p*. com. cl. of stars not very rich.	6756
63	1793 Mar. 3	ζ Pixidis Naut. L. C. 777	*p*	2 25	*s*	0 24	2	A L. cl. of scattered S*st*. *i*F. considerably rich.	2627
64	4	— —	*p*	20 55	*s*	1 9	1	A L. cl. of *st*. of a middling size. *i*E. considerably rich. The stars are chiefly in rows.	2567
65	8	2 Navis	*p*	16 10	*n*	0 38	1	A S. cl. of *v*S *st*. considerably rich and compressed.	2401
66	1794 Oct. 18	7 Cephei	*f*	16 45	*s*	1 7	2	A cl. of cons. com. *v*S. and L. stars about 12′ diam. considerably rich.	7142
67	1799 Jan. 30	15 (π′) Canis	*f*	42 33	*s*	0 14	2	A cl. of com. stars, considerably rich.	2421

Eighth Class. Coarsely scattered Clusters of Stars.

VIII.	1788.	Stars.		M. S.		D. M.	Ob.	Description.	N.G.C.
79	Dec. 16	11 (β) Cassiop	*f*	20 35	*n*	1 5	1	A coarsely sc. cl. of L*st*. mixed with smaller ones, not very rich.	129
80	18	1 Camelopar.	*p*	41 36	*s*	1 29	1	A cl. of S. stars, containing one large one, 10; 9m. 2 or 3′ diam. not rich.	1444
81	1789 July 18	5 Vulpeculæ	*p*	2 46	*n*	2 4	1	A sc. cl. of *c*L. *st*. *i*F. pretty rich, above 15′ in extent.	6793
82	1790 Sept. 11	57 Cygni	*f*	1 0	*n*	0 52	1	A L. cl. of *p*S. stars of several sizes.	6989
83	30	51 Cygni	*p*	25 24	*s*	0 1	1	A cl. of sc. stars, above 15′ diam. pretty rich, joining to the milky-way, or a projecting part of it.	6895

VIII.	1790	Stars.		M. S.		D. M.	Ob.	Description.	N.G.C.
84	Dec. 28	33 (*a*) Persei	*f*	9 14	*n*	1 36	1	A cl. of S*st*. not very rich.	1348
85	—	41 Persei Hev.	*f*	2 42	*s*	0 2	1	A coarsely sc. cl. of L*st*. pretty rich.	1545
86	1792 Sept. 15	34 Cygni	*p*	9 43	*n*	0 15	1	A coarsely sc. cl. of L. stars, of a right-angled triangular shape.	6874
87	1793 Mar. 8	2 Navis	*p*	7 10	*s*	0 15	1	A small cl. of S. stars, not very rich.	2425
88	1799 Dec. 28	46 (*ξ*) Persei	*p*	27 13	*n*	1 29	2	A cl. of coarsely sc. L*st*. about 15′ diam.	1342

[*Notes to the Third Catalogue of Nebulæ.*

I. 246. Second obs., Sw. 1038, Apr. 8, 1793, 39 Ursæ, f. 44^m 5^s, s. 0° 2′. Both these obs. give the P.D. 4′ or 5′ too small. In the first (Sw. 951) there is at the end a note to the effect that the line was contracted by moisture, 16′ by the quadrant at beginning and end. In Sw. 1038 is I. 271, the P.D. of which is also 5′ too small. Yet the zero is the same by three stars, but they are all at the beginning of the sweep.

I. 262. R.A. is 92^s too great. Probably a reduction to the meridian has been forgotten, as in the case of II. 879 and several stars. Other nebulæ in this sweep (1036) are all right, viz. II. 880, I. 263 and V. 51.

I. 264. R.A. is 60^s too small, which is apparently caused by a reduction to the meridian of -63^s.

I. 271. See I. 246.

I. 272. Second obs., Mar. 9., 1796, 10^h 41^m S.T. "The neb. of March 4 is about 7′ or 8′ s. of the Geor. Planet and a few degrees more p. in position than the second satellite, which almost points to it." The P.D. from the first obs. is 6′ too small. As the P.D. of Uranus on March 9 was 79° 41′·0, the P.D. of the neb. on that day comes out about 8′ greater than on March 4, and is much nearer the truth.

I. 278, 279. *P.T.* has followed Sw. 1068. But the P.D.'s seem to have been interchanged, as the obs. makes the f. one 1° 53′ n. of the p. one instead of *vice versâ*. In Sw. 1074, Dec. 20, 1797, only I. 279 was seen, 4 Drac. Hev., p. 4^m 38^s, s. 0° 46′, which is correct.

I. 282, 283, 284. The three brightest of the fifteen nebulæ observed by H. in Sw. 1096, April 2, 1801. He referred them all to "208 (N) Camelop. of BODE's Cat.," which is 4 H. Draconis = B. 1634. The following four stars were observed:—

9^h 37^m 24^s	11° 9′	* 8 m.	[G. 1561 = P. IX. 112, magn. 6·3]
10 28 40	13 22	* 6·7 m	[B. 1446 = G. 1650, magn. 4·6]
12 6 22	11 57	* 7 m	[B. 1633 = F. 2024, magn. 6·2]
12 6 38	11 49	* 6 m	[B. 1634 = G. 1859, magn. 4·6].

To most of the transits corrections to centre of field have been applied, which adds to the uncertainty of the resulting places due to the small Polar Distance. All the objects have been identified on plates taken at Greenwich in 1911 (*M.N.* vol. 71, p. 509), the places for 1860 being—

I. 282	9^h 29^m 42^s	14° 30′·5	
I. 283	10 9 42	15 7·4	= N.G.C. 3183, d'A.
I. 284	10 32 44	12 27·6	= N.G.C. 3329, h. 733, d'A.

See below, under II. 903–905 and III. 963–971.

II. 781. Also observed in Sw. 929, Apr. 26, 1789, 5 Canum, p. 20^m 31^s, n. 1° 0′, in good accordance with the first obs. No modern obs. known.

II. 794, 795, 796; III. 778–783. These groups were observed in Sw. 921 (Apr. 14 1789) and in Sw. 1001 (Mar. 24, 1791). In the former there are four stars, 77 (ε), 79 (ξ), 81 and 83 Ursæ maj. There is *not* any error in the obs. of P.D. of 77 Ursæ (as asserted by h. in G.C. p. 31); it agrees perfectly with the others, though the transit seems to require a correction of $+20^s$. All the details in *P.T.* are from Sw. 921, and the places derived from them by Auwers are distinctly better than those from Sw. 1001. But h. is right in thinking that the two observations supposed to belong to II. 794 refer to different objects. The second one has here been called II. 910. What put him out was probably that he did not know of N.G.C. 4669 (d'A.) which is =III. 778. The R.A.'s of the second group are very inaccurate, but there are five objects within 72^s. In Sw. 1001 there are three stars, Harv. 4038, 73 Ursæ and G. 1903; in the last there is an error of 1^m. In the following table the places are for 1860.

	Sweep 921, Auwers.	Sweep 1001.	Modern Observations.	Observer.
II. 794	$12^h\ 36^m\ 26^s$ 34° 6′	....	$12^h\ 36^m\ 19^s$ 34° 4′·7	d'A.
II. 910	....	$12^h\ 36^m\ 18^s$ 34° 18′	36 27 34 22·5	d'A.
III. 778	12 38 46 34 21	12 38 55 34 25	38 23 34 21·9	d'A.
II. 795	12 39 31 34 30	....	39 10 34 29·5	d'A.
II. 796	12 40 36 34 42	12 40 6 34 37	40 20 34 42·3	d'A.
III. 985	....	12 40 42 34 47	41 14 34 51·5	d'A.
III. 779	12 59 18 32 57	....	59 28 32 56·0	h. 1532
III. 781	13 0 16: 35 39:	....	59 32 35 33·9	Rümker / Howe
III. 783	13 0 18 35 45	....	59 36 35 40·6	h. 1533
III. 782	13 0 30 35 37	....	59 56 35 35·4	Rümker
III. 780	13 0 19 33 36	....	13 0 9 33 34	Bigourdan

II. 797. There must be an error of about 100^s in the transit. A second obs. in Sw. 929, Apr. 26, 1789, 82 Ursæ p. $10^m\ 36^s$, n. 0° 12′ agrees perfectly with Bigourdan ($13^h\ 23^m\ 34^s$, 36° 10″·5 for 1860).

II. 801. This is 14^s f., 6′ n. of I. 232. A second obs. supposed to be of it (Sw. 1003, Apr. 2, 1791) belongs to I. 232.

II. 803. Second obs., Sw. 951, Mar. 18, 1790, 69 δ Ursæ p. $3^m\ 52^s$, n. 0° 44′, agrees perfectly with the first.

II. 810. Not found four times by Bigourdan. In the sweep (928) it comes between III. 812 and II. 811, both of which are nearly correct. It is no doubt =N.G.C. 6127 (Swift IV.) with an error of 20′ in P.D., that read off being 31° 59′ instead of 31° 39′, the true Δ P.D. being 3° 43′.

II. 818. R.A. in G.C. and N.G.C. is 2^m too great (therefore not found by Bigourdan). The comp. star is "12 Drac. Hev. Woll. Cat.," *i.e.* G. 2280; G. 2182 was also observed and both agree. The neb. is probably =I.C. 1100 (Swift IX.), vF. pS. lE.

II. 824. Is =h. 994; C. H. and Auwers both agree with h's place.

II. 825. Is identical with III. 716 (*q.v.* in Second Cat.).

II. 827. In Sw. 948 (the last one that night) there was no Flamsteed star, for which reason 77 Ursæ was taken from the previous sweep. In Sw. 948, II. 827 was $5^m\ 34^s$ f., 10′ s. of I. 238, and $21^m\ 19^s$ f., 2° 10′ n. of G. 2013.

II. 862. Identification difficult, as it is one of a group. In Sw. 990, 57 Aurigæ is the only comparison star and the neb. is 2^s p., 2′ n. of II. 736. Auwers gives for 1860 $7^h\ 0^m\ 8^s$, 39° 37′ It is probably one of Kobold's nebulæ in the I.C.

II. 868. Not seen by d'A., and h. only observed the f. one (II. 869).

II. 872. Is 21^s p., 2′ s. of III. 881, while h. gives it as 24^s f., 2′ s. of it. H − h = -53^s. Sweep (1004) examined.

II. 878. P.D. is 5′ too great (* D.M. +56° ·2331, $12^m\ 35^s$ f., 34′ n., gives 2′ less) owing to a change in the P.D. cord noted in the sweep.

II. 879. Sw. 1036, the only one that night (Apr. 6, 1793); the stars disagree badly in R.A. The resulting R.A. of the neb. is $1^m\ 47^s$ too great. There is a * 7m. $10^m\ 25^s$ p., 3′ s. of the neb., which must be +68°·632, but though right in P.D., it gives the R.A. 35^s too small. Something has been erased in the transit column between this star and the neb. II. 880 comes next, the place being correct; then comes I. 262 (*q.v.*).

II. 881. Not found by d'A. and Bigourdan. Sweep examined.

II. 903, 904, 905. See above, under I. 282. The Greenwich places are for 1860—

9^h	42^m	10^s	$13°$	$28'{\cdot}7$
10	52	43	14	8·0
11	23	46	14	36·0

II. 908. Omitted in the *P.T.*

II. 909. Omitted in the *P.T.* "Three, the place is that of the last, which is F. pL. R. The sp. one eF. vS., about 1′ more south and 1 F. = 20″ preceding [*i.e.* 20^s]. The np. one pB., stellar, about 3′ more north than that of which the place is taken and 1·5 F. = 30″ preceding." These three are N.G.C. 3063 = II. 333, 3065 = II. 334, and 3066 = II. 909.

II. 910. See above, under II. 794.

III. 751. Place is from the second obs. in Sw. 908, Feb. 22, 1789. It agrees well with the first (Sw. 902), * 7 [Lund 4808], p. 20^m 18^s, s. 1° 52′. The R.A. of h. (one obs.) is $\frac{1}{2}^m$ less.

III. 753. Sw. 907. R.A. is 37^s too small; so is the R.A. of the only other neb. in this short sweep (a second obs. of III. 606). "Extremely windy," and clock error from 33 Cancri differs 35^s from that of the previous sweep.

III. 769. Not found by Bigourdan.

III. 773. This is certainly = II. 830 (40^s f. the place of III. 773) which has a * 13 on the p. edge. In the sweep (920) III. 773 is 1^m 28^s f. I. 227.

III. 774. A second obs. in Sw. 946, Mar. 17, 1790, has * 6m. [G. 1807], f. 5^m 17^s, n. 2° 11′.

III. 776. Not found by Bigourdan.

III. 778–783. See above, under II. 794.

III. 791. The description is ambiguous: "Two, cB. R. vgmbM., has another p. vF. R. S., nearly in the mer. 3′ or 4′ dist. prec.," with a note added afterwards to the word "mer.": "By the description it should be perhaps nearly in the parallel." In a second obs. of April 2, 1791, H. saw only I. 232. Bigourdan has a neb. 4^s f. II. 801 on the parallel, but 3′ or 4′ p. I. 232 neither he nor d'A. saw any neb.

III. 794. Not found by Bigourdan.

III. 797. Also observed in Sw. 953, Mar. 19, 1790, 76 Ursæ p. 27^m 59^s, s. 2° 50′, or I. 253, f. 13^m 47^s, s. 1° 59′. This agrees well with the place of Bigourdan, which is = N.G.C. -37^s +4′.

III. 800–801. Very probably the word "two" refers to III. 799 and III. 800, as nobody seems to have seen three nebulæ in the place.

III. 803. Observed twice. Sw. 924, Apr. 17, 1789, "eF. vS. I was too late to verify with 300. I had, however, a single glimpse which seemed to confirm it. 12 ι Draconis p. 1^h 50^m 41^s, s. 1° 32′." Sw. 926, Apr. 24, 1789. Suspected eF. vS., but may be a deception, probably 2 S. close stars. 69 Ursæ Hev. Woll. Cat. [= G. 2002] f. 9^m 33^s, s. 2° 53′." In Sw. 924 it is 3^m 19^s p., 4′ s. of a star 6m. which is G. 2030; this gives for 1860 13^h 31^m 57^s, 32° 8′·8, or 1^m less than the result from 926. G. C. has taken the mean of the two, and Bigourdan could not see anything in that place.

III. 808. Is no doubt identical with II. 826, both observed once only and in different sweeps.

III. 812. In the sweep (928) there is a nearer star, G. 2296, f. 13^m 6^s, n. 2° 40′.

III. 821. According to the sweep (929) it is 1^m 7^s p., 1′ n. of the star L.L. 24969 (+53°·1622). Not seen by Bigourdan.

III. 824. There is an error of reduction of 6^m in the G.C. and the nebula is identical with h. 3316.

III. 835. Is = III. 804.

III. 839. A nearer comparison star is G. 1429, p. 20^m 8^s, n. 0° 12′.

III. 840. The P.D. is 9′ too great, probably caused by an error of 10′ in reading off the quadrant.

III. 842. R.A. is 40^s too great. Reduction to centre of field -40^s, evidently underestimated.

III. 845. Not found by Bigourdan.

III. 848. A better star is G. 1965, p. 23^s, s. 0° 4′.

III. 882. A better star is G. 2091, f. 16^m 20′, n. 0° 13′, which agrees perfectly with Bigourdan's place (14^h 25^m 39^s, 19° 40′·8), while that derived from 9 Ursæ is 44^s out.

III. 883. G. 2091 is nearer in P.D.; in the sweep (1005) the neb. f. 1^h 54^m 53^s, n. 0° 38′, which gives for 1860 16^h 2^m 45^s, 19° 8′ in good agreement with Bigourdan.

III. 884. In the same sweep as III. 883, and its R.A. is also nearly 1^m too great. The neb. followed same star G. 2091, 1^h 57^m 3^s, s. 8′, which gives 16^h 5^m 4^s, 19° 54′, agreeing well with Bigourdan.

III. 907. Bigourdan's R.A. is 1^m 13^s greater. In the sweep (1037) there is not any star nearer in P.D. than 6 Draconis, but I. 264 is 35^m 26^s p., 47′ s. of III. 907, which gives 12^h 45^m 36^s, 17° 38′, much nearer to B.'s place.

III. 909. In same sweep. Bigourdan's R.A. is 1^m greater than that of Auwers. IV. 70 followed 11^m 17^s, 8′ s., which gives 13^h 7^m 47^s, 18° 38′ or Auwers $+36^s$ +1.

III. 912. Not found by Bigourdan. In the sweep (1038) it precedes III. 913 16^m 0^s, 5′ north, so it is no doubt identical with either III. 917 or 918, which were observed the following night (Sw. 1039) without any mention of III. 912.

III. 937. In the sweep (1064) the observation of I. 274 seems to be inaccurate, but III. 937 is between two well determined stars (Kasan 2331 and 2388).

	* 6·7, f. 8^m 52^s, s. 2′.	
	* 7·8, p. 25 40, s. 31.	
These give respectively,	12^h 58^m 29^s	13° 52′
	58 4	49

and III. 937 is therefore = h. 1527.

III. 940. This is the same as III. 971. R.A. in N.G.C. is 1^m too small (clerical error).

III. 946. A better star is G. 2066, p. 11^m 6^s n. 30′, which gives 13^h 39^m 35^s, 9° 49′, agreeing much better with Bigourdan's place, 13^h 38^m 49^s, 9° 50′.

III. 949. A better star is Kasan 2528, p. 16^m 8^s, n. 30′, which gives 14^h 25^m 17^s, 9° 15′, differing nearly 2^m from the place in N.G.C., in which Bigourdan twice searched in vain.

III. 954. R.A. 28^s too great. Doubtless a correction to centre of field was forgotten.

III. 959. This is I.C. 324, 11^s f., 1′·2 south of I. 60. N.G.C. 1331 to be struck out. The place given by H. no doubt refers to I. 60.

III. 963–971. See above, under I. 282. The places from the Greenwich plates are, for 1860:—

	h	m	s	°	′	
III. 963	9	23	17	13	3·6	
964	10	3	7	15	5·4	= N.G.C. 3144, d'A.
965	10	5	18	14	57·5	3155, h. 676, d'A.
966	10	0	16	11	29·6	
967	10	48	56	14	3·6	3465, h. 795, d'A.
968	10	51	23	14	2·9	
969	11	23	46	14	51·0	
970	11	34	19	11	51·0	
971	11	41	26	14	55·2	3890, III. 940, d'A.

III. 977. In the sweep (1111) is G. 1562, p. 5^m 0^s, n. 34′, which gives 9^h 25^m 48^s, 9° 39′, agreeing much better with Bigourdan's place.

III. 978. Not found by Bigourdan. The place is sufficiently correct, as appears from the star B. 1439 = G. 1643, p. 33^m 41^s, s. 13′, which gives for 1860 9^h 48^m 31^s, 8° 59′.

III. 979, 980, 981. These are not given in the *P.T.* The R.A. of N.G.C. 3210 requires a correction of $+1^m$; d'A. observed the 2nd and 3rd, Bigourdan all three.

III. 982, 983. Not in the *P.T.* In the Cape Observations, where h. gives these "omitted nebulæ," ΔP.D. is misprinted 2° 30′, and the P.D. of Auwers is therefore 9′ too great. Bigourdan's places ($\Delta\alpha = 48^s$) agree well with H's. d'Arrest's R.A. of III. 983, adopted in the N.G.C., to be diminished by 1^m (one obs.).

III. 984. Not in *P.T.* The place is correct, the two stars are np. distant 0′·5 and 1′·5.

III. 985. See above, under II. 794.

IV. 65. Sweep 935. There was some uncertainty as to whether the P.D. was 90° or 91°. A star 6·5 m. following 2^m 29^s, 1° 20′ n., was supposed to be 22 Monocerotis, but the observed P.D. of this star must be 1° wrong, as the "P.D. piece" was immediately afterwards set from 88° 50′ to 89° 48′, "supposing it to have been set upon the wrong degree or changed by some accident." Then comes a star 7·8 m. 20^m 3^s f., 1° 2′ s. of the neb., which agrees with D.M. −1°·1738, and the P.D. of the neb. is 90° and not 91°, as stated in the *Phil. Trans.*, 1791, p. 82.

J. L. E. D.]

L.

Observations of the Transit of Mercury over the Disk of the Sun; to which is added, an Investigation of the Causes which often prevent the proper Action of Mirrors.

[*Phil. Trans.*, 1803, pp. 214–232.]

Read February 10, 1803.

THE following observations were made with a view to attend particularly to every phenomenon that might occur during the passage of the planet Mercury over the sun's body. My solar apparatus, on account of the numerous observations I have lately been in the habit of making, was in great order for viewing the sun in the highest perfection; and, very fortunately, the weather proved to be as favourable as I could possibly have wished it.

The time at which the observations were made, not being an object of my investigation, is only to be considered as denoting the order of their succession.

November 9, 1802. About 40′ after seven o'clock in the morning, I directed a telescope, with a glass mirror of 7 feet focal length, and 6·3 inches in diameter, to the sun; and perceived the planet Mercury. It was easily to be distinguished from the openings in the luminous clouds, generally called spots, of which there were more than forty in number. Its perfect roundness would have been sufficient to point it out, had I not already known where to look for it.

10^h 0′. When the sun was come to a sufficient altitude to show objects on its surface with distinctness, I directed my attention to the contour of the mercurial disk, and found its termination perfectly sharp.

With a 10-feet reflector, and magnifying power of 130, I saw the corrugations of the luminous solar surface, up to the very edge of the whole periphery of the disk of Mercury.

10^h 27′. When the planet was sufficiently advanced towards the largest opening of the northern zone, I compared the intensity of the blackness of the two objects; and found the disk of Mercury considerably darker, and of a more uniform black tint, than the area of the large opening.

10^h 32′. The preceding limb of Mercury cuts the luminous solar clouds with the most perfect sharpness; whereas, in the great opening, the descending parapet, down the preceding side, was plainly visible.

It should be remarked, that the instrument here applied to the sun, with the moderate power of 130, is the same 10-feet reflector which, in fine nights, when directed to very minute double stars, will show them distinctly with a magnifier of 1000.

Having often attempted to use high magnifiers in viewing the sun, I wished to make another trial ; though pretty well assured I should not succeed, for reasons which will appear hereafter.

With two small double convex lenses, both made of dark green glass, and one of them having the side which is nearest the eye thinly smoked, in order to take off some light, I viewed the sun. Their magnifying power was about 300 ; and I saw Mercury very well defined ; but that complete distinctness, which enables us to judge with confidence of the condition of the object in view, was wanting.

With a single eye-glass, smoked on the side towards the eye, and magnifying 460 times, I also saw Mercury pretty well defined ; but here the sun appeared ruddy, and no very minute objects could be perceived.

11^h 28′. The planet having advanced towards the preceding limb of the sun, it was now time to attend to the appearances of the interior and exterior contacts.

11^h 32′. 10-feet reflector. The whole disk of Mercury is as sharply defined as possible ; there is not the least appearance of any atmospheric ring, or different tinge of light, visible about the planet.

11^h 37′. Appearances remain exactly as before.

11^h 42′. The sharp termination of the whole mercurial disk, appears to be even more striking than before. This may be owing to its contrast with the bright limb of the sun, which, having many luminous ridges in the northern zone, is remarkably brilliant about the place of the planet.

11^h 44′. I was a few moments longer writing down the above than I should have been, to see the interior contact so completely as I could have wished ; however, the thread of light on the sun's limb was but just breaking, or broken ; but no kind of distortion, either of the limb or of the disk of Mercury, took place.

The appearance of the planet, during the whole time of its emerging from the sun, remained well defined, to the very last.

The following limb of Mercury remained sharp, till it reached the very edge of the sun's disk ; and vanished without occasioning the smallest distortion of the sun's limb, in going off, or suffering the least alteration in its own figure.

As soon as the planet had quitted the sun, the usual appearance of its limb was so instantly and perfectly restored, that not the least trace remained whereby the place of its disappearance could have been distinguished from any other adjacent part of the solar disk.

It will not be amiss to add, that very often, during the transit, I examined the appearance of Mercury with a view to its figure, but could not perceive the least deviation from a spherical form ; so that, unless its polar axis should have happened to be situated, at the time of observation, in a line drawn from the eye to the sun the planet cannot be materially flattened at its poles.

OBSERVATIONS AND EXPERIMENTS RELATING TO THE CAUSES WHICH OFTEN AFFECT MIRRORS, SO AS TO PREVENT THEIR SHOWING OBJECTS DISTINCTLY.

It is well known to astronomers, that telescopes will act very differently at different times. The cause of the many disappointments they may have met with in their observations, is however not so well understood.

Sometimes we have seen the failure ascribed to certain tremors, as belonging to specula; and remedies have been pointed out for preventing them. Not unfrequently again, the telescope itself has been condemned; or, if its goodness could not admit of a doubt, the weather in general has been declared bad, though possibly it might be as proper for distinct vision as any we can expect in this changeable climate.

The experience acquired by many years of observation, will however, I believe, enable me now to assign the principal cause of disappointments to which we are so often exposed. Unwilling to hazard any opinion that is not properly supported by facts, I shall have recourse to a collection of occasional observations. They have been made with specula of undoubted goodness, so that every cause which impeded their proper action must be looked upon as extrinsic. I shall arrange these observations under different heads, that, when they have been related, there may remain no difficulty to draw a few general conclusions from them, which will be found to throw a considerable light upon our subject.

Moisture in the Air.

(1.) October 5, 1781. I see double stars, with 460, completely well. The air is very damp.

(2.) Nov. 23, 1781. 15^h 30′. The morning is uncommonly favourable, and I see the treble star ζ Cancri, with 460, in high perfection. The air is very moist, and intermixed with passing clouds.

(3.) Sept. 7, 1782. I viewed the double star preceding 12 Camelopardalis,* with 932. In this, and several other fine nights which I have lately had, the condensing moisture on the tube of my telescope has been running down in streams; which proves that damp air is no enemy to good vision.

(4.) Dec. 28, 1782. 17^h 30′. The water condensing on my tube keeps running down; yet I have seen very well all night. I was obliged to wipe the object-glass of my finder almost continually. The specula, however, are not in the least affected with the damp. The ground was so wet that, in the morning, several people believed there had been much rain in the night, and were surprised when I assured them there had not been a drop.

(5.) Feb. 19, 1783. I have seen perfectly well till now † that a frost is coming

* See *Phil. Trans.* Vol. LXXV. Part I. page 68; II. 53. [This edition, Vol. I. p. 184.]

† The time is not marked in the journal; but, from the number of the observations that had been made during the night, it must have been towards morning.

on ; though Datchet Common, which is just before my garden, is all under water ; and the grass on which I stand with my telescope is as wet as possible.

(6.) Feb. 26, 1783. All the ground is covered with snow ; yet I see remarkably well.

(7.) March 8, 1783. The common before my garden is all under water ; my telescope is running with condensed vapour ; not a breath of air stirring. I never saw better.

(8.) August 25, 1783. My telescope ran with water all the night. The small speculum, which sometimes gathers moisture, was never affected in the 7-feet tube, but was a little so in the 20-feet. The large eye-glasses and object-glasses of the finders required wiping very often. I saw all night remarkably well.

Fogs.

(9.) Oct. 30, 1779. It grows very foggy, and the moon is surrounded with strong nebulosity ; nevertheless, the stars are very distinct, and the telescope will bear a considerable power.

(10.) August 20, 1781. It is so foggy that I cannot see an object at the distance of 40 feet ; yet the stars are very distinct in the telescope. By an increase of the fog, α Piscium can no longer be seen by the eye ; yet, in the telescope, it being double, I see both the stars with perfect distinctness.

(11.) Sept. 6, 1781. A fog is come on ; yet I see very well.

(12.) Sept. 9, 1781. There is so strong a fog, that hardly a star less than 30° high is to be seen ; and yet, in the telescope, at great elevations, I see extremely well.

(13.) March 9, 1783. It is very foggy ; yet in the telescope I see the stars without aberration, and they are very bright. α Serpentarii is without a single ray.

(14.) April 6, 1783. A very thick fog settles upon all my glasses ; but the specula, even the 20-feet, which has so large a surface, remain untouched. I see perfectly well.

Frost.

(15.) Nov. 15, 1780 ; 5 o'clock in the morning. An excellent speculum, No. 2, will not act properly ; the frosty morning probably occasions an alteration in its figure. Another speculum, No. 1, acts but indifferently, though I have known it to shew very well formerly in a very hard frost : for instance, November 23, 1779, I saw with the same mirror, and a power of 460, the vacancy between the two stars of the double star Castor, without the least aberration.

(16.) Oct. 22, 1781. Frost seems to be no hindrance to perfect vision. The tube of my 7-feet telescope is covered with ice ; yet I see very well.

(17.) Nov. 19, 1781. It freezes very hard, and the stars, even those which are 50° high, are very tremulous. I suspect their apparent diameters to be diminished ;

and, if I recollect right, this is not the first time that such a suspicion has occurred to me.

(18.) Jan. 10, 1782. My telescope would not act well, even at an altitude of 70 or 80 degrees. There is a strong frost.

(19.) Jan. 31, 1782. I cannot see with a power of 460, the stars seem to dance so unaccountably, and yet the air is perfectly calm : even at 60 or 70 degrees of altitude, vision is impaired.

(20.) Feb. 9, 1782. That frost is no hindrance to seeing well is evident ; for, not only my breath freezes upon the side of the tube, but more than once have I found my feet fastened to the ground, when I have looked long at the same star.

(21.) Oct. 4, 1782. It froze very severely this night. At first, when the frost came on, I saw very badly, every object being tremulous ; but, after some time, and at proper altitudes, I saw as well as ever. Between 5 and 6 o'clock in the morning, objects began to be tremulous again ; occasioned, I suppose, by the coming on of a thaw.

(22.) Jan. 1, 1783. I made a number of delicate observations this night, notwithstanding, at 4 o'clock in the morning, my ink was frozen in the room; and, about 5 o'clock, a 20-feet speculum, in the tube, went off with a crack, and broke into two pieces. On looking at FAHRENHEIT'S thermometer, I found it to stand at 11°

(23.) May 6, 1783. It freezes, and in the telescope the stars seem to dance extremely.

Hoar-frost.

(24.) Nov. 6, 1782. There is a thick hoar-frost ; yet I see extremely well. It seems to enlarge the diameters of the stars ; but, as I see the minutest double stars well, the apparent enlargement of the diameters must be a deception.

(25.) Dec. 22, 1782. There is a strong hoar-frost gathering upon the tubes of my telescopes ; but I see very well.

Dry Air.

(26.) Dec. 21, 1782. The tube of my telescope is dry, and I do not see well.

(27.) April 30, 1783. The stars are extremely tremulous and confused ; the outside of the tube of my telescope is quite dry.

Northern Lights.

(28.) Sept. 25, 1781. There are very strong northern lights ; their flashing does not seem to interfere with telescopic vision ; but all objects appear tremulous, and indifferently defined.

(29.) Aug. 30, 1782. There are very bright northern lights, in broad arches, with white streaks ; yet I see perfectly well.

(30.) March 26, 1783. An Aurora Borealis is so bright, that η Herculis, which it covers, can hardly be seen ; yet, in the telescope, and with a power of 460, I find no difference. I compared that star with γ Coronæ, which was in a bright part of the heavens, and in the telescope they appeared nearly alike. I suspected η Herculis to be somewhat more tinged with red than it should be; and examined it afterwards, when clear of the Aurora: it was indeed less red; but, as it had gained more altitude, the experiment was not decisive.

Windy Weather.

(31.) Jan. 8, 1783. It is very windy. The diameters of the stars are strangely increased, even those at 60 and 70° of altitude. Every star seems to be a little planet.

(32.) Jan. 9, 1783. Wind increases the apparent diameters of the stars.

(33.) Sept. 20, 1783. The night has been very windy; and I do not remember ever to have seen so ill, with such a beautiful appearance of brilliant star-light.

Fine in Appearance.

(34.) May 28, 1781. The evening, though fine in appearance, is not favourable. No instrument I have will act properly. The wind is in the east.

(35.) August 30, 1781. The stars appear fine to the naked eye, so that I can see ε Lyræ very distinctly to be two stars; yet my telescope will show nothing well. There are flying clouds, which, by their rapid motion, indicate a disturbance in the upper regions of the air; though, excepting now and then a few gusts of wind, it is in general very calm. At a distance there are continual flashes of lightning, but I can hardly hear any thunder.

(36.) Sept. 14, 1781. I see very small stars with the naked eye; but the telescope will not act so well as it should.

(37.) Sept. 24, 1781. The evening is apparently fine; but, with the telescope, I can see neither η Coronæ nor μ Bootis double; nor indeed can I see any other stars well.

Over a Building.

(38.) August 24, 1780. I viewed ε Bootis with 449, 737, and 910, but saw it very indifferently. The star was over a house.

(39.) Oct. 26, 1780. ε Bootis being near the roof of a house, I saw it not so distinctly as I could wish.

The Telescope lately brought out.

(40.) Oct. 10, 1780. 6^h 30′. Having but just brought out my telescope, it will not act well.

6^h 45′. The tube and specula are now in order, and perform very well.

(41.) Jan. 11, 1782. To all appearance, the morning was very fine, but still the telescope, when first brought out, would not act well. After half an hour's exposure, it performed better.

Confined Place.

(42.) July 19, 1781. 13^h 15′. My telescope would not act well; and, supposing the exhalations from the grass in my garden to affect vision, I carried the telescope into the street, (the observation was made at Bath,) and found it to perform to admiration.

(43.) July 19, 1781. My telescope acted very well; but a slight field-breeze springing up, and brushing through the street where my instrument was placed, it would no longer bear a magnifying power of 460.

Haziness and Clouds.

(44.) Sept. 22, 1783. The weather is now so hazy, that the double star δ Cygni is but barely visible to the naked eye. This has taken off the rays of the large star, so that I now see the small one extremely well, which at other times it is so difficult to perceive, even with a magnifying power of 932.

(45.) August 13, 1781. A cloud coming on very gradually upon fixed stars, has this remarkable effect, that their apparent diameters diminish gradually to nothing.

(46.) July 7, 1780. The air was very hazy, but extremely calm. I had Arcturus in the field of view of the telescope, and, the haziness increasing, it had a very beautiful effect on the apparent diameter of this star. For, supposing the first of the points

No. 1 2 3 4 5 6 7 8 9 10
• • • • • • • • • •

to represent its magnitude when brightest, I saw it gradually decrease, and assume, with equal distinctness, the form of all the succeeding points, from No. 1 to No. 10, in the order of the numbers placed over them. The last magnitude I saw it under, could certainly not exceed two-tenths of a second; but was perhaps less than one. This leads to the discovery of one of the causes of the apparent magnitude of the fixt stars.

Focal Length.

(47.) Nov. 14, 1801. The focal length of my 10-feet mirror increases by the heat of the sun. I have often observed this before; the difference, by several trials, amounts to 8 hundredths of an inch.

(48.) Dec. 13, 1801. The focal length of my 10-feet mirror, while I was looking at the sun, became shorter, contrary to what it used to do; but, there being a strong frost, I guess that the object metal grows colder, notwithstanding its exposure to the sun's rays.

(49.) Nov. 9, 1802. 10^h 50′. The focus of my 7-feet glass mirror became 18 hundredths of an inch shorter, on being exposed for about a minute to the sun. The figure of the speculum was also distorted; the foci of the inside and outside rays differing considerably, though its curvature, by observations on the stars, has been ascertained to be strictly parabolical.

12^h 0′. The same mirror, exposed one minute to the action of the sun, became 21 hundredths shorter in focal length.

The focus of a 10-feet metalline mirror, when exposed one minute to the sun's rays, became 15 hundredths of an inch longer than it was before.

(50.) January 9, 1803. When I looked with the glass 7-feet mirror, several times, a minute or two at the sun, it shortened generally, ·24, ·26, and ·30 of an inch, in focal length.

The observations which are now before us, appear to be sufficient to establish the following principle ; namely,

"That in order to see well with telescopes, it is required that the temperature of the atmosphere and mirror should be uniform, and the air fraught with moisture."

This being admitted, we shall find no difficulty in accounting for every one of the foregoing observations.

If an uniform temperature be necessary, a frost after mild weather, or a thaw after frost, will derange the performance of our mirrors, till either the frost or the mild weather are sufficiently settled, that the temperature of the mirror may accommodate itself to that of the air. For, till such an uniformity with the open air, in the temperature of the mirror, the tube, the eye-glasses, and I would almost add the observer, be obtained, we cannot expect to see well. See observation 15, 17, 18, 19, and 23.

But, when a frost, though very severe, becomes settled, the mirror will soon accommodate itself to the temperature ; and we shall find our telescopes to act well. See obs. 16, 20, 21, 22, 24, and 25.

This explains, with equal facility, why no telescope just brought out of a warm room can act properly. See obs. 40 and 41.

Nor can we ever expect to make a delicate observation, with high magnifying powers, when looking through a door, window, or slit in the roof of an observatory ; even a confined place, though in the open air, will be detrimental. See obs. 42 and 43.

It equally shows, that windy weather in general, which must occasion a mixture of airs of different temperatures, cannot be favourable to distinct vision. See obs. 31, 32, and 33.

The same remark will apply to Auroræ Boreales, when they induce, as they often do, a considerable change in the temperature of the different regions of air. See obs. 28.

But, should they not be accompanied by such a change, there seems to be no reason why they should injure vision. See obs. 29 and 30.

The warm exhalations from the roof of a house in a cold night, must disturb the uniformity of the temperature of a small portion of air; so that stars which are over the house, and at no considerable distance, may be affected by it. See obs. 38 and 39.

Sometimes the weather appears to be fine, and yet our telescopes will not act well. This may be owing to dryness occasioned by an easterly wind; or to a change of temperature, arising from an agitation of the upper regions of the atmosphere. See obs. 34 and 35.

Or, possibly, to both these causes combined together. See obs. 36 and 37.

If moisture in the atmosphere be necessary, dry air cannot be proper for vision. See obs. 26 and 27.

And therefore, on the contrary, dampness, and haziness of the atmosphere, must be favourable to distinct vision. See obs. 1, 2, 3, 4, 6, and 8.

Fogs also, which certainly denote abundance of moisture, must be very favourable to distinct vision. See obs. 9, 10, 11, 12, 13, and 14.

Nay, if the observatory should be surrounded by water, we need be under no apprehension on that account. Perhaps, were we to erect a building for astronomical purposes only, we ought not to object to grounds which are occasionally flooded; the neighbourhood of a river, a lake, or other generally called damp situations. See obs. 5 and 7.

It is however possible, that fogs and haziness may increase to such a degree as, at last, to take away, by their interposition, all the light which comes from celestial objects; in which case, they must of course put an end to observation; but they will nevertheless be accompanied with distinct vision to the very last. See obs. 44, 45, and 46.

We have now only the four last observations to account for. They relate to the change of the focal length of mirrors in solar observations, and its attendant derangement of the foci of the different parts of the reflecting surface; and, as simplicity is one of the marks of the truth of a principle, I believe we need not have recourse to any other cause than the change of temperature produced by the action of the solar rays that occasion heat; which will be quite sufficient to explain all the phenomena. But, in order to show this in its proper light, I shall relate the following experiments.

1st Experiment.

I placed a glass mirror, of 7-feet focal length, in the tube belonging to the telescope; and, having laid it open at the back, I prepared a stand, on which the iron used in my experiments on the terrestrial Rays that occasion Heat (see *Phil. Trans.* for 1800, Plate XVI. Fig. 1)* might be placed, so as to heat the mirror from behind, while I kept a certain object in the field of view of the telescope. Having

* [See above, p. 91.]

measured the focal length, and also examined the figure of the mirror, which was parabolical, the heated iron was applied so as to be about 2¼ inches from the back of the glass mirror. The consequence of this was, that a total confusion in all the foci took place, so that the letters on a printed card in view, which before had been extremely distinct, became instantly illegible. In 15 seconds, the focus of the mirror was shortened 2·3 inches; in half a minute, 3·47 inches; and, at the end of the minute, I found it no less than 4·59 inches shorter than it had been before the application of the hot iron.

On repeating the experiment, but placing the heated iron no more than ¾ of an inch from the back of the mirror, its focal length, in 1½ minute, became 5·33 inches shorter.

I tried also a more moderate heat; and, placing the iron at 3 inches from the back, the focus of the mirror shortened in one minute 2·83 inches.

A thermometer placed in contact with the reflecting surface of the mirror, could hardly be perceived to have risen, during the time in which the hot iron produced the alteration of the focal length.

2d Experiment.

Every thing remaining as before, I suspended a small globe of heated iron in front of the mirror, at one inch and a half from its vertex; and, in two minutes, the focus was lengthened 5·3 inches. The figure of the mirror was also deranged; so that the letters on the card could not be distinguished.

I made a second trial, with the suspended iron a little more heated, and brought it as near the surface of the mirror as I judged it to be safe; since a contact would probably have cracked the mirror. In consequence of this arrangement, the focus lengthened, in one minute, 1·64 inch.

On removing the heated iron, the mirror returned, in one minute, to within 18 inch of its former focal length; and, at the end of the second minute seemed to be nearly restored. But the disagreement of the foci of the different parts of the reflecting surface might be perceived for a long time afterwards, and caused an indistinctness of vision, which plainly indicated that, under such circumstances, the magnifying power of the telescope, 225, was more than it ought to be, in order to see well.

3d Experiment.

I now changed the glass mirror for a metalline one; and, on placing the heater near the back of it, the focus of the speculum, in 30 seconds, became ·77 inch shorter. But, continuing the observation, instead of shortening still farther in the next 30 seconds, it became ·3 inch longer, so that, at the end of a minute, it was only ·47 shorter than before the approach of the hot iron.

4th Experiment.

When the small heated globe of the 2d experiment was suspended in front of the mirror, the focus lengthened ·27 inch in one minute ; nor would the lengthening increase by leaving the hot iron longer in its position. The foci in this, as well as in the 3d experiment, were so much injured that they could not be measured with any precision ; and it was evident, that high magnifying powers ought not to be used with a mirror of which the temperature is undergoing a continual change.

I repeated the experiment with the iron nearly red hot ; and found the focus lengthened 1·48 inch in 30 seconds. Five minutes after the removal of the iron, the regularity of the figure of the mirror was pretty well restored.

With a moderate heat, I had, in 30 seconds, a lengthening of the focus, of ·57 inch ; and, in about $1\frac{1}{2}$ minute after the removal of the heated iron, distinct vision was nearly restored.

These four experiments show, that a change in the temperature of mirrors, occasioned by heat, is attended with an alteration of their focal length ; and also prove, that the figure of the reflecting surface is considerably injured, during the time that such a change takes place. We are consequently authorised to believe, that the small alteration in the focus of a mirror exposed to the rays of the sun, arises from the same cause. For, since a thermometer, when the sun is shining upon it, will show that its temperature is altered, the action of the solar rays upon a mirror must be attended with a similar effect in its temperature. See obs. 47, 48, 49, and 50.

The same experiments will now also explain why the observations of the sun, related in our transit of Mercury, between $10^h\ 32'$ and $11^h\ 28'$, were not attended with success ; for we have seen that heat occasions a derangement in the action of the reflecting surface ; and it follows that, under such circumstances, high magnifying powers cannot be expected to show objects very distinctly.

If it should be remarked, that I have not explained why the focus of a glass mirror should shorten by the same rays of the sun which lengthen that of a metalline speculum, I confess that this at present does not appear ; and, as it is not material to our purpose, I might pass it over in silence. We are however pretty well assured, that the alterations of the focal length must be owing to a dilatation of the glass or metal of which mirrors are made, and must be greatest where most heat is applied. Our experiments therefore cannot agree perfectly with solar observations ; for, in the glass mirror, the application of partial heat in front, must undoubtedly have been much stronger about the middle of the mirror (though the centre of it was sometimes guarded by a brass plate equal to the size of the small speculum) than at the circumference. But when, on the contrary, a mirror is exposed to the sun, every part of the surface will receive an equal portion of heat.

It may also be said, that I have pointed out a defect in telescopes used for solar observations, without assigning a cure for it. It will however be allowed, that tracing an evil to its cause must be the first step towards a remedy Had the imperfection of the figure brought on by the heat of the solar rays been of a regular nature, an elliptical speculum might have been used to counteract the assumed hyperbolical form ; or *vice versâ*.

And now, as, properly speaking, the derangement of the figure of a mirror used in observing the sun, is not so much caused by the heat of its rays as by their partial application to the reflecting surface only, which produces a greater dilatation in front than at the back, there may be a possibility of counteracting this effect, by a contrary application of heat against the back, or by an interception of it on the front. But this we leave to future experiments.

LI.

Account of the Changes that have happened, during the last Twenty-five Years, in the relative Situation of Double-stars; with an Investigation of the Cause to which they are owing.

[*Phil. Trans.*, 1803, pp. 339–382.]

Read June 9, 1803.

In the Remarks on the Construction of the Heavens, contained in my last Paper on this subject,* I have divided the various objects which astronomy has hitherto brought to our view, into twelve classes. The first comprehends insulated stars.

As the solar system presents us with all the particulars that may be known, respecting the arrangement of the various subordinate celestial bodies that are under the influence of stars which I have called insulated, such as planets and satellites, asteroids and comets, I shall here say but little on that subject. It will, however, not be amiss to remark, that the late addition of two new celestial bodies, has undoubtedly enlarged our knowledge of the construction of the system of insulated stars. Whatever may be the nature of these two new bodies, we know that they move in regular elliptical orbits round the sun. It is not in the least material whether we call them asteroids, as I have proposed; or planetoids, as an eminent astronomer, in a letter to me, suggested; or whether we admit them at once into the class of our old seven large planets. In the latter case, however, we must recollect, that if we would speak with precision, they should be called very small, and exzodiacal; for, the great inclination of the orbit of one of them to the ecliptic, amounting to 35 degrees, is certainly remarkable. That of the other is also considerable; its latitude, the last time I saw it, being more than 15 degrees north. These circumstances, added to their smallness, show that there exists a greater variety of arrangement and size among the bodies which our sun holds in subordination, than we had formerly been acquainted with, and extend our knowledge of the construction of the solar, or insulated sidereal system. It will not be required that I should add any thing farther on the subject of this first article of my classification; I may therefore immediately go to the second, which treats of binary sidereal systems, or real double stars.

We have already shewn the possibility that two stars, whatsoever be their relative magnitudes, may revolve, either in circles or ellipses, round their common

* See *Phil. Trans.* for 1802, p. 477 [above, p. 199].

centre of gravity ; and that, among the multitude of the stars of the heavens, there should be many sufficiently near each other to occasion this mutual revolution, must also appear highly probable. But neither of these considerations can be admitted in proof of the actual existence of such binary combinations. I shall therefore now proceed to give an account of a series of observations on double stars, comprehending a period of about 25 years, which, if I am not mistaken, will go to prove, that many of them are not merely double in appearance, but must be allowed to be real binary combinations of two stars, intimately held together by the bond of mutual attraction.

It will be necessary to enter into a certain theory, by which these observations ought to be examined, that we may find to what cause we should attribute such

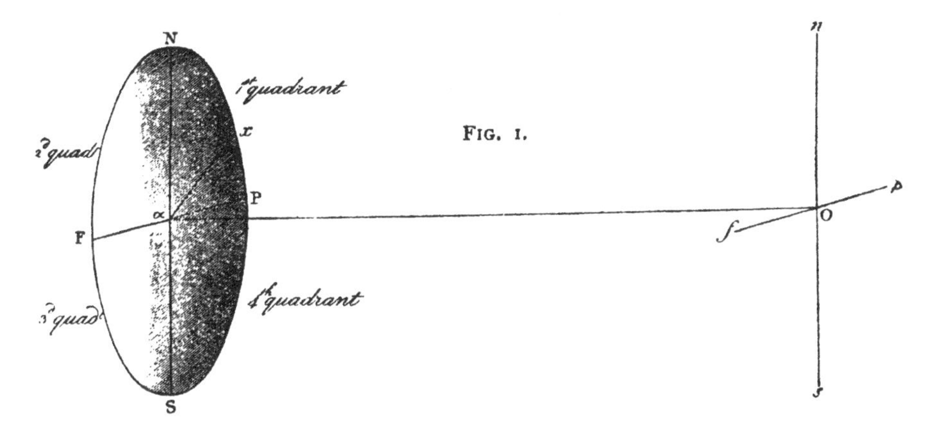

FIG. 1.

changes in the position, or distance, of double stars, as will be reported ; and, in order to make the required principles very clear, I shall give them in a few short and numbered sentences, that they may be referred to hereafter.

In Fig. 1, let us call the place of the sun, which may also be taken for that of the observer, O. In the centre of an orbit or plane NFSP is α Geminorum ; and, if any other star is to be examined, we have only to exchange the letter α for that by which such double star is known. This letter is always understood to represent the largest of the two stars which make up the double star ; and a general expression for its smaller companion will be x. N, F, S, P, represents the positions of the different parts of the heavens, with respect to α, north, following, south, and preceding ; and the small letters n, f, s, p, stand for the same directions with respect to O. $x\alpha$P, is the angle of position of the two stars x and α, with the parallel FP.

As the motion of an observer affects the relative situation of objects, we have three bodies to consider, in our investigation of the cause of the changes which will be pointed out ; the sun, the large star, and the small star, or, as we have shortly called them, O, α, x. This admits of three cases : a motion of one of the three bodies ; another, of two ; and a third, of all the three bodies together. We shall now point out the consequences that will arise in each of the cases.

Single Motions.

No. 1. Motion of x. When a and O are at rest, the motion of x may be assumed, so as perfectly to explain any change of the distance of the two stars, and of their angle of position.

No. 2. Motion of a. When x and O are at rest, and a has a motion, either towards P, N, F, or S, then the effect of it, whatever may be the angle PaO, will be had by entering the following Table, with the direction of the given motion.

Motion.	Distance.	Angle.	Quadrants.
aP	− +	+ −	1st and 4th 2 — 3
aF	+ −	− +	1 — 4 2 — 3
aN	− +	− +	1 — 2 3 — 4
aS	+ −	+ −	1 — 2 3 — 4

No. 3. Motion of O. 1st case. When a and x are at rest, and the angle PaO is 90 degrees, a proper motion of O, towards either p, f, n, or s, which will be extremely small when compared with the distance of O from a, can have no effect on the apparent distance, or angle of position, of the two stars; and therefore no other motion, composed of the directions we have mentioned, will induce a change in the comparative situation of a and x.

2d case. When the plane PNFS is oblique to the ray aO, and the angle PaO more than 90 degrees, the effect of the motion of O will be had by the following Table.

Motion.	Distance.	Angle.	Quadrants.
Op	+	−	1st and 2d 3 — 4
Of	−	+	1 — 2 3 — 4
On	+ −	+ −	1 — 3 2 — 4
Os	− +	− +	1 — 3 2 — 4

3d case. When the angle PαO is less than 90 degrees, the following Table must be used.

Motion.	Distance.	Angle.	Quadrants.
Op	–	+	1st and 2d 3 — 4
Of	+	–	1 — 2 3 — 4
On	– +	– +	1 — 3 2 — 4
Os	+ –	+ –	1 — 3 2 — 4

Double Motions.

No. 4. If we admit different motions in two of our three bodies, and if the ratio of the velocities, the directions of the motions, and the ratio of the distances of the bodies be given quantities, a supposition in which we admit their concurrence, may explain the phenomena of a double star, but can never be probable.

Motions of the three Bodies.

No. 5. If we admit different motions in every one of the three bodies, O, α, x, and if the velocities and directions of the motions, as well as the relative distances of the three bodies are determined, an hypothesis which admits the existence of such motions and situations, may resolve the phenomena of a double star, but cannot have any pretension to probability.

The compass of this Paper will not allow me to give the observations of my double stars at full length; I shall therefore, in the examination of every one of them, only state those particulars which will be required for the purpose of investigating the cause of the changes that have taken place, either in the distance, or angle of position, of the two stars of which the double star is composed.

As the arguments in the case of most of these stars will be nearly the same, it may be expected, that the first two or three which are to be examined will take up a considerable space; and the number of double stars, in which I have already ascertained a change, amounting to more than fifty, it will not be possible to give them all in one paper; I shall therefore confine the present one to a moderate length, and leave it open for a continuation at a future opportunity.

α Geminorum.

From my earliest observations on the distance of the two stars which make up the double star in the head of Castor, given in the first of my catalogues of double stars, we find, that about 23 years and a half ago, they were nearly two diameters of the large star asunder. These observations have been regularly continued, from the year 1778 to the present time, and no alteration in the distance has been perceived: the stars are now still nearly 2 diameters of the large one asunder.

It will be necessary to enter a little into the practicability of ascertaining distances by a method of estimation apparently so little capable of precision. From a number of observations and experiments I have made on the subject, it is certain that the apparent diameter of a star, in a reflecting telescope, depends chiefly upon the four following circumstances: the aperture of the mirror with respect to its focal length; the distinctness of the mirror; the magnifying power; and the state of the atmosphere at the time of observation. By a contraction of the aperture, we can increase the apparent diameter of a star, so as to make it resemble a small planetary disk. If distinctness should be wanting, it is evident that the image of objects will not be sharp and well defined, and that they will consequently appear larger than they ought. The effect of magnifying power is, to occasion a relative increase of the vacancy between two stars that are very near each other; but the ratio of the increase of the distance is not proportional to that of the power, and sooner or later comes to a maximum. The state of the atmosphere is perhaps the most material of the four conditions, as we have it not in our power to alter it. The effects of moisture, damp air, and haziness, (which have been related in a paper where the causes that often prevent the proper action of mirrors were discussed,) show the reason why the apparent distance of a double star should be affected by a change in the atmosphere. The alteration in the diameter of Arcturus, extending from the first to the last of the ten images of that star, figured in the above-mentioned paper,* shows a sufficient cause for an increase of the distance of two stars, by a contraction of their apparent disks. A skilful observer, however, will soon know what state of the air is most proper for estimations of this kind. I have occasionally seen the two stars of Castor, from 1½ to 2 and 2½ diameters asunder; but, in a regular settled temperature and clear air, their distance was always the same. The other three causes which affect these estimations, are at our own disposal; an instance of this will be seen in the following trial. I took ten different mirrors of 7 feet focal length, each having an aperture of 6·3 inches, and being charged with an eye-glass which gave the telescope a magnifying power of 460. With these mirrors, one after another, the same evening, I viewed the two stars of our double star; and the result was, that with every one of them, the stars were precisely at an equal distance from each other. These

* See *Phil. Trans.* for 1803, page 232 [above, p. 244].

mirrors were all sufficiently good to show minute double stars well; and such a trial will consequently furnish us with a proper criterion, by which we may ascertain the goodness of our telescope, and the clearness of the atmosphere required for these observations. To those who have not been long in the habit of observing double stars, it will be necessary to mention, that, when first seen, they will appear nearer together than after a certain time; nor is it so soon as might be expected, that we see them at their greatest distance. I have known it to take up two or three months, before the eye was sufficiently acquainted with the object, to judge with the requisite precision.

Whatever may be the difficulties, or uncertainties, attending the method of determining the distance of two close stars by an estimation of the apparent diameter, it must however be confessed, that we have no other way of obtaining the same end with so much precision. Our present instance of α Geminorum, will show the degree of accuracy of which such estimations are capable, and at the same time prove, that the purpose for which I shall use the estimated interval between the two stars will be sufficiently answered. By an observation of the 10th of May, 1781, we have the diameter of the largest of the two stars to that of the smallest as 6 to 5; and, according to several measures I have taken with the micrometer, we may admit their distance, diameters included, to be five seconds. Then, as the vacancy between the two stars is nearly, but not quite, 2 diameters of the large one, I shall value it at $1\frac{7}{8}$. From this we calculate, that the diameter of the large star, under the circumstances of our estimation, is nearly 1″·35: so that an error of one quarter of such a diameter, which is the most we can admit, will not exceed 0″·34. Nor is it of much consequence, if the measure of 5″ should not be extremely correct; as a small mistake in that quantity will not materially affect the error of estimation by the diameter, which, from what has been said, if the measure was faulty to a second, would not amount to more than one-fifteenth part of it.

Having thus ascertained that no perceptible change in the distance of the stars has taken place, we are now to examine the angle of position. In the year 1779, it was 32° 47′ north preceding; and, by a mean of the three last measures I have taken, it is now only 10° 53′. In the space of about 23 years and a half, therefore, the angle of position has manifestly undergone a diminution, of no less than 21° 54′; and, that this change has been brought on by a regular and gradual decrease of the angle, will be seen when the rest of the measures come to be examined.

The accuracy of the micrometer which has been used, when the angles of position were taken, being of the utmost importance, it becomes necessary to ascertain how far it will be safe to rely on the result of the measures. It might be easily shown that, in the day time, a given angle, delineated on a card, and stuck up at a convenient distance, may be full as accurately measured by a telescope furnished with this micrometer, as it can be done by any known method, when the card is laid on a table before us; but this would not answer my purpose. For, objects in motion, like the stars, especially when at a distance from the pole, cannot be measured with such steadiness as those which are near us, and at rest. The method of illuminating the wires, and other circumstances, will likewise affect the accuracy of the angles that are measured, especially when the distance of the stars is very small. I shall therefore have recourse to astronomical observations, in order to see what the micrometer has actually done.

January 22, 1802. The position of A Orionis was taken. 1st measure, 52° 38′ south preceding; 2d measure, 54° 14′. Mean of the two measures, 53° 26′. Deviation of the measures from the mean, 48′.

March 4, 1802. 11 Monocerotis. 1st measure, 28° 18′ south following; 2d measure, 26° 49′. Mean of the two, 27° 34′. Deviation from the mean, 45′.

February 9, 1803. α Geminorum. 1st measure, 6° 11′ north preceding; 2d measure, 4° 48′. Mean of the two, 5° 29′. Deviation from the mean, 41′.

September 6, 1802. η Coronæ. 1st measure, 89° 42′ north following; 2d measure, 89° 38′. Mean of the two, 89° 40′. Deviation from the mean, 2′.

When these observations are considered, we shall not err much if we admit that, in favourable circumstances, and with proper care, the micrometer, by a mean of two measures, will give the position of a double star true to nearly one degree; but, as the opportunities of taking very accurate measures are scarce, it will be necessary to have recourse to some more discordant observations.

February 18, 1803. β Orionis. 1st measure, 72° 58′ south preceding; 2d measure, 67° 24′. Mean of the two, 70° 11′. Deviation from the mean, 2° 47′.

But a memorandum to the observation says, that the evening was not favourable. We may therefore admit, that in the worst circumstances which can be judged proper for measuring at all, an error in the angle of position by two measures will not amount to three degrees.

It will be remarked, when we come to compare single measures which have been taken on different nights, that they are somewhat more discordant; but I have not ventured to reject them on that account, except in cases where it was pretty evident that some mistake in reading off, or other accident to which all astronomical observations are liable, was to be apprehended. Nor can such disagreements materially affect the conclusions I have drawn, when it appears that the deviations happen sometimes to be on one side, and sometimes on the other side, of the true angle of position. For, since that angle is not a thing that will change in the course of a few nights, the excess of one measure will serve to correct the defect of another; and we are not to think it extraordinary, when stars are so near together, and their motion through the field of view (in consequence of the high magnifying power we are obliged to use) so quick, that we should now and then even fall short of that general accuracy which may be had by a careful use of the micrometer.

I shall now enter into an examination of the cause of the change in the angle of position of the small star near Castor.

A revolving star, it is evident, would explain in a most satisfactory manner, a continual change in the angle of position, without an alteration of the distance. But this, being a circumstance of which we have no precedent, ought not to be admitted without the fullest evidence. It will therefore be right to examine, whether the related phenomena cannot be satisfactorily explained by the proper motions of the stars, or of the sun.

Single Motions.

(*a*) The three bodies we have to consider, are O, α, and x; and, supposing them to be placed as they were observed to be in the year 1779; the angle $x\alpha$P, in Fig. 1, will be 32° 47′ north preceding. We are at liberty to let the angle PαO be what will best answer the purpose. Then, in order to examine the various hypotheses that may be formed, according to the arrangement of the principles we have given, we shall begin with No. 1; and, as this admits that all phenomena may be resolved by a proper motion of x, let us suppose this star to be placed any where far beyond α, but so as to have been seen, in the year 1779, where the angle of position, 32° 47′ north preceding, and the observed distance, near 2 diameters of the large star, required it. With a proper velocity, let it be in motion towards the place where it may now be seen at the same distance from Castor, but under an angle of position only 10° 53′ north preceding. It may then be admitted, that a small decrease of the distance which would happen at the time when the angle of position was 21° 50′, could not have been perceived; so that the gradual change in the observed angle of position, as well as the equality of the distance of the two stars, will be sufficiently accounted for. But the admission of this hypothesis requires, that α Geminorum and the solar system should be at rest; and, by the observations of astronomers, which I shall soon have occasion to mention, neither of these conditions can be conceded.

(*b*) If, according to No. 2, we admit the motion of α, we shall certainly be more consistent with the observations which astronomers have made on the proper motion of this star; * and, as a motion of the solar system, which I shall have occasion to mention hereafter, has not been rigidly proved, it may, for the sake of argument, be set aside; nor has a proper motion of the star x been any where ascertained. The retrograde annual proper motion of Castor, in right ascension, according to Dr. MASKELYNE, is 0″·105. This, in about 23½ years, during which time I have taken notice of the angle of position and distance of the small star, will amount to a change of nearly 2″·47. Then, if we enter the short Table I have given in No. 2, with the motion αP, we find, that in the first quadrant, where the small star is placed,

* See TOBIÆ MAYERI *Opera inedita. De motu fixarum proprio*, page 80. Also Dr. MASKELYNE'S first Volume of *Observations*. Explanation and Use of the Tables, page iv. Or Mr. WOLLASTON'S *Astronomical Catalogue*, end of the Preface. Likewise *Connoissance des Temps pour l'Année* VI. page 203. *Sur le Mouvement particulier propre à différentes Etoiles*; par Mons. DE LA LANDE.

the distance between the two stars will be diminished, and the angle of position increased. But since it appears, by my observations, that the distance of the stars is not less now than it was in 1780; and that, instead of an increase in the angle of position, it has actually undergone a diminution of nearly 22 degrees; it follows, that the motion of α Geminorum in right ascension will not explain the observed alterations in the situation of this double star. If, according to Mr. DE LA LANDE'S account,* we should also consider the annual proper motion of α in declination, which is given $0''\cdot12$ towards the north, we shall find, by entering our Table with the motion αN, amounting to $2''\cdot82$, that the distance of the two stars will be still more diminished; but that, on the contrary, the angle of position will be much lessened; and, by combining the two motions together, the apparent disks of the two stars should now be a little more than one-tenth of a second from each other, and the angle of position 35 degrees south preceding. But, since neither of these effects have taken place, the hypothesis cannot be admitted.

(*c*) That the sun has a proper motion in space, I have shown with a very high degree of evidence, in a paper which was read at the Royal Society about twenty years ago.† The same opinion was before, but only from theoretical principles, hinted at by Mr. DE LA LANDE, and also by the late Dr. WILSON, of Glasgow;‡ and has, since the publication of my paper, been taken up by several astronomers,§ who agree that such a motion exists. In consequence of this, let us now, according to No. 3, assign to the sun a motion in space, of a certain velocity and direction. Admitting therefore α and x to be at rest, let the angle PαO be 90 degrees; then, by the 1st case of No. 3, we find that none of the observed changes of the angles of position will admit of an explanation. There is moreover an evident concession of the point in question, in the very supposition of the above angle of 90 degrees; for, if x be at the same distance as α from the sun, and no more than 5″ from that star, its real distance, compared to that of the sun from the star, will be known; and, since that must be less than the 40 thousandth part of our distance from Castor, these two stars must necessarily be within the reach of each other's attraction, and form a binary system.

(*d*) Let us now take the advantage held out by the 2d case of No. 3, which allows us to place x far behind α; in which situation, the angle PαO will be more than 90 degrees. The star x being less than α, renders this hypothesis the more plausible. Now, as a motion of Castor, be it real or apparent, has actually been ascertained, we cannot set it aside; the real motion of O, therefore, in order to account for the apparent one of α, must be of equal velocity, and in a contrary direction; that is, when decomposed, $0''\cdot105$ towards *f*, and $0''\cdot12$, towards *s*.

* See page 211 of the treatise before referred to.

† See *Phil. Trans.* Vol. LXXIII. page 247. [This Ed., Vol I. p. 108.]

‡ See my note in *Phil. Trans.* Vol. LXXIII. page 283. [Vol. I. p. 130.]

§ See *Astronomisches Jahrbuch für das Jahr* 1786; *Seite* 259. *Uber die Fortrückung unseres Sonnen-Systems, von* HERRN *Professor* PREVOST. Also 1805; *Seite* 113.

The effect of the sun's moving from O towards *f*, according to the 1st Table in No. 3, is, that the distance between the two stars will be diminished, and the angle of position increased. But these are both contrary to the observations I have given. The motion of O in declination towards *s*, according to the same Table, will still diminish the distance of the two stars, but will also diminish the angle of position. Then, since a motion in right ascension increases the angle, while that in declination diminishes it, the small star may be placed at such a distance that the difference in the parallax, arising from the solar motion, shall bring the angle of position, in $23\frac{1}{2}$ years, from 32° 47′ to 10° 53′; which will explain the observed change of that angle. The distance of the star *x*, for this purpose, must be above $2\frac{1}{3}$ times as much as that of α from us. But, after having in this manner accounted for the alteration of the angle of position, we are, in the next place, to examine the effect which such a difference of parallax must produce in the apparent distance of the two stars from each other. By a graphical method, which is quite sufficient for our purpose, it appears, that the union of the two motions in right ascension and declination, must have brought the two stars so near, as to be only about half a diameter of the large star from each other; or, to express the same in measures, the centres of the stars must now be 1″·8 nearer than they were $23\frac{1}{2}$ years ago. But this my observations cannot allow; for we have already shown, that any change of more than 3 or 4-tenths of a second must have been perceived.

If, on the other hand, we place the star *x* at such a distance that the solar parallax may only bring it about 4-tenths of a second nearer to α, which is a quantity we may suppose to have escaped our notice in estimating the apparent distance of the two stars, then will the angle of position be above 20 degrees too large. This shows, that no distance, beyond Castor, at which we can place the star, will explain the given observations.

(*e*) The last remaining trial we have to examine, is to suppose *x* to be nearer than α; the angle PαO will then be less than 90 degrees; and the effect of a motion of O towards *f*, by the 2d Table in No. 3, will be an increase of the distance of the two stars, and a diminution of their angle of position. But the motion O*s*, which is also to be considered, will add to the increase of the distance, and counteract the diminution of the angle. It is therefore to be examined, whether such an increase of distance as we can allow to have escaped observation, will explain the change which we know to have happened in the angle, during the last $23\frac{1}{2}$ years. By the same method of compounding the two motions as before, it immediately appears, that we cannot place the small star more than about 1-tenth of the distance Oα on this side of Castor, without occasioning such an increase of the apparent distance of the two stars as cannot possibly be admitted; and that, even then, the angle of position, instead of being less, will be a few degrees larger, at the end of $23\frac{1}{2}$ years, than it was at the beginning. This hypothesis, therefore, like all the foregoing ones, must also be given up, as inconsistent with my observations.

It is moreover evident, that the observations of astronomers on the proper motion of the stars in general, will not permit us to assume the solar motion at pleasure, merely for the sake of accounting for the changes which have happened in the appearances of a double star. The proper motion of Castor, therefore, cannot be intirely ascribed to a contrary motion of the sun. For we can assign no reason why the proper motion of this star alone, in preference, for instance, to that of Arcturus, of Sirius, and of many others, should be supposed to arise from a motion of the solar system. Now, if they are all equally intitled to partake of this motion, we can only admit it in such a direction, and of such a velocity, as will satisfy the mean direction and velocity of the general proper motions of the stars ; and place all deviations to the account of a real proper motion in each star separately.

Double Motion.

(*f*) In order to explain the phenomena of our double star, according to No. 4, by the motion of two bodies, for instance α and x, it will be required that they both should move in given directions ; that the velocities of their motions should be in a given ratio to each other ; and that this ratio should be compounded with the ratio of their distances from O ; a supposition which must certainly be highly improbable. To show this with sufficient evidence, let us admit that, according to the best authorities, the annual proper motion of Castor is $-0''\cdot105$ in right ascension, and $0''\cdot12$ in declination towards the north. Then, as the small star, without changing its distance, has moved through an angle of 21° 54′, the only difference in the two motions of these stars, will be expressed by the extent of the chord of that angle. To produce the required effect, it is therefore necessary that the motion of α, which is given, should regulate that of the small star, whose relative place at the end of $23\frac{1}{2}$ years is also given. Then, as α moves in an angle of 53° 31′ north preceding, and with a velocity which, being expressed by the space it would describe in $23\frac{1}{2}$ years, will be $3''\cdot51$, it is required that x shall move in an angle of 29° 25′, likewise north preceding, and with a velocity of $3''\cdot02$. The ratio of the velocities, therefore, and the directions of the motions, are equally given. But this will not be sufficient for the purpose : their distance from O must also be taken into consideration. It has been shown, that the two stars cannot be at an equal distance from us, without an evident connection ; it will therefore be necessary for those who will not allow this connection, to place one of them nearer to us than the other. But, as the motions which have been assumed, when seen from different distances, will subtend lines whose apparent magnitudes will be in the inverse ratio of the assumed distances, it is evident that this ratio, if the motions are given, must also be a given one ; or that, if the distances be assumed, the ratio of the motions must be compounded with the ratio of the distances. How then can it be expected that such precise conditions should be made good, by a concurrence of circumstances owing to mere chance ? Indeed, if we were inclined to pass by the difficulties we have considered, there is still a point left which cannot be set aside. The motion of the solar system, although its precise direction and velocity may still be unknown, can hardly admit of a doubt ; we have therefore a third motion to add to the former two, which consequently will bring the case under the statement contained in our 7th number, and will be considered hereafter.

(*g*) If we should intend to change our ground, and place the two motions in O and *x*, it will then be conceded, that the motion of *a* is only an apparent one, which owes its existence to the real motion of the sun. By this, the effect of the solar parallax on any star at the same distance will be given ; and it cannot be difficult to assume a motion in *x*, which shall, with the effect of this given parallax, produce the apparent motion, in the direction of a chord from the first to the last angle of position pointed out by my observations ; taking care, however, not to place the stars *a* and *x* at the same distance from us ; and using the inverse ratio of the solar parallax as a multiple in the assigned motion. For instance, let the sun have a motion of the velocity expressed as before by 3″·51, and in a direction which makes an angle of 53° 31′ south following with the parallel of *a* Geminorum ; and let the small star *x* have a real motion in an angle of 18° 40′ south preceding from the parallel of its situation, and with a real velocity which, were it at the distance of *a*, would carry it through 1″·89. Then, if the distance of the small star be to that of the large one as 3 to 2, the effect of the solar parallax upon it will be $\frac{2}{3}$ of its effect upon *a* ; that is, while *a*, which is at rest, appears to move over a space of 3″·51, in an angle of 53° 31′ north preceding, the parallactic change of place in *x* will be 2″·34 in the same direction. This, though only an apparent motion, will be compounded with the real motion we have assigned to it, but which, at the distance of *a*, will only appear as 1″·26 ; and the joint effect of both will bring the star from the place in which it was seen 23½ years ago, to that where now we find it situated. *a*, in the same time, will appear to have had an annual proper motion of −0″·105 in right ascension, and 0″·12 in declination towards the north; and thus all phenomena will be explained.

From this statement, we may draw a consequence of considerable importance. If we succeed, in this manner, in accounting for the changes observed in the relative situation of the two stars of a double star, we shall fail in proving them to form a binary system ; but, in lieu of it, we shall gain two other points, of equal value to astronomers. For, as *a* Geminorum, according to the foregoing hypothesis, is a star that has no real motion, its apparent motion will give us the velocity and direction of the motion of the solar system ; and, this being obtained, we shall also have the relative parallax of every star, not having a proper motion, which is affected by the solar motion. Astronomical observations on the proper motion of many different stars, however, will not allow us to account for the motion of *a* Geminorum in the manner which the foregoing instance requires ; the hypothesis, therefore, of its being at rest, must be rejected.

(*h*) If we place our two motions in O and α, we shall be led to the same conclusion as in the last hypothesis. The known proper motion of α, and the situations of the small star in 1779 and 1803, given by my observations, will ascertain the apparent motion of x, now supposed to be at rest. Then, since the change in the place of x must be intirely owing to the effect of parallax, it will consequently give us, in the same manner as before, the quantity and direction of the motion of the solar system, and the relative distances of all such stars as are affected by it. But, here again, the solar motion required for the purpose is such as cannot be admitted; and the hypothesis is not maintainable.

Motion of the three Bodies.

(*i*) There is now but one case more to consider, which is, according to No. 5, to assign real motions to all our three bodies; and this may be done as follows. Suppose the sun to move towards λ Herculis, with the annual velocity 1.

Let the apparent motion of α Geminorum be as it is stated in the astronomical tables before mentioned; but suppose it to arise from a composition of its real motion with the effect of the systematical parallax, as we may call that apparent change of place of stars which is owing to the motion of the solar system. Let the real motion of x, aided by the effect of the same parallax, be the cause of the changes in the angle of position which my observations have given. We may admit the largest of the two stars of our double star to be of the second magnitude; and, as we are not to place x too near α, we may suppose its distance from O to be to that of α from the same as 3 to 2. In this case, O will move from the parallel of α, in an angle of 60° 37′ north following, with an apparent annual velocity of ·4536. The motion of α in right ascension, may be intirely ascribed to solar parallax; but its change of declination cannot be accounted for in the same manner. Let us therefore admit that the solar velocity, in the direction we have calculated, will produce an apparent retrograde motion in α, which, in $23\frac{1}{2}$ years will amount to 2″·085 in right ascension. But the same parallax will also occasion a change in declination, towards the south preceding, of 3″·701; and, as this will not agree with the observed motion of α, we must account for it by a proper motion of this star directly towards the north. The real annual velocity required for this purpose, must be 1·3925.

The apparent motion of x, by parallax, at the distance we have placed this star, will be 2″·832 towards the south preceding; and, by assigning to it an annual proper motion of the velocity 1·3354, in the direction of 73° 10′ north preceding its own parallel, the effect of the solar parallax and this proper motion together, will have caused the small star, in appearance, to revolve round α, so as to have produced all the changes in the angle of position which my observations have given; and, at the same time, α will have been seen to move from its former place, at the annual rate of 0″·105 in right ascension, and 0″·12 in declination towards the north.

In this manner, we may certainly account for the phenomena of the changes which have taken place with the two stars of α Geminorum. But the complicated requisites of the motions which have been exposed to our view, must surely compel every one who considers them to acknowledge, that such a combination of circumstances involves the highest degree of improbability in the accomplishment of its conditions. On the other hand, when a most simple and satisfactory explanation of the same phenomena may be had by the effects of mutual attraction, which will support the moving bodies in a permanent system of revolution round a common centre of gravity, while at the same time they follow the direction of a proper motion which this centre may have in space, it will hardly be possible to entertain a doubt to which hypothesis we ought to give the preference.

As I have now allowed, and even shown, the possibility that the phenomena of the double star Castor may be explained by proper motions, it will appear that, notwithstanding my foregoing arguments in favour of binary systems, it was necessary, on a former occasion, to express myself in a conditional manner,* when, after having announced the contents of this Paper, I added, "*should these observations be found sufficiently conclusive*"; for, if there should be astronomers who would rather explain the phenomena of a small star appearing to revolve round Castor by the hypothesis we have last examined, they may certainly claim the right of assenting to what appears to them most probable.

I shall now enter into a more detailed exainmation of the several angles of position I have taken at different times, and show that they agree perfectly well with the appearances which must arise from the revolution of a small star round Castor. A calculation of these angles may be had, by finding the annual motion of the small star, from the change of 21° 54′, which has been shown to have taken place in 23 years and 142 days. Accordingly, I have given, in the 1st column of the following Table [p. 264], the time when the angles were taken. In the 2d, are the angles as they were found by measure; they are all in the north-preceding quadrant. The 3d column contains a calculation from the annual motion of 56′·18, obtained as before mentioned: it shows what these angles should have been, according to our present supposition of a revolving star. And the last column gives the difference between the observed and calculated angles.

On looking over the 4th column of this table, it will be found, that the differences between the observed and calculated angles are not greater than may be expected, considering that most of the early measures are single, and cannot have the accuracy which may be obtained by repetition. Even as they are, we must acknowledge them sufficient to ascertain the gradual change in the angle of position of the two stars. In one place, the difference amounts to six degrees; but it will soon

* See *Phil. Trans.* for 1802, page 486 [above, p. 204].

appear, that a more accurate annual motion gives a calculated position which takes off much of the error of this measure.

Times of the observations.	Observed angles.	Calculated angles.	Differences.
Nov. 5, 1779	32° 47′	32° 47′	0° 0′
Feb. 23, 1791	22 57	22 11	+0 46
Feb. 26, 1792	27 16	21 16	+6 0
Dec. 15, 1795	13 52	17 42	−3 50
March 26, 1800	18 8	13 41	+4 27
April 23, 1800	10 30	13 37	−3 7
Dec. 31, 1801	7 58	12 2	−4 4
Jan. 10, 1802	10 53	12 1	−1 8
Jan. 23, 1802	10 28	11 59	−1 31
Feb. 28, 1802	13 0	11 53	+1 7
Feb. 11, 1803	7 53	11 0	−3 7
March 23, 1803	13 23	10 54	+2 29
March 27, 1803	10 53	10 53	0 0

In a conversation with my highly esteemed friend the Astronomer Royal, he happened some time ago accidentally to mention, that Dr. BRADLEY had formerly observed the two stars of α Geminorum to stand in the same direction with Castor and Pollux. It occurred to me immediately, that if the time of this observation could be nearly ascertained, it would be of the greatest importance to the subject at present under consideration. For, should Dr. BRADLEY'S position be very different from a calculated one, it would induce us at once to give up the idea of a revolving star. The observation was made by Dr. BRADLEY with a view to see whether any change could be perceived in the course of the year, by which the annual parallax of the stars might be discovered. Dr. MASKELYNE, who had this information from Dr. BRADLEY in conversation, had made a memorandum of it in his papers. He has been so kind as to look for it; and, as soon as he found the note, he sent me the following copy, which I have his permission to transcribe.

"*Double star Castor. No change of position in the two Stars: the line joining them, at all times of the year, parallel to the line joining Castor and Pollux in the heavens, seen by the naked eye.*"

Dr. MASKELYNE informs me, that the observation must have been made about the year 1759; and also mentions, that he himself verified the fact, as to the line joining the two stars appearing through the telescope parallel to the line joining Castor and Pollux, in 1760 or 1761; but that he did not examine it at various times of the year.

The advantage of having an angle of position observed in 1759 by Dr. BRADLEY, and so soon after verified by Dr. MASKELYNE, will give us an addition of 20 years to our period. On calculating the right ascension and polar distance of Castor and Pollux for November 5, 1759, it appears, that a line drawn from Pollux through

Castor, must have made an angle of 56° 32′ north preceding with the parallel of that star ; and, this being also the position of our double star, we have an interval of 43 years and 142 days, for a change of 45° 39′, from the time of Dr. BRADLEY'S observation to that of my last measure of the angle. By this we are now enabled to correct our former calculation, which was founded upon a supposition that the first angle of position I had taken was perfect ; but this could hardly be expected, and on examination it appears that the measure was 2° 40′ too little. The annual motion, by our increased period, is 1° 3′·1 ; and the computation of the angles of position in the 3d column of the following Table, as well as the differences contained in the 4th, are made according to this motion.

When the result of this Table is compared with that of the former, it will be seen that my observations agree not only very well with Dr. BRADLEY'S position, but even give more equally divided differences than before, so that the excess and differences counteract each other better than in the first Table.

The time of a periodical revolution may now be calculated from the arch of 45° 39′, which has been described in 43 years and 142 days. The regularity of the motion gives us great reason to conclude, that the orbit in which the small star moves about Castor, or rather, the orbits in which they both move round their common centre of gravity, are nearly circular, and at right angles to the line in which we see them. If this should be nearly true, it follows, that the time of a whole apparent revolution of the small star round Castor, will be about 342 years and two months.

Times of the observations.	Observed angles.	Calculated angles.	Differences.
Nov. 5, 1759	56° 32′	56° 32′	0° 0′
Nov. 5, 1779	32 47	35 29	−2 42
Feb. 23, 1791	22 57	23 36	−0 39
Feb. 26, 1792	27 16	22 32	+4 44
Dec. 15, 1795	13 52	18 32	4 40
March 26, 1800	18 8	14 3	+4 5
April 23, 1800	10 30	13 58	−3 28
Dec. 31, 1801	7 58	12 12	−4 14
Jan. 10, 1802	10 53	12 10	−1 17
Jan. 23, 1802	10 28	12 7	−1 39
Feb. 28, 1802	13 0	12 1	+0 59
Feb. 11, 1803	7 53	11 1	−3 8
March 23, 1803	13 23	10 54	+2 29
March 27, 1803	10 53	10 53	0 0

γ Leonis.

Our foregoing discussions will greatly abridge the arguments which may be used, to show that this star and its small companion are also probably united in forming a binary system. But, in order to give more clearness to our disquisition, we shall follow the arrangement which has been used with α Geminorum, and prefix the same letters to our paragraphs. Then, if any one article should appear to be not sufficiently explained, we need but turn back to our first double star, where the same letter will point out what has already been said more at large on the subject; and an application of it may easily be made.

The distance of the stars γ and x, as I shall again call the small one, has undergone a visible alteration in the last 21 years. The result of a great number of observations on the vacancy between the two stars, made with the magnifying powers of 278, 460, 657, 840, 932, 1504, 2010, 2589, 3168, 4294, 5489, and 6652, is, that with the standard power and aperture of the 7-feet telescope, the interval in 1782 was $\frac{1}{4}$ of a diameter of the small star, and is now $\frac{3}{4}$. With the same telescope, and a power of 2010, it was formerly $\frac{1}{2}$ of a diameter of the small star, and is now full 1 diameter. In the years 1795, 1796, and 1798, the interval was found to have gradually increased; and all observations conspire to prove, that the stars are now $\frac{1}{2}$ a diameter of the small one farther asunder than they were formerly. The proportion of the diameter of γ to that of x, I have, by many observations, estimated as 5 to 4.

The first measured angle in 1782, is 7° 37′ north following;* and the last, which has been lately taken, is 6° 21′ south following. The sum of these angles gives 13° 58′, for the change that has taken place in 21 years and 38 days. To account for this, we are to have recourse, as before, to the various motions of the three bodies.

Single Motions.

(*a*) The motion of x alone cannot be admitted, since it is known that γ Leonis is not at rest. The annual proper motion of this star, according to M. DE LA LANDE, is +0″·38 in right ascension, and 0″·04 in declination towards the south.

(*b*) γ cannot be the only moving body; because its motion in right ascension only, which, in 21·1 years, at the parallel of γ, amounts to 7″·49, would have long ago taken it away from the small star.

(*c, d, e*) The sun cannot be the only moving body; because its motion in right ascension will not account for that of γ Leonis, which star therefore cannot be at rest. And, if we were willing to give up the former assumed solar motion, in order to fix upon such a one as would explain the motion of γ, we should be under a necessity to contradict the united evidence of the proper motions of many principal stars which are in opposition to it.

* In my second Catalogue of double stars, (*Phil. Trans.* for 1785, page 48,) [Vol. I. p. 173] the angle of position is 5° 24′. This was taken April 18, 1783; and, not being acquainted with the motion of the small star, I supposed it to be more accurate than the former measure.

Double Motions.

(*f*) When two motions are proposed, we cannot fix upon γ and x for the moving bodies, unless we should set aside the solar motion, and this, we know, cannot properly admit of a doubt.

(*g*) That we cannot allow O and x to be the two bodies in motion, follows from the insufficiency of the solar motion to account for that of γ, which must be real, or at least partly so.

(*h*) If O and γ are the moving bodies, the given situations of x, in the years 1782 and 1783, point out an apparent motion of x, which must be intirely owing to the solar parallax; and, therefore, those who will admit this hypothesis, must grant the discovery of the motion of the solar system, and of the proportional parallax of the two stars γ and x. Let us however examine whether any motion of the sun, such as we can admit, will account for the change of position and distance pointed out by my observations of the small star near γ Leonis.

The joint effect of proper motion and parallax, has carried γ from its situation in 1782 to that where we now find it. The small star, having all this time, in appearance, accompanied γ, must have gone through a space of $7''\cdot98$, in a direction which makes an angle of $8^\circ\ 30'$ south following with the parallel of γ, in order to be at its present distance from it, and at the same time to have undergone the required change of its angle of position. Now, as the supposition we are examining requires this small star to be actually at rest, it will be necessary to assign to the sun an opposite motion of the same velocity, in order to make that of x only an apparent one. The consequence of this will be a retrograde motion of the sun, which it is well known cannot be admitted.

Motion of the three Bodies.

(*i*) A motion of all the three bodies, is the only way left to explain the phenomena of our double star; and I shall now again point out the very particular circumstances which it is requisite should all happen together, to produce the intended effect.

Let the motion of the sun, with the same annual velocity 1, as in the case of α Geminorum, be directed towards λ Herculis. Then the effect of this motion will show itself at the place of γ Leonis, in the annual velocity of $\cdot3314$, and in a direction which makes an angle of $31^\circ\ 11'$ south preceding with the parallel of that star. In this calculation, I have admitted the distance of the largest of the two stars of γ from the sun to be 3, that of α Geminorum being 2. But, if any other distance should hereafter be considered as more probable, the calculation may be easily adapted to it. The consequence of the parallax thus produced on γ Leonis in 21 1 years, will be an apparent motion of $2''\cdot788$ south preceding, in the abovementioned direction; and, on x, it will be in the same time, and in the same direction, $1''\cdot091$.

As the small star must not be too near γ, we have, in the calculation, supposed it to be at the distance of 4 from O.

The real annual proper motion of γ is required to be 3·5202 ; and its direction must make an angle of 3° 40′ north following with the parallel. By this motion alone, γ would have passed over a space of 9″·87 in 21·1 years; but, when it is combined with the apparent motion arising from parallax, the star will come into its present situation.

The real annual motion of x must be 4·6294, in a direction 0° 20′ south following This will carry it over 9″·74, in 21·1 years; and, when combined with the apparent motion which the solar parallax will occasion, both together will bring it to its proper distance from γ Leonis, and to a situation which will agree with the last observed angle of position.

From what has been said, it is again evident, that not only as many particular circumstances must concur in explaining the phenomena of γ Leonis as we have pointed out with α Geminorum, but that a very marked condition is added in our second double star, which requires an adjustment of velocities in γ and x, which shall also fit the same solar motion that was used in α Geminorum. And this proves, that every additional double star which requires the same condition in order to have its appearances explained, will inforce the arguments which have been used, in a compound ratio.

If, on the other hand, we have recourse to the simplicity of the known effects of attraction, and admit the two stars of our present double star to be united in one system, all the foregoing difficulties of accounting for the observed phenomena will vanish. Whatever may be the proper motion of the sun, the parallax arising from that cause will affect both stars equally, on account of their equal distance from the sun. The proper motion of γ Leonis also may be in any direction, and of any given velocity, such as will agree best with astronomical observations ; since the motion of a system of bodies will not interfere with the particular motion of the bodies that belong to it, so that our secondary star will continue its revolution round the primary one without disturbance.

It will now be necessary to examine the observed angles of position, and to compare them with calculated ones ; but, as there has been a change in the distance of the two stars, it is evident that, if they revolve in circular orbits, the situation of the plane of their revolution must be considerably inclined to the line in which we see the principal star.

Let N F S P, Fig. 2, be the orbit in which x revolves about γ placed in the centre. Suppose a perpendicular to be erected at γ leading to O, not expressed in the figure. By an observation of Feb. 16, 1782, we have the angle $F\gamma x = 7° \ 37'$ north following ; and the proportion of the apparent diameter of γ to that of x has been given as 5 to 4. It has also been ascertained, that the vacancy between the apparent diameters, when the first angle of position was taken, was $\frac{1}{4}$ diameter of the small star; and the last angle of position being 6° 21′ south following, with a distance between the stars of $\frac{3}{4}$ diameter of the small star, we obtain the two points or centres of the small stars *x x′*, through which an ellipsis *a b x x′ c d* may be drawn about γ. This will be the apparent orbit in which the small star will be seen to move about γ, by an eye placed at O. And the inclination of the orbit to the line in which we see the double star, will be had sufficiently accurate to enable us to give a calculation of the several angles of position that have been taken. The ellipsis we have delineated shows that the small star, in its first situation *x*, could not be much past its

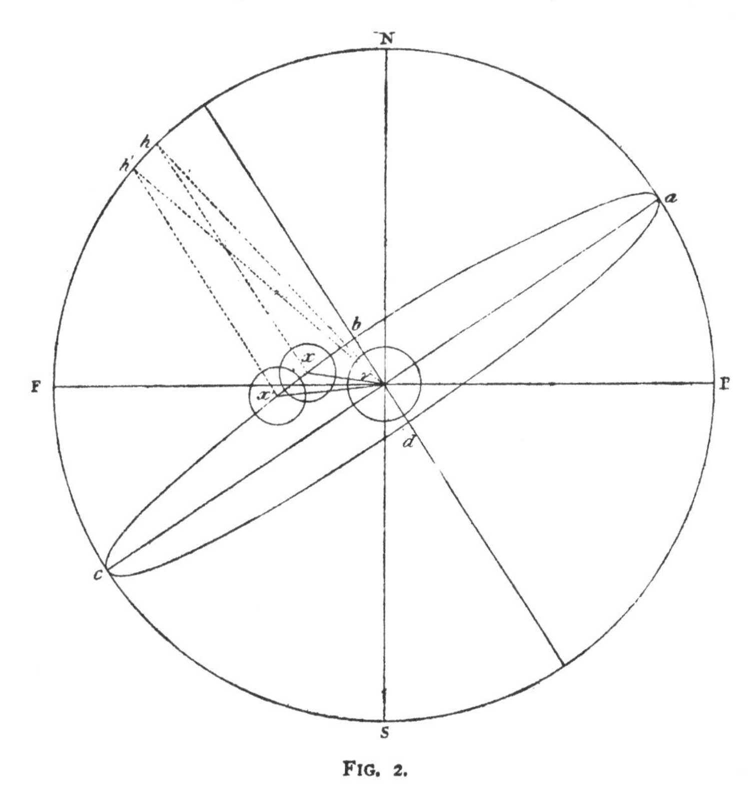

Fig. 2.

conjunction at *b*, and that, consequently, in passing from *x* to *x′*, the parts of the apparent elliptical arch, which are projections of the real circular arch *h h′*, would be described in times nearly proportional to the time in which the whole arch has been described. Upon these principles, the 3d column of the Table [see page 270] has been calculated.

The difference between the calculated and observed angles, contained in the 4th column of the following Table, is so little, that we may look upon the gradual change of these angles as established by observation; and we may form a calculated estimate of the time which will be taken up by the mutual revolution of the two stars. The apparent places *x x′*, being referred to their real ones, give the arch *h h′*, which has been described in 21 years and 38 days; and this arch, seen from the centre γ, is about 6° 20′: it follows, that the length of a whole revolution of our small star round γ Leonis, will be about 1200 years.

Times of the observations.	Observed angles.	Calculated angles.	Differences.
Feb. 16, 1782	7° 37′ *nf*	7° 37′	0° 0′
April 18, 1783	5 24 *nf*	6 51	−1 27
Jan. 24, 1800	3 16 *sf*	4 15	−0 59
Feb. 19, 1800	3 23 *sf*	4 18	−0 55
March 26, 1800	3 47 *sf*	4 22	−0 35
Jan. 26, 1802	6 4 *sf*	5 35	+0 29
Feb. 10, 1803	3 33 *sf*	6 16	−2 43
March 22, 1803	6 32 *sf*	6 20	+0 12
March 26, 1803	6 21 *sf*	6 21	0 0

ε Bootis.

This beautiful double star, on account of the different colours of the stars of which it is composed, has much the appearance of a planet and its satellite, both shining with innate but differently coloured light.

There has been a very gradual change in the distance of the two stars; and the result of more than 120 observations, with different powers, is, that with the standard magnifier, 460, and the aperture of 6·3 inches, the vacancy between the two stars, in the year 1781, was 1½ diameter of the large star, and that it now is 1¾. By some earlier observations, the vacancy was found to be considerably less in 1779 and 1780; but the 7-feet mirror then in use was not so perfect as it should have been, for the purpose of such delicate observations. By many estimations of the apparent size of the stars, I have fixed the proportion of the diameter of ε to that of x, as 3 to 2. August 31, 1780, the first angle of position measured 32° 19′ north preceding; * and, March 16, 1803, I found it 44° 52′, also north preceding: the motion, therefore, in 22 years and 207 days, is 12° 33′. It should also be noticed, that while the apparent motion of α Geminorum, and of γ Leonis, is retrograde, that of ε Bootis is direct.

A proper motion in this star, if it has any, is still unknown; our former arguments, therefore, cannot be applied to it, without some additional considerations; and, as many others of my double stars will stand in the same predicament, I shall give an outline of what may be said, to show that this, and probably many of the rest, are also binary systems.

Single Motions.

(*a—e*) If ε Bootis is a star in which no proper motion can be perceived, we may infer, from the highly probable motion of the solar system, that this star, which is of the 3d magnitude, and on that account within the reach of parallax, must have

* The angle of position, in my first Catalogue of double Stars, *Phil. Trans.* for 1782, page 115, [Vol. I. p. 60] is 31° 34′ (it should be 54′) north preceding. This will be found to be a mean of the three first measures hereafter given in a Table of positions.

a real motion, to keep up with the sun, in order to prevent an apparent change of place, which must otherwise have happened. In this case, no single motion can be admitted to explain the phenomena of our double star. But, if a real proper motion of ϵ Bootis should hereafter be ascertained, the arguments we have used in the case of γ Leonis, will lead to the same conclusion.

Double Motions.

(*f*) ϵ and x cannot be the moving bodies; and our former argument (*f*) will apply to every double star whatsoever.

(*g*) O and x cannot be alone in motion; for, if no motion in ϵ can be perceived, it must move in a similar manner with the sun, and none of the three bodies will be at rest. But, if its proper motion shall hereafter be found out, it must either be exactly the reverse of the solar motion, and therefore only an apparent one, or it will be more or less different. In the latter case, all the three bodies must be in motion; in the former, the exact quantity of the solar motion will be discovered, and the relative parallax of many stars may be had by observation.

(*h*) If O and ϵ are the two bodies in motion, and if at the same time no motion in ϵ can be perceived, then the apparent motion of x must be intirely owing to the different effect of the solar parallax on ϵ and x; but the effect of the solar parallax on x, can only be in a direction contrary to the motion of the sun, which, being north following the small star, whether it be nearer or farther from us than ϵ, must have an apparent motion towards the south preceding part of the heavens. But this is directly in opposition to my observation of the motion of the small star, which, these last 23 years, has been directed towards the north following.

Motion of the three Bodies.

(*i*) Let the motion of the sun be again towards λ Herculis; then, if no motion in ϵ Bootis be perceivable, it must move exactly like O. Highly improbable as it is, let it be admitted. Then, in addition to this extraordinary supposition, a third motion is also required for x, which, aided by the solar parallax, is to carry it likewise within a quarter of a diameter of ϵ, into the same place where, though unperceived, the large star has been carried by its own motion; that is, in order to be apparently at rest, the sun, ϵ Bootis, and its small companion, must all move exactly alike, setting aside the very little difference in the position and distance of the small star, which, in the whole, amounts to little more than 6-tenths of a second; than which, certainly nothing can be more improbable.

But, if ϵ shall hereafter be found not to have been at rest during the time of my observations upon it, then its place will be given; and, since also the situation of x, with respect to ϵ, is to be had from my angles of position and distances of the two stars, the case will be similar to that which has already been considered, in the paragraph (*i*), under the head of γ Leonis.

I may here add a remark with regard to ϵ Bootis, which will be applicable to several more of my double stars. In the milky-way, a multitude of small stars are

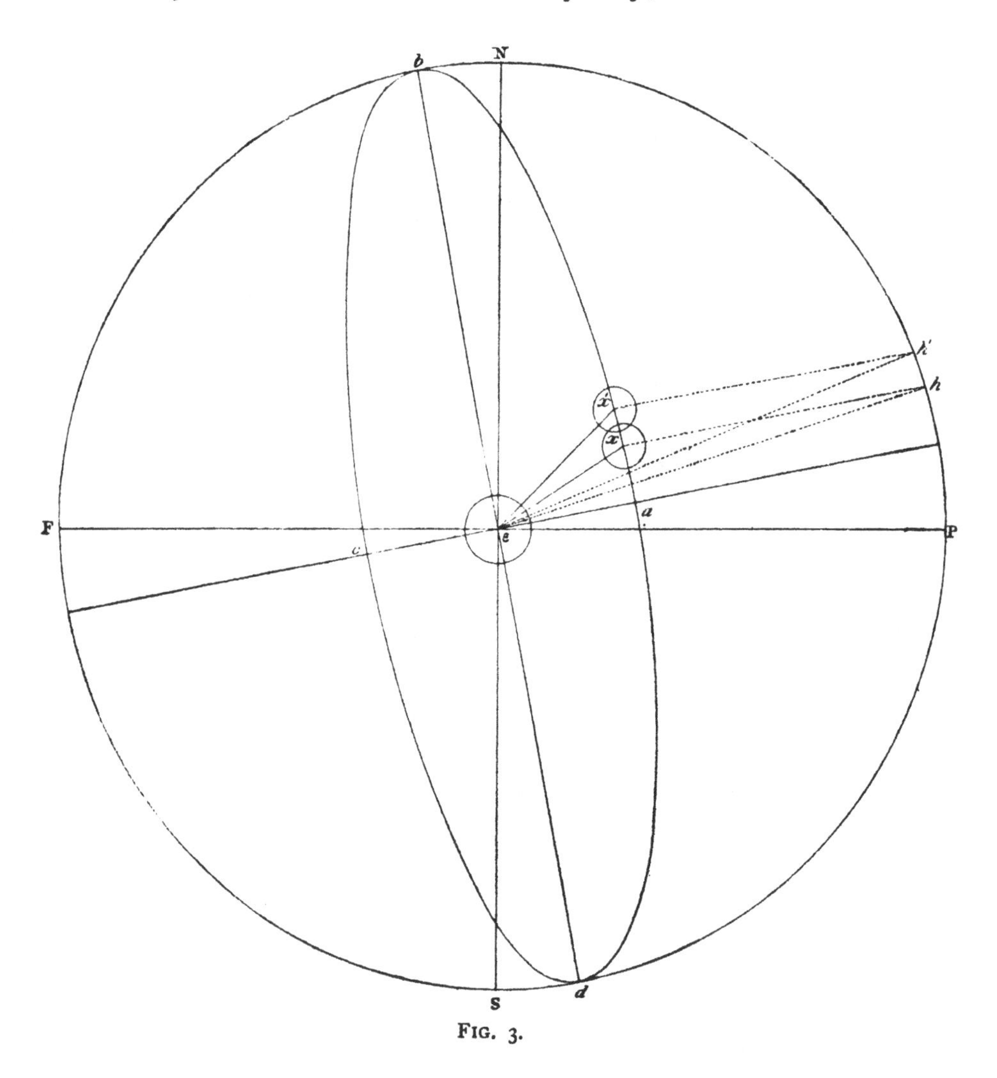

FIG. 3.

profusely scattered, and their arrangement is very different from what we perceive in those parts of the heavens which are at a considerable distance from it. About ϵ Bootis, which is situated in what I have formerly called figuratively a nebulous part of the heavens,* there are, comparatively speaking, hardly any stars; and, that so remarkable a star as ϵ should have a companion, seems almost to amount to a proof that this very companion is, as it appears to be, a connected star. The *onus probandi*, therefore, ought in justice to fall to the share of those who would deny the truth of what we may call a fact; and I believe the utmost they could do, would be

* See *Phil. Trans.* for 1784, page 449 [Vol. I. p. 165].

to prove that we may be deceived; but they cannot show that this star has no connection with ε Bootis.

This argument will be much supported, when we consider that many of the double stars in the milky-way are probably such as have one of the scattered stars, nearly in the same line, at a great distance behind them. In this case, the two stars of the double star have no connection with each other; and the great number of them in the milky-way, is itself an indication of this effect of the scattered multitude of small stars. In the single constellation of Orion, for instance, we have no less than 43, pointed out by my catalogues; ten of which are of the first class, and yet have undergone no change of distance or position since I first perceived them. But, with apparently insulated stars, such as ε Bootis, the case is just the reverse.

If, in consequence of our former arguments, and the present remarks, we place ε Bootis among the stars which hold a smaller one in combination, we may delineate its orbit as in Fig. 3.

Let P N F S represent a circle, projected into the elliptical orbit $a\,x\,x'\,b\,c\,d$. ε is the large star; and $x\;x'$ are the first and last measured north preceding situations of the small one, as given in the following Table.

Times of the observations.	Observed angles.	Calculated angles.	Differences.
August 31, 1780	32° 19′	33° 58′	−1° 39′
March 13, 1781	30 21	34 13	−3 52
May 10, 1781	33 1	34 18	−1 17
Feb. 17, 1782	38 26	34 40	+3 46
August 18, 1796	45 32	41 40	+3 52
Jan. 28, 1802	49 18	44 19	+4 59
August 31, 1802	46 47	44 36	+2 11
March 23, 1803	43 43	44 52	−1 9
March 26, 1803	44 52	44 52	0 0

The real motion from h to h' is projected into that from x to x'; and, while the elliptical arch subtends an angle of 12° 33′, the circular one will be about 4° 50′.

From the figure of the orbit, we may conclude that the small star, in its first position, at x or h, was not more than between 30 and 40 years past its conjunction; and that, consequently, the parts of the arch $x\;x'$, were nearly proportional to the times of their being described. The positions have been calculated upon this principle; but with some allowance for the first observed angle, which I suppose to have been a little too small; and, though the differences of the observed and calculated angles are pretty considerable, the observations are still sufficiently consistent to prove the gradual change of the situation of the small star.

The quantity of the change in 22 years and 207 days, will show that a periodical revolution cannot take up less than 1681 years. The real figure and situation of the orbit, with many other particulars, are still unknown; it is, therefore, unnecessary to point out the uncertainties in which the investigation of the periodical time of the small star about ε Bootis must long remain involved.

ζ *Herculis.*

My observations of this star furnish us with a phenomenon which is new in astronomy; it is, the occultation of one star by another. This epoch, whatever be the cause of it, will be equally remarkable, whether owing to solar parallax, proper motion, or motion in an orbit whose plane is nearly coincident with the visual ray. My first view of this star, as being double, was July 18, 1782. With 460, the stars were then ½ diameter of the small star asunder. The large star is of a beautiful bluish white; and the small one ash-coloured.

July 21, of the same year, I measured the angle of position, 20° 42′ north following. With the standard power, the distance of the stars remained as before. With 987, they were one full diameter of the small one asunder.

In the year 1795, I found it difficult to perceive the small star; however, in October of the same year, I saw it plainly double, with 460; and its position was north following.

Other business prevented my attending to this star till the year 1802, when I could no longer perceive the small star. Sometimes, however, I suspected it to be still partly visible; and, in September of the same year, with 460, the night being very clear, the apparent disk of ζ Herculis seemed to be a little lengthened one way. With the 10-feet telescope, and a power of 600, I saw the two stars of η Coronæ very distinctly; and, having in this manner proved the instrument to act well, I directed it to ζ Herculis, and found it to have the appearance of a lengthened, or rather wedge-formed star; after which, I took a measure of the position of the wedge.

Our temperature is seldom uniform enough to permit the use of very high powers; however, on the 11th of April, 1803, I examined the apparent disk, with a magnifier of 2140, and found it, as before, a little distorted; but there could not be more than about ⅜ of the apparent diameter of the small star wanting to a complete occultation. Most probably, the path of the motion is not quite central; if so, the disk will remain a little distorted, during the whole time of the conjunction. Our present observations cannot determine which of the stars is at the greatest distance; but this will occasion no difference in the appearance; for, if the small star should be the nearest, its light will be equally lost in the brightness of the large one.

The observations I have made on this star, are not sufficient to direct us in the investigation of the nature of the motion by which this change is occasioned.

We may however be certain, that with regard to

Single Motions,

(*a, b*) Neither *x* nor ζ can be supposed to be the only moving bodies, without contradicting the highly probable arguments for the sun's motion.

(*c, d*) If we admit the sun to be the moving body, the stars ζ and *x* being at rest, we may calculate the effect of the solar parallax upon them, as follows. Let O move towards λ Herculis, with the annual velocity 1, as in the case of α Geminorum; then, from the situation and magnitude of the large star of ζ Herculis, which we will suppose 4m, the effect of the solar motion at ζ will be only ·0522; and, at *x*, supposed to be at the distance 5m, it will be ·0418. This will show itself at the parallel of ζ in a direction of 25° 5′ north preceding, the solar motion being in the opposite direction south following. But this parallax will only produce, in 20 years and 10 months, an apparent change of 0″·444 in ζ, and of 0″·355 in *x*; and will separate the stars, instead of bringing them to a conjunction.

(*e*) A considerable advantage may be gained, by placing *x* at a little more than $\frac{1}{3}$ the distance of ζ from O. For as, in the abovementioned time, this would make the effect of parallax upon it 1″·18, a conjunction should now take place. But then the stars, though very near each other, would not be quite in contact; much less could one of them occasion an occultation of the other. The supposition also, that the small star should be only $\frac{1}{3}$ of the distance of the large one from us, is not very favourable to the hypothesis.

δ Serpentis.

This double star has undergone a very considerable change in the angle of position, but none in the distance of the two stars. The 5th of September, 1782, an accurate measure of the position was 42° 48′ south preceding; and February 7, 1802, it measured 61° 27′ south preceding. In 19 years and 155 days, therefore, the small star has moved, in a retrograde order, over an arch of 18° 39′.

Every argument, to examine the cause of this motion, which has been used with ε Bootis, in the paragraphs from (*a*) to (*i*), will completely apply to this star; from this we may conclude, that the most natural way of accounting for the observed changes, is to admit the two stars to form a binary system. In this case we calculate, with considerable probability, that the periodical time of a revolution of the small star round δ Serpentis, must be about 375 years.

γ Virginis.

This double star, which has long been known to astronomers,* has undergone a visible change since the year 1780, when I first began my observations of it. The 21st of November, 1781, I measured the position of the two stars, which was 40° 44′ south following. The stars are so nearly equal, that I have but lately ascertained

* *Mémoires de l'Académie des Sciences.* Ann. 1720.

the following one to be rather larger than its companion; the position, therefore, ought now to be called north preceding. By a mean of three measures, that were taken on the 15th of April, 1803, the angle was 30° 20′ *np*.

The distance, as far as estimations by the diameter can determine, when the stars are so far asunder as these are, remains without alteration. May 21, 1781, they were 2½ diameters asunder; and, by estimations lately made, with the same instrument and power as were used 21 years ago, the stars are still at the same distance of 2½ diameters.

A very small proper motion in declination, of 0″·02 towards the south, has been assigned to this double star; * but the quantity is hardly sufficient for us to rely much upon the accuracy of the determination. I shall therefore rather consider γ Virginis as one of the stars of which we have no proper motion ascertained; and the arguments to which I shall refer, will consequently be those which have been given with ϵ Bootis.

The change of the angle of position, in the time of 21 years and 145 days, amounts to 10° 24′; from which we obtain the annual motion of 29′·16. The observed and calculated angles, with their differences, on which it will not be necessary to make any remarks, are in the following Table.

Times of the observations.	Observed angles.	Calculated angles.	Differences.
Nov. 21, 1781	40° 44′	40° 44′	0° 0′
Jan. 29, 1802	28 22	30 51	−2 29
April 15, 1803	30 20	30 20	0 0
May 28, 1803	32 2	30 17	+1 45

As a confirmation of the accuracy of these observations, we may have recourse to a position of the same stars, deduced from the places of them, as they are given in MAYER's *Zodiacal Catalogue*. By two observations, reduced to the beginning of the year 1756, the preceding one was 3″·8 before the other in right ascension, and 5″·3 more north than that star. From this we calculate the position, which was 54° 21′ 37″ north preceding. The interval from the 1st of January, 1756, to the 21st of November, 1781, is 25 years and 325 days. When this is added to the period I have given, we have 47 years and 105 days, for a motion of 24° 2′. The annual motion, deduced from this lengthened period, which is 30′·5, differs less than 1½ minute from that which has been calculated from my observations. With the assistance, therefore, of MAYER's observation, which greatly supports our calculation, we may conclude, that the two stars of γ Virginis revolve round each other in about 708 years.

* *Connoissance des Temps*, Année VI. page 213.

LII.

Continuation of an Account of the Changes that have happened in the relative Situation of double Stars.

[*Phil. Trans.*, 1804, pp. 353–384.]

Read June 7, 1804.

In my former Paper,* I have given the changes which have happened in the situation of six double stars. When the causes of these observed changes in the double star Castor were investigated, I had recourse to the most authentic observations I could find, of the motions in right-ascension and polar distance of this star. But the Tables which have been lately published, in the last volume of the observations made by the Astronomer Royal at Greenwich, give us now the proper motions of 36 principal stars, of which α Geminorum is one; and, as the motion of this star, especially in north polar distance, is very different from what it has been supposed in my former examination, it will be necessary to review the arguments which have been used, in order to ascertain what will be the result of this new motion. We shall here again follow the order of the paragraphs of the former Paper, and denote those which treat of the same motions, with the same letters, that they may be readily compared.

Single Motions.

(*a*) The small star x cannot be alone in motion, as we have now, in the new Tables I have mentioned, an evident proof that the large star α is not at rest.

(*b*) As the observations of the Astronomer Royal have ascertained the motion of Castor, so it is no less evident, from the series of observations which has been given in my Paper, that its smaller companion has also changed, if not its real situation, at least its relative one with respect to the large star. Let us therefore examine, whether the motion of α can be the cause of the apparent change that has taken place in the relative situation of these two stars.

The annual proper motion of α Geminorum, in right ascension, by the new Tables, is $0''\cdot15$; which, in $23\frac{1}{2}$ years, will amount to $3''\cdot525$. The annual proper motion in polar distance, by the same Tables, is $0''\cdot04$; which, in the same time, will amount to $0''\cdot94$; the former motion being retrograde, and the latter towards

* See *Phil. Trans.* for 1803, p. 339 [above, p. 250].

the south. Let FP, in Figure 1, be the parallel of Castor, and make $\alpha\,\alpha'$ equal to 2978·5; which will be the quantity of its motion in right ascension in the parallel, when it is 3525 in the equator. At right angles to αP, make $\alpha'\,\alpha''$ equal to 940; and this will represent the motion of the star in polar distance towards the south. Draw the line αx so as to make an angle of 32° 47′ with the parallel FαP on the north preceding side, and place x at the distance of 3765 from α. Then will α and x be the situation in which these two stars were observed in the year 1779; their apparent distance, estimated in diameters of the large star, being $1\frac{7}{8}$; and the angle of position, as has been stated, 32° 47′ north-preceding.

If the star x had been at rest while α moved towards α'', the relative situation of the two stars in the year 1803 would have been represented by $\alpha''\,x$; that is to say, the apparent disks of these two stars would have been hardly $1\frac{1}{3}$ diameter of the largest asunder, and the angle of position $x\,\alpha''$ P′ must have been 86° 25′ north-preceding. But this is quite inconsistent with the observations that have been given; according to which, the small star, in the year 1803, was situated at x'. It is therefore proved, that the motion of α alone cannot account for the change which has taken place.

(*c*) If the motion of Castor should be only an apparent one, arising from the motion of the solar system, then the proper motion of the sun must be just the reverse of that which the new Tables assign to α Geminorum. This being admitted, let us examine what will be the result with regard to the relative situation of the small star, which, since only the sun is supposed to be in motion, must now remain at rest, as well as α. The effect of the parallax, which we are now considering, is inversely as the distances of the stars which are affected by it. Hence arise the three cases which have been examined in my first Paper.

When a line from the sun to Castor, Oα,* is perpendicular to the line αx, joining the two stars, no change in their relative situation can take place, arising from parallax, which will act equally on both. For, let α, α'' and x, in Fig. 2, be placed as they were in Fig. 1; and the real motion of the sun from O to O′, will produce the parallactic motion of Castor from α to α''. It will also occasion an apparent motion of x, equal to that of α, and in a parallel direction with it. This star will therefore appear to have moved from x to x', in the same time that the large star has moved from α to α'', so that their relative situation will remain unchanged.

(*d*) If x be placed beyond α, the effect of parallax, exerted in the direction $x\,x''$, parallel to $\alpha\,\alpha''$, will be less upon this star than on Castor; and its apparent motion must fall short of the situation x'. The consequence of this will be an increase of the angle of position; but, as we know, from the observations which have been given, that this angle has been decreasing, it follows, that the small star cannot be admitted to have been at rest, if we place it farther from us than α.

* See Figure 1 of the former Paper.

(*e*) When the smallest of the two stars of our double star is supposed to be much nearer than the largest, the effect of parallax will carry it beyond x'. Let its distance from us to that of a be, for instance, as 3123 to 6076. In this case, while a appears to move as far as a'', x will be seen to move to x''; where its angle of position $x'' a''$ P′, will be just 10° 53′ north-preceding, as by observation it was found to be in the year 1803. But, according to this hypothesis, the distance $a'' x''$ of the two stars, ought now to be nearly double what it was in 1779; and, since this is contrary to observation, we must also give up this last supposition.

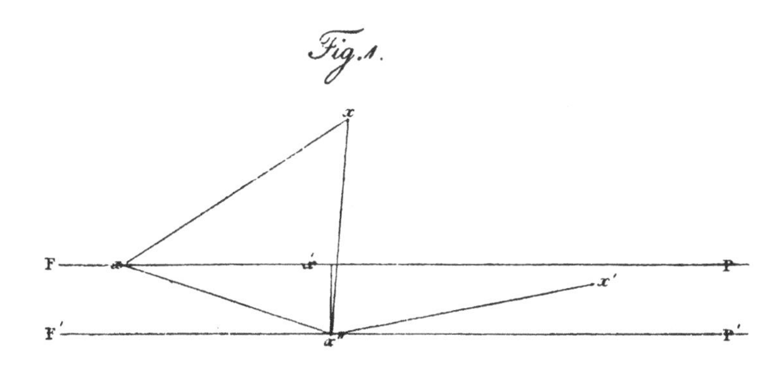

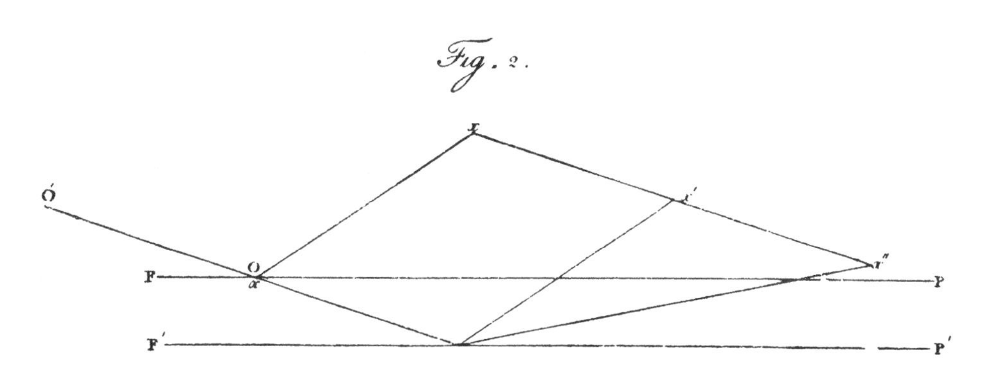

Double Motions.

(*f*) Let us now suppose a and x to be in motion, while the solar system remains at rest. Then, since there will be no parallax that can affect the appearance of these motions, they must be real, and proper to each of the stars. But the circumstances that must take place, in order to produce the phenomena which have been observed, are so particular, that we shall soon find the great improbability of such an accidental arrangement of them as would answer the end. It has already been shown, in the paragraph (*c*) of the former Paper, that we cannot place the two stars at an equal distance from us; and it would be the height of improbability to suppose them to move in parallel planes. But, whatever may be the directions and velocities of the motions of the stars, or at whatsoever different distances we may place them, the effect which is to arise from these combined circumstances is positively determined ; for the star α must appear to move at the rate of 3″·123 of an arch of a great circle in 23½ years, and in the direction of 17° 31′ south-preceding its parallel ; while the star x, in the same time, must seem to move over an arch of 4″·179, in a direction of 32° 52′ south-preceding the same parallel. When these quantities, resulting from the proper motions of our new Tables, are substituted for those which have been used in the paragraph (*f*) of my former Paper, the arguments which it contains will remain in full force, and need not be here repeated.

(*g*) The same argument which has been used in the first Paper, when the sun and the small star only were supposed to be in motion, will perfectly apply to the proper motion of Castor, as given in the new Tables. For, as this motion is now to be accounted for by the motion of the sun, we have only to substitute the velocity of 3″·123, in a direction which makes an angle of 17° 30′ 56″ north-following with the parallel of α Geminorum, for the quantity of the solar motion before used ; and to assign a proper motion to the small star, having a direction of 68° 10′ south-preceding the parallel of α Geminorum, with a velocity which, if the star was at the distance of α from us, would carry it in 23½ years through 1″·4303.

(*h*) When the sun and Castor only are supposed to be in motion, the former statement of the case will in every respect remain conclusive.

Motion of the three Bodies.

(*i*) It remains now only to be shown, that the arguments which are contained in my first Paper, against the probability of a supposition which ascribes all the observed phenomena to three real motions, will not be affected by the given alteration in the proper motion of Castor. Without repeating any part of the discussion of the former paragraph (*i*), it will be sufficient if I point out three motions, such as will answer the required purpose.

Let the solar motion, as before, be towards λ Herculis, with such a velocity as will in 23½ years produce a parallactic motion, at the distance and situation of α Geminorum, amounting to 2″·2805, in a direction of 60° 36′ 57″ south-preceding the parallel of that star. Let Castor have a real motion, which in 23½ years would carry it over an arch of 2″·1341, in a direction of 29° 23′ 3″ north-preceding its parallel ; and let the real motion of the small star be such that in 23½ years, at its distance from us, supposed to be to that of Castor as 3 to 2, it would describe an arch of 2″·9212, in a direction of 18° 50′ 13″ south-preceding. Then would the parallactic motion of $\alpha = 2''{\cdot}2805$, compounded with the real motion we have mentioned, give us an apparent annual motion equal to that which, in Dr. MASKELYNE's Table, is called the proper motion in right ascension and polar distance of this star. And the parallactic motion of $x = 1''{\cdot}5203$, compounded with the real motion we have assigned, would also produce an apparent annual motion which would correspond with my series of observed situations of this small star. But, for the high improbability of such an hypothesis, I refer to the paragraph (*i*) of my former Paper.

What has been said of Castor, will apply to every other double star of which the proper motion may hereafter be assigned; for, unless the parallactic motion arising from the motion of the solar system should completely explain the observed changes, the same arguments will still remain in full force.

I shall now proceed to a continuation of my account of the changes that have happened in the relative situation of double stars, either in their position or their mutual distance; and, in the following list of them, it will be seen that, of 50 changeable double stars which are given, 28 have undergone only moderate alterations, such as do not amount to an angle of 10 degrees. None of them however have been admitted, except where the change was at least so considerable, that the micrometer which was used on this occasion could ascertain the change with a proper degree of accuracy. Two of the stars, indeed, have hardly suffered any alteration in the angle of position; but, with them it will be found, that a change in their distance has been so ascertained as not to admit of any doubt. Thirteen of the stars have altered their situation above 10 degrees, but less than 20. Three stars have undergone a change in the angle of position, of more than 20, and as far as 30 degrees. The six remaining stars afford instances of a still greater change, which, in the angle of position of some of them, amounts to more than 30 degrees; in others, to near 40, 50, 60, and upwards, to 130 degrees.

α *Herculis*. II, 2.*

The two stars of this double star have undergone a considerable change in their angle of position. By a measure taken May 20, 1781, it was 21° 28′ south-following.† April 3, 1783, two measures gave 25° 29′. A mean of two measures, taken Feb. 21 and March 4, 1802, was 31° 38′. By five measures, taken in 1803 and the beginning of 1804, it was 31° 54′; and, June 3, 1804, by a very accurate measure, with an improved illumination of the wires, it was 32° 50′. This gives a change of 11° 22′, in 23 years and 14 days.

It does not appear that the distance has undergone any perceptible alteration.

As we have now the proper motion of this star in Dr. MASKELYNE's new Tables, we are enabled to enter upon an examination of the cause of the observed change; but first it will be necessary to mention, that in this and all the following stars, I have no longer supposed the solar motion to be directed towards λ Herculis. A point at no very great distance from this star has been chosen, for reasons which it would lead us too far from our present subject to assign, and which are of no absolute

* The numbers after the name of the star, refer to my Catalogues of double Stars, published in the *Philosophical Transactions*. For instance, II, 2, denotes that α Herculis is the 2d star in the 2d class.

† By mistake, the first angle of position in my Catalogue is given 30° 35′, instead of 21° 28′, and should be corrected. See *Phil. Trans.* Vol. LXXII. Part I. p. 122 [Vol. I. p. 64].

consequence to it. The motion of the solar system, towards this assumed point, will produce an opposite parallactic motion, in every star that is not too far from us to be sensibly affected by it.

That change of place which astronomers have established by observation, and which is called the proper motion of a star, either may agree with this parallactic motion, (in which case it will be only an apparent one, the star being really at rest,) or it may be directed to another part of the heavens, so as to differ from our parallactic motion. Whenever this happens, the star will have the following three motions: a real, a parallactic, and an apparent one; the latter being a composition of the former two.

That α Herculis is one of those stars which has these three motions, will appear thus: the parallactic motion which this star, from its magnitude and consequent proximity, must be allowed to have, will carry it, in an angle of about 58½ degrees, towards the south-preceding part of the heavens; but the motion assigned to it in the new Tables, has a direction towards the north. Hence it follows, that α Herculis has also a real motion, which, by its composition with the parallactic one, produces the tabular apparent one.

We are now to examine the effect of these three motions, on the position of the two stars of our double star, in order to see how far they will account for the observed change. The two stars are sufficiently different in magnitude, for us to expect a difference of parallax, on a supposition that their distances from us are inversely as their apparent magnitudes. The change of the angle of position, arising from a superior parallactic motion of the large star, would have occasioned a retrograde motion of the small one; but this, by my observations, has moved according to the order of signs; its change of situation, therefore, will admit of no explanation from the effect of parallax.

The real motion of α Herculis, being such as, with the union of the parallactic one, will produce an apparent motion towards the north, is determined by the velocities and directions of the other two motions. It must however be towards the north-following part of the heavens, and of a velocity considerably greater than the proper motion given in the new Tables; but, since it is known to be compounded with the parallactic one, we are now only to consult the direction and velocity of that composition, which is such that the large star, in 23 years and 14 days, must have been carried $5''\cdot299$ towards the north. If the stars are not connected, the most favourable case we can put, will be to suppose the small one at rest, and at such a distance from us as to be intirely free from sensible parallax. This being admitted, the large star, by its motion, should now have left the small one so far behind, that the distance of the centres of the two stars, (which Sept. 25, 1781, by a measure with my lamp micrometer, was $4''\ 34'''$,) should now be $7''\cdot92$; while, at the same time, the angle of position ought to have increased to 52½ degrees. My last observations, however, give so different a result, that this hypothesis cannot be admitted.

If the small star, which is not so much less than the large one that we can justly place it at the above mentioned distance, should partake of some parallactic motion, it will then increase the objections we have stated ; for, if the effect of it should be only one quarter of what it is upon the large star, it will add to the magnitude of the angle of position, and increase the distance of the two stars.

Hence it follows, that, unless we should admit the supposition of three independent motions, the high improbability of which has been sufficiently shown, we have good reason to believe that the large star has, during the 23 last years, carried the small one along with itself, in the path it describes in space ; both being equally affected by parallax and real motion. If this be admitted, a mutual revolution of the two stars will be the immediate consequence, when the laws of gravitation are taken into consideration ; and the change of position they have undergone, will be a necessary consequence of it.

γ *Arietis*. III, 9.

This star being only of the 4th magnitude, and of the third class as a double star, we have no reason to expect a great change in the angle of position ; and yet, with the assistance of a very distant observation, which we have in MAYER's *Zodiacal Catalogue*, a considerable change may be proved to have taken place. The position, Nov. 2, 1779, was 84° 0′ south-following.* Oct. 10, 1780, it was 86° 5′ ; and, Feb. 7, 1802, it was 89° 10′. The change, in 22 years and 97 days, is 5° 10′. From the given right ascension and declination of the two stars, in MAYER's Catalogue, we compute, that their position in 1756 was 78° 46′ south-following ; which gives a change of 5° 14′, in 23 years and 306 days, up to the time of my first observation. The two periods, which are nearly equal, give a change of 10° 24′, for 46 years and 38 days. A motion of γ Arietis, arising from systematical parallax, by which we may admit the smallest of the two stars (on account of its supposed greater distance) not to be so much affected as the large one, will perfectly account for the change ; unless, hereafter, the proper motion of this star, when known, should lead to a different conclusion.

ξ *Ursæ*. I, 2.

This double star has undergone a very extraordinary change in the angle of position. Dec. 19, 1781, the smallest of the two stars was 53° 47′ south-following. Feb. 4, 1802, it was 7° 31′ ; and, January 29, 1804, the position was only 2° 38′. This gives a motion of 51° 9′, for 22 years 41 days, and amounts to 2° 19′ *per* year. If an annual alteration to this amount should continue to take place for the future, a very few years would be sufficient to ascertain the cause of this change, as no motion but a revolving one could possibly explain the phenomenon. If, on the contrary, the parallactic motion of the largest star should have occasioned the change

* This position, for reasons explained in the note to *p* Serpentarii [*infra*, p. 290], has not been given in my Catalogue.

of situation, which is not impossible, it will soon be verified by an increased distance of the two stars, accompanied with very little angular change in their position. The little difference in the magnitude of the two stars, however, does not well agree with a supposition which gives a parallactic motion to one of them only.

γ *Andromedæ*. III, 5.

It has already been noticed, on a former occasion, that this double star is one of the most beautiful objects in the heavens. The striking difference in the colour of the two stars, suggests the idea of a sun and its planet, to which the contrast of their unequal size contributes not a little. The position of the small star, when we consider that this double star is one of the third class, has undergone a sufficient change to deserve notice. In the year 1781, Oct. 15, it was 19° 37′ north-following. Feb. 3, 1802, 26° 34′. Feb. 11, 1803, 26° 5′; and, Feb. 5, 1804, 27° 39′. The difference, in 22 years and 113 days, is 8° 2′. The distance of the two stars is too great to be accurately estimated by their apparent diameters; and measures taken with a micrometer, unless fractions of a second of space could be strictly ascertained, would be useless. If we suppose the small star sufficiently removed not to partake of the systematical parallax of the large one, the change of the angle of position may be accounted for, upon the principle of the solar motion. The stars, however, are hardly so different in magnitude as would be required for that purpose. We ought also to know, whether a proper motion has been observed in this star.

μ *Draconis*. II, 13.

The change in the relative situation of the two stars of this double star is pretty considerable. The position, Sept. 24, 1781, was 37° 38′. This may stand either for south-preceding or north-following, because the stars were then regarded as being equal. March 4, 1802, a measure of the position gave 50° 32′. Feb. 5, 1804, position 49° 0′ south-preceding; and, Feb. 6, 1804, 50° 4′. A memorandum annexed to the observation says, that the preceding star is the smallest, but that the difference is so little as to require much attention to be perceived. The alteration, in 22 years and 135 days, is 12° 26′. The two stars being nearly of an equal magnitude, we can have no inducement to suppose them to be at very different distances from us. This makes it not probable that the difference of their parallactic motion should be the cause of the change of the angle of position; otherwise, the direction of that motion would be sufficiently favourable.

δ *Geminorum*. II, 27.

The measures of the position of the two stars of this double star are attended with great difficulty, on account of the faintness of the smallest; a considerable disagreement will therefore be excuseable. The position, Nov. 18, 1781, was 85° 51′ south-preceding. Jan. 28, 1802, it was 76° 21′. Feb. 4, 1802, 73° 5′; and, Feb. 6,

1804, 69° 52′. The difference, in 22 years and 80 days, is 15° 59′. We can have no assistance from observations made on the distance of the two stars, which is too great for estimation. A parallactic motion, which, on account of the great difference in the magnitude of the stars, might be admitted, would lessen their distance, and make the angle of position retrograde, which, by my observation, has moved in a contrary direction. A connection between the two stars is also rendered improbable, on account of the great number of small ones that are scattered in this neighbourhood, of which our small star may be one; so that we have good reason to ascribe the change which has happened in the situation of our two stars, to a proper motion of δ.

ε *Draconis.* I, 8.

In this star, we have to notice a great change of the angle of position, but none in the distance. In the year 1782, Sept. 4, with 460, I found the stars to be 1½ diameter of L. asunder. May 22, 1804, they were still at the same distance of 1½ diameter of L. Oct. 20, 1781, the position was 63° 14′ north-preceding; and, May 22, 1804, it was 84° 29′; which proves a change of 21° 15′, in 22 years and 214 days. This cannot be owing to a parallactic motion of the large star; for the effect arising from such a motion, would have been directly contrary to the change which has taken place: the angle of position would have undergone a direct, instead of a retrograde alteration. We are consequently assured that ε Draconis cannot be at rest. If future observations on the proper motion of the stars should furnish us with that of ε, and if this motion should also fail to explain my observed change of the angle of position, without a change of distance, we shall then have good reason to admit this star into the list of those that have a small one revolving about it. For, to ascribe an additional and independent motion to the small star, would be to have recourse to three separate motions, of given velocities, in given directions, and at given distances; the improbability of which has been sufficiently pointed out.

ζ *Aquarii.* II, 7.

The position, Nov. 26, 1779, was 71° 5′ north-following. Sept. 24, 1781, it was 71° 39′. June 19, 1782, 72° 7′. Jan. 3, 1802, 78° 3′. The change is 6° 58′, in 22 years and 38 days. As the equality of the two stars gives little room for admitting a difference in their parallactic motions, we cannot reasonably ascribe the change of situation to that cause; though, otherwise, the direction of such a motion in the largest of the two, would be sufficiently favourable. The situation of the stars being much insulated, a connection between them may be admitted, with a high degree of probability.*

* The calculation of the probability of a connection, which has been given in the *Phil. Trans.* for 1802, page 484 [above, p. 203], makes it above 75 millions to 1, that these two stars are not situated as they are, by a mere casual scattering of them in space.

ξ *Bootis.* II, 18.

The change in the situation of the two stars of this double star is very remarkable. The small star, April 15, 1782, was 65° 53′ north-following the large one. In one of my *sweeps*, April 20, 1792, I perceived the small star in the 20-feet reflector; and estimated its position, as it passed the field of view, to be about 85° north-preceding. When the sweep was finished, I found that this star could not be in the situation I had just seen it, unless it had undergone a considerable change since the year 1782; and, that no mistake had been made in the estimation of this evening, appeared very clearly, by a measure taken of its position, which actually gave 85° 43′·5 north-preceding. This pointed out a retrograde motion of the small star. March 22, 1795, the position was 84° 56′. April 1, 1802, 82° 57′; and, April 2, 1804, I found it 83° 54′. A mean of the two last measures, will give the present situation 83° 26′ north-preceding; and the total change of the angle of position, in 21 years and 352 days, will be 30° 41′.

If it should be remarked, that the measure taken in 1795 appears to be inconsistent, it ought to be recollected, that the cause of this apparent motion remains to be investigated. If the largest of the two stars should pass closely by the smallest, which, on account of its supposed great distance from us, may appear fixed, a very great and quick alteration in the angle of position will take place; but, in a short time the change will become very moderate, and not long after insensible. The same appearances may also happen, although the small star should not be fixed, but revolve about the large one; for, if its orbit were in a plane with the line of sight, it would be seen to move with great velocity, about the opposition, and soon after appear to be almost stationary. That either one or the other of the stars has really had a motion approaching to a straight line, is ascertained from an alteration of the distance; for, in the year 1781, the vacancy between the two stars, with 460, was 3 diameters of the large one. But, April 2, 1804, with 527, their distance was greater than estimations by diameters can determine; and, comparing ξ with π Bootis, I found that the stars of ξ were farther asunder than those of π; notwithstanding, in the year 1782, the former was placed in the 2d class, and the latter in the 3d. The change of the angle of position, if it were owing to a parallactic motion, would have been direct, instead of retrograde.

ω *Leonis.* I, 26.

In a note added to this star, which is the 26th in my second Catalogue, a suspicion is expressed, that the two stars which compose this very minute double star, were receding from each other.* This has since been completely verified; for, having seen the two stars close upon one another, and afterwards by degrees disengaged, as related in my second Catalogue, the separation between them kept on

* See *Phil. Trans.* Vol. LXXV. Part I. page 48 [Vol. I. p. 172].

increasing, and, on the 21st of April, 1795, they were ½ diameter of the small star asunder. Feb. 5, 1804, with a power of 527 the vacancy between them was nearly 1 diameter of the small one. The position has likewise undergone a sensible alteration. Nov. 13, 1782, it was 20° 54′ south-following. Feb. 4, 1802, 41° 28′. Feb. 5, 1804, 40° 17′. A mean of the two last measures, is 40° 53′. The change, therefore, amounts to 19° 59′, in 21 years and 84 days, and is probably owing to a real motion of ω Leonis; for the effect of a parallactic motion would have shown itself in a contrary alteration of the angle of position.

π *Arietis*. I, 64.

This star is marked as being treble; and the third star, as it happens, is now of use, in verifying the measures which have ascertained the relative change in the situation of the other two. The position of π and its adjacent star, Oct. 29, 1782, was 19° 9′ south-following; and the third star was in the same line of that angle continued. Oct. 17, 1802, the position was 34° 11′; and, Feb. 6, 1804, by a mean of two measures, 31° 15′; which gives a change of 12° 6′, in 21 years and 100 days.

That this change has taken place gradually, is confirmed by two observations of the third star. Jan. 15, 1795, the distant star was observed to have remained a little behind, while the near one had advanced; and, Oct. 17, 1802, it was again remarked, that the three stars were no longer in a line, and that the nearest small star had advanced according to the order of the signs, which had increased its angle of position.

The multitude of small stars in this neighbourhood, and the minuteness of the two that have been observed with π, as well as the distance of the farthest, render a connection between the three stars very improbable; nor can the change of situation be owing to parallax, as this would have occasioned a retrograde motion of the small star, which, on the contrary, has been direct. From these considerations we may conclude, that π Arietis has a proper motion, to which we must look for the cause of the observed change.

η *Coronæ*. I, 16.

This very minute double star has undergone a great alteration in the relative situation of the two stars. Sept. 9, 1781, their position was 59° 19′ north-following; and, Sept. 6, 1802, by a mean of two very accurate measures, it was 89° 40′ north-preceding; which amounts to a change of 31° 1′, in 20 years and 362 days.* The distance of the two stars has not been subject to any sensible alteration. Sept. 9, 1781, a very small division might be seen, with 460. August 30, 1794, they were so close that, with a 10-feet reflector, and power of 600, a very minute division could but just be perceived. April 15, 1803, with a 10-feet reflector, a very small division was also visible, with 400, though better with 600. And, May 15, 1803, I saw the separation between the two stars, with the same 7-feet reflector, and

* [The position angle in 1802 was 89° *sf*, not *np*, and the change was 149°.—ED.]

magnifying power of 460, with which I had seen it 22 years before. The stars differ very little in magnitude; so that we have no reason to expect any effect from a difference of parallax. Besides, if the small one were out of the reach of it, a parallactic motion of the largest alone, would have occasioned the small one to move apparently according to the order of the signs; but the motion has been retrograde.

Fl. 21 *Ursæ*. II, 73.

Nov. 17, 1782, the two stars were in the position of 36° 45′ north-preceding; and, May 20, 1802, I found them 47° 37′; which gives a change of 10° 52′, in 19 years and 184 days. A parallactic motion will account for it; unless, hereafter, a proper motion of the large star should be found to have a different tendency.

Fl. 4 *Aquarii*. I, 44.

The position of the two stars, July 23, 1783, was 81° 30′ north-preceding; and, by a mean of two observations, August 28 and 29, 1802, it was 61° 5′ north-following. Both the last measures are positive, with regard to the position being following, and not preceding, as it certainly was in the year 1783. This proves a change of 37° 25′, in 19 years and 37 days. The distance is perhaps a little increased. Sept. 5, 1782, it was ⅙ diameter of S. August 29, 1802, less than ½ diameter of S. A parallactic motion of the large star, would have brought on a retrograde motion of the small one, which, on the contrary, we find has been direct. This proves a real motion, the nature of which cannot remain many years unknown; its velocity, hitherto, having been at the rate of nearly 2 degrees *per* year, of angular change.

South-preceding π Serpentis. I, 81.*

The position, March 7, 1783, was 49° 48′ south-preceding. August 30, 1802, it was 59° 5′. The change is 9° 17′, in 19 years and 176 days. If the stars were a little more different in magnitude, a parallactic motion of the largest would account for the change of position.

Near μ Bootis. I, 17.

There is a considerable change in the relative situation of the two stars of this double star; and, by the assistance of μ Bootis, it is remarkably well ascertained. This star is so near, that it may be brought into the same field of view with our double star. Sept. 3, 1782, the position was 87° 14′ north-preceding; and, about a year before, the situation of μ Bootis had been determined, so that it appeared, from the two measures, that the three stars were almost in a line, the small star being, however, 6° 49′ on the *following* side. August 30, 1802, the position of the small star was 76° 14′ north-preceding; which, in 19 years and 361 days, gives a change of 11° 0′; and it was at the same time observed, that when all the three stars were seen

* We now have the place of this double star in BODE's Catalogue, where it is called Serpentis 112.

together, the small one was on the *preceding* side of the line which joins this double star and μ Bootis. A parallactic motion of the large star, would have occasioned the small one to go in a direct order; but it has had a retrograde motion.

North-preceding Fl. 18 *Persei*. I, 38.*

The two stars, August 20, 1782, were situated in a direction 8° 24′ north-preceding; and, by a mean of two measures, taken March 7, 1804, the position was 20° 34′. This gives a change amounting to 12° 10′, in 21 years and 199 days. There is probably a little increase in the distance of the stars. The first observations, with 460, give ½ diameter of either of them, supposing the stars to be equal; and the last, with 527, make it a diameter of the smallest; the stars being then considered as pretty unequal. If the difference of the parallactic motion of the two stars should be sufficiently considerable, that motion would account, not only for the change of the angle of position, but also for a small increase of the distance of the two stars.

σ *Coronæ*. I, 3.

This star has undergone a great change. The position of the two stars, Oct. 15, 1781, was 77° 32′ north-preceding; but, Sept. 6, 1802, it was 78° 36′ north-following; which gives an alteration of 23° 52′, in 20 years and 326 days. The great number of small stars in this neighbourhood, is not favourable to a supposed connection between any of them and σ Coronæ. As the two stars are considerably unequal, we may suppose the large one to be affected by a parallactic motion, which will sufficiently account for the angular change.

ε *Lyræ*. II, 5 and 6.

This remarkable double-double star has undergone a change of situation in each double star separately, which is not very considerable, but deserves our notice, on account of a certain similarity in the directions of the alteration. The position of II, 5, Nov. 2, 1779, was 56° 5′ north-following; and, by a mean of three observations, taken Sept. 20, 1802, May 26, and 29, 1804, it was 59° 14′; which gives a change of 3° 9′; the motion of the angle being retrograde. The position of II, 6, on the same days, was 83° 28′, and 75° 35′, south-following. This gives a difference of 7° 53′; the motion being also retrograde. Now, from the position of the apex of the translation of the solar system, it follows, that the parallax arising from this principle, cannot account for the motion of both the sets of double stars: it may explain the change of the preceding, but not of the following one. The situation of both, however, is in a part of the heavens which is so rich in scattered small stars, that a variety of casual, and merely apparent combinations may be expected.

* The place of this star is now given in BODE's Catalogue, where it is the Persei 85.

p Serpentarii Fl. 70. II, 4.

The alteration of the angle of position, that has taken place in the situation of this double star, is very remarkable. Oct. 7, 1779, the stars were exactly in the parallel, the preceding star being the largest; the position therefore was 0° 0′ following.* Sept. 24, 1781, it was 9° 14′ north-following; † and, May 29, 1804, it was 48° 1′ north-preceding; which gives a change of 131° 59′, in 24 years and 234 days. This cannot be owing to the effect of systematical parallax, which could never bring the small star to the preceding side of the large one.

λ *Ophiuchi*. I, 83.

The position, March 9, 1783, was 14° 30′ north-following. May 20, 1802, it was 20° 41′. The difference, in 19 years and 72 days, is 6° 11′. March 9, 1783, the distance, with 460, was $\frac{1}{4}$ or $\frac{1}{3}$ diameter of the small star. May 1 and 2, 1802, I could not perceive the small star, though the last of the two evenings was very fine. May 20, 1802, with 527, I saw it very well, but with great difficulty. The object is uncommonly beautiful; but it requires a most excellent telescope to see it well, and the focus ought to be adjusted upon ε of the same constellation, so as to make that perfectly round. The appearance of the two stars is much like that of a planet with a large satellite or small companion, and strongly suggests the idea of a connection between the two bodies, especially as they are much insulated. The change of the angle of position, might be explained by a parallactic motion of the large star; but the observations on the distance of the two stars, can hardly agree with an increase of it, which would have been the consequence of that motion.

North-preceding Fl. 29 *Capricorni*. I, 47.

The position, July 23, 1783, was 84° 48′ north-preceding. Sept. 1, 1802, it was 66° 50′. This gives a change of 17° 58′, in 19 years and 40 days. The effect of a parallactic motion would fall chiefly on the distance; it will, however, account for the change of the angle.

Near Fl. 3 *Pegasi*. II, 62.

The position, May 3, 1783, was 88° 24′ north-preceding. August 31, 1802, it was 79° 38′ south-following. The change is 8° 46′, in 19 years and 120 days. The stars are so nearly equal, that in 1783 I supposed the preceding one to be the smallest, and in 1802 the following one; which occasions the different denomination of the

* The first position was not given in my Catalogue, as I had no reason to suppose, at the time of its publication, that the positions of the stars were liable to any progressive change. It may be remembered, that my principal aim was, if possible, to find out some small annual variation, or libration of position, which might lead to a discovery of the parallax of the fixed stars.

† [The original observation had *sf*, which agrees with a diagram in the Journal. But the com panion must have been *nf* in 1781.—ED.]

angles of position. If the distance of the preceding star should be much greater than that of the following one, a parallactic motion would explain the change of the angle, but not otherwise.

Fl. 49 *Serpentis.* I, 82.

In the year 1783, March 7, the position of the two stars of this double star, was 21° 33′ north-preceding. May 20, 1802, 32° 52′; and, April 2, 1804, 35° 10′; which gives a change of 13° 37′, in 21 years and 26 days. The stars are now a little farther asunder than they were formerly. A parallactic motion would account for the change of the angle, but not for the increased distance.

Preceding Fl. 11 *Serpentarii.* II, 23.

The position of the stars, May 18, 1782, was 46° 24′ north-preceding. May 20, 1802, it was 66° 56′; which gives a change of 20° 32′, in 20 years and 2 days.* A parallactic motion, if the small star should be sufficiently distant from us, will account for it.

Fl. 38 *Piscium.* II, 50.

The position, June 30, 1783, was 25° 3′ south-preceding, and, August 31, 1802, it was 34° 43′. The change is 9° 40′, in 19 years and 62 days. The small star has been retrograde. If the change had been owing to the systematical parallax, the motion would have been direct.

Near Fl. 64 *Aquarii.* III, 69.†

The position, August 21, 1783, was 20° 3′ north-following. Oct. 16, 1802, it was 31° 34′.‡ The change, in 19 years and 56 days, is 11° 31′; and may be accounted for by a parallactic motion of the large star, especially as the stars are extremely unequal in apparent magnitude.

Fl. 46 *Herculis.* I, 79.

There is a small change in the distance of the two stars of this double star. Feb. 5, 1783, the interval between them, with 227,§ was nearly 1 diameter of L, and with 460, 1¾ diameter of L. Sept. 29, 1802, it was 2½ or 3 diameters of L. The position, Feb. 5, 1783, was 66° 36′ south-following. Sept. 29, 1802, it was 76° 18′. The alteration is 9° 42′, in 19 years and 236 days; but cannot be owing to parallactic motion.

* [There must be an error of 20° in the angle of 1802, as the Position Angle is still 316° or 46° *np*.—Ed.]

† In Bode's Catalogue, it is now called Aquarii 222.

‡ [The angle of 1783 must be 10° in error. No change.—Ed.]

§ In my Catalogue, the power is called 460, instead of 227, as it should have been; and the rest of the observation, with 460, was by mistake omitted [Vol. I. p. 180].

δ Cygni. I, 94.

This double star, I believe, has furnished us with a second instance of a conjunction, resembling that of ζ Herculis. The position, Sept. 22, 1783, was 18° 21′ north-following. Jan. 3, 10, and 11, 1802, I could no longer perceive the small star; which must have been at least so near the large one as to be lost in its brightness. Jan. 29, 1804, I examined this star with powers from 527 to 1500, and saw it as a lengthened star, but not with sufficient clearness to take a measure of its position. May 22, 1804, in a very clear evening, I tried 527 and 1500, with the 10-feet reflector, which acted remarkably well on other double stars, but I could not perceive the small star of δ Cygni. In hopes that the superior light of a 20-feet reflector would show it, I examined the star, May 29, 1804, with the powers 157 and 360, but could not perceive the small one. A parallactic motion of δ will perfectly account for this occultation; for the situation of the two stars, in 1783, was such, that this motion must have carried the large star, by this time, nearly upon the small one.

b Draconis. I, 7.

The position, Oct. 10, 1780, was 77° 19′ north-following; and, Oct. 30, 1802, it was 83° 41′. The change is 6° 22′, in 22 years and 20 days. The effect of a parallactic motion of the largest star, would have shown itself in a direction contrary to the observed one; a proper motion of one of the stars, at least, must be admitted.

South-preceding Fl. 30 *Orionis.* I, 75.

The position, Jan. 9, 1783, was 89° 36′ north-preceding; and, Jan. 22, 1802, it was 79° 12′ north-following; which gives a change of 11° 12′, in 19 years and 13 days. A parallactic motion of either of the stars, for they are nearly equal, would chiefly affect their distance; besides, the stars are so numerous in this part of the heavens, that we can only look upon this as a casual double star; a proper motion therefore must be recurred to.

η Cassiopeæ. III, 3.

The situation of the two stars of this beautiful double star, June 14, 1782, was 27° 56′ north-following; and, Feb. 11, 1803, it was 19° 14′; which gives a change of 8° 42′, in 20 years and 242 days. This arises probably from a real motion of η in space; for parallax would have had a contrary effect.

d Serpentis. I, 12.

This star has not altered its angle of position sufficiently to be certain of the change, which only amounts to 2° 8′; this quantity being too small for the precision of the micrometer, when only two measures are taken; but the alteration in the distance of the two stars is well ascertained. Oct. 22, 1781, with 278, it was 1⅓ diameter of L. April 28, 1783, with 460, it was 2½ diameters; and, May 4, 1802, it was not less than 4 or 5 diameters of L. If this change had arisen from a parallactic motion, there must have been a considerable alteration in the angle of position, which cannot be admitted ; it may, therefore, more properly be ascribed to a real motion of *d* Serpentis.

North of 105 *Herculis.* I, 86.

The alteration in the angle of position of this star is uncommonly great. April 27, 1783, it was 79° 24′ north-preceding ; and, Sept. 29, 1802, it measured only 22° 27′ ; which denotes a change of 56° 57′, in 19 years and 155 days.* The distance has undergone very little alteration, but is rather less now than it was formerly. A real motion of the largest star, in a north-following direction, may explain this change, which cannot be ascribed to a parallactic motion of the stars.

Rigel. II, 33.

This bright star has undergone a change of situation with regard to its distance from the small one, which is near it ; but, in the angle of position, very little difference can be perceived. By eleven measures, taken between Jan. 1, 1802, and Feb. 18, 1803, the mean position is 69° 5′ south-preceding ; which is but little more than 68° 12′, the measure of Oct. 1, 1781, given in my Catalogue.

The distance was estimated, Oct. 1, 1781, with 460, to be more than 3 diameters of Rigel ; and, as I supposed it to be one of those double stars of which I might ascertain the vacancy between the two stars, by estimating the number of diameters of the large one that would fill it up, I placed the star in the second class. However, by a measure taken with a micrometer, Oct. 22, 1781, the stars were found to be far enough asunder to come into the third class. By a mean of six measures, which were taken the first 18 months of my observing the star, their distance was 9″ 32‴ ; and, by a repetition of estimations, it appeared, Dec. 22, 1781, that the vacancy between the two stars was not less than 4 diameters, and, when the air was tremulous, 4 or 5. After an interval of more than 21 years, having omitted estimations by the diameter, as not very proper to be used with these stars, I wished to compare their distance with the former estimations ; and, with the same instrument and same magnifying power that had been used before, the vacancy, Feb. 22, 1803, amounted to 5 or 6 diameters of the large star ; so that, certainly, an increase of distance must be admitted.

The number of scattered stars in this neighbourhood, and the smallness of the star to which the relative situation of Rigel has been referred, render it probable that there is only a casual proximity, and no real connection, between these two stars. Nor can the change of their relative situation be accounted for by a parallactic motion of Rigel, although we should admit the small star to be without the reach of solar parallax ; for the effect arising from parallactic motion, would not

* [The star observed in 1802 cannot have been I. 86, in which system there has been no change. It is described thus : " It is a star about 1¾° from the 2 st of 107 Herc., the largest of several."]

only lessen the distance of the two stars, but would occasion a considerable diminution in the angle of position, neither of which have taken place.

As we have now the proper motion of Rigel, in Dr. MASKELYNE'S new Tables, we can no longer be at a loss for the cause of the change; for, by a composition of the tabular motions in right ascension and polar distance, this star, in 21 years and 144 days, must have moved about 3″·481, in an angle of 79° 29′ 33″, towards the north-following part of the heavens. This would consequently remove it from the small star, which is placed almost in an opposite direction, and would occasion hardly any change in its angle of position; and these are the very phenomena which have been established by my observations.

ζ *Cancri*. III, 19.

The position of the stars, Nov. 21, 1781, was 88° 16′ south-preceding; and, Feb. 7, 1802, it was 81° 47′ south-following. The change is 9° 57′, in 20 years and 78 days; and may be ascribed to a parallactic motion of the large star, which is in favour of the observed alteration.

ρ *Capricorni*. II, 51.

The position, July 4, 1783, was 84° 0′ south-following; and, August 29, 1802, it was 86° 55′ south-preceding.* This gives a change of 9° 5′, in 19 years and 56 days; and a motion arising from parallax will sufficiently account for it.

North-preceding Fl. 56 *Andromedæ*. I, 89.†

The position, July 28, 1783, was 75° 30′ south-following; and, Sept. 19, 1802, it was 67° 4′. The change is 8° 26′, in 19 years and 53 days. A parallactic motion of the large star would have occasioned the change of the angle to be direct, instead of retrograde.

Near 37 *Aquilæ*. I, 13.‡

The position, Oct. 6, 1782, by a mean of two measures, was 37° 15′ north-preceding; and, Oct. 2, 1802, it was 44° 45′. The change is 7° 30′, in 19 years and 361 days; and may be owing to a parallactic motion.

α *Ursæ minoris*. IV, 1.

There has been a small alteration in the relative situation of the pole star; but, when we consider that this double star is of the fourth class, we cannot expect that any great change in the angle of position should have taken place, in the course of 20 years. The position, Dec. 19, 1781, was 66° 42′ south-preceding; and, June 17,

* [This should be *sf*, as there has not been any change. No diagram in original observation.—ED.]

† The Andromedæ 241 of BODE'S Catalogue, gives us now the place of this star.

‡ The place of this star is now given in BODE'S Catalogue, where it is Aquilæ 136.

1782, it was 67° 23'. A mean of both measures, is 67° 3'. March 4, 1802, the position was 61° 43'; which gives a difference of 5° 20', in 19 years and 350 days. A parallactic motion of the large star, which, considering the great difference of size between the two, may well be admitted, will account for the angular change; especially as the distance of the two stars exceeds the limits which probability points out for connected stars, when the large one is of the third magnitude.

North-preceding Fl. 62 *Aquilæ*. I, 93.

The position, Sept. 12, 1783, was 19° 9' north-preceding; and, Oct. 2, 1802, it was 13° 21'. The change is 5° 48', in 19 years and 20 days. A parallactic motion of the largest of the two stars, would have occasioned a contrary apparent motion of the small one.

Preceding τ Orionis. I, 54.

The position, January 22, 1783, was 35° 42' north-preceding; and, Jan. 25, 1802, it was 41° 27'. The change is 5° 45', in 19 years and 3 days; and may be owing to the effect of parallax.

ζ Ursæ majoris. III, 2.

The position, Nov. 18, 1781, was 56° 46' south-following; and, Oct. 3, 1802, it was 51° 14'. The change is 5° 32', in 20 years and 319 days; but this cannot be accounted for by a parallactic motion of ζ, which would have occasioned a contrary change of the angle.

North-following φ Herculis. I, 37.

The position, Oct. 6, 1782, was 59° 48' south-following; and, Sept. 20, 1802, it was 65° 0'; which gives a change of 5° 12', in 19 years and 349 days. It cannot be ascribed to a parallactic motion of the largest star.

North-following ν Aquarii. I, 46.

The position, July 31, 1783, was 62° 27' north-preceding; and, August 29, 1802, it was 67° 27'. The change is 5° 0', in 19 years and 29 days. The distance of these stars is now greater than it was formerly. July 31, 1783, with 460, they were rather more than 1 diameter asunder. August 29, 1802, I found them too far distant to be put into the first class. If any effect of parallax can reach such small stars, it is so far in favour, that it will account for an increase of the distance, but not for the change of the angle of position.

α Piscium. II, 12.

The position of the stars, Oct. 19, 1781, was 67° 23' north-preceding; and, by a mean of three measures, taken Jan. 28 and Feb. 4, 1802, it was 63° 0'. This gives a change of 4° 23', in 20 years and 105 days. The parallactic motion of α will account for the alteration, unless a proper motion should hereafter lead to a different conclusion, which, from the insulated situation of this double star, is not improbable.

Fl. 11 *Monocerotis*. II, 17.

The position, Oct. 20, 1781, was 31° 38′ south-following; and, by a mean of two measures, taken Feb. 4, and March 4, 1802, it was 27° 34′. The change, which is 4° 4′, in 20 years and 121 days, may be accounted for by a parallactic motion.

North-preceding γ Aquilæ. I, 91.

The position, August 7, 1783, was 8° 18′ north-preceding; and, Sept. 20, 1802, it was 12° 23′. This gives a change of 4° 5′, in 19 years and 44 days; and may be accounted for upon the principles of parallax.

e Geminorum. III, 47.

The position, Oct. 2, 1782, was 89° 54′ south-following; and, April 6, 1802, it was 86° 6′ south-preceding; which gives a change of 4° 0′, in 19 years and 186 days. This cannot be ascribed to parallactic motion.

Fl. 32 *Eridani*. II, 36.

The position, Oct. 22, 1781, was 73° 23′ north-preceding; and, Feb. 6, 1804, it was 77° 19′. The change is 3° 56′, in 22 years and 107 days. It cannot be owing to a parallactic motion, which would have produced a different effect.

LIII.

Experiments for ascertaining how far Telescopes will enable us to determine very small Angles, and to distinguish the real from the spurious Diameters of celestial and terrestrial Objects: with an Application of the Result of these Experiments to a Series of Observations on the Nature and Magnitude of Mr. HARDING'S *lately discovered Star.*

[*Phil. Trans.*, 1805, pp. 31–64.]

Read December 6, 1804.

THE discovery of Mr. HARDING having added a moving celestial body to the list of those that were known before, I was desirous of ascertaining its magnitude; and as in the observations which it was necessary to make I intended chiefly to use a ten-feet reflector, it appeared to me a desideratum highly worthy of investigation to determine how small a diameter of an object might be seen by this instrument. We know that a very thin line may be perceived, and that objects may be seen when they subtend a very small angle; but the case I wanted to determine relates to a visible disk, a round, well defined appearance, which we may without hesitation affirm to be circular, if not spherical.

In April of the year 1774, I determined a similar question relating to the natural eye: and found that a square area could not be distinguished from an equal circular one till the diameter of the latter came to subtend an angle of 2′ 17″. I did not think it right to apply the same conclusions to a telescopic view of an object, and therefore had recourse to the following experiments.

1st *Experiment, with the Heads of Pins.*

I selected a set of pins with round heads, and deprived them of their polish by tarnishing them in the flame of a candle. The diameters of the heads were measured by a microscopic projection, with a magnifying power of 80. These measures are so exact, that when repeated they will seldom differ more than a few ten thousandths parts of an inch from each other. Their sizes were as follows: ·1375, ·0863, ·0821, ·0602, ·0425. I placed the pins in a regular order upon a small post erected in my garden, at 2407·85 inches from the centre of the object mirror of my ten-feet reflecting telescope. The focal length of the mirror on Arcturus is 119·64 inches, but on these objects 125·9. The distance was measured with deal rods.

When I looked at these objects in the telescope, I found immediately that only the smallest of them, at this distance could be of any use ; for with an eye-glass of 4 inches, which gives the telescope a magnifying power of no more than 31·5, this pin's head appeared to be a round body, and the view left no doubt upon the subject. It subtended an angle of 3″·64 at the centre of the mirror, and the magnified angle under which I saw it was 1′ 54″·6. This low power however required great attention.

With a lens 3·3, power 38·15, I saw it instantly round and globular. The magnified angle was 2′ 18″·9.

With a magnifying power of 231·8,* I saw it so plainly that the little notch in the pin's head between the coils of the wire making the head, appeared like a narrow black belt surrounding the pin in the manner of the belts of Jupiter. This notch by the microscopic projection measured ·00475 inch ; and subtended an angle, at the centre of the mirror, of 0″·407.

With 303·5 I saw the belt still better, and could follow it easily in its contour.

With 432·0 I could see down into the notch, and saw it well defined within.

With 522·3 the pin's head was a very striking globular object, whose diameter might easily be divided by estimation into ten parts, each of which would be equal to 0″·364.

With 925·6 I saw all the same phenomena still plainer.

The result of this experiment is, that an object having a diameter ·0425 may be easily seen in my telescope to be a round body, when the magnified angle under which it appears is 2′ 18″·9, and that with a high power a part of it, subtending an angle of 0″·364, may be conveniently perceived.

When I considered the purpose of this experiment, I found the result not sufficient to answer my intention ; for as the size of the object I viewed obliged me to use a low power, a doubt arose whether the instrument would be equally distinct when a higher should be required. To resolve this question, it was necessary either to remove my objects to a greater distance, or to make them smaller.

2d Experiment, with small Globules of Sealing-wax.

I melted some sealing-wax thinly spread on a broad knife, and dipt the point of a fine needle, a little heated, into it, which took up a small globule. With some practice I soon acquired the art of making them perfectly round and extremely small. To prevent my seeing them at a distance in a different aspect from that in which they were measured under the microscope, I fixed the needles with sealing-wax on small slips of cards before the measures were taken.

Eight of these globules of the following dimensions ·0466, ·0325, ·0290, ·02194, ·0210, ·0169, ·0144, ·00763 were placed upon the post in my garden, and I viewed them in the telescope.

* The powers have been strictly ascertained as they are at the distance where these objects were viewed.

With a power of 231·8 I saw all the first seven numbers well defined, and round, and could see their gradual decrease very precisely from No. 1 to No. 7.

With 303·5 I saw them better, and had a glimpse of No. 8, but could not be sure that I saw it distinctly round; though the magnified angle was 3′ 18″·2.

With 432·0 they are all very palpable objects, and, as a solid body, No. 8 may be seen without difficulty; at the centre of the mirror it subtends an angle of 0″·653. With attention we may also be sure of its roundness; but here the magnified angle is not less than 4′ 42″·1.

With 522·3 I see them all in great perfection as spherical bodies, and the magnitude of No. 7 may be estimated in quarters of its diameters. The angle is 1″·253, and one quarter of it is 0″·313. No. 8 may be divided into two halves with ease; each of which is 0″·327.

With 925·4 I saw No. 8 still better; but sealing-wax is not bright enough for so high a power.

By this experiment it appears, that with a globule so small as ·00763 of a substance not reflecting much light, the magnified angle must be between 4 and 5 minutes before we can see it round. But it also appears that a telescope with a sufficient power, will show the disk of a faint object when the angle it subtends at the naked eye is no more than 0″·653.

3d Experiment, with Globules of Silver.

As the objects made of sealing-wax, on account of their colour, did not appear to be fairly selected for these investigations, I made a set of silver ones. They were formed by running the end of silver wires, the 305th and 340th part of an inch in diameter, into the flame of a candle. It requires some practice to get them globular, as they are very apt to assume the shape of a pear; but they are so easily made that we have only to reject those which do not succeed.

Thirteen of them, in a pretty regular succession of magnitude, were selected and placed upon the post. Their dimensions were ·03956, ·0371, ·0329, ·0317, ·0272, ·0260, ·0187 ·0178, ·0164, ·0125, ·01137, ·00800, ·00556.

For the sake of more conveniency I had removed my telescope from its station in the library to a work-room. The distance of the objects from the mirror of the telescope, measured with deal rods, was here only 2370·5 inches; and the focal length of the mirror, the magnifying powers of the telescope, and the angles subtended by the objects have been calculated accordingly.

With 522·7 I see all the globules, from No. 1 to No. 13, perfectly well, and can estimate the latter in quarters of its diameter. The angle it subtends at the centre of the mirror is 0″·484; and one quarter of it is 0″·121.

With the same power I see the wires which hold the balls, so well that even the smallest of them may be divided into half its thickness. It measures ·00237; the angle is 0″·206; and half of it 0″·103.

With 433·0 I see all the globules of a round form, and can by estimation divide No. 13 into two halves. The magnified angle is here 3′ 29″·0, but as its diameter could by estimation be divided into two parts, the round form of a globule somewhat less might probably have been perceived, so that the magnified angle would perhaps not have much exceeded the quantity 2′ 18″·9 that has been assigned before.

After some time the weather became much overcast, and as the globules were placed over a cut hedge, the leaves and interstices of which did not reflect much light, they received the greatest part of their illumination from above. This made them gradually assume the shape of half moons placed horizontally. The dark part of these little lunes, however, did not appear sensibly less than the enlightened part, so that there could not be any thing spurious about them.

By this experiment we find that the telescope acts very well with a high power, and will show an object subtending only 0″·484 so large that we may divide it into quarters of its diameter.

4th Experiment, with Globules of Pitch, Bee's-wax, and Brimstone.

I had before objected to sealing-wax globules on account of their dingy-red colour; in the last experiment a doubt was raised with regard to the silver ones, because they were perhaps too glossy. In order to compare the effect of different substances together in the same atmosphere, I put up three globules, No. 1 of silver, diameter ·01137; No. 2 of sealing-wax, ·01125; No. 3 of pitch, ·00653.

With 522·7 I saw No. 1 round, and could estimate $\frac{1}{4}$ of its diameter. The angle is 0″·989; $\frac{1}{4}$ of it is 0″·247.

I saw No. 2 round, but of a dusky-red colour. It is not nearly so bright as No. 1; nor does it appear quite so large as the proportional measure of the globules would require. I can estimate $\frac{1}{3}$ of its diameter. The angle is 0″·979; and $\frac{1}{3}$ of it is 0″·326.

No. 3 reflects so little light that I can barely perceive the globule, but not its form; and yet it subtends an angle of 0″·568.

To discover whether this ought to be ascribed intirely to the want of reflection of the pitch, I took up some white melted bee's-wax, by dipping the fine point of a needle perpendicularly into it. This happened to be only half a globule, and its diameter was ·0105.

When I examined the object with 523 I saw it with great ease, and could estimate $\frac{1}{4}$ of its diameter. The angle is 0″·914; and $\frac{1}{4}$ of it is 0″·228. I saw also that it was but half a globule.

I took up another, that I might have a round one; but found that again I had only half a globule. It was so perfectly bisected, that art and care united could not have done it better. Its diameter was ·0108. In the telescope I saw its semi-globular form, and could estimate $\frac{1}{4}$ of its diameter.

By some further trials it appeared, that a perfect globule of this substance could not be taken up, the reason of which it is not difficult to perceive ; for as it melts with very little heat, it will cool the moment the needle is lifted up ; and the surface, which cools first, will be flat.

The roundness of the objects being a material circumstance, I melted a small quantity of the powder of brimstone, and dipping the point of a needle into it, I found that globules, perfectly spherical and extremely small, might be taken up. I had one of them that did not exceed the 640th part of an inch in diameter.

When four of the following sizes, ·00962, ·009125, ·00475, ·002375 were placed on the post in the garden and viewed from the work-room station with 522·7 I saw No. 1, 2, and 3, round, but No. 4 was invisible.

These globules reflect but little light, so that they are not easily to be distinguished from the surrounding illumination of the atmosphere ; but when I placed some dark blue paper a few inches behind them, I then could also perceive No. 4 as a round body. The angle it subtends is 0″·207

5th Experiment with Objects at a greater Distance.

Having carried the minuteness of the globules as far as appeared to be proper, I considered that a valuable advantage would be gained by increasing the distance of the objects. The experiments might here be made upon a larger scale, and the body of air through which it would be necessary to view the globules would bring the action of the telescope more upon a part with an application of it to celestial objects.

On a tree, at 9620·4 inches from the object mirror of the telescope, I fixed the sealing-wax globules of the 2d experiment. The distance was measured by a chain compared with deal rods, and by calculation the altitude of the objects has been properly taken into the account.

With 502·6 No. 1 is a very large object ; so that were I to see a celestial body under the same angle, I could never mistake it for a small star. The angle it subtends is 0″·999.

I see the diameters of No. 2 and 3 very clearly, and can divide them by estimation into two parts, half of No. 3 is 0″·311.

I see No. 4 and 5 as round bodies, but cannot divide them by estimation. The diameter of No. 5 is 0″·45. No. 6 may also be seen, but 7 and 8 are invisible.

These objects reflecting too little light, the silver globules of the 3d experiment were placed on the tree. It will be right to mention that they were all so far tarnished by having been out in the open air for more than a fortnight, that no improper reflection was to be apprehended.

The air being uncommonly clear, I saw with 502·6 the globules No. 1, 2, 3, 4, 5, and 6, as well defined black balls. I could easily distinguish $\frac{1}{4}$ of the diameter of No. 6 ; which is 0″·139.

With 415·7 I saw them all round as far as No. 10 included.

With 502·6 I saw No. 9 and 10 very sharp and black, and could divide No. 10 into two parts, each of which would be 0″·134.

With a new lens, power 759·7, I saw No. 10 better than with 502·6, and could with ease distinguish it into halves, or even third parts of its diameter. $\frac{1}{3}$ of it is 0″·089.

With 223·1 I saw them all as far as No. 10 included as visible objects, but the smallest of them were mere points. No. 6 might be divided with this power into two parts; each being 0″·279.

With 292·1 I saw No. 10 sharp and round. The magnified angle is only 1′ 18″·3. One half of No. 6 may be perceived with great ease.

The weather being as favourable as possible, I saw with 415·7 the globule No. 10 round at first sight; the magnified angle is 1′ 51″·2. I can see No. 12 steadily round; the angle is 0″·172. It is however a mere point, and divisions of it cannot be made.

With a new 10-feet reflector, power 540, the globule No. 10 is beautifully well defined, and $\frac{1}{2}$ of it may be estimated; the angle is 0″·268; $\frac{1}{2}$ of it is 0″·134.

With the old reflector, and 502·6, I see No. 12 steadily round. No. 7, 11, and 13, have met with an accident, and could not be observed.

6th Experiment with illuminated Globules.

The night being very dark, 8 silver globules, from ·0291 to ·00596 in diameter were placed on the post, and illuminated by a lantern held up against them.

With 522·7 I saw them all perfectly well, but the small quantity of light thrown on them was not sufficient to make angular experiments upon them. As objects I saw them as easily as in the day time. Probably the phases of the illuminated disks I saw might be such as the moon would show when about 9 or 10 days old. The angle of No. 8, had it been full, would have been 0″·519. A better way of illumination might be contrived.

SPURIOUS DIAMETERS OF CELESTIAL OBJECTS.

Observations and Experiments, with Remarks.

July 17 1779. With a 7-feet reflector, power 280, I saw the body of Arcturus, very round and well defined. I saw also ζ Ursæ majoris and other stars equally round, and as well defined.

REMARKS.

(1.) As these diameters are undoubtedly spurious, it follows that, with the stars, the spurious diameters are larger than the real ones, which are too small to be seen.

Sept. 9, 1779. The two stars of ε Bootis are of unequal diameters; one of them being about three times as large as the other.

(2.) From this and many estimations of the spurious diameters of the stars * it follows, not only that they are of different sizes, but also that under the same circumstances, their dimensions are of a permanent nature.

August 25, 1780. The large star of γ Andromedæ is of a very fine reddish colour, and the small one blue.

(3.) By this and many other observations it appears, that the spurious diameters of the stars are differently coloured, and that these colours are permanent when circumstances are the same.

Nov. 23, 1779. I viewed α Geminorum with a power of 449, and saw the two stars in the utmost perfection. The vacancy between them was about 1½ diameter of the largest. I found when I looked with a lower power, that the proportion between the distance and magnitude of the stars underwent an alteration. With 222, the vacancy was 1¼ diameter, and with 112, it was no more than 1 diameter of the smallest of the two stars, or less.

(4.) By many observations, a number of instances of which may be seen in my catalogues of double stars, their spurious diameters are lessened by increasing the magnifying power, and increase when the power is lowered.

(5.) It is also proved by the same observations, that the increase and decrease of the spurious diameters, is not inversely as the increase and decrease of the magnifying power, but in a much less ratio.

Nov. 13, 1782. The two stars of the double star 40 Lyncis, with a power of 460 are very unequal; and with 227 they are extremely unequal.

(6.) From this we find, that the magnifying power acts unequally on spurious diameters of different magnitudes; less on the large diameters, and more on the small ones.

Aug. 20, 1781. I saw ϵ Bootis with 460, and the vacancy between the two stars was 1¼ diameter of the large one. I then reduced the aperture of the telescope by a circle of pasteboard from 6·3 inches to 3·5, and the vacancy between the two stars became only ½ diameter of the small star.

The proportion of the diameters of the two stars to each other was also changed considerably; for the small one was now at least ⅔ if not ¾ of the large one.

(7.) This shows that when the aperture of the telescope is lessened, it will occasion an increase of the spurious diameters, and when increased will reduce them.

(8.) It also shows that the increase and decrease of the unequal spurious diameters, by an alteration of the aperture of the telescope, is not proportional to the diameters of the stars:

(9.) But that this alteration acts more upon small spurious diameters, and less upon large ones.

Aug. 7, 1783. I tried some excessively small stars near γ Aquilæ. When γ was perfectly distinct and round, the extremely small stars were dusky and ill

* See Catalogues of double Stars, *Phil. Trans.* for 1782, p. 115; and for 1785, p. 40 [Vol. I. pp. 60 and 167].

defined ; the *excessively* small ones were still less defined. As there are stars of all sizes in this neighbourhood, I saw some so very minute, that they only had the appearance of a small dusky spot, approaching to mere nebulosity. By very long attention I perceived many small dusky nebulous spots, which had it not been for this attention might have been in the field of view without the least suspicion.

(10.) From this we find that stars, when they are extremely small, lose their spurious diameters, and become nebulous.

July 7, 1780. I saw the spurious diameter of Arcturus gradually diminished by a haziness of the atmosphere till it vanished intirely.

A more circumstantial account of this observation has already been given ; and some other causes that affect the spurious diameter of the stars, have been pointed out in the same paper, such as tremulous air, wind, and hoar-frost.*

January 31, 1783. The star in the back of Columba makes a spectrum, about 5 or 6″ long, and about 2″ broad, finely coloured by the prismatic power of the atmosphere at this altitude.

July 28, 1783. Fomalhaut gives a beautiful prismatic spectrum, on account of its low situation.

July 17, 1781. With a new lens, power between 5 and 6 hundred, I saw ζ Aquarii, and found the vacancy between the two stars exactly 2 diameters. With my old one, power only 460, it was full 2 diameters. As it should have been larger with the high power than with the low one, it shows that the best eye-lens will give the least spurious diameter.

Oct. 12, 1782. I tried a new plain speculum, made by a very good workman, and found that when I viewed α Geminorum with 460, the vacancy between the two stars was barely 1½ diameter, but the same telescope and power with my own small speculum, made the distance 2 diameters, so that the figure of this mirror affects the spurious diameters of the stars.

(11.) Hence we may conclude that many causes will have an influence on the apparent diameter of the spurious disks of the stars ; but they are so far within the reach of our knowledge, that with a proper regard to them, the conclusion we have drawn in Rem. (2.) "that under the same circumstances their dimensions are permanent," will still remain good.

SPURIOUS DIAMETERS OF TERRESTRIAL OBJECTS, WITH SIMILAR REMARKS.

7th Experiment with Silver Globules.

A number of silver globules were put on the post, before they had been tarnished; and the sun shone upon them. When I viewed them in the telescope, there was on each of them a lucid appearance resembling the spurious disk of a star. I could distinguish this bright spot from the real diameters of the globules perfectly well, and found it much less than they were.

* See *Phil. Trans.* for 1803, page 224 [above, p. 244].

REM. (1.) Hence we conclude that the terrestrial, spurious disks of globules are less than the real disks; whereas we have seen, in Remark (1.) of the celestial spurious disks, that these are larger than the real ones.

8th Experiment.

The luminous spots, or spurious disks of the globules were of unequal diameters. The globule No. 1 had the largest disk, and the smaller ones the least; and the gradation of the sizes followed the order of the numbers.

(2.) This agrees with the spurious disks of celestial objects: the stars of the first, second, and third magnitude having a larger spurious disk than those that are of inferior magnitudes.

9th Experiment.

I found that there was a considerable difference in the colour of the spurious disks; one of them was of a beautiful purple colour, another was inclined to orange, a large one was straw coloured, a small one pale-ash coloured, and most of them were bluish-white.

(3.) With respect to colours, therefore, the terrestrial also agree with the celestial spurious disks.

10th Experiment.

I made two globules of different diameters, and placed them very near each other, so that their spurious disks might resemble those of a double star; this succeeded perfectly well. I viewed them with different powers.

With 177, the vacancy between them is $\frac{3}{4}$ diameter of the large star.

With 232, it is $1\frac{1}{4}$ diameter.

With 303·8, it is $1\frac{5}{8}$ diameter.

With 432·3, it is $1\frac{3}{4}$.

(4.) This experiment proves that the spurious diameters of the globules are also in this respect like the spurious disks of the stars; for they are proportionally lessened by increasing the magnifying power, and increased when the power is lowered.

(5.) When the estimations are compared with the powers, it will also be seen that the increase and decrease of the spurious disks of the globules is not inversely as the powers, but in a much less ratio.

11th Experiment.

Two other globules of different sizes were examined; and

With 706·3 they were pretty unequal.

With 522·7 they were considerably unequal.

With 303·8 they were very unequal.

(6.) This proves that the effect of magnifying power is unequally exerted on spurious diameters; and that, as with celestial objects, so with terrestrial, this power acts more on the small spurious disks than on the large ones.

12th Experiment.

I viewed a different artificial double star with 522·7, and keeping always the same power, changed the aperture of the telescope.

With the inside rays I found them considerably unequal, and 2½ diameters of the largest asunder. The spurious disks are perfectly well defined, round, and of a planetary aspect.

With all the mirror open, they are also round and well defined.

With the outside rays, they are near 4 diameters of the largest asunder, and are also round and distinct, but surrounded with flashing rays and bright nodules in continual motion.

(7.) This shows that the spurious terrestrial disks, in this respect again resemble those of the stars ; increasing when the aperture is lessened, and decreasing when it is enlarged.

13th Experiment.

With the same magnifying power 432·3, but a change of aperture, I viewed two equal globules, and two unequal ones.

With the inside rays the equal globules were 1 diameter asunder.

With all the mirror open, they were 1½ diameter asunder.

And with the outside rays they were 2 diameters asunder.

The unequal globules, with the inside rays, were a little unequal, and 1 diameter of the large one asunder.

With the outside rays they were considerably unequal, and 2 diameters of the large one asunder.

(8.) By these experiments it is proved, that the increase and decrease of the diameters occasioned by different apertures is not proportional to the diameters of the spurious disks.

(9.) But that the change of the apertures acts more on the small, and less on the large ones.

14th Experiment.

No. 1 of a set of globules, has the largest spurious diameter. No. 3 is larger than No. 2 ; whereas No. 2 has the largest real diameter. It is inclined to a greenish colour. No. 3 is now reddish, and is larger than No. 1, which is at present less than No. 2. No. 1 grows bigger, and is now the largest.

The sun which had been shining, was obscured by some clouds, but the spurious diameters of the globules I was viewing remained visible, and were almost as bright as when the sun shone upon them.

I saw one of the globules lose its spurious diameter while the sun continued to shine. After some time the spurious diameter came on again, and very gradually grew brighter, but not larger. The colour of one of the globules being of a beautiful purple, changed soon after to a brilliant white.

The sun being obscured by some clouds, a globule lost its spurious diameter, and acquired the shape of an half moon, of the size of the real disk or diameter of the globule. I saw the sun break out again, and the half moon was gradually transformed into a much smaller spurious disk.

(10.) The spurious disks of globules are lost for want of proper illumination, but do not change their magnitude on that account. The brightness of the atmosphere in a fine day is sufficient to produce them ; though the illumination of the sun is generally the principal cause of them.

(11.) The diameters of spurious disks are liable to change from various causes ; an alteration in the direction of the illumination will make the reflection come from a different part of the globule, which can hardly be expected to be equally polished in its surface, or of equal convexity every where, being very seldom perfectly spherical ; but as upon the whole the figure of them is pretty regular, the apparent diameter of the spurious disks will generally return to its former size.

15th Experiment, with Drops of Quicksilver.

At a time of the year when bright sun-shine is not very frequent, I found that my silver globules would seldom give me an opportunity for experiments on spurious disks ; to obviate this inconvenience, I used small drops of quicksilver. They are more lucid, and will give a bright spot with very little sunshine. Many of these drops of all sizes were exposed upon a plate of glass, and some on slips of steel. The management of them is a little different from that of the globules. For in order to represent a double star these must be placed one almost behind the other, as otherwise they cannot be brought near enough without running together. The following general observation will include all the necessary particulars.

The bright spots on drops of quicksilver are very small compared to the size of the drops.

They are not proportional to the magnitude of the drops, though less on the small ones and greater on large ones.

In some of the large ones the bright spot is about $\frac{1}{30}$ or $\frac{1}{40}$ of the diameter of the drop.

The magnitude of the luminous spots is liable to changes, but is rather more permanent than with the silver globules.

There is a little difference in the colour of the luminous spots ; they are generally of a brilliant white, but sometimes they incline to yellow, and the small ones to ash-colour.

With high magnifying powers they are very well defined, and, on account of their brightness, will bear these powers better than the silver globules.

If M and m, stand for the diameters of the large and small mirror of my telescope, then will an aperture $= \sqrt{\frac{M^2 - m^2}{2} + m^2}$ give half the light of the telescope.

With this I examined two of the drops, and found the luminous spots upon them with

925·4 nearly equal, and 2½ diameters of the largest asunder.

706·3 nearly equal, and above 2 diameters of the largest asunder.

432·3 pretty unequal, and 2 diameters of the largest asunder.

177·0 considerably unequal, and 1¼ diameter of the largest asunder.

I examined also two other drops, with different apertures, without changing the power, which was 706·3.

With the inside rays they were very little unequal, and ¾ diameter of the largest asunder.

With the outside rays they were considerably unequal, and 1¼ diameter of the largest asunder.

From what has been said, it appears that all the remarks which have been made with regard to the spurious disks of the silver globules are confirmed by the luminous spots on the drops of quicksilver. There is a difference in the proportion which the spurious disks on quicksilver bear to the drops, and that on the silver globules to the size of the globules; the latter also give a greater variety of colours and magnitudes than those on quicksilver; these are circumstances of which it would be easy to assign the cause, but they can be of no consequence to the result we have drawn from the experiment.

16th Experiment, with black and white Circles.

I tried to measure some of the spurious disks by projecting them on a scale with a moveable index, but found their diameters were too small for accuracy by this method; for this reason I had recourse to artificial measuring-disks, and prepared a set of eleven white circleson a black ground, and eight black ones on a white ground. In order to guard against deceptions, I fixed them up against a tablet 154 inches from the eye, where it was intended to project the spurious disks of the globules, and examined them at that distance with the naked eye. Comparing then the size of the black to the white, I judged No. 1 of the black to be a little larger than No. 6 of the white circles. By a measure taken afterwards, it appeared that the black one was ·40 and the white ·39. Without supposing that every estimation may be made at this distance with equal accuracy, to the hundredth part of an inch, it is sufficiently evident that no material deception can take place in estimating by either of the sets of circles on account of their colour.

17th Experiment, with different Illumination.

A similar experiment was made in the microscope, by which the globules were measured. Two of them were placed on the measuring stand, and with an illumination from below, they appeared black, and were projected on white paper. The diameter of each globule and the distance between them were then measured. fter this, I caused the illumination to come from above, and the globules being now of a silvery white, were projected on a slate. In this situation, when I repeated the former measures, no difference could be perceived.

18th Experiment. Measures of spurious Disks.

The spurious disk of a globule was then projected on the tablet where the white circles were placed. While I was comparing it with No. 4, which is ·31 in diameter and estimated it to be a little less than the circle, the spurious disk grew brighter; but it remained still of the same size; so that a variation in the quantity of the illumination will make no difference.

Every thing being now arranged for the measurement, I viewed the spurious diameter, with a magnifying power of 522·7, and compared it to the circles which succeeded each other by small differences of magnitude.

With all the mirror, from the centre to 8·8 inches open, the diameter of the spurious disk was ·31 inches.

With 6·3 inches open, it was less than ·40 and larger than ·355.

With 5 inches open, it was ·40.

With 4 inches open, it was ·42.

With 3 inches open, it was ·465 nearly.

From these measures it might be supposed that by lessening the quantity of light, we bring on a certain indistinctness which gives more diameter to the spurious object; to prove that this is not the cause of the increase, I used the following apertures.

With an annular opening from 6·5 to 8·8 inches, the spurious disk was rather less than ·18.

With another from 5 to 8·8 it was exactly ·18.

With an opening from 4 to 6·5 it was ·22.

With another from 1·6 to 4 it was ·42.

(12.) Now since the outside rim from 6·5 to 8·8, which reflected less than half the light of the mirror, produced a spurious disk less than ·18 in diameter, and the whole light as we have seen gave a disk of ·31, it is evidently not the quantity of the light, but the part of the mirror from which it is reflected, that we are to look upon as the cause of the magnitude of the spurious disks of objects.

(13.) These measures therefore point out an improvement in my former method of putting any terrestrial disk we suspect to be spurious to the test. For the inside rays of a mirror, as before, will increase the diameter of these disks, but the outside rays alone will have a greater effect in reducing it, than when the inside rays are left to join with them.

19th Experiment. Trial of Estimations.

I placed two silver globules at a small distance from each other upon the post, but without measuring either the globules or their distance. When I viewed them with 522·7 they appeared in the shape of two half moons in an horizontal situation. The unenlightened parts of them were also pretty distinctly visible. I estimated the vacancy between the cusps of the lunes to be $\frac{1}{4}$ diameter of the largest.

On measuring the diameters and distance under the microscope, it appeared that the largest was ·0312; a quarter of which is ·0078. The distance of the globules from each other measured ·0111. The difference in the estimation ·0033 is less than $\frac{1}{300}$ part of an inch.

The experiment was repeated with a change of the distance of the globules from each other. They were then estimated to be less than the diameter of the large one asunder, but full that of the small one. When they were measured it was found that their distance was ·02608, and the diameter of the small one was ·0247, which estimation is still more accurate than the former.

20th Experiment. Use of the Criterion.

It remained now to be ascertained whether these half moons were spurious or real; for although I could also imperfectly perceive the dark part of the disks of the globules, yet a doubt would arise whether the two halves were really of equal magnitude; to resolve this question, I viewed them first with the inside rays of the mirror, then with the outside, and found that in both cases the distance of the lunes remained without the least alteration. I viewed them also with the whole mirror open, but it occasioned no change.

21st Experiment. Measures of the comparative Amount of the spurious Diameters, produced by the Inside and Outside Rays.

I divided the aperture of the mirror into two parts, one from 0 to 4·4 and the other from 4·4 to 8·8 inches. When I measured the spurious diameter of a globule, the inside rays made it ·40; with all the mirror open it was ·31; and with the outside rays it was ·22.

(14.) From this we may conclude that the diameters given by the inside rays, by all the mirror open, and by the outside rays, are in an arithmetical progression; and that the inside rays will nearly double the diameter given by the outside. It remains however to be ascertained whether this will hold good with spurious disks of various magnitudes.

It will not be necessary to carry the divisions of the aperture farther; for as the application of these experiments is chiefly intended for astronomical purposes, we can hardly do with less than half the mirror open; and on the other hand with a very narrow rim of reflection from the outside of the mirror, distinctness would be apt to fail.

22d Experiment. Trial of the Criterion on celestial Objects.

I viewed α Lyræ with the outside rays, and found its spurious disk to be small; with all the mirror open it was larger, and with the inside rays it was largest.

As far as the imagination will enable us to compare objects we see in succession, the magnitudes appeared to be in an arithmetical progression.

23d Experiment.

I examined α Geminorum with 410·5, and with the outside rays the stars were considerably unequal, and 1¼ diameter of the largest asunder. With all the mirror open they were more unequal, and 1½ diameter of the largest. With the inside rays they were very unequal, and 1⅞ of the largest asunder.

These experiments show that, if it had not been known that the apparent disks of the stars were spurious, the application of the improved criterion of the apertures would have discovered them to be so; and that consequently the same improvement is perfectly applicable to celestial objects.

OBSERVATIONS ON THE NATURE AND MAGNITUDE OF MR. HARDING'S LATELY DISCOVERED STAR.

It will be remembered that in a former Paper, where I investigated the nature of the two asteroids discovered by Signior PIAZZI and Dr. OLBERS, I suggested the probability that more of them would soon be found out; it may therefore be easily supposed that I was not much surprised when I was informed of Mr. HARDING'S valuable discovery.

On the day I received an account of it, which was the 24th of September, I directed my telescope to the calculated place of the new object, and noted all the small stars within a limited compass about it. They were then examined with a distinct high magnifying power; and since no difference in their appearance was perceivable, it became necessary to attend to the changes that might happen in the situation of any one of them. They were delineated as in Fig. 1, which is a mere eye-draught, to serve as an elucidation to a description given with it in the journal; and the star marked *k*, as will be seen hereafter, was the new object.

Sept. 25. The moon was too bright to see minute objects well, and my description the night before, for the same reason, had not been sufficiently particular; nor did I expect, from the account received, that the star had retrograded so far in its orbit.

Sept. 26. The weather being very hazy, no regular observations could be made; but as I noticed very particularly a star not seen before, it was marked *l* in Fig. 2, and proved afterwards to have been the lately discovered one, though still unknown this evening, for want of fixed instruments.

Sept. 27. I was favoured with Dr. MASKELYNE'S account of the place of the star, taken at the Royal Observatory, by which communication I soon found out the object I was looking for.

Sept. 29. Being the first clear night, I begun a regular series of observations; and as the power of determining small angles, and distinctness in showing minute disks, whether spurious or real, of the instrument I used on this occasion, has been sufficiently investigated by the foregoing experiments, there could be no difficulty in the observation, with resources that were then so well understood, and have now been so fully ascertained.

" Mr. HARDING'S new celestial body precedes the very small star in Fig. 3, between 29 and 33 Piscium, and is a little larger than that star; it is marked A. *f g h* are taken from Fig. 1. I suppose *g* to be of about the 9th magnitude, so that the new star may be called a small one of the 8th."

With the 10-feet reflector, power 496·3, I viewed it attentively, and comparing

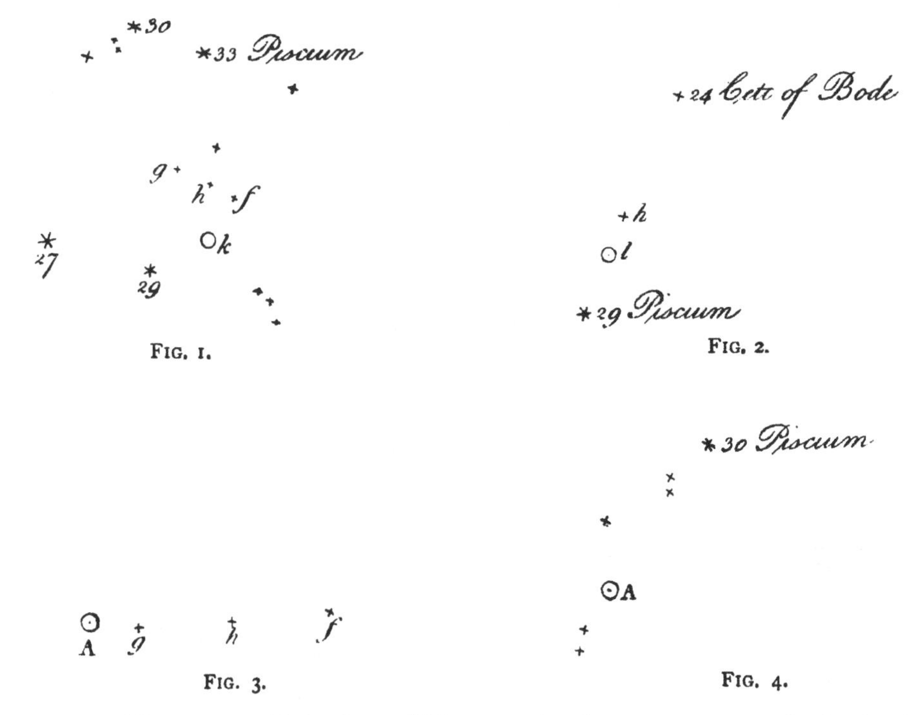

FIG. 1. FIG. 2. FIG. 3. FIG. 4.

it with *g* and *h*, Fig. 3, could find no difference in the appearance but what might be owing to its being a larger star.

By way of putting this to a trial, I changed the power to 879·4, but could not find that it magnified the new one more than it did the stars *g* and *h*.

" I cannot perceive any disk; its apparent magnitude with this power is greater than that of the star *g*, and also a very little greater than that of *h*; but in the finder and the night-glass *g* is considerably smaller than the new star, and *h* is also a very little smaller."

I compared it now with a star which in the finder appeared to be a very little larger; and in the telescope with 879·4 the apparent magnitude of this star was also larger than that of the new one.

" As far as I can judge without seeing the asteroids of Mr. PIAZZI and Dr. OLBERS at the same time with Mr. HARDING'S, the last must be at least as small as the smallest of the former, which is that of Dr. OLBERS."

"The star *k*, Fig. 1, observed Sept. 24, is wanting, and was therefore the object I was in search of, which by computation must have been that day in the place where I saw it."

"The new star being now in the meridian with all those to which I am comparing it, and the air at this altitude being very clear, I still find appearances as before described: the new object cannot be distinguished from the stars by magnifying power, so that this celestial body is a true ASTEROID."

Mr. BODE's stars 19, 25 and 27 Ceti are marked 7m, and by comparing the asteroid, which I find is to be called Juno, with these stars, it has the appearance of a small one of the 8th magnitude.

With regard to the diameter of Juno, which name it will at present be convenient to use, leaving it still to astronomers to adopt any other they may fix upon, it is evident that, had it been half a second, I must have instantly perceived a visible disk. Such a diameter, when I saw it magnified 879·4 times, would have appeared to me under an angle of 7′ 19″·7, one half of which, it will be allowed, from the experiments that have been detailed, could not have escaped my notice.

Oct. 1. Between flying clouds, I saw the asteroid, which in its true starry form has left the place where I saw it Sept. 29. It has taken the path in which by calculation I expected it would move. This ascertains that no mistake in the star was made when I observed it last.

Oct. 2, 7^h Mr. HARDING's asteroid is again removed, but is too low for high powers.

8^h 30′. I viewed it now with 220·3, 288·4, 410·5, 496·3, and 879·4. No other disk was visible than that spurious one which such small stars have, and which is not proportionally magnified by power.

With 288·4, the asteroid had a larger spurious disk than a star which was a little less bright, and a smaller spurious disk than another star that was a little more bright.

Oct. 5, with 410·5. The situation of the asteroid is now as in Fig. 4. I compared its disk, which is probably the spurious appearance of stars of that magnitude, with a larger, an equal, and a smaller star. It is less than the spurious disk of the larger, equal to that of the equal, and larger than that of the smaller star. The gradual difference between the three stars is exceedingly small.

"With 496·3, and the air uncommonly pure and calm, I see so well that I am certain the disk, if it be not a spurious one, is less than one of the smallest globules I saw this morning in the tree."

The diameter of this globule was ·02. It subtended an angle of 0″·429, and was of sealing-wax; had it been a silver one, it would have been still more visible.

With 879·4. All comparative magnitudes of the asteroid and stars, remain as with 496·3.

I see the minute double star *q* Ophiuchi * in high perfection, which proves that the air is clear, and the telescope in good order.

The asteroid being now in the meridian, and the air very pure, I think the comparative diameter is a little larger than that of an equal star, and its light also differs from star-light. Its apparent magnitude, however, can hardly be equal to that of the smallest globule I saw this morning. This globule measured ·01358, and at the distance of 9620·4 inches subtended an angle of 0″·214.

When I viewed the asteroid with 879·4 I found more haziness than an equal star would have given: but this I ascribe to want of light. What I call an equal star, is one that in an achromatic finder appears of equal light.

Oct. 7. Mr. HARDING's asteroid has continued its retrograde motion. The weather is not clear enough to allow the use of high powers.

Oct. 8. If the appearance resembling the spurious disks of small stars, which I see with 410·5 in Mr. HARDING's asteroid, should be a real diameter, its quantity then by estimation may amount to about 0″·3. This judgment is founded on the facility with which I can see two globules often viewed for this purpose.

The angle of the first is 0″·429, and of the other 0″·214; and the asteroid might be larger than the latter, but certainly was not equal to the former.

With 496·3, there is an ill defined hazy appearance, but nothing that may be called a disk visible. When there is a glimpse of more condensed light to be seen in the centre, it is so small that it must be less than two-tenths of a second.

To decide whether this apparent condensed light was a real or spurious disk, I applied different limitations to the aperture of the telescope, but found that the light of the new star was too feeble to permit the use of them. From this I concluded that an increase of light might now be of great use, and viewed the asteroid with a fine 10-feet mirror of 24 inches diameter, but found that nothing was gained by the change. The temperature indeed of these large mirrors is very seldom the same as that of the air in which they are to act, and till a perfect uniformity takes place, no high powers can be used.

The asteroid in the meridian, and the night beautiful. After many repeated comparisons of equal stars with the asteroid, I think it shows more of a disk than they do, but it is so small that it cannot amount to so much as 3-tenths of a second, or at least to no more.

It is accompanied with rather more nebulosity than stars of the same size.

The night is so clear, that I cannot suppose vision at this altitude to be less perfect on the stars, than it is on day objects at the distance of 800 feet in a direction almost horizontal.

Oct. 11. By comparing the asteroid alternately and often with equal stars, its disk, if it be a real one, cannot exceed 2, or at most 3-tenths of a second. This estimation is founded on the comparative readiness with which every fine day I have

* See Cat. of double Stars, I. 87.

seen globules subtending such angles in the same telescope, and with the same magnifying power.

" The asteroid is in the meridian, and in high perfection. I perceive a well defined disk that may amount to 2 or 3-tenths of a second ; but an equal star shows exactly the same appearance, and has a disk as well defined and as large as that of the asteroid."

RESULT AND APPLICATION OF THE EXPERIMENTS AND OBSERVATIONS.

We may now proceed to draw a few very useful conclusions from the experiments that have been given, and apply them to the observations of the star discovered by Mr. HARDING ; and also to the similar stars of Mr. PIAZZI and Dr. OLBERS. These kind of corollaries may be expressed as follows.

(1.) A 10-feet reflector will show the spurious or real disks, of celestial and terrestrial objects, when their diameter is $\frac{1}{4}$ of a second of a degree ; and when every circumstance is favourable, such a diameter may be perceived so distinctly, that it can be divided by estimation into two or three parts.

(2.) A disk of $\frac{1}{4}$ of a second in diameter, whether spurious or real, in order to be seen as a round, well defined body, requires a distinct magnifying power of 5 or 6 hundred, and must be sufficiently bright to bear that power.

(3.) A real disk of half a second in diameter will become so much larger by the application of a magnifying power of 5 or 6 hundred, that it will be easily distinguished from an equal spurious one, the latter not being affected by power in the same proportion as the former.

(4.) The different effects of the inside and outside rays of a mirror, with regard to the appearance of a disk, are a criterion that will show whether it is real or spurious, provided its diameter is more than $\frac{1}{4}$ of a second.

(5.) When disks, either spurious or real, are less than $\frac{1}{4}$ of a second in diameter, they cannot be distinguished from each other ; because the magnifying power will not be sufficient to make them appear round and well defined.

(6.) The same kind of experiments are applicable to telescopes of different sorts and sizes, but will give a different result for the quantity which has been stated at $\frac{1}{4}$ of a second of a degree. This will be more when the instrument is less perfect, and less when it is more so. It will also differ even with the same instrument, according to the clearness of the air, the condition, and adjustment of the mirrors, and the practical habits of the observer.

With regard to Mr. HARDING'S new starry celestial body, we have shown, by observation, that it resembles, in every respect, the two other lately discovered ones of Mr. PIAZZI and Dr. OLBERS ; so that Ceres, Pallas, and Juno, are certainly three individuals of the same species.

That they are beyond comparison smaller than any of the seven planets cannot be questioned, when a telescope that will show a diameter of $\frac{1}{4}$ of a second of a degree, leaves it undecided whether the disk we perceive is a real or a spurious one.

A distinct magnifying power, of more than 5 or 6 hundred, has been applied to Ceres, Pallas, and Juno, but has either left us in the dark, or at least has not fully removed every doubt upon this subject.

The criterion of the apertures of the mirror, on account of the smallness of these objects, has been as little successful; and every method we have tried has ended in proving their resemblance to small stars.

It will appear, that when I used the name asteroid to denote the condition of Ceres and Pallas, the definition I then gave of this term * will equally express the nature of Juno, which, by its similar situation between Mars and Jupiter, as well as by the smallness of its disk, added to the considerable inclination and excentricity of its orbit, departs from the general condition of planets. The propriety therefore of using the same appellation for the lately discovered celestial body cannot be doubted.

Had Juno presented us with a link of a chain, uniting it to those great bodies, whose rank in the solar system I have also defined,† by some approximation of a motion in the zodiac, or by a magnitude not very different from a planetary one, it might have been an inducement for us to suspend our judgment with respect to a classification; but the specific difference between planets and asteroids appears now by the addition of a third individual of the latter species to be more fully established, and that circumstance, in my opinion, has added more to the ornament of our system than the discovery of another planet could have done.

Slough, near Windsor,
Dec. 1, 1804.

* See *Phil. Trans.* for 1802, p. 229 [above, p. 196].

† *Ibid.* page 224 of the same Paper [above, p. 194].

LIV.

On the Direction and Velocity of the Motion of the Sun, and Solar System.

[*Phil. Trans.*, 1805, pp. 233–256.]

Read May 16, 1805.

OUR attention has lately been directed again to the construction of the heavens, on which I have already delivered several detached papers. The changes which have taken place in the relative position of double stars, have ascertained motions in many of them, which are probably of the same nature with those that have hitherto been called proper motions. It is well known that many of the principal stars have been found to have changed their situation, and we have lately had a most valuable acquisition in Dr. MASKELYNE'S Table of proper motions of six and thirty of them. If this Table affords us a proof of the motion of the stars of the first brightness, such as are probably in our immediate neighbourhood, the changes of the position of minute double stars that I have ascertained, many of which can only be seen by the best telescopes, likewise prove that motions are equally carried on in the remotest parts of space which hitherto we have been able to penetrate.

The proper motions of the stars have long engaged the attention of astronomers, and in the year 1783, I deduced from them, with a high degree of probability, a motion of the sun and solar system towards λ Herculis. The reasons which were then pointed out for introducing a solar motion, will now be much strengthened by additional considerations; and the above mentioned Table of well ascertained proper motions will also enable us to enter rigorously into the necessary calculations for ascertaining its direction, and discovering its velocity. When these points are established, we shall be prepared to draw some consequences from them that will account for many phenomena which otherwise cannot be explained.

The scope of this Paper, wherein it is intended to assign not only the direction, but also the velocity of the solar motion, embraces an extensive field of observation and calculation; but as to give the whole of it would exceed the compass of the present sheets, I shall reserve the velocity of the solar motion for an early future opportunity, and proceed now to a disquisition of the first part of my subject, which is the direction of the motion of the sun and solar system.

Reasons for admitting a solar Motion.

It may appear singular that, after having already long ago pointed out a solar motion, and even fixed upon a star towards which I supposed it to be directed, I should again think it necessary to show that we have many substantial reasons for admitting such a motion at all. What has induced me to enter into this inquiry is, that some of the consequences hereafter to be drawn from a solar motion when established, seem to contradict the very intention for which it is to be introduced. The chief object in view, when a solar motion was proposed to be deduced from observations of the proper motions of stars, was to take away many of these motions by investing the sun with a contrary one. But the solar motion, when its existence has been proved, will reveal so many concealed real motions, that we shall have a greater sum of them than it would be necessary to admit, if the sun were at rest; and, to remove this objection, the necessity for admitting its motion ought to be well established.

Theoretical Considerations.

A view of the motion of the moons, or secondary planets, round their primary ones, and of these again round the sun, may suggest the idea of an additional motion of the latter round some other unknown center; and those who like to indulge in fanciful reviews of the heavens, might easily build a system upon hypotheses not altogether without some plausibility in their favour. Accordingly we find that Mr. LAMBERT, in a work which is full of the most fantastic imaginations, has framed a system wherein the sun is supposed to move about the nebula in Orion.* But, setting aside the extravagant idea of making this luminous spot a center of motion, it must certainly be admitted that the solar motion itself is at least a very possible event.

I have already mentioned, in a note to my former Paper,† that the possibility of a solar motion has also been shown from theoretical principles by the late Dr. WILSON of Glasgow; and its probability afterwards, from reasons of the same nature, by Mr. DE LA LANDE. The rotatory motion of the sun, from which he concludes a displacing of the solar center, must certainly be allowed to indicate a motion of translation in space; for though it may be possible, it does not appear probable, that any mechanical impression should have given the former, without occasioning the latter. But, as we are intirely unacquainted with the cause of the rotatory motion, the solar translation in space from theoretical reasons, can only be admitted as a very plausible hypothesis.

It would be worth while for those who have fixed instruments, to strengthen this argument by observing the stars which are known to change their magnitudes periodically. For as we have great reason to ascribe these regular changes to a

* See *Système du Monde de* Mr. LAMBERT, page 152 and 158.

† See *Phil. Trans.* for the year 1783, page 283. [This Ed., Vol. I. p. 130.]

rotatory motion of the stars,* a real motion in space may be expected to attend it ; and the number of these stars is so considerable that their concurring testimony would be very desirable.

Perhaps Algol, which according to these ideas must have a very quick rotatory motion, may be found to have also a considerable progressive one ; and if that should be ascertained, the position of the axis of the rotation of this star will be in a great measure thereby discovered.

An argument from the real motion to a rotatory one is nearly of equal validity, and therefore all the stars that have a motion in space may be surmised to have also a rotation on their axes.

Symptoms of parallactic Motions.

But, setting aside theoretical arguments, I shall now proceed to such as may be drawn from observation ; and, as all parallactic motions are evident indications that the observer of them is not at rest, it will be necessary to explain three sorts of motions, of which the parallactic is one ; they will often engage our attention in the following discussion.

Let the sun be supposed to move towards a certain part of the heavens, and since the whole solar system will have the same motion, the stars must appear to an inhabitant of the earth to move in an opposite direction. In the triangle *spa*, Fig. 1, let *sp* represent the parallactic motion of a star ; then, if this star is one that has no real motion, *sp* will also be its apparent motion ; but if the star in the same time, that by its parallactic motion it would have gone from *s* to *p*, should have a real motion which would have carried it from *s* to *r*, then will it be seen to move along the diagonal *sa*, of the parallelogram *srpa* ; and *pa*, which is parallel and equal to *sr*, will represent its real motion. Therefore, in the above mentioned triangle *spa*, which I suppose to be formed in the concave part of the heavens by three arches of great circles, the eye of the observer being in the center, the three sides will represent, or stand for, the three motions I have named : *sp* the parallactic, *pa* the real, and *sa* the apparent motion of the star. The situation and length of these arches, in seconds of a degree, will express, or rather represent, not only the direction but also the quantity of each motion, such as it must appear to an eye in the above mentioned central situation. And calling the solar motion S, the distance of the star from the sun d, and the sine of the star's distance from the point towards which the sun is moving ϕ, the parallactic motion, when these are given, will be had by the expression $\frac{\phi \,.\, S}{r \,.\, d} = sp$. This theorem, and its corollaries, of which frequent use will be made hereafter, it will not be necessary here to demonstrate.

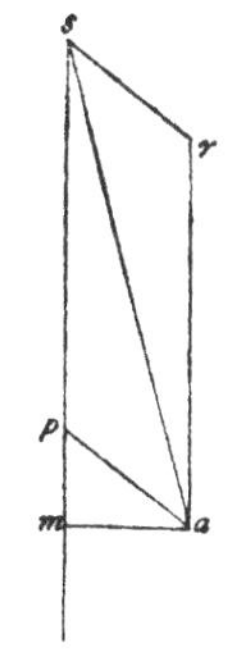

FIG. 1.

When I call the arch *pa* the real motion, it should be understood that I only

* See *Phil. Trans.* for the year 1795, page 68. [Vol. I. p. 482.]

mean its representative ; for it must be evident that the absolute motion of a star in space, as well as its intrinsic velocity, will still remain unknown, because the inclination of that motion on which also its real velocity will depend, admits of the greatest variety of directions. We are only acquainted with the plane in which the motion must be performed, and with the length of the arch in seconds by which that motion may be measured. We may add that the chords of the arches representing the three motions are the smallest velocities of these motions that can be admitted ; for in every other direction but at right angles to the line of sight, the actual space over which the star will move must be greater than the arch or chord by which its motion is represented.

Now, since a motion of the sun will occasion parallactic motions of the stars, it follows that these again must indicate a solar motion ; but in order to ascertain whether parallactic motions exist, we ought to examine those stars which are most liable to be visibly affected by solar motion. This requisite points out the brightest stars as the most proper for our purpose ; for any star may have a great real motion, but in order to have a great parallactic one, it must be in the neighbourhood of the sun. And as we can only judge of the distance of the stars by their splendour we ought to choose the brightest, on account of a probability that, being nearer than faint ones, they may be more within the reach of parallax, and thus better qualified to show its effects.

We are also to look out for a criterion whereby parallactic may be distinguished from real motions ; and this we find in their directions. For if a solar motion exists, all parallactic motions will tend to a point in opposition to the direction of that motion ; whereas real motions will be dispersed indiscriminately to all parts of space.

With these distinctions in view, we may examine the proper motions of the principal stars ; for these, if the sun is not at rest, must either be intirely parallactic, or at least composed of real and parallactic motions ; in the latter case they will fall under the denomination of one of the three motions we have defined, namely *sa*, the apparent motion of the star.

In consequence of this principle I have delineated the meeting of the arches arising from a calculation of the proper motions of the 36 stars in Dr. MASKELYNE'S catalogue, on a celestial globe ; and, as all great circles of a sphere intersect each other in two opposite points, it will be necessary to distinguish them both : for, if the sun moves to one of them, it may be called the apex of its motion, and as the stars will then have a parallactic motion to the opposite one, the appellation of a parallactic center may very properly be given to it. The latter falling into the southern hemisphere, among constellations not visible to us, I shall only mention their opposite intersections ; and of these I find no less than ten that are made by stars of the first magnitude, in a very limited part of the heavens, about the constellation of Hercules. Upon all the remaining surface of the same globe there is not the least appearance of any other than a promiscuous situation of intersections; and of these only a single one is made by arches of principal stars.

The ten intersecting points made by the brightest stars are as follows. The 1st is by Sirius and Arcturus, in the mouth of the Dragon. The 2d by Sirius and Capella, near the following hand of Hercules. The 3d by Sirius and Lyra, between the hand and knee of Hercules. The 4th by Sirius and Aldebaran, in the following leg of Hercules. The 5th by Arcturus and Capella, north of the preceding wing of the Swan. The 6th by Arcturus and Aldebaran, in the neck of the Dragon. The 7th by Arcturus and Procyon, in the preceding foot of Hercules. The 8th by Capella and Procyon, south of the following hand of Hercules. The 9th by Lyra and Procyon, preceding the following shoulder of Hercules. And the 10th is made by Aldebaran and Procyon, in the breast of Hercules.

The following Table gives the calculated situation of these ten intersections in right ascension and north polar distance.

TABLE I.

No.	Right Ascension.	Polar Distance.
1	255° 39′ 50″	36° 41′ 34″
2	275 9 32	64 21 48
3	272 23 58	58 23 24
4	263 25 38	44 39 47
5	290 0 58	32 7 23
6	267 2 19	33 57 20
7	235 3 13	46 21 34
8	272 51 49	73 7 56
9	266 46 49	66 48 11
10	260 1 29	60 59 34

We might rest satisfied with having shown that the parallactic effect of which we are in search is plainly to be perceived in the motion of the brightest stars; however, by way of further confirmation, we may take in some large stars of the next order, in whose motions evident marks of the influence of parallax may likewise be perceived. When the intersections made by their proper motions and the arches in which the stars of the first magnitude are moving, are examined, we find no less than fifteen which unite with the former ten, in pointing out the same part of the heavens as a parallactic center. It will be sufficient only to mention the opposite points of the situation of these intersections, and the stars by which they are made, without giving a calculated table of them.

The 1st is the following leg of Hercules, and is made by Sirius and β Tauri. The 2d is also in the following leg of Hercules, by Sirius and α Andromedæ. The 3d is in the following hand of Hercules, by Sirius and α Arietis. The 4th in the neck of the Dragon, by Arcturus and β Tauri. The 5th between the Lyre and the northern wing of the Swan, by Capella and α Andromedæ. The 6th near the following hand of Hercules, by Capella and α Arietis. The 7th preceding the head of Hercules, by Lyra and β Tauri. The 8th between the Lyre and northern wing of the Swan, by Lyra and α Andromedæ. The 9th in the following arm of Hercules, by Lyra and α Arietis. The 10th in the following leg of Hercules, by Aldebaran and β Tauri. The 11th in the following leg of Hercules, by Aldebaran and α Andromedæ. The 12th in the head of Hercules, by Aldebaran and α Arietis. The 13th in the following arm of Hercules, by Procyon and β Tauri. The 14th in the back of Hercules, by Procyon and α Andromedæ. And the 15th near the following arm of Hercules, is made by Procyon and α Arietis.

An argument like this, founded upon the most authentic observations, and supported by the strictest calculations, can hardly fail of being convincing. And though only the ten principal apices of the twenty-five that are given have been calculated, the other fifteen may nevertheless be depended upon as true to less than one degree of the sphere.

Changes in the Position of double Stars.

We have lately seen that the alterations in the relative situation of a great number of double stars may be accounted for by a parallactic motion. Among the 56 stars which I have given, the changes of more than half of them appear to be of this nature; and it will certainly be more eligible to ascribe them to the effect of parallax than to admit so many separate motions in the different stars; especially when it is considered that if the alterations of the angle of position were owing to a motion of the largest star of each set, the direction of such motions must, in contradiction to all probability, tend nearly to one particular part of the heavens.

This argument, drawn from the change of the position of double stars, may be considered as deriving its validity from the same source with the former, namely, the parallactic motions of at least 28 more stars, pointing out the same apex of a solar motion by their direction to its opposite parallactic center.

Incongruity of proper Motions.

It may be remarked that the proper motions of the stars, if they were in reality such as they appear to be, would contain a certain incongruous mixture of great velocity and extreme slowness. Arcturus alone describes annually an arch of more than two seconds: Aldebaran hardly one-tenth and a quarter of a second: Rigel little more than one-tenth and a half: even Lyra moves barely three and a quarter tenths of a second, while Procyon has almost four times that velocity. Out of 36 stars whose proper motion we have examined, there are 15 that do not reach two-tenths of a second: β Virginis moves seventy-seven hundredths, and α Cygni only six. But it will be shown, when the direction and velocity of the solar motion come to be explained, that these kind of incongruities are mere parallactic appearances; and that there is so general a consistency among the real motions of the stars, that Arcturus is in no respect singled out as a star whose motion is far beyond the rest.

By giving this remark a place among the reasons for admitting a solar motion, it is not intended to lay any particular stress upon it; for it may be objected that our idea of the congruence or harmony of the celestial motions can be no criterion of their real fitness and symmetry. But when such discordant proper motions as those I have mentioned in stars of no very different lustre are under consideration, and may be easily shown to be only parallactic phenomena, the method by which this can be done must certainly appear eligible, and when added to many other inducements, will throw some share of weight into the scale.

Sidereal Occultation of a small Star.

Of nearly the same importance with the former argument is the account of the occultation of a small star by a large one, which I have given in my last Paper. When the solar motion has been established, we shall prove that the vanishing of the small star near δ Cygni, as far as we can judge at present, is only a parallactic disappearance. It must be granted that a real motion of the large star would also explain the same phenomenon; but then again, this star must be supposed to move towards the very same parallactic center which the changes in the position of other double stars point out, and this cannot be probable.

Direction of the solar Motion.

From what has been said, I believe the expedience of admitting a solar motion will not be called in question; our next endeavour therefore must be to investigate its direction.

To return to the before mentioned intersections of the arches, in which the proper motions of the stars are performed, I shall begin by proving that when the proper motions of two stars are given, an apex may be found, to which, if the sun be supposed to move with a certain velocity, the two given motions may then be resolved into apparent changes, arising from sidereal parallax, the stars remaining perfectly at rest.

Let the stars be Arcturus and Sirius, and their annual proper motions as given in the Astronomer Royal's Tables.

When the annual proper motion of Arcturus, which is $-1''{\cdot}26$ in right ascension, and $+1''{\cdot}72$ in north polar distance, is reduced by a composition of motions to a single one, it will be in a direction which makes an angle of $55^\circ\ 29'\ 42''$ south-preceding with the parallel of Arcturus, and of a velocity so as to describe annually $2''{\cdot}08718$ of a great circle.

The annual proper motion of Sirius, $-0''{\cdot}42$ in right ascension, and $+1''{\cdot}04$ in north polar distance, by the same method of composition, becomes a motion of

1″·11528, in a direction which makes an angle of 68° 49′ 41″ south-preceding with the parallel of Sirius.

By calculation, the arches in which these two stars move, when continued, will meet in what I have called their parallactic center, whose right ascension is 75° 39′ 50″, and south polar distance is 36° 41′ 34″. The opposite of this, or right ascension 255° 39′ 50″, and north polar distance 36° 41′ 34″, is what we are to assume for the required apex of the solar motion.

When a star is situated at a certain distance from the sun, which we shall call 1; and 90° from the apex of the solar motion, its parallactic motion will be a maximum. Let us now suppose the velocity of the sun to be such that its motion, to a person situated on this star, would appear to describe annually an arch of 2″·84825, or, which is the same thing, that the star would appear to us, from the effect of parallax, to move over the above mentioned arch in the same time.

To apply this to Arcturus, we find by calculation that its distance from the apex of the solar motion is 47° 7′ 6″; its parallactic motion therefore, which is as the sine of that distance, will be 2″·08718; and this, as has been shown, is the apparent motion which observation has established as the proper motion of Arcturus.

In the next place, if we admit Sirius to be a very large star situated at the distance 1·6809 from us, and compute its elongation from the apex of the solar motion, we shall find it 138° 50′ 14″·5. With these two data we calculate that its parallactic motion will be $\frac{\phi \, . \, S}{r \, . \, d} = sp = 1''\cdot11528$; and this also agrees with the apparent motion which has been ascertained by observation as the proper motion of Sirius.

Now since, according to the rules of philosophising, we ought not to admit more motions than will account for the observed changes in the situation of the stars, it would be wrong to have recourse to the motions of Arcturus and Sirius, when that of the sun alone will account for them both; and this consideration would be a sufficient inducement for us to fix at once on the calculated apex, as well as on the relative distances that have been assigned to these stars, if other proper motions could with equal facility be resolved into similar parallactic appearances. But from the nature of proper motions, it follows, that when a third star does not lead us to the same apex as the other two, its apparent motion cannot be resolved by the effect of parallax alone. And to enhance our difficulties, the number of apices, that would be required to solve all proper motions into parallactic ones, increases not as the number of stars admitted to have proper motions, but, when their situation happens to be favourable, as the sum of an arithmetical series of natural numbers, beginning at 0, continued to as many terms as there are stars admitted: so that if two stars give only one apex, one star added to it will give three apices; and ten, for instance, will give no less than 45, and so on.

The method of reasoning which, on this subject, I have adopted, is so closely connected with astronomical observations that I shall keep them constantly in view;

and therefore shall illustrate what has been advanced, by taking in Capella as a third star. The three apices which then are pointed out will be that in the mouth of the Dragon, by Arcturus and Sirius ; a second under the northern wing of Cygnus, by Arcturus and Capella ; and a third in the following hand of Hercules, by Sirius and Capella. The calculation of them is in Table I.

The annual proper motions of our third star in Dr. MASKELYNE'S Tables are +0″·21 in right ascension, and +0″·44 in north polar distance ; and by calculation these quantities give an annual motion of 0″·46374 to Capella, in a direction which makes an angle of 71° 35′ 22″·4 south-following with the parallel of this star.

The distance of Capella from the same calculated apex of the solar motion, by which we have already explained the apparent motions of the other two stars, is 80° 54′ 46″ ; and, admitting again the velocity of the sun towards the same point as stated before, it will occasion a parallactic motion of Capella, in a direction 89° 54′ 48″ south-following its parallel, amounting to 2″·8125. In this calculation Capella has been taken for a star of the first magnitude, supposing its distance from us to be equal to that of Arcturus.

By constructing then a triangle, the three sides of which will represent the three motions which every star must have that is not at rest in space ; we have one of the sides, representing the apparent motion of the star, equal to 0″·4637 ; the other side, being the parallactic motion of the star 2″·8125 ; and the included angle 18° 19′ 27″. From these data we obtain the third side, representing the real motion of the star, which will be 2″·3757. By the given situation of this triangle with respect to the parallel of declination of Capella, the angle of the real motion will also be had, which is 86° 34′ 11″ north-following the parallel of this star. A composition of the parallactic and the real motion in the directions we have assigned, will produce the annual apparent motion which has been established by observation.

But to apply what has been said to our present purpose, it may be observed, that although we have accounted for the proper motion of our third star by retaining the same apex of the solar motion, which has given us an explanation of the apparent motions of the other two, yet in doing this we have been obliged to assign a great degree of real motion to Capella ; and to this it may be objected, that we can have no authority to deprive Arcturus and Sirius of real motions, in order to give one of the same nature to our third star : and indeed to every star that has a proper motion which does not tend to the same parallactic center as the motions of Arcturus and Sirius.

This objection is perfectly well founded, and I have given the above calculation on purpose to show that, when we are in search of an apex for the solar motion, it ought to be so fixed upon as to be equally favourable to every star which is proper for directing our choice. Hence a problem will arise, in our present case, how to find a point whose situation among three given apices shall be so that, if the sun's motion be directed towards it, there may be taken away the greatest quantity of proper motion possible from the given three stars. The intricacy of the problem is greater than at first it may appear, because by a change of the distance of the apex from any one of the stars, its parallactic motion, which is as the sine of that distance, will be affected ; so that it is not the mere alteration of the angle of direction, which is concerned. However, it will not be necessary to enter into a solution of the problem ; for it must be very evident that a much more complex one would immediately succeed it, since three stars would certainly not be sufficient to direct us in our present endeavour to find the best situation of an apex for the solar motion ; I shall therefore now leave these stars, and the apices pointed out by them, in order to proceed to a more general view of the subject.

We have already seen that the brightest stars are most proper for showing the effect of parallax, and that in our search after the direction of the solar motion, our aim must be to reduce the proper motions of the stars to their lowest quantities. The six principal stars, whose intersecting arches have been given, when their proper motions in right ascension and polar distance are brought into one direction, will have the apparent motions contained in the following Table.

TABLE II.

Names of the Stars.	Direction of the apparent Motions.	Quantities of the apparent Motions.
	° ′ ″	″
Sirius	68 49 40·7 south-preceding	1·11528 per year
Arcturus	55 29 42·0 south-preceding	2·08718 ———
Capella . .	71 35 22·4 south-following	0·46374 ———
Lyra	56 20 57·3 north-following	0·32435 ———
Aldebaran	76 29 37·3 south-following	0·12341 ———
Procyon	50 2 24·5 south-preceding	1·23941 ———
	Sum of the apparent motions	5·35337

We must now recur to what has been said, when the construction of the triangle expressing the three motions of a star, that is not at rest, was explained ; and, as we are to find out a solar motion which will require the least real motion in our six stars, an attention to this triangle will be of considerable use ; for when the line pa, Fig. 1, which represents the real motion, is brought into the situation ma, where it is perpendicular to sp, the real motion which is required will then be a minimum. It also follows, from the construction of the same triangle, that if by the choice of an apex for the solar motion we can lessen the angle made at s by the lines sp and sa, we shall lessen the quantity of real motion required to bring the star from the parallactic line spm to the observed position a.

It has already been shown, in the case of Sirius and Arcturus, that when two stars only are given, the line sp may be made to coincide with the lines sa, of both the stars, whereby their real motions will be reduced to nothing. It has also been proved, by adding Capella to the former two, that when three stars are concerned, some real motion must be admitted in one of them. Now, since all parallactic motions are directed to the same center, a single line may represent the direction of the effect of the parallax, not only of these three stars but of every star in the heavens. According to this theory let the line sP or sS, in Fig. 2, stand for the direction of the parallactic motion of the stars; and as in the foregoing Table we have the angles of the apparent motion of six stars with the parallel of each star, we must now also compute the direction of the line sP or sS with the parallels of the same stars. This may be done as soon as an apex for the solar motion is fixed upon. The difference between these angles and the former will give the several parallactic angles Psa or Ssa, required for an investigation of the least quantity ma, belonging to every star.

For instance, let the point towards which we may suppose the sun to move, be λ Herculis; and calculating the required angles of the direction in which the effect of parallax will be exerted, with the six stars we have selected for the purpose of our investigation, we find them as in the following Table.

TABLE III.

Angles of the parallactic Motion with the Parallel.

	°	′	″	
Sirius	32	54	8·5	south-preceding.
Arcturus	17	23	45·7	south-preceding.
Capella	85	10	3·9	south-following.
Lyra	35	59	49·5	north-following.
Aldebaran	71	21	35·4	south-following.
Procyon	47	43	44·6	south-preceding.

The difference between these parallactic, and the former apparent angles, with the parallel of each star, will give the required angles for our second figure. They will be as follows.

TABLE IV.

Angles of the apparent with the parallactic Motion.

	°	′	″	
Sirius	35	55	32·2	south-following.
Arcturus	38	5	56·3	south-following.
Capella	13	34	41·5	south-following.
Lyra	20	21	7·8	north-preceding.
Aldebaran	5	8	1·9	south-preceding.
Procyon	2	18	39·9	south-following.

By these angles, with the assistance of the lines sa, whose lengths represent the annual quantity of the apparent motions as given in our former Table, the Figure No. 2 has been constructed. When the situation of these angles is regulated as in that figure, we may draw the several lines ma perpendicular to SP, and, by computation, their value and sum will be obtained as follows.

TABLE V.

Quantities and Sum of the least real Motions.

	″
Sirius	0·65437
Arcturus	1·28784
Capella	0·10887
Lyra	0·11281
Aldebaran	0·01104
Procyon	0·04998
Sum	2·22491.

The result of this investigation is, that by admitting a motion of the sun towards λ Herculis, the annual proper motions of our six stars, of which the sum is 5″·3537, may be reduced to real motions of no more than 2″·2249.

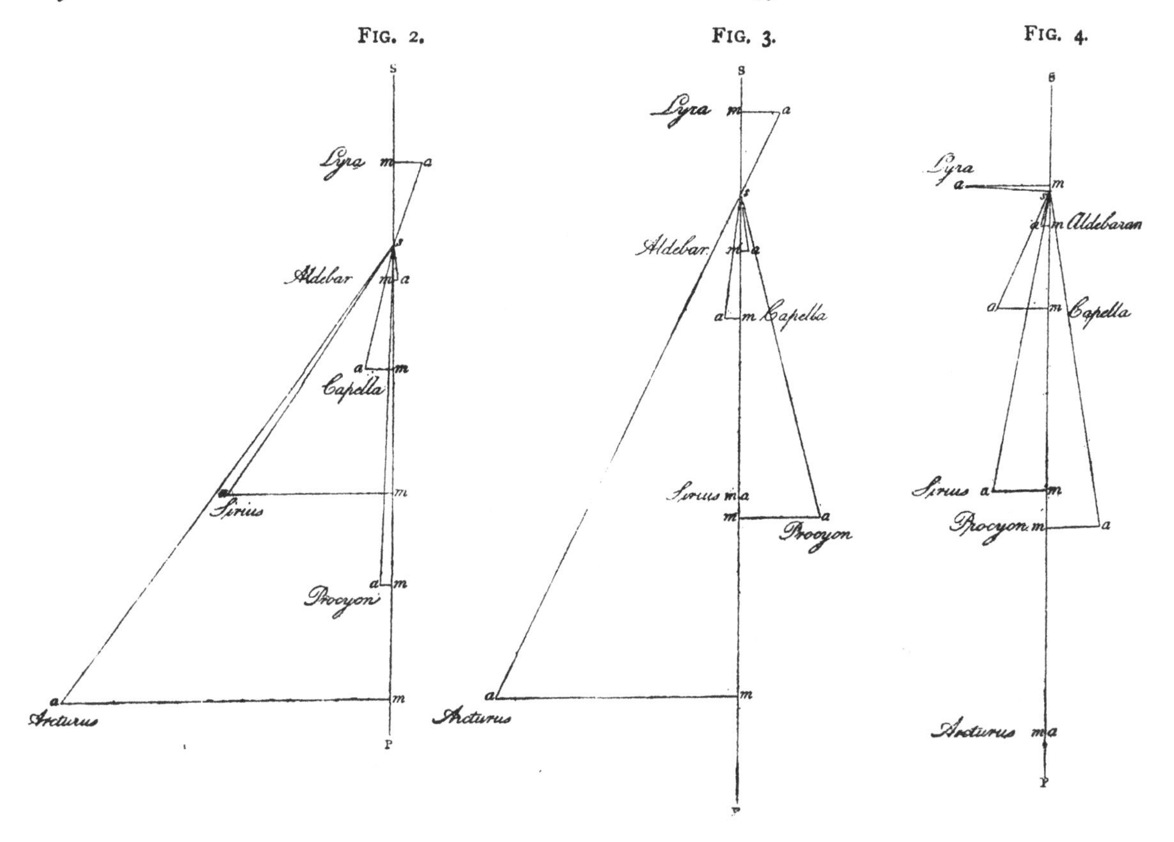

When first I proposed λ Herculis as an apex for the solar motion, it may be remembered that a reference to future observations was made for obtaining greater accuracy.* Such observations we have now before us, in the valuable Tables from which I have taken the proper motions of the six stars; and I shall prove that, with their assistance, we may fix on a solar motion that will be considerably more favourable.

* See *Phil. Trans.* for 1783, pp. 273, 274 [Vol. I. pp. 124, 125].

We have already shown, that to ascertain the precise place of the best apex is attended with some difficulty ; but from the inspection of the figure which represents the quantities of real motion required when λ Herculis is fixed upon, it will be seen that, by a regular method of approximation, we may turn the line SP into a situation, where all the angles of the apparent motion of the six stars will be much reduced. The quantities which are required for constructing another figure to represent the threefold motions of our six stars, when a different apex is fixed upon, are to be found by the same method we have pursued in the instance of λ Herculis ; and the figure that has been given with respect to that star, shows evidently that the parallactic line SP should be turned more towards the line *sa*, representing the apparent motion of Sirius. We shall accordingly try a point near the following knee of Hercules, whose right ascension is 270° 15′, and north polar distance 54° 45′.

The result of a calculation of the angles and the least quantities of real motion of our six stars, according to this apex, is collected in the following Table, and represented in Fig. 3.

TABLE VI.

Stars.	Angles of the parallactic Motion with the Parallel.	Angles of the apparent with the parallactic Motion.	Least Quantities ot the real Motion.
	° ′ ″	° ′ ″	″
Sirius . .	68 51 5 south-preceding	0 1 25 south-preceding	0·0004561
Arcturus .	29 30 32 south-preceding	25 59 10 south-following	0·9145072
Capella .	77 54 0 south-following	6 18 38 south-following	0·0509727
Lyra . .	27 38 47 north-following	28 42 9 north-preceding	0·1557761
Aldebaran .	66 20 17 south-following	10 9 21 south-preceding	0·0217607
Procyon .	64 48 27 south-following	14 46 1 south-preceding	0·3159051
			Sum 1·4593779

By this Table it appears that the annual proper motion of our six stars may be reduced to 1″·4594, which is 0″·7655 less than the sum in the 5th Table, where the apex was λ Herculis.

In the approximation to this point it appeared, that when the line of the parallactic motion of Sirius is made to coincide with its apparent motion, we may soon obtain a certain minimum of the other parallactic motions ; but as Sirius is not the star which has the greatest proper motion, it occurred to me that another minimum, obtained from the line in which Arcturus appears to move would be more accurate ; for, on account of its great proper motion, we have reason to suppose it more affected than other stars, by the parallax arising from the motion of the sun ; and, with a view to this, I soon was led to a point not only in the line of the apparent motion of Arcturus, but equally favourable to Sirius and Procyon, the remaining two stars that have the greatest motions.

If the principle of determining the direction of the solar motion by the stars which have the greatest proper motion be admitted, the following apex must be extremely near the truth; for, an alteration of a few minutes in right ascension or polar distance either way, will immediately increase the required real motion of our stars. Its place is: right ascension 245° 52′ 30″, and north polar distance 40° 22′.

The calculated motions of the same stars by this apex will be as in the following Table, and are delineated in Fig. 4.

TABLE VII.

Stars.	Angles of the parallactic Motion with the Parallel.	Angles of the apparent with the parallactic Motion.	Least Quantities of the real Motion.
	° ′ ″	° ′ ″	″
Sirius . .	58 24 56 south-preceding	10 24 44 following	0·20157
Arcturus .	55 29 45 south-preceding	0 0 3 preceding	0·00003
Capella .	83 44 17 south-preceding	24 40 21 following	0·19358
Lyra . .	36 28 33 south-following	92 49 30 following	0·32396
Aldebaran .	89 48 35 south-preceding	13 41 48 following	0·02922
Procyon .	59 43 10 south-preceding	9 40 46 preceding	0·20839
			Sum 0·95675

The sum of the real motions required, with the apex of the solar motion above mentioned, is less in this Table than that in the former by 0″·50263.

In these calculations we have proceeded upon the principle of obtaining the least possible quantity of real motion, by way of coming at the most favourable situation of a solar apex, and have proved that the sum of the observed proper motions of the six principal stars, amounting to 5″·3534, may be the result of a composition of two other motions, and that the real motions of these stars, if they could be reduced to their smallest possible quantities, would not exceed 0″·9568.

But as I do not intend to assert that these real motions can be actually brought down to the low quantities that have been mentioned, it will be necessary to show that the validity of the arguments for establishing the method I have pursued will not be affected by that circumstance. In the first place then, we should consider that although the great proper motions of Arcturus, Procyon, and Sirius, are strong indications of their being affected by parallax, it does not follow, nor is it probable, that the apparent changes of the situation of these stars should be intirely owing to solar motion; on the contrary, we may reasonably expect that their own real motions will have a great share in them. Next to this, it is evident that in the case of parallactic motions the distance of a star from the sun is of material consequence; and as this cannot be assumed at pleasure, we are consequently not at liberty to make the parallactic motion *sp* in Fig. 1, equal to the line *sm* of the same figure;

hence it follows, that the real motion of the star cannot be from *m* to *a*, as the foregoing calculations have supposed ; but will be from *p* to *a*. It is however very evident, that if *ma* be a minimum, the line *pa*, when *sp* is given, will also be a minimum ; and if all the *ma*'s in Fig. 4 are minima, it follows also that all the *sp*'s, whatever they may be, will give the *pa*'s as small as possible : and this is the point that was to be established.

Whatever therefore may be the sum of real motions required to account for the phenomena of proper motions, our foregoing arguments cannot be affected by the result ; for, as by observation it is known that proper motions do exist, and since no solar motion can resolve them intirely into parallactic ones, we ought to give the preference to that direction of the motion of the sun which will take away more real motion than any other, and this, as we have shown, will be done when the right ascension of the apex is 245° 52′ 30″, and its north polar distance 40° 22′.

LV.

Observations on the singular Figure of the Planet Saturn.

[*Phil. Trans.*, 1805, pp. 272–280.]

Read June 20, 1805.

THERE is not perhaps another object in the heavens that presents us with such a variety of extraordinary phenomena as the planet Saturn: a magnificent globe, encompassed by a stupendous double ring: attended by seven satellites: ornamented with equatorial belts: compressed at the poles: turning upon its axis: mutually eclipsing its ring and satellites, and eclipsed by them: the most distant of the rings also turning upon its axis, and the same taking place with the farthest of the satellites: all the parts of the system of Saturn occasionally reflecting light to each other: the rings and moons illuminating the nights of the Saturnian: the globe and satellites enlightening the dark parts of the rings: and the planet and rings throwing back the sun's beams upon the moons, when they are deprived of them at the time of their conjunctions.

It must be confessed that a detail of circumstances like these, appears to leave hardly any room for addition, and yet the following observations will prove that there is a singularity left, which distinguishes the figure of Saturn from that of all the other planets.

It has already been mentioned on a former occasion, that so far back as the year 1776 I perceived that the body of Saturn was not exactly round; and when I found in the year 1781 that it was flattened at the poles at least as much as Jupiter, I was insensibly diverted from a more critical attention to the rest of the figure. Prepossessed with its being spheroidical, I measured the equatorial and polar diameters in the year 1789, and supposed there could be no other particularity to remark in the figure of the planet. When I perceived a certain irregularity in other parts of the body, it was generally ascribed to the interference of the ring, which prevents a complete view of its whole contour; and in this error I might still have remained, had not a late examination of the powers of my 10-feet telescope convinced me that I ought to rely with the greatest confidence upon the truth of its representations of the most minute objects I inspected.

The following observations, in which the singular figure of Saturn is fully investigated, contain many remarks on the rest of the appearances that may be seen when this beautiful planet is examined with attention; and though they are not immediately necessary to my present subject, I thought it right to retain them, as they show the degree of distinctness and precision of the action of the telescope, and the clearness of the atmosphere at the time of observation.

April 12, 1805. With a new 7-feet mirror of extraordinary distinctness, I examined the planet Saturn. The ring reflects more light than the body, and with a power of 570 the colour of the body becomes yellowish, while that of the ring remains more white. This gives us an opportunity to distinguish the ring from the body, in that part where it crosses the disk, by means of the difference in the colour

of the reflected light. I saw the quintuple belt, and the flattening of the body at the polar regions; I could also perceive the vacant space between the two rings.

The flattening of the polar regions is not in that gradual manner as with Jupiter, it seems not to begin till at a high latitude, and there to be more sudden than it is towards the poles of Jupiter. I have often made the same observation before, but do not remember to have recorded it any where.

April 18; 10-feet reflector, power 300. The air is very favourable, and I see the planet extremely well defined. The shadow of the ring is very black in its extent over the disk south of the ring, where I see it all the way with great distinctness.

The usual belts are on the body of Saturn; they cover a much larger zone than the belts on Jupiter generally take up, as may be seen in the figure I have given here; and also in a former representation of the same belts in 1794.*

* See *Phil. Trans.* for 1794, Table VI. page 32 [Vol. I. Plate XIII. p. 452].

The figure of the body of Saturn, as I see it at present, is certainly different from the spheroidical figure of Jupiter. The curvature is greatest in a high latitude.

I took a measure of the situation of the four points of the greatest curvature, with my angular micrometer, and power 527. When the cross of the micrometer passed through all the four points, the angle which gives the double latitude of two of the points, one being north the other south of the ring, or equator, was 93° 16′. The latitude therefore of the four points is 46° 38′ ; it is there the greatest curvature takes place. As neither of the cross wires can be in the parallel, it makes the measure so difficult to take, that very great accuracy cannot be expected.

The most northern belt comes up to the place where the ring of Saturn passes behind the body, but the belt is bent in a contrary direction being concave to the north, on account of its crossing the body on the side turned towards us, and the north pole being in view.

There is a very dark, but narrow shadow of the body upon the following part of the ring, which as it were cuts off the ring from the body.

The shadow of the ring on the body, which I see south of the ring, grows a little broader on both sides near the margin of the disk.

The division between the two rings is dark, like the vacant space between the ansæ, but not black like the shadow I have described.

There are four satellites on the preceding side near the ring ; the largest and another are north-preceding ; the other two are nearly preceding.

April 19. I viewed the planet Saturn with a new 7-feet telescope, both mirrors of which are very perfect. I saw all the phenomena as described last night, except the satellites, which had changed their situation ; four of them being on the following side. This telescope however is not equal to the 10-feet one.

The remarkable figure of Saturn admits of no doubt : when our particular attention is once drawn to an object, we see things at first sight that would otherwise have escaped our notice.

10-feet reflector, power 400. The night is beautifully clear, and the planet near the meridian. The figure of Saturn is somewhat like a square or rather parallelogram, with the four corners rounded off deeply, but not so much as to bring it to a spheroid. I see it in perfection.

The four satellites that were last night on the preceding, are now on the following side, and are very bright.

I took a measure of the position of the four points of the greatest curvature, and found it 91° 29′. This gives their latitude 45° 44′·5. I believe this measure to be pretty accurate. I set first the fixed thread to one of the lines, by keeping the north-preceding and south-following two points in the thread ; then adjusted the other thread in the same manner to the south-preceding and north-following points.

May 5, 1805. I directed my 20-feet telescope to Saturn, and, with a power of about 300, saw the planet perfectly well defined, the evening being remarkably clear. The shadow of the ring on the body is quite black. All the other phenomena are very distinct.

The figure of the planet is certainly not spheroidical, like that of Mars and Jupiter. The curvature is less on the equator and on the poles than at the latitude of about 45 degrees. The equatorial diameter is however considerably greater than the polar.

In order to have the testimony of all my instruments, on the subject of the structure of the planet Saturn, I had prepared the 40-feet reflector for observing it in the meridian. I used a magnifying power of 360, and saw its form exactly as I had seen it in the 10 and 20-feet instruments. The planet is flattened at the poles, but the spheroid that would arise from this flattening is modified by some other cause, which I suppose to be the attraction of the ring. It resembles a parallelogram, one side whereof is the equatorial, the other the polar diameter, with the four corners rounded off so as to leave both the equatorial and polar regions flatter than they would be in a regular spheroidical figure.

The planet Jupiter being by this time got up to a considerable altitude, I viewed it alternately with Saturn in the 10-feet reflector, with a power of 500. The outlines of the figure of Saturn are as described in the observation of the 40-feet telescope; but those of Jupiter are such as to give a greater curvature both to the polar and equatorial regions than takes place at the poles or equator of Saturn which are comparatively much flatter.

May 12. I viewed Saturn and Jupiter alternately with my large 10-feet telescope of 24 inches aperture; and saw plainly that the former planet differs much in figure from the latter.

The temperature of the air is so changeable that no large mirror can act well.

May 13. 10-feet reflector, power 300. The shadows of the ring upon the body, and of the body upon the ring, are very black, and not of the dusky colour of the heavens about the planet, or of the space between the ring and planet, and between the two rings. The north-following part of the ring, close to the planet, is as it were cut off by the shadow of the body; and the shadow of the ring lies south of it, but close to the projection of the ring.

The planet is of the form described in the observation of the 40-feet telescope; I see it so distinctly that there can be no doubt of it. By the appearance, I should think the points of the greatest curvature not to be so far north as 45 degrees.

The evening being very calm and clear, I took a measure of their situation, which gives the latitude of the greatest curvature 45° 21′. A second measure gives 45° 41′.

Jupiter being now at a considerable altitude, I have viewed it alternately with Saturn. The figure of the two planets is decidedly different. The flattening at the poles and on the equator of Saturn is much greater than it is on Jupiter, but the curvature at the latitude of from 40 to 48° on Jupiter is less than on Saturn.

I repeated these alternate observations many times, and the oftener I compared the two planets together, the more striking was their different structure.

May 26. 10-feet reflector. With a parallel thread micrometer and a magnifying power of 400, I took two measures of the diameter of the points of greatest curvature. A mean of them gave 64·3 divisions $=11''\cdot98$. After this, I took also two measures of the equatorial diameter, and a mean of them gave 60·5 divisions $=11''\cdot27$; but the equatorial measures are probably too small.

To judge by a view of the planet, I should suppose the latitude of the greatest curvature to be less than 45 degrees. The eye will also distinguish the difference in the three diameters of Saturn. That which passes through the points of the greatest curvature is the largest; the equatorial the next, and the polar diameter is the smallest.

May 27. The evening being very favourable, I took again two measures of the diameter between the points of greatest curvature, a mean of which was 63·8 divisions $=11''\cdot88$. Two measures of the equatorial diameter gave 61·3 divisions $=11''\cdot44$.

June 1. It occurred to me that a more accurate measure might be had of the latitude in which the greatest curvature takes place, by setting the fixed thread of the micrometer to the direction of the ring of Saturn, which may be done with great accuracy. The two following measures were taken in this manner, and are more satisfactory than I had taken before. The first gave the latitude of the south-preceding point of greatest curvature $43^\circ\ 26'$; and the second $43^\circ\ 13'$. A mean of the two will be $43^\circ\ 20'$.

June 2. I viewed Jupiter and Saturn alternately with a magnifying power of only 300, that the convexity of the eye-glass might occasion no deception, and found the form of the two planets to differ in the manner that has been described.

With 200 I saw the difference very plainly; and even with 160 it was sufficiently visible to admit of no doubt. These low powers show the figure of the planets perfectly well, for as the field of view is enlarged, and the motion of the objects in passing it lessened, we are more at liberty to fix our attention upon them.

I compared the telescopic appearance of Saturn with a figure drawn by the measures I have taken, combined with the proportion between the equatorial and polar diameters determined in the year 1789; * and found that, in order to be a perfect resemblance, my figure required some small reduction of the longest diameter, so as to bring it nearly to agree with the measures taken the 27th of May When I had made the necessary alteration, my artificial Saturn was again compared with the telescopic representation of the planet, and I was then satisfied

* See *Phil. Trans.* for 1790, page 17 [Vol. I. p. 380].

that it had all the correctness of which a judgment of the eye is capable. An exact copy of it is given in the figure [p. 333]. The dimensions of it in proportional parts are,

The diameter of the greatest curvature	36
The equatorial diameter	35
The polar diameter	32
Latitude of the longest diameter	43° 20′.

The foregoing observations of the figure of the body of Saturn will lead to some intricate researches, by which the quantity of matter in the ring, and its solidity, may be in some measure ascertained. They also afford a new instance of the effect of gravitation on the figure of planets ; for in the case of Saturn, we shall have to consider the opposite influence of two centripetal and two centrifugal forces : the rotation of both the ring and planet having been ascertained in some of my former Papers.

LVI.

On the Quantity and Velocity of the Solar Motion.

[*Phil. Trans.*, 1806, pp. 205–237.]

Read February 27, 1806.

THE direction of the solar motion having been sufficiently ascertained in the first part of this Paper,* we shall now resume the subject, and proceed to an inquiry about its velocity.

The proper motions, when reduced to one direction, have been called quantities, to distinguish them from the velocities required in the moving stars to produce those motions. It will be necessary to keep up the same distinction with respect to the velocity of the solar motion; for till we are better acquainted with the parallax of the earth's orbit, we can only come to a knowledge of the extent of the arch which this motion would be seen to describe in a given time, when seen from a star of the first magnitude placed at right angles to the motion. There is, however, a considerable difference between the velocity of the solar motion and that of a star; for at a given distance, when the quantity of the solar motion is known its velocity will also be known, and every approximation towards a knowledge of the distance of a star of the first magnitude will be an approximation towards the knowledge of the real solar velocity; but with a star it will be otherwise; for though the situation of the plane in which it moves is given, the angle of the direction of its motion with the visual ray will still remain unknown.

As hitherto we have consulted only those proper motions which have a marked tendency to a parallactic centre, we ought now, when the question is to determine the velocity of the solar motion, to have in view the real motion of every star whose apparent motion we know; for as it would not be proper to assign a motion to the sun, either much greater or much less than any real motion which may be found to exist in some star or other, it follows that a general review of proper motions ought to be made before we can impartially fix on the solar velocity; but as trials with a number of stars would be attended with considerable inconvenience, I shall use only our former six in laying down the method that will be followed with all the rest.

* *Phil. Trans.* for 1805, page 233 [above, p. 317].

Proportional Distance of the Stars.

We are now come to a point no less difficult than essential to be determined. Neither the parallactic nor real motion of a star can be ascertained till its relative distance is fixed upon. In attempting to do this it will not be satisfactory to divide the stars into a few magnitudes, and suppose *these* to represent the relative distances we require. There are not perhaps among all the stars of the heavens any two that are exactly at the same distance from us; much less can we admit that the stars which we call of the first magnitude are equally distant from the sun. And indeed, if the brightness of the stars is admitted as a criterion by which we are to arrange them, it is perfectly evident that all those of the first magnitude must differ as much in distance as they certainly do in lustre; yet imperfect as this may be, it is at present the only rule we have to go by.

The relative brightness of our six stars, may be expressed as follows: Sirius – – – Arcturus – Capella -, Lyra – – Aldebaran . Procyon.

The notations here used are those which have been explained in my first Catalogue of the relative Brightness of the Stars; * but to denominate the magnitudes of these six stars so that they may with some probability represent the distances at which we should place them according to their relative brightness, I must introduce a more minute subdivision than has been commonly admitted, by using fractional distinctions, and propose the following arrangement.

TABLE VIII.

Proportional Distances of Stars.

Star	Distance	Star	Distance
Sirius	1·00	Lyra	1·30
Arcturus	1·20	Aldebaran	1·40
Capella	1·25	Procyon	1·40

The interval between Sirius and Arcturus is here made very considerable; but whoever will attentively compare together the lustre of these two stars, when they are at an equal altitude, must allow that the difference in their brightness is fully sufficient to justify the above arrangement.

The order of the other four stars is partly a consequence of the distance at which Arcturus is placed, and of the comparative lustre of these stars such as it has been estimated by observations. But if it should hereafter appear that other more exact estimations ought to be substituted for them, the method I have pursued will equally stand good with such alterations. I have tried all the known, and many new ways of measuring the comparative light of the stars, and though I have not yet found one that will give a satisfactory result, it may still be possible to discover some method of mensuration preferable to the foregoing estimations, which are only the result of repeated and accurate comparisons by the eye. When-

* *Phil. Trans.* for 1796, page 189 [Vol. I. p. 543].

ever we are furnished with more authentic data the calculations may then be repeated with improved accuracy.

Effect of the Increase and Decrease of the Solar Motion, and Conditions to be observed in the Investigation of its Quantity.

The following Table, in which the 2d, 4th, and 5th columns contain the sides of the parallactic triangle, is calculated with a view to show that an increase or decrease of the solar motion will have a contrary effect upon the required real motions of different stars; and as we are to regulate the solar velocity by these real motions, an attention to this circumstance will point out the stars which are to be selected for our purpose.

TABLE IX.

Stars and relative Distances.	Apparent Motion.	Solar Motion.	Parallactic Motion.	Real Motion.	Velocities.
Sirius 1·00	″ 1·11528	1·0 1·5 2·0	″ 0·67768 1·01652 1·35536	″ +0·46518 +0·21701 −0·32776	465175 217007 327755
Arcturus 1·20	2·08718	1·0 1·5 2·0	0·53579 0·80368 1·07158	+1·57389 +1·30478 +1·01561	1888670 1565735 1218736
Capella 1·25	0·46374	1·0 1·5 2·0	0·79593 1·19390 1·59186	−0·42159 −0·79637 −1·18662	526987 995465 1483270
Lyra 1·30	0·32435	1·0 1·5 2·0	0·32542 0·48812 0·65083	−0·47065 −0·59923 −0·74135	611839 778995 963750
Aldebaran 1·40	0·12341	1·0 1·5 2·0	0·65117 0·97676 1·30234	−0·53208 −0·85737 −1·18283	744913 1200324 1655967
Procyon 1·40	1·23941	1·0 1·5 2·0	0·66394 0·99591 1·32788	+0·59548 +0·30731 −0·23385	833665 430227 327390

The real motion of Arcturus contained in the 5th column compared with that of Aldebaran, shows that when the solar motion is increased from 1·0 to 1·5 and to 2″·0 the real motion of Arcturus will be gradually diminished from 1·57 to 1·30 and to 1″·02, while that of Aldebaran undergoes a contrary change from 0·53 to 0·86 and to 1″·18. We may also notice that Capella and Aldebaran, which have a negative sign prefixed to their real motions when the solar motion is 1″·0 are affected differently from Arcturus, Sirius, and Procyon, which have a positive sign; and that even the motions of the two last become negative when the solar motion is increased beyond a certain point. It may be easily understood that the motion of Arcturus itself would become negative were we to increase the solar motion till the parallactic motion of this star should exceed its apparent motion.

From these considerations it appears, that a certain equalization, or approach to equality may be obtained between the motions of the stars, or between that of the sun and any one of them selected for the purpose; for instance, the motions of Arcturus and Aldebaran being contrary to each other, may be made perfectly equal by supposing the sun's annual motion to be 1″·85925. For then we shall have the real annual motion of Arcturus towards the parallactic centre 1″·091, and that of Aldebaran towards the opposite part of the heavens, in which the solar apex is placed, will be 1″·091 likewise; the first in a direction 55° 29′ 39″ south-preceding, the latter 88° 16′ 31″ north-following their respective parallels; and a composition of these motions with the parallactic ones arising from the given solar motion, will produce the apparent motions of these stars which have been established by observation. But since Arcturus, by the hypothesis which has been adopted in Table VIII. is a nearer star than Aldebaran, the velocities of the real motions, describing these equal arches will be 1309109 in the former and 1527780 in the latter. And it is not the arches but these velocities that must be equalized. Therefore, in order to have this required equality, let the solar motion be 1″·718865, then will a velocity of 1399478 in Arcturus, and 1399842 in Aldebaran, which are sufficiently equal, occasion such angular real motions in the two stars as will bring them, when compounded with their parallactic motions, to the apparent places in which we find them by observation.

Before we proceed, it will be proper to obviate a remark that may be made against this way of equalization or approach to equality. We have said that the calculated velocities are such as would be true if the stars were at the assumed distances, and if their real motions were performed in lines at right angles to the visual ray; to which it may here be objected that the last of these assumptions is so far from having any proof in its favour that even the highest probability is against it. We may admit the truth of what the objection states, without apprehending that any error could arise on that account, if the solar motion were determined by this method. For if the stars do not move at right angles to the visual ray, their real velocity will exceed the calculated one; so that in the first place we should certainly have the minimum of their velocities: and if we were obliged, for want of data to leave the other limit of the motion unascertained, it must be allowed to be a considerable point gained if we could show what is likely to be the least velocity of the solar motion; but a more satisfactory defence of the method is, that if we were to assume a mean of all the angular deviations from the perpendicular to the visual ray that may take place in the directions of the real motions of the stars, the only position we could fix upon as a mean would be an inclination of 45 degrees. For in this case the chance of a greater or smaller deviation would be equal; and when a number of stars are taken, the deviations either way might then be supposed to compensate each other; but what is chiefly to our purpose, not only the angle of 45 degrees, but also any other, that might be fixed upon as a proper one to represent the mean quantity of sidereal motions, would lead exactly to the same result of the solar velocity to be investigated. For if the velocities of any two stars were equalized, when their motions are supposed to be perpendicular to the visual ray, they would be as much so when they make any other given angle with it; and it is the equalization or approach to equality and not the quantity of the velocities that is the spirit of this method. I have only to add, that an equalization of the solar motion with that of any star selected for the purpose may be had by a direct method of calculation, and will therefore be of great use in settling the rate of the motion to be determined.

It must be evident from what has been said, that a certain mean rate, or middle rank, should be assigned to the motion of the sun, unless very sufficient reasons should induce us to depart from this condition. To obtain this end must consequently be our principal aim; and if we can at the same time bring the sidereal motions to a greater equality among each other, it will certainly be a very proper secondary consideration.

There are two ways of taking a mean of the sidereal motions, one of them may be called the rate and the other the rank. For instance, a number equal to the mean rate of the six numbers, 2, 6, 13, 15, 17, 19, would be 12; but one that should hold a middle rank between the three highest and three lowest of the six would be 14. In assigning the rate of the solar motion it appears to be most eligible that it should hold a middle rank among the sidereal velocities. We shall however find that nearly the same result will be obtained from either of the methods.

With respect to our second consideration, we may see that it also admits of a certain modification by the choice of the solar motion; for in Table IX. when this motion is 1″·5 the velocity of Arcturus 1565735, will exceed that of Sirius, 217007, more than seven times; whereas a solar motion of 1″ will give us the proportional velocities of these stars as 188867 to 465174; and the former will then exceed the latter only four times.

Calculations for drawing Figures that will represent the observed Motions of the Stars.

The necessary calculations for investigating the solar motion are of considerable extent, and may be divided into two classes, the first of which will remain unaltered whatsoever be the solar motion under examination, while the other must be adjusted to every change that may be required.

The direction of the sun remaining as it has been settled in the first part of this Paper, the permanent computation of each star will contain the annual quantity of the observed or apparent motion, its direction with the parallel of the star, its direction with the parallactic motion, and its velocity. The changeable part will consist of the angular quantity of the real motion, the parallactic direction of this motion, and its velocity.

Before we can make a calculation of the required velocities, we must fix upon the probable relative distance of the rest of the stars, in the same manner as we have done with the first six. In this I have thought it advisable to distinguish the stars that, from their lustre, may be called principal, and have limited their extent to the brightest of the second magnitude, on account of the uncertainty which still remains about their progressive distances. For though it appears reasonable to allow that the bright stars of the second magnitude may be twice as far from us as those of the first, it will admit of some doubt whether this rule ought to be strictly followed up to the 3d, 4th, 5th, and 6th magnitude; especially when it is not easy to ascertain the boundaries which should limit the magnitudes of very small stars.

The number of these principal stars is 24. The remaining 12 are also arranged by admitting that their magnitudes express their relative distances; and notwithstanding the doubtfulness we have noticed their testimony with respect to the proper quantity of a solar motion, though it should be received with some diffidence, must not be neglected; some considerable alteration in their supposed distances, however, would have but little effect upon the conclusions intended to be drawn from their velocities.

The following Table contains the result of the calculations that relate to the permanent quantities. In the first and second columns, we have the names of the stars, and their assigned relative distances. The third gives the apparent angular motions, and the fourth their direction. The fifth contains the direction of the same motions, with respect to the parallactic motions arising from the given solar direction; and the sixth gives the velocity of the stars which produce the quantity of the apparent motions.

The contents of this Table will enable us to examine the motions of the stars in different points of view. For instance, by the apparent motions in the third column, and their directions in the fourth, a figure may be drawn which will represent the actual state of the heavens, with respect to those annual changes in the situations of our 36 stars, which in astronomical tables are called their proper motions.

Fig. 1 gives us these motions brought into one view, so that by supposing successively every one of the stars to be represented by the central point of the figure, we may see the angular quantity and direction of the several annual proper motions represented by the line which is drawn from the centre to each star. By this means we have the comparative arrangement and quantity of these movements with respect to their directions.

TABLE X.

Names of the Stars.	Proportional Distances.	Apparent Motions.	Direction with the Parallel.	Direction with the parallactic Motion.	Velocity of the Stars.
		"	° ′ "	° ′ "	
Sirius . . .	1·00	1·11528	68.49.40·7 *sp*	10.24.44·3 *sf*	1115281
Arcturus . . .	1·20	2·08718	55.29.42·0 *sp*	0. 0. 3 *sp*	2504621
Capella . . .	1·25	0·46374	71.35.22·4 *sf*	24.40.21 *sf*	579668
Lyra . . .	1·30	0·32435	56.20.57·3 *nf*	92.49.30 *nf*	421657
Rigel . . .	1·35	0·16273	79.29.33·9 *np*	159.28. 1 *np*	219684
α Orionis . . .	1·35	0·13038	85.38.14·6 *nf*	169.18.58 *np*	176010
Procyon . . .	1·40	1·23941	50. 2.24·5 *sp*	9.40.46 *sp*	1735172
Aldebaran . .	1·40	0·12341	76.29.37·3 *sf*	13.41.48 *sf*	172778
Pollux . . .	1·42	0·65037	0. 0. 0 *prec.*	61.30.34 *sp*	923523
Spica . . .	1·44	0·19102	84. 5. 1·8 *np*	144.13.16 *np*	275065
Antares . . .	1·46	0·26000	90. 0. 0 *north*	178.57.44 *np*	379600
Altair . . .	1·47	0·71912	48.40.12·0 *nf*	103.17.29 *nf*	1057105
Regulus . . .	1·48	0·22886	20.27.37·5 *np*	70. 9.20 *sp*	338711
β Leonis . . .	1·50	0·55324	7.16. 8·4 *sp*	40.34.31 *sp*	829856
β Tauri . . .	1·50	0·10039	84.58.27·1 *sf*	13.17.11 *sf*	150579
Fomalhaut . .	1·50	0·30698	11.16.16·3 *nf*	16.47. 5 *sf*	460469
α Cygni . . .	1·60	0·06440	27.45.56·3 *np*	177.31.39 *np*	103036
Castor . . .	2·00	0·13294	17.30.40·6 *sp*	45.25.43 *sp*	265869
α Ophiuchi . .	2·00	0·07698	40.30.24·8 *sf*	33.29.28 *sf*	153955
α Coronæ . . .	2·00	0·23279	7.24.15·4 *sf*	105. 0.43 *nf*	465587
α Aquarii . . .	2·00	0·20615	67.10.17·1 *np*	162.43.46 *nf*	412295
α Andromedæ . .	2·00	0·09268	40.20.48·2 *sf*	12.55.11 *sf*	185360
α Serpentis . .	2·00	0·21913	60. 7.12·5 *nf*	161.34. 4 *nf*	438257
α Pegasi . . .	2·00	0·18917	72. 5.16·0 *np*	157.45.25 *nf*	378338
α Hydræ . . .	2·30	0·16598	57.30.24·8 *np*	107. 6.24 *np*	381763
α^2 Libræ . . .	2·40	0·18376	54.42.52·9 *np*	127. 3. 7 *np*	441022
γ Pegasi . . .	2·50	0·17355	59.48. 7·9 *np*	174. 5.15 *nf*	433880
α Arietis . . .	2·50	0·11587	37. 9.15·9 *sf*	29.32.47 *sf*	289685
α Ceti . . .	2·80	0·14406	33.44. 2·9 *np*	141.18.55 *np*	403356
α Herculis . .	3·00	0·23000	90. 0. 0 *north*	168.23.41 *nf*	690000
β Virginis . .	3·00	0·77706	17.59.25·5 *sf*	111.11.44 *nf*	2331169
γ Aquilæ . . .	3·00	0·19320	55.54.41·7 *np*	178.25.20 *nf*	579589
α^2 Capricorni . .	3·50	0·26452	79.23.35·3 *nf*	136.21.18 *nf*	925819
β Aquilæ . . .	4·00	0·35127	85. 7.37·0 *sp*	39.49.15 *sp*	1405079
α′ Capricorni . .	4·20	0·28000	90. 0. 0 *north*	146.59.44 *nf*	1176000
α′ Libræ . . .	6·00	0·20898	59.27.58·4 *np*	131.46. 7 *np*	1253875

Fig. 3 represents the same motions, but instead of being drawn so as to show their directions with regard to the several meridians and parallels of the stars, they are laid down by the angles contained in the fifth column ; and will therefore indicate their arrangement with respect to a line drawn from the solar apex towards the parallactic centre. These directions will remain the same, whatever may be the velocity of the solar motion upon which we shall ultimately fix, provided no change be made in the situation of the apex towards which the sun has been admitted to move.

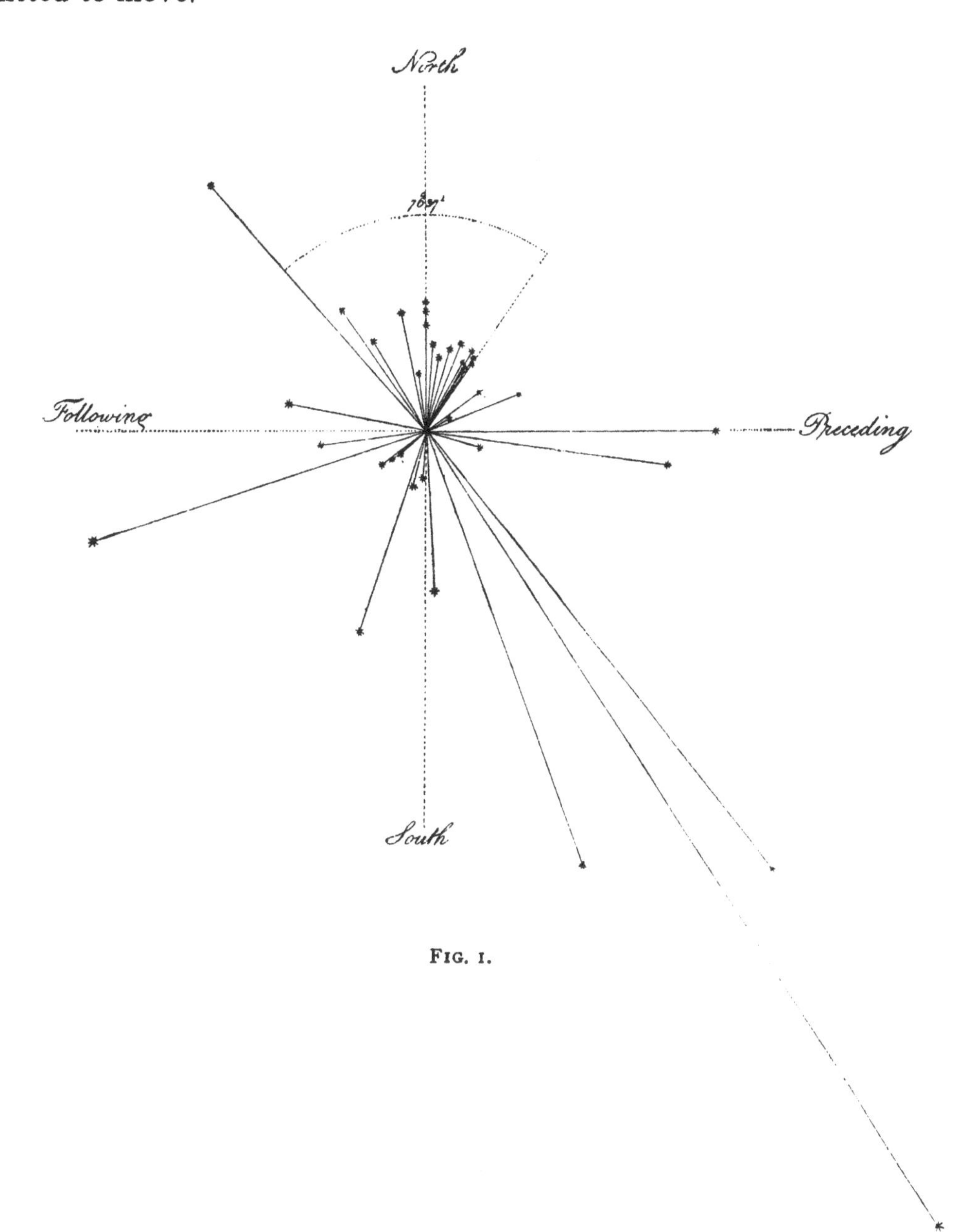

FIG. I.

In these two figures, the lines drawn from the centre give us only the angular changes of the places that have been either observed or calculated, and not the velocities which are required in the stars to produce them. It will therefore be necessary to represent the velocities by two other figures, in which the same directions are preserved, but where the extent of each line is made proportional to the distance of the stars in the second column.

Fig. 2 is drawn according to this plan; the angles of the directions remain as in the fourth column, but the lines are lengthened so as to give us the velocities contained in the sixth.

In Fig. 4, the angles of the 3d figure are preserved, but the lines are again lengthened as in Fig. 2.

N.B. These two last figures would have been of an inconvenient size if they had been drawn on the same scale with the two foregoing ones, for which reason,

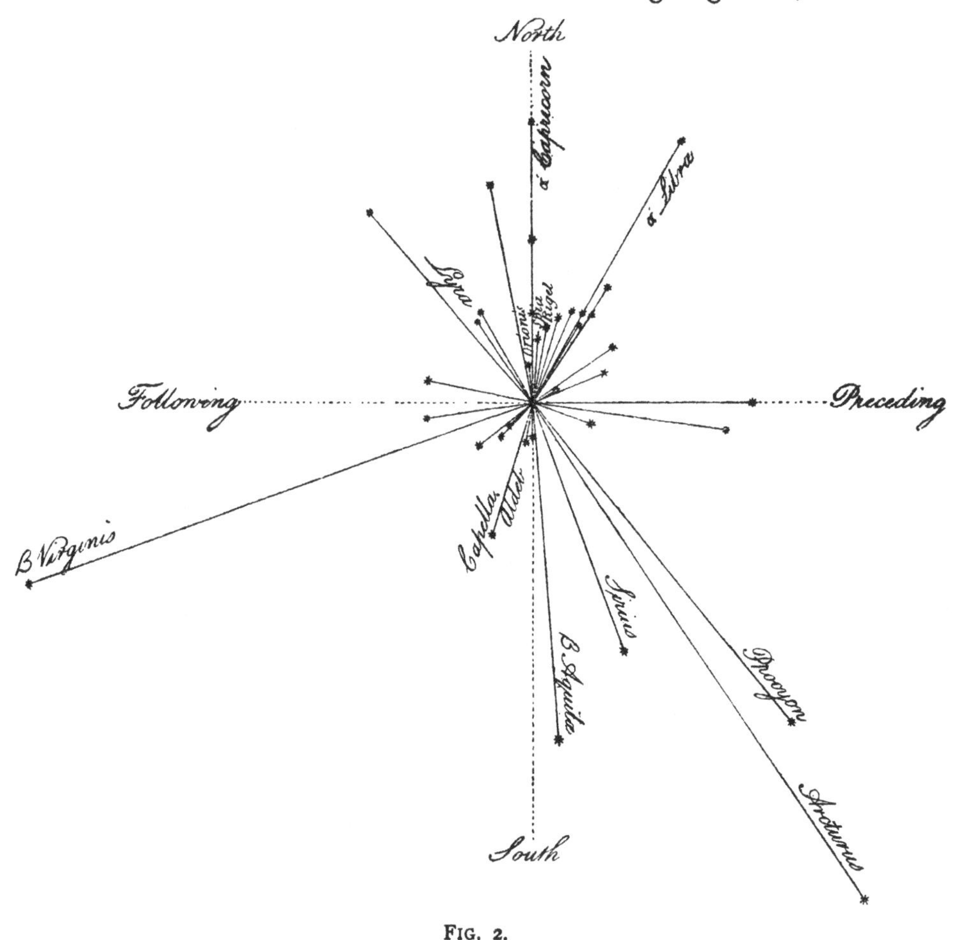

FIG. 2.

in comparing the 2d and 4th with the 1st and 3d, it must be remembered that the former are reduced to one half of the dimensions of the latter.

Remarks on the sidereal Motions as they are represented from Observation.

As we have now before us a set of figures which give a complete view of the result of the calculations contained in the Xth Table, we may examine the arrangements of the stars, and draw a few conclusions from them, that will throw some light upon the subject of our present inquiry.

In the first place, then, we have to observe in Fig. 1, that 17 out of the 21 stars, whose motions are directed towards the north, are crowded together into a compass of little more than $76\frac{1}{2}$ degrees. But this figure, as we have shown, is drawn from observation. We are consequently obliged to conclude, that, if these motions are the real ones, there must be some physical cause which gives a bias to the directions in which the stars are moving; if so, it would not be improbable that the sun, being situated among this group of stars, should partake of a motion towards the same part of the heavens.

Our next remark concerns the velocity of the sidereal motions; and therefore we must have recourse to Fig. 2, where we perceive that the greatest motions are not confined to the brightest stars. For instance, the velocity of β Virginis is but little inferior to that of Arcturus, and exceeds the velocity of Procyon. Likewise the velocities of β Aquilæ, α′ Libræ, and α′ Capricorni, surpass that of Sirius; and an inspection of the rest of the figure will be sufficient to show how very far the velocities of Capella, Lyra, Rigel, α Orionis, Aldebaran, and Spica, are exceeded by those of many other stars.

If we look at the arrangement of the stars with respect to the direction of the solar motion, we find in Fig. 3, that a somewhat different scattering of them has taken place; but still most of the stars appear to be affected by some cause which tends to lead them to the same part of the heavens, towards which the sun is moving; and the directions of the greatest number of them are not very distant from the line of the solar motion.

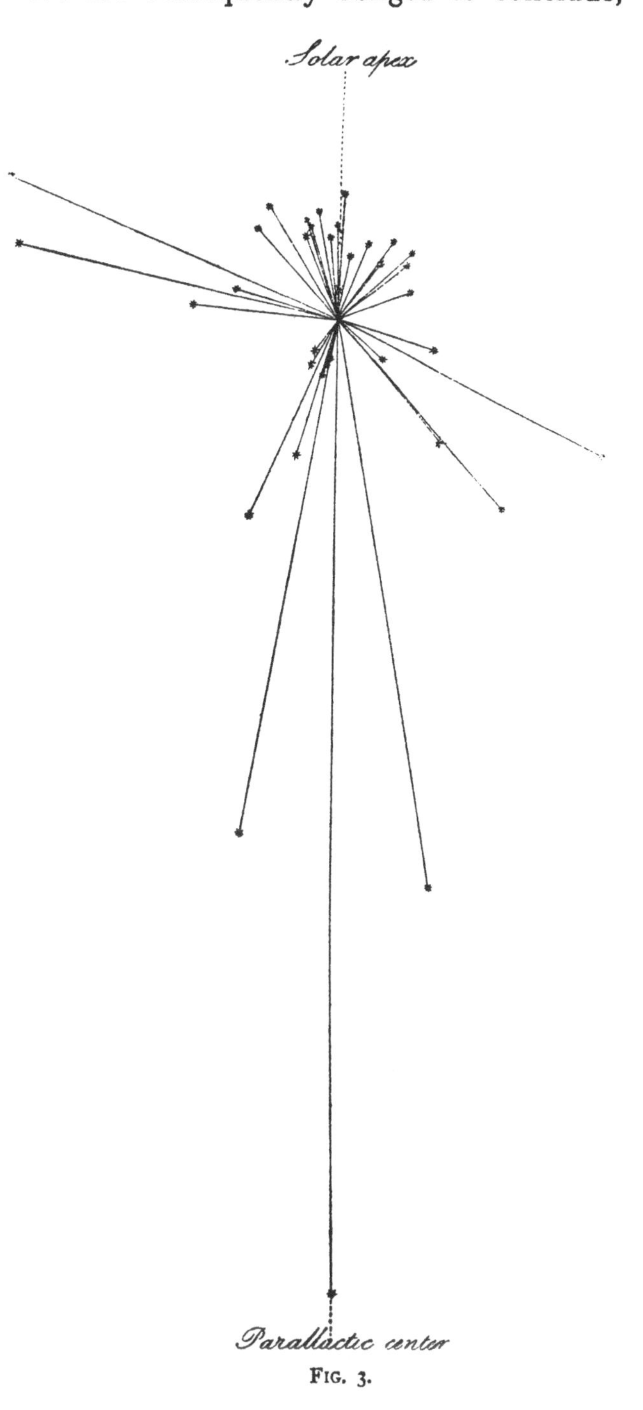

FIG. 3.

The whole appearance of this figure presents us with the idea of a great compression above the centre, arising from some general cause, and a still greater expansion in the lower part of it. The considerable projection of a few stars on both sides, is however a plain indication that the compressing or dilating cause does not act in their directions.

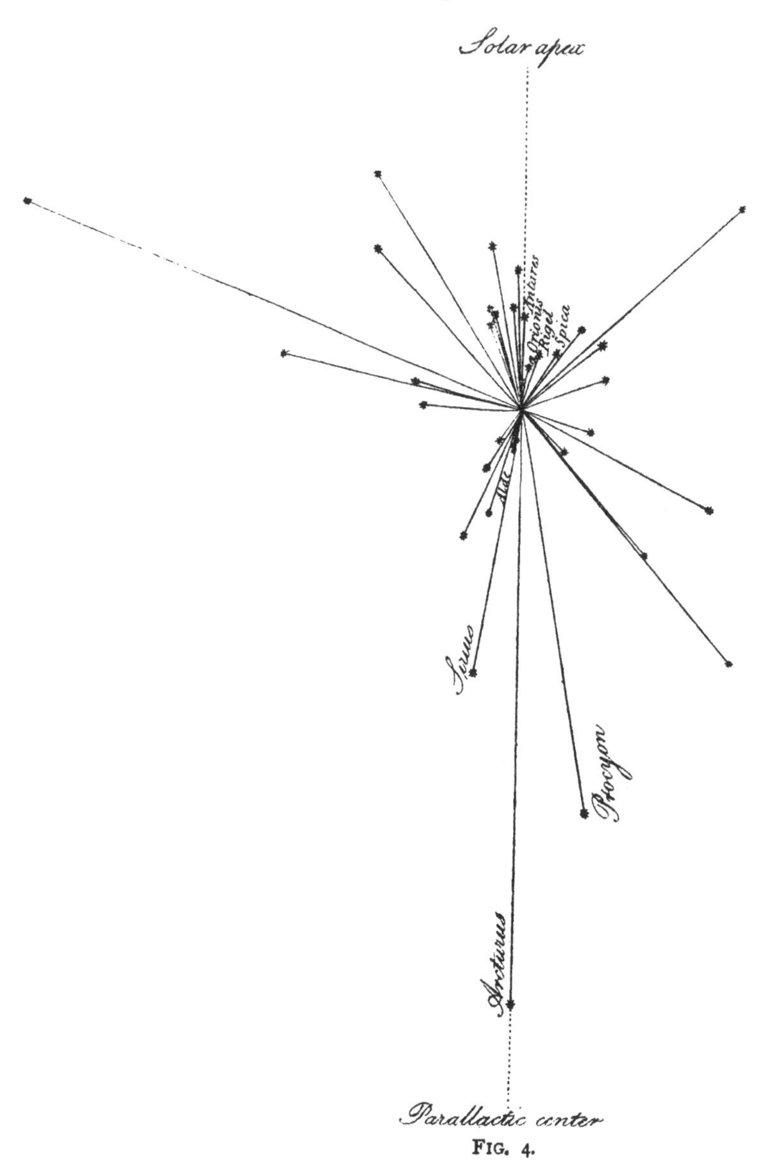

Fig. 4.

When the velocity of the stars, represented in the same point of view in Fig. 4, is examined, we find a particularity in the direction and comparative velocities in the largest stars that must not be overlooked. Four of them, Rigel, α Orionis, Spica, and Antares, have a motion towards that part of the heavens in which the solar apex is placed, and their motions are very slow. Three other stars of the 1st magnitude, Arcturus, Procyon, and Sirius, move towards the opposite part of the heavens, and their motions, on the contrary, are very quick.

The direction of the motion of Aldebaran, compared with its small velocity, is no less remarkable; and seems to be contrary to what has been pointed out with the three last mentioned stars; we shall however soon have an opportunity of showing that it is perfectly consistent with the principles of the solar motion.

The Solar Motion and its Direction assigned in the first Part of this Paper are confirmed by the Phenomena attending the observed Motions of the 36 Stars.

An application of some of the foregoing remarks will be our next subject; and I believe it will be found, that in the first place they point out the expediency of a solar motion. That next to this they also direct us to the situation of the apex of this motion: and lastly that they will assist us in finding out the quantity requisite for giving us the most satisfactory explanation of the phenomena of the observed proper motions of the stars.

In examining the second figure, it has been shown that no less than six stars of the first magnitude, namely, Capella, Lyra, Rigel, α Orionis, Aldebaran, and Spica, have less velocity than nine or ten much smaller stars. Aldebaran and α Orionis indeed have so little motion that there are but three stars in all the 36 that have less. But the situation of these bright stars, from their nearness, must be favourable to our perceiving their real motions if they had any, unless they were counteracted by some general cause that might render them less conspicuous. Now to suppose that the largest stars should really have the smallest motions, is too singular an opinion to be maintained ; it follows, therefore, that the apparently small motions of these large stars is owing to some general cause, which renders at least some part of their real motion invisible to us. But when a solar motion is introduced, the parallax arising from that cause will completely account for the singularity of these slow motions.

If the foregoing argument proves the expediency of a solar motion, its direction is no less evidently pointed out by it. For if the parallax occasioned by the motion of the sun is to explain the appearances that have been remarked, it will follow, that a direction in opposition to the motion of Arcturus, will answer that end in the most satisfactory manner. That compression, for instance, which has been remarked in the motions of the stars moving toward the solar apex in Fig. 3, and which is so completely accounted for by a parallactic motion arising from the motion of the sun, points out the direction in which the sun should move, in order to produce this required parallactic motion. The expansion of the motions that are in opposition to the former is evidently owing to the same parallactic motions, which in this direction unite with the real motions of the stars ; and as, in the former case, the observed motions are the differences between the parallactic and real motions, so here they are the sum of them.

The remark that stars having a side motion, are not affected by the cause of the compression or expansion, which acts upon the rest, is perfectly explained ; for a parallactic motion, in the direction of the motion of Arcturus, can have no effect in lengthening or shortening the perpendicular distance to which a star may move in a side direction.

I have only to add, that the small velocities of Rigel, α Orionis, Spica, and Antares, in Fig. 4, as well as the great velocities of Arcturus, Procyon, and Sirius, point out the same apex which in the first part of this Paper has already been established by more extended computations.

The case of Aldebaran, though seemingly contrary to what has been shown, confirms the same conclusions. This will appear by considering that a star, moving towards the solar apex with a greater real motion than its parallactic one, must continue apparently to move in its real direction ; but should a star, such as

Aldebaran, move towards the apex with less velocity than the parallactic motion which opposes it, there will arise a change of direction, and the star will be seen moving towards the opposite part of the heavens.

Trial of the Method to obtain the Quantity of the Solar Motion by its Rank among the sidereal Velocities.

According to the conditions that have been explained, a calculation may be made with a view of equalizing the velocities of the sun and the star α Orionis; and the result of it will show that the proposed equality will be obtained when the solar motion is 1″·266230. It will moreover be found that so small an increase of this motion as 0″·01 would give us 19 stars with less, and 17 with more velocity than that which the calculation assigns to the sun; this consequently fixes one of the limits to which the solar motion ought not to come up, if we intend it should hold a middle rank among the sidereal velocities.

On the other hand, by a similar calculation of the velocities of the star Pollux and the sun, it appears that a solar motion of 0″·967754 will make them equal; and that a diminution of this motion not exceeding 0″·01 would give us 19 stars moving at a greater rate than the sun, and only 17 falling short of its velocity. This consequently fixes the other limit to which the solar motion ought not to be depressed. And thus it appears by this method, that the quantity we are desirous of ascertaining, is confined within very narrow bounds, and that by fixing upon a mean of the two limits, we may have the rank of the solar motion true to less than 0″·15.

Calculations for investigating the Consequences arising from any proposed Quantity of Solar Motion, and for delineating them by proper Figures.

Before we can justly examine the real motions of stars which it will be necessary to admit in consequence of a given solar motion, it will be convenient to have them represented in two figures that we may see their arrangement and extent; and as a calculation of the required particulars will oblige us to fix upon a certain quantity, we shall take the motion that has been ascertained to belong to the middle rank of the sidereal velocities for a pattern. The result of the necessary calculations will be found in Table XI.

By the contents of this Table, Fig. 5 is drawn with the lines contained in the third column and the angles of the fourth; the scale of it is that of the 1st and 3d figures; and it represents the directions and angular quantities of the real motions that are required to compound with the parallactic effects of the second column, so as to produce those annual proper motions which are established by observation.

Fig. 6 is drawn on the reduced scale of the 2d and 4th figures. The lines make the same angles with the direction of the solar motion as before, but their lengths are in the proportion of the velocities contained in the last column.

TABLE XI.

Names.	Parallactic Motion.	Real Motion.	Parallactic Angle.	Velocity.
	"	"	° ′ ″	
Sun	0·00000	1·116992	00.00.00	1116992
Sirius	0·75697	0·395212	149.20. 6 *sf*	395212
Arcturus	0·59847	1·488713	179.59.55·7 *sp*	1786455
Capella	0·88905	0·506123	22.29.12·5 *nf*	632654
Lyra	0·36349	0·498949	40.29.14 *nf*	648634
Rigel	0·55470	0·709381	4.36.52 *np*	957665
α Orionis	0·71410	0·842559	1.38.38 *np*	1137455
Procyon	0·74161	0·523428	156.32.21 *sp*	732799
Aldebaran . . .	0·72736	0·608148	2.45.15 *nf*	851407
Pollux	0·78643	0·743971	50.12.11 *np*	1056439
Spica	0·74009	0·902004	7. 6.44 *np*	1298886
Antares	0·74110	1·000835	0.16.10·5 *np*	1461219
Altair	0·64544	1·071042	40.48. 4 *nf*	1574431
Regulus	0·75095	0·706833	17.43.53 *np*	1046113
β Leonis	0·68003	0·443842	54.10.14·5 *np*	665763
β Tauri	0·73063	0·633317	2. 5.15·5 *nf*	949976
Fomalhaut . . .	0·66693	0·383414	13.22. 5·5 *nf*	575121
α Cygni	0·46516	0·529503	0.18. 2·2 *np*	847204
Castor	0·55841	0·474647	11.30.32 *np*	949293
α Ophiuchi . . .	0·35202	0·290934	8.23.43 *nf*	581869
α Coronæ	0·23427	0·370580	37.21.17 *nf*	741160
α Aquarii	0·55743	0·756754	4.38.19·5 *nf*	1513508
α Andromedæ . . .	0·55389	0·464035	2.33.34 *nf*	928071
α Serpentis . . .	0·38655	0·598458	6.38.54 *nf*	1196917
α Pegasi	0·55567	0·734265	5.35.47·5 *nf*	1468530
α Hydræ	0·46554	0·538281	17. 8.26 *np*	1238046
α² Libræ	0·43377	0·563892	15. 4.29 *np*	1353342
γ Pegasi	0·44540	0·618272	1.39.27 *nf*	1545679
α Arietis	0·43893	0·342934	9.35.29·5 *nf*	857336
α Ceti	0·33271	0·454165	11.26. 5·5 *np*	1271662
α Herculis . . .	0·21909	0·446795	5.56.38·5 *nf*	1340388
β Virginis . . .	0·36039	0·967572	48.29. 2·5 *nf*	2902716
γ Aquilæ	0·30898	0·502168	0.36.25 *nf*	1506503
α² Capricorni . . .	0·31390	0·537285	19.51.52·5 *nf*	1880497
β Aquilæ	0·24370	0·226458	96.36.59·5 *sp*	905830
α′ Capricorni . . .	0·26151	0·519230	17. 4.54·5 *nf*	2180769
α′ Libræ	0·17347	0·349371	26.29.44·5 *np*	2096229

Remarks that lead to a necessary Examination of the Cause of the sidereal Motions.

The first particular that will strike us when we cast our eye on Fig. 5, is the uncommon arrangement of the stars. It seems to be a most unaccountable circumstance that their real motions should be as represented in that figure; indeed, if we except only ten of the stars, all the rest appear to be actuated by the same influence, and, like faithful companions of the sun, to join in directing their motions towards a similarly situated part of the heavens.

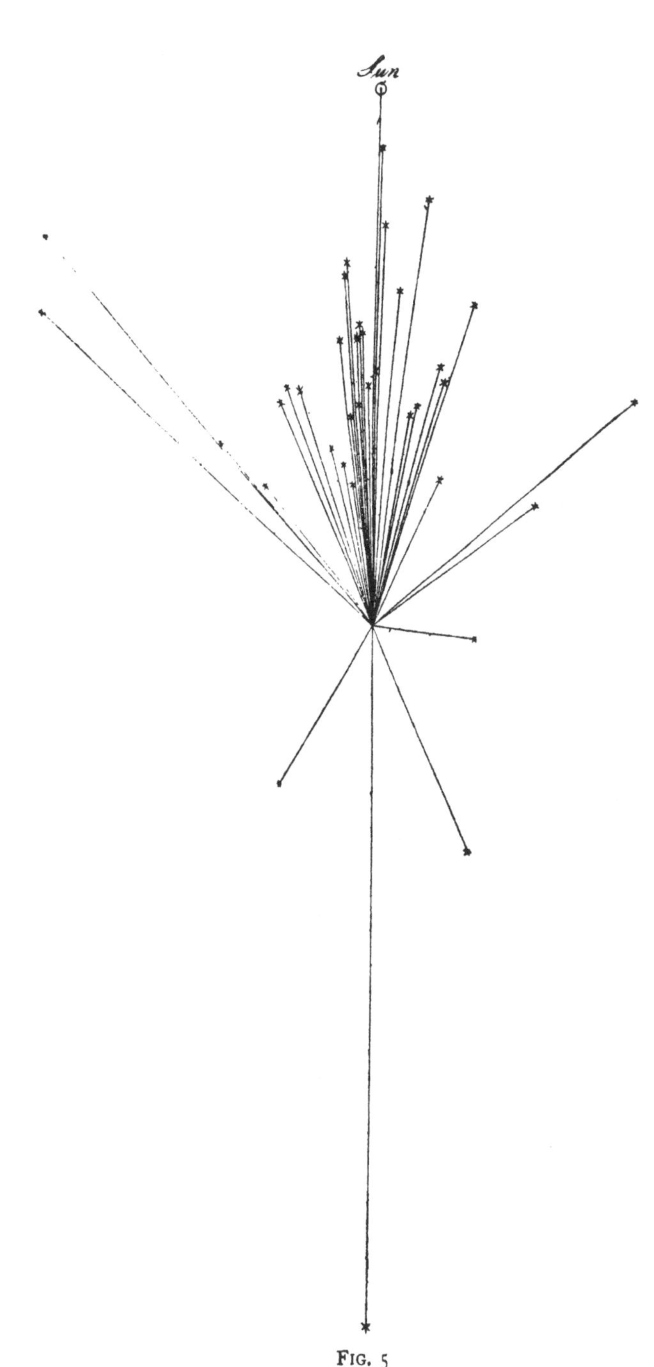

FIG. 5

This singularity is too marked not to deserve an examination; for unless a cause for such peculiar directions can be shown to exist, I do not see how we can reconcile them with a certain equal distribution of situations, quantities, and motions, which our present investigation seems to demand. In order to penetrate as far as we can into this intricate subject, we shall take a general view of the causes of the motions of celestial bodies.

A motion of the stars may arise either from their mutual gravitation towards each other, or from an original projectile force impressed upon them. These two causes are known to act on all the bodies belonging to the solar system, and we may therefore reasonably admit them to exert their influence likewise on the stars. But it will not be sufficient to know a general cause for their motions, unless we can show that its influence will tend to make them go towards a certain part of the heavens rather than to any other. Let us examine how these causes are acting in the solar system.

The projectile motions of the planets, the asteroids, and the satellites, excepting those of the Georgium Sidus, are all decidedly in favour of a marked singularity of direction. We may add to them the comet of the year 1682, whose regular periodical return in 1759 has sufficiently proved its permanent connection with the solar system. Here then we have not less than 23 various bodies belonging to the solar system to show that this cause not only can, but in the only case of which we have a complete knowledge, actually does influence the celestial motions, so as to give them a very particular appropriate direction. Even the exception of the Georgian satellites may be brought in confirmation of the same peculiarity; for though they do not unite with the rest of the bodies of our system, they still conform among each other to establish the same tendency of a similar direction in their motion round the primary planet. And thus it is proved that the similar direction of the motion of a group of stars may be ascribed to their similar projectile motions without incurring the censure of improbability.

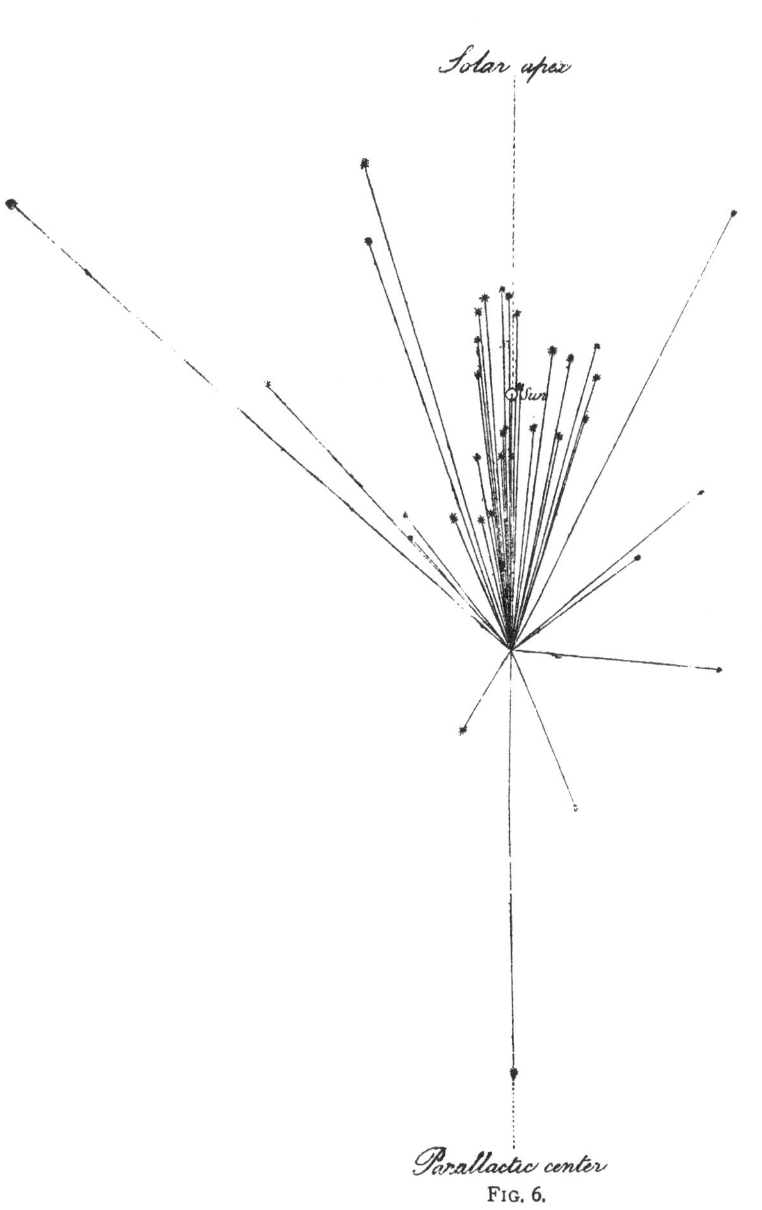

FIG. 6.

Let us however pursue the objection a little farther, and as we have shown that the celestial bodies of the solar system actually have these similar projectile motions, it may be required that we should also prove that the stars have them likewise; since the appearances in Fig. 5 may otherwise be looked upon as merely the consequence of the assumed solar motion. To this I answer, that setting aside the solar motion, and allowing the observations of astronomers on the proper motion of the stars to give us the real direction and angular quantity of these motions, even then the same similarity will equally remain to be accounted for. In my examination of Fig. 1 and 3, it has been shown that we ought to ascribe the similar directions of the sidereal motions to some physical cause, which probably exerts its influence also on the solar motion; therefore in reverting to those figures I may be said to appeal to the actual state of the heavens, for a proof of what has been advanced, with respect to the similarity of the directions of projectile motions.

Having thus examined one cause of the sidereal motions, and shown that as far as we are acquainted with its mode of acting in the solar system, it is favourable to a similarity of direction; and that moreover, if we ascribe the motion of the stars to it, we have also good reason, from observation, to believe it to be in favour of the same similarity; we may in the next place proceed to consider the mutual gravitation of the stars towards each other. This is an acknowledged principle of motion, and the laws of its exertion being perfectly known, we shall in this inquiry meet with no difficulty relating to its direction, which is always towards the attracting body.

Considerations of the attractive Power required for a sufficient Velocity of the sidereal Motions.

As attraction is a power that acts at all distances, we ought to begin by examining whether the motions of our stars can be accounted for by the mutual gravitation of neighbouring stars towards each other, or by a periodical binal revolution of them about a common centre of gravity; or whether we ought not rather to have recourse to some very distant attractive centre. This may be decided by a calculation of the effects arising from the laws according to which the principle of attraction is known to act. For instance, let the sun and Sirius be two equal bodies placed in the most favourable situation to permit a mutual approach by attraction: that is, let them be without projectile motions, and removed from all other stars which might impede their progress towards each other, by opposite attractions. Then, by calculation, the space over which one of them would move in a year, were the matter of both collected in the other as an attractive centre, would be less than a five thousand millionth part of a second; supposing that motion to be seen by an eye at the distance of Sirius, and admitting the parallax of the whole orbit of the earth on this star to be one second.

This proves evidently that the mere attraction of neighbouring stars acting upon each other cannot be the cause of the sidereal motions that have been observed.

In the case of supposed periodical binal revolutions of stars about a common centre of gravity, where consequently projectile motions must be admitted, the united power of the connected stars, provided the mass of either of them did not greatly exceed that of the sun, would fall very short of the attraction required to give a sufficient velocity to their motions. The star Arcturus for example, which happens to move, as is required, in an opposite direction to the proposed solar motion, were it connected with the sun, and had the proper degree of necessary projectile motion, could not describe an arch of 1″ of its orbit, about their common centre, in less than 102 years; and though the opposite motion of the sun, by a parallactic effect would double that quantity, it still would fall short of the change we observe in this star in the course of a single year.

Other considerations are still more against the admission of such partial connections: they would intirely oppose the similarity of the directions of the sidereal motions that have been proved to exist, and which we are now endeavouring to explain.

Let us then examine in what manner a distant centre of attraction may be the cause of the required motions. By admitting this centre to be at a great distance, we shall have its influence extended over a space that will take in a whole group of stars, and thus the similar directions of their motions will be accounted for. Their velocities also may be ascribed to the energy of the centre, which may be sufficiently great for all the purposes of the required motions. A circumstance, however, attends the directions of the motions to be explained, which shows that a distant centre of attraction alone will not be sufficient; for these motions, as we may see in Fig. 3, though pretty similar in their directions, still are diverging; whereas if they were solely caused by attraction, they would converge towards the attracting centre, and point out its situation. It is therefore evident that projectile motions must be combined with attraction, and that the motions of the stars when regulated in this manner, are not unlike the disposition by which the bodies of the solar system are governed. If we pursue this arrangement, it will be proper to consider the condition, and probable existence of such a centre of attraction.

There are two ways in which a centre of attraction, so powerful as the present occasion would require, may be constructed; the most simple of them would be a single body of great magnitude; this may exist, though we should not be able to perceive it by any superiority of lustre; for notwithstanding it might have the usual starry brightness, the decrease of its light arising from its great distance would hardly be compensated by the size of its diameter; but to have recourse to an invisible centre, or at least to one that cannot be distinguished from a star, would be intirely hypothetical, and, as such, cannot be admitted in a discussion, the avowed object of which is to prove its existence.

The second way of the construction of a very powerful centre, may be joint attraction of a great number of stars united into one condensed group.

The actual existence of such groups of stars has already been proved by observations made with my large instruments; many of those objects, which were looked upon as nebulous patches, having been completely resolved into stars by my 40 and 20-feet telescopes. For instance, the nebula discovered by Dr. HALLEY in the year 1714, in which the discoverer, and other observers after him, have seen no star, I have ascertained to be a globular cluster, containing, by a rough calculation, probably not less than fourteen thousand stars. From the known laws of gravitation, we are assured that this cluster must have a very powerful attractive centre of gravity, which may be able to keep many far distant celestial bodies in control.

But the composition of an attractive centre is not limited to one such cluster. An union of many of them will form a still more powerful centre of gravitation, whose influence may extend to a whole region of scattered stars. To prove that I argue intirely from observations, I shall mention that another nebula, discovered by Mr. MESSIER in the year 1781, is, by the same instruments, also proved to consist of stars; and though they are seemingly compressed into a much smaller space, and have also the appearance of smaller stars, we may fairly presume that these circumstances are only indications of a greater distance, and that, being a globular cluster, perfectly resembling the former, the distance being allowed for, it is probably not less rich in the number of its component stars. The distance of these two clusters from each other is less than 12 degrees, and we are certain that somewhere in the line joining these two groups there must be a centre of gravitation, far superior in energy to the single power of attraction that can be lodged in either of the clusters.

I have selected these two remarkable objects merely for their situation, which is very near the line of the direction of the solar motion; but were it necessary to bring farther proof of the existence of combined attractions, the numerous objects of which I have given catalogues * would amply furnish me with arguments.

If a still more powerful but more diffused exertion of attraction should be required than what may be found in the union of clusters, we have hundreds of thousands of stars, not to say millions, contained in very compressed parts of the milky way, some of which have already been pointed out in a former Paper.† Many of these immense regions may well occasion the sidereal motions we are required to account for; and a similarity in the direction of these motions will want no illustration.

With regard to the situation of the condensed parts of the milky way, and of the two clusters that have been mentioned, we must remark, that the seat of attraction may be in any part of the heavens whatsoever; for when projectile motions are given to bodies that are retained by an attractive centre, they may have any direction, even that at right angles to its situation not excepted.

It will give additional force to the arguments I have used for the admission of far distant centres of attraction, as well as projectile motions in the stars that are connected with them, when we take notice that, independent of the solar motion, and setting that intirely aside, the action of these causes will be equally required to explain the acknowledged proper motions of the stars. For if the sun be at rest, then Arcturus must actually change its place more than 2″ a year, and

* *Phil. Trans.* for 1786, page 457; for 1789, page 212; for 1802, page 477 [Vol. I. pp. 260 and 329; Vol. II. p. 199].

† *Ibid.* for 1802, page 495 [above, p. 211].

consequently this and many other stars, which are well known to change their situation, must be supposed to have projectile motions, and to be subject to the attraction of far distant centres.

Determination of the Quantity of the Solar Motion.

If I am not mistaken, it will now be allowed that no objection can arise against any solar velocity we may fix upon, for want of a cause that may be assigned to act upon the sun, and many stars, so as to account for their motions, and similar tendency towards a certain part of the heavens; we may consequently proceed in examining whether the quantity that has been assumed for calculating the contents of the XIth Table, will sufficiently come up to the conditions we have adopted for directing our determination.

In Fig. 6 we have the velocities of the 36 stars delineated, and by examining the last column of the Table from which they are taken, we find that the parallactic effects arising from the proposed solar motion require the velocity of 18 stars to exceed that of the sun, and exactly the same number to be inferior to it; so far then the rank which has been assigned to the solar motion is a perfect medium among the sidereal velocities.

If we examine in the next place how this motion will agree with a mean rate deduced from the velocities in the above mentioned column, we find a 36th part of their sum to be 1196550. A solar motion, therefore, which agrees with this mean rate will differ from one assigned by the middle rank no more than 0″·079558; and, on account of the smallness of this quantity, the calculations required to lessen it, by some little increase of the solar motion, might well be dispensed with; but if we were desirous of greater precision, the secondary purpose, next to be considered, would rather incline us to an opposite alteration.

The great disparity of the sidereal motions, which has been mentioned as an incongruity in the first part of this Paper, and has more evidently been shown to exist when we examined the representations of these motions in the 3d figure, is the next point we have to consider in the effect of the solar motion. Let us see how far we have been successful in lessening the ratio these velocities bear to each other. The last column of the Xth Table contains them as they must have been admitted if the sun had been at rest. The proportion of the quickest motion to the slowest is there as 2504621 to 103036; and the velocity of one is therefore above 24 times greater than that of the other. But in consequence of the solar motion we have used, the two extreme velocities are reduced to 2902716 and 395212; which gives a proportion of less than 7½ to 1.

If the quantity of the solar motion were lessened to 1″, we might bring the ratio of the extreme velocities so low as 6 to 1; but as the middle rank has already given it a little below the mean rate, I do not think that we ought to lower it still more; so that when all circumstances are properly considered, there is a great probability that the quantity assumed in the last calculation may not be far from the truth. It appears, therefore, that in the present state of our knowledge of the observed proper motions of the stars, we have sufficient reason to fix upon the quantity of the solar motion to be such as by an eye placed at right angles to its direction, and at the distance of Sirius from us, would be seen to describe annually an arch of 1″·116992 of a degree; and its velocity, till we are acquainted with the real distance of this star, can therefore only be expressed by the proportional number of 1116992.

Concluding Remarks and Inferences.

We have now only to notice a few remarks that may be made, by way of objection to the solar motion I have fixed upon. If the quantity of this motion is to be assigned by the mean rank of sidereal velocities, it may be asked, will not the addition of every star, whose proper motion shall be ascertained, destroy that middle rank, which has been established? To this I shall answer, that future observations may certainly afford us more extensive information on the subject, and even show that the solar motion should not exactly hold that middle rank, which from various motives we have been induced to assign to it; but at present it appears, that according to the doctrine of chances, a middle rank among the sidereal velocities must be the fairest choice, and will remain so, unless, what is now a secondary consideration, should hereafter become of more importance than the first. That this should happen is not impossible, when a general knowledge of the proper motions of all the stars of the 1st, 2d, and 3d magnitudes can be obtained; but then the method of calculation that has been traced out in this and the former Paper, is so perfectly applicable to any new lights observation may throw upon the subject, that a more precise and unobjectionable solar motion can be ascertained by it with great facility. Hitherto we find that a mean rank agrees sufficiently with the phenomena that were to be explained: the apparent velocities of Arcturus and Aldebaran, without a solar motion for instance, were to each other, in the IXth Table, as 208 to 12; our present solar motion has shown, that when the deception arising from its parallactic effect is removed by calculation, these velocities are to each other only as 179 to 85, or as 2 to 1. And though Arcturus still remains a star that moves with great velocity, yet in the XIth Table we have 4 or 5 stars with nearly as much motion; and 4 with more.

Our solar motion also removes the deception by which the motion of a star of the consequence of α Orionis is so concealed as hardly to show any velocity; whereas by computation we find that it really moves at a rate which is fully equal to the motion of the sun.

I must now observe, that the result of calculations founded upon facts, such as we must admit the proper motions of the stars to be, should give us some useful information, either to satisfy the inquisitive mind, or to lead us on to new discoveries. The establishment of the solar motion answers both these ends. We have already seen that it resolves many difficulties relating to the proper motions of the stars, and reconciles apparent contradictions; but our inquiries should not terminate here. We are now in the possession of many concealed motions, and to bring them still more to light, and to add new ones by future observations should become the constant aim of every astronomer.

This leads me to a subject, which though not new in itself, will henceforth assume a new and promising aspect. An elegant outline of it has long ago been laid before the public in a most valuable paper on general Gravitation, under the form of "Thoughts" on the subject; * but I believe, from what has been said in this Paper, it will now be found that we are within the reach of a link of the chain which connects the principles of the solar and sidereal motions with those that are the cause of orbital ones.

A discovery of so many hitherto concealed motions, presents us with an interesting view of the construction of that part of the heavens which is immediately around us. The similarity of the directions of the sidereal motions is a strong indication that the stars, having such motions, as well as the sun, are acted upon by some connecting cause, which can only be attraction; and as it has been proved that attraction will not explain the observed phenomena without the existence of projectile motions, it must be allowed to be a necessary inference, that the motions of the stars we have examined are governed by the same two ruling principles which regulate the orbital motions of the bodies of the solar system. It will also be admitted that we may justly invert the inference, and from the operation of these causes in our system, conclude that their influence upon the sidereal motions will tend to produce a similar effect; by which means the probable motion of the sun, and of the stars in orbits, becomes a subject that may receive the assistance of arguments supported by observation.

What has been said in a paragraph of a former Paper, where the sun is placed among the insulated stars,† does not contradict the present idea of its making one of a very extensive system. On the contrary, a connection of this nature has been alluded to in the same Paper.‡ The insulation ascribed to the sun relates merely to a supposed binary combination with some neighbouring star; and it has now been proved by an example of Arcturus, that the solar motion cannot be occasioned or accounted for by a periodical revolution of the sun and this or any other star about their common centre of gravity.

* See the note *Phil. Trans.* for 1783, page 283 [Vol. I. p. 130].

† *Ibid.* for 1802, page 478 [above, p. 200].

‡ *Ibid.*

LVII.

Observations and Remarks on the Figure, the Climate, and the Atmosphere of Saturn, and its Ring.

[*Phil. Trans.*, 1806, pp. 455–467.]

Read June 26, 1806.

My last year's observations on the singular figure of Saturn having drawn the attention of astronomers to this subject, it may be easily supposed that a farther investigation of it will be necessary. We see this planet in the course of its revolution round the sun in so many various aspects, that the change occasioned by the different situations in which it is viewed, as far as relates to the ring, has long ago been noticed; and HUYGENS has given us a very full explanation of the cause of these changes.*

As the axis of the planet's equator, as well as that of the ring, keeps its parallelism during the time of its revolution about the sun, it follows that the same change of situation, by which the ring is affected, must also produce similar alterations in the appearance of the planet; but since the shape of Saturn, though not strictly spherical, is very different from that of the ring, the changes occasioned by its different aspects will be so minute that only they can expect to perceive them who have been in the habit of seeing very small objects, and are furnished with instruments that will show them distinctly, with a very high and luminous magnifying power.

If the equator of the planet Jupiter were inclined to the ecliptic like that of Saturn, I have no doubt but that we should see a considerable change in its figure during the time of a synodical revolution; notwithstanding the spheroidical figure occasioned by the rotation on its axis has not the extended flattening of the polar regions that I have remarked in Saturn. But since not only the position of the Saturnian equator is such that it brings on a periodical change in its aspect, amounting to more than 62 degrees in the course of each revolution, but that moreover in the shape of this planet there is an additional deviation from the usual spheroidical figure arising from the attraction of the ring, we may reasonably expect that our present telescopes will enable us to observe a visible alteration in its appearance, especially as our attention is now drawn to this circumstance.

* See *Systema Saturnium*, page 55, where the changes of the ring are represented by a plate.

In the year 1789 I ascertained the proportion of the equatorial to the polar diameter of Saturn to be 22·81 to 20·61,* and in this measure was undoubtedly included the effect of the ring on the figure of the planet, though its influence had not been investigated by direct observation. The rotation of the planet was determined afterwards by changes observed in the configuration of the belts, and proper figures to represent the different situation of the spots in these belts were delineated.† In drawing them it was understood that the shape of the planet was not the subject of my consideration, and that consequently a circular disk, which may be described without trouble, would be sufficient to show the configurations of the changeable belts.

Those who compare these figures, and others I have occasionally given, in which the particular shape of the body of the planet was not intended to be represented, with the figure which is contained in my last Paper, of which the sole

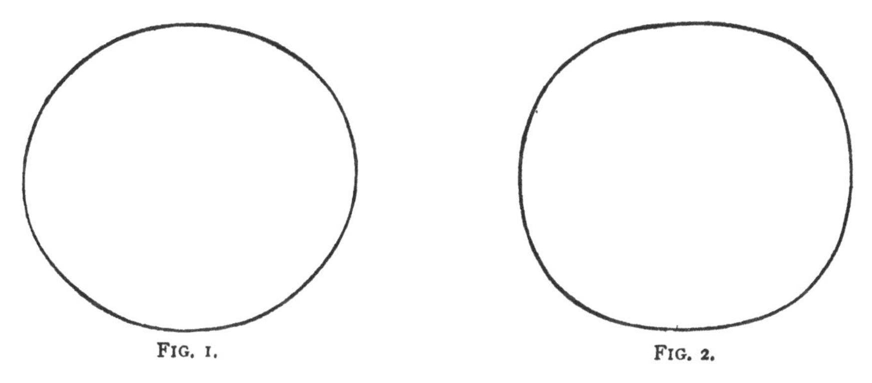

FIG. 1. FIG. 2.

purpose was to express that figure, and wonder at the great difference, have probably not read the measures I have given of the equatorial and polar diameters of this planet; and as it may be some satisfaction to compare the appearance of Saturn in 1789 with the critical examination of it in 1805, I have now drawn them from the two papers which treat of the subject; Fig. 1 represents the spheroidical form of the planet as observed in 1789, at which time the singularity of the shape since discovered was unknown; and Fig. 2 represents the same as it appeared the 5th of May 1805. The equatorial and polar diameters that were established in 1789 are strictly preserved in both figures, and the last differs from the first only in having the flattening at the poles a little more extended on both sides towards the equatorial parts. It is in consequence of the increase of the length of this flattening, or from some other cause, that a somewhat greater curvature in the latitudes of 40 or 45 degrees north and south has taken place; and as these differences are very minute, it will not appear extraordinary that they should have been overlooked in 1789, when my attention was intirely taken up with an examination of the two principal diameters of the planet.

The use of various magnifying powers in observing minute objects is not

* *Phil. Trans.* for 1790, page 17 [Vol. I. p. 380].

† *Ibid.* for 1792, page 22 [Vol. I. p. 430].

generally understood. A low power, such as 200 or 160, with which I have seen the figure of Saturn, is not sufficient to show it to one who has not already seen it perfectly well with an adequate high power ; an observer, therefore, who has not an instrument that will bear a very distinct magnifying power of 500, ought not to expect to see the outlines of Saturn so sharp and well defined as to have a right conception of its figure. The quintuple belt is generally a very good criterion ; for if that cannot be seen the telescope is not sufficient for the purpose ; but when we have intirely convinced ourselves of the reality of the phenomena I have pointed out, we may then gradually lower the power, in order to be assured that the great curvature of the eyeglasses giving these high powers, has not occasioned any deceptions in the figure to be investigated, and this was the only reason why I mentioned that I had also seen the remarkable figure of Saturn with low powers.

In very critical cases it becomes necessary to calculate every cause of an appearance that falls under the province of mathematical investigation. For this reason I have always looked upon an astronomical observation without a date as imperfect, and the journal-method of communicating them is undoubtedly what ought to be used. For instance, when it is known that my last year's most decisive observation, relating to the singular figure of Saturn, was made the 5th of May, astronomers may then calculate by this date the place of Saturn and of the earth ; their distances from each other, and the angle of illumination of the Saturnian disk ; by these means we find the gibbosity of the planet in the given situation, and ascertain that the defalcation of light could not then amount to the one hundredth part of a second of a degree, and that consequently no error could arise from that cause.

I have divided the following observations into two heads, one relating intirely to the figure of the body of Saturn, the other concerning the physical condition or climate and atmosphere of the planet.

Observations of the Figure of Saturn.

In the collection of my observations on the planet Saturn, I have met with one made 18 years ago, which is perfectly applicable to the present subject, and is as follows :

August 2, 1788, 21^{h} 58′. 20-feet reflector, power 300. Admitting the equatorial diameter of Saturn to lie in the direction of the ring, the planet is evidently flattened at the poles. I have often before, and again this evening, supposed the shape of Saturn not to be spheroidical, (like that of Mars and Jupiter,) but much flattened at the poles, and also a very little flattened at the equator, but this wants more exact observations.

April 16, 1806. I examined the figure of the body of Saturn with the 7 and 10-feet telescopes, but they acted very indifferently, and, were I to judge by present appearances, I should suppose the planet to have undergone a considerable change; should this be the case, it will then be necessary to trace out the cause of such alterations.

April 19. 10-feet, power 300. The polar regions are much flattened. The figure of the planet differs a little from what it appeared last year. This may be owing to the increased opening of the ring, which in four places obstructs now the view of the curvature in a higher latitude than it did last year. The equatorial regions on the contrary are more exposed to view than they have been for some time past.

May 2. 10-feet, power 375. The polar regions are much flatter than the equatorial: the latter being more disengaged from the ring appear rather more curved than last year, so that the figure of the planet seems to have undergone some small alteration, which may be easily accounted for from our viewing it now in a different aspect. The planet Jupiter not being visible, we cannot compare the figure of Saturn with it; but from memory I am quite certain that the flattening of the Saturnian polar regions is considerably more extended than those of Jupiter.

May 4. 10-feet, power 527. The equatorial region of Saturn appears to be a little more elevated than last year. This part of the Saturnian figure could not be examined so well then as it may at present, the ring interfering with our view of it in four places, which are now visible. The flattening on both sides of the pole is continued to a greater extent than in a figure merely spheroidical, such as that of Jupiter; and this makes the planet more curved in high latitudes. The planet being in the meridian, the equatorial shape of Saturn appears a little more curved than last year; but the air is not sufficiently pure to bear high powers well.

May 5. 10-feet, power 527. The air is very favourable, and I see the planet well with this power; its figure is very little different from what it was last year. The polar regions are more extendedly flat than I suppose they would have been if the planet had received its form only from the effect of the centrifugal force arising from its rotatory motion. The equatorial region is a little more elevated than it appeared last year. The diameter which intersects the equator in an angle of about 40 or 45 degrees is apparently a little longer than the equatorial, and the curvature is greatest in that latitude. The planet being in the meridian and the night beautiful, I have had a complete view of its figure. It has undergone no change since last year, except what arises from its different situation, and a greater opening of the ring.

May 9. Power 527. The air being very clear, I see the figure of Saturn nearly the same as last year; the flattening at the poles appears at present somewhat less; the equatorial and other regions are still the same.

May 15, 10^h $30'$. I examined the appearance of Saturn, and compared it with the engraving representing its figure in last year's volume of the *Phil. Trans.* The outlines and all the other features of this engraving are far more distinct than we can ever see them in the telescope at one view ; but it is the very intention of a copper-plate to collect together all that has been successfully discovered by repeated and occasional perfect glimpses, and to represent it united and distinctly to our inspection. Indeed by looking at the drawings contained in books of astronomy this will be found to be the case with them all.*

The equatorial diameter of my last year's figure is however a very little too short ; it should have been to the polar diameter as 35·41 to 32, which is the proportion that was ascertained in 1789, from which I have hitherto found no reason to depart.

The following particulars remain as my last year's observations have established them.

The flattening at the poles of Saturn is more extensive than it is on the planet Jupiter. The curvature in high latitudes is also greater than on that planet. At the equator, on the contrary, the curvature is rather less than it is on Jupiter. Upon the whole, therefore, the shape of the globe of Saturn is not such as a rotatory motion alone could have given it. I see the quintuple belt, the division of the ring, a very narrow shadow of the ring across the body, and another broader shadow of the body upon the following part of the ring ; and unless all these particulars are very distinctly visible we cannot expect that our instrument should show the outlines of the planet sufficiently well to perceive its peculiar formation.

May 16, 10^h 10′. The greatest curvature on the disk of Saturn seems to be in a latitude of about 40 degrees.

May 18. The difference between the equatorial and polar diameters appears to be a little less than the measures taken September 14, 1789, give it ; but as the eye was then in the plane of the equator, and is now about 16 degrees elevated above it, we cannot expect to see it quite so much flattened at present.

June 3. The shadow of the ring falls upon the body of the planet southwards of the ring, towards the limb ; it grows a little broader at both ends where it is upon the turn round the globe.

June 5. The planet Jupiter is not sufficiently high for distinct vision, and Saturn is already too low to use a proper magnifying power ; but nevertheless the difference in the formation of the two planets is evident. The equatorial as well as polar regions on Jupiter are more curved than those of Saturn.

June 9. The air is beautifully clear, and proper for critical observations. The breadth of the ring is to the space between the ring and the body of Saturn as about 5 to 4. See Fig. 3. The ring appears to be sloping towards the body of the planet, and the inside edge of it is probably of a spherical or perhaps hyperbolical form. The shadow of the ring on the planet is broader on both sides than

* For an instance of this, see TOBIÆ MAYERI *Opera inedita. Appendix Observationum. Ad Tabulam Selenographicam Animadversiones*, where the annexed accurate and valuable plate represents the moon such as it never can be seen in a telescope.

in the middle; this is partly a consequence of the curvature of the ring which in the middle of its passage across the body hides more of the shadow in that place than at the sides. The shadow of the body upon the ring is a little broader at the north than the south, so as not to be parallel with the outline of the body; nor is it so broad at the north as to become square with the direction of the ring. The most northern dusky belt comes northwards on both sides as far as the middle of the breadth of the ring where it passes behind the body. It is curved towards the south in the middle.

I viewed Jupiter, and compared its figure with that of Saturn. An evident difference in the formation of the two planets is visible. To distinguish the figure of Jupiter properly it may be called an ellipsoid, and that of Saturn a spheroid.

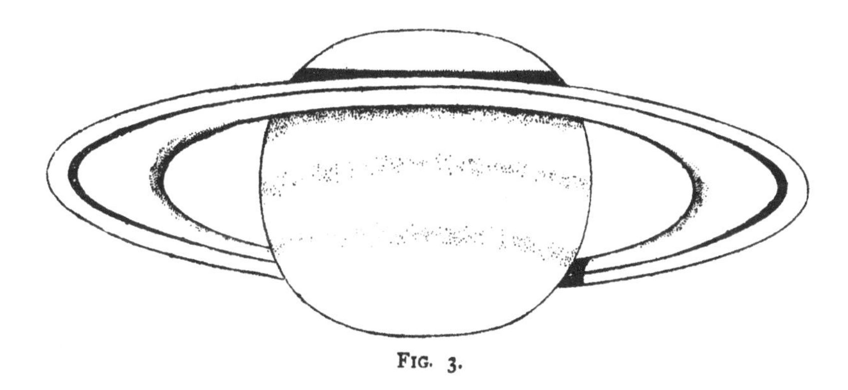

FIG. 3.

Observations on the periodical Changes of the Colour of the polar Regions of Saturn.

In the observations I have given on the planet Mars, it has been shown that an alternate periodical change takes place in the extent and brightness of the north and south polar spots; * and I have there suggested an idea that the cause of the brightness might be a vivid reflection of light from frozen regions, and that the reduction of the spots might be ascribed to their being exposed to the sun.

The following observations, I believe, will either lead us to similar conclusions with respect to the appearance of the polar regions of Saturn, or will at least draw the attention of future observers to a farther investigation of the subject.

With high magnifying powers the objects we observe require more light than when the power is lower; this affords us a good method of determining the relative brightness of the different parts of a planet. The less bright object will be found deficient in illumination when the power exceeds what it will bear with ease. I have availed myself of this assistance in the observations that follow.

June 25, 1781. With an aperture of 6·3 inches I used a magnifying power of 460. This gave a kind of yellowish colour to the planet Saturn, while the ring still retained its full white illumination.

* *Phil. Trans.* for 1784, page 260 [Vol. I. p. 149].

November 11, 1793. From the quintuple belt towards the south pole the whole distance is of a pale whitish colour; less bright than the white belts, and much less bright than the ring. This has been represented in a figure which was given in the volume of the *Phil. Trans.* for 1794, page 32.* It is to be noticed that the south pole of the planet had been long exposed to the influence of the sun, and the former polar whitishness was no longer to be seen.

Jan. 1, 1794. The south polar regions are a little less bright than the equatorial belt.

Nov. 5, 1796. The space between the quintuple belt and the northern part of the ring is of a bright white colour.

This seems to indicate that the whiteness of the northern hemisphere of Saturn increases when there is less illumination from the sun.

May 6, 1806. The north pole of Saturn being now exposed to the sun, its regions have lost much of their brightness; the space about the south pole has regained its former colour, and is brighter and whiter than the equatorial parts.

May 15. The south polar regions of Saturn are white; those of the north retain also some whitishness still.

May 18. With a magnifying power of 527, the south polar regions remain very white. The equatorial parts become of a yellowish tinge, and about the north pole there is still a faint dusky white colour to be seen.

June 3. The south polar regions are considerably brighter than those of the north.

These observations contrasted with those which were made when the south pole was in view complete nearly half a Saturnian year, and the gradual change of the colour of the polar regions seems to be in a great measure ascertained. Should this be still more confirmed, there will then be some foundation for admitting these changes to be the consequence of an alteration of the temperature in the Saturnian climates. And if we do not ascribe the whiteness of the poles in their winter seasons immediately to frost or snow, we may at least attribute the different appearance to the greater suspension of vapours in clouds, which, it is well known, reflect more light than a clear atmosphere through which the opaque body of the planet is more visible. The regularity of the alternate changes at the poles ought however to be observed for at least two or three of the Saturnian years, and this, on account of their extraordinary length, can only be expected from the successive attention of astronomers.

On the Atmosphere of Saturn.

June 9, 1806. The brightness which remains on the north polar regions, is not uniform, but is here and there tinged with large dusky looking spaces of a cloudy atmospheric appearance.

* [Vol. I. p. 452, Pl. XIII. fig. 1.]

From this and the foregoing observations on the change of the colour at the polar regions of Saturn arising most probably from a periodical alteration of temperature, we may infer the existence of a Saturnian atmosphere; as certainly we cannot ascribe such frequent changes to alterations of the surface of the planet itself: and if we add to this consideration the changes I have observed in the appearance of the belts, or even the belts themselves, we can hardly require a greater confirmation of the existence of such an atmosphere.

A probability that the ring of Saturn has also its atmosphere has already been pointed out in a former Paper.

Slough, near Windsor,
June 12, 1806.

LVIII.

Experiments for investigating the Cause of the coloured concentric Rings, discovered by Sir ISAAC NEWTON, *between two Object-glasses laid upon one another.*

[*Phil. Trans.*, 1807, pp. 180–233.]

Read February 5, 1807.

THE account given by Sir I. NEWTON, of the coloured arcs and rings which he discovered by laying two prisms or object-glasses upon each other, is highly interesting. He very justly remarks, that these phenomena are "of difficult consideration," but that "they may conduce to farther discoveries for completing the theory of light, especially as to the constitution of the parts of natural bodies on which their colours or transparency depend."*

With regard to the explanation of the appearance of these coloured rings, which is given by Sir I. NEWTON, I must confess that it has never been satisfactory to me. He accounts for the production of the rings, by ascribing to the rays of light certain fits of easy reflection and easy transmission alternately returning and taking place with each ray at certain stated intervals.† But this, without mentioning particular objections, seems to be an hypothesis which cannot be easily reconciled with the minuteness and extreme velocity of the particles of which these rays, according to the NEWTONIAN theory, are composed.

The great beauty of the coloured rings, and the pleasing appearances arising from the different degrees of pressure of the two surfaces of the glasses against each other when they are formed, and especially the importance of the subject, have often excited my desire of enquiring farther into the cause of such interesting phenomena; and with a view to examine them properly I obtained, in the year 1792, the two object-glasses of HUYGENS, in the possession of the Royal Society, one of 122, the other of 170 feet focal length, and began a series of experiments with them, which, though many times interrupted by astronomical pursuits, has often been taken up again, and has lately been carried to a very considerable extent. The conclusions that may be drawn from them, though they may not perfectly account for all the phenomena of the rings, are yet sufficiently well supported, and of such a nature as to point out several modifications of light that have been totally overlooked, and others that have never been properly discrimin-

* NEWTON'S *Optics*, 4th ed., p. 169.

† *Ibid.*, p. 256.

ated. It will, therefore, be the aim of this paper to arrange and distinguish the various modifications of light in a clear and perspicuous order, and afterwards to give my sentiments upon the cause of the formation of the concentric rings. The avowed intricacy of the subject,* however, requires, in the first place, a minute detail of experiments, and afterwards a very gradual developement of the consequences to be deduced from them.

As the word modification will frequently be used, it may not be amiss to say, that when applied to light, it is intended to stand for a general expression of all the changes that are made in its colours, direction, or motion : thus, by the modification of reflection, light is thrown back ; by that of refraction, it is bent from its former course ; by the modification of dispersion, it is divided into colours, and so of the rest.

I. *Of different Methods to make one set of concentric Rings visible.*

In the beginning of my experiments I followed the NEWTONIAN example, and having laid the two object-glasses of HUYGENS upon one another I soon perceived the concentric rings. It is almost needless to say that I found all the NEWTONIAN observations of these rings completely verified ; but as his experiments seemed to be too much confined for drawing general conclusions, I endeavoured to extend them : and by way of rendering the methods I point out very clear, I have given one easy particular instance of each, with the addition of a generalization of it, as follows :

First Method. On a table placed before a window I laid down a slip of glass the sides of which were perfectly plain, parallel, and highly polished. Upon this I laid a double convex lens of 26 inches focal length, and found that this arrangement gave me a set of beautiful concentric rings.

I viewed them with a double convex eye lens of 2½ inches focus mounted upon an adjustable stand, by which simple apparatus I could examine them with great ease ; and as it was not material to my present purpose by what obliquity of incidence of light I saw the rings, I received the rays from the window most conveniently when they fell upon the lens in an angle of about 30 degrees from the perpendicular, the eye being placed on the opposite side at an equal angle of elevation to receive the reflected rays.

Generalization. Instead of a plain slip of glass, the plain side of a plano-concave, or plano-convex lens of any focal length whatsoever may be used : and when the convex side of any lens is laid upon it, whatever may be the figure of the other surface, whether plain, concave, or convex, and whatever may be its focal length, a set of concentric rings will always be obtained. I have seen rings with lenses of all varieties of focus, from 170 feet down to one quarter of an inch. Even a common watch glass laid upon the same plain surface will give them.

* NEWTON'S *Optics*, 4th ed. p. 288 ; end of Obs. 12.

To insure success, it is necessary that the glasses should be perfectly well cleaned from any adhering dust or soil, especially about the point of contact; and in laying them upon each other a little pressure should be used, accompanied at first with a little side motion, after which they must be left at rest.

If the surface of the incumbent lens, especially when it is of a very short focal length, is free from all imperfection and highly polished, the adjustment of the focus of the above mentioned eye-glass, which I always use for viewing the rings, is rather troublesome, in which case a small spot of ink made upon the lens will serve as an object for a sufficient adjustment to find the rings.

Second Method. Instead of the slip of glass, I laid down a well polished plain metalline mirror; and placing upon it the same 26-inch double convex lens, I saw again a complete set of concentric rings.

It is singular that, in this case, the rings reflected from a bright metalline surface will appear fainter than when the same lens is laid on a surface of glass reflecting but little light; this may however be accounted for by the brilliancy of the metalline ground on which these faint rings are seen, the contrast of which will offuscate their feeble appearance.

Generalization. On the same metalline surface every variety of lenses may be laid, whatever be the figure of their upper surface, whether plain, concave, or convex, and whatever be their focal lengths, provided the lowest surface remains convex, and concentric rings will always be obtained; but for the reason mentioned in the preceding paragraph, very small lenses should not be used till the experimentalist has been familiarized with the method of seeing these rings, after which lenses of two inches focus, and gradually less, may be tried.

Third Method. Hitherto we have only used a plain surface upon which many sorts of glasses have been placed; in order therefore to obtain a still greater variety, I laid down a plano-convex lens of 15 inches focal length, and upon the convex surface of it I placed the 26-inch double convex lens, which produced a complete set of rings.

Fourth Method. The same lens placed upon a convex metalline mirror of about 15 inches focal length gave also a complete set of rings.

Generalization. These two cases admit of a much greater variety than the first and second methods; for here the incumbent glass may have not only one, but both its surfaces of any figure whatsoever; whether plain, concave, or convex; provided the radius of concavity, when concave lenses are laid upon the convex surface of glass or metal, is greater than that of the convexity on which they are laid.

The figure of the lowest surface of the subjacent substance, when it is glass, may also be plain, concave, or convex; and the curvature of its upper surface, as well as of the mirror, may be such as to give them any focal length, provided the radius of their convexities is less than that of the concavity of an incumbent lens; in all which cases complete sets of concentric rings will be obtained.

Fifth Method. Into the concavity of a double concave glass of 8 inches focal length I placed a 7-inch double convex lens, and saw a very beautiful set of rings.

Sixth Method. Upon a 7 feet concave metalline mirror I placed the double convex 26-inch lens, and had a very fine set of rings.

Generalization. With these two last methods, whatever may be the radius of the concavity of the subjacent surface, provided it be greater than that of the convexity of the incumbent glass; and whatever may be the figure of the upper surface of the lenses that are placed upon the former, there will be produced concentric rings. The figure of the lowest surface of the subjacent glass may also be varied at pleasure, and still concentric rings will be obtained.

II. *Of seeing Rings by Transmission.*

The great variety of the different combinations of these differently figured glasses and mirrors will still admit of further addition, by using a different way of viewing the rings. Hitherto, the arrangement of the apparatus has been such as to make them visible only by reflection, which is evident, because all the experiments that have been pointed out may be made by the light of a candle placed so that the angle of incidence and of reflection towards the eye of the observer, may be equal. But Sir I. NEWTON has given us also an observation where he saw these rings by transmission, in consequence of which I have again multiplied and varied the method of producing them that way, as follows:

First Method. On a slip of plain glass highly polished on both sides place the same double convex lens of 26 inches, which had already been used when the rings were seen by reflection. Take them both up together and hold them against the light of a window, in which position the concentric rings will be seen with great ease by transmitted light. But as the use of an eye-glass will not be convenient in this situation, it will be necessary to put on a pair of spectacles with glasses of 5, 6, or 7 inches focus, to magnify the rings in order to see them more readily.

Second Method. It would be easy to construct an apparatus for viewing the rings by transmission fitted with a proper eye-glass; but other methods of effecting the same purpose are preferable. Thus, if the two glasses that are to give the rings be laid upon a hollow stand, a candle placed at a proper angle and distance under them will show the rings conveniently by transmitted light, while the observer and the apparatus remain in the same situation as if they were to be seen by reflection.

Third Method. A still more eligible way is to use daylight received upon a plain metalline mirror reflecting it upwards to the glasses placed over it, as practised in the construction of the common double microscope; but I forbear entering into a farther detail of this last and most useful way of seeing rings by transmission, as I shall soon have occasion to say more on the same subject.

Generalization. Every combination of glasses that has been explained in the first, third, and fifth methods of seeing rings by reflection will also give them by transmission, when exposed to the light in any of the three ways that have now been pointed out. When these are added to the former, it will be allowed that we have an extensive variety of arrangements for every desirable purpose of making experiments upon rings, as far as single sets of them are concerned.

III. *Of Shadows.*

When two or more sets of rings are to be seen, it will require some artificial means, not only to examine them critically, but even to perceive them ; and here the shadow of some slender opaque body will be of eminent service. To cast shadows of a proper size and upon places where they are wanted, a pointed penknife may be used as follows.

When a plain slip of glass or convex lens is laid down, and the point of a penknife is brought over either of them, it will cast two shadows, one of which may be seen on the first surface of the glass or lens, and the other on the lowest.

When two slips of glass are laid upon each other, or a convex lens upon one slip, so that both are in contact, the penknife will give three shadows ; but if the convex lens should be of a very short focus, or the slips of glass a little separated, four of them may be perceived ; for in that case there will be one formed on the lowest surface of the incumbent glass or lens ; but in my distinction of shadows this will not be noticed. Of the three shadows thus formed the second will be darker than the first, but the third will be faint. When a piece of looking glass is substituted for the lowest slip the third shadow will be the strongest.

Three slips of glass in contact, or two slips with a lens upon them, or also a looking glass, a slip and a lens put together, will give four shadows, one from each upper surface and one from the bottom of the lowest of them.

In all these cases a metalline mirror may be laid under the same arrangement without adding to the number of shadows, its effect being only to render them more intense and distinct.

The shadows may be distinguished by the following method. When the point of the penknife is made to touch the surface of the uppermost glass or lens, it will touch the point of its own shadow, which may thus at any time be easily ascertained : and this in all cases I call the first shadow ; that which is next to it, the second ; after which follows the third, and so on.

In receding from the point, the shadows will mix together, and thus become more intense ; but which, or how many of them are united together, may always be known by the points of the shadows.

When a shadow is to be thrown upon any required place, hold the penknife nearly half an inch above the glasses, and advance its edge foremost gradually towards the incident light. The front should be held a little downwards to keep the light from the underside of the penknife, and the shadows to be used should be obtained from a narrow part of it.

With this preparatory information it will be easy to point out the use that is to be made of the shadows when they are wanted.

IV. *Of two sets of Rings.*

I shall now proceed to describe a somewhat more complicated way of observation, by which two complete sets of concentric rings may be seen at once. The new or additional set will furnish us with an opportunity of examining rings in situations where they have never been seen before, which will be of eminent service for investigating the cause of their origin, and with the assistance of the shadows to be formed, as has been explained, we shall not find it difficult to see them in these situations.

First Method. Upon a well polished piece of good looking-glass lay down a double convex lens of about 20 inches focus. When the eye glass has been adjusted as usual for seeing one set of rings, make the shadow of the penknife in the order which has been described, pass over the lens ; then, as it sometimes happens in this arrangement that no rings are easily to be seen, the shadow will, in its passage over the surface, show where they are situated. When a set of them is perceived, which is generally the primary one, bring the third shadow of the penknife over it, in which situation it will be seen to the greatest advantage.

Then, if at the same time a secondary set of rings has not yet been discovered, it will certainly be perceived when the second shadow of the penknife is brought upon the primary set. As soon as it has been found out, the compound shadow, consisting of all the three shadows united, may then be thrown upon this secondary set, in order to view it at leisure and in perfection. But this compound shadow should be taken no farther from the point than is necessary to cover it ; nor should the third shadow touch the primary set. The two sets are so near together, that many of the rings of one set intersect some of the other.

When a sight of the secondary set has been once obtained, it will be very easy to view it alternately with the primary one by a slight motion of the penknife, so as to make the third shadow of it go from one set to the other.

Besides the use of the shadows, there is another way to make rings visible when they cannot be easily perceived, which is to take hold of the lens with both hands, to press it alternately a little more with one than with the other ; a tilting motion, given to the lens in this manner, will move the two sets of rings from side to side ; and as it is well known that a faint object in motion may be sooner perceived than when it is at rest, both sets of rings will by these means be generally detected together.

It will also contribute much to facilitate the method of seeing two sets of rings. if we receive the light in a more oblique angle of incidence, such as 40, 50, or even

60 degrees. This will increase the distance between the centers of the primary and secondary sets, and at the same time occasion a more copious reflection of light.

Instead of a common looking-glass a convex glass mirror may be used, on which may be placed either a plain, a concave, or a convex surface of any lens or glass, and two sets of rings will be obtained.

In the same manner, by laying upon a concave glass mirror a convex lens, we shall also have two sets of rings.

The generalizations that have been mentioned when one set of rings was proposed to be obtained, may be easily applied with proper regulations, according to the circumstances of the case, not only to the method by glass mirrors already mentioned, but likewise to all those that follow hereafter, and need not be particularized for the future. In the choice of the surfaces to be joined, we have only to select such as will form a central contact, the focal length of the lenses and the figure of the upper surface being variable at pleasure.

Second Method. On a plain metalline mirror I laid a parallel slip of glass, and placed upon it the convex surface of a 17-inch plano-convex lens, by which means two sets of rings were produced.

Upon the same mirror the plain side of the plano-convex glass may be laid instead of the plain slip, and any plain, convex, or concave surface being placed upon the convexity of the subjacent lens, will give two sets of rings.

The plain side of a plano-concave glass may also be placed upon the same mirror, and into the concavity may be laid any lens that will make a central contact with it, by which arrangement two sets of rings will be obtained.

Third Method. Upon a small well polished slip of glass place another slip of the same size, and upon them lay a 39-inch double convex lens. This will produce two sets of rings ; one of them reflected from the upper surface of the first slip of glass, and the other from that of the second.

Instead of the uppermost plain slip of glass we may place upon the lowest slip the plain side of a plano-convex or plano-concave lens, and the same variety which has been explained in the third method, by using any incumbent lens that will make a central contact, either with the convexity or concavity of the subjacent glass, will always produce two sets of rings.

Fourth Method. A more refined but rather more difficult way of seeing two sets of rings, is to lay a plain slip of glass on a piece of black paper, and when a convex lens is placed upon the slip, there may be perceived, but not without particular attention, not only the first set, which has already been pointed out as reflected from the first surface of the slip, but also a faint secondary set from the lowest surface of the same slip of glass.

It will be less difficult to see two sets of rings by a reflection from both surfaces of the same glass, if we use, for instance, a double concave of 8 inches focus with a double convex of $7\frac{1}{2}$ inches placed upon it. For, as it is well known that glass will reflect more light from the farthest surface when air rather than a denser medium is in contact with it, the hollow space of the 8-inch concave will give a pretty strong reflection of the secondary set.

Fifth Method. The use that is intended to be made of two sets of rings requires that one of them should be dependent upon the other : this is a circumstance that will be explained hereafter, but the following instance, where two independent sets of rings are given, will partly anticipate the subject. When a double convex lens of 50 inches is laid down with a slip of glass placed upon it, and another double convex one of 26 inches is then placed upon the slip, we get two sets of rings of different sizes ; the large rings are from the 50-inch glass, the small rings from the 26-inch one. They are to be seen with great ease, because they are each of them primary. By tilting the incumbent lens or the slip of glass these two sets of rings may be made to cross each other in any direction ; the small set may be laid upon the large one, or either of them may be separately removed towards any part of the glass. This will be sufficient to show that they have no connection with each other. The phenomena of the motions, and of the various colours and sizes assumed by these rings, when different pressures and tiltings of the glasses are used, will afford some entertainment. With the assistance of the shadow of the penknife the secondary set belonging to the rings from the 26-inch lens will be added to the other two sets ; but in tilting the glasses this set will never leave its primary one, while that from the 50-inch lens may be made to go any where across the other two.

V. *Of three Sets of Rings.*

To see three sets of concentric rings at once is attended with some difficulty, but by the assistance of the methods of tilting the glasses and making use of the multiplied shadows of a penknife we may see them very well, when there is a sufficient illumination of bright daylight.

First Method. A 26-inch double convex lens placed upon three slips of plain glass will give three sets of rings. The slips of glass should be nearly 2-tenths of an inch thick, otherwise the different sets will not be sufficiently separated. When all the glasses are in full contact the first and second sets may be seen with a little pressure and a small motion, and, if circumstances are favourable the third, which is the faintest, will also appear. If it cannot be seen, some of the compound shadows of the penknife must be thrown upon it ; for in this case there will be five shadows visible, several of which will fall together and give different intensity to their mixture.

Second Method. When a single slip of glass, with a 34-inch lens upon it, is placed upon a piece of good looking-glass, three sets of rings may be seen : the first and third sets are pretty bright, and will be perceived by only pressing the lens a little upon the slip of glass ; after which it will be easy to find the second set with the assistance of the proper shadow. In this case four shadows will be seen ; and when the third shadow is upon the first set, the fourth will be over the second set and render it visible.

Third Method. When two slips of glass are laid upon a plain metalline mirror, then a 26-inch lens placed upon the slips will produce three sets of rings ; but it is not very easy to perceive them. By a tilting motion the third set will generally appear like a small white circle, which at a proper distance will follow the movement of the first set. As soon as the first and third sets are in view the third shadow of the penknife may be brought over the first set, by which means the fourth shadow will come upon the second set, and in this position of the apparatus it will become visible.

Fourth Method. On a plain metalline mirror lay one slip of glass, but with a small piece of wood at one end under it, so that it may be kept about one-tenth of an inch from the mirror, and form an inclined plane. A 26-inch lens laid upon the slip of glass will give three sets of rings. Two of them will easily be seen ; and when the shadow of the penknife is held between them the third set will also be perceived. There is but one shadow visible in this arrangement, which is the third ; the first and second shadows being lost in the bright reflection from the mirror.

Fifth Method. I placed a $6\frac{3}{4}$-inch double convex upon an 8-inch double concave, and laid both together upon a plain slip of glass. This arrangement gave three sets of rings. They may be seen without the assistance of shadows, by using only pressure and tilting. The first had a black and the other two had white centers.

VI. *Of four Sets of Rings.*

The difficulty of seeing many sets of rings increases with their number, yet by a proper attention to the directions that are given four sets of concentric rings may be seen.

First Method. Let a slip of glass, with a 26-inch lens laid upon it, be placed upon a piece of looking-glass. Under one end of the slip, a small piece of wood one-tenth of an inch thick must be put to keep it from touching the looking-glass. This arrangement will give us four sets of rings. The first, third, and fourth may easily be seen, but the second set will require some management. Of the three shadows, which this apparatus gives, the second and third must be brought between the first and fourth sets of rings, in which situation the second set of rings will become visible.

Second Method. When three slips of glass are laid upon a metalline mirror, and a plano-convex lens of about 17 inches focus is placed with its convex side upon them, four sets of rings may be seen ; but this experiment requires a very bright day, and very clean, highly polished slips of plain glass. Nor can it be successful unless all the foregoing methods of seeing multiplied sets of rings are become familiar and easy.

I have seen occasionally, not only four and five, but even six sets of concentric rings, from a very simple arrangement of glasses: they arise from reiterated internal reflections; but it will not be necessary to carry this account of seeing multiplied sets of rings to a greater length.

VII. *Of the Size of the Rings.*

The diameter of the concentric rings depends upon the radius of the curvature of the surfaces between which they are formed. Curvatures of a short radius, *cæteris paribus*, give smaller rings than those of a longer; but Sir I. NEWTON having already treated on this part of the subject at large, it will not be necessary to enter farther into it.

I should however remark, that when two curves are concerned, it is the application of them to each other that will determine the size of the rings, so that large ones may be produced from curvatures of a very short radius. A double convex lens of 2¼-inches focus, for instance, when it is laid upon a double concave which is but little more in focal length, gives rings that are larger than those from a lens of 26 inches laid upon a plain slip of glass.

VIII. *Of Contact.*

The size of the rings is considerably affected by pressure. They grow larger when the two surfaces that form them are pressed closer together, and diminish when the pressure is gradually removed. The smallest ring of a set may be increased by this means to double and treble its former diameter; but as the common or natural pressure of glasses laid upon any flat or curved surface is occasioned by their weight, the variations of pressure will not be very considerable when they are left to assume their own distance or contact. To produce that situation, however, which is generally called contact, it will always be necessary to give a little motion backwards and forwards to the incumbent lens or glass, accompanied with some moderate pressure, after which it may be left to settle properly by its own weight.

IX. *Of measuring Rings.*

It may be supposed from what has been said concerning the kind of contact, which is required for glasses to produce rings, that an attempt to take absolute measures must be liable to great inaccuracy. This was fully proved to me when I wanted to ascertain, in the year 1792, whether a lens laid upon a metalline surface would give rings of an equal diameter with those it gave when placed on glass. The measures differed so much that I was at first deceived; but on proper consideration it appeared that the HUYGENIAN object glass, of 122 feet focus, which I used for the experiment, could not so easily be brought to the same contact on metal as on glass; nor can we ever be well assured that an equal distance between the two surfaces in both cases has been actually obtained. The colour of the central point, as will be shown hereafter, may serve as a direction; but even that cannot be easily made equal in both cases. By taking a sufficient number of measures of any given ring of a set, when a glass of a sufficient focal length is used, we may however determine its diameter to about the 25th or 30th part of its dimension.

Relative measures, for ascertaining the proportion of the different rings in the same set to each other, may be more accurately taken, for in that case the contact with them all will remain the same, if we do not disturb the glasses during the time of measuring.

X. *Of the Number of Rings.*

When there is a sufficient illumination, many concentric rings in every set will be perceived; in the primary set we see generally 8, 9, or 10, very conveniently. By holding the eye in the most favourable situation I have often counted near 20, and the number of them is generally lost when they grow too narrow and minute to be perceived, so that we can never be said fairly to have counted them to their full extent. In the second set I have seen as many as in the first, and they are full as bright. The third set, when it is seen by a metalline mirror under two slips, will be brighter than the second, and almost as bright as the first: I have easily counted 7, 8, and 9 rings.

XI. *Of the Effect of Pressure on the Colour of the Rings.*

When a double convex object glass of 14 or 15 feet focus is laid on a plain slip of glass, the first colours that make their faintest appearance will be red surrounded by green; the smallest pressure will turn the center into green surrounded by red: an additional pressure will give a red center again, and so on till there have been so many successive alterations as to give us six or seven times a red center, after which the greatest pressure will only produce a very large black one surrounded by white.

When the rings are seen by transmission, the colours are in the same manner subject to a gradual alternate change occasioned by pressure; but when that is carried to its full extent, the center of the rings will be a large white spot surrounded by black.

The succession and addition of the other prismatic colours after the first or second change, in both cases is extremely beautiful; but as the experiment may be so easily made, a description, which certainly would fall short of an actual view of these phenomena, will not be necessary.

When the rings are produced by curves of a very short radius, and the incumbent lens is in full contact with the slip of glass, they will be alternately black and white; but by lessening the contact, I have seen, even with a double convex lens of no more than two-tenths of an inch focus, the center of the rings white, red, green, yellow, and black, at pleasure. In this case I used an eye-glass of one inch focus; but as it requires much practice to manage such small glasses, the experiment may be more conveniently made by placing a double convex lens of 2-inch focus on a plain slip of glass, and viewing the rings by an eye glass of $2\frac{1}{2}$ inches; then having first brought the lens into full contact, the rings will be only black and white, but by gently lifting up or tilting the lens, the center of the rings will assume various colours at pleasure.

XII. *Of diluting and concentrating the Colours.*

Lifting up or tilting a lens being subject to great uncertainty, a surer way of acting upon the colours of the rings is by dilution and concentration. After having seen that very small lenses give only black and white when in full contact, we may gradually take others of a longer focus. With a double convex lens of four inches the outward rings will begin to assume a faint red colour. With 5, 6, and 7, this appearance will increase; and proceeding with lenses of a larger focus, when we come to about 16, 18, or 20 inches, green rings will gradually make their appearance.

This and other colours come on much sooner if the center of the lens is not kept in a black contact, which, in these experiments must be attended to.

A lens of 26 inches not only shows black, white, red, and green rings, but the central black begins already to be diluted so as to incline to violet, indigo, or blue. With one of 34, the white about the dark center begins to be diluted, and shows a kind of gray inclining to yellow. With 42 and 48, yellow rings begin to become visible. With 55 and 59, blue rings show themselves very plainly. With a focal length of 9 and 11 feet, orange may be distinguished from the yellow, and indigo from the blue. With 14 feet, some violet becomes visible. When the 122 feet HUYGENIAN glass is laid on a plain slip, and well settled upon it, the central colour is then sufficiently diluted to show that the dark spot, which in small lenses, when concentrated, had the appearance of black, is now drawn out into violet, indigo, and blue, with little admixture of green; and that the white ring, which used to be about the central spot, is turned partly green with a surrounding yellow, orange, and red-coloured space or ring; by which means we seem to have a fair analysis of our former compound black and white center.

One of my slips of glass, which is probably a little concave, gave the rings still larger when the 122 feet glass was firmly pressed against it. I used a little side motion at the same time, and brought the glasses into such contact that they adhered sufficiently to be lifted up together. With this adhesion I perceived a colour surrounding a dark center which I have never seen in any prismatic spectrum. It is a kind of light brown, resembling the colour of a certain sort of Spanish snuff. The 170 feet object glass showed the same colour also very clearly.

XIII. *Of the Order of the Colours.*

The arrangement of the colours in each compound ring or alternation, seen by reflection is, that the most refrangible rays are nearest the center; and the same order takes place when seen by transmission. We have already shown that when a full dilution of the colours was obtained their arrangement was violet, indigo, blue, green, yellow, orange, and red; and the same order will hold good when the colours are gradually concentrated again; for though some of them should vanish before others, those that remain will always be found to agree with the same arrangement.

If the rings should chance to be red and green alternately, a doubt might arise which of them is nearest the center; but by the method of dilution, a little pressure, or some small increase of the focal length of the incumbent lens, there will be introduced an orange tint between them, which will immediately ascertain the order of the colours.

In the second set of rings the same order is still preserved as in the first; and the same arrangement takes place in the third set as well as in the fourth. In all of them the most refrangible rays produce the smallest rings.

XIV. *Of the alternate Colour and Size of the Rings belonging to the primary and dependent Sets.*

When two sets of rings are seen at once, and the colour of the center of the primary set is black, that of the secondary will be white; if the former is white, the latter will be black. The same alternation will take place if the colour of the center of the primary set should be red or orange; for then the center of the secondary one will be green; or if the former happens to be green, the latter will be red or orange. At the same time there will be a similar alternation in the size of rings; for the white rings in one set will be of the diameter of the black in the other; or the orange rings of the former will be of equal magnitude with the green of the latter.

When three sets of rings are to be seen, the second and third sets will be alike in colour and size, but alternate in both particulars with the primary set.

The same thing will happen when four sets are visible; for all the sets that are formed from the primary one will resemble each other, but will be alternate in the colour and dimensions of their rings with those of the primary set.

XV. *Of the sudden Change of the Size and Colour of the Rings in different Sets.*

When two sets of rings are viewed which are dependent upon each other, the colour of their centers and of all the rings in each set, may be made to undergo a sudden change by the approach of the shadow of the point of a penknife or other opaque slender body. To view this phenomenon properly, let a 16-inch double convex lens be laid upon a piece of looking-glass, and when the contact between them has been made to give the primary set with a black center, that of the secondary will be white. To keep the lens in this contact, a pretty heavy plate of lead with a circular hole in it of nearly the diameter of the lens should be laid upon it. The margin of the hole must be tapering, that no obstruction may be made to either the incident or reflected light. When this is properly arranged, bring the third shadow of the penknife upon the primary set, which is that towards the light. The real colours of this and the secondary set will then be seen to the greatest advantage. When the third shadow is advanced till it covers the second set, the second shadow will at the same time fall upon the first set, and the colour of the centers, and of all the rings in both sets, will undergo a sudden transformation from black to white and white to black.

The alternation of the colour is accompanied with a change of size, for as the white rings before the change were of a different diameter from the black ones, these latter, having now assumed a black colour, will be of a different size from the former black ones.

When the weight is taken from the lens the black contact will be changed into some other. In the present experiment it happened that the primary set got an orange coloured center and the secondary a green one. The same way of proceeding with the direction of the shadow being then pursued, the orange center was instantly changed to a green one, while at the same moment the green center was turned into orange. With a different contact I have had the primary set with a blue center and the secondary with a deep yellow one ; and by bringing the second and third shadows alternately over the primary set, the blue center was changed to a yellow, and the yellow center to a blue one ; and all the rings of both sets had their share in the transformation of colour and size.

If there are three sets of rings, and the primary set has a black center, the other two will have a white one ; and when the lowest shadow is made to fall on the third set, the central colour of all the three sets will be suddenly changed, the first from black to white, the other two from white to black.

A full explanation of these changes, which at first sight have the appearance of a magical delusion, will be found in a future article.

XVI. *Of the Course of the Rays by which different Sets of Rings are seen.*

In order to determine the course of the rays, which give the rings both by reflection and by transmission, we should begin from the place whence the light proceeds that forms them. In figure 1, we have a plano-convex lens laid upon three slips of glass, under which a metalline mirror is placed. An incident ray, 1, 2, is transmitted through the first and second surface of the lens, and comes to the point of contact at 3. Here the rings are formed, and are both reflected and transmitted : they are reflected from the upper surface of the first slip, and pass from 3 to the eye at 4 : they are also transmitted through the first slip of glass from 3 to 5 ; and at 5 they are again both reflected and transmitted ; reflected from 5 to 6, and transmitted from 5 to 7 ; from 7 they are reflected to 8, and transmitted to 9 ;

and lastly they are reflected from 9 to 10. And thus four complete sets of rings will be seen at 4, 6, 8, and 10.

The most convenient way of viewing the same rings by transmission, is that which has been mentioned in the second article of this paper, when light is conveyed upwards by reflection. In figure 2, consisting of the same arrangement of glasses as before, the light by which the rings are to be seen comes either from 1, 2, or 3, or from all these places together, and being reflected at 4, 5, and 6, rises up by transmission to the point of contact at 7, where the rings are formed. Here they are both transmitted up to the eye at 8, and reflected down to 9; from 9 they are reflected up to 10 and transmitted down to 11; from 11 they are reflected to 12 and

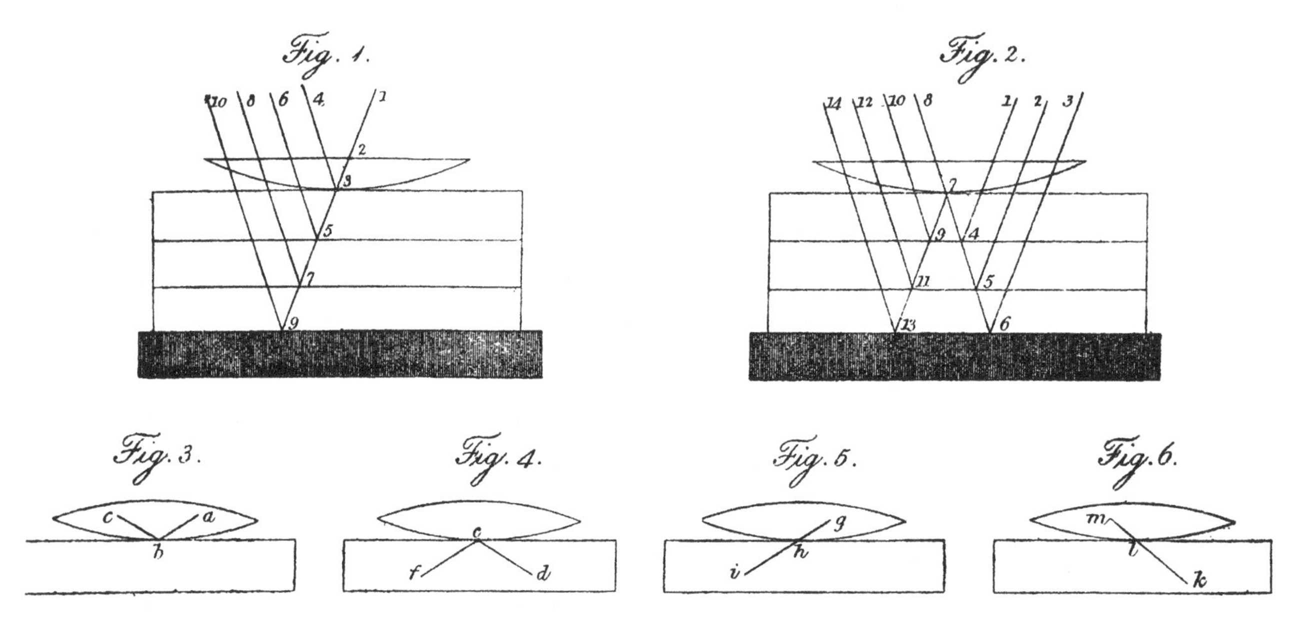

transmitted to 13; and lastly, from 13 they are reflected to 14; so, that again four sets of rings will be seen at 8, 10, 12, and 14.

This being a theoretical way of conceiving how the rays of light may produce the effects, it will be required to show by experiments that this is the actual progress of the rays, and that all the sets of rings we perceive are really reflected or transmitted in the manner that has been pointed out; but as we have so many reflections and transmissions before us, it will be necessary to confine these expressions to one particular signification when they are applied to a set of rings.

When the center of the rings is seen at the point of contact, it is a primary set; and I call it reflected, when the rays which come to that point and form the rings undergo an immediate reflection. But I call it transmitted, when the rays after having formed the rings about the point of contact are immediately transmitted.

Thus in figure 3 and 4 the rays *a b c*, *d e f*, give reflected sets of rings; and the rays *g h i*, *k l m*, in figure 5 and 6, give transmitted sets.

In this denomination, no account is taken of the course of the rays before they come to *a, d, g, k* ; nor of what becomes of them after their arrival at *c, f, i, m* : they may either come to those places or go from them by one or more transmissions or reflections, as the case may require ; but our denomination will relate only to their course immediately after the formation of the rings between the glasses.

The secondary and other dependent sets will also be called reflected or transmitted by the same definition : and as a set of these rings formed originally by reflection may come to the eye by one or more subsequent transmissions ; or being formed by transmission, may at last be seen by a reflection from some interposed surface, these subsequent transmissions or reflections are to be regarded only as convenient ways to get a good sight of them.

With this definition in view, and with the assistance of a principle which has already been proved by experiments, we may explain some very intricate phenomena ; and the satisfactory manner of accounting for them will establish the truth of the theory relating to the course of rays that has been described.

The principle to which I refer is, that when the pressure is such as to give a black center to a set of rings seen by reflection, the center of the same set, with the same pressure of the glasses seen by transmission will be white.*

I have only mentioned black and white, but any other alternate colours, which the rings or centers of the two sets may assume, are included in the same predicament.

XVII. *Why two connected Sets of Rings are of alternate Colours.*

It has already been shown, when two sets of rings are seen, that their colours are alternate, and that the approach of the shadow of a penknife will cause a sudden change of them to take place. I shall now prove that this is a very obvious consequence of the course of rays that has been proposed. Let figures 7 and 8 represent the arrangement given in a preceding article, where a 16-inch lens was laid upon a looking-glass, and gave two sets of rings with centers of different colours : but let figure 7 give them by one set of rays, and figure 8 by another. Then, if the incident rays come in the direction which is represented in figure 7, it is evident that we see the primary set with its center at 2 by reflection, and the secondary one at 4 by transmission. Hence it follows, in consequence of the admitted principle, that if the contact is such as to give us the primary set with a black center, the secondary set must have a white one ; and thus the reason of the alternation is explained.

But if the rays come as represented in figure 8, we see the primary set by transmission, and the secondary one by reflection ; therefore, with an equal pressure of the glasses, the primary center must now be white, and the secondary one black.

Without being well acquainted with this double course of rays, we shall be liable to frequent mistakes in our estimation of the colour of the centers of two sets

* See Article XI. of this Paper.

of rings; for by a certain position of the light, or of the eye, we may see one set by one light and the other set by the other.

XVIII. *Of the Cause of the sudden Change of the Colours.*

Having thus accounted for the alternation of the central colours, we may easily conceive that the interposition of the penknife must have an instantaneous effect upon them. When it stops the rays of figure 7, which will happen when its second shadow falls upon the primary set, the rings will then be seen by the rays 1, 2, 3, 4, and 1, 2, 3, 5, 6, of figure 8. When it stops the rays of figure 8, which must happen when the third shadow falls upon the primary set, we then see both sets by the rays

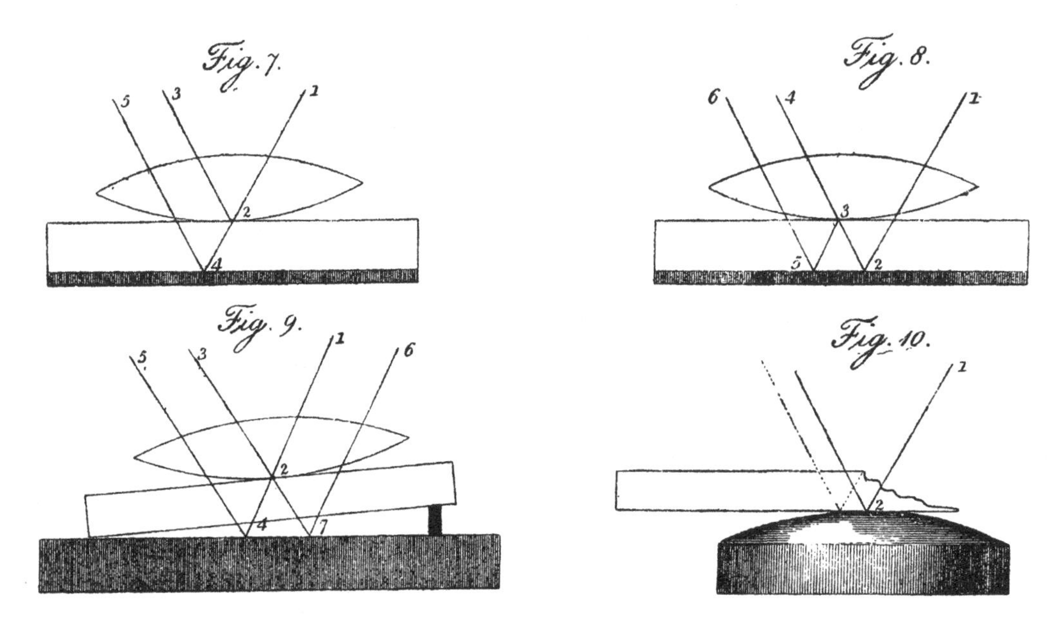

1, 2, 3, and 1, 2, 4, 5, of figure 7. When the penknife is quite removed both sets of rays will come to the point of contact, and in some respects interfere with each other; but the strongest of the two, which is generally the direct light of figure 7, will prevail. This affords a complete explanation of all the observed phenomena: by the rays of figure 7 the centers will be black and white; by those of figure 8 they will be white and black; and by both we shall not see the first set so well as when the third shadow being upon it, has taken away the rays of figure 8: indeed we can hardly see the secondary set at all, till the shadow of the penknife has covered either the rays of figure 7 or of figure 8.

As soon as we are a little practised in the management of the rays, by knowing their course, we may change the colour so gradually as to have half the center white while the other half shall still remain black; and the same may be done with green and orange, or blue and yellow centers. The rings of both sets will also participate in the gradual change; and thus what has been said of the course of rays in the 16th article will again be confirmed.

XIX. *Of the Place where the different Sets of Rings are to be seen.*

By an application of the same course of the rays, we may now also determine the situation of the place where the different sets of rings are seen : for according to what has been said in the foregoing article, the situation of the primary set should be between the lens and the surface of the looking-glass : and the place of the secondary one at the metalline coating of the lowest surface. To try whether this be actually as represented, let us substitute a metalline mirror with a slip of glass laid upon it in the room of the piece of looking-glass ; and let there be interposed a short bit of wood, one-tenth of an inch thick, between the slip of glass and the mirror, so as to keep up that end of the slip which is towards the light. This arrangement is represented in figure 9, where both sets of rays are delineated. Then if we interpose a narrow tapering strip of card, discoloured with japan ink, between the slip of glass and the mirror, so as to cover it at 7, we do not only still perceive the primary set, but see it better than before : which proves that being situated above the slip of glass the card below cannot cover it. If on the contrary we insert the strip of card far enough, that it may at the same time cover the mirror both at 4 and at 7, we shall lose the secondary set, which proves that its situation was on the face of the mirror.

When several sets of rings are to be perceived by the same eye-glass, and they are placed at different distances, a particular adjustment of it will be required for each set, in order to see it well defined. This will be very sensible when we attempt to see three or four sets, each of them situated lower than the preceding ; for without a previous adjustment to the distance of the set intended to be viewed we shall be seldom successful ; and this is therefore a corroborating proof of the situation that has been assigned to different sets of rings.

XX. *Of the Connection between different Sets of Rings.*

It will now be easy to explain in what manner different sets of rings are connected, and why they have been called primary and dependent. When the incident rays come to the point of contact and form a set of rings, I call it the primary one : when this is formed some of the same rays are continued by transmission or reflection, but modified so as to convey an image of the primary set with opposite colours forward through any number of successive transmissions or reflections ; whenever this image comes to the eye, a set of rings will again be seen, which is a dependent one. Many proofs of the dependency of second, third, and fourth sets of rings upon their primary one may be given ; I shall only mention a few.

When two sets of rings are seen by a lens placed upon a looking-glass, the center of the secondary set will always remain in the same plane with the incident and reflected rays passing through the center of the primary one. If the point of contact by tilting is changed, the secondary set will follow the motion of the primary set ;

and if the looking-glass is turned about, the secondary will be made to describe a circle upon that part of the looking-glass which surrounds the primary one as a center. If there is a defect in the center or in the rings of the primary set there will be exactly the same defect in the secondary one; and if the rays that cause the primary set are eclipsed, both sets will be lost together. If the colour of the primary one is changed, that of the secondary will also undergo its alternate change, and the same thing will hold good of all the dependent rings when three or four sets of them are seen that have the same primary one.

The dependency of all the sets on their primary one may also be perceived when we change the obliquity of the incident light; for the centers of the rings will recede from one another when that is increased and draw together when we lessen it, which may go so far that by an incidence nearly perpendicular we shall bring the dependent sets of rings almost under the primary one.

XXI. *To account for the Appearance of several Sets of Rings with the same coloured Centers.*

It has often happened that the colour of the centers of different sets was not what the theory of the alternation of the central colours would have induced me to expect: I have seen two, three, and even four sets of rings, all of which had a white center. We are however now sufficiently prepared to account for every appearance relating to the colour of rings and their centers.

Let an arrangement of glasses be as in figure 9. When this is laid down so as to receive an illumination of day light, which should not be strong, nor should it be very oblique, the reflection from the mirror will then exceed that from the surface of glass; therefore the primary set will be seen by the rays 6, 7, coming to the mirror at 7, and going through the point of contact in the direction 7, 2, 3, which proves it to be a set that is seen by transmission, and it will therefore have a white center. The rays 1, 2, 4, passing through the point of contact, will also form a transmitted set with a white center, which will be seen when the reflection from 4 to 5 conveys it to the eye. But these two sets have no connection with each other; and as primary sets are independent of all other sets, I have only to prove that this secondary set belongs not to the primary one which is seen, but to another invisible one. This may be done as follows.

Introduce the black strip of card that has been mentioned before, till it covers the mirror at 7; this will take away the strong reflection of light which overpowers the feeble illumination of the rays 1, 2, 3; and the real hitherto eclipsed primary set belonging to the secondary one with a white center, will instantly make its appearance with a black one. We may alternately withdraw and introduce again the strip of card, and the center of the primary set will be as often changed from one colour to its opposite; but the secondary set, not being dependent on the rays 6, 7, will not be in the least affected by the change.

If the contact should have been such as to give both sets with orange centers, the introduction of the strip of card will prove that the set which is primary to the other has really a green center.

Another way of destroying the illusion is to expose the same arrangement to a brighter light, and at the same time to increase the obliquity of the angle of incidence; this will give a sufficient reflection from the surface of the glass to be no longer subject to the former deceptive appearance; for now the center of the primary set will be black, as it ought to be.

XXII. *Of the reflecting Surfaces.*

The rays of light that form rings between glasses, must undergo certain modifications by some of the surfaces through which they pass, or from which they are reflected; and to find out the nature of these modifications, it will be necessary to examine which surfaces are efficient. As we see rings by reflection and also by transmission, I shall begin with the most simple, and show experimentally the situation of the surface that reflects, not only the primary, but also the secondary sets of rings.

Upon a slip of glass, the lowest surface of which was deprived of its polish by emery, I laid an object-glass of 21 feet focal length, and saw a very complete set of rings. I then put the same glass upon a plain metalline mirror, and saw likewise a set of them. They were consequently not reflected from the lowest surface of the subjacent glass or metal.

It will easily be understood, that were we to lay the same object glass upon a slip of glass emeried on both sides, or upon an unpolished metal, no rings would be seen. It is therefore neither from the first surface of the incumbent object-glass, nor from its lowest, that they are reflected; for if they could be formed without the modification of reflection from the upper surface of a subjacent glass or metal, they would still be seen when laid on rough surfaces; and consequently, the efficient reflecting surface, by which we see primary sets of rings, is that which is immediately under the point of contact.

To see a secondary set of rings by reflection, is only an inversion of the method of seeing a primary one. For instance, when a lens is laid upon a looking-glass, the course of the rays represented in figure 8, will show that the rays 1, 2, 3, 5, 6, by which a secondary set is seen, are reflected about the point of contact at 3, and that the lowest surface of the incumbent lens is therefore the efficient reflecting one; and thus it is proved, that in either case of seeing reflected rings, one of the surfaces that are joined at the point of contact contributes to their formation by a certain modification of reflection.

XXIII. *Of the transmitting Surfaces.*

It would seem to be almost self-evident, that when a set of rings is seen by transmission, the light which occasions them must come through all the four surfaces of the two glasses which are employed; and yet it may be shown that this is not necessary. We may, for instance, convey light into the body of the subjacent glass through its first surface, and let it be reflected within the glass at a proper angle, so that it may come up through the point of contact, and reach the eye, having been transmitted through no more than three surfaces. To prove this I used a small box, blackened on the inside, and covered with a piece of black pasteboard, which had a hole of about half an inch in the middle. Over this hole I laid a slip of glass with a 56-inch lens upon it; and viewed a set of rings given by this arrangement very obliquely, that the reflection from the slip of glass might be copious. Then guarding the point of contact between the lens and the slip of glass from the direct incident light, I saw the rings, after the colour of their center had been changed, by means of an internal reflection from the lowest surface of the slip of glass; by which it rose up through the point of contact, and formed the primary set of rings, without having been transmitted through the lowest surface of the subjacent glass. The number of transmitted surfaces is therefore by this experiment reduced to three; but I shall soon have an opportunity of showing that so many are not required for the purpose of forming the rings.

XXIV. *Of the Action of the first Surface.*

We have already shown that two sets of rings may be seen by using a lens laid upon a slip of glass; in which case, therefore, whether we see the rings by reflection or by transmission, no more than four surfaces can be essential to their formation. In the following experiments for investigating the action of these surfaces I have preferred metalline reflection, when glass was not required, that the apparatus might be more simple.

Upon a plain metalline mirror I laid a double convex lens, having a strong emery scratch on its upper surface. When I saw the rings through the scratch, they appeared to have a black mark across them. By tilting the lens, I brought the center of the rings upon the projection of the scratch, so that the incident light was obliged to come through the scratch to the rings, and the black mark was again visible upon them, but much stronger than before. In neither of the situations were the rings disfigured. The stronger mark was owing to the interception of the incident light, but when the rings had received their full illumination the mark was weaker, because in the latter case the rings themselves were probably complete, but in the former deficient.

I placed a lens that had a very scabrous polish on one side, but was highly polished on the other, upon a metalline mirror. The defective side being uppermost, I did not find that its scabrousness had any distorting effect upon the rings.

I splintered off the edge of a plain slip of glass; it broke as it usually does with a waving striated, curved slope coming to an edge. The splintered part was placed upon a convex metalline mirror of 2-inch focus, as in figure 10. The irregularity of the striated surface through which the incident ray 1, 2, was made to pass had very little effect upon the form of the rings; the striæ appearing only like fine dark lines, with hardly any visible distortion; but when, by tilting the returning ray, 2, 3, was also brought over the striated surface, the rings were much disfigured. This experiment therefore seems to prove that a very regular refraction of light by the first surface is not necessary; for though the rings were much disfigured when the returning light came through the splintered defect, this is no more than what must happen to the appearance of every object which is seen through a distorting medium.

I laid the convex side of a plano-convex lens of 2·8-inch focus with a diameter of 1·5 upon a plain mirror, and when I saw a set of rings I tilted the lens so as to bring the point of contact to the very edge of the lens, both towards the light and from the light, which, on account of the large diameter of the lens, gave a great variety in the angle of incidence to the rays which formed the rings; but no difference in their size or appearance could be perceived. This seems to prove that no modification of the first surface in which the angle of incidence is concerned, such as refraction and dispersion, has any share in the production of the rings, and that it acts merely by the intromission of light; and though even this is not without being influenced by a change of the angle, it can only produce a small difference in the brightness of the rings.

A more forcible argument, that leads to the same conclusion, is as follows. Laying down three 54-inch double convex lenses, I placed upon the first the plain side of a plano-convex lens of $\frac{6}{8}$ inch focus; upon the second, a plain slip of glass; and upon the third, the plain side of a plano-concave lens also $\frac{6}{8}$ inch focus. I had before tried the same experiment with glasses of a greater focal length, but selected these to strengthen the argument. Then, as nothing could be more different than the refraction of the upper surfaces of these glasses, I examined the three sets of rings that were formed by these three combinations, and found them so perfectly alike that it was not possible to perceive any difference in their size and colour. This shows that the first surface of the incumbent glasses merely acts as an inlet to the rays that afterwards form the rings.

To confirm the idea that the mere admission of light would be sufficient, I used a slip of glass polished on one side but roughened with emery on the other; this being laid upon a 21-feet object-glass, I saw a set of rings through the rough surface; and though they appeared hazy, they were otherwise complete in figure and colour. The slip of glass when laid in the same manner upon the letters of a book made them appear equally hazy; so that the rings were probably as sharply formed as the letters.

Having now already great reason to believe that no modification, that can be given by the first surface to the incident rays of light, is essential to the formation of the rings, I made the following decisive experiment.

Upon a small piece of looking-glass I laid half a double-convex lens of 16-inches focus, with the fracture exposed to the light, as represented in figure 11. Under the edge of the perfect part of the lens was put a small lump of wax, soft enough to allow a gentle pressure to bring the point of contact towards the fractured edge, and to keep it there. In this arrangement it has already been shown that there are two different ways of seeing two sets of rings: by the rays 1, 2, 3, we see a primary set; and by 1, 2, 4, 5, the secondary set belonging to it: by the rays 6, 7, 2, 3 we see a different primary set; and by 6, 7, 2, 4, 5, we see its secondary one. That this theory is well founded has already been proved; but if we should have a doubt

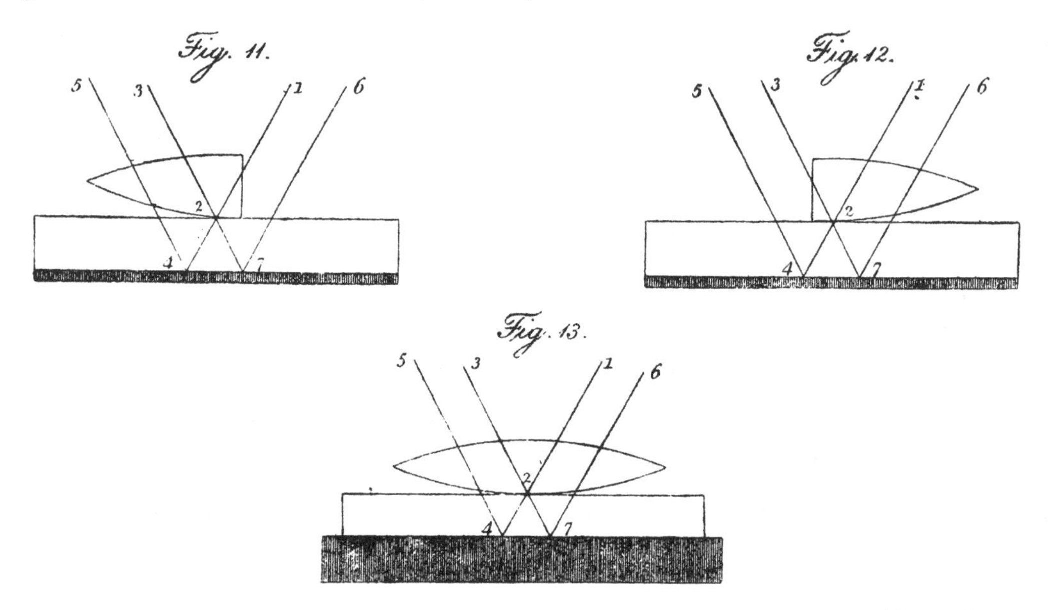

remaining, the interposition of any small opaque object upon the looking-glass near the fracture will instantly stop the latter two sets of rings, and show the alternate colours of the two sets that will then be seen by the rays 1, 2, 3, and 1, 2, 4, 5. Remove in the next place the stop from the looking-glass, and bring the second shadow of the penknife over the primary set, and there will then only remain the two sets of rings formed by incident rays which come from 6, and which have never passed through the upper surface of the lens. Now, as both sets of rings in this case are completely formed by rays transmitted upwards from the coated part of the looking-glass without passing through the first surface of the incumbent lens, the proof that the modifying power of that surface is not required to the formation of the rings is established.

It can hardly be supposed that the first surface of the lens should have any concern in the formation of the rings when the rays are reflected from the looking-glass towards the eye; but the same experiment, that has proved that this surface was not required to be used with incident rays, will show that we may do without it when they are on their return. We need only invert the fractured lens, as in figure 12, when either the rays 1, 2, 4, 5, or 6, 7, 2, 4, 5, will convey the image of the rings after their formation to the eye without passing through any part of the lens.

XXV. *Of the Action of the second Surface.*

As rings are formed when two glasses are laid upon each other, it is but reasonable to expect that the two surfaces at least which are placed together should have an immediate effect upon them, and so much the more, as it has been ascertained that the first surface assists only by permitting light to pass into the body of the glass. Some of the experiments that have been instituted for examining the action of the first surface will equally serve for investigating that of the second.

The lens already used with a strong emery scratch being again placed on the mirror, but with the injured side downwards, I found that the rings, when brought under the scratch, were not distorted ; they had only a black mark of the same shape as the scratch across them.

The lens with a scabrous side was also placed again upon the mirror, but with the highly polished side upwards. In this position the scabrousness of the lowest surface occasioned great irregularity among the rings, which were indented and broken wherever the little polished holes that make up a scabrous surface came near them; and if by gently lifting the lens a strong contact was prevented, the colours of the rings were likewise extremely disfigured and changed.

As we have now seen that a polished defect upon the second surface will affect the figure of the rings that are under them, it will remain to be determined whether such defects do really distort them by some modification they give to the rays of light in their passage through them, or whether they only represent the rings as deformed, because we see them through a distorted medium. For although the scabrousness did not sensibly affect the figure of the rings when it was on the first surface, we may suppose the little polished holes to have a much stronger effect in distorting the appearance of the rings when they are close to them. The following experiment will entirely clear up this point.

Over the middle of a 22-inch double-convex lens I drew a strong line with a diamond, and gave it a polish afterwards that it might occasion an irregular refraction. This being prepared, I laid a slip of glass upon a plain metalline mirror, and placed the lens with the polished line downwards upon the slip of glass. This arrangement has been shown to give two sets of rings. When I examined the primary set, a strong disfiguring of the rings was visible ; they had the appearance of having been forced asunder, or swelled out, so as to be much broader one way than another. The rings of the secondary set had exactly the same defects, which being strongly marked, could not be mistaken. The centers of the two sets, as usual, were of opposite colours, the first being black, the second white ; and all those defects that were of one colour in the first set, were of the opposite colour in the second. When, by the usual method, I changed the colours of the centers of the rings, making that of the primary white and of the secondary black, the defects in both sets were still exactly alike, and as before; except that they had also undergone the like transformation of colour, each having assumed its opposite. It remains now only to show that this experiment is decisive; for by the established course of the rays we saw the secondary set of rings when it had a white center by the transmitted rays marked 1, 2, 4, 5, in figure 13; and when it had a black one, by the reflected rays 6, 7, 2, 4, 5, of the same figure; but in neither of these two cases did the rays come through the defective part of the lens in their return to the eye.

This experiment proves more than we might at first be aware of; for it does not only establish that the second surface, when properly combined with a third surface, has a modifying power whereby it can interrupt the regularity of the rings, but also one whereby it contributes to their formation; for, if it can give an irregular figure to them by transmitting its irregularly modified rays, it follows, that when these rays are regularly modified it will be the cause of the regular figure of the rings. Nay, it proves more; for if it modifies the figure of the rings by transmission, it modifies them no less by reflection; which may be seen by following the course of the rays 6, 7, 2, 4, 5; for as they do not pass through the defective place of the lens, they can only receive their modification from it by reflection. This opens a field of view to us that leads to the cause of all these intricate phenomena, of which in a second part of this paper I shall avail myself.

XXVI. *Of the Action of the third Surface.*

When a double-convex lens is laid upon a plain metalline mirror that happens to have an emery scratch in its surface, we see it as a black line under the rings that are formed over them. This shows, that when a defect from want of polish has not a power to reflect light in an irregular manner, it cannot distort the rings that are formed upon it.

When I laid a good 21-feet object glass upon a plain slip that had some defects in its surface, the rings, in every part of the object glass that was brought over them, were always disfigured; which proves that a reflection from a defective third surface has a power of forming distorted rings, and that consequently a reflection from one that is perfect must have a power of forming rings without distortion, when it is combined with a proper second surface.

When the defective slip of glass, with a perfect lens upon it, was placed upon a metalline mirror, I saw the secondary set affected by distortions of the rings that were perfectly like those in the primary set; which proves that a polished defect in the third surface will give modifications to the rays that form the rings by transmission as well as by reflection.

XXVII. *The Colour of the reflecting and transmitting Surfaces is of no consequence.*

I laid seven 54-inch double-convex lenses upon seven coloured pieces of plain glass. The colours of the glasses were those which are given by a prism, namely, violet, indigo, blue, green, yellow, orange, and red. The rings reflected from each of these glasses were in every respect alike ; at least so far that I could have a black, a white, a red, an orange, a yellow, a green, or a blue center with every one of them, according to the degree of pressure I used. The lenses being very transparent, it may be admitted that the colours of the glasses seen through them would in some degree mix with the colours of the rings ; but the action of the cause that gives the rings was not in the least affected by that circumstance.

I saw the rings also by direct transmission through all the coloured glasses except a dark red, which stopped so much light that I could not perceive them. The colour of the glasses, in this way, coming directly to the eye, gave a strong tinge to the centers of the rings, so that instead of a pure white I had a bluish-white, a greenish-white, and so of the rest ; but the form of the rings was no less perfect on that account.

XXVIII. *Of the Action of the fourth Surface.*

We have already seen that a set of rings may be completely formed by reflection from a third surface, without the introduction of a fourth ; this, at all events, must prove that such a surface is not essential to the formation of rings, but as not only in direct transmission, but also when two sets of rings are to be seen, one of which may be formed by transmission, this fourth surface must be introduced, I have ascertained by the following experiments how far the same has any share in the formation of rings.

In direct transmission, where the light comes from below, the fourth surface will take the part which is acted by the first, when rings are seen reflected from a metalline mirror. Its office therefore will be merely to afford an entrance to the rays of light into the substance of the subjacent glass ; but when that light is admitted through the first, second, and third surfaces, the fourth takes the office of a reflector, and sends it back towards the point of contact. It will not be required to examine this reflection, since the light thus turned back again is, with respect to the point of contact, in the same situation in which it was after its entrance through the first surface when it proceeded to the same point ; but when two sets of rings are to be formed by rays, either coming through this point directly towards the fourth surface, or by reflection from the same point towards the place where the secondary rings are to be seen, it will then be necessary to examine whether this surface has any share in their formation, or whether these rings, being already completely formed, are only reflected by it to the eye. With a view to this, I selected a certain polished defect in the surface of a piece of coach-glass, and when a 26-inch lens was laid upon it, the rings of the set it produced were much distorted. The lens was then put upon a perfect slip of glass, and both together were laid upon the defective place of the coach-glass. The rings of the secondary set reflected by it were nevertheless as perfect as those of the primary set. It occurred to me that these rings might possibly be reflected from the lowest surface of the perfect slip of glass, especially as by lifting it up from the coach-glass I still continued to see both sets. To clear up this point, therefore, I took away the slip, and turning the defective place of the coach-glass downwards, produced a set of perfect rings between the lens and the upper surface of the coach-glass, and brought it into such a situation that a secondary set must be reflected from the defective place of the lowest surface. This being obtained, the rings of this set were again as well formed and as free from distortions as those of the primary set.

Upon a plain metalline mirror I laid down two lenses, one a plano-convex, the other a plano-concave, both of 2·9 inches focus, and having the plain side upwards. When two 21-inch double-convex glasses were laid upon them, the secondary sets of both the combinations were of equal size, and perfectly like their primary sets; which proves that the refraction of the fourth surface is either not at all concerned, or at least has so little an effect in altering the size of the rings that it cannot be perceived.

The result of the foregoing experiments, relating to the action of the several surfaces, is,

I. That only two of them are essential to the formation of concentric rings.

II. That these two must be of a certain regular construction, and so as to form a central contact.

III. That the rays from one side or the other, must either pass through the point of contact, or through one of the surfaces about the same point to the other to be reflected from it.

IV. And that in all these cases a set of rings will be formed, having their common center in the place where the two surfaces touch each other.

XXIX. *Considerations that relate to the Cause of the formation of concentric Rings.*

It is perfectly evident that the phenomena of concentric rings must have an adequate cause, either in the very nature or motion of the rays of light, or in the modifications that are given to them by the two essential surfaces that act upon them at the time of the formation of the rings.

This seems to reduce the cause we are looking for to an alternative that may be determined; for if it can be shown that a disposition of the rays of light to be alternately reflected and transmitted cannot account for the phenomena which this hypothesis is to explain, a proposition of accounting for them by modifications that may be proved, even on the very principles of Sir I. NEWTON to have an existence, will find a ready admittance. I propose, therefore, now to give some arguments, which will remove an obstacle to the investigation of the real cause of the formation of the concentric rings; for after the very plausible supposition of the alternate fits, which agrees so wonderfully well with a number of facts that have been related, it will hardly be attempted, if these should be set aside, to ascribe some other inherent property to the rays of light, whereby we might account for them ; and thus we shall be at liberty to turn our thoughts to a cause that may be found in the modifications arising from the action of the surfaces which have been proved to be the only essential ones in the formation of rings.

XXX. *Concentric Rings cannot be formed by an alternate Reflection and Transmission of the Rays of Light.*

One of the most simple methods of obtaining a set of concentric rings is to lay a convex lens on a plain metalline mirror ; but in this case we can have no transmission of rays, and therefore we cannot have an alternate reflection and transmission of them. If to get over this objection it should be said that, instead of transmission, we ought to substitute absorption ; since those rays which in glass would have been transmitted will be absorbed by the metal, we may admit the elusion ; it ought however to have been made a part of the hypothesis.

XXXI. *Alternate Fits of easy Reflection and easy Transmission, if they exist, do not exert themselves according to various Thicknesses of thin Plates of Air.*

In the following experiment, I placed a plain well polished piece of glass 5·6 inches long, and 2·3 thick, upon a plain metalline mirror of the same length with the glass ; and in order to keep the mirror and glass at a distance from each other, I laid between them, at one end, a narrow strip of such paper as we commonly put between prints. The thickness of that which I used was the 640th part of an inch ; for 128 folds of them laid together would hardly make up two-tenths. Upon the glass I put a 39-inch double-convex lens ; and having exposed this combination to a proper light, I saw two complete sets of coloured rings.

In this arrangement, the rays which convey the secondary set of rings to the eye must pass through a thin wedge of air, and if these rays are endowed with permanent fits of easy reflection, and easy transmission, or absorption, their exertion, according to Sir I. NEWTON, should be repeated at every different thickness of the plate of air, which amounts to the $\frac{1}{88952}$ part of an inch, of which he says "Hæc est crassitudo aeris in primo annulo obscuro radiis ad perpendiculum incidentibus exibito, qua parte is annulus obscurissimus est." The length of the thin wedge of air, reckoned from the line of contact, to the beginning of the interposed strip of paper, is 5·2 inches, from which we calculate that it will have the above mentioned thickness at $\frac{1}{27}$ of an inch from the contact ; and therefore at $\frac{1}{54}$, $\frac{3}{54}$, $\frac{5}{54}$, $\frac{7}{54}$, $\frac{9}{54}$, $\frac{11}{54}$, &c. we shall have the thickness of air between the mirror and glass, equal to $\frac{1}{178000}$, $\frac{3}{178000}$, $\frac{5}{178000}$, $\frac{7}{178000}$, &c. of which the same author says that they give "crassitudines Aeris in omnibus Annulis lucidis, qua parte illi lucidissimi sunt." Hence it follows that, according to the above hypothesis, the rings of the secondary set which extended over a space of ·14 of an inch, should suffer more than seven interruptions of shape and colour in the direction of the wedge of air.

In order to ascertain whether such an effect had any existence, I viewed the secondary set of rings upon every part of the glass-plate, by moving the convex lens from one end of it gradually to the other; and my attention being particularly directed to the 3d, 4th, and 5th rings, which were extremely distinct, I saw them retain their shape and colour all the time without the smallest alteration.

The same experiment was repeated with a piece of plain glass instead of the metalline mirror, in order to give room for the fits of easy transmission, if they existed, to exert themselves; but the result was still the same; and the constancy of the brightness and colours of the rings of the secondary set, plainly proved that the rays of light were not affected by the thickness of the plate of air through which they passed.

XXXII. *Alternate Fits of easy Reflection and easy Transmission, if they exist, do not exert themselves according to various Thicknesses of thin Plates of Glass.*

I selected a well polished plate of coach glass 17 inches long, and about 9 broad. Its thickness at one end was 33, and at the other 31 two-hundredths of an inch; so that in its whole length it differed $\frac{1}{100}$ of an inch in thickness. By measuring many other parts of the plate I found that it was very regularly tapering from one end to the other. This plate, with a double convex lens of 55 inches laid upon it, being placed upon a small metalline mirror, and properly exposed to the light, gave me the usual two sets of rings. In the secondary set, which was the object of my attention, I counted twelve rings, and estimated the central space between them to be about $1\frac{1}{3}$ times as broad as the space taken up by the 12 rings on either side; the whole of the space taken up may therefore be reckoned equal to the breadth of 40 rings of a mean size: for the 12 rings, as usual, were gradually contracted in breadth as they receded from the center, and, by a measure of the whole space thus taken up, I found that the breadth of a ring of a mean size was about the 308th part of an inch.

Now, according to Sir I. NEWTON'S calculation of the action of the fits of easy reflection and easy transmission in thick glass plates, an alternation from a reflecting to a transmitting fit requires a difference of $\frac{1}{137545}$ part of an inch in thickness; * and by calculation this difference took place in the glass plate that was used at every 80th part of an inch of its whole length; the 12 rings, as well as the central colour of the secondary set, should consequently have been broken by the exertion of the fits at every 80th part of an inch; and from the space over which these rings extended, which was about ·13 inch, we find that there must have been more than ten such interruptions or breaks in a set of which the 308th part was plainly to be distinguished. But when I drew the glass plate gently over the small mirror, keeping the secondary set of rings in view, I found their shape and colour always completely well formed.

* NEWTON'S *Optics*, p. 277.

This experiment was also repeated with a small plain glass instead of the metalline mirror put under the large plate. In this manner it still gave the same result, with no other difference but that only six rings could be distinctly seen in the secondary set, on account of the inferior reflection of the subjacent glass.

XXXIII. *Coloured Rings may be completely formed without the Assistance of any thin or thick Plates, either of Glass or of Air.*

The experiment I am now to relate was at first intended to be reserved for the second part of this paper, because it properly belongs to the subject of the flection of the rays of light, which is not at present under consideration ; but as it particularly opposes the admission of alternate fits of easy reflection and easy transmission of these rays in their passage through plates of air or glass, by proving that their assistance in the formation of rings is not required, and also throws light upon a subject that has at different times been considered by some of our most acute experimentalists, I have used it at present, though only in one of the various arrangements, in which I shall have occasion to recur to it hereafter.

Sir I. NEWTON placed a concave glass mirror at double its focal length from a chart, and observed that the reflection of a beam of light admitted into a dark room, when thrown upon this mirror, gave " four or five concentric irises or rings of colours like " rainbows." * He accounts for them by alternate fits of easy reflection and easy transmission exerted in their passage through the glass-plate of the concave mirror.†

The Duke De CHAULNES concluded from his own experiments of the same phenomena, " that these coloured rings depended upon the first surface of the mirror, and that the second surface, or that which reflects them after they had passed the first, only served to collect them and throw them upon the pasteboard, in a quantity sufficient to make them visible."‡

Mr. BROUGHAM, after having considered what the two authors I have mentioned had done, says, " that upon the whole there appears every reason to believe that the rings are formed by the first surface out of the light which, after reflection from the second surface, is scattered, and passes on to the chart." §

My own experiment is as follows. I placed a highly polished 7 feet mirror, but of metal instead of glass, that I might not have two surfaces, at the distance of 14 feet from a white screen, and through a hole in the middle of it one-tenth of an inch in diameter I admitted a beam of the sun into my dark room, directed so as to fall perpendicularly on the mirror. In this arrangement the whole screen remained perfectly free from light, because the focus of all the rays which came to the mirror was by reflection thrown back into the hole through which they entered. When all was duly prepared, I made an assistant strew some hair-powder with a puff into

* NEWTON'S *Optics*, p. 265. † *Ibid.*, p. 277.
‡ PRIESTLEY'S *History, &c.* on the Colours of thin Plates, p. 515.
§ *Phil. Trans.* for 1796, p. 216.

the beam of light, while I kept my attention fixed upon the screen. As soon as the hair-powder reached the beam of light the screen was suddenly covered with the most beautiful arrangement of concentric circles displaying all the brilliant colours of the rainbow. A great variety in the size of the rings was obtained by making the assistant strew the powder into the beam at a greater distance from the mirror; for the rings contract by an increase of the distance, and dilate on a nearer approach of the powder.

This experiment is so simple, and points out the general causes of the rings which are here produced in so plain a manner, that we may confidently say they arise from the flection of the rays of light on the particles of the floating powder, modified by the curvature of the reflecting surface of the mirror.

Here we have no interposed plate of glass of a given thickness between one surface and another, that might produce the colours by reflecting some rays of light and transmitting others; and if we were inclined to look upon the distance of the particles of the floating powder from the mirror as plates of air, it would not be possible to assign any certain thickness to them, since these particles may be spread in the beam of light over a considerable space, and perhaps none of them will be exactly at the same distance from the mirror.

I shall not enter into a further analysis of this experiment, as the only purpose for which it is given in this place is to show that the principle of thin or thick plates, either of air or glass, on which the rays might alternately exert their fits of easy reflection and easy transmission, must be given up, and that the fits themselves of course cannot be shown to have any existence.

XXXIV. *Conclusion.*

It will hardly be necessary to say, that all the theory relating to the size of the parts of natural bodies and their interstices, which Sir I. NEWTON has founded upon the existence of fits of easy reflection and easy transmission, exerted differently, according to the different thickness of the thin plates of which he supposes the parts of natural bodies to consist, will remain unsupported; for if the above mentioned fits have no existence, the whole foundation on which the theory of the size of such parts is placed, will be taken away, and we shall consequently have to look out for a more firm basis on which a similar edifice may be placed. That there is such a one we cannot doubt, and what I have already said will lead us to look for it in the modifying power which the two surfaces, that have been proved to be essential to the formation of rings, exert upon the rays of light. The Second Part of this Paper, therefore, will enter into an examination of the various modifications that light receives in its approach to, entrance into, or passage by, differently disposed surfaces or bodies; in order to discover, if possible, which of them may be the immediate cause of the coloured rings that are formed between glasses.

LIX.

Observations on the Nature of the new celestial Body discovered by Dr. OLBERS, *and of the Comet which was expected to appear last January in its return from the Sun.*

[*Phil. Trans.*, 1807, pp. 260–266.]

Read June 4, 1807.

THE late discovery of an additional body belonging to the solar system, by Dr. OLBERS, having been communicated to me the 20th of April, an event of such consequence engaged my immediate attention. In the evening of the same day I tried to discover its situation by the information I had obtained of its motion; but the brightness of the moon, which was near the full, and at no great distance from the object for which I looked, would not permit a star of even the 5th magnitude to be seen, and it was not till the 24th that a tolerable view could be obtained of that space of the heavens in which our new wanderer was pursuing its hitherto unknown path.

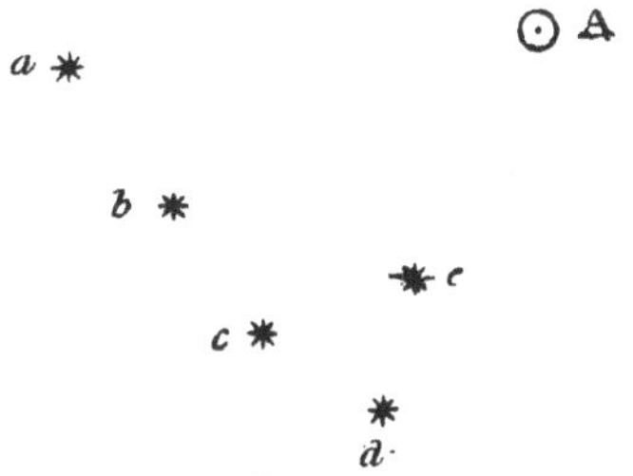

⊙ A

a ✳

b ✳

✳ e

c ✳

✳
d

FIG. 1.

As soon as I found that small stars might be perceived, I made several delineations of certain telescopic constellations, the first of which was as represented in figure 1, and I fixed upon the star A, as most likely, from its expected situation and brightness, to be the one I was looking for. The stars in this figure, as well as in all the other delineations I had made, were carefully examined with several magnifying powers, that in case any one of them should hereafter appear to have been the lately discovered object, I might not lose the opportunity of an early acquaintance with its condition. An observation of the star marked A, in particular, was made with a very distinct magnifying power of 460, and says, that it had nothing in its appearance that differed from what we see in other stars of the same size; indeed Dr. OLBERS, by mentioning in the communication which I received, that with such magnifying powers as he could use it was not to be distinguished from a fixed star,* had already prepared me to expect the newly discovered heavenly body to be a valuable addition to our increasing catalogue of asteroids.

* Der neue Planet zeigt sich als ein Stern zwischen der 5ten und 6ten Grösse und ist im Fernrohr, wenigstens mit den Vergrösserungen die ich anwenden kann, von einem Fixstern nicht zu unterscheiden.

The 25th of April I looked over my delineations of the preceding evening and found no material difference in the situation of the stars I had marked for examination; and in addition to them new asterisms were prepared, but on account of the retarded motion of the new star, which was drawing towards a period of its retrogradation, the small change of its situation was not sufficiently marked to be readily perceived the next day when these asterisms were again examined, which it is well known can only be done with night-glasses of a very low magnifying power.

A long interruption of bad weather would not permit any regular examination of the situation of small stars; and it was only when I had obtained a more precise information from the Astronomer Royal, who, by means of fixed instruments, was already in possession of the place and rate of motion of the new star, that I could direct my telescope with greater accuracy by an application of higher magnifying

FIG. 2. FIG. 3.

powers. My observations on the nature of this second new star discovered by Dr. OLBERS are as follows.

April 24. This day, as we have already seen, the new celestial object was examined with a high power; and since a magnifier of 460 would not show it to be different from the stars of an equal apparent brightness, its diameter must be extremely small, and we may reasonably expect it to be an asteroid.

May 21. With a double eye-piece magnifying only 75 times the supposed asteroid A makes a right-angled triangle with two small stars *f g*. See fig. 2. With a very distinct magnifier of 460 there is no appearance of any planetary disk.

May 22. The new star has moved away from *f g*, and is now situated as in fig. 3. The star A of figure 1 is no longer in the place where I observed it the 24th of April, and was therefore the asteroid. I examined it now with gradually increased magnifying powers, and the air being remarkably clear, I saw it very distinctly with 460, 577, and 636. On comparing its appearance with these powers alternately to that of equal stars, among which was the 463d of BODE'S Catalogue of the stars in the Lion of the 7th magnitude, I could not find any difference in the visible size of their disks.

By the estimations of the distances of double stars, contained in the first and second classes of the catalogues I have given of them, it will be seen that I have always considered every star as having a visible, though spurious, disk or diameter; and in a late paper I have entered at large into the method of detecting real disks from spurious ones ; it may therefore be supposed that I proceeded now with Vesta (which name I understand Dr. OLBERS has given the asteroid), as I did before in the investigation of the magnitudes of Ceres, Pallas, and Juno.

The same telescopes, the same comparative views, by which the smallness of the latter three had been proved, convinced me now that I had before me a similar fourth celestial body. The disk of the asteroid which I saw was clear, well defined, and free from nebulosity. At the first view I was inclined to believe it a real one ; and the Georgian planet being conveniently situated so that a telescope might without loss of time be turned alternately either to this or to the asteroid, I found that the disk of the latter, if it were real, would be about one-sixth of the former, when viewed with a magnifying power of 460. The spurious nature of the asteroidal disk, however, was soon manifested by an increase of the magnifying power, which would not proportionally increase its diameter as it increased that of the planet ; and a real disk of the asteroid still remains unseen with a power of 636.

May 23. The new star has advanced, and its motion is direct ; its situation with respect to the two small stars *f g*, is given in figure 4. Its apparent disk with a magnifier of 460 is about 5 or 6-tenths of a second ; but this is evidently a spurious appearance, because higher powers destroy the proportion it bears to a real disk when equally magnified. The air is not sufficiently pure this evening to use large telescopes.

* * ⊙ A
f g

FIG. 4.

May 24. With a magnifying power of 577 I compared the appearance of the Georgian planet to that of the asteroid, and with this power the diameter of the visible disk of the latter was about one 9th or 10th part of the former. The apparent disk of the small star near β Leonis, which has been mentioned before, had an equal comparative magnitude, and probably the disks of the asteroid and of the star it resembles are equally spurious.

The 20 feet reflector, with many different magnifying powers, gave still the same result ; and being already convinced of the impossibility, in the present situation of the asteroid, which is above two months past the opposition, to obtain a better view of its diameter, I used this instrument chiefly to ascertain whether any nebulosity or atmosphere might be seen about it. For this purpose the valuable quantity of light collected by an aperture of 18¾ inches directly received by an eye-glass of the front-view without a second reflection, proved of eminent use, and gave me the diameter of this asteroid intirely free from all nebulous or atmospheric appearances.

The result of these observations is, that we now are in possession of a formerly unknown species of celestial bodies, which by their smallness and considerable deviation from the path in which the planets move, are in no danger of disturbing, or being disturbed by them ; and the great success that has already attended the pursuit of the celebrated discoverers of Ceres, Pallas, Juno, and Vesta, will induce us to hope that some further light may soon be thrown upon this new and most interesting branch of astronomy.

*Observations of the expected Comet.**

The comet which has been seen descending to the sun, and from the motion of which it was concluded that we should probably see it again on its return from the perihelion, was expected to make its reappearance about the middle of last January, near the southern parts of the constellation of the whale.

January 27. Towards the evening, on my return from Bath, where I had been a few days, I gave my sister CAROLINA the place where this comet might be looked for, and between flying clouds, the same evening about 6^h $49'$ she saw it just long enough to make a short sketch of its situation.

January 31. Clouds having obscured the sky till this time, I obtained a transitory view of the comet, and perceived that it was within a few degrees of the place which had been assigned to it; the unfavourable state of the atmosphere, however, would not permit the use of any instrument proper for examining it minutely.

There will be no occasion for my giving a more particular account of its place, than that it was very near the electrometer of the constellation, which in Mr. BODE'S maps is called *machina electrica*; the only intention I had in looking for it, being to make a few observations upon its physical condition.

February 1. The comet had moved but very little from the place where it was last night; and as the air was pretty clear, I used a 10-feet reflector with a low power to examine it. There was no visible nucleus, nor did the light which is called the coma increase suddenly towards the centre, but was of an irregular round form, and with this low power extended to about 5, 6, or 7 minutes in diameter. When I magnified 169 times it was considerably reduced in size, which plainly indicated that a farther increase of magnifying power would be of no service for discovering a nucleus. On account of cloudy weather I never had an opportunity of seeing the comet afterwards.

When I compare these observations with my former ones of 15 other telescopic comets, I find that out of the 16 which I have examined, 14 have been without any visible solid body in their centre, and that the other two had a very ill defined small central light, which might perhaps be called a nucleus, but did not deserve the name of a disk.

* [Comet 1806, II. It passed the perihelion on Dec. 28.—ED.]

LX.

Observations of a Comet, made with a View to investigate its Magnitude and the Nature of its Illumination. To which is added, an Account of a new Irregularity lately perceived in the apparent Figure of the Planet Saturn.

[*Phil. Trans.*, 1808, pp. 145–163.]

Read April 7, 1808.

THE comet which we have lately observed, was pointed out to me by Mr. PIGOTT, who discovered it at Bath the 28th of September,* and the first time I had an opportunity of examining it was the 4th of October, when its brightness to the naked eye gave me great hopes to find it of a different construction from many I have seen before, in which no solid body could be discovered with any of my telescopes.

In the following observations, my attention has been directed to such phenomena only, as were likely to give us some information relating to the physical condition of the comet; it will therefore not be expected that I should give an account of its motion, which I was well assured would be most accurately ascertained at the Royal Observatory at Greenwich.

The different parts of a comet have been generally expressed by terms that may be liable to misapprehension, such as the head, the tail, the coma, and the nucleus; for in reading what some authors say of the head, when they speak of the size of the comet, it is evident that they take it for what is often called the nucleus. The truth is, that inferior telescopes, which cannot show the real nucleus, will give a certain magnitude of the comet, which may be called its head; it includes all the very bright surrounding light; nor is the name of the head badly applied, if we keep it to this meaning; and since, with proper restriction, the terms which have been used may be retained, I shall give a short account of my observations of the comet, as they relate to the above-mentioned particulars, namely, the nucleus, the head, the coma, and the tail, without regarding the order of the time when they were made; the date of each observation, however, will be added, that any person who may hereafter be in possession of more accurate elements of the comet's orbit, than those which I have at present, may repeat the calculations in order to obtain a more correct result.

* [The great comet of 1807, first seen in Sicily on Sept. 9.—ED.]

Of the Nucleus.

From what has already been said, it will easily be understood that by the nucleus of the comet, I mean that part of the head which appears to be a condensed or solid body, and in which none of the very bright coma is included. It should be remarked, that from this definition it follows, that when the nucleus is very small, no telescope, but what has light and power in an eminent degree, will show it distinctly.

Observations.

Oct. 4, 1807. 10-feet reflector. The comet has a nucleus, the disk of which is plainly to be seen.

Oct. 6. I examined the disk of the comet with a proper set of diaphragms, such as described in a former paper,* in order to see whether any part of it were spurious; but when the exterior light was excluded, so far from appearing larger, as would have been the case with a spurious disk, it appeared rather diminished for want of light; nor was its diameter lessened when I used only the outside rays of the mirror. The visible disk of the comet therefore is a real one.

Oct. 4. I viewed the comet with different magnifying powers, but found that its light was not sufficiently intense to bear very high ones. As far as 200 and 300, my 10-feet reflector acted very well, but with 400 and 500 there was nothing gained, because the exertion of a power depending on the quantity of light was obstructed,† which I found was here of greater consequence than the increase of magnitude.

Illumination of the Nucleus.

Oct. 4, 6^h 15′. The nucleus is apparently round, and equally bright all over its disk. I attended particularly to its roundness.

Oct. 18. The nucleus is not only round, but also every where of equal brightness.

Oct. 19. I see the nucleus again, perfectly round, well defined, and equally luminous. Its brilliant colour in my 10-feet telescope is a little tinged with red; but less so than that of Arcturus to the naked eye.

Magnitude of the Nucleus.

Oct. 26. In order to see the nucleus as small as it really is, we should look at it a long while, that the eye may gradually lose the impression of the bright coma which surrounds it. This impression will diminish gradually, and when the eye has got the better of it, the nucleus will then be seen most distinctly, and of a determined magnitude.

Oct. 4. With a 7-feet reflector I estimated the diameter of the nucleus of the

* See *Phil. Trans.* for 1805, page 53. Use of the Criterion [above, p. 310].

† See *Phil. Trans.* for 1800, p. 78 [above, p. 47].

comet at first to be about five seconds, but soon after I called it four, and by looking at it longer, I supposed it could not exceed three seconds.

Oct. 6. 10-feet reflector, power 221. The apparent disk of the comet is much less than that of the GEORGIAN planet, which being an object I have seen so often with the same instrument and magnifying power, this estimation from memory cannot be very erroneous.

Oct. 5. Micrometers for measuring very small diameters, when high magnifying powers cannot be used, being very little to be depended upon, I erected a set of sealing wax globules upon a post at 2422 inches from the object mirror of my 10-feet reflector, and viewed them with an eye-glass, which gives the instrument a power of 221, this being the same which I had found last night to show the nucleus of the comet well. I kept them in their place all the day, and reviewed them from time to time, that their magnitudes might be more precisely remembered in the evening, when I intended to compare the appearance of the nucleus with them.

On examining the comet, I found the diameter of its nucleus to be certainly less than the largest of my globules, which being ·0466 inch, subtended an angle of 3″·97 at the distance of the telescope in the day time.

Comparing the nucleus also with the impressions, which the view of the second and third had left in my memory, and of which the real diameters were ·0325 and ·0290 inch ; and magnitudes at the station of the mirror 2″·77 and 2″·47, I found that the comet was almost as large as the second, and a little larger than the third.

Oct. 18. The nucleus is less than the globule which subtends 2″·77.

Oct. 19. The air being uncommonly clear, I saw the comet at 40 minutes after five, and being now at a considerable altitude, I examined it with 289, and having but very lately reviewed my globules, I judged its diameter to be not only less than my second globule, but also less than the third ; that is, less than 2″·47.

Oct. 6. The 20-feet reflector, notwithstanding its great light, does not show the nucleus of the comet larger than the 10-feet, with an equal magnifier, makes it.

Oct. 28. My large 10-feet telescope, with the mirror of 24 inches in diameter, does not increase the size of the nucleus.

Oct. 6. Being fully aware of the objections that may be made against the method of comparing the magnitude of the nucleus of the comet with objects that cannot be seen together, I had recourse to the satellites of Jupiter for a more decisive result, and with my 7-feet telescope, power 202, I viewed the disk of the third satellite and of the nucleus of the comet alternately. They were both already too low to be seen very distinctly ; the diameter of the nucleus however appeared to be less than twice that of the satellite.

Oct. 18. With the 10-feet reflector, and the power 221, a similar estimation was made ; but the light of the moon would not permit a fair comparison.

Oct. 19. I had prepared a new 10-feet mirror, the delicate polish of my former one having suffered a little from being exposed to damp air in nocturnal observations.

This new one being uncommonly distinct, and the air also remarkably clear, I turned the telescope from the comet to Jupiter's third satellite, and saw its diameter very distinctly larger than the nucleus of the comet. I turned the telescope again to the comet, and as soon as I saw it distinctly round and well defined, I was assured that its diameter was less than that of the satellite.

6^h 20′. I repeated these alternate observations, and always found the same result. The night is beautifully clear, and the moon is not yet risen to interfere with the light of the comet.

Nov. 20. With a 7-feet reflector, and power only 75, I can also see the nucleus; it is extremely small, being little more than a mere point.

Of the Head of the Comet.

When the comet is viewed with an inferior telescope, or if the magnifying power, with a pretty good one, is either much too low, or much too high, the very bright rays immediately contiguous to the nucleus will seem to belong to it, and form what may be called the head.

Oct. 19. I examined the head of the comet with an indifferent telescope, in the manner I have described, and found it apparently of the size of the planet Jupiter, when it is viewed with the same telescope and magnifying power.

With a good telescope, I saw in the centre of the head a very small well defined round point.

Nov. 20. The head of the comet is now less brilliant than it has been.

Of the Coma of the Comet.

The coma is the nebulous appearance surrounding the head.

Oct. 19. By the field of view of my reflector, I estimate the coma of this comet to be about 6 minutes in diameter.

Dec. 6. The extent of the coma, with a mirror of 24 inches diameter, is now about 4′ 45″.

Of the Tail of the Comet.

Oct. 18. 7^h. With a night glass, which has a field of view of nearly 5°, I estimated the length of the tail to be $3°\frac{3}{4}$; but twilight is still very strong, which may prevent my seeing the whole of it.

Nov. 20. The tail of the comet is still of a considerable length, certainly not less than $2\frac{1}{2}$ degrees.

Oct. 26. The tail of the comet is considerably longer on the south-preceding, than on the north-following side.

It is not bifid, as I have seen the comet of 1769 delineated, by a gentleman who carefully observed it.*

Oct. 28. 7-feet reflector. The south-preceding side of the tail in all its length,

* Dr. LIND of Windsor.

except towards the end, is very well defined ; but the north-following side is every where hazy and irregular, especially towards the end ; it is also shorter than the south-preceding one.

The shape of the unequal length of the sides of the tail, when attentively viewed, is visible in a night glass, and even to the naked eye.

Oct. 31. 10-feet reflector. The tail continues to be better defined on the south-preceding than on the north-following side.

Dec. 6. The length of the tail is now reduced to about 23′ of a degree.

Of the Density of the Coma and Tail of the Comet.

Many authors have said, that the tails of comets are of so rare a texture, as not to affect the light of the smallest stars that are seen through them. Unwilling to take any thing upon trust that may be brought to the test of observation, I took notice of many small stars that were occasionally covered by the coma and the tail, and the result is as follows.

Oct. 26, 6^h 15′. Large 10-feet reflector, 24 inches aperture. A small star within the coma is equally faint with two other stars that are on the north-following side of the comet, but without the coma.

7^h 30′. The coma being partly removed from the star, it is now brighter than it was before.

Oct. 31, 6^h 5′. 10-feet reflector. A star in the tail of the comet, which we will call *a*, is much less bright than two others, *b* and *c*, without the tail.

Two other stars, *d* and *e*, towards the south of *b* and *c*, are in the following skirts of the tail, and are extremely faint.

7^h 20′. The star *c* is now considerably bright, the tail having left it, while *d*, which is rather more involved than it was before, is hardly to be seen.

7^h 50′. The star *a*, towards which the comet moves, is involved in denser nebulosity than before, and is grown fainter.

d is involved in brighter nebulosity than before, but being near the margin, it will soon emerge.

8^h 35′. Being still more involved, the star *a* is now hardly visible.

e is quite clear of the tail, and is a considerable star ; *d* remains involved.

9^h 10′. The star *d* is also emerged, but the comet is now too low to estimate the brightness of stars properly.

Nov. 25, 7^h 35′. There is a star *a* within the light of the tail, near the head of the comet, equal to a star *b* situated without the tail, but near enough to be seen in the field of view with *a*. The path of the head of the comet leads towards *a*, and a more intense brightness will come upon it.

8^h 46′. The star *a* is now involved in the brightness near the head of the comet, and is no longer visible, except now and then very faintly, by occasional imperfect glimpses ; but the star *b* retains its former light.

Nebulous appearance of the Comet.

Dec. 6. The head of the comet, viewed with a mirror of 24 inches diameter, resembles now one of those nebulæ which in my catalogues would have been described, "a very large, brilliant, round nebula, suddenly much brighter in the middle."

Dec. 16. 7-feet reflector. The night being fine, and the moon not risen, the comet resembles "a very bright, large, irregular, round nebula, very gradually much brighter in the middle, with a faint nebulosity on the south preceding side."

Jan. 1. 1808. 7-feet. "Very bright, very large, very gradually much brighter in the middle."

If I had not known this to be a comet, I should have added to my description of it as a nebula, that the center of it might consist of very small stars, but this being impossible, I directed my 10-feet telescope with a high power to the comet, in order to ascertain the cause of this appearance; in consequence of which I perceived several small stars shining through the nebulosity of the coma.

Jan. 14. 7-feet. "Bright, pretty large, irregular round, brighter in the middle."

Feb. 2. 10-feet, 24-inch aperture. "Very bright, large, irregular round, very gradually much brighter in the middle." There is a very faint diffused nebulosity on the north-preceding side; I take it to be the vanishing remains of the comet's tail.

Feb. 19. "Considerably bright; about $\frac{1}{7}$ of the field $= 3' \ 26''$ in diameter, gradually brighter in the middle." The faint nebulosity in the place where the tail used to be, still projects a little farther from the center than in other directions.

Feb. 21. Less bright than on the 19th; nearly of the same size; gradually brighter in the middle. The nebulosity still a little projecting on the side where the tail used to be.

Result of the foregoing Observations.

From the observations which are now before us, we may draw some inferences, which will be of considerable importance with regard to the information they give us, not only of the size of the comet, but also of the nature of its illumination.

A visible, round and well defined disk, shining in every part of it with equal brightness, elucidates two material circumstances; for since the nucleus of this comet, like the body of a planet, appeared in the shape of a disk, which was experimentally found to be a real one, we have good reason to believe that it consists of some condensed or solid body, the magnitude of which may be ascertained by calculation. For instance, we have seen that its apparent diameter, the 19th of October, at $6^h \ 20'$, was not quite so large as that of the 3d satellite of Jupiter. In order therefore to have some idea of the real magnitude of our comet, we may admit that its diameter at the time of observation was about $1''$, which certainly cannot be far from truth. The diameter of the 3d satellite of Jupiter, however, is known to have a permanent disk, such as may at any convenient time be measured with all the accuracy that can be used; and when the result of such a measure has given us the diameter of this satellite, it may by calculation be brought to the distance from the earth at which, in my observation, it was compared with the diameter of the comet, and thus more accuracy, if it should be required, may be obtained. The following result of my calculation however appears to me quite sufficient for the purpose of a general information. From the perihelion distance 0·647491, and the rest of the given elements of the comet, we find that its distance from the ascending node on its orbit at the time of observation was 73° 45′ 44″; and having also the earth's distance from the same node, and the inclination of the comet's orbit, we compute by these data the angle at the sun. Then by calculating in the next place the radius vector of the comet, and having likewise the distance of the earth from the sun, we find by computation that the distance of the comet from the earth at the time of observation was 1·169192, the mean distance of the earth being

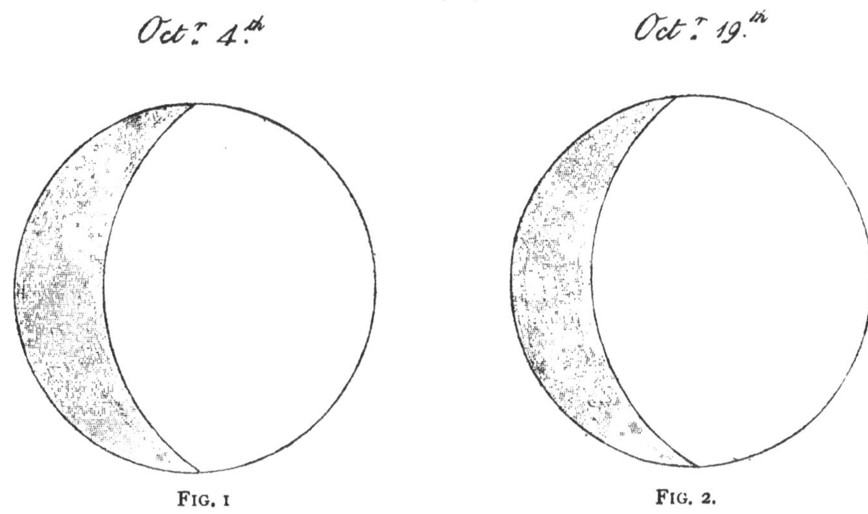

FIG. 1 FIG. 2.

1. Now since the disk of the comet was observed to subtend an angle of 1″, which brought to the mean distance of the earth gives 1″·169, and since we also know that the earth's diameter, which, according to Mr. DALBY, is 7913·2 miles,* subtends at the same distance an angle of 17″·2 we deduce from these principles the real diameter of the comet, which is 538 miles.

Having thus investigated the magnitude of our comet, we may in the next place also apply calculation to its illumination. The observations relating to the light of the comet were made, from the 4th of October to the 19th. In all which time the comet uniformly preserved the appearance of a planetary disk fully enlightened by the sun: it was every where equally bright, round, and well defined on its borders. Now as that part of the disk which was then visible to us, could not possibly have a full illumination from the sun, I have calculated the phases of the comet for the 4th and for the 19th, the result of which is, that on the 4th the illumination was 119° 45′ 9″ as represented in figure 1, and that on the 19th it had gradually increased to 124° 22′ 40″, of which a representation is given in figure 2. Both

* See *Phil. Trans.* for 1791, page 239, Mr. DALBY gives the two semi-axes of the earth, from a mean of which the above diameter 7913·1682 is obtained.

phases appear to me sufficiently defalcated, to prove that the comet did not shine by light reflected from the sun only; for had this been the case, the deficiency I think would have been perceived, notwithstanding the smallness of the object. Those who are acquainted with my experiments on small silver globules,* will easily admit, that the same telescope, which could shew the spherical form of balls, which subtended only a few tenths of a second in diameter, would surely not have represented a cometary disk as circular, if it had been as deficient as are the figures which give the calculated appearances.

If these remarks are well founded, we are authorised to conclude, that the body of the comet on its surface is self-luminous, from whatever cause this quality may be derived. The vivacity of the light of the comet also, had a much greater resemblance to the radiance of the stars, than to the mild reflection of the sun's beams from the moon, which is an additional support of our former inference.

The changes in the brightness of the small stars, when they are successively immerged in the tail or coma of the comet, or cleared from them, prove evidently, that they are sufficiently dense to obstruct the free passage of star-light. Indeed if the tail or coma were composed of particles that reflect the light of the sun to make them visible, we ought rather to expect, that the number of solid reflecting particles, required for this purpose, would entirely prevent our seeing any stars through them. But the brightness of the head, coma, and tail alone, will sufficiently account for the observed changes, if we admit that they shine not by reflection, but by their own radiance; for a faint object projected on a bright ground, or seen through it, will certainly appear somewhat fainter, although its rays should meet with no obstruction in coming to the eye. Now, as in this case, we are sure of the bright interposition of the parts of the comet, but have no knowledge of floating particles, we ought certainly, not to ascribe an effect to an hypothetical cause, when the existence of one, quite sufficient to explain the phenomena, is evident.

If we admit that the observed full illumination of the disk of the comet cannot be accounted for from reflection, we may draw the same conclusion, with respect to the brightness of the head, coma, and tail, from the following consideration. The observation of the 2d of February mentions that not only the head and coma were still very bright, but that also the faint remains of the tail were still visible; but the distance of the comet from the earth, at the time of observation, was nearly 240 millions of miles,† which proves, I think, that no light reflected from floating particles could possibly have reached the eye, without supposing the number, extent, and density of these particles, far greater than what can be admitted.

My last observation of the comet, on the 21st of February, gives additional support to what has been said; for at the time of this observation, the comet was almost 2·9 times the mean distance of the sun from the earth.‡ It was also nearly

* See *Phil. Trans.* for 1805, page 38, the 5th experiment [above, p. 301].

† 239894939.

‡ The sun's mean distance being 1, that of the comet was 2·89797.

2·7 from the sun.* What chance then could rays going to the comet from the sun, at such a distance, have to be seen after reflection, by an eye placed at more than 275 millions of miles † from the comet? And yet the instant the comet made its appearance in the telescope, it struck the eye as a very conspicuous object.

The immense tails also of some comets that have been observed, and even that of the present one, whose tail, on the 18th of October, was expanded over a space of more than 9 million of miles,‡ may be accounted for more satisfactorily, by admitting them to consist of radiant matter, such as, for instance, the aurora borealis, than when we unnecessarily ascribe their light to a reflection of the sun's illumination thrown upon vapours supposed to arise from the body of the comet.

By the gradual increase of the distance of our comet, we have seen that it assumed the resemblance of a Nebula; and it is certain, that had I met with it in one of my sweeps of the zones of the heavens, as it appeared on either of the days between the 6th of December, and the 21st of February, it would have been put down in the list I have given of nebulæ. This remark cannot but raise a suspicion that some comets may have actually been seen under a nebulous form, and as such have been recorded in my catalogues; and were it not a task of many years' labour, I should undertake a review of all my nebulæ, in order to see whether any of them were wanting, or had changed their place, which certainly would be an investigation that might lead to very interesting conclusions.

Account of a new irregularity lately perceived in the apparent Figure of the Planet Saturn.

The singular figure of Saturn, of which I have given an account in two papers, has continued, for several reasons, to claim my attention. When I saw the uncommon flattening of the polar regions of this planet, in the 40-feet telescope, I ascribed it to the attractive matter in the ring,§ and of its tendency to produce such an effect we can have no doubt; but as another circumstance, which was also noticed, namely, an apparent small flattening of the equatorial parts, cannot be explained on the same principles, I wished to ascertain what physical cause might be assigned for this effect, and with a view to an investigation of this point, I have continued my observations. The position of the ring, at the last appearance of the planet, however, proved to be quite unfavourable for the intended purpose; for the very parts, which I was desirous of inspecting, were covered by the passage of the ring over the disk of the planet in front, or were projected on the ring, where it passed behind the body.

In my attempts to pursue this object, I perceived a new irregularity in the Saturnian figure, which, I am perfectly assured, had no existence the last time I examined the planet, and the following observations contain an account of it.

* The comet's distance from the sun was 2,683196. † 275077889.
‡ 9160542. § See *Phil. Trans.* for 1805, page 276 [above, p. 335].

Observations.

June 16, 1807. The two polar regions of Saturn are at present of a very different apparent shape. The northern regions, as in former observations are flattened; but the southern are more curved or bulged outwards.

I asked my son JOHN HERSCHEL, who after me looked at Saturn while I was writing down the above observation, if he perceived that there was a difference in the curvature of the north and south pole, and if he did, to mark on a slate how it appeared to him. When I examined the slate, I found that he had exactly delineated the appearance I have described.

In a letter to a very intelligent astronomical friend,* who has one of my 7-feet reflectors, I requested the favour of him to examine both the polar regions of Saturn, and to let me know whether he could perceive any difference in the appearance of their curvature; in answer to which I received, the 23d of June, a letter inclosing a drawing, in which also the southern regions were marked as more protuberant, with a greater falling off close to the irregularity. My friend, with his usual precaution, called this an illusion; and it will be seen by and by, that we shall have no occasion to ascribe this irregularity to a real want of due proportion, or settled figure of the polar regions of Saturn.

June 22, 9^h 24′. I see the same curved appearance at the south pole of Saturn, which was observed the 16th.

June 24. The air is very clear, and all the most critical phenomena are very distinctly to be seen; the shadow of the ring towards the south upon the planet; the shadow of the body towards the north-following side upon the ring; the belts upon the body; the division of the two rings; and with the same distinctness, I also see the protuberance of the south pole.

My seeing this appearance, at present, is a proof that it is not a physical irregularity or distortion of only some particular spot on the polar regions; for, in that case, it could not have been seen this evening, as from the rotation of the planet on its axis, which is 10^h 16′, the space of the polar circle which is now exposed to our view, must have been very different from what I saw the 16th and 22d.

Many observations were made afterwards, which all confirm the reality of this appearance.

It is so natural for us to reflect upon the cause of a new phenomenon, that I cannot forbear giving an opinion on this subject. To suppose a real change in the whole zone of the planet, cannot be probable; it seems therefore that this appearance must be, as my friend calls it, an illusion. But since the reality of this illusion, if I may use the expression, has been ascertained by observation, it is certain that there must be some extrinsic cause for its appearance; and also that the same cause must not act upon the northern hemisphere. Now the only difference in the circum-

* Dr. WILSON of Hampstead, late Professor of Astronomy at Glasgow. [Patrick, son of Alexander Wilson.—ED.]

stances under which the two polar regions of Saturn were seen in the foregoing observations is the situation of its ring, which passes before the planet at the south, but behind at the north. The rays of light therefore which come to the eye from the very small remaining southern zone of the Saturnian globe, pass at no great distance by the edge of the ring, while those from the north traverse a space clear of every object that might disturb their course. If therefore we are in the right to ascribe the observed illusion to an approximate interposition of the ring, we have, in the case under consideration, only two known causes that can modify light so as to turn it out of its course, which are inflection and refraction. The insufficiency of the first to account for the lifting up of the protuberant small segment of the northern regions will not require a proof. The effects of refraction on the contrary are known to be very considerable. Let us therefore examine a few of the particulars of the case. The greatest elevation of the visible segment above the ring did not amount to more than one second and three or four tenths. Then supposing the ring, the edge of which is probably of an elliptical figure, to have a surrounding atmosphere, it will most likely partake of the same form, and the rays which pass over its edge will undergo a double refraction : the first on their entrance into this atmosphere, and the second at their leaving it, and these refractions seem to be sufficient to produce the observed elevation. For should they raise the protuberant appearance only half a second, or even less, the segment could no longer range with the rest of the globe of Saturn, but must assume the appearance of a different curvature or bulge outwards.

The refractive power of an atmosphere of the ring has been mentioned in a former paper,* when the smallest satellites of Saturn were seen as it were bisected by the narrow luminous line under which form the ring appeared when the earth was nearly in the plane of it ; and the phenomenon, of which the particulars have now been described, appears to be a second instance in support of the former.

* See *Phil. Trans.* for 1790, page 7 [Vol. I. p. 373].

LXI.

Continuation of Experiments for investigating the Cause of coloured concentric Rings, and other Appearances of a similar Nature.

[*Phil. Trans.*, 1809, pp. 259–302.]

Read March 23, 1809.

In the first part of this paper, I have pointed out a variety of methods that will give us coloured concentric rings between two glasses of a proper figure applied to each other, and it has been proved that only two surfaces, namely, those that are in contact with each other, are essential to their formation ; it will now be necessary to enlarge the field of prismatic phenomena, by showing that their appearance in the shape of rings has been owing to our having only used spherical curves to produce them.

35. *Cylindrical Curves produce Streaks.*

As soon as it occurred to me, that the cause of the figure of any certain prismatic appearance must be looked for in the nature of the curvature of one or both of the surfaces, that are essential to its production, I was prepared to expect that if a spherical curve, when applied to a plain surface of glass, produces coloured rings, a cylindrical one applied to the same would give coloured lines or streaks. To put this to the proof of an experiment, I ground one side of a plate of glass into a cylindrical curve, and after having given it a polish, I laid a slip of plain glass upon it, and soon perceived a beautiful set of coloured streaks. The broadest of them was at the line of contact, and on each side they were gradually narrower and less bright. The colours in the streaks were similar to those in the rings, and they were in the same manner changeable by pressure as in them. Their order was likewise the same, if we reckon from the line of contact, as with rings we do from the center ; so that these streaks differed in no respect from rings, except in their linear instead of circular arrangement.

When the cylindrical surface was laid upon a plain slip of glass, the same streaks were seen as in the former experiment. They were of a lively red and green colour, and I saw at least ten, eleven, or twelve on each side of the line of contact.

Metalline surfaces had the same effect, for when the cylindrical surface of glass was laid on a plain metalline mirror, I had red, orange, yellow, green, and blue streaks. In the same manner a plain slip of glass placed upon a polished part of a brass cylinder of 3½-inch in diameter, produced also coloured streaks.

The combination of two cylindrical surfaces has an effect on the streaks, which is similar to that which the contact of two spherical ones has on the rings; for when I placed the cylindrical surface of glass longitudinally upon the polished part of the brass cylinder, the streaks were contracted as rings would have been by the application of two spherical curves to each other.

36. *Cylindrical and spherical Surfaces combined produce coloured elliptical Rings.*

The theory which suggests to us that the particular figure of every prismatic appearance between glasses depends on the curvature of the surfaces which are in contact, is still farther confirmed when spherical and cylindrical curves are applied to each other; for these, accordingly should give elliptical rings; and when I tried the experiment, by laying a 26-inch double-convex lens upon the cylindrical surface of my plate of glass, it produced a coloured elliptical central part, encompassed with gradually vanishing rings of the same figure. By changing the focal length of the lens, I could alter the proportion of the conjugate to the transverse axes of these elliptical rings at pleasure. A lens of 55 inches gave ellipses that were much flattened, and one of 5 inches gave them nearly circular.

37. *Irregular Curves produce irregular Figures.*

The modifying power of surfaces may be further established by such as have no regular figure; for these ought to give irregular prismatic phenomena, and this was fully proved by the following experiment.

I took a large piece of mica which had a very glossy but irregular surface, and when a 34-inch double-convex lens was placed upon a small ridge of it, several pretty straight streaks might be seen, but wherever the ridge was waving the streaks were following the same direction. In some places the mica gave irregular, coloured arcs, that were concave to some distant centre; and in others, the various contorted figures, that were to be seen, exceeded all the imaginary forms which the most inventive fancy can paint. The flexibility of mica also gave room for using different degrees of pressure, by which means a continual change of figure and succession of prismatic colours was produced.

When I laid a piece of this mica upon a cylinder, and placed a plain slip of glass or double convex lens upon it, all its irregularities were modified into disfigured streaks with the former, and distorted ellipses with the latter.

Experiments of a similar nature were made upon the irregular surface of Iceland crystal and other substances, which all gave the same result.

38. *Curved Surfaces are required for producing the coloured Appearances at present under Consideration.*

It has already been seen, in the first part of this paper, that spherical curves give circular rings, and I have now shown that cylindrical forms produce streaks; that a combination of spherical and cylindrical curvatures give elliptical rings, and that all sorts of variegated coloured phenomena are made visible by surfaces, which are irregularly and variously curved; these experiments prove in the fullest manner that the curvature of surfaces is the cause of the appearance, as well as of the shape of the coloured phenomena which are produced. For if we can invariably predict, from the nature of the curves we employ in an experiment, what will be the appearance and form of the colours that will be seen, it certainly must prove the efficacy of these curvatures in the production of such phenomena. This will receive additional confirmation in the following article, which shows that

39. *Coloured Appearances cannot be produced between the plain Surfaces of two parallel Pieces of Glass applied to one another.*

As the production and modification of the figure of the coloured appearances, that have hitherto been considered, has in the last article been ascribed to curved surfaces, it will be necessary to examine whether such phenomena may not also be seen between the plain surfaces of two parallel pieces of glass applied to each other directly in contact, or inclined towards each other in some certain extremely small angle.

The latter of these cases has already been considered in the 31st article of the first part of this paper, where I have shown that two plain surfaces, let the angle of the wedge of air between them be as small as you please, will not give coloured streaks. I have indeed seen two thin plain pieces of glass, with a slip of platina of an extraordinary thinness between them at one end tied together, which showed some streaks near the place where the glasses were in contact, but when I removed the thread that bound them together, the streaks vanished, which proves that the glasses had been constrained, and thus had probably assumed some curvature at the point of contact.

I have also tried two flat surfaces of glass, which were so perfect that no colour could be perceived unless they were by unequal pressure somewhat disfigured, and when that was the case large flashy coloured appearances became visible, and their configuration followed very evidently the stress which I laid upon the different parts of the glasses.

It is however unnecessary to dwell on proofs, that streaks cannot be seen when two plain parallel pieces of glass are applied to each other, as it will hereafter be shown that when the incumbent plain glass is not of a parallel thickness, coloured phenomena may be rendered visible between two perfectly plain surfaces, although no force or strain should be used to produce a fallacious, curved, contact.

40. *Of the Production of coloured Appearances.*

Hitherto I have only considered the coloured rings which Sir ISAAC NEWTON has pointed out, and have shown, at the end of the 28th article, that no more than two surfaces are essential to their formation. It has now also been proved, that the configuration of the coloured phenomena arises from the curvature of one or both of the two essential surfaces. From these principles it will be seen, that we are to distinguish between the production of the colours and that of their configuration when produced. By the experiments that have been given, the cause of the configuration is laid open to our view; but the production and arrangement of the colours remain to be investigated.

The leading feature of the arrangement of the colours of the rings is prismatic; that is to say their order is red, orange, yellow, green, blue, indigo, and violet; in order, therefore, to enter minutely into the subject, I shall have recourse to some prismatic experiments.

It will be necessary here to mention, that the proposed enumeration of the modifications of light, which was intended to have been given in this part of my paper, is grown to such an extent by the number of experiments I have made upon the subject, that its introduction would occasion a long interruption of the present subject; and although undoubtedly the action of bodies and surfaces on light would be better understood, if all the modifications wherein colours are produced had been before us, yet as the experiments I have to relate may be made plain, either by referring to modifications that are sufficiently known, or by explaining what is not already familiar, I shall postpone the intended enumeration to some future opportunity, and confine myself at present to a few remarks relating to them.

The colours contained in white light may be separated by reflection, as well as by refraction, and what is perfectly to my present purpose, the order, in which the colours thus produced are arranged, is the same in both cases; each of these principles therefore may cause coloured appearances, which the particular figure of the surfaces we use will mould into different configurations.

Sir ISAAC NEWTON, for instance, has shown that the rays of light will be separated, by what he calls a different reflexibility, when they fall on the base of a prism; the violet being reflected first, and the red last.* By this property of the differently coloured rays, he has explained a very remarkable phenomenon, which is that in a prism, when exposed in the open air, and when the eye is properly placed "the spectator will see a bow of a blue colour." † From the little the author has said of this bow, it may be supposed that he did not examine it farther than was required for his purpose; it will therefore be necessary to enter more fully into the subject.

* See the illustration of the 9th experiment in the first book of NEWTON'S *Optics*, page 46.

† See the 16th experiment in the second part of the first book, page 145.

41. *Particulars relating to the Newtonian prismatic blue Bow.*

The Newtonian blue bow may very conveniently be examined, when a right-angled prism is laid down on a table before an open window. The eye being then brought to a convenient altitude, and pretty near the side of the prism, we see in it a bow, which from the predominant colour may be called blue. It contains some green followed by blue, indigo, and violet. A very faint red, orange, and yellow may also be perceived above the greenish colour; but these belong not to the blue bow, and have not been noticed by the author. Their appearance will hereafter be accounted for.

To analyse this blue bow more particularly, let us admit that the colours, which give it the general appearance of what may be called blue, consist of half the green, and of all the blue, indigo, and violet rays, which are reflected while the other half of the green, the yellow, orange, and red are transmitted. Then the angle of obliquity, at which this separation of the colours will happen, in consequence of the different refrangibility of the differently coloured rays assigned by NEWTON, will be 49° 46′ 12″·5.

Let A B C D E, Fig. 1, be rays of light moving within glass in such directions, as to fall on the interior base F G upon the points α β γ δ ϵ. Then, if it be required that these rays after reflection from the base should meet in the point H, and form the blue bow, the angles $A\alpha G$, $B\beta G$, $C\gamma G$, $D\delta G$, and $E\epsilon G$ must be respectively equal to 49° 46′ 12″·5; 49° 49′ 20″; 49° 55′ 33″·6; 49° 59′ 41″·4; and 50° 7′ 54″; which will give the angles $\alpha H\beta$, $\beta H\gamma$, $\gamma H\delta$, $\delta H\epsilon$, equal to 3′ 7″·5, 6′ 13″·6, 4′ 7″·8, and 8′ 12″·6, making in the whole the angle subtended by the bow $\alpha H\epsilon$ 21′ 41″·5.* For in consequence of the different reflexibility of the differently coloured rays the violet, indigo, blue, and faintest half of the green rays will be reflected between α and β, if they fall on that space in any angle between the above mentioned ones contained between $A\alpha G$ and $B\beta G$; and will therefore meet at H, and form the greenish blue part of the bow. The red, orange, yellow, and the brightest half of the green rays, on the contrary being less reflexible, will be transmitted through the base between α and β, and by refraction pass in proper angles into the air. The letters *v i b* $\frac{1}{2}$*g*, which in the figure are placed within the space α H β, denote the reflected colours, and $\frac{1}{2}$*g y o r* put under the base between α and β are the initials of the transmitted colours; and in the same manner the reflections and transmissions which must happen between β γ, γ δ, and δ ϵ are expressed by the letters over the base for the former, and under it for the latter. The order of the colours of the blue bow when it is seen at H, is perfectly explained by the letters in

* There is a mistake in one of the angles given by NEWTON, when in his *Optics*, page 145, he explains the blue bow; for 49 deg. $\frac{1}{28}$ taken from 50 deg. $\frac{1}{9}$, makes the breadth of the bow 1° 4′ 31″·4, which contradicts the refractions he has given, page 112. As he only takes in the blue, indigo, and violet colours, instead of $49\frac{1}{28}$ degrees, it should rather be $49\frac{28}{28}$.

the reflected part; and the eye must be placed, for seeing it, at the mean obliquity between the angles A α G and E ϵ G, which is 49° 57′ 3″·3.

In order to conform this account of the blue bow, to the manner in which it was viewed by NEWTON, I have preserved his way of ascribing the separation of the rays to their different reflexibility, which however is merely the effect of their different

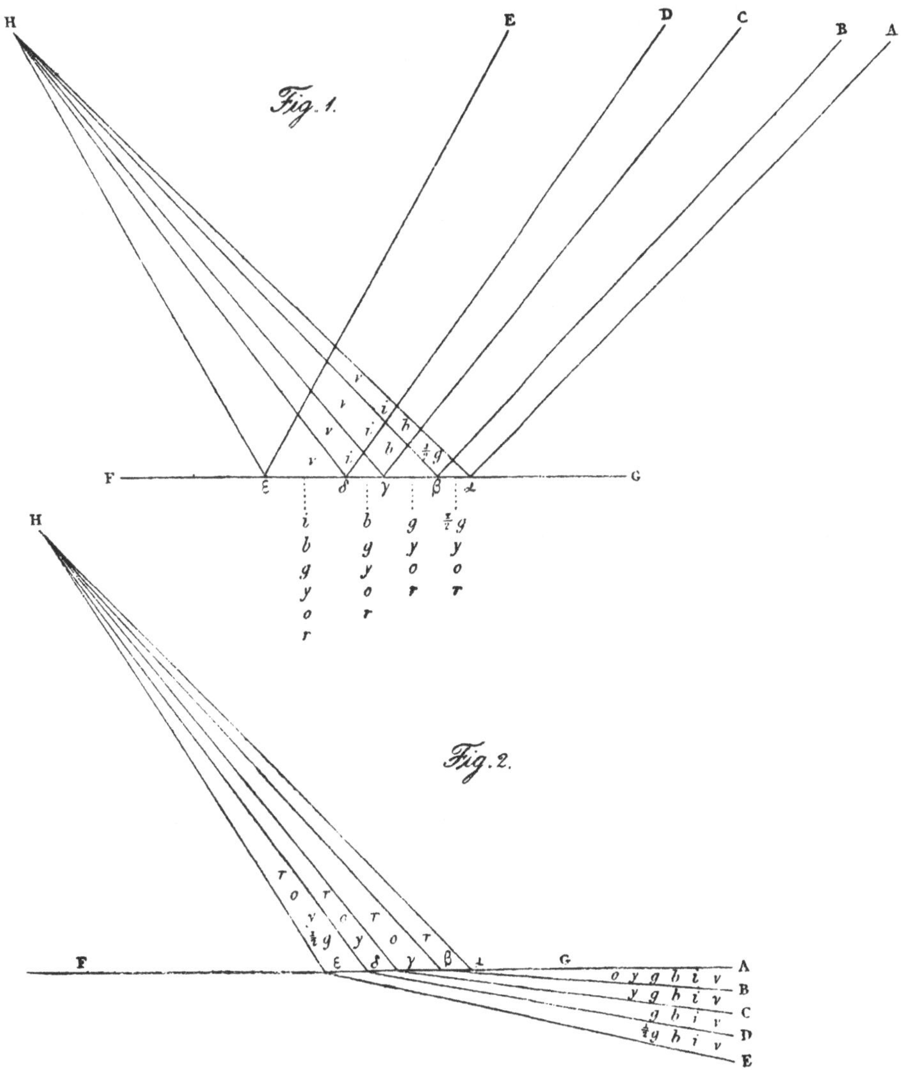

refrangibility. The angles at which the rays that constitute the blue bow are separated from the rest, may very properly be called *critical*, and the effect, which is the consequence of the oblique incidences that have been given, may with equal propriety be called a *critical separation* of the differently coloured rays of light.

42. *Account of a prismatic red Bow.*

I must now introduce a prismatic appearance, which on account of its similarity with the Newtonian blue bow, from which it only differs in colour, I have called a prismatic red bow. It consists of red, orange, yellow, and some green rays; and the red colour being upon the whole very predominant, it may not improperly be called a red bow. It is not produced by the Newtonian different reflexibility of the differently coloured rays of light, but owes its origin to a modification which takes effect at the outside of the prism at very oblique angles of incidence, and may be called a different intromissibility; but this, like the Newtonian different reflexibility, is only the consequence of the different refrangibility of light.

To see the red bow, an observer should place himself in the open air, and standing with his back within a few feet of some wall or building, hold the side of an equilateral prism flat over his eyes, and look upwards to an altitude of about 30° at the heavens; he will then see a beautiful arch of a deep red colour, succeeded by a bright orange and yellow, with a considerable portion of green on the inside. The comparative darkness of the building behind will show the light in front to the best advantage. It is also to be observed, that all experiments on prismatic bows succeed best, when the heavens are totally overcast with an uniform cloudiness.

To analyze the production of this bow, let A B C D E, Fig. 2, be rays of light moving in air, in such directions as to fall on the exterior base F G, of a piece of glass, upon the points $\alpha\ \beta\ \gamma\ \delta\ \epsilon$; then, if it be required that these rays, after their intromission into the glass should meet in the point H and form the red bow, the angles A α H, B β H, C γ H, D δ H, and E ϵ H, must be respectively equal to 130° 29′ 33″·6; 133° 40′ 33″·2; 134° 29′ 28″·2; 135° 36′ 13″·2; and 136° 10′ 38″·0; from which we have the angles A β B, A γ C, A δ D, and A ϵ E, which a red ray would make were it to pass out of glass into air, equal to 3° 15′ 45″·5; 4° 7′ 30″·5; 5° 19′ 17″·5; and 5° 56′ 50″·5. Now by the laws of the different refrangibility of light, the red rays are intromissible at α, when by refraction they make the angle H α F = 49° 30′ 26″·4; but the orange cannot be intromitted any where between α and β with any effect on the red bow, since it is only at β, where the angle H β F is 49° 35′ 12″·3, that they can enter the glass so as to come to the eye at H. The yellow rays will, for the same reason, be efficiently intromitted only at γ, where they will make the angle H γ F 49° 38′ 2″·3, and the brightest half of the green rays will find an efficient entrance from δ to ϵ, since the smallest angle of their intromission H δ F is 49° 43′ 4″·3, and the angle H ϵ F, which terminates the red bow, is 49° 46′ 12″·5. The arrangement of the colours of this bow will be seen, as it was in the blue bow, from the letters placed above the base, which denote those that are intromitted so as to come to the eye; the rest of the colour-making rays, which cannot come in that direction, being marked by letters placed under the base. The whole angle of the red bow α H ϵ is 15′ 46″·1, and the mean obliquity of the eye at H is 49° 38′ 19″·5.

In the calculation of both the bows, the situation of the eye at H has been determined, as it would be, were the rays to remain in glass; but as they will be refracted by the side of a prism, when they come out of it, proper computations must be made not only of the place of the eye in air, but also of the angle which the bow will subtend; for this will be found to be considerably different in different prisms; those that have large refracting angles will magnify the bows more, and require the eye to be nearer than others that have smaller angles.

These bows may be examined at leisure, by projecting them upon a white ground in the following manner:

In a dark room, by a reflecting apparatus, I admitted a horizontal beam of the solar light through an opening of about an inch and a half in diameter. The formation of the bows requiring scattered light,* I covered the opening with a piece of glass evenly roughened on both sides. Then, with an intention to obtain a projection of the blue bow, I placed a prism having one angle of 91° and the other two nearly equal, close to the emeried surface, and turned it upon its axis till the angle of obliquity of the scattered rays, that fell on one side of the prism, was proper for the required critical separation of the coloured rays. The obliquity of the middle ray with the base, for this purpose, it has been shown, must be 49° 57′ 3″·3. In this position the interior critical separation of the prismatic colours taking place, the blue part, namely the violet, indigo, blue, and about one half of the green rays were reflected, and passing through the opposite side of the prism projected the blue bow upon the ceiling of the room. The colours may there be conveniently seen; but as this bow is composed of the least luminous rays of the prismatic spectrum, it requires considerable attention to perceive the faintest of them. The green and blue are most visible, and by receiving the bow upon a screen of white paper held at the most favourable distance, the fainter colours, when the illumination is very bright may also be perceived.

In order then to project also the red bow, I turned the prism upon its axis till the scattered light fell with a proper obliquity on the base of it; the angle required for this purpose, it has been shown, must be from 0° 0′ 0″ to 5° 56′ 50″·5; the side of the prism, which is turned towards the opening, should be covered with a slip of pasteboard to prevent any light from entering it. In this situation, I saw a very bright arch containing red, orange, and yellow projected at some distance backwards upon the ceiling; that part of the green which no doubt was also transmitted, was lost in the brightness which is to be seen within the bow, for the same reason that the faint colours of the blue bow can only with great difficulty, if at all, be perceived; namely, that they join the dark inside of the bow. For NEWTON has proved that the space beyond the convex part of the blue bow must be bright, and that beyond the concave dark; but in the red bow, as my theory will show, we have that on the convex dark, and on the concave bright. This experiment therefore proves, that here, by the gradual intromission of the differently coloured rays, a critical separation takes place on the outside of the prism, similar to that which by reflection happens in the blue bow at the inside; and by which, in the present case,

* See the first paragraph of the 46th article of this paper.

the red part of the prismatic spectrum, that is, the red, orange, yellow, and some of the green, can only reach the eye.

43. *Of a sudden Change of the Colours of the Bows.*

It has been shown that the red bow should be seen nearly in the same place where the Newtonian blue bow is visible. For in the 41st article the place of the eye, for seeing the blue bow in the prism of 100 degrees, was determined to be at an obliquity of 49° 57′ 3″·3 ; and with the red bow, and in the same prism, it has been shown that the eye must be placed at the obliquity of 49° 38′ 19″·5. The difference is only 18′ 43″·8, and by the following experiments it will be found, that both the bows may actually be seen nearly in the same part of every prism ; and that the direction of the light, by which we see either the blue or the red bow, determines which of the two will be visible. To prove this, let a right angled prism be laid down on a sheet of white paper before a window, and when the eye is placed in the proper situation for seeing a reflected blue bow, we may instantly transform it into a transmitted red one, by covering the side of the prism which is towards the incident light with a slip of pasteboard; for by stopping the direct light, which before fell on the base of the prism, and was there reflected, we then see the bow by light intromitted from the paper through the base, which, as has been explained, will be red.

With proper management we may have the bow half red and half blue ; blue in the middle with red sides, or red in the middle with blue sides ; which appearances it will not be required to explain any farther, especially after what has already been said in the 18th article of the first part of this paper of the change of the colours of rings.

When we have before us a bow that is half blue and half red, it will be seen that both taken together contain all the prismatic colours in their regular order of refrangibility. It will now also appear that the faint red, orange, and yellow, which I have said are to be perceived above the blue bow,* may arise either from an imperfectly transmitted red bow, which always lies concealed under the Newtonian blue one, or perhaps more probably from the partial reflection of the red, orange, and yellow rays, many of which will come to the eye notwithstanding they are also copiously transmitted.†

According to my account of the red bow, it ought to be seen in the prism a little above the blue one, and this is also further confirmed by any one of the experiments in which we have some part of each bow in view at the same time, for then the relative situation of the two bows will be visible.

* See the first paragraph of the 41st article.

† In my modifications of light I have proved, by undeniable experiments, that within a prism as well as on the outside of it the rays of all the colours are equally reflexible, and that a critical separation of them only takes place at those angles where by refraction a ray cannot be transmitted.

Similar experiments may be made by candle light upon either of the bows; for when a sheet of white paper is pinned against a wall, that it may reflect the light of a candle placed upon a table about three or four inches from the paper, we may then see the blue bow in a prism placed upon a dark ground before the reflecting paper; and the green colour, which it is not very easy to perceive distinctly in day-light, will here be very visible, and the more so if we use an equilateral prism instead of a right angled one. When the reflecting paper is removed from the wall and laid under the prism, that the light may then be thrown upwards and transmitted through the base, we see a bow of a lively red colour.

Before I can introduce more intricate phenomena, it will be necessary to advert to some other particulars relating to these bows.

44. *Of Streaks and other Phenomena produced from the prismatic blue and red Bows.*

It has been remarked in the 40th article, that the production of colours and their configuration when produced are owing to different causes; this will now be confirmed by an experiment.

Scattered rays, when they fall on a prism will by a critical separation of the colours, produce both the blue and the red bows, and these coloured appearances when produced may be modified into streaks, circular rings, and other forms, by the configurating power of surfaces. When a plain glass or metalline mirror is laid under the base of a right angled prism in which we see the blue bow, the contact of the two plain surfaces will immediately produce a great number of coloured streaks. They will be found to be parallel to the bow, most of them within and some just under it. They may be seen without any lens, merely by looking into the prism with the eye pretty close to the surface through which we see the blue bow. This experiment proves that plain surfaces, though they cannot produce colours, have a power of modifying and multiplying them when produced. As I shall have occasion hereafter to be more particular, I shall now only mention that when we lay a spherical surface, such as an object glass, under the prism it will immediately give us several sets of innumerable concentric coloured rings; and, as will now be readily expected, a cylindrical surface placed under the prism will give a number of lenticular appearances, such as are contained between the intersections of two circular arches drawn concave towards each other. The irregular surface of mica will in like manner produce multiplications of appearances, that may be seen much better than they can be described.

When the same surfaces are applied to the red bow, phenomena that are perfectly of the same form will be made visible within and just under the bow; and the streaks will also be in a parallel direction.

The side of the prism, to which a plain glass must be applied, is of singular use in the explanation of many appearances of the coloured phenomena, which are to be seen, and it is on this account that the formation of the generated colours into all sorts of configurations has been noticed before I come to that part of this paper, wherein this subject must find a further discussion ; for by the application of a slip of plain glass, we can decisively ascertain the nature of any coloured appearance in the prism. Thus, when we see a common coloured red or blue arch, occasioned by the mere different refrangibility of light, the plain glass any how applied to the prism will give no streaks. If we apply the plain glass to a transmitting side, we can have no streaks from a critical blue bow, because it is occasioned by reflection ; and for the same reason, when the plain glass is applied to a reflecting side, we can have no streaks that belong to a critical red bow, because it originates at the intromitting surface. With the assistance of this criterion I may now proceed to a review of more complicated phenomena.

45. *Explanation of various Appearances relating to prismatic Bows.*

If, in the open air, we look into the zenith with a right angled prism held across the eyes, we shall see two red bows convex towards each other. They are caused by the bright transmissions of the light of the heavens through the sides in which the bows appear ; for when to either of these sides the criterion of the plain glass is applied, we shall have coloured streaks. The course of the rays which produce the two bows is delineated in Fig. 3. A B C represents the prism, and the rays that can enter the eye when they fall on A B within the limits *a b* A from 0° 0′ 0″ to 5° 56′ 50″·5, which are the red, orange, yellow, and the brightest part of the green, will form the red bow ; and the situation of the eye at E will be had by the mean refrangibility of the rays which give the bow ; for as the angle B *c d* must be 49° 38′ 19″·5, we have the obliquity B *d c* = 85° 21′ 40″·5 and the angle C *e* E that conveys the ray to the eye will be 82° 49′ 34″·2. The same thing will happen on the other side of the prism, where the rays *m n o p q* will come to the eye at E, in an equal but differently directed angle B *q* E, and cause an inverted red bow to be seen in the side A C.

When we look down into the side of an equilateral prism we see a blue bow, but on lifting the eye and prism gently up together towards the zenith, the bow, at a certain altitude, will be changed from blue to red ; and by the application of the criterion, it is proved that we see the first by reflection, and the last by transmission. For, suppose *a b*, Fig. 4, to be a ray of a mean refrangibility between the violet, indigo, blue, and half the green ; when this falls on the side A C of the equilateral prism A B C with an obliquity *a b* A of 57° 58′ 28″·5, it will be refracted so as to make the angle C *c d* 70° 2′ 56″·8 which gives 49° 57′ 3″·3 for the angle C *d c* ; and consequently the ray *d e f* E will come to the eye by the same angles of reflection and refraction as it entered the prism, and make A *f* E equal to A *b a*. The eye at E will therefore see a blue bow. Then if a plain glass be applied to the transmitting side A C there can be no streaks ; for blue bows being caused by the critical separation of the rays occasioned by the Newtonian reflexibility, the plain glass must be in contact with the reflecting side ; and as soon as we hold it against B C, the coloured streaks will make their appearance. The change of the colour of the bow, on lifting the prism and eye together towards the zenith, is represented in figure 5 ; for the light from the sky, which will enter the prism on the side A B, will eclipse the blue bow which was seen before by light entering from the ground through the side A C in figure 4 ; then if *a b* fig. 5 is a ray of the mean refrangibility

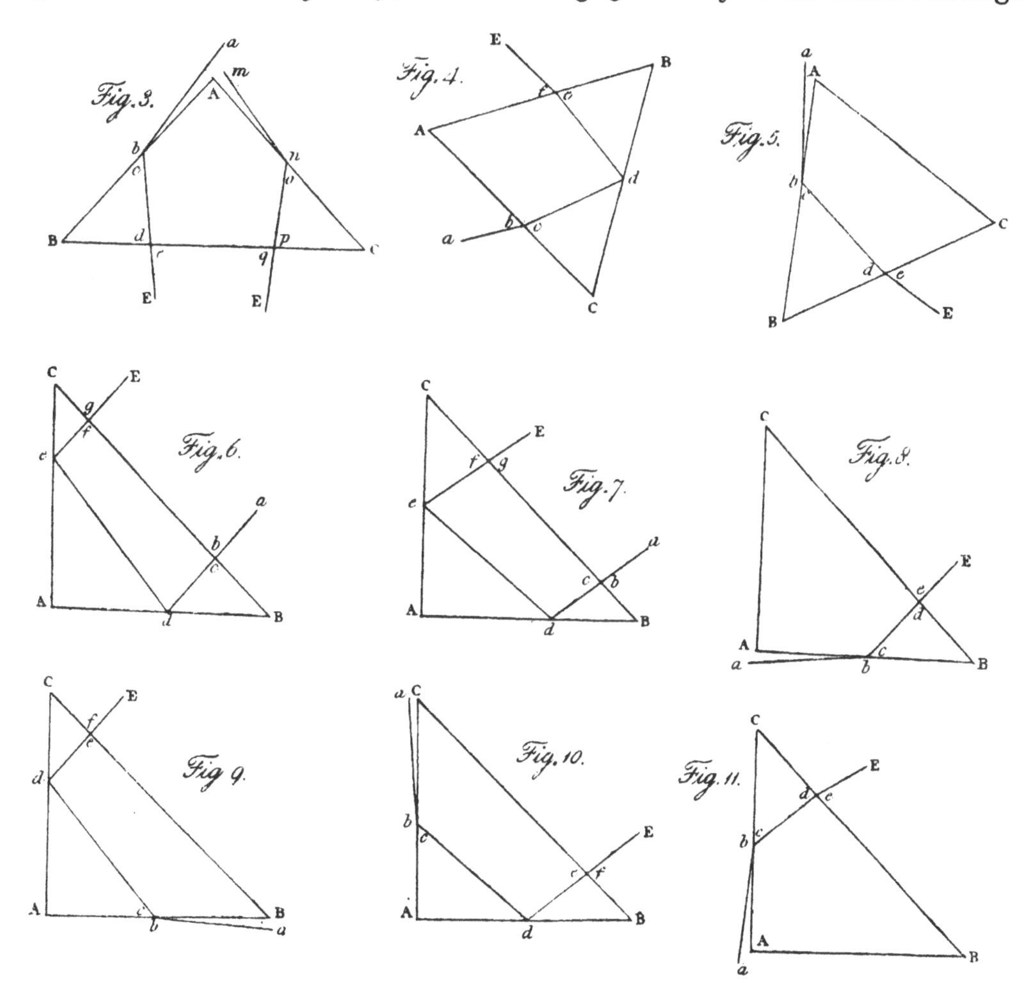

of the red bow, it will by refraction give the angle B *c d* 49° 38′ 19″·5, from which we obtain B *d b* equal to 70° 21′ 40″·5, and the ray will, by a second refraction, come to the eye in an angle C *e* E of 58° 44′ 12″·4, where the red bow will be seen ; but in order to produce coloured streaks, the plain glass must now be applied to the transmitting side A B.

When a right angled prism is held in the hand, so that the light of the sky through an open window may fall upon the base, if then an observer with his back to the light looks through the base into the side A C of the prism A B C fig. 6, he will see an erect blue bow by two reflections, only one of which however is the cause of the critical separation of the coloured rays, the other being a common one. For when a mean refrangible blue-bow-ray falls with an obliquity *a b* C of 82° 17′ 31″ on B C, it will by refraction give the angle B *c d* =85° 2′ 56″·8, from which we obtain A *d e* =49° 57′ 3″·3, which being the mean angle of the critically separated rays, they will by reflection pass to the side A C, where the angle of the common reflection C *e f* will be 40° 2′ 56″·8; this gives *e f* B 85° 2′ 56″·8, and by refraction the middle of the blue bow will be seen by an eye at E in an angle E *g* B equal to the angle *a b* C. From the construction of the figure, it is evident that the eye may be drawn from E towards *a*, and always keep the blue bow in view, which will still remain erect; for when the eye comes to *a*, the rays by which the bow is seen will then enter at E, and the critical reflection will still remain at *d*, as may be satisfactorily proved by an application of the plain glass to A C, which will cause no streaks, whereas they will immediately appear when it is held under the side A B.

When the eye looks into the side B C with the same obliquity of 82° 17′ 31″, but differently directed, so that in fig. 7 the angle may be *a b* B, instead of *a b* C a blue bow will again be seen, but in an inverted position. This also may be drawn over into the other side of the prism without an alteration of its appearance, the reason of which is sufficiently evident from the construction of the figure; but in this case the critical reflection will be at *e*, and the common one at *d*.

It will be proper to shew that like appearances of the red bow may be seen; for this purpose let the prism be laid with one side upon a sheet of white paper placed in a window, with the base towards the observer, as represented in fig. 8. In this position, the light from without reflected by the paper under the prism will be brighter than that from within the room, and the very oblique incident rays *a b* will be refracted by the horizontal side A B, so as to make the angle B *c d* equal to 49° 38′ 19″·5, from which we have B *d c* =85° 21′ 40″·5, and by refraction C *e* E =82° 49′ 34″·2, the eye placed at E will therefore see an erect red bow in the horizontal side A B, which may be drawn over into the perpendicular side without change of position; for the scattered rays reflected from the paper will also enter the prism in the same oblique angle of incidence from the opposite direction *a b* fig. 9; where having caused the red bow by an intromissive critical separation at *c*, they will come to the eye after a common reflection from the side A C, in the same angle as before.

When an inverted red bow is to be seen the eye must be placed a little lower, and the calculation of the angles in the 10th and 11th figures, which represent the course of the rays, being similar though differently directed, will be sufficiently understood by an inspection of them; but as in fig. 8 and 9, the intromissive separation was produced by the horizontal side, so it is, in these figures, effected by the vertical one; all which may be proved by a proper application of the criterion.

There are many other phenomena attending the bows, but as they are more intricate, and not necessary for my present purpose, I leave them to the ingenuity of those who have entered into the preceding calculations, which are quite sufficient to point out the method that should be taken for explaining them.

46. *The first Surface of a Prism is not concerned in the Formation of the blue Bow, nor of the Streaks that are produced by a plain Glass applied to the efficient Surface.*

It has already been mentioned that the bows are formed by scattered light; but to have a direct experimental proof that such light, if not absolutely necessary to the formation of the bows, is at least equally efficient with regularly refracted light, I took a prism with one side of it roughened on emery, and receiving the light through it when the eye was in the situation required for seeing the blue bow, I saw it as completely formed by scattered light, as it could have been by light regularly refracted through a polished side.

A natural consequence of this experiment seems to be, that the form of the surface through which light enters can be of no consequence; this will however admit of a more convincing proof, as follows: upon the middle of the side of a right angled prism, through which the rays entered that caused the blue bow, I laid a plano-convex lens of an inch and a half focus; the result was, that not the least alteration could be perceived either in the form or in the colour of the bow, both which remained as perfect under the place where the incident rays passed through the lens as they were on each side of it. When I changed the convex lens for a plano-concave glass of the same focus, appearances were still the same; and when by a critical application of a plain glass I produced coloured streaks from the base of the prism, the interposition of either the convex or concave glass was equally immaterial. A scattering glass applied to the incident ray, had no other effect than to diminish the brightness of the bow.

The same experiment may be repeated with the red bow; but as here the first surface is essential to the formation of the bow, the plain side of the convex lens or concave glass, when placed against the prism, as before, will produce streaks; neither the bow, nor its streaks however will be in the least affected by the convexity or concavity of the outward surface of the glass applied, through which the light is admitted. A scattering glass will have no effect to disturb the bow or its streaks, and when this glass is emeried on both sides, we have again the bow complete, but without streaks; and by this fact it is proved, that unless a polished plain reflecting surface is applied to the prism, streaks cannot be formed.

47. *The Streaks which may be seen in the blue Bow contain the Colours of both the Parts of the prismatic Spectrum, by the critical separation of which the Bow is formed.*

The most favourable way of observing the colours of the blue bow streaks that are formed when a plain glass is laid under the base of a right angled prism, is to place a screen of white paper before an open window, and to let the direct solar light shine through it upon the side of the prism. This scattered light will be bright and uniform, and cause no adventitious colours to mix with the streaks. The eye should be within six or seven inches of the prism. A streak consists of a certain principal colour and the intermediate tint which separates it from the next; and in the following memorandum of fourteen streaks, which I saw in the manner above described, the principal colours are placed in front, and the dividing tints at the side between them.

Principal colour	Dividing tint
1. Very faint blue,	
. . . . .	Pale red.
2. Faint blue,	
. . . . .	Pale red.
3. Blue,	
. . . . .	Pale red.
4. Bright blue,	
. . . . .	Faint red.
5. Purple blue,	
. . . . .	Whitish red.
6. Bluish red,	
. . . . .	Whitish red.
7. Deep red,	
. . . . .	Greenish white.
8. Red,	
. . . . .	Greenish white.
9. Red,	
. . . . .	Pale bluish green.
10. Red,	
. . . . .	Pale bluish green.
11. Pale red,	
. . . . .	Pale bluish green.
12. Paler red,	
. . . . .	Dirty white.
13. Dingy yellow,	
. . . . .	Dirty white.
14. Dingy yellow.	

To ascertain whether the second surface of the subjacent glass, which by other experiments I know to have a multiplying power of at least six or seven reiterated interior reflections, all of which may be seen through the side of the prism, had any share in the production of these streaks, I fixed on one side of it a glass, of which the lowest surface was emeried, and on the other a metalline plain mirror, but found that the streaks were both in number and colour perfectly alike in them all.

By this account it is evident that the streaks derived from the blue bow contain not only the colours of the blue reflected, but also those of the red transmitted part of the spectrum. This fact is a clear indication of the office which is performed by the surface of the subjacent plain glass, which is simply that of reflecting back the rays of the transmitted red part of the spectrum, which being mixed with the blue part, both together, by their intersections, produce the observed streaks, as will be explained hereafter.

That the colours of the transmitted part of the spectrum are reflected back into the prism, is a point which I suppose will be admitted; but if it should be imagined that the red rays in the streaks of the blue bow might come into the prism by a scattered reflection of the light which falls on the plain glass under its base, then I say that a sheet of white paper or double emeried glass, ought to give the brightest streaks; whereas, on the contrary, neither of them produces any; * it is therefore evident, that a regular reflecting surface is necessary to their formation; but such a surface, be it glass or metal, can only reflect red rays when it receives them; and since we know that the red part of the spectrum is transmitted, and must fall on the reflecting surface, it is but fair to conclude that the rays, of which that part is composed, are those which by reflection re-enter the prism.

* See the last paragraph of the preceding article.

48. *On the Formation of Streaks.*

As I have now ascertained that the streaks we see when a plain glass is laid under a prism, which shows the blue bow, are formed by the principle of reflection, which throws back the transmitted rays, it will be a considerable satisfaction if we can trace the course of these rays far enough to have some idea of the arrangement, whereby such appearances may be produced. To show, by calculation, the complete formation of the streaks in a case that is liable to such variation, on account of the different contact between the modifying surfaces, the position of the light and the inclination of the eye, would be a most laborious, if not endless, undertaking; it will therefore be sufficient, if I can make it appear, that streaks must unavoidably be produced by the rays which after transmission are reflected back again, and mix with those that form the bow; and this I believe will not be difficult. For instance, let F G, fig. 12, be the base of a solid piece of glass, in which a compound ray of light is moving from A to α, with an obliquity A α G $=49^\circ\ 46'\ 12''\cdot 5$; and let I K be the plain surface of a reflecting substance placed under the base; then will the violet, indigo, blue, and the faintest part of the green of this ray be reflected at α, and the remaining green, the yellow, orange, and red will be transmitted. Now, in order to understand the intention of this figure, it will be necessary to observe that on account of the minuteness of the operations of light, all the lines and distances are represented upon a scale one thousand times larger than what the calculation gives them.* The real dimensions of several lines therefore cannot find room in the figure, and must be supplied by imagination. The distance of the eye from the base F G, for instance, which in the calculation has been assumed to be only three inches, will be 3000; the diameter of the pupil of the eye 200; the breadth of the base not less than 2160; and the subtense of the whole blue bow will be twenty-four inches eight tenths. The distance between the reflecting surface I K, and base F G, I have supposed to be the ten thousandth part of an inch; it is therefore in this figure represented by one tenth of an inch, and the space $\alpha\beta$, in which the colours that have been mentioned are transmitted, and which by calculation is ·003588 is expressed by 3·59 inches.

The rays of the different colours which are transmitted at α will be refracted in different angles, and when they come to the reflecting plane will be returned to the base in such a direction, as to come to it again in the same angle in which by refraction they left it; but their distance from the point α, when they reach the base, will differ considerably. If we call the angle of refraction ϕ, and the distance of the reflecting plane from the base x, then $2x \times \dfrac{\text{rad.}}{\text{tan.}\ \phi}$ will be an expression for the intervals at which the several rays will re-enter the base, which for red will be $\alpha r = \cdot 0019198$, for orange $\alpha o = \cdot 0022974$, for yellow $\alpha y = \cdot 0026675$, and for green

* [In this edition figures 12 and 13 are on a scale $= \frac{6}{10}$ of the original.—ED.]

$a\,g = \cdot0043053$. At these places the rays will be a second time refracted, and rise towards the eye in parallel directions, and with an obliquity of 49° 46′ 12″·5 equal to that of their incidence A a G. Their course is represented in the figure by the letters $a\,r\,r'\,r''$, $a\,o\,o'\,o''$, $a\,y\,y'\,y''$, and $a\,g\,g'\,g''$.

These things being premised, I proceed to explain the consequences that must arise from the mixture of the transmitted with the originally reflected rays. The first is, that the rays which after transmission re-enter the prism at different points, and are the cause of the streaks, will not proceed in a parallel direction with those

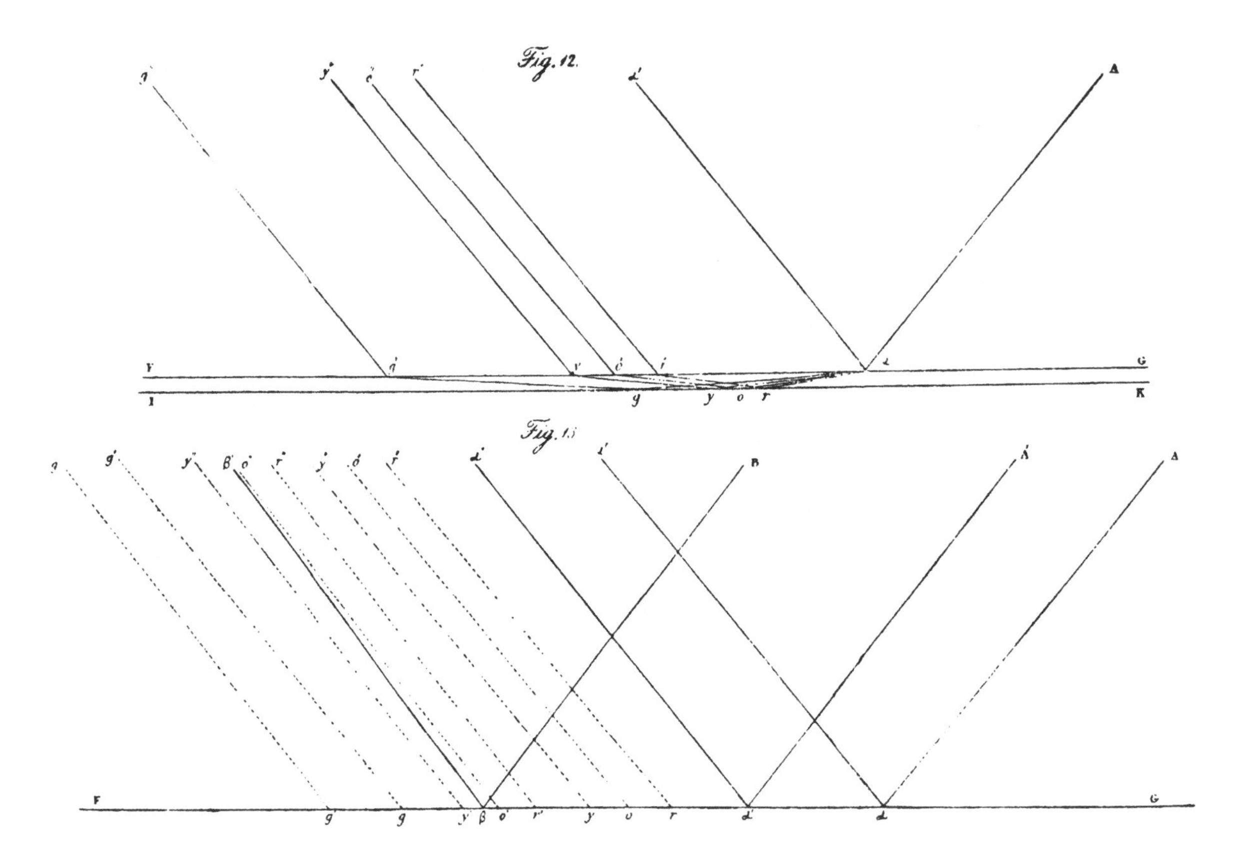

that by reflection from the same or neighbouring points form the blue bow. For instance, let A $a\,a'$, and B $\beta\,\beta'$, fig. 13, be two incident and reflected rays of the blue bow; then if the yellow ray transmitted at a after two refractions, and one reflection, not expressed in this figure, re-enters the prism at y, it will make the angle $y'\,y$ F equal to the angle A a G. But from the construction of the blue bow, it has been shown that B β G is greater than A a G; $\beta'\,\beta$ F is therefore greater than $y'\,y$ F, and the rays $\beta\,\beta'$ and $y\,y'$ will meet somewhere in the line $\beta\,\beta'$ produced. If we call the greatest of the two angles m, the smallest n, and the distance of the angular points d, then $d \times \frac{\sin. n}{\sin. \overline{m - n}}$ will give us the length of the line $\beta\,\beta'$, at which the two rays will meet and intersect each other, which according to the enlarged size of

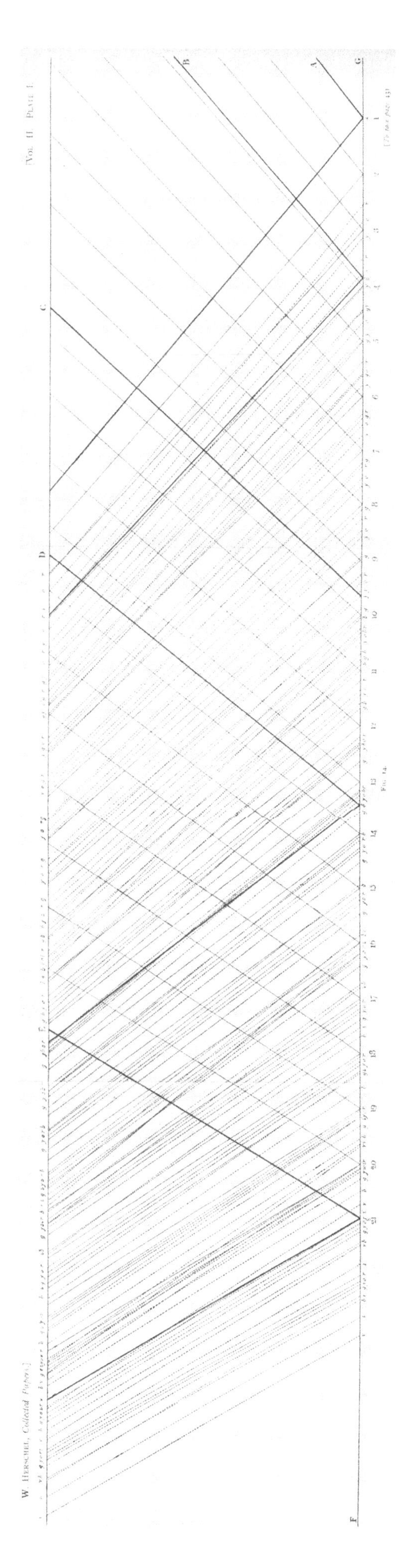

The material originally positioned here is too large for reproduction in this reissue. A PDF can be downloaded from the web address given on page iv

this figure, will be at 773 inches from β.* For the same reason the orange ray $o\,o'$ will meet $\beta\,\beta'$ at 1084 inches, and the red ray $r\,r'$ at 1401 inches from β.* It follows also from the same construction, that some of the transmitted rays will diverge from the reflected ones; for instance, the green ray transmitted at α, which re-enters the prism at g, will make the angle $g'\,g$ F less than the angle $\beta'\,\beta$ F; the rays $\beta\,\beta'$ and $g'\,g$ will therefore diverge. To this may be added, that $g\,g'$ $y\,y'$ $o\,o'$ $r\,r'$ and $\alpha\,\alpha'$ will be parallel.

If such difference between the directions of the transmitted and reflected rays takes place, it will be seen that the rays transmitted through different points are among themselves subject to the same variety in the direction of their course; $r'\,r''$, $o'\,o''$, $y'\,y''$, $g'\,g''$, for instance, which passed through the point α, are parallel to each other; but all of them converge respectively to $r\,r'$, $o\,o'$, $y\,y'$, $g\,g'$ transmitted through c; and on the other hand $y'\,y''$, $o'\,o''$, $r'\,r''$, diverge from $g\,g'$.

Fig. 14, Plate I., is a general representation of the course of the rays of the blue bow, and of those that produce the streaks. The base of the bow is divided into twenty equal parts, and one ray of the bow reflected from each of the points of the division is marked by a line. Twenty-one sets of rays of the different colours transmitted through the same points re-enter the base at their calculated places, and are represented by dotted lines drawn at proper angles; but here it should be noticed, that the difference of the twenty angles being much too small to give any idea of their converging or diverging condition, the difference between each set has been expressed by one degree less towards the right, and one degree more towards the left; the angle of the middle ray being of its proper magnitude. The strong lines marked A α, B β, C γ, D δ, E ϵ, show the division of the colours, and are the same which in fig. 1 were used to explain the construction of the blue bow. The rays incident on the base F G, in the direction of these lines, which are reflected in the same angles, and are also marked with strong lines, meet at the point where the eye is supposed to be placed.

The figure has been drawn by the result of a strict calculation contained in the following table. In the first column are the angles of the obliquity of the incident rays; in the second we have the distances of the reflecting points on the base from α. The remaining columns contain the distances also reckoned from α, at which the transmitted rays of the several colours re-enter the base, after two refractions and one reflection.

From the complex nature of this figure, it will immediately be seen that we cannot attempt an investigation of the particular streaks, that will be formed by the mixture of the transmitted with the reflected rays. An inspection of it, however, will be sufficient to show that streaky appearances must be produced. For instance, between α and the first red ray which re-enters the base, a narrow blue streak should be seen; this will be broken in upon by the mixture of two sets of red, orange, and yellow rays, which together with the reflected colours of the bow, the

* [These figures should be multiplied by 0·6 on account of the reduced scale.—Ed.]

Table of Calculations.

No.	Obliquity.	Distances.	Red.	Orange.	Yellow.	Green.	Blue.	Indigo.	Violet.
	° ′ ″								
1	α 49 46 12·50	·0000000	·0019198	·0022974	·0026675	·0043053			
2	49 47 17·58	·0012397	·0030962	·0034310	·0037464	·0049503			
3	49 48 22·65	·0024794	·0042782	·0045780	·0048491	·0057881			
	β 49 49 20·00	·0035880	.. ..	.. ..	.. ..	.. ..			
4	49 49 27·73	·0037191	·0054653	·0057356	·0059726	·0067327			
5	49 50 32·80	·0049588	·0066566	·0069021	·0071117	·0077443	·0118750		
6	49 51 37·88	·0061985	·0078517	·0080759	·0082630	·0088005	·0112212		
7	49 52 42·95	·0074382	·0090501	·0092559	·0094243	·0098887	·0115764		
8	49 53 48·03	·0086779	·0102514	·0104413	·0105938	·0110005	·0122780		
9	49 54 53·10	·0099176	·0114553	·0116311	·0117701	·0121302	·0131457		
	γ 49 55 34·00	·010721	.. ..	.. ..	.. ..	.. ..	.. ..		
10	49 55 58·18	·0111573	·0126614	·0128249	·0129523	·0132741	·0141088		
11	49 57 3·25	·012397	·0138695	·0140221	·0141395	·0144294	·0151325	·0186142	
12	49 58 8·33	·0136367	·0150797	·0152224	·0153309	·0155939	·0161973	·0183685	
13	49 59 13·40	·0148764	·0162915	·0164254	·0165261	·0167661	·0172916	·0188450	
	δ 49 59 41·00	·0154406	.. ..	.. ..	.. ..	.. ..	.. ..	.. ..	.. ..
14	50 0 18·48	·0161161	·0175048	·0176307	·0177246	·0179447	·0184081	·0196009	
15	50 1 23·55	·0173558	·0187194	·0188382	·0189259	·0191288	·0195415	·0204987	·0231720
16	50 2 28·63	·0185955	·0199354	·0200476	·0201299	·0203177	·0206884	·0214808	·0231409
17	50 3 33·70	·0198352	·0211525	·0212588	·0213361	·0215106	·0218460	·0225168	·0236904
18	50 4 38·78	·0210749	·0223707	·0224715	·0225444	·0227070	·0230125	·0235906	·0244812
19	50 5 43·85	·0223146	·0235899	·0236857	·0237545	·0239066	·0241864	·0246915	·0253990
20	50 6 48·93	·0235543	·0248100	·0249013	·0249663	·0251090	·0253664	·0258130	·0263933
21	ε 50 7 54·00	·024794	·0260309	·0261180	·0261797	·0263138	·0265518	·0269504	·0274380

green being still wanting, must give a pale red division immediately joining the blue streak. When we advance farther into the figure, the great mixture of the colours and the different directions of the rays are so various, that nothing particular can be determined without entering into a very complicated calculation of the meeting and intersections of the rays; we see, however, that these mixtures will produce a condensation of rays in some parts, and vacancies in others, so that no uniform tinge can remain, and consequently streaky appearances must be seen. The same conclusion may be drawn from an inspection of the places where the transmitted colours re-enter the base; for the green, which is transmitted between α and β does not enter again till after the fourth division of the base; the blue which begins to be transmitted at β cannot find admittance again till after the tenth; the indigo transmitted from γ to δ does not re-enter into the composition till after the sixteenth division; and the violet transmitted between δ and ϵ will only come in again after the nineteenth. There will consequently be a considerable space without green, another without blue, a third without indigo, and a fourth without violet; from which it follows, that streaky appearances must every where be seen in the composition of the rays that come to the eye. We should also notice that towards δ all colours but violet will be transmitted, for which reason when they rise again a compound of them will produce streaks that approach to white, such as pale red, pale bluish green, dingy yellow, and dirty white; so that both at the beginning and end of the bow-streaks all observations * of them agree perfectly with what is pointed out by the foregoing remarks; and though we have not analysed the particular construction of the streaks in the middle of the bow, yet what has been said will sufficiently prove that various successive changes of the colours must also take place.

It will be understood that I have only attempted to give some idea of the action of surfaces, in giving configuration to colours that are already produced; but that the principle of reflection is the cause of streaks will remain evident, even if the method of its action should not have been explained so much to our satisfaction as we might wish. It will also remain to be proved, that streaks are only the effect of one of those modifications which depend on the figure of the reflecting surface; † and having got thus far in this research, I may advance towards a final consideration of my subject.

49. *Prismatic Bows when seen at a Distance are straight Lines.*

The next point to be shown, in order to approach gradually to a solution of my problem, is that the apparently arched figure of the blue and red bows, which may be seen in a prism, is merely the consequence of the position of the eye, and the modifying power of the surface through which it sees them. For a proof of this,

* See the first paragraph of the last article.

† See the second paragraph of the 44th article.

it would be sufficient to refer to the principles of the formation of the bows, from which it must be evident that the critical separation of the rays will be exerted in every direction, and that the extent of the bows we see would consequently be parallel to the sides and base of the prism, if the eye could receive the rays which form them, every where in the same angle from a line drawn parallel to that side of the prism through which they pass. An experimental confirmation of this we have by laying down a prism, and keeping either of the bows in view while we gradually draw the eye away; it will then be seen that the curvature, which the bows had assumed, will continually be diminished, and nearly vanish at a very moderate distance.

50. *The Colours of the Bow-streaks owe their Production to the Principle of the critical Separation of the different Parts of the prismatic Spectrum.*

That streaks will be produced when a plain glass is laid under the side of a prism which forms either of the coloured bows, has already been sufficiently shown; but that these streaks, as well as the rest of the phenomena which have been mentioned in the 44th article, are exclusively to be deduced from the same principle by which the bows have been explained will require some proof. With regard to streaks, the following experiment, I believe, will remove every doubt upon the subject.

Let a plain glass be laid under the base of a right angled prism; then, if the eye at first be placed very low, no streaks will be seen; but when afterwards the eye is gradually elevated, till by the appearance of the blue bow we find that the principle of the critical separation of colours is exerted, the streaks will become visible, and not before; nor will they remain in view when the eye is lifted higher than the situation in which the effects of the critical separation are visible. It is therefore evident, not only that the colours are furnished by the same cause which produces the bow but also that they are modified into streaks by the plain surface under the prism.

In addition to this, it must be remarked that the criterion, which has been successfully used in the explanation of several prismatic phenomena, proves that no other colours, but those which arise from the same source, can be modified so as to give streaks. The following experiment will show the foundation on which this criterion is established.

Let there be an horizontal opening in the upper part of a window-shutter, of about three feet long and one foot high; then, if we look at it through one side of a right angled prism, we shall see a red bow from the highest margin of the opening, and a blue one from the lowest; but when a plain glass is applied to either of the sides of the prism through which we see these bows, neither of them will give any coloured streaks. The experimenter must carefully keep the critical bows out of the way; for should either of them fall upon those which are under examination, streaks must of course be seen to pass over them.

When a spherical surface is placed under the prism, it has likewise been shown that coloured rings will be seen ; but these, like the streaks, will not be visible when the eye is below the place where the bows can be seen, which would not have happened had a plain glass been used instead of the prism ; for with such an arrangement, coloured rings may be seen at the most oblique as well as perpendicular stations of the eye.—As soon as the blue bow is perceived, the rings begin to be formed first partly then half, and lastly, we see them completed ; and what is remarkable, these coloured rings are of such a magnitude and brightness, that they cannot be a moment mistaken for those we see when a plain glass is laid upon the same spherical curve.—The eye being then gradually elevated above the range in which the bows may be seen, these rings will pretty suddenly shrink in their dimensions and lose much of their brilliancy ; till at last, when the eye comes to a perpendicular situation, we find them dwindled away to the size and appearance of such as may be seen when a plain glass is substituted for the prism.

Irregular surfaces are no less decisive in the phenomena they exhibit ; for when an equilateral prism is laid upon red mica in a strong illumination of scattered light, we may see a most admirable variety of very minute coloured appearances, whenever the eye is brought to the blue bow place ; but as soon as it is in the least elevated above, or depressed below that situation, these fantastical figures are sure to vanish.

51. *A Lens may be looked upon as a Prism bent round in a circular Form.*

Those who have followed me in the analysis of the blue and red bows, will readily enter into the application I shall make of this theory to the generation of coloured rings by lenses.

It has been proved, that the different refrangibility of the prismatic colours, at certain critical angles, will cause the violet, indigo, blue, and part of the green rays to be separately reflected, and that, according to what has been said in the 49th article, this will produce an extended straight-lined appearance tinged with the above-mentioned colours. It has also been shown that the same principle, at certain critical angles, will cause the red, orange, yellow, and part of the green rays to be exclusively intromitted, in such directions as will produce a similarly extended straight-lined appearance tinged with these latter colours. From the angle in which the eye must receive these appearances in a prism, they are converted into the blue and red bows ; but, since they would appear to be straight lines, if they were seen in directions perpendicular to a line drawn parallel to the edges of the prism, it follows, that were a long prism bent round into a circular form so that its two ends might meet, these lines would then be changed into rings, one of which would be formed by reflection, the other by transmission.

A lens may be said to be such a prism, from which indeed it differs only in one respect, which is, that an angle contained between two lines applied as tangents to different parts of its surface is changeable, whereas the refracting angle of a given prism is constant.

If it should be remarked that in consequence of considering a lens in this light, a plano-convex one, for instance, ought to present us, in certain situations, with a ring of the colours of the blue bow, and in others with a similar ring containing those of the red one, I must observe that the reason why such rings or bows can never be seen by the eye, though the physical separation of the rays should actually take place, is owing to that particular circumstance in which, we have remarked, the lens differs from a prism, namely, the curvature of the refracting surface; for although it has been proved that the figure of the first surface of a prism is not concerned in the formation of the blue bow, yet that of the surface through which it is seen by the eye is of material consequence, as will appear by the following experiment.

An equilateral prism, one side of which I had made cylindrical, was exposed so as to receive the incident light through the convex surface. In this situation, the eye being about three or four inches from the prism, a bow was formed which in every respect was like one I saw in another equilateral prism, whose three sides were flat; but when the convexity of the first prism was turned towards the eye, the bow could no longer be seen, although the critical separation of the rays would undoubtedly form it in this, as well as in the other prism; the two sides and angles of each exposed to the light being perfectly equal. By much attention to what may be perceived when the eye is placed at various distances, I found that the curvature of the surface through which I tried to see the bow, produced a focal contraction and subsequent inversion of the rays in their passage to the eye, and thus occasioned a total change of appearances. Now, since a ring or bow would not be visible in a prism bent round, if the side through which it must be seen were curved, we cannot expect to see such appearances in a lens, which every where presents us with a spherical surface.

The effect upon the appearance of the bows, produced by the surface through which the rays must pass to come to the eye, may be still better examined by laying the plain side of a plano-convex glass of a short focus upon the flat side of a prism, through which we see either of the bows; for when the eye is near the focus of the lens, they will be entirely effaced as far as they are covered by the lens.

A consequence of great importance may be drawn from these experiments; for since the cause of the coloured appearances, which have been called bows when seen in a prism, is now perfectly understood to be the critical separation of the colours of the incident light, it must be admitted that such a separation will certainly take place whenever a beam of light can find an entrance into glass, so as to make the required angles either with an interior or exterior surface, be it in the shape of a prism, lens, or solid of any kind, although the figure of the last transmitting surface should not permit such coloured-appearance-making-rays to reach the eye. A plano-convex lens will consequently by its construction separate the rays of light which enter at the convex surface in such a manner, as by reflection to produce what, if it could be seen, would be called a blue bow, and by rays that come in at the plain side, separate them by intromission so as to produce a red one.

To remove all doubt about the truth of this theory, I ground a small part of a plano-convex lens flat, that I might look into it, as it were, through a window, to see what passed within. The flat made an angle with the base of about thirty-four degrees, and I saw through it very plainly, in different directions of the illumination, a blue bow by light entering at the convex surface, and a red bow by light coming in at the plain one.

With regard to a plain glass contained between parallel surfaces, it may be remembered that when in the last paragraph of the 39th article I said that streaks could not be seen by laying another plain glass under it, I intimated at the same time the formation of colours; this will now admit of a satisfactory explanation. Scattered rays will enter into a parallel piece of glass, and by reflection the critical separation of colours will take place on its interior surface, so that if this effect could be seen, a blue bow would appear; and in the same manner a red bow might be seen by rays intromitted through the lowest surface. In consequence of the course of these rays, streaks would also appear from each of the bows when another plain glass is laid under the parallel piece; but from a calculation made according to the principles that have been established in the preceding part of this paper, the reflection of a mean ray of the blue bow from the interior surface being at the angle 49° 57′ 3″·3; and this being also the oblique incidence on the upper surface, a ray which comes in that direction with the mean refrangibility of the rays of the blue bow cannot come out of glass. The angle of obliquity of the mean intromitted ray for the red bow is 49° 38′ 19″·5, and on computing its direction by the mean refrangibility of the red bow, it will also be found that it cannot clear the glass. I have seen the bows and their streaks when the upper surface of the glass was inclined only nine degrees to the lower one; and possibly a much smaller angle would have been sufficient to permit the emergence of the coloured rays. The strong reflection from the outside of the glass, and the contraction of the dimensions of the bows are however much against perceiving them at a great obliquity.

52. *The critical Separation of the Colours, which takes place at certain Angles of Incidence, is the primary Cause of the Newtonian coloured Rings between Object-glasses.*

It has been proved that streaks, concentric rings, lenticular figures, and all sorts of irregular coloured phenomena may be seen by means of the prism; and in the 35th, 36th, and 37th articles, it has already been sufficiently explained that the cause of the great variety of these appearances is to be found in the configurating power of surfaces. I have also remarked in the 40th article, that in order completely to account for the Newtonian rings, it remained only to be shown how the colours thus modified are produced.

The prismatic experiments contained in this paper have explained in what manner a critical separation of the colours, which takes place at certain angles of incidence, is the cause of the appearance of the blue and red bows; since the different reflexibility of the rays of light, by which NEWTON has accounted for the blue bow, brings on a critical separation of the blue colours, and since also the different intromissibility by which I have explained the red bow, occasions an equally critical separation of the red ones.

In the 50th article I have not only proved that all the above described various appearances, which in the first part of this paper were produced by convex glasses, may be equally well obtained by the use of a prism, but have also shown that the great simplicity of this valuable optical instrument has cleared up great difficulties, by pointing out to us that the colours which are modified into such various shapes, are in all prismatic experiments exclusively produced by the critical separation of the rays of light. Now, as this must be admitted; it will certainly not be philosophical to look for a different cause of the same or similar effects, when convex glasses, which have all the required prismatic properties are used to produce them.*

To show the great similarity, or rather the identity of these effects, let us examine them in different points of view, and since the variety of the configurations is no longer an object that wants explaining, I shall only take the most simple case of each, namely, the coloured rings, that are produced when a plano-convex lens is laid with its convex side upon a plain reflecting surface; and the coloured streaks which are produced when the base of a right angled prism is in the same manner placed upon such a surface.

The form of rings arises from the spherical figure of the lens.†

The right-lined appearance of the streaks is owing to the straight figure of the plain surface of the prism.‡

The colour of the rings may suddenly be changed.§

The colour of the blue bow-streak may as instantly be converted into those of the red bow.||

The cause of the sudden change of the rings has been shown to be that the sets of one colour are seen by reflection, and those of other by transmission.¶

*It has also been shown that the blue bow-streaks are seen by reflection, and those of the red bow by transmission.***

* By this it will be understood that if any case should occur, in which the critical separation cannot account for the observed phenomena, we are then authorised to look out for some other cause to explain them.

† See the first paragraph of this paper.

‡ See the 49th article.

§ See the 15th article of the first part of this paper.

|| See the 43d article.

¶ See the 8th article of the first part of this paper.

** See the 43d article.

In a lens we may at the same time see, in half the set, the colours of the reflected and in the other half, the colours of the transmitted rings.*

And in a prism held before an open window, when the eye is close to it, and when half the bow falls on the side of the room, we may see blue streaks by reflection from half the blue bow, and green streaks by transmission from half the red bow.†

When deep convex, or concave glasses, are laid upon the first surface of a lens, the rings are not affected by it.‡

And when the same glasses are laid upon the first surface of a prism the streaks remain unaltered.§

When the convexity of the lens, which is placed on the reflecting surface, is changed, the size of the rings is also changed.||

And when the angle of the prism is increased or diminished, the distance of the streaks undergoes a proportional alteration.¶

When the lens is pressed upon the plain glass, the rings increase in diameter.**

And by a pressure of the plain glass against the prism the distance of the streaks grows larger.

To form rings by a lens, scattered light is only required.††

And the same light is best for the production of streaks by a prism.‡‡

Many other instances of similarity might be adduced, but those that have been recited will surely be sufficient to show that the same operations, which will produce these prismatic phenomena, will equally account for those that are formed by the lens; now, as it has been clearly proved, that the critical separation of the colours, which takes place at certain angles of incidence, occasions all the phenomena of the blue and red bows, and of the streaks, rings, and other regular or irregular appearances, that may be seen in a prism, there cannot remain a doubt but that the Newtonian rings observed between object glasses, are owing to the same cause.

53. *Remarks relating to the Newtonian alternate Fits of easy Reflection and easy Transmission.*

In attempting to rescue the science of optics, from what has been so long considered as unsatisfactory for explaining the great question about the cause of the coloured rings, I have made use of a principle, the effects of which have so near a resemblance to those of the supposititious fits of easy reflection and easy transmission,

* See the second paragraph in the 18th article.
† The experiment has been made, though not mentioned in this paper.
‡ See the sixth paragraph of the 24th article.
§ See the second paragraph of the 46th article.
|| See the first paragraph of the 7th article.
¶ See the fourth paragraph of the 42d article.
** See the 8th article.
†† See the seventh paragraph of the 24th article.
‡‡ See the third paragraph of the 46th article.

that the author of them might easily be misled by appearances. But although the principle of a critical separation of the colours substituted for these fits, admits the reflection of some rays at the same angles of incidence at which others are transmitted, yet since the Newtonian different refrangibility of light will account for these critical reflections within glass, and the equally critical intromissions from without, we can have no longer any reason to ascribe original fits to the rays of light, which in the first part of this paper, they have already been proved not to possess, and which now in all prismatic experiments, I have shown are not necessary for explaining appearances that may be accounted for without them.

Slough near Windsor,
Dec. 9, 1808.

[The three papers in the *Phil. Trans.* for 1807, 1809, 1810 have been reprinted here for the sake of completeness, although the ideas set forth in them, and Herschel's arguments against Newton's "fits," were entirely against the exact experimental evidence in the *Opticks*. For an account of the strong opposition with which this paper met on the part of the committee of publication of the Royal Society, see the Introduction to Vol. I of this edition.—ED.]

LXII.

Supplement to the First and Second Part of the Paper of Experiments for Investigating the Cause of Coloured Concentric Rings between Object Glasses, and other Appearances of a similar Nature.

[*Phil. Trans.*, 1810, pp. 149–177.]

Read March 15, 1810.

WHEN the intricacy of the subject, on which my two last papers have been treating, is considered, it will not appear singular that a few supplementary articles should be given. The compression of the account of the experiments into a small compass, where many material circumstances must be left unnoticed, may throw some obscurity on the results, which can only be removed by examining the subject in a fuller extent, and from various points of view. I hope the following illustration and additional explanations will have the effect of clearing up what may possibly to some appear obscure or doubtful, in either the first or second part of my paper, and serve also to make the conclusions, which in the second have been chiefly supported by prismatic experiments, directly applicable to such as have in the first been made by convex glasses.

That the colours in all prismatic phenomena, which have been examined in the 44th, 45th, 46th, 47th, and 48th articles of my paper, are produced either by the interior critical separation arising from the different reflexibility of the rays which cause the blue bow, or by the exterior critical separation arising from the different intromissibility of the rays which cause the red bow, has been so clearly and circumstantially proved that it can admit of no doubt; it may even be conceived by some that I have been too particular in giving the precise angles, *when we see in the Lectiones Opticæ,* Sect. II. *Par.* 2, *page* 257, 258, *how far Sir* I. NEWTON *has explained the blue bow*; but a sufficient reason for this minuteness was to give greater clearness to my explanation of the new phenomenon of a red bow, which I have with equal precision described, and which by this means may be, step by step, compared with the production of the blue bow. By this precaution I hoped to anticipate any objection that might occur, such as, for instance, that Sir I. NEWTON has *also explained the red bow which* (it may be supposed) *is merely the converse of the blue bow.* This conception, although NEWTON no where speaks of a red bow, seems to be countenanced by what is said after he has shown that the blue bow is caused by the different reflexibility of the rays of light; for as he affirms that the red, orange, and yellow colours are transmitted, he contrives a method of proving it experimentally, by adding a second prism, placed under that which gives the blue bow, and thus making the transmitted rays visible. The full import of this NEWTONIAN experiment will be considered in the following articles.

LIV. *Supplemental Considerations, which prove that there are two primary prismatic Bows, a blue one and a red one.*

As it will be admitted that we have a primary blue bow, I shall only repeat that by the use of the criterion, which has been indicated at the conclusion of my 44th article, we find that when a plain surface of glass is brought into contact with that side of the prism by which the reflective, or intromissive critical separation is performed, the bow will be turned into streaks, and that the blue bow, which NEWTON has explained, will stand the test of this criterion.

It will now be necessary to prove that the red bow which I have introduced in my 42d article is a phenomenon of equal originality with the NEWTONIAN blue bow. That it will stand the test of the criterion, has already been proved in the 44th article, since by the contact of a plain surface of glass with the efficient surface of the prism this bow is also turned into streaks; but as we find that the NEWTONIAN experiment, by the addition of a second prism, has made the red, orange, and yellow colours, which are the residue of the blue bow, visible, it will be necessary to show that the phenomenon which may thus be viewed is not the red bow I have described.

First consideration. It is a necessary consequence from NEWTON'S explanation of the 49th figure in his *Lectiones Opticæ*, page 260, for a copy of which see fig. 1; that the angle tsp subtended by the transmitted colours must be exactly equal to the angular breadth of the blue bow tSp and that also the angular position Cts, when the eye sees the transmitted colours, must be exactly equal to CtS, in which it is to be placed, that the blue bow may be seen. But these angles in my red bow, are not the same as they are in the NEWTONIAN blue one; for, in figure 2, the angle tsp, which the red bow subtends, is 5′ 55″·4 less than the angle tSp in figure 3, subtended by the NEWTONIAN blue one; and the position of the eye Cts, in figure 2, for seeing the vertex of the red bow, is 15′ 46″ less than CtS figure 3, in which position the eye sees the vertex of the blue bow. Now as these angles arise immediately out of the critical separation of the rays, it is evident that one of these bows cannot be the converse of the other, but that we have two critical separations essentially different, namely the reflective and the intromissive.

Second consideration. The transmitted colours, which NEWTON makes visible by the addition of a second prism, cannot be seen without it. For, if the red, orange, and yellow rays, were not intercepted by the additional prism, they would be refracted at prt, figure 1, and pass into the air, scattered in such a manner, as to be totally unfit for giving a distinct image. My red bow, on the contrary, may be seen in one prism, laid down in open daylight, just as we see the blue bow explained by NEWTON.

Third consideration. The residuary colours of the NEWTONIAN blue bow, being transmitted at *p r t*, the interposition of a second prism will refract them downwards to *s*, for which reason, they can never be seen in the form of a red bow, by an eye placed above the prism at S, where the blue bow is visible; whence it follows that, if we were not now acquainted with an original red bow, all the phenomena of the sudden change of the colours of the bows, which have been explained in my 43d and 45th articles, would still remain unaccounted for.

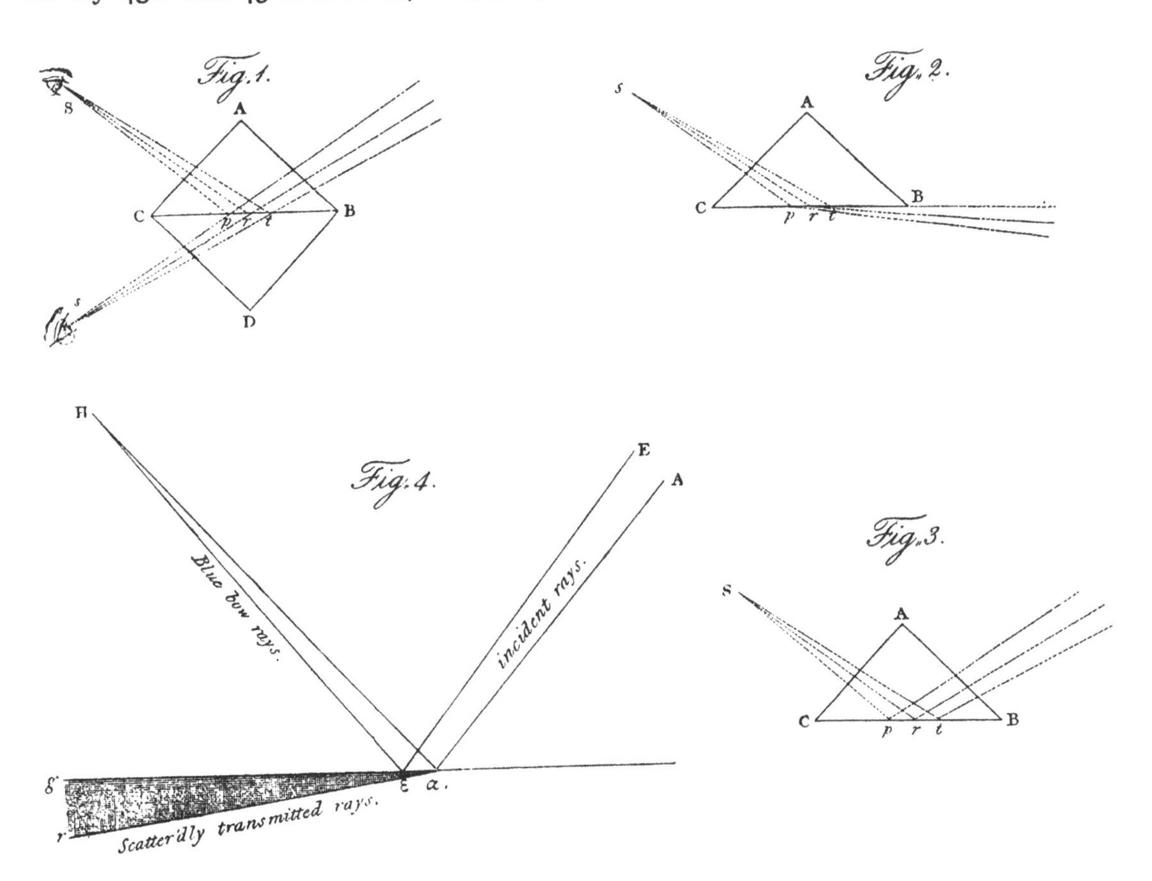

Fourth consideration. The course of the rays which produce the red bow, is so essentially different from the course of those which form the blue bow, that a mere inspection of the figures which represent them, proves, that one cannot be the converse of the other. The incident rays between A E, α ϵ, figure 4, are critically separated by the action of the *interior* base; those which are reflected from the space between α and ϵ, go to H, and form the blue bow; those which are transmitted through the same space, are by refraction scattered over an expansion of 9° 11′ 4″·3 contained between the mean refrangible green, passing from α to *g*, and the least refrangible red, going from ϵ to *r*. The incident rays in figure 5, which enter between A E and $\alpha\epsilon$, are likewise critically separated, but the cause of this separation is the action of the *exterior* base of the prism ; and those which are intromitted between *a* and *ε* so as to come to H, produce the red bow ; the rest being also intromitted, but not coming to H, are by refraction scattered over a space not exceeding 37′ 7″ ; the most refrangible violet going from *a* to *v*, and the mean refrangible green, from *ε* to *g*. As the smallness of some of the angles cannot be accurately expressed in the figures, they may be more correctly compared by the calculated particulars, which are as follows.

	Blue bow.	Red bow.
Convergency of the incident rays	0° 21′ 41″·5	5° 56′ 50″·5
Elevation of the vertices	49 57 3·3	49 38 19·5
Divergency of the scattered rays	9 11 4·3	0 37 7

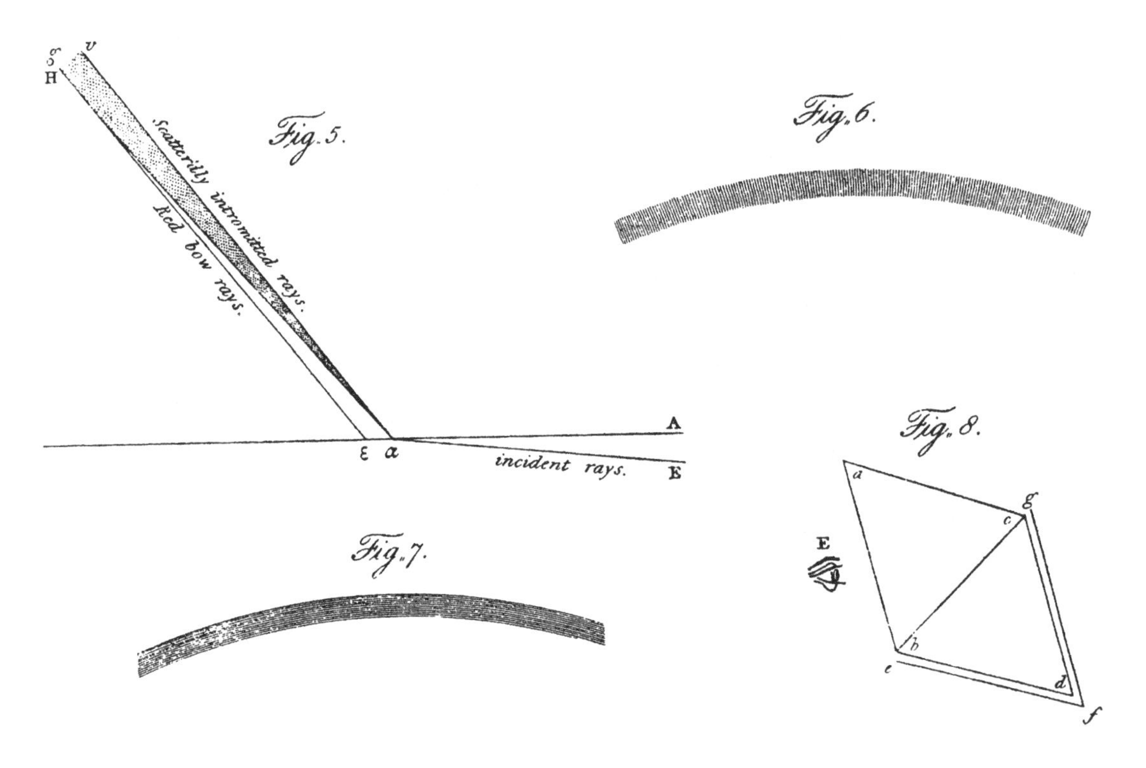

LV. *Illustration of the Dependence of the Streaks of both the Bows, upon the critical Separation.*

One of the reasons, which in the 44th article have been given, for ascribing the colours of the bow-streaks to the critical separation which causes the bows, is their being always in a direction parallel to the bows. With respect to this it may be thought by some, who are still inclined to believe in the fits of easy reflection, and easy transmission, *that streaks parallel to the bows, (though not dependent on critical separation) will in that situation be seen most easily, and most distinctly, because the visual ray in that position passes most obliquely through the stratum of air between the surfaces.* This observation, however, it will be found, cannot be applied to the streaks of either of the bows ; for in the 50th article it has already been proved, not that these streaks can most easily, and most distinctly be seen in the place where the bows are, but that they can absolutely not be seen any where else.

First illustration. To enter more minutely into a subject which is so essential to the support of the arguments contained in the 52d article, let us see whether it will be possible to assign any other reason why the streaks should be parallel to the bows, but their dependence on critical separation ? What is there in two plain surfaces that can determine the direction of streaks, supposing they could possibly be formed without depending on critical separation ? Why, for instance, should they not be as in figure 6, rather than in figure 7, since in both cases, their arrangement in the shape of a bow, would, according to the objection, be still in the position where the visual ray passes most obliquely through the stratum of air ? The necessity of some cause for the direction of the streaks, may be inferred from the experiment which has been given in the 35th article. For when a plain slip of glass is laid upon a cylindrical curvature, the line of contact, which way soever it be turned, will determine the direction of the streaks that are to be seen ; but when two plain surfaces touch all over equally, no bias of this sort can be given to the direction of streaks : the same cause, therefore, which determines the direction of the bow, must also determine that of its streaks, and this establishes their dependence on critical separation.

Second illustration. In what has been said, the possibility that streaks might be formed between two plain surfaces independent of critical separation, has been admitted, but this I cannot allow. The advocates for the colours by thin plates, themselves, must confess, that an uniformly thin plate of air between two plain surfaces, ought not to produce streaks, which contain a variety of colours, so that the very existence of streaks already proves the action of some principle that will produce different colours ; but when the plain side of a prism is laid upon a plain surface of glass, in which situation it has been proved, by the appearance of the bows, that either the reflective or intromissive colour making principle, may be made to exert itself at the interior or exterior base of the prism, and when in either case, streaks are immediately produced, their dependence on the same cause that produces the bows, namely, the critical separation cannot be doubted.

Third illustration. That all the bow-streaks are not only dependent on critical separation, but that each collection depends in particular, on the very principle which forms the bow to which it belongs, is proved by the characteristic colours of the streaks. In the blue bow-streaks, the blue colour is greatly predominant ; and in the streaks of the red bow, the red and green are most abundant, which could never happen if the reflective separation did not copiously furnish the blue, and the intromissive separation as copiously the red and green colours ; and what must put this dependence past all doubt, is the sudden change which may be made in the colour of the streaks ; for by the mere interposition of a screen, or by lifting up the prism towards the light, we may not only change one bow into the other, as has been proved in the 43d article, but when a plain surface is held under the base of the prism, in order to turn the bow into streaks, we may then change the colours of the streaks belonging to one bow, into those which belong to the streaks of the other, with as much certainty as we can change one bow into the other.

A beautiful experiment to prove this, is as follows. Let two equilateral prisms be tied together as in figure 8. Then standing at the distance of five or six feet from an open window, with the prism held in the situation as represented, the sides *bd, dc* being covered with a pasteboard screen *efg*, look into the side *ab*, straight forward to the window, and you will see beautiful blue bow-streaks. The rays which produce them enter through *ac*, are critically separated *by reflection* from the surface *bc*, and cause the blue bow, which by the plain surface *bc*, of the prism *bdc*, is converted into streaks that go to the eye at E.

Without altering either the position of the eye, or of the prism, drop the screen *efg*, and the blue bow-streaks will instantly be changed into those of the red bow. The rays which cause them, enter through the side *bc*, are critically separated *by intromission*, and form the red bow, but are at the same time turned into streaks by the side *bc*, of the prism *bcd*, and go to the eye at E.

The experiment will succeed equally well, if, instead of the prism *bcd*, a highly polished plain slip of glass of the size of the base of the prism *abc* is tied to it.

LVI. *Illustration of the Dependence of Rings, seen in a Prism upon the critical Separation.*

If it should now be granted, that streaks which may be seen by applying a plain glass to the side of a prism, depend entirely upon critical separation, it may still be doubted, whether the rings which are produced, when a prism is laid upon a spherical surface, are likewise to be ascribed to the same cause, but this may also be decided by a very satisfactory experiment as follows.

Upon a small board, lay a sheet of white paper to reflect light upwards, and through the paper fasten three short tacks into the board; then place an object glass upon the tacks, and put a right angled prism across its surface, which should be of the convexity of a globe of about 30 or 40 feet diameter. A pasteboard screen, formed as in figure 9, must be hung over the vertex to darken the exposed side, that only the scattered light which comes from the paper, may enter the prism through the base, and cause a red bow. The board should be placed upon a stand, near a door which admits the unconfined light of the heavens, where no adventitious colours will disturb the experiment. As soon as the eye comes to the altitude of the bow, a set of rings will be seen, whose colours, when the bow goes across their center, will be red and green. Some motion of the eye to bring the bow a very little above or below the center, will show the colours to advantage, and in this position of the eye, we are sure to see the rings precisely in the range of the bow which is turned into rings, but remains visible at both sides, where the critical separation is known to take place. When they have been sufficiently viewed, let the screen be removed, that the brighter light of the heavens from above, may transform the red bow into a blue one, which will at the same time instantly change the colours of the rings from red and green to blue.

If it should now be alleged, that streaks or rings may still be independent of critical separations, notwithstanding their taking the colours of the bows, because they must necessarily appear blue, red, or green, when they are seen in rays of these colours, we may answer this objection by proving experimentally, that any adventitious colours that may occasionally mix with streaks or rings, can only tinge them in the places where they pass to the eye in the same direction, but can themselves not produce either blue, red, or green streaks or rings. Place yourself before a window, and holding in your hand a right angled prism, with a plain slip of glass under the base, look in at one side, and turn the prism upon its axis, till

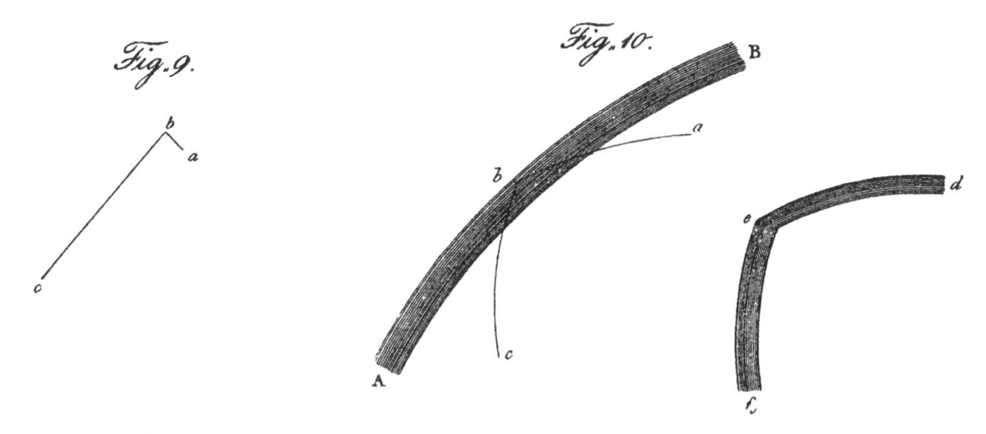

you see the horizontal bars of the window tinged with blue above, and red below; bring the red bow-streaks upon one of these bars, and lowering the streaks gradually, you will find that the colours of the bar merely affect only those parts of the streaks over which they pass, but do not cause any additional streaks of their own colour. To see this still better, take an equilateral prism with a slip of plain glass held under it as before, and turn the axis of the prism gradually down at the left, till the vertical separations of the panes of glass in the window, are equally tinged with the horizontal ones. Let *ab* and *bc*, figure 10, represent one of the angles made by the meeting of the divisions between the panes of glass; then bring the bow-streaks into the direction A B, and draw them gradually over the corner *b*, this motion will cause the streaks to be successively cut by the adventitious colours, but you will at the same time observe these colours to remain confined to the individual place over which they pass, and to produce no other effect than what must result from a mixture of their tinge, with the particular colour of the streaks at the place of their meeting. The corner *abc* will remain perfectly single, which plainly proves that the adventitious colours, not being caused by critical separation, cannot produce streaky phenomena, whereas if they could diffuse themselves, we ought to see at least 5, 6, or 7 coloured angular figures, parallel to each other, as represented at *def*.

LVII. *Remarks on Colours supposed to be produced by thin Plates or Wedges of Air.*

First remark. In the 39th article of my paper it has been shown, that coloured appearances, such as streaks, cannot be seen between the plain surfaces of two parallel pieces of glass applied to each other; if an objection should however be made to this, by showing an experiment with two supposed plain surfaces of glass in contact, where irregular streaks, or flashy appearances may be seen, I shall be authorized to avail myself of what has been proved in the 37th article; for it has been shown, that irregular surfaces will cause irregular figures; for which reason such appearances will only prove, not that plain surfaces can produce them, but that the surfaces between which we see them, are not strictly plain.

Second remark. If it should still farther be conceived that by means of a wedge-formed plate of air, *straight bands of colour would be produced between plain surfaces slightly inclined*, the following experiment will show, that the objection cannot be well founded. I selected two plates of glass, their surfaces being as perfectly plain and parallel as I could possibly find them, and the event shows that they were sufficiently so. The plates were applied to each other in such a manner, that the end of one touched the surface of the other, in a very sharp straight line, while at the opposite end they were kept from contact, by a very fine single thread of the silk worm placed between them, which would produce the required *slight inclination*. No streaks were then visible. I pressed the line of contact strongly together, and streaks became visible; but they were disfigured by pressure, and most disfigured where I pressed most. As soon as the pressure was removed, these coloured appearances vanished. My plain slips were cut with a diamond out of a parallel plate of glass, polished by an optician for optical purposes, and the incumbent slip had its tangent edge finely ground in an angle of about 70 or 80 degrees, to make it a straight line without injuring its plain figure.

Third remark. It will be proper also to take notice of an objection that may be made to the foregoing experiment, by appealing to one of an opposite result; for possibly two plates of glass, supposed to be plain, may be shown, which, when put together slightly inclined, as the experiment requires, will produce streaks near the line of contact; but, if this should be the case, my 35th article accounts so well for the appearance of such streaks, that it would not be philosophical to ascribe them to plain surfaces, when it has been shown, that cylindrical curves of any figure, will invariably produce them; for which reason, I should think myself justified in concluding, that one or other of the plates, which were supposed to be plain, had a cylindrical termination; the figure of which might be circular, elliptical, parabolical, hyperbolical, or indeed of any other variety of cylindrical curvature.

LVIII. *Illustrating Remarks on the Intention of the 14th Figure, explained in the 48th Article of my Paper.*

The great difficulty of representing rays of light, which are compressed beyond all conception, is such, that even a figure one thousand times magnified, which gives a delineation of them, is hardly less inadequate to give a tolerable idea of what is to be expressed, than if it had not been at all increased in its dimensions. This being the case, it might be expected that some objections would arise, such as that *in my figure constructed for explanation of the streaks the vacancies are observed to correspond with, and to depend upon the intervals between the rays* 1, 2, 3, 4, *&c. originally assumed as separated by blank intervals.* There may appear to be some plausibility in this objection, but still if some notion of this kind should be entertained, it may be shown, not only that such a remark would not be quite correct, but also, that the supposed force of it, is founded on a misconception of the figure.

First remark. When we look upon this figure in a cursory way, it may seem as if the vacancies corresponded with the assumed distances of the rays, but this is partly erroneous; for even in the short extent of the figure, the vacancy between 5 and 6 at the bottom, does not correspond with that at the top. The lower single vacancy at 18, does not agree with the two vacancies above. Between 16 and 17 is a vacancy at the bottom, but none at the top. The lower vacancy at 11, has no corresponding one above; nor has that between 4 and 5 below, a corresponding vacancy at the top.

Second remark. The rays 1, 2, 3, 4, &c. are by no means assumed as separated by blank intervals, but as rays at a certain distance from each other, one thousand times greater in the figure, than in their compressed natural state.*

Third remark. The appearance of the rays in my figure, was not intended to represent streaks such as will be seen, but to denote their incipient course in passing from the base of the prism, to an eye whose distance from that base we are to suppose not less than 3000 inches. The visible arrangement and colour of the streaks, erroneously conceived to be expressed in my figure, can only be deduced from the mixture of rays at the place where they enter the eye, the pupil of which it should be remembered, must have a proportional diameter of 200 inches. The angles also, it has been explained, could not possibly be drawn of just dimensions, and from what has been said, we may conclude, that to make a calculation of the mixture and colour of all the rays when they reach the eye, even with the spare quantity of them which has been drawn in the figure, would be extremely laborious, and that a thorough investigation of this particular point would really, as I have before said, be an endless undertaking. The only fault, therefore, that may be found with my figure, I believe is, that it enters perhaps too particularly into this circumstance. It is well known, that similar discussions about irises, fringes, or halos, have generally been dispatched without giving us the least intimation about

* [Plate I. is on a scale = $\frac{2}{3}$ of the original.—ED.]

the real angular course of the rays that produce them. It was indeed, not incumbent on me to go so far, but where angles and distances and intersections fell in my way, that could be determined, I was unwilling to pass them by unnoticed, as such considerations certainly tend to facilitate our conception of the ultimate production of the streaks.

Fourth remark. That such delineations may be used, to show in what manner we may conceive intricate optical phenomena to be produced, I have the authority of eminent writers in my favour ; thus NEWTON, in his first figure of the 3d book of *Optics*, assumes four rays on each side, to illustrate in what manner we may conceive that a hair can give a proportionally larger shadow near its body, than at a distance ; but no one will affirm, that these four rays can give an idea of the actual quantity of light, and the real angles in which it falls on the different parts of the paper ; all which it would be necessary to show, in order to prove, that the appearance of it on paper, agrees with the hypothesis ; and yet we may nevertheless perfectly well conceive the author's meaning, and can make no serious objection to his explanation, on account of his having taken but four rays, at four arbitrary distances, moving in four arbitrary angles.

LIX. *Experiments on the multiplying Power of Surfaces in contact, which modify the Form of prismatic Appearances.*

The simplicity of the following experiments is such as will ensure them a ready admittance, even by those who may not have an opportunity of repeating them ; their application also to some of the most intricate phenomena of modifying the form of the prismatic appearances, must render them of considerable value.

First experiment. Upon a plain metalline mirror $5\frac{3}{4}$ inches long, and 4 broad, I laid the base ABCD, figure 11, of a right-angled prism ; and having darkened the room, a candle was placed so as to throw its light upon one side of the prism, the reflection of which from the base, I saw through the other. The eye was then gradually lifted up to such an altitude, that a blue bow, if it were made visible by the admission of an uniform scattered light, would extend from *a* to *b* ; the candle was then withdrawn, till only the inverted flame of it remained visible at *c*. A small pasteboard screen as long as the prism, and bent at the top as in figure 9, must be hung by the end *a* upon the vertex of the prism, to cover the reflected image of the candle, and the side *bc* must be short enough to leave about one or two tenths of an inch open for scattered light to enter, so as by reflection from the mirror, in proper angles to make the blue bow visible. Every thing being in this arrangement, and the room properly darkened, as well as the eye guarded from the direct light of the candle, place the pasteboard screen on the prism, and you will then perceive a very bright small spectrum of red and green light at *d*, which consists of those rays that in the blue bow place are transmitted through the base of the prism, and are reflected by the mirror in such a direction, as to come to the eye. I do not mention the blue bow streaks which may be seen with some attention; they are very faint, and are not the object of this experiment, serving only to prove, that the coloured spectrum is formed in the bow place.

Second experiment. Every thing remaining arranged as before, lay a narrow slip of thin pasteboard *ef* under the end of the prism at AC, figure 12, but leave BD in contact; the eye must also be elevated till the bow place comes up to *ab*. In this position you will see the small image multiplied, so that, according to the brightness of the candle, and clearness of the prism and reflector, 6, 7, or 8 coloured spectra may be perceived, arranged from *d* to *g* as expressed in the figure. They are not perfect images of the candle, but so many reflections of the red, yellow, and green light transmitted at the blue bow place, and every one of them will

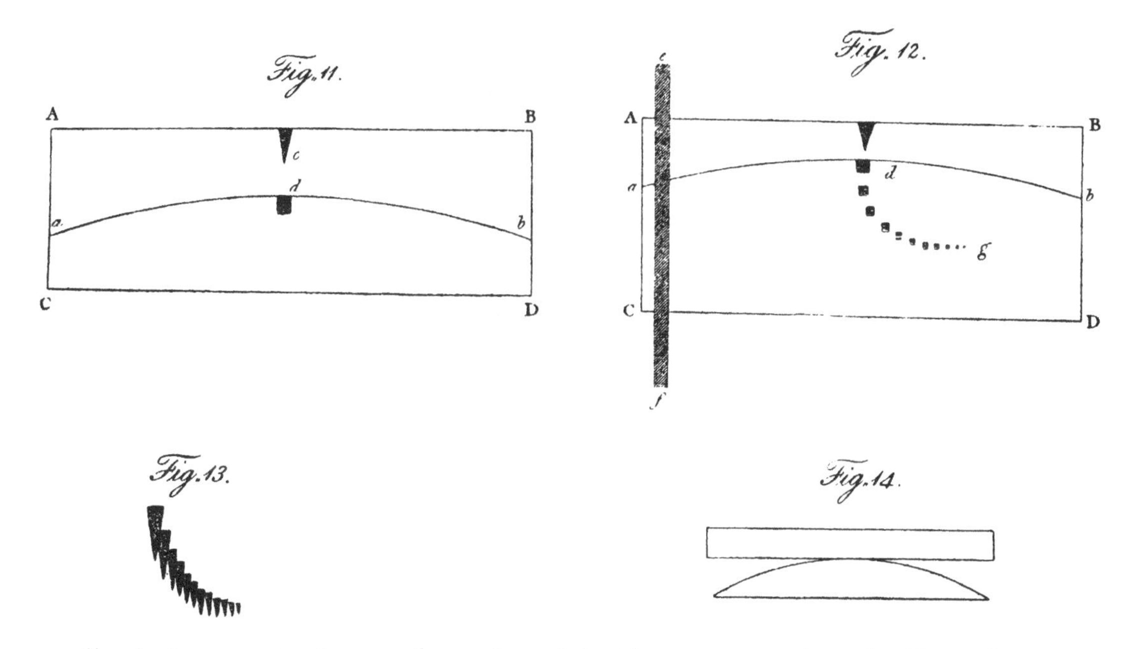

accordingly be seen to be nearly as broad in the green part, as in the red; none of them coming to a tapering point, like the white image of the candle. The spectra are consequently occasioned by a reiterated reflection of the critically separated rays between the subjacent mirror, and the exterior surface of the base of the prism. Indeed, nothing can be more evident than this reiteration of reflections, which is so well known, that opposite parallel mirrors are often put up in rooms, to produce a multiplied extent.

Third experiment. When the eye is lifted high enough to have the line of the critical reflection quite above the prism, and the small screen is also taken away, we may repeat lifting and depressing the end AC, and the two reflecting surfaces will then, by reiteration in the common way, give us a set of complete tapering images of the white flame of the candle, arranged as in figure 13. And in consequence of the various distance of the prism from the mirror, the distance of the several images of the candle will increase and decrease, so that when at last, the end AC is again set down on the mirror, they will apparently coalesce into one single bright image.

The use that may be made of these experiments is as follows. From the laws of reflection we know, that the extent of the multiplied images perceived in re-iterated reflections, between two surfaces, may at all times be ascertained when their distance is given. It is also well known, that when two plates are in what is called contact, we can in fact, only suppose them to be extremely near each other. The production of streaks, when a plain glass is laid under a prism, is a sufficient proof that, even when they are in the closest contact which can be made, the subjacent surface still acts by reflection upon the rays that pass through the base of the prism; for if the contact of the two surfaces were so complete as to make one solid mass of glass, no reflection would take place within its substance.

LX. *Of the breadth of the Streaks compared to that of the Bows, and the Cause why they must take up a broader Space than the Bows from which they are derived.*

It must have been noticed by those who have examined the beautiful streaks, which in my paper it has been shown, will be produced when a plain surface is held under the base of the prism, that they take up a broader space than the bows; and notwithstanding what has been said in explanation of their production, by the reflection of the transmitted rays in the 14th figure of the second part of my paper, it may probably be objected, *that since the space occupied by the streaks adjacent to the bows, is much broader than either of the bows, it is conceived, that critical separation will not account for them.* At a first view of this remark, it may in some measure appear to be justified; for if there were no other cause than merely critical separation, the increase of the breadth of the streaks could not well be accounted for. It must however be recollected, that according to what has been proved in the 47th and 48th articles, the modifying power of surfaces in the production of streaks, is added to the principle of the critical separation which produces the colours.

First cause. In consequence of the reflection of the transmitted rays, from the plain surface held under the prism, it may already be seen in the above-mentioned 14th figure, that the streaks must take up a greater extent than the bows; for instance, the last reflected colour which is marked, enters the prism again at *v*, 2·67 inches beyond the faintest part of the bow. This gives a magnified extent of 27·44 to the streaks, that of the bow being 24·8.

Second cause. But in order to find a greater correspondence between calculation and actual observation, I must repeat that we are not to suppose the twenty intervals between the assumed rays to be blanks. The reason why more rays were not introduced, was to avoid crowding the figure unnecessarily; but let us take, for instance, blue rays falling on the interior base at ·003 of an inch from β towards No. 4; then, in order to be reflected so as to reach the eye, they must have the oblique incidence of 49° 49′ 21″; and since we know that many of the rays which fall on the same spot, will be transmitted while others are reflected, we find by computation, that blue rays with the above incidence being transmitted, will be refracted in such a manner, as to arrive at the subjacent reflecting surface in an angle of 11′ 32″, and will, therefore, re-enter the prism at the magnified distance of 59·6 inches from the place at which they left it, or 63·2 from a; and this will give an extent to streaks, amounting to nearly 2½ times the breadth of the bow. The same will take place with indigo and violet rays to an indefinite extent, which it will not be necessary to particularise.

Third cause. In the foregoing article it has been shown, that beside the single reflection, which in the 14th figure is delineated as taking place between the base of the prism, and the subjacent reflecting plane, 6, 7, and 8 succeeding reiterations of the same effect will carry on the reflected rays to a certain extent which is assignable. These rays would have crowded the figure so much, that they could not be inserted; but let us see how far they may extend the breadth of the streaks. I have already shown, that the first reflection, on the magnified dimensions of the figure, will carry the transmitted rays 2·64 inches beyond the bow; these rays, by reiterated reflections, may therefore be extended to 14·84; 18·48, or 21·12 inches beyond it, which alone will be sufficient; but if moreover, the intermediate rays are here also taken into consideration, there cannot remain a doubt, but that the breadth of the streaks is sufficiently accounted for.

LXI. *Of the Manner in which Rays that are Separated by critical Reflection or Intromission come to the Eye.*

The subject of vision in general, affords so many intricate phenomena, that we must not be surprised if some things occur, that are of difficult conception. By means of the principle of the intromissive separation of the colours, I have already accounted for several appearances, that no other principle, not even the NEWTONIAN fits of easy reflection and easy transmission can possibly reach. If therefore, it should be objected to my ascribing the generation of the colours of the NEWTONIAN rings, likewise to critical separation *that rings must arise from some other cause than critical separation, because they can be seen at the under surface of a glass terminated by parallel planes* (*as in figure* 14), *and in other situations in which critical separation cannot reach the eye*, it will be necessary to examine how far this observation is well founded.

The objection seems to convey a double assertion; the first is, that in the situation of a plain glass laid upon a convex surface, no critical separation can reach the eye; and the next, that if I cannot show how the rays come to the eye, the rings cannot be caused by critical separation. The first of these positions contains something taken for granted, which cannot be admitted. It supposes that I affirm the critical separation to be the sole cause of the rings, whereas I have plainly shown, that this separation furnishes only the colours, and that the modifying power of the subjacent spherical reflecting surface, turns these colours into rings. Now to show how very different an effect may be produced, when the critically separated rays are acted upon by the modifying power of a surface upon which they fall after their separation, I need but refer to the experiment which has been mentioned in the 5th paragraph of the 42d article, where the rays of the bows, which can only be seen in one particular situation, when they come directly to the eye, are effectually rendered visible in every direction, by the scattering power of the surface on which they are thrown.

I proceed now, by different decisive experiments to prove, that the objection in neither of the senses it may be taken, can affect the validity of the theory I have explained.

First set of experiments. Having ground and polished a metalline mirror, to the convexity of a sphere of 40 feet diameter, I laid upon it a right angled prism, and when they were properly exposed to the light, I lifted the eye gradually up to the blue bow place, and saw the rings that were formed of the colours critically separated by the base of the prism. That these rings owe their formation to the joint effect of the critical separation, and modifying power of the spherical metalline reflecting surface, cannot, after what has been proved in the 56th article, admit of a doubt. I then lifted the eye very slowly higher and higher, till it was brought to the vertex of the prism, and attending minutely to the rings all this time, I could no where perceive the least interruption in their uniform visibility. I do not take notice of the gradual changes in the colour and size of the rings, because such gradual changes equally happen to those that are seen between object glasses. When the eye is over the vertex, the prism being equiangular at the base, we see in the opposite side an equal set of rings. We may then advance the eye still farther, and keep the first set in view, or, what will be more convenient, take up the opposite set of rings, and confining our attention to it, may draw the eye down again till we lose it: which will not be, till when the eye is nearly brought down to the level of the side exposed to the light. Here then we have an instance of rings composed of the colours furnished by critical separation, which may be seen from the obliquity of 82° 17′ 31″, down to I suppose about 5 degrees, which gives an angular space of at least 77 degrees.

To extend this range farther, I used several prisms with different refracting angles; first, one where that angle was 30 degrees; then one with 25; another with 20; and also one of 9 degrees. By this successive change of prisms, it was ascertained, that the range of visibility increased, when a smaller refracting angle was used. In the last prism, the rings became visible at an elevation of 36° 43′ 6″; this up to 90 on one side, and down again at the other to 5 degrees, gives a range of more than 138 degrees, in which these rings may be seen. To manage this experiment it is necessary, when the eye is vertical, gently to turn the mirror with the prism upon it half round, that the eye may then be depressed gradually, without interfering with the incident light.

From these experiments it may be presumed, that were the refracting angle still farther diminished, it would increase the range at last to that of a plain glass, which therefore, I am authorized to look upon, as a prism with a vanishing refracting angle. It will be seen presently, that even this has been completely verified.

As the modifying power of spherical surfaces, to render critically separated colours visible in every direction, is by these experiments established, we might take it for granted, that a similar power will be exerted by all sorts of curvatures and irregularities of reflecting surfaces; but in order to take nothing upon trust that may be proved, I had recourse to the following experiments with different curvatures.

Second set of experiments. Upon a ridge of glass ground and polished to a cylindrical form, I laid the base of a prism, with one angle of 96; and two of 42 degrees each. In this I saw beautifully coloured streaks, or rather very narrow lenticular configurations, on one side as low as the angle of 42° will allow the critical separation to be seen, and on the other down to within about 5° of the level of the plane, which gives a range of visibility of nearly 82 degrees.

Having also provided a refracting angle of 3° 43′ 8″, and lain it upon the same cylindrical ridge, the visibility of these phenomena was extended to 152 degrees.

A plain glass laid upon the same ridge, extended this range to 170 degrees.

That these effects of extending the range of the angular space, in which the narrow coloured lenticular forms are visible, is owing to the modifying power of the cylindrical surface, is particularly evident from the parallelism of the coloured phenomena with the line of contact, and from the direction of their extended visibility at right angles to this line.

Having laid a right angled prism upon the same ridge, I perceived not only the coloured primary figure, but also a similar, magnified secondary one, which I ascribed to a reflection from the flat base of the glass, the upper side of which, contained the polished cylindrical ridge. To take off this secondary image, I deprived the base of its reflecting power, by rubbing it on emery, which had the required effect.

A doubt might occur, with regard to the cause which produced the colours, because in these experiments the subjacent medium was of glass; and although on account of the emeried base, we may be certain that the curved ridge only acted by reflecting the critically separated transmitted colours in a variety of angles, yet to prove in the most satisfactory manner, that these colours were exclusively from the prism, I ground and polished two metalline cylindrical pieces to different curvatures, and laying upon them my small prismatic angle of 3° 43′, the minute lenticular figures became again visible over an extent of 152 degrees.

To vary the experiment, I placed a cylindrical ridge in contact with a plain metalline mirror, and saw with great facility through the plain side of the glass which was towards the eye, and was highly polished, the extended range of the modified colours, which in this case amounted to 170 degrees.

From this variation a very important consequence may be drawn, which is, that a radiation from curved surfaces reflected by plain ones, will effect the same extension of visibility, as a radiation from plain surfaces reflected by curved ones. How much more then must this effect be produced, when two curved surfaces are applied to each, as for instance, when two double convex object glasses are laid together ?

Third set of experiments. Upon a piece of mica tied over a cylinder, I placed a right angled prism, and having brought the eye to the altitude where critically separated rays are visible, I perceived a number of irregular forms of beautiful colours. When the position of the eye was gradually changed, I found that the modifying power of the irregularly curved surface of the mica, made these configurations visible over the same angular space in which I had already seen the rings, and lenticular figures.

These appearances are often so delicate, that they may easily escape our notice; although we should follow them with great attention when the eye is moved. This will happen, especially when they are extremely minute, and in the experiment of the last paragraph of the 50th article, I had actually overlooked them; but on repeating the same afterwards with a magnifier, I perceived them without much difficulty.

I tried not only the smaller angles of the former experiments, with the same result of a gradually increased range of visibility, but had also recourse to the plate of glass of unequal thickness mentioned in the 32d article, the sides of which, when produced, would meet in an angle of 2′ 2″. Its surfaces approach so nearly to parallelism, that the inference I had drawn from the trial of smaller angles, when rings on the convex mirror were examined, was now verified by an application of this plate to the surface of mica; for with the assistance of a magnifier, I saw the coloured forms over an angular space amounting to 171° 18′ 28″, which is full as much as we could have seen with a plain glass. From this range, in which the actual angles of elevation of the eye above the plane of the glass at each extreme were measured, it appears that 5 degrees, which I have before allowed for this purpose, is more than sufficient.

The foregoing three sets of experiments prove, that the first of the assertions, into which I have divided the objection, is not well founded, because the modification of the subjacent reflecting surface, so essential to the formation of the phenomena under consideration, has not been attended to. In addition to this, the inference I have drawn from a foregoing experiment proves, that not only the modification of the reflecting surface, but also that through which the rays are transmitted and radiate upon the subjacent one, must be equally taken into consideration.

It remains therefore established, that by combination, the figure of either of the surfaces in contact, be it the reflecting or the radiating one, will make these appearances visible over an extended space, in the shape of rings, ellipses, lenticular figures, and all sorts of irregular configurations, except in the only case, where both reflection and radiation happen between two plain surfaces in contact, and where consequently no change in the angle of seeing the critically separated rays can take place. The uniformity of this modification will then produce streaks, only visible in the bow place.

If the objection should now assume the second form, which is, that unless I can show how the rays of the critical separation thus modified can reach the eye, the rings must arise from some other cause, I may then fairly say, it is sufficient to have proved two very essential points, the first of which is, that these rings are formed from the colours of the critical separation, modified by the subjacent reflecting surface ; and the next, that when this modification is caused by subjacent, or even by incumbent surfaces of any curvature whatsoever, that can be brought into proper contact, their modification of reflection or radiation will then increase the field of visibility of all the various coloured phenomena that can be produced, to whatever extent the circumstances of the combination of the two essential surfaces may allow. The very case proposed in the objection, of a slip of plain glass laid on a spherical surface, has been examined in the most simple form possible, by using for the spherical curvature, a piece of polished metal, to exclude all adventitious source of the generation of colours, and employing for the contact of a plain surface, the base of a prism, on the inside of which, according to the NEWTONIAN doctrine of the different reflexibility of light, it must be admitted, the colours will be critically separated. Then, without the least change of form, or contact of the two essential surfaces, it has been proved, that a diminution of the prismatic angle, will gradually extend the visibility of the rings, till, even before that angle comes to a vanishing state, where the prism would be converted into a plain slip of glass, the rings may be seen in every direction, and at any elevation of the eye, in which they can be seen, through a slip of glass such as the objection supposes. As this then has been proved not only of rings, but of all other possible configurations, that may be caused by the rays of the critically separated colours, modified by reflection or radiation from curved surfaces, it is evident, that an objection which asserts that such colours cannot be seen, contradicts the plainest and best established facts.

With regard to the actual course of the rays from the very moment of their critical separation into colours, till they produce the required effect, it cannot surely be expected that I should trace them through a most intricate complication of reflections from curve to curve, when it has been shown, in the second part of my paper, that even with streaks, which are produced by the contact of two plain surfaces, it would be an endless undertaking to follow them till they enter the eye. Enough has already been said upon this subject to convince every intelligent reader, that all the phenomena of coloured rings, which have been ascribed to the effect of certain fits of easy reflection and easy transmission of the rays of light, as well as the great variety of other coloured appearances of which I have treated, admit of a most satisfactory solution, by substituting the solid principle of the critical separation of the different colours in the room of these fits.

LXIII.

Astronomical Observations relating to the Construction of the Heavens, arranged for the Purpose of a critical Examination, the Result of which appears to throw some new Light upon the Organization of the celestial Bodies.

[*Phil. Trans.*, 1811, pp. 269–336.]

Read June 20, 1811.

A KNOWLEDGE of the construction of the heavens has always been the ultimate object of my observations, and having been many years engaged in applying my forty, twenty, and large ten feet telescopes, on account of their great space-penetrating power to review the most interesting objects discovered in my sweeps, as well as those which had before been communicated to the public in the *Connoissance des Temps*, for 1784, I find that by arranging these objects in a certain successive regular order, they may be viewed in a new light, and, if I am not mistaken, an examination of them will lead to consequences which cannot be indifferent to an inquiring mind.

If it should be remarked that in this new arrangement I am not entirely consistent with what I have already in former papers said on the nature of some objects that have come under my observation, I must freely confess that by continuing my sweeps of the heavens my opinion of the arrangement of the stars and their magnitudes, and of some other particulars, has undergone a gradual change; and indeed when the novelty of the subject is considered, we cannot be surprised that many things formerly taken for granted, should on examination prove to be different from what they were generally, but incautiously, supposed to be.

For instance, an equal scattering of the stars may be admitted in certain calculations; but when we examine the milky way, or the closely compressed clusters of stars, of which my catalogues have recorded so many instances, this supposed equality of scattering must be given up. We may also have surmised nebulæ to be no other than clusters of stars disguised by their very great distance, but a longer experience and better acquaintance with the nature of nebulæ, will not allow a general admission of such a principle, although undoubtedly a cluster of stars may assume a nebulous appearance when it is too remote for us to discern the stars of which it is composed.

Impressed with an idea that nebulæ properly speaking were clusters of stars, I used to call the nebulosity of which some were composed, when it was of a certain appearance, *resolvable*; but when I perceived that additional light, so far from resolving these nebulæ into stars, seemed to prove that their nebulosity was not different from what I had called milky, this conception was set aside as erroneous. In consequence of this, such nebulæ as afterwards were suspected to consist of stars, or in which a few might be seen, were called *easily resolvable* ; but even this expression must be received with caution, because an object may not only contain stars, but also nebulosity not composed of them.

It will be necessary to explain the spirit of the method of arranging the observed astronomical objects under consideration in such a manner, that one shall assist us to understand the nature and construction of the other. This end I propose to obtain by assorting them into as many classes as will be required to produce the most gradual affinity between the individuals contained in any one class with those contained in that which precedes and that which follows it : and it will certainly contribute to the perfection of this method, if this connection between the various classes can be made to appear so clearly as not to admit of a doubt. This consideration will be a sufficient apology for the great number of assortments into which I have thrown the objects under consideration ; and it will be found that those contained in one article, are so closely allied to those in the next, that there is perhaps not so much difference between them, if I may use the comparison, as there would be in an annual description of the human figure, were it given from the birth of a child till he comes to be a man in his prime.

The similarity of the objects contained in each class will seldom require the description of more than one of them, and for this purpose, out of the number referred to, the selected one will be that which has been most circumstantially observed ; however, those who wish either to review any other of the objects, or to read a short description of them, will find their place in the heavens, or the account of their appearance either in the catalogues I have given of them in the *Philos. Trans.* or in the *Connoissance des Temps* for 1784, to which in every article proper references will be given for the objects under consideration.

If the description I give should sometimes differ a little from that which belongs to some number referred to, it must be remembered that objects which had been observed many times, could not be so particularly and comprehensively detailed in the confined space of the catalogues as I now may describe them ; additional observations have also now and then given me a better view of the objects than I had before. This remark will always apply to the numbers which refer to the *Connoissance des Temps* ; for the nebulæ and clusters of stars are there so imperfectly described, that my own observation of them with large instruments may well be supposed to differ entirely from what is said of them. But if any astronomer should review them, with such high space-penetrating-powers, as are absolutely required, it will be found that I have classed them very properly.

It will be necessary to mention that the nebulous delineations in the figures are not intended to represent any of the individuals of the objects which are described otherwise than in the circumstances which are common to the nebulæ of each assortment: the irregularity of a figure, for instance, must stand for every other irregularity; and the delineated size for every other size. It will however be seen, that in the figure referred to there is a sufficient resemblance to the described nebula to show the essential features of shape and brightness then under consideration.*

1. *Of extensive diffused Nebulosity.*

The first article of my series will begin with extensive diffused nebulosity, which is a phenomenon that hitherto has not been much noticed, and can indeed only be perceived by instruments that collect a great quantity of light. Its existence, when some part of it is pointed out by objects that are within the reach of common telescopes, has nevertheless obtruded itself already on the knowledge of astronomers, as will be seen in my third article.

The widely diffused nebulosity under consideration has already been partially mentioned in my catalogues.†

The description of the object I shall select is of No. 14 in the 5th class, and is as follows: "Extremely faint branching nebulosity; its whitishness is entirely of the milky kind, and it is brighter in three or four places than in the rest; the stars of the milky way are scattered over it in the same manner as over the rest of the heavens. Its extent in the parallel is nearly $1\frac{1}{2}$ degree, and in the meridional direction about 52 minutes. The following part of it is divided into several streams and windings, which after separating, meet each other again towards the south." See figure 1 [Plate II.].

This account, which agrees with what will be found in all the other numbers referred to, with regard to the subject under consideration, namely, a diffused milky nebulosity, will give us already some idea of its great abundance in the heavens; my next article however will far extend our conception of its quantity.

2. *Observations of Nebulosities that have not been published before.*

It may be easily supposed that in my sweeps of the heavens I was not inattentive to extensive diffusions of nebulosity, which occasionally fell under my observation. They can only be seen when the air is perfectly clear, and when the observer has been in the dark long enough for the eye to recover from the impression of having been in the light.

I have collected fifty-two such observations into a table, and have arranged them in the order of right ascension. In the first column they are numbered; in

* [About the figures accompanying this paper, Sir John Herschel wrote in the *Phil. Trans.*, 1864, p. 41: "They do not profess to be resemblances, and are given rather as types of certain classes of objects into which he there considers the nebulæ to be distributable. At least they are made from very rude diagrams."—ED.]

† See *Phil. Trans.* 1786, p. 471; 1789, p. 226 [Vol. I. pp. 268 and 337]; and 1802, p. 503 [above, p. 215]. The following ten nebulosities are in the Vth class, No. 13, 14, 15, 17, 28, 30, 31, 33, 34, 38.

the second and third columns are the right ascension and north polar distance of a place which is the central point of a parallelogram comprehending the space which the nebulosity was observed to fill. They are calculated for the year 1800.

The length and breadth of the parallelograms are set down in the 4th and 5th columns in degrees and minutes of a great circle. The time taken up in the transit of each parallelogram having been properly reduced to space by the polar distance given in the 3d column, in order to make it agree with the space contained in the breadth of the zone described by the telescope; the dimensions of the former space therefore is in the parallel, and that of the latter in the meridian. My field of view, being fifteen minutes in diameter, its extent has been properly considered in the assigned dimensions of the parallelograms. It is however evident that the limits of the sweeping zone leave the extent of the nebulosity in the meridian unascertained. The beginning of it is equally uncertain, since the nebulous state of the heavens could only be noticed when its appearance became remarkable enough to attract attention. The ending is always left undetermined; for, as the right ascension was only taken once, I have allowed but a single minute of time for the extent of the nebulosity in that direction, except where the time was repeatedly taken with a view to ascertain how far it went in the parallel; or when the circumstances of its brightness pointed out a longer duration.

The sixth column of the table contains the size of the observed nebulosity reduced to square degrees and decimals, computed from the two preceding columns; and in the last I have given the account of these nebulosities as recorded in my sweeps at the time they were made; namely within a period of nineteen years, beginning in 1783 and ending in 1802.

*Table of extensive diffused Nebulosity.**

No.	R.A.	P.D.	Paral.	Merid.	Size.	Account of the Nebulosity.
	h ′ ″	° ′	° ′	° ′	Deg.	
1	0 5 2	81 7	1 44	1 55	3·3	Much affected with nebulosity.
2	0 12 31	86 34	3 0	2 34	7·7	Much affected.
3	0 17 17	61 24	0 41	2 40	1·8	Affected.
4	0 20 31	86 34	1 30	2 34	3·6	Much affected.
5	0 25 5	67 8	0 29	2 34	1·2	Much affected.
6	0 31 22	90 4	2 30	2 19	5·7	Appeared to be affected with very faint nebulosity.
7	0 32 54	49 23	1 33	3 1	4·7	Affected with nebulosity.
8	0 34 21	51 17	1 17	2 49	3·6	Unequally affected.
9	0 36 13	47 3	2 37	3 18	8·6	Suspected faint nebulosity.
10	0 43 32	46 58	0 26	3 18	1·4	Suspected faint nebulosity.
11	1 35 32	60 42	0 28	2 40	1·3	Suspected to be tinged with milky nebulosity.

* [About these nebulous regions see Roberts, *Astr. Nachr.*, No. 3836, and *Month. Not. R.A.S.*, vol. 62, p. 26; Barnard, *Astrophys. Journal*, vol. 17, p. 77, and M. Wolf, *Month. Not.*, vol. 63, p. 303.—Ed.]

No.	R.A.	P.D.	Paral.	Merid.	Size.	Account of the Nebulosity.
	h ′ ″	° ′	° ′	° ′	Deg.	
12	2 22 19	71 27	0 29	2 29	1·2	Much affected with nebulosity.
13	3 56 14	65 6	0 29	2 27	1·7	Much affected.
14	4 17 21	55 7	1 4	2 38	2·8	Suspected pretty strong nebulosity.
15	4 18 21	55 6	1 53	2 38	5·0	Suspected nebulosity.
16	4 21 35	97 44	0 30	2 15	1·1	Strong milky nebulosity.
17	4 23 14	69 23	0 29	2 36	1·3	Much affected.
18	4 38 17	69 23	0 29	2 36	1·3	Much affected.
19	4 46 17	63 25	1 46	2 31	4·4	Strong suspicion of very faint milky nebulosity.
20	5 9 44	65 6	1 23	2 27	3·4	Very much affected.
21	5 13 14	65 6	0 29	2 27	1·7	Affected.
22	5 23 59	97 1	2 31	2 31	6·3	Affected with milky nebulosity.
23	5 25 16	92 48	0 30	2 40	1·3	Affected.
24	5 27 2	94 23	1 48	2 32	4·6	Visible and unequally bright nebulosity. I am pretty sure this joins to the great nebula in Orion.
25	5 30 40	92 35	2 45	2 33	7·0	Diffused milky nebulosity.
26	5 31 58	97 1	1 56	2 31	4·9	A pretty strong suspicion of nebulosity.
27	5 38 5	88 55	1 6	2 37	2·9	Affected with milky nebulosity.
28	5 55 55	86 17	0 30	2 34	1·3	Much affected.
29	5 56 36	110 28	1 48	2 48	5·0	Affected.
30	6 33 7	48 39	0 26	3 4	1·3	Affected.
31	9 22 56	108 3	0 29	2 30	1·2	Affected.
32	9 27 19	18 21	0 24	4 4	1·6	Much affected with very faint whitish nebulosity.
33	10 6 56	98 33	3 58	2 17	9·1	Very faint whitish nebulosity.
34	10 16 1	37 58	0 24	4 9	1·7	Much affected.
35	10 34 29	26 44	0 29	3 15	1·6	Affected with very faint nebulosity.
36	10 58 24	26 44	0 42	3 15	2·3	Affected.
37	11 56 59	58 50	0 41	2 54	2·0	Affected with whitish nebulosity.
38	12 7 34	58 50	0 41	2 54	2·0	Affected with whitish nebulosity.
39	13 7 33	55 20	0 27	2 17	1·0	Much affected.
40	13 58 0	55 20	0 42	2 17	1·6	Very much affected; and many faint nebulæ suspected.
41	15 5 7	70 40	1 52	2 31	4·7	Affected with very faint nebulosity.
42	20 58 20	92 17	1 45	2 21	4·1	Much affected with whitish nebulosity.
43	20 48 50	73 38	0 29	2 52	1·4	A good deal affected.
44	20 51 4	46 51	0 59	2 53	2·8	Faint milky nebulosity scattered over this space, in some places pretty bright.
45	20 52 28	91 57	0 49	0 56	0·8	Much affected with whitish nebulosity.
46	20 53 31	47 7	1 8	3 18	3·7	Suspected nebulosity joining to plainly visible diffused nebulosity.
47	21 0 26	76 3	0 44	2 46	2·0	Affected.
48	21 29 27	80 8	0 30	2 15	1·1	Much affected.
49	21 42 16	68 57	0 29	2 36	1·2	Affected.
50	22 52 36	64 47	0 29	2 47	1·3	Much affected.
51	22 53 6	64 47	0 42	2 47	1·9	Affected.
52	22 55 29	61 15	0 28	2 37	1·2	A little affected.

When this account says *affected*, it is intended to mean that the ground upon which, or through which we see, or may see stars, is affected with nebulosity.

In looking over this table, it may be noticed that I have inserted several nebulosities that were only suspected. Had I been less scrupulous at the time of observation the word suspected would generally have been omitted; for with this nebulosity, as well as with the great number of nebulæ that in my catalogues are marked suspected, I have almost without exception found, in a second review, that the entertained suspicion was either fully confirmed, or that, without having had any previous notice of the former observation, the same suspicion was renewed when I came to the same place again.

When these observations are examined with a view to improve our knowledge of the construction of the heavens, we see in the first place that extensive diffused nebulosity is exceedingly great indeed; for, the account of it, as stated in the table, is 151·7 square degrees; but this, it must be remembered, gives us by no means the real limits of it, neither in the parallel nor in the meridian; moreover the dimensions in the table give only its superficial extent; the depth or third dimension of it may be far beyond the reach of our telescopes; and when these considerations together are added to what has been said in the foregoing article, it will be evident that the abundance of nebulous matter diffused through such an expansion of the heavens must exceed all imagination.

By nebulous matter I mean to denote that substance, or rather those substances which give out light, whatsoever may be their nature, or of whatever different powers they may be possessed.

Another remark of equal importance arises from the consideration of the observed nebulosities. By the account of the table we find that extreme faintness is predominant in most of them; which renders it probable that our best instruments will not reach so far into the profundity of space, as to see more distant diffusions of it. In No. 44 of the table, we have an instance of faint milky nebulosity, which, though pretty bright in some places, was completely lost from faintness in others; and No. 46 confirms the same remark. It has also been already mentioned in the first article, that the nebulosity in V. 14 was brighter in three or four places than in the rest. The stars also of the milky way which were scattered over it, and were generally very small, appeared with a brilliancy that will admit of no comparison with the dimness of the brightest nebulosity. In consequence of this, we may already surmise that the range of the visibility of the nebulous matter is confined to very moderate limits.

3. *Of Nebulosities joined to Nebulæ.*

The nature of diffused nebulosity is such that we often see it joined to real nebulæ; for instances of this kind we have the fourteen following objects.*

* See I. 81, 207, 214. IV. 41. V. 32, 35, 37, 44, 51, 52. *Connoissance des Temps* 17, 42, 64, 78.

The account of the three first nebulæ being shortened in the catalogue, I give it here more at length.

No. 81 in the first class * is "A considerable bright and large nebula. Its nebulosity is of the milky kind, and a small part of it is considerably brighter than the rest. The greatest extent of the milkiness is preceding the bright part, and the termination of it is imperceptible." To No. 207 † should be added, "It seems to join to imperceptible nebulosity on the south preceding side"; and to No. 214, ‡ "It terminates abruptly to the north and is diffused to the south." See fig. 2.

No. 42 of the *Connoissance* is the great nebula in the constellation of Orion discovered by HUYGHENS. This highly interesting object engaged my attention already in the beginning of the year 1774, when viewing it with a NEWTONIAN reflector I made a drawing of it, to which I shall have occasion hereafter to refer; and having from time to time reviewed it with my large instruments, it may easily be supposed that it was the very first object to which, in February 1787, I directed my forty feet telescope. The superior light of this instrument shewed it of such a magnitude and brilliancy that, judging from these circumstances, we can hardly have a doubt of its being the nearest of all the nebulæ in the heavens, and as such will afford us many valuable informations. I shall however now only notice that I have placed it in the present order because it connects in one object the brightest and faintest of all nebulosities, and thereby enables us to draw several conclusions from its various appearance.

The first is that the extensive diffused nebulosities contained in the objects of the preceding articles are of the same nature with the nebulosity in this great nebula; for when we pursue it in its extensive course it assumes precisely the same appearance as the before-mentioned diffused nebulosities.

The second consequence we may draw from the circumstance of its containing both the brightest and faintest nebulosity joined in one object is a confirmation of an opinion already conceived in the second article, namely, that the range of the visibility of nebulous matter is what may be called very limited. The depth of the nebula may undoubtedly be exceedingly great, but when we consider that its greatest brightness does not equal that of small telescopic stars, as may be seen by comparing four of them situated within the inclosed darkness of the nebula, and several within its brightest appearance, with the intensity of the nebulous light; it cannot be expected that such nebulosities will remain visible when exceedingly farther from us than this prime nebula: the ratio of the known decrease of light will not admit of a great range of visibility within the narrow limits whereby this shining substance can affect the eye.

From this argument a secondary conclusion may be drawn, which adds to what has already been said in the foregoing article, namely, that if our best tele-

* [N.G.C. 3344.] † [N.G.C. 4096.] ‡ [N.G.C. 5474.]

scopes cannot be expected to reach the nebulous matter, which by analogy we may suppose to be lodged among the very small stars plainly to be seen by them; the actual quantity of its diffusion may still farther exceed even the vast abundance of it already proved to exist. A nebulous matter, diffused in such exuberance throughout the regions of space, must surely draw our attention to the purpose for which it probably may exist; and it must be the business of a critical inquirer to attend to all the appearances under which it will be exposed to his view in the following observations.

4. *Of detached Nebulosities.*

The nebulosities of the preceding articles are not restricted to an extensive diffusion; we meet with them equally in detached collections; I shall only mention the following six.*

V. 21 † consists of "A broad faint nebulosity extended in the form of a parallelogram with a short ray from the preceding corner towards the south. The nebulosity is nearly of an equal brightness throughout the parallelogram, which is about 8′ long and 5 or 6′ broad, but ill defined." See fig. 3, *a*, *b*, *c*.

5. *Of milky Nebulæ.*

When detached nebulosities are small we are used to call them nebulæ, and it is already known from my catalogues that their number is very great. It will therefore be sufficient to refer only to a few, of which the nebulosity is of the milky kind. ‡

No. 9 in the 5th class § is "A large, extended, broad, faint nebula; its nebulosity, like that of the preceding one (which is DE LA CAILLE's last but one in the *Catalogue des Nébuleuses du Ciel Austral* ||) is of the milky kind."

The only purpose for which the nebulæ of these two classes have been placed in this connection, is to show that large detached nebulosities, whatever may be their appearance, as well as those nebulæ expressly called milky, partake of the general nature of the diffused nebulous matter, pointed out in the preceding articles.

6. *Of milky Nebulæ with Condensation.*

In looking at the beautiful nebula in Orion; to which I refer, because every common good telescope will shew it sufficiently well for the present purpose, we perceive that it is not equally bright in all its parts, but that its light is more condensed in some places than in others. The idea of condensation occurs so naturally to us when we see a gradual increase of light, that we can hardly find a more intelligible mode of expressing ourselves than by calling it condensed. The numerous

* See I. 92. V. 21, 26, 36, 41, 42.
† [N.G.C. 2359.]
‡ See I. 204. III. 1, 116. IV. 7, 20, 30. V. 9, 25.
§ [N.G.C. 6526 = I.C. 1271. The N.P.D. is 114°.]
|| See *Connoissance des Temps* for 1784, page 272. [= N.G.C. 6523.]

instances that will be given hereafter of nebulæ that have this kind of condensation, renders it unnecessary to refer to more than the following four.*

The first of these, No. 11 in the first class,† is "A bright nebula of some extent, although not very large. It is of an irregular figure, and the greatest brightness lies towards the middle. The whitishness of this nebula is of the milky kind." See figure 4.

By attending to the circumstances of the size and figure of this nebula, we find that we can account for its greater brightness towards the middle in the most simple manner by supposing the nebulous matter of which it is composed to fill an irregular kind of solid space, and that it is either a little deeper in the brightest place, or that the nebulosity is perhaps a little more compressed. It is not necessary for us to determine at present to which of these causes the increase of brightness may be owing; at all events it cannot be probable that the nebulous matter should have different powers of shining such as would be required independent of depth or compression.

7. *Of Nebulæ which are brighter in more than one Place.*

It is not an uncommon circumstance that the same nebula is brighter in several different places than in the rest of its compass. The following six are of this sort.‡

No. 213 in the first class § is "A very brilliant and considerably large nebula, extended in a direction from south preceding to north following. It seems to have three or four bright nuclei." See fig. 5.

From this construction of the nebula, we may draw some additional information concerning the point which was left undetermined in my last article; for since there it was proposed as an alternative, that the nebulous matter might either be of a greater depth or more compressed in the brightest part of the nebula then under consideration, we have now an opportunity to examine the probability of each case. If here the appearance of several bright nuclei is to be explained by the depth of the nebulous matter, we must have recourse to three or four separate very slender and deep projections, all situated exactly in the line of sight; but such a very uncommon arrangement of nebulous matter cannot pretend to probability; whereas a moderate condensation, which may indeed be also accompanied with some little general swelling of the nebulous matter about the places which appear like nuclei, will satisfactorily account for their superior brightness.

The same method of reasoning may be as successfully applied to explain the number of unequally bright places in the diffused nebulosities which have been described in the 1st, 2d, and 3d articles. For instance, in the branching nebulosity V. 14,|| we find three or four places brighter than the rest—in the nebulosity

* See I. 11, 84. III. 457. IV. 12.

† [N.G.C. 4153.]

‡ See I. 165, 213, 261. II. 297, 406. III. 49.

§ [N.G.C. 4449.]

|| [N.G.C. 6992.]

No. 44 of the table we have places of different brightness. In the nebula of Orion, there are many parts that differ much in lustre; and in V. 37 * of the same article I found, by an observation in the year 1790, the same variety of appearance. In all these cases a proportional condensation of the nebulous matter in the brighter places will sufficiently account for their different degree of shining.

This way of explaining the observed appearances being admitted, it will be proper to enter into an examination of the probable cause of the condensation of the nebulous matter. Should the necessity for such a condensing cause be thought to be admitted upon too slight an induction, a more detailed support of it will hereafter be found in the condition of such a copious collection of objects, as will establish its existence beyond all possibility of doubt.†

Instead of inquiring after the nature of the cause of the condensation of nebulous matter, it would indeed be sufficient for the present purpose to call it merely a condensing principle; but since we are already acquainted with the centripetal force of attraction which gives a globular figure to planets, keeps them from flying out of their orbits in tangents, and makes one star revolve around another, why should we not look up to the universal gravitation of matter as the cause of every condensation, accumulation, compression, and concentration of the nebulous matter? Facts are not wanting to prove that such a power has been exerted; and as I shall point out a series of phenomena in the heavens where astronomers may read in legible characters the manifest vestiges of such an exertion, I need not hesitate to proceed in a few additional remarks on the consequences that must arise from the admission of this attractive principle.

The nebula, for instance, which has been described at the beginning of this article, as containing several bright nuclei, has probably so many predominant seats of attraction, arising from a superior preponderance of the nebulous matter in those places; but attraction being a principle which never ceases to act, the consequence of its continual exertion upon this nebula will probably be a division of it, from which will arise three or four distinct nebulæ. In the same manner its operation on the diffused nebulosities that have many different bright places, will possibly occasion a breaking up of them into smaller diffusions and detached nebulæ; but before I proceed with conjectures, let us see what observations we have to give countenance to such expectations.

8. *Of double Nebulæ with joined Nebulosity.*

In addition to the instances referred to in the preceding article, of nebulæ that have more than one centre of attraction I give the following list of what may be called double nebulæ.‡

The 316th nebula in the second class to which in the catalogue is joined the

* [N.G.C. 7000.]

† See Article 24.

‡ See I. 56, 176, 178, 193. II. 80, 271, 309, 316, 832. III. 45, 644. IV. 8, 28. *Connoiss.* 27, 51.

317th,* consists of "two small faint nebulæ of an equal size within 1′ of each other. Each has a seeming nucleus, and their apparent nebulosities run into each other. Their relative position is in a direction from south preceding to north following." See fig. 6, *a* and *b*.

Each of the fifteen objects referred to contains two nuclei or centers of attraction, and if the active principle of condensation carries on its operation, a division of their at present united nebulosities must, in the end, be the consequence. I have given two figures for the same double nebulæ. For, although the nebulosities of figure *b*, when seen in the direction of the dotted lines will appear to run together, they may nevertheless be at some small distance from each other; but the same cause which will bring on a separation of it in figure *a* will also make two distinct nebulæ of figure *b*.

With regard to their being double nebulæ, it may be objected that this double appearance may be a deception; and indeed if this were a double star, instead of a double nebula, there might be some room for such a surmise. But on two accounts the case is very different. In the first place, we have not nebulæ without number at all distances to which we might have recourse, in supposing one to be far behind the other, as we have stars behind stars to produce an appearance of their being double. In the next; if what has been said of the confined range of the visibility of the nebulous matter be recollected, especially where it is so faint as in the double nebula which has been described, we cannot harbour an idea that the two objects of which it is composed are very far asunder. Add to this their great resemblance in size, in faintness, in nucleus, and in their nebulous appearance; from all which I believe it must be evident that their nebulosity has originally belonged to one common stock.

9. *Of double Nebulæ that are not more than two Minutes from each other.*

To add to the probability of the separation of nebulæ, we ought to have a considerable number of them already separated. The following twenty-three are completely divided although not more than two minutes from one another.†

A description of II. 714 ‡ is "Two pretty bright nebulæ; they are both round, small, and about 2′ from each other, in a meridional direction."

Of III. 755 § is "Two very faint, very small extended nebulæ within 1½′ from each other."

That all these nebulæ are really double, is founded on the reason already assigned in the last article. Then if we would enter into some kind of examination how they came to be arranged into their binary order, we cannot have recourse to a promiscuous scattering, which by a calculation of chances can never account

* [N.G.C. 2371–72.]

† See I. 116, 190, 197. II. 8, 28, 57, 111, 178, 450, 714. III. 92, 228, 280, 591, 687, 719, 755, 855, 886, 943, 952, 959, 967.

‡ [N.G.C. 5353.]

§ [N.G.C. 4403–04.]

for such a peculiar distribution of them. If, on the contrary, we look to a division of nebulous matter by the condensing principle, then every parcel of it, which had more than one preponderating seat of attraction in its extent, must in the progress of time have been divided.

No doubt can be suggested on account of the great length of time such a division must have taken up, when we have an eternity of past duration to recur to.

10. *Of double Nebulæ at a greater Distance than 2′ from each other.*

It may well be supposed that more than one attractive center would not be so frequent a case in small distances, as in nebulosities of a more extended compass; accordingly we find that separated nebulæ at more than 2′ from each other are much more numerous. The following 101 double nebulæ referred to will confirm this statement.*

No. 36 and 37 in the first class † are "Two small bright nebulæ, both a little extended."

No. 74 and 75 in the second class ‡ are "Two pretty bright nebulæ; the preceding of them is almost round; the following very much extended in length; they are not far from the same parallel, and about 8 or 10′ distant from each other."

No. 127 and 128 in the third class § are "Two extremely faint nebulæ, about 3′ from each other, and nearly in the same parallel. The second is a very little brighter than the first, and is of an irregular round figure."

It is remarkable that in the description of all these 101 nebulæ, there are not more than five or six which differ so much in brightness from one another, that we could suppose them to be at any considerable different distance from us; and equal brightness or faintness runs through them all in general; but supposing that any two nebulæ should even differ as much from one another, as the set of the first class which has been described, is different from the faintness of the last described set, yet this would not nearly amount to the difference in the brightness of one part of the nebula in Orion from that of another of the same nebula.

11. *Of treble, quadruple, and sextuple Nebulæ.*

If it was supposed that double nebulæ at some distance from each other would frequently be seen, it will now on the contrary be admitted that an expectation of finding a great number of attracting centers in a nebulosity of no great extent is not so probable; and accordingly observation has shewn that greater combinations of nebulæ than those of the foregoing article are less frequently to be seen.

* See I. 28, 36, 90, 145. II. 17, 44, 55, 61, 74, 84, 85, 155, 118, 121, 139, 153, 167, 219, 228, 233, 333, 388, 426, 429, 455, 518, 546, 550, 580, 614, 679, 684, 692, 751, 764, 787, 789, 841, 842, 865, 868. III. 9, 15, 35, 44, 51, 62, 97, 117, 121, 127, 129, 138, 154, 159, 162, 166, 167, 172, 196, 199, 210, 216, 231, 250, 277, 306, 323, 335, 344, 351, 377, 402, 404, 407, 416, 422, 431, 511, 546, 551, 572, 574, 592, 629, 635, 657, 678, 707, 758, 781, 798, 800, 802, 807, 869, 897, 917, 957, 959, 974.

† [N.G.C. 4550–51.] ‡ [N.G.C. 4754–62.] § [N.G.C. 5706–09.]

The following list however contains 20 treble, 5 quadruple, and 1 sextuple nebulæ of this sort.*

Among the treble nebulæ there is one, namely V. 10,† of which the nebulosity is not yet separated. "Three nebulæ seem to join faintly together, forming a kind of triangle; the middle of which is less nebulous, or perhaps free from nebulosity; in the middle of the triangle is a double star of the 2d or 3d class; more faint nebulosities are following."

Among the quadruple nebulæ we have III. 358.‡ "Four nebulæ, all within three minutes. The largest is faint and small; the other three are less and fainter. They form a small quartile, the largest being the most north of the preceding side."

"The nebulæ which form the sextuple one are all very faint and very small; they take up a space of more than 10 or 12 minutes."

12. *Of the remarkable Situation of Nebulæ.*

The number of compound nebulæ that have been noticed in the foregoing three articles being so considerable, it will follow, that if they owe their origin to the breaking up of some former extensive nebulosities of the same nature with those which have been shewn to exist at present, we might expect that the number of separate nebulæ should far exceed the former, and that moreover these scattered nebulæ should be found not only in great abundance, but also in proximity or continuity with each other, according to the different extents and situations of the former diffusions of such nebulous matter. Now this is exactly what by observation, we find to be the state of the heavens.

In the following seven assortments we have not less than 424 nebulæ; some of them of unascertained size, figure, or condensation; and the rest with only the first of these three essential features recorded.

The reason for not having a more circumstantial account of such a number of objects, is that they crowded upon me at the time of sweeping in such quick succession, that of sixty-one I could but just secure the place in the heavens, and of the remaining three hundred and sixty-three, I had only time to add the relative size.§

* See *treble nebulæ.* I. 17. II. 50, 123, 141, 171, 215, 392, 447. III. 85, 94, 117, 156, 300, 358, 382, 592, 873, 900, 945. V. 10.
Quadruple. II. 482, 568. III. 356, 358, 562.
Sextuple. III. 391.

† [N.G.C. 6514, the "trifid nebula."]

‡ [N.G.C. 4169, 73, 74, 75.]

§ See *sixty-one nebulæ.* II. 30, 66, 68, 70, 109, 114, 117, 125, 138, 170, 174, 176, 345, 361, 390, 391, 496, 499, 541, 542, 543, 572, 573, 629, 631, 806, 898. III. 20, 26, 31, 33, 39, 41, 42, 89, 103, 189, 193, 205, 332, 353, 363, 364, 365, 390, 413, 432, 481, 482, 483, 484, 485, 669, 670, 705, 796, 819, 930, 934, 936. *Connoiss.* 84.
Ten extremely small nebulæ. III. 98, 108, 194, 195, 230, 238, 297, 526, 545, 639.
One hundred and thirty-six very small nebulæ. II. 22, 64, 67, 72, 91, 93, 287, 354, 367, 464, 497, 527, 544, 640, 641, 675, 720, 724, 739, 876. III. 6, 13, 22, 24, 34, 37, 38, 104, 111, 140, 164, 166, 186, 190, 237, 247, 255, 283, 285, 302, 303, 304, 309, 315, 317, 319, 325, 326, 333, 338, 339, 343, 354, 385, 386, 387, 389, 398, 411, 412, 421, 425, 430, 433, 435, 437, 443, 444, 453, 459, 460, 467, 470, 501, 507,

Neither of the nebulæ in these seven divisions will require a description, as the title of each assortment contains all that has been ascertained about them; but their number and situation, especially when added to those that will be contained in the following articles, completely supports what has been asserted, namely, that the present state of the heavens presents us with several extensive collections of scattered nebulæ, plainly indicating by their very remarkable arrangement, that they owe their origin to some former common stock of nebulous matter.

To refer astronomers to the heavens for an inspection of these and the following nebulæ, would be to propose a repetition of more than eleven hundred sweeps to them, but those who wish to have some idea of the nebulous arrangements may consult Mr. BODE's excellent *Atlas Cœlestis*. A succession of places where the nebulæ of my catalogues are uncommonly crowded, will there be seen beginning over the tail of Hydra and proceeding to the southern wing, the body and the northern wing of Virgo, Plate 14. Then to Coma Berenices, Canes venatici, and the preceding arm of Bootes, Plate 7. A different branch goes from Coma Berenices to the hind legs of Ursa major. Another branch passes from the wing of Virgo to the tail and body of Leo, Plate 8.

It will not be necessary to point out many other smaller collections which may be found in several plates of the same Atlas.

On the other hand, a very different aspect of the heavens will be perceived when we examine the following constellations. Beginning from the head of Capricorn, Plate 16, thence proceeding to Antinous, to the tail of Aquila, Plate 9, to Ramus Cerberus, and the body of Hercules, Plate 8, to Quadrans Muralis, Plate 7 and to the head of Draco, Plate 3. We may also examine the constellations of Auriga, Lynx, and Camelopardalus, Plate 5.

In this second review, it will be found that here the absence of nebulæ is as remarkable, as the great multitude of them in the first mentioned series of constellations.

509, 525, 539, 544, 578, 579, 607, 618, 623, 625, 634, 638, 640, 641, 645, 650, 652, 659, 666, 702, 704, 708, 716, 718, 731, 733, 738, 762, 766, 775, 787, 788, 789, 799, 803, 809, 827, 831, 833, 836, 837, 838, 839, 848, 849, 866, 875, 883, 884, 894, 895, 905, 912, 913, 919, 956, 960, 961, 962, 965, 966.

Forty-two not very small nebulæ. I. 119. II. 65, 73, 100, 163, 248, 327, 352, 375, 382, 472, 606, 639, 765, 821, 838. III. 17, 30, 249, 281, 321, 327, 366, 375, 504, 548, 615, 628, 647, 660, 667, 698, 712, 715, 734, 751, 773, 774, 840, 850, 941. *Connois.* 89.

One hundred and seven small nebulæ. I. 25, 123. II. 18, 42, 46, 60, 71, 92, 94, 169, 264, 294, 324, 343, 350, 351, 356, 363, 374, 379, 381, 395, 396, 397, 398, 441, 493, 512, 529, 530, 559, 577, 578, 678, 710, 743, 778, 779, 794, 800. III. 25, 48, 57, 59, 60, 69, 74, 192, 206, 235, 243, 308, 328, 329, 334, 337, 350, 380, 420, 446, 458, 462, 464, 475, 478, 502, 516, 517, 529, 550, 588, 611, 651, 661, 664, 668, 721, 722, 723, 729, 761, 763, 769, 779, 780, 794, 797, 814, 826, 833, 841, 843, 861, 880, 881, 894, 915, 924, 925, 926, 927, 928, 939, 950, 951, 954, 969.

Fifty-eight pretty large nebulæ. I. 22, 24, 85, 169, 283. II. 34, 83, 107, 119, 137, 146, 296, 342, 358, 362, 366, 380, 383, 384, 385, 386, 387, 419, 498, 630, 652, 670, 713, 748, 801, 844, 862, 903, 905. III. 14, 18, 40, 70, 75, 76, 102, 213, 261, 279, 318, 340, 367, 372, 374, 415, 454, 473, 503, 543, 599, 662, 790, 970.

Ten large nebulæ. II. 106, 120, 175, 176. III. 28, 361, 440, 480. V. 6. *Connoiss.* 58.

13. *Of very narrow long Nebulæ.*

In order to advance in our knowledge of the condition of the nebulous matter, we may investigate the form of its expansion by the figure of the nebulæ that have been observed. The following five are particular instances of some that were much extended in length, but very little in breadth.*

No. 254 in the 3d class † is "A very faint nebula, extended from north-preceding to south following. It is about 5′ long and less than ¼ minute broad." See fig. 7.

The expansion of the nebulous matter in general may be considered as consisting of three dimensions; these may all be either nearly equal, or one of them may be much less than the other two; or the extent of two of them may be very inferior to that of the third. The nebulæ which have now been referred to exclude a nebulosity of three nearly equal dimensions, which can never be seen under less than two of them. When two of the dimensions of the nebulous matter are nearly equal, one of them may indeed be only visible; but then the chance that the other should be exactly parallel to the line of sight, is by no means favourable. The most plausible way of accounting for the apparent figure of these nebulæ is, therefore, to admit that the expansion of the nebulosity consists indeed of a very narrow length, and not much depth. This form when ascribed to nebulous matter, is sufficiently uncommon for us to expect to see many nebulæ of the figure of extended rays.

14. *Of extended Nebulæ.*

This class of nebulæ, which are chiefly extended in length, but at the same time have a considerable breadth, is very numerous. I have divided the nebulæ it contains, which are 284, into five assortments as follows.‡

* See I. 23, 206. III. 254. IV. 72. V. 20.

† [N.G.C. 3044.]

‡ See *one hundred and sixty-one extended nebulæ of various small sizes.* I. 80, 89, 194, 202, 234, II. 14, 53, 72, 82, 108, 133, 145, 164, 206, 260, 262, 278, 280, 305, 348, 414, 436, 437, 486, 507, 520, 522, 574, 585, 611, 627, 638, 642, 649, 668, 682, 696, 700, 723, 731, 742, 772, 785, 786, 802, 809, 810, 826, 830, 831, 835, 837, 844, 847, 853, 859, 885. III. 4, 23, 56, 58, 65, 66, 73, 79, 82, 100, 110, 132, 183, 218, 225, 236, 241, 242, 244, 248, 258, 265, 305, 313, 314, 316, 342, 347, 348, 355, 369, 370, 406, 410, 419, 427, 429, 441, 442, 445, 450, 479, 487, 490, 494, 496, 499, 510, 514, 515, 520, 521, 528, 554, 557, 567, 569, 570, 586, 598, 599, 601, 612, 613, 619, 646, 649, 653, 677, 681, 682, 713, 714, 727, 730, 732, 752, 767, 771, 778, 783, 792, 804, 806, 808, 811, 812, 813, 816, 832, 845, 846, 874, 885, 892, 904, 914, 920, 929, 932, 942, 948, 949, 973.

Sixty-two extended nebulæ of various large sizes. I. 14, 20, 76, 141, 189, 212, 215, 220, 253. II. 3, 17, 23, 63, 113, 126, 134, 147, 152, 156, 165, 188, 221, 235, 251, 300, 326, 335, 344, 355, 378, 407, 453, 492, 525, 548, 566, 579, 595, 607, 619, 628, 671, 687, 703, 750, 755, 762, 799. III. 253, 282, 290, 346, 414, 492, 498, 508, 610, 689, 740, 766, 776, 921.

Thirty-one extended nebulæ from ¾ to 2′ long. II. 150, 181, 222, 237, 365, 479, 510, 514, 535, 582, 624, 654, 655, 674, 763, 798, 807, 829, 881, 897, 899, 901. III. 203, 368, 506, 556, 620, 648, 692, 906, 907.

Twenty-four extended nebulæ from 2 to 5′ long. I. 94, 174, 201. II. 227, 284, 291, 402, 432, 490, 536, 558, 600, 664, 747, 784, 900. III. 362, 523, 524, 553, 603, 710, 711, 717.

Six extended nebulæ from 5 to 15′ long. I. 134, 153, 285. II. 824. V. 5, 23.

II. 514 * is " A faint nebula extended from south-preceding to north-following; it is about 2′ long and 1′ broad." See fig. 8.

III. 523 † is " A very faint nebula extended from south-preceding to north-following; it is 3 or 4′ long and nearly 3′ broad."

I. 134 ‡ is " A considerably bright nebula, 7 or 8 minutes long and about 3′ broad."

The considerable breadth of these nebulæ, although chiefly extended in length, proves that two of the dimensions of the nebulous matter, namely, the breadth and depth, are probably not very different; for if the depth, which is the dimension we do not see, should be equal to the length, the chance of its being out of sight is not sufficiently probable to happen very frequently. It is therefore to be supposed that the extension in length is really the greatest; for as we actually see it under this form, we are assured that it is at least as long as it appears, whereas one of the other dimensions, if not both, must certainly be less than the length. This kind of expansion admits of the utmost variety of lengthened form and position; and from the great number of nebulæ to which I have referred, the existence of such nebulosities is fairly to be deduced.

15. *Of Nebulæ that are of an irregular Figure.*

Among the various figures that may be seen in nebulæ we have a great many that are of an irregular appearance; I have divided the following ninety-three into two assortments.§

I. 61 ‖ is " A very bright small nebula north-following a star of the 9th magnitude. It is of an irregular figure." See fig. 9.

II. 289 ¶ is " A faint pretty large nebula; it is of an irregular triangular figure."

By calling the figure of a nebula irregular, it must be understood that I saw no particular dimension of it sufficiently marked to deserve the name of length; for had there been such a distinction, its extension in the longitudinal direction would have been recorded, or, as it frequently happened, for want of time, the nebula would shortly have been called extended. From this consideration it follows, that the nebulous matter which assumes an irregular figure when seen in a telescope, cannot be very different in two of its dimensions; and this leaving

* [N.G.C. 1620.] † [N.G.C. 4682.] ‡ [N.G.C. 4781.]

§ See *sixty-one irregular nebulæ of various small sizes.* I. 61, 284. II. 185, 242, 259, 274, 281, 306, 339, 415, 445, 586, 597, 601, 605, 647, 744, 761, 834, 886, 893, 907. III. 12, 83, 191, 259, 273, 287, 301, 310, 456, 465, 485, 486, 493, 495, 533, 535, 537, 555, 581, 582, 605, 642, 663, 675, 699, 701, 724, 735, 795, 817, 834, 847, 851, 868, 879, 893, 963, 976, 977.

Thirty-two irregular nebulæ of various large sizes. I. 138, 246, 248, 282. II. 43, 81, 149, 289, 346, 349, 360, 421, 467, 468, 495, 587, 651, 681, 711, 749, 756, 804, 877. III. 137, 257, 274, 463, 683, 695, 765, 911, 938.

‖ [N.G.C. 2974.] ¶ [N.G.C. 1888.]

the third entirely undetermined, it may be of greater, equal, or less extent than either of the other two. But to be greater or less than the dimensions that were seen it would require the particular situation of the third dimension in either case to be in the direction of the line of sight, which is so far at least improbable, that we may fairly suppose the unseen dimension not to differ much from either of the former two.

16. *Of Nebulæ that are of an irregular round Figure.*

The apparent figure of the nebulæ contained in the foregoing articles has already assisted me in a great measure to assign the expanded form of the nebulous matter of which they consist. The irregular round appearance of the following fifty-five nebulæ however, being of a much more marked description than the former, will lead to more decisive conclusions. I have divided them into three assortments.*

No. 177 in the third class † is "A very faint nebula of an irregular round figure, about 2 or 3 minutes in diameter." See fig. 10.

The appearance of an irregular round figure necessarily requires that the extent of two dimensions of the nebulous matter should be nearly equal in every direction at right angles to each other. The unseen dimensions may certainly be longer or shorter than the visible irregular diameter ; but then it must be absolutely extended centrally in the line of sight, which is a condition that has no probability in its favour ; and the greater the number is, of such nebulæ, the less is the probability that the form of the nebulous matter should be irregularly cylindrical, or conical. For, except an irregular cylinder or cone, placed in the particular required situation, no expansion of the nebulous matter but an irregular globular one can be the cause of the irregular round figure of the above-mentioned nebulæ. Then since the irregular globular form has this advantage above the cylindrical and conical figure, that it will answer the required end in any situation whatsoever, it is certainly that which ought to be admitted as the cause of the observed appearance.

This method of reasoning upon the form of the nebulous matter from the observed figure of nebulæ, will lead us a step farther than it might have been supposed. For granting it to be highly probable, that the appearance of irregular round nebulæ are owing to so many irregular globular expansions of nebulous matter, it will be necessary to direct our attention to the cause which has

* See *twenty-eight nebulæ of an irregular round figure of various small sizes.* I. 231. II. 97, 191, 243, 254, 273, 336, 560, 758, 895, 896. III. 208, 224, 311, 474, 566, 600, 614, 621, 673, 674, 688, 728, 784, 813, 835, 931, 955.

Twenty-one nebulæ of an irregular round figure of various large sizes. I. 69, 108, 161. II. 197, 240, 494, 513, 537, 538, 552, 685, 727, 872, 890. III. 426, 447, 558, 862, 876. V. 7. *Connoiss.* 70.

Six nebulæ of an irregular round figure of a mean diameter from 1 *to* 5′. III. 131, 177, 223, 261, 542, 617.

† [N.G.C. 925.]

formed this matter into such masses. To ascribe an highly improbable event to chance is not philosophical; especially as a forming cause offers itself to our view, when we direct an eye to the globular figure of the planets and satellites of the solar system.

17. *Of round Nebulæ.*

From what has been said, it appears that the figure of nebulæ is a subject of more interest than mere curiosity. The following fifty-seven were observed to be round, and I give them here in four assortments.*

As the title of each sort gives all that is necessary for the present purpose relating to the various sizes of round nebulæ, a description of one of the last will be sufficient. The observation of I. 269 † says, that it is "A considerably bright round nebula of about one minute in diameter." See fig. 11.

The arguments which I have given in the foregoing article, where only nebulæ of an irregular round figure were considered, need not be repeated when a regular circular form is presented to our view; for the additional number of nebulæ, and the regularity of their figure are both greatly in favour of a conclusion, that the mass of the nebulous matter which occasions their appearance must be of a globular form.

In the last article I have only directed our attention to the cause of this very particular construction, but from the observations of the nebulæ above referred to, we may now more confidently assign the attraction of gravitation as the principle which has drawn the nebulous matter towards a center, and collected it into a spherical compass.

I have already shewn that the same principle appears to be the cause of the condensation of the nebulous matter in the bright places of nebulæ that shine with unequal degrees of light in the different parts of their extent, ‡ and a concurrence of arguments established upon very different foundations cannot fail to give additional weight to the reasonings by which they are supported.

18. *Of Nebulæ that are remarkable for some particularity in Figure or Brightness.*

Among the nebulæ, which I have described as of an irregular figure, the following might have been inserted; but the real form of the nebulous matter of which they consist is probably as irregular as the figure or brightness of the nebulæ themselves. I have arranged thirty-five of them into three assortments.§

* See *three round nebulæ.* III. 381, 511, 754.

Forty-one round nebulæ of various small sizes. I. 275. II. 54, 218, 223, 225, 329, 659, 760, 803. III. 11, 50, 78, 94, 95, 96, 149, 150, 180, 181, 209, 221, 222, 295, 371, 451, 477, 505, 622, 631, 671, 684, 726, 760, 800, 801, 810, 842, 888, 909, 946, 971.

Ten round nebulæ of various large sizes. I. 7, 124, 252. II. 19, 481, 889. III. 54, 77, 112, 452.

Three round nebulæ from 1 to 6′ in diameter. I. 269. II. 593. V. 16.

† [N.G.C. 3488.]

‡ See Article 7.

§ See *two nebulæ of remarkable figure.* I. 286. V. 19.

Ten unequally bright nebulæ. I. 254. II. 200, 210, 422, 557, 591, 646. III. 142, 245, 534.

Twenty-three nebulæ that are brightest on one side. I. 113, 162. II. 26, 27, 136, 155, 313, 332, 364, 369, 370, 442, 506, 531, 555, 589, 623. III. 120, 153, 286, 676, 700. V. 22.

V. 19 * is "A considerably bright nebula about 15′ long and 3′ broad; its length is divided in the middle by a black division at least three or four minutes long." See fig. 12.

The nebulous matter of this nebula is probably a ring in a very oblique position with respect to the line of sight.

II. 646 † is "A pretty bright, large nebula, of an irregular figure; it is unequally bright."

The inequality of its brightness in different parts may arise from unequal condensation or from greater depth of nebulous matter.

II. 313 ‡ is "A pretty bright nebula, a little extended in the parallel. The greatest brightness is towards the following side, which is also the broadest; the preceding part being more like a ray proceeding from it."

The irregular figure of these latter kind of nebulæ may be admitted to arise from the as yet imperfect concentration of a nebulous mass, in which the preponderating matter of it is not in the center.

19. *Of Nebulæ that are gradually a little brighter in the middle.*

The investigation of the form of the nebulous matter in the 13, 14, 15, and 16th articles has been founded only upon the observed figure of nebulæ; and in the 17th article the globular form of this matter deduced from the round appearance of nebulæ, has been ascribed to the action of the gravitating principle. I am now entering upon an examination of nebulæ of which, besides their figure, I have also recorded the different degrees of light, and the situation of the greatest brightness with respect to their figure. These observations will establish the former conclusions by an additional number of objects, and by the decisive argument of their brightness, which points out a seat of attraction.

In the following four assortments are one hundred and fifty nebulæ, which all agree in being a little brighter in the middle. This increase of brightness must be understood to be always very gradual from the outside towards the middle of the nebula, whatever be its figure; and although this circumstance, for want of time, has often been left unnoticed in the observation, I am very sure that had the gradation of brightness been otherwise, it would certainly not have been overlooked.§

* [N.G.C. 891.] † [N.G.C. 5112.] ‡ [N.G.C. 5084.]

§ See *thirty-two nebulæ, the particular figure of which has not been ascertained, gradually a little brighter in the middle.* II. 201, 401, 424, 444, 457, 528, 532, 616, 617, 648, 673, 677, 736, 904. III. 90, 106, 148, 331, 436, 472, 489, 519, 596, 633, 654, 655, 656, 686, 853, 860, 896, 978.

Twenty-four extended nebulæ, gradually a little brighter in the middle. II. 184, 192, 252, 285, 412, 478, 480, 565, 621, 688, 906. III. 141, 233, 449, 461, 468, 488, 532, 577, 736, 890. V. 8, 40, 50.

Twenty nebulæ of an irregular figure, gradually a little brighter in the middle. II. 213, 357, 403, 471, 487, 491, 524, 533, 594, 717, 729. III. 272, 428, 434, 626, 690, 857, 903, 947. V. 29.

Seventy-four round or nearly round nebulæ, gradually a little brighter in the middle. II. 7, 40, 102, 129, 131, 162, 190, 249, 258, 267, 276, 286, 290, 308, 320, 338, 428, 459, 474, 476, 477, 509, 516, 526,

III. 853 * is " A very faint small nebula ; it is very gradually a little brighter in the middle."

III. 488 † is " A very faint extended nebula, near 3′ long, and above 2′ broad ; it is gradually a little brighter in the middle." Fig. 13.

II. 549 ‡ is " A very large and pretty bright nebula of an irregular figure ; it is a little brighter in the middle." Fig. 14.

II. 812 § is " A faint, small, round nebula ; it is very gradually a little brighter in the middle, and the increase of brightness begins at a distance from the center." Fig. 15.

It is hardly necessary to say that the united testimony of so many objects can leave no doubt about the central seat of attraction, which in every instance of figure is pointed out to be in the middle.

The only remark I have to make, relates to the exertion of the condensing power, which in the case of these nebulæ appears to have produced but a very moderate effect. This may be ascribed either to the unshapen mass of nebulous matter which would require much time before it could come to some central arrangement of form either in length, or in length and breadth, or lastly in all its three dimensions. It may also be ascribed to the small quantity of the preponderating central attractive matter ; or even to the shortness of its time of acting : for in this case millions of years, perhaps are but moments.

20. *Of Nebulæ which are gradually brighter in the middle.*

By the general description of a nebula, when it is said to be gradually brighter in the middle, we are to understand that its light was observed to be obviously brighter about the center than in other parts. Had the nebulæ of this class been only a little brighter, or had they been much brighter in the middle, such additional expressions would certainly have been used ; except where time would not allow to be more particular. I have sorted two hundred and twenty-three of these nebulæ like the foregoing, according to their figure, into four classes.||

602, 637, 699, 726, 737, 770, 780, 797, 811, 812. III. 62, 63, 94, 105, 121, 122, 123, 133, 162, 163, 252, 292, 296, 298, 330, 388, 409, 448, 466, 497, 522, 597, 608, 665, 680, 746, 750, 753, 818, 822, 823, 824, 858, 867, 889, 891, 908, 917, 918, 923.

* [N.G.C. 3073.] † [N.G.C. 2848.] ‡ [N.G.C. 4818.] § [N.G.C. 6338.]

|| See *Thirty-nine nebulæ of an unascertained figure, gradually brighter in the middle.* I. 19, 49, 264. II. 24, 49, 87, 88, 89, 90, 319, 337, 347, 368, 373, 409, 440, 515, 534, 590, 610, 634, 636, 672, 783, 830, 840, 856, 857, 858, 860, 861, 863. III. 275, 584, 587, 602, 872, 892, 935.

Fifty extended nebulæ gradually brighter in the middle. I. 1, 55, 62, 131, 199, 241, 259, 263, 279. II. 1, 10, 52, 77, 95, 132, 135, 157, 203, 205, 211, 253, 266, 302, 325, 405, 417, 508, 539, 545, 583, 592, 613, 625, 643, 656, 667, 697, 709, 730, 773, 880, 882. III. 246, 267, 589, 594, 864, 902. V. 4, 39.

Twenty-nine nebulæ of an irregular figure, gradually brighter in the middle. I. 95, 196, 227, 266. II. 36, 56, 96, 130, 226, 265, 295, 314, 353, 423, 433, 434, 475, 488, 553, 596, 657, 663, 690, 793, 819, 825, 887. III. 397, 500.

One hundred and five round, or nearly round nebulæ, gradually brighter in the middle. I. 5, 12, 54, 70, 98, 106, 120, 148, 168, 186, 211, 222, 229, 243, 245, 274. II. 50, 51, 128, 151, 158, 160, 161, 196,

II. 409 * is "A pretty bright and pretty large nebula; it is very gradually brighter in the middle."

I. 55 † is "A considerably bright, extended nebula about 4′ long and 2′ broad, in a meridional direction; it is gradually brighter in the middle." Fig. 16.

I. 266 ‡ is "A considerably bright, and pretty large nebula, of an irregular figure; it is gradually brighter in the middle." Fig. 17.

I. 98 § is "A considerably bright, and pretty large round nebula; it is brighter in the middle, the brightness diminishing very gradually from the center towards the circumference." Fig. 18.

From the account of these nebulæ, we find again that all what has been said concerning the seat of the forming and condensing power of the nebulous matter, is abundantly confirmed by observation.

I have only to remark that, the exertion of the gravitating principle in these nebulæ, is in a more advanced state than with those of the last article; and that the same conceptions which have already been suggested, namely, the original form of the nebulous matter, its quantity in the seat of the attracting principle, and the length of the time of its action, when properly considered, will sufficiently account for the present state of these nebulæ.

21. *Of Nebulæ that are gradually much brighter in the middle.*

The nebulous matter which appears under the various forms of the following four assortments, containing two hundred and two nebulæ, assumes now a more condensed aspect, than that under which it was seen in either of the two foregoing collections; and thus by its gradually greater compression, gives us a still more decisive indication of the central seat of attraction.‖

208, 224, 247, 255, 256, 263, 275, 293, 307, 312, 330, 331, 333, 359, 376, 399, 408, 411, 435, 458, 461, 465, 511, 517, 523, 562, 567, 580, 588, 594, 614, 615, 622, 632, 633, 635, 662, 712, 719, 741, 769, 777, 792, 817, 818, 845, 851, 852, 865, 866, 873, 879, 883, 884, 888, 902. III. 2, 88, 107, 138, 139, 220, 491, 527, 541, 609, 694, 739, 749, 825, 829, 865, 870, 871, 882, 899, 900, 933, 937, 940, 972.

* [N.G.C. 4190.] † [N.G.C. 7479.] ‡ [N.G.C. 3206.] § [N.G.C. 5273.]

‖ See *Twenty-five nebulæ of unascertained figure, gradually much brighter in the middle.* I. 73, 121, 127, 140, 155, 181, 287. II. 35, 177, 187, 299, 439, 452, 540, 653, 658, 669, 686, 694, 795, 828, 855, 871. III. 863. *Connoiss.* 99.

Fifty-four extended nebulæ, gradually much brighter in the middle. I. 29, 31, 33, 35, 38, 53, 58, 64, 72, 82, 86, 93, 97, 101, 104, 125, 154, 157, 164, 184, 209, 233, 239, 247, 271, 277. II. 12, 13, 31, 37, 182, 212, 231, 282, 318, 416, 431, 463, 504, 604, 612, 626, 691, 701, 702, 704, 725, 753, 775, 875. III. 179, 198. V. 47. *Connoiss.* 49.

Nineteen nebulæ of an irregular figure, gradually much brighter in the middle. I. 10, 26, 59, 66, 109, 110, 114, 115, 219, 235, 237, 276. II. 2, 20, 438, 503, 734, 827. III. 299.

One hundred and four round or nearly round nebulæ, gradually much brighter in the middle. I. 8, 16, 21, 30, 42, 63, 65, 67, 68, 74, 79, 83, 87, 88, 100, 102, 105, 111, 112, 118, 129, 135, 136, 142, 144, 147, 150, 158, 159, 166, 171, 175, 182, 185, 216, 218, 221, 232, 238, 244, 257, 260, 265, 273, 278. II. 5, 11, 38, 69, 98, 148, 230, 236, 245, 250, 257, 269, 270, 277, 288, 292, 301, 303, 309, 311, 328, 418, 420, 446, 462, 466, 556, 561, 564, 575, 598, 632, 644, 645, 660, 666, 695, 707, 728, 738, 757, 767, 774, 782, 816, 823, 839, 854, 874. III. 250, 284, 512, 531, 624, 744, 859, 878. *Connoiss.* 59, 96.

II. 828 * is " A pretty bright small nebula, very gradually much brighter in the middle."

I. 101 † is " A considerably bright pretty large nebula, extended in the meridional direction, about 4′ or 5′ long ; much brighter in the middle." In the 40 feet telescope I saw the very gradual increase of brightness towards the middle of its length ; a longer extent of the nebula was also visible. Fig. 19.

I. 219 ‡ is " A very bright considerably large nebula of an irregular figure, very gradually much brighter in the middle." Fig. 20.

I. 63 § is " A bright round nebula of about one minute in diameter ; it is much brighter in the middle, and very faint towards the border." Fig. 21.

The greater difference between the comparative brightness of the center, and the outward parts of these nebulæ, may certainly be ascribed to the same causes that have been considered in the two foregoing articles ; but in the present case, and taking into the account that this is already a third step of condensation from a little brighter to brighter, then, to much brighter, there appears to be some foundation for supposing rather that this greater effect is produced by a longer time of the action of the attractive principle, than that it should arise merely from an original more favourable expansion of the nebulous matter.

22. *Of Nebulæ that have a Cometic appearance.*

Among the numerous nebulæ I have seen, there are many that have the appearance of telescopic comets. The following are of that sort.||

I. 4 ¶ is "A pretty large cometic nebula of considerable brightness ; it is much brighter in the middle, and the very faint chevelure is pretty extensive." Fig. 22.

By the appellation of cometic, it was my intention to express a gradual and strong increase of brightness towards the center of a nebulous object of a round figure ; having also a faint chevelure or coma of some extent, beyond the faintest part of the light, gradually decreasing from the center.

It seems that this species of nebulæ contains a somewhat greater degree of condensation than that of the round nebulæ of the last article, and might perhaps not very improperly have been included in their number. Their great resemblance to telescopic comets, however, is very apt to suggest the idea, that possibly such small telescopic comets as often visit our neighbourhood may be composed of nebulous matter, or may in fact be such highly condensed nebulæ.

* [N.G.C. 2756.] † [N.G.C. 779.] ‡ [N.G.C. 3665.] § [N.G.C. 1052.]

|| See *Seventeen cometic nebulæ.* I. 3, 4, 34, 217. II. 6, 15, 33, 59, 104, 153, 154, 241, 315, 404. III. 5, 21. *Connoiss.* 95.

¶ [N.G.C. 3169.]

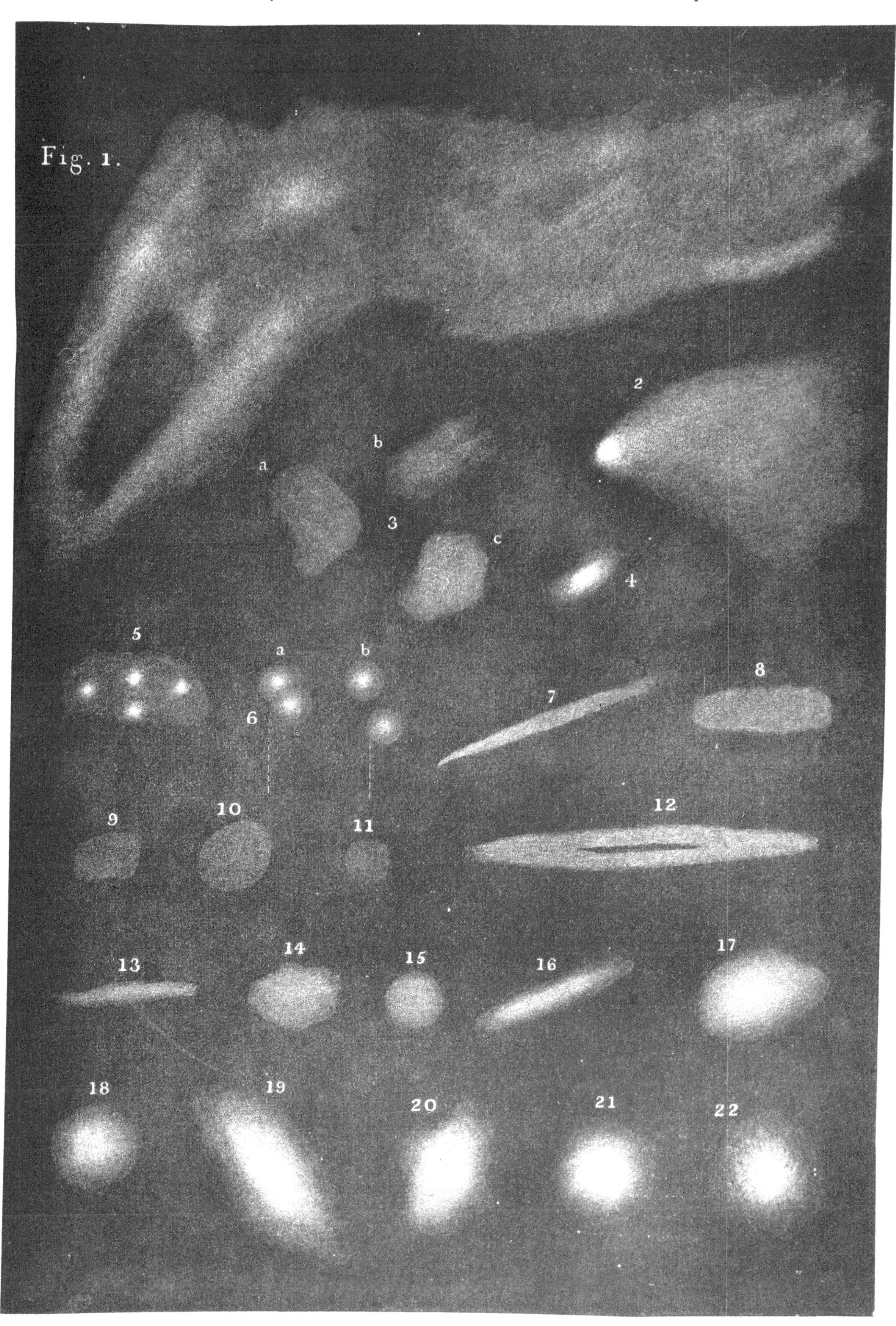
Fig. 1.
2
b
a
3
c
4
5
a
b
6
7
8
9
10
11
12
13
14
15
16
17
18
19
20
21
22

23. *Of Nebulæ that are suddenly much brighter in the middle.*

From the third degree of visible condensation, I have in the 21st article intimated, that the length of the time of the action of the attracting principle, would explain the observed gradual accumulation of the nebulous matter. In the following eighteen nebulæ we may see a still more advanced compression of it, amounting almost to the appearance of a nucleus.*

II. 814 † is "A small faint nebula, very suddenly much brighter in the middle."

I. 39 ‡ is "A very bright nebula, extended from south-preceding to north-following, about 4′ or 5′ long, and 3′ broad; it is much brighter in the middle, but the brightness breaks off abruptly, so as almost to resemble a nucleus." Fig. 23.

I. 256 § is "A very bright pretty large nebula of an irregular figure; it is suddenly much brighter in the middle." Fig. 24.

I. 99 ‖ is "A very bright, small, round nebula; it is very suddenly much brighter in the middle." Fig. 25.

From the appearance of these nebulæ, we see plainly that a progressive concentration of the nebulous matter has an existence; it is also remarkable that the condensation in long nebulæ inclines towards the shape of a nucleus, as well as in round ones, which can be ascribed only to the continued action of the attracting principle, tending to draw the nebulous extended expansion into a globular form.

A nucleus, to which these nebulæ seem to approach, is an indication of consolidation; and should we have reason to conclude that a solid body can be formed of condensed nebulous matter, the nature of which has hitherto been chiefly deduced from its shining quality, we may possibly be able to view it with respect to some other of its properties.

24. *Of round Nebulæ increasing gradually in brightness up to a Nucleus in the middle.*

It has already been proved, from the figure and central brightness of round nebulæ, that the nebulous matter of which they consist must be admitted to be of a globular form; but the following thirteen nebulæ lead me to a remark which not only applies to them, but to all the round nebulæ of the last five articles, which added to these amount to three hundred and twenty-one. They are not only round, but the gradual condensation from the circumference to the very center being of equal density of light at equal central distances, every ring or circle drawn round the center, bears witness to the existence of a central attraction. For whatever

* See *One nebula of unascertained figure, suddenly much brighter in the middle.* II. 814.
Seven extended nebulæ, suddenly much brighter in the middle. I. 39, 91, 96, 200. II. 183, 505. *Connoiss.* 66.
Two nebulæ of an irregular figure, suddenly much brighter in the middle. I. 256. II. 521.
Eight round or nearly round nebulæ, suddenly much brighter in the middle. I. 99, 138. II. 410, 413, 698. III. 251, 685. *Connoiss.* 54.

† [N.G.C. 4732.] ‡ [N.G.C. 4697.] § [N.G.C. 5322.] ‖ [N.G.C. 5557.]

may be the intensity or ratio of the concentration at any given central distance, it follows, from the equality of brightness at the assigned distance, that no figure but a globular one can with any kind of probability explain the appearance ; and that the concentration, as well as the figure, is produced by a general gravitation of the nebulous matter.*

I. 151 † is "A considerably bright and considerably large, round cometic nebula; it is very gradually much brighter in the middle, with a nucleus in the center." Fig. 26.

From the description of these nebulæ, we find that an actual nucleus has been formed in the attracting center ; and that consequently a certain degree of consolidation of the nebulous matter is highly probable ; for, although the quality of shining only points out the existence of something that is luminous, yet from analogy we have reason to conclude that certain material substances must be present to produce the light we perceive ; and that they must be opaque, may be inferred from every thing we know about shining substances.

25. *Of Nebulæ that have a Nucleus.*

It may be expected that some considerable change will take place in the appearance of a nebula after it has come to a certain degree of continued gradual condensation. We are as yet so little acquainted with the nature and distribution of this matter, that an application of mathematical calculations, founded on the attraction of gravitation, for want of data, cannot be applied in order to suggest to us what appearance might next be expected ; I shall therefore proceed in a regular manner to give the observations, which shew what these appearances are, without entering into any theoretical discussions.

In the following two assortments we have forty nebulæ.‡

Number 63 of the *Connoissance des Temps* § is "A very bright nebula, extending from north-preceding to south-following, 9 or 10′ long, and near 4′ broad ; it has a very brilliant nucleus." Fig. 27.

I. 107 ‖ is "A very bright round nebula, about 1½ minutes in diameter ; it has a bright nucleus in the middle." Fig. 28.

The nuclei of these nebulæ, after what has been proved, of the existence of a condensing power, I need not hesitate to ascribe to the longer continuance of its action, which appears to bring on a consolidation ; and that this may be the consequence we may conclude, not only from the power of condensing, which argues

* See I. 2, 6, 132, 151, 173, 236, 272. II. 25, 189, 716, 864. III. 518. IV. 6.

† [N.G.C. 524.]

‡ See *Twenty-seven extended nebulæ, with a nucleus.* I. 43, 77, 126, 156, 170, 180, 208, 224, 240, 250, 255, 270, 280, 281. II. 238, 460, 759, 768, 796, 846, 849, 891. V. 18, 24, 48. *Connoiss.* 63, 101.
Thirteen round or nearly round nebulæ, with a nucleus. I. 107, 133, 139, 152, 167, 203, 225. II 99, 501, 746, 754. III. 178. *Connoiss.* 90.

§ [N.G.C. 5055.]

‖ [N.G.C. 1407.]

a sufficient quantity of matter, but also from the quality of shining ; for this proves that the substance which throws out the nebulous light is endowed with some other of the general qualities of matter besides that of gravitation.

A second remark I have to make is, that the opaque nature of the nebulous matter which was before inferred from analogy, is here supported by observation ; for these consolidated nuclei have a considerable resemblance to the disks of planets ; and if this matter consisted only of a luminous substance, the increase of light would probably far exceed their observed lustre : this being the case, the power of arresting light in its passage is an additional quality, very different from those which have already been mentioned, and seems to be analogous to properties which we know to belong to hard and solid bodies.

26. *Of extended Nebulæ that shew the Progress of Condensation.*

When the nebulous matter is much extended in length, it appears from the following nebulæ, that with those which have a nucleus completely formed, the nebulosity on each side of it is comparatively reduced to a fainter state than it is in nebulæ of which the nucleus is apparently still in an incipient state. These faint opposite appendages to the nucleus I have in my observations called branches.

In some nebulæ there is also an additional small faint nebulosity of a circular form about the nucleus, and this I have called the chevelure. The following two assortments contain twenty-eight nebulæ of this kind.*

Number 65 of the *Connoissance* † is " A very brilliant nebula extended in the meridian, about 12′ long. It has a bright nucleus, the light of which suddenly diminishes on its border, and two opposite very faint branches." Fig. 29.

I. 205 ‡ is " A very brilliant nebula, 5′ or 6′ long and 3 or 4′ broad ; it has a small bright nucleus with a faint chevelure about it, and two opposite very extensive branches." Fig. 30.

The construction of these nebulæ is certainly complicated and mysterious, and in our present state of knowledge it would be presumptuous to attempt an explanation of it ; we can only form a few distant surmises, which however may lead to the following queries. May not the faintness of the branches arise from a gradual diminution, of the length and density of the nebulous matter contained in them, occasioned by its gravitation towards the nucleus into which it probably subsides ? Are not these faint nebulous branches joining to a nucleus, upon an immense scale, somewhat like what the zodiacal light is to our sun in miniature ? Does not the chevelure denote that perhaps some of the nebulous matter still

* See *Twenty-three extended nebulæ with a nucleus and two opposite faint branches.* I. 9, 13, 15, 27, 32, 75, 130, 160, 163, 187, 188, 195, 223, 228, 230. II. 101, 650, 733. IV. 61. V. 43. *Connoiss.* 65, 83, 98.
Five with a nucleus, chevelure and branches. I. 194, 205, 210. V. 45. *Connoiss.* 94.

† [N.G.C. 3623.]

‡ [N.G.C. 2841.]

remaining in the branches, before it subsides into the nucleus, begins to take a spherical form, and thus assumes the semblance of a faint chevelure surrounding it in a concentric arrangement? And, if we may venture to extend these queries a little farther—will not the matter of these branches in their gradual fall towards the nucleus, when discharging their substance into the chevelure, produce a kind of vortex or rotatory motion? Must not such an effect take place, unless we suppose, contrary to observation, that one branch is exactly like the other; that both are exactly in a line passing through the center of the nucleus, by way of causing exactly an equal stream of it from each branch to enter the chevelure at opposite sides; and, this not being probable, do we not see some natural cause which may give a rotatory motion to a celestial body in its very formation?

27. *Of round Nebulæ that shew the Progression of Condensation.*

When round nebulæ have a nucleus, it is an indication that they have already undergone a high degree of condensation. From their figure we are assured that the form of the nebulosity of which they are composed is now spherical, whatever may have been its original shape; and being surrounded by a chelevure, we may look upon its different evanescent degrees of faintness as a sign whereby to judge of the gradual progress of the consolidation of the nucleus. The following seventeen nebulæ are given in two assortments.*

IV. 23 † is "A considerably bright nebula with a very bright nucleus, and a chevelure about 3 or 4′ in diameter." Fig. 31.

III. 99 ‡ is "A small nebula with a pretty bright nucleus and very faint chevelure; it is almost like a nebulous star." Fig. 32.

The chevelure of these nebulæ consists probably of the rarest nebulous matter, which not having as yet been consolidated with the rest, remains expanded about the nucleus in the shape of a very extended atmosphere; or it may be of an elastic nature, and be kept from uniting with the nucleus, as their elasticity causes the atmospheres of the planets to be expanded about them. In this case we have another property of the nebulous substance to add to the former qualities of its matter.

With those nebulæ where this chevelure is uncommonly faint, and the nucleus very bright, the consolidation appears to have reached a still higher degree, and their resemblance to nebulous stars may lead to very interesting consequences.

28. *Of round Nebulæ that are of an almost uniform Light.*

The argument that the nebulous matter is in some degree opaque which is given in the 25th article, will receive considerable support from the appearance of

* See *fifteen round or nearly round nebulæ, with a nucleus and faint chevelure.* I. 40, 137, 226, 242, 251, 262. II. 321. III. 291, 373. IV. 23, 54, 56, 59, 76. *Connoiss.* 32.
Two nebulæ with a nucleus and chevelure resembling nebulous stars. II. 32. III. 99.

† [N.G.C. 936.] ‡ [N.G.C. 5210.]

the following nebulæ; for they are not only round, that is to say the nebulous matter of which they are composed is collected into a globular compass, but they are also of a light which is nearly of an uniform intensity except just on the borders. I give these nebulæ in two assortments.*

Number 97 of the *Connoissance* † is "A very bright, round nebula of about 3′ in diameter; it is nearly of equal light throughout, with an ill defined margin of no great extent."

IV. 13 ‡ is "A pretty faint nebula of about 1′ diameter; it is perfectly round, and of an equal light throughout; and the edges of it are pretty well defined." Fig. 33.

Admitting that these sixteen nebulæ are globular collections of nebulous matter, they could not appear equally bright, if the nebulosity of which they are composed consisted only of a luminous substance perfectly penetrable to light; at least this could not happen unless a certain artificial condensation of it were introduced, which can have no pretension to probability in its favour. Is it not rather to be supposed, that a certain high degree of condensation has already brought on a sufficient consolidation to prevent the penetration of light, which by this means is reduced to a superficial planetary appearance?

29. *Of Nebulæ that draw progressively towards a Period of final Condensation.*

In the course of the gradual condensation of the nebulous matter, it may be expected that a time must come when it can no longer be compressed, and the only cause which we may suppose to put an end to the compression is, when the consolidated matter assumes hardness. It remains therefore to be examined, how far my observations will go to ascertain the intensity of its consolidation.

The following two assortments contain seven nebulæ, from whose appearance a considerable degree of solidity may be inferred.§

IV. 55 ‖ is "A pretty bright round nebula, almost of an even light throughout approaching to a planetary appearance, but ill defined, and a little fainter on the edges; it is about ¾ or 1 minute in diameter." Fig. 34.

IV. 37 ¶ is "A very bright planetary disk of about 35″ in diameter, but ill defined on the edges; the center of it is rather more luminous than the rest, and with long attention a very bright well defined round center becomes visible." Fig. 35.

In these nebulæ we have three different indications of the compression of the

* See *Four from 2′ to 4′ in diameter.* IV. 50, 62, 67. *Connoiss.* 97.
Twelve nebulæ from ¾ of a minute to 2′ in diameter. I. 267. II. 186, 209, 705, 836, 870. III. 152, 877. IV. 13, 14, 16, 39.

† [N.G.C. 3587.]

‡ [N.G.C. 6894.]

§ See *Four nebulæ of a planetary appearance.* IV. 55, 60, 68, 78.
Three planetary disks with a bright central point. II. 268. IV. 37, 73.

‖ [N.G.C. 2537.]

¶ [N.G.C. 6543.]

nebulous matter of which they are composed: their figure, their light, and the small compass into which it is reduced. The round figure is a proof that the nebulous mass is collected into a globular form, which cannot have been effected without a certain degree of condensation.

Their planetary appearance shews that we only see a superficial lustre such as opaque bodies exhibit, and which could not happen if the nebulous matter had no other quality than that of shining, or had so little solidity as to be perfectly transparent. That there is a certain maximum of brightness occasioned by condensation, is to be inferred from the different degrees of light of round nebulæ that are in a much less advanced state of compression; for these are gradually more bright towards the center; which proves that brightness keeps up with condensation till the increase of it brings on a consolidation which will no longer permit the internal penetration of light, and thus a planetary appearance must in the end be the consequence; for planets are solid opaque bodies, shining only by superficial light, whether it be innate or reflected.

From the size of the nebulæ as we see them at present, we cannot form an idea of the original bulk of the nebulous matter they contain; but let us admit, for the sake of computation, that the nebulosity of the above described nebula IV. 55, when it was in a state of diffusion, took up a space of 10′ in every cubical direction of its expansion; then, as we now see it collected into a globular compass of less than one minute, it must of course be more than nineteen hundred times denser than it was in its original state. This proportion of density is more than double that of water to air.

With regard to planetary disks, which have bright central points, we may surmise that their original diffused nebulosity was more unequally scattered, and that they passed through the different stages of extended nebulæ, gradually acquiring a nucleus, chevelure, and branches. For in nebulæ of this construction, the consolidation of a nucleus is already much advanced at the time when a considerable quantity of nebulous matter, on account of its greater central distance, remains still unformed in the branches; and if the condensation of the nucleus should keep the lead, it will come to a state of great solidity and maximum of brightness by the time that the rest of the nebulosity is drawn into a planetary appearance.

30. *Of Planetary Nebulæ.*

The objects of which I shall give an account in this article have so near a resemblance to planets, that the name of planetary nebulæ very justly expresses their appearance; for notwithstanding their planetary aspect, some small remaining haziness, by which they still are more or less surrounded, evinces their nebulous origin. In my catalogues the places of the following ten have been given.*

* See *Planetary nebulæ* IV. 1, 11, 18, 26, 27, 34, 51, 53, 64, 70.

IV. 18 * is "A beautiful bright round nebula, having a pretty well defined planetary disk of about 10 or 12″ in diameter. It is a little elliptical, and has a very small star following, which gives us the idea of a small satellite accompanying its planet. It is visible in a common finder as a small star." Fig. 36.

IV. 27 † is "A beautiful very brilliant globe of light, hazy on the edges, but the haziness going off suddenly. I suppose it to be from 30 to 40″ in diameter, and perhaps a very little elliptical. The light of it seems to be all over of the uniform lustre of a star of the 9th magnitude. The haziness on the edges does not exceed the 20th part of the diameter."

IV. 51 ‡ is "A small beautiful planetary nebula, but considerably hazy upon the edges; it is of a uniform light, and considerably bright, perfectly round, and about 10 or 15″ in diameter."

IV. 53 § is "A pretty bright planetary nebula of nearly 1′ in diameter; it is round, or a little elliptical; its light is uniform, and pretty well defined on the borders."

IV. 64 ‖ is "A beautiful planetary nebula of a considerable degree of brightness, but not very well defined, about 12 or 15″ in diameter."

The remarks which have been made on the nebulæ of the foregoing article, will here apply with additional propriety; for the light of these planetary nebulæ must be considerably more condensed than that of the foregoing sets. The diameter of four of them does not exceed 15″, so that if we again suppose the original diffused nebulosity from which they sprang of 10′ in cubical dimensions, we shall have a condensation, which has reduced the nebulous matter to less than the one-hundred and twenty-two thousandth part of its former bulk.

One of them, number 34 in the 4th class, appeared even in the 20 feet telescope, with the sweeping power, like a star with a large diameter, and it was only when magnified 240 times that it resembled a small planetary nebula; nor can any of these nebulæ be distinguished from the neighbouring small stars in a good common telescope, night glass, or finder.

When we reflect upon these circumstances, we may conceive that, perhaps in progress of time these nebulæ which are already in such a state of compression, may be still farther condensed so as actually to become stars.

It may be thought that solid bodies, such as we suppose the stars to be from the analogy of their light with that of our sun when seen at the distance of the stars, can hardly be formed from a condensation of nebulous matter; but if the immensity of it required to fill a cubical space, which will measure ten minutes when seen at the distance of a star of the 8th or 9th magnitude, is well considered, and properly compared with the very small angle our sun would subtend at the same distance, no degree of rarity of the nebulous matter, to which we may have recourse, can be

* [N.G.C. 7662.] † [N.G.C. 3242.] ‡ [N.G.C. 6818.] § [N.G.C. 1501.] ‖ [N.G.C. 2440.]

any objection to the solidity required for the construction of a body of equal magnitude with our sun.*

A circumstance which allies these very compressed nebulæ to the character of many of our well known celestial bodies, such as some of the planets and their satellites, the sun and all periodical stars, is that very probably most, if not all of them, turn on their axes. Seven of the ten I have mentioned are not perfectly round, but a very little elliptical. Ought we not to ascribe this figure to the same cause which has flattened the polar diameter of the planets, namely, a rotatory motion ?

At the end of the 26th article I have already pointed out one configuration of the nebulous matter, of which the final condensation seems to be properly disposed for bringing on a rotatory motion of the nucleus ; but, if we consider this matter in a general light, it appears that every figure which is not already globular must have eccentric nebulous matter, which in its endeavour to come to the center, will either dislodge some of the nebulosity which is already deposited, or slide upon it sideways, and in both cases produce a circular motion ; so that in fact we can hardly suppose a possibility of the production of a globular form without a consequent revolution of the nebulous matter, which in the end may settle in a regular rotation about some fixed axis. Many of the extended and irregular nebulæ are considerably elliptical, and the irregular round ones shew a general approach to the oval form ; now these figures are all favourable to a surmise, that a rotatory motion may often take place even before the nucleus of a nebula can have arrived to a state of consolidation. An objection, that this remarkable form of planetary nebulæ may be owing to chance, will hardly deserve to be mentioned, because the improbability of such a supposition must exclude it from all claim to refutation.

31. *Of the Distance of the Nebula in the Constellation of Orion.*

In my 3d article I concluded, from the appearance of the great nebula in Orion, that the range of the visibility of the diffused nebulous matter cannot be great, because we may there see in one and the same object, both the brightest and faintest appearance of nebulosities that can be seen any where. It will therefore be a case of some interest, if we can form any conception of the place among the fixed stars to which we ought to refer the situation of this nebula ; and this I believe my observation of it will enable us to determine pretty nearly.

In the year 1774, the 4th of March, I observed the nebulous star, which is the 43d of the *Connoissance des Temps*,† and is not many minutes north of the great nebula ; but at the same time I also took notice of two similar, but much smaller

* A cubical space, the side of which at the distance of a star of the 8th magnitude is seen under an angle of 10′, exceeds the bulk of the sun (2208600000000000000) two trillion and 208 thousand billion times.

† [N.G.C. 1982, " Mairan's nebula."]

nebulous stars ; one on each side of the large one, and at nearly equal distances from it. Fig. 37 is a copy of a drawing which was made at the time of observation.

In 1783, I examined the nebulous star, and found it to be faintly surrounded with a circular glory of whitish nebulosity, faintly joining to the great nebula.

About the latter end of the same year I remarked that it was not equally surrounded, but most nebulous towards the south.

In 1784 I began to entertain an opinion that the star was not connected with the nebulosity of the great nebula of Orion, but was one of those which are scattered over that part of the heavens.

In 1801, 1806, and 1810 this opinion was fully confirmed, by the gradual change which happened in the great nebula, to which the nebulosity surrounding this star belongs. For the intensity of the light about the nebulous star had by this time been considerably reduced, by the attenuation or dissipation of the nebulous matter ; and it seemed now to be pretty evident that the star is far behind the nebulous matter, and that consequently its light in passing through it is scattered and deflected, so as to produce the appearance of a nebulous star. A similar phenomenon may be seen whenever a planet or a star of the 1st or 2nd magnitude happens to be involved in haziness ; for a diffused circular light will then be seen, to which, but in a much inferior degree, that which surrounds this nebulous star bears a great resemblance.

When I reviewed this interesting object in December 1810, I directed my attention particularly to the two small nebulous stars, by the sides of the large one, and found that they were perfectly free from every nebulous appearance ; which confirmed not only my former surmise of the great attenuation of the nebulosity, but also proved that their former nebulous appearance had been entirely the effect of the passage of their feeble light through the nebulous matter spread out before them.

The 19th of January 1811, I had another critical examination of the same object in a very clear view through the 40-feet telescope ; but notwithstanding the superior light of this instrument, I could not perceive any remains of nebulosity about the two small stars, which were perfectly clear, and in the same situation, where about thirty-seven years before I had seen them involved in nebulosity.

If then the light of these three stars is thus proved to have undergone a visible modification in its passage through the nebulous matter, it follows that its situation among the stars is less distant from us than the largest of the three, which I suppose to be of the 8th or 9th magnitude. The farthest distance therefore, at which we can place the faintest part of the great nebula in Orion, to which the nebulosity surrounding the star belongs, cannot well exceed the region of the stars of the 7th or 8th magnitude, but may be much nearer ; perhaps it may not amount to the distance of the stars of the 3d or 2nd order ; and consequently the most luminous appearance of this nebula must be supposed to be still nearer to us. From the very considerable changes I have observed in the arrangement of its nebulosity, as well as from its great extent, this inference seems to have the support of observation ; for in very distant objects we cannot so easily perceive changes as in near ones, on account of the smaller angles which both the objects and its changes subtend at the eye. The following memorandum was made when I viewed it in 1774 ; "its shape is not like that which Dr. SMITH has delineated in his *Optics*, though somewhat resembling it, being nearly as in fig. 37 : from this we may infer that there are undoubtedly changes among the regions of the fixed stars ; and perhaps from a careful observation of this lucid spot, something may be concluded concerning the nature of it."

In January 1783, the nebulous appearance differed much from what it was in 1780, and in September it had again undergone a change in its shape since January.

March 13, 1811. With a view to ascertain such obvious alterations in the disposition of the nebulous matter as may be depended on, I selected a telescope that had the same light and power which thirty-seven years ago I used, when I made the above-mentioned drawing ; and the relative situation of the stars remaining as before, I found that the arrangement of the nebulosity differs considerably. The northern branch *N* still remains nearly parallel to the direction of the stars *ab* ; but the southern branch *S* is no longer extended towards the star *d* : its direction is now towards *e*, which is very faintly involved in it. The figure of the branch is also different ; the nebulosity in the parallel *PF* of the three stars being more advanced towards the following side than it was formerly.

I compared also the present appearance of this nebula with the delineation which HUYGENS has given of it in his *Systema Saturnium*, page 8, of which fig. 38 is a copy. The twelve stars which he has marked are sufficient to point out the arrangement of the nebulous matter at the time of his observation. By their situation we find that the nebula had no southern branch, nor indeed any to the north, unless we call the nebulosity in the direction of the parallel a branch ; but then this branch is not parallel to a line drawn from *a* to the star *b* ; moreover the star *f* is now involved in faint nebulosity, which also reaches nearly up to *g*, and quite incloses *h*. The star *b* which is now nebulous, is represented as perfectly out of all nebulosity, and can hardly be supposed to have been affected when HUYGENS observed it.

The changes that are thus proved to have already happened, prepare us for those that may be expected hereafter to take place, by the gradual condensation of the nebulous matter ; for had we no where an instance of any alteration in the appearance of nebulæ, they might be looked upon as permanent celestial bodies, and the successive changes, to which by the action of an attracting principle they have been conceived to be subject, might be rejected as being unsupported by observation.

The various appearances of this nebula are so instructive, that I shall apply them to the subject of the partial opacity of the nebulous matter, which has already been inferred from its planetary appearance, when extremely condensed in globular masses ; but which now may be supported by more direct arguments. For when I formerly saw three fictitious nebulous stars, it will not be contended that there were three small shining nebulosities, just in the three lines in which I saw them, of which two are now gone and only one remaining. As well might we ascribe the light surrounding a star, which is seen through a mist, to a quality of shining belonging to that particular part of the mist, which by chance happened to be situated where the star is seen. If then the former nebulosity of the two stars which have ceased to be nebulous can only be ascribed to an effect of the transit or penetration of their light through nebulous matter which deflected and scattered it, we have now a direct proof that this matter can exist in a state of opacity, and may possibly be diffused in many parts of the heavens without our being able to perceive it.

That there has been shining as well as opaque nebulous matter about the large star, appears from several observations I have made upon the light which surrounded it. In 1783 the nebulosity about it was so considerable in brightness, and so much on one side of it, that the star did not appear to have any connection with it. The reason of which is plainly, that the shining quality of the nebulous matter then overpowered the feeble scattering of the light of the star in the nebulosity.

32. *Of Stellar Nebulæ.*

It has been remarked that diffused nebulosities may exist unknown to us, among the more distant regions of the fixed stars ; and though we may not be able to see a nebulous diffusion that is farther from us than the moderate distance at which we now have reason to suppose the faintest visible nebulosity of the nebula in Orion to be placed ; yet if some former diffusion of the nebulous matter should be already reduced into separate and much condensed nebulæ, they might then come within the reach of telescopes that have a great power of collecting light : this being admitted, there is a probability that some of the various diffusions of the nebulous matter, from which our present nebulæ derive their origin, may have been much farther from us than others. For, in every description of figure, size and condensation, of which I have given instances, we find not only very bright and very large, but also faint and small, as well as extremely faint and extremely small nebulæ ; and the same gradations will now be found to run through that class which I have called stellar nebulæ. This classification was introduced in my sweeps when the objects to be recorded came in so quick a succession that I found it expedient to express as much as I could in as few words as possible, and by calling a nebula stellar, I intended to denote that the object to which I gave this name was, in the first place as small, or almost as small, as a star ; and in the next, that notwithstanding its smallness, and starlike appearance, it bore evident marks of not being one of those objects which we call stars, and of which I saw many at the same time in the telescope.

The following three collections contain one hundred and seventeen stellar nebulæ, which have been assorted by their brightness, that their comparative condensation might be estimated according to the different distances at which we may suppose other nebulæ of the same degree of light to be placed.*

I. 71 † is "A considerably bright, very small, almost stellar nebula; the brightness diminishing insensibly and breaking off pretty abruptly. The whole together is not more than about 7 or 8″ in diameter." A second observation, made in a remarkably clear morning, says, that "the greatest brightness is towards the following side, and that the very faint nebulosity extends to near a minute."

This is probably a condensation of a former nucleus with surrounding chevelure.

I. 268 ‡ is "A very bright, very small, round stellar nebula." Fig. 39.

This may be a former planetary nebula in a higher state of condensation.

II. 110 § is "A very bright small stellar nebula or star with a bur all-around." Fig. 40.

This star with a bur is probably one that was formerly a planetary nebula with a pretty strong haziness on the borders.

II. 603 || is "A pretty bright stellar nebula, or a pretty considerable star with a very faint chevelure." Fig. 41.

This may have been a planetary nebula with a faint haziness about the margin.

IV. 46 ¶ is "A very small pretty bright, or considerably bright stellar nebula, like a star with burs."

It may have been a pretty well defined planetary nebula.

If it should be deemed singular that we have not a greater number of bright stellar nebulæ, I must remark that, if the stellar is a succession of the planetary state, the number of bright stellar is sufficiently proportionable to that of the planetary nebulæ; and as the faint nebulæ are far more numerous than the bright ones, so it will be seen by the references in the two next assortments, that in proportion as brightness decreases, we have a much more copious collection of stellar nebulæ.**

II. 663 †† is "A pretty bright very small stellar nebula."

This nebula and the rest of them, which are all of the same description, must

* See *First assortment containing six of the brightest stellar nebulæ.* I. 71, 268. II. 110, 603. IV. 32, 46.

† [N.G.C. 5812.] ‡ [N.G.C. 3458.] § [N.G.C. 4262.]

|| [N.G.C. 1278.] ¶ [N.G.C. 5493.]

** See *Second assortment containing eleven stellar nebulæ of the next degree of brightness.* II. 159, 178, 179, 204, 232, 663, 676, 689, 708, 820, 867.

†† [N.G.C. 4963.]

be looked upon as condensations of distant nebulæ that had nuclei, or were nearly about the planetary condition.*

In this collection of nebulæ we have many of a different description. In some, the mark whereby they were distinguished from stars was their figure, the object not being so small but that its figure might still be perceived. Of others, some difference in the brightness between the center and outside was visible; and many of them were only called stellar, because by some deficiency or other in the appearance it was evident they were not perfect stars. Instances of every sort will be seen in the following descriptions.

II. 424 † is "A very faint stellar nebula, or a little larger."

II. 805 ‡ is "An extremely faint very small round stellar nebula."

II. 425 § is "A faint very small stellar nebula, of an irregular figure."

III. 145 || is "A very faint stellar nebula; a little extended."

III. 691 ¶ is "A considerably faint stellar nebula, suddenly much brighter in the middle."

33. *Of Stellar Nebulæ nearly approaching to the Appearance of Stars.*

The starlike appearance of the following six nebulæ is so considerable that the best description, which at the time of observation I could give of them, was to compare them to stars with certain deficiencies.**

IV. 49 †† is "A pretty bright stellar nebula, like a star with a small bur all around."

The other two are of the same nature.

IV. 15 ‡‡ is "A stellar nebula, or rather like a faint star with a small chevelure and two burs."

The other two are nearly of the same description.§§

34. *Of doubtful Nebulæ.*

It may have been remarked, that many stellar nebulæ of my catalogues have the memorandum added to their descriptions that they were confirmed with a higher magnifying power, and that this was sometimes attended with difficulty, and sometimes could not be successfully done.

* See *Third assortment containing one hundred stellar nebulæ of several degrees of faintness.* II. 127, 194, 244, 340, 341, 425, 443, 448, 449, 454, 550, 551, 576, 618, 620, 692, 693, 718, 721, 722, 735, 740, 781, 815, 848. III. 81, 109, 114, 119, 125, 136, 145, 151, 161, 167, 168, 169, 170, 171, 172, 173, 175, 188, 215, 231, 232, 234, 240, 260, 276, 277, 278, 289, 294, 320, 322, 341, 400, 401, 418, 422, 423, 424, 438, 439, 469, 476, 530, 536, 561, 562, 563, 564, 565, 571, 576, 590, 606, 627, 672, 691, 706, 737, 741, 764, 768, 770, 772, 777, 786, 793, 805, 815, 821, 828, 843, 852, 855, 856, 916.

† [N.G.C. 5347.] ‡ [N.G.C. 4290.] § [N.G.C. 5990.]

|| [N.G.C. 7052.] ¶ [N.G.C. 5791.]

** See *Three stars with burs.* II. 655. IV. 47, 49. †† [N.G.C. 5507.]

‡‡ [N.G.C. 16.] §§ See *Three stars with a faint chevelure.* IV. 15, 21, 31.

A collection of thirty-four nebulæ that come under this description is as follows:*

II. 470 † is "A small stellar nebula." By a second observation a doubt entertained in the first was removed with 240, which shewed it "pretty bright, but hardly to be distinguished from a star."

III. 29 ‡ is "A very faint extremely small stellar nebula or rather nebulous star." The sweeping power left me rather doubtful; 240 verified it.

It must be noticed, that in these nebulæ the doubt which was entertained did not relate to the existence of the objects, but merely to their nature; and when the suspected nebula was so faint that even its existence was doubtful, a higher power was applied only with a view to ascertain whether the object existed as nebula or as star; for had the suspicion of its existence not been accompanied with the expectation of its being a nebula, it could never have been attempted to be verified.§

III. 270 || is "A very faint extremely small stellar nebula; 240 verified it with difficulty, and considerable attention, the night being uncommonly clear."

When difficulty is mentioned, it is always to be understood that a considerable time as well as attention was required in the examination before a decisive opinion could be formed.¶

III. 7 ** is "A nebulous star, but doubtful of the nebulosity. With 240 the same doubtful appearance continues." Fig. 42.

With this object the doubt which remained could only relate to the nature of it; for being at first sight taken to be a nebulous star, its existence could not be a subject for examination; but the unresolved doubt, whether an object is a nebula, or a star, must certainly be allowed to be as great a proof of identity as we can possibly expect to see.

35. *Concluding Remarks.*

The total dissimilitude between the appearance of a diffusion of the nebulous matter and of a star, is so striking, that an idea of the conversion of the one into the other can hardly occur to any one who has not before him the result of the critical examination of the nebulous system which has been displayed in this paper. The end I have had in view, by arranging my observations in the order in which they have been placed, has been to shew, that the above mentioned extremes may be connected by such nearly allied intermediate steps, as will make it highly probable that every succeeding state of the nebulous matter is the result of the action of

* See *First assortment containing twenty-five verified stellar nebulæ.* II. 470, 502, 661. III. 29, 80, 84, 124, 135, 174, 184, 187, 202, 207, 214, 226, 264, 266, 268, 269, 513, 549, 604, 742, 748, 964.

† [N.G.C. 1140.] ‡ [N.G.C. 3768.]

§ See *Second assortment, containing five stellar nebulæ verified with difficulty.* III. 115, 212, 219, 262, 270.

|| [N.G.C. 2089.]

¶ See *Third assortment, containing four objects that could not be verified.* III. 7, 176, 263, 293.

** [N.G.C. 2508.]

gravitation upon it while in a foregoing one, and by such steps the successive condensation of it has been brought up to the planetary condition. From this the transit to the stellar form, it has been shown, requires but a very small additional compression of the nebulous matter, and several instances have been given which connect the planetary to the stellar appearance.

The faint stellar nebulæ have also been well connected with all sorts of faint nebulæ of a larger size ; and in a number of the smaller sort, their approach to the starry appearance is so advanced, that in my observations of many of them it became doubtful whether they were not stars already.

It must have been noticed, that I have confined myself in every one of the preceding articles to a few remarks upon the appearance of the nebulous matter in the state in which my observations represented it ; they seemed to be the natural result of the observations under consideration, and were not given with a view to establish a systematic opinion, such as will admit of complete demonstration. The observations themselves are arranged so conveniently that any astronomer, chemist, or philosopher, after having considered my critical remarks, may form what judgment appears most probable to him. At all events, the subject is of such a nature as cannot fail to attract the notice of every inquisitive mind to a contemplation of the stupendous construction of the heavens ; and what I have said may at least serve to throw some new light upon the organization of the celestial bodies.

Synopsis of the Contents of this Paper.

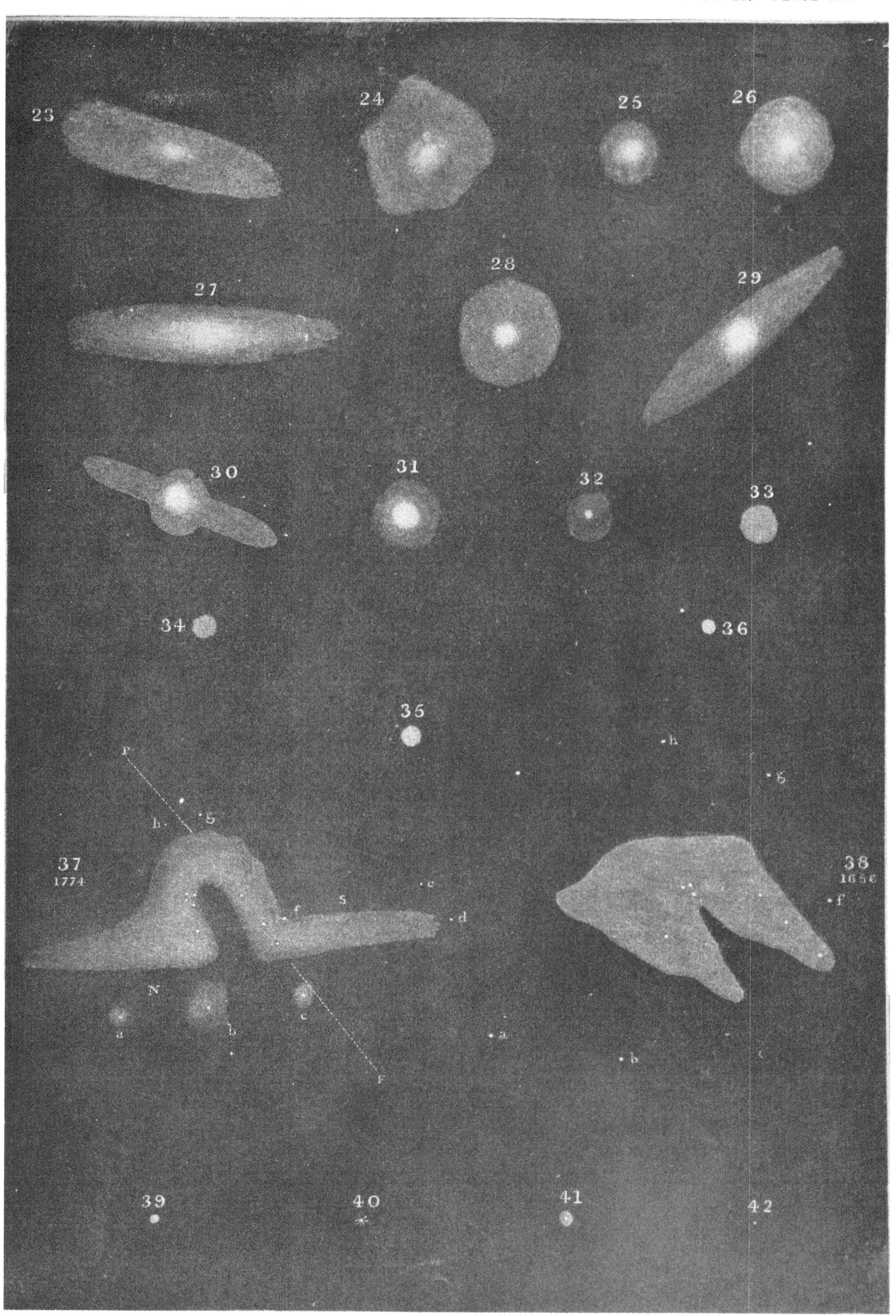

[*To face page* 496.

POSTSCRIPT.

It will be seen that in this paper I have only considered the nebulous part of the construction of the heavens, and have taken a star for the limit of my researches. The rich collection of clusters of stars contained in the 6th, 7th, and 8th classes of my Catalogues, and many of the *Connoissance des Temps*, have as yet been left unnoticed. Several other objects, in which stars and nebulosity are mixed, such as nebulous stars, nebulæ containing stars, or suspected clusters of stars which yet may be nebulæ, have not been introduced, as they appeared to belong to the sidereal part of the construction of the heavens, into a critical examination of which it was not my intention to enter in this Paper.

WILLIAM HERSCHEL.

Slough, near Windsor,
May 26, 1811.

LXIV.

Observations of a Comet, with Remarks on the Construction of its different Parts.

[*Phil. Trans.*, 1812, pp. 115–143.]

Read December 19, 1811.

THE comet which has lately visited the solar system * has moved in an orbit very favourably situated for astronomical observations. I have availed myself of this circumstance, and have examined all the parts of it with a scrutinizing attention, by telescopes of every degree of requisite light, distinctness, and power.

The observations I have made have been so numerous, and so often repeated, that I shall only give a selection of such as were made under the most favourable circumstances, and which will serve to ascertain the most interesting particulars relating to the construction of the comet.

As my attention in these observations was every night directed to as many particulars as could be investigated, it will be most convenient to assort together those which belong to the same object; and in the following arrangement I shall begin with the principal part, which is

The planetary Body in the Head of the Comet.

By directing a telescope to that part of the head where with the naked eye I saw a luminous appearance not unlike a star; I found that this spot, which perhaps some astronomers may call a nucleus, was only the head of the comet; but that within its densest light there was an extremely small bright point, entirely distinct from the surrounding glare. I examined this point with my 20 feet, large 10 feet, common 10 feet, and also with a 7 feet telescope; and with every one of these instruments I ascertained the reality of its existence.

At the very first sight of it, I judged it to be much smaller than the little planetary disk in the head of the comet of the year 1807; but as we are well assured that if any solidity resembling that of the planets be contained in the comet, it must be looked for in this bright point; I have called it the planetary body; in order to distinguish it from what to the naked eye or in small telescopes appeared to be a nucleus, but which in fact was this little body with its surrounding light or head seen together as one object.

* [The great comet, 1811 I.—ED.]

With a new 10 feet mirror of extraordinary distinctness, I examined the bright point every fine evening, and found that although its contour was certainly not otherwise than round, I could but very seldom perceive it definedly to be so.

As hitherto I had only used moderate magnifiers from 100 to 160, because they gave a considerable brightness to the point, it occurred to me that higher powers might be required to increase its apparent magnitude; accordingly the 19th of October, having prepared magnifiers of 169, 240, 300, 400, and 600, I viewed the bright point successively with these powers.

With 169 it appeared to be about the size of a globule which in the morning I had seen in the same telescope and with the same magnifier, and which by geometrical calculation subtended an angle of 1″·39.

I suspected that this apparent size of the bright point was only such as will spuriously arise from every small star-like appearance; and this was fully confirmed when I examined it with 240; for by this its magnitude was not increased; which not only proved that my power was not sufficient to reach the real diameter of the object, but that the light of this point was, like that of small stars, sufficiently intense to bear being much magnified.

I viewed it next with 300, and here again I could perceive no increase of size.

When I examined the point with 400, it appeared to me somewhat larger than with 300; I saw it indeed rather better than with a lower power, and had reason to believe that its real diameter was now within reach of my magnifiers. Curiosity induced me to view it in the 7 feet telescope with a power of 460; and notwithstanding the inferior quantity of light of this instrument, the magnitude was fully sufficient to show that the increase of size in this telescope agreed with that in the 10 feet.

Returning again to the latter I examined the bright point with 600, and saw it now so much better than with 400, that I could keep it steadily in sight while it passed the field of view of the eye-glass.

With this power I compared its appearance to the size of several globules, that have been examined with the same telescope and magnifier, and by estimation I judged it to be visibly smaller than one of 1″·06 in diameter, and rather larger than another of 0″·68.

It should be noticed that I viewed the globules, which were of sealing wax, without sunshine, in the morning after the observation as well as the morning before; referring in one case the bright point to the globules, and in the other the globules to the bright point.*

* A similar method was used with the comet of 1807. See *Phil. Trans.* for 1808, page 145. [Above, p. 403.]

The apparent and real Magnitudes of the planetary Body.

The size of the bright point being much more like the smallest of the two globules, I shall add one quarter of their difference to 0″·68, and assume the sum, which is 0″·775 as the apparent diameter of the planetary disk.

Then by a calculation from some corrected elements of the comet's orbit, which, though not very accurate, are however sufficiently so for my purpose, I find that the distance of the comet from the earth, at the time of observation, was nearly 114 millions of miles; from which it follows that the bright point, or what we may admit to be the solid or planetary body of the comet, is about 428 miles in diameter.

The Eccentricity and Colour of the planetary Body.

The situation of the bright point was not in the middle of the head, but was more or less eccentric at different times.

The 16th of October that part of the head, which was towards the sun, was a little brighter and broader than that towards the tail, so that the planetary disk or point was a little eccentric.

The 17th I found its situation to be a little beyond the centre, reckoning the distance in the direction of a line drawn from the sun through the centre of the head.

The 4th of November it was more eccentric than I had ever seen it before.

Nov. 10, I found no alteration in the eccentricity since the last observation.

The colour of the planetary disk was of a pale ruddy tint, like that of such equally small stars as are inclined to red.

The Illumination of the planetary Body.

The smallness of the disk, even when most magnified, rendered any determination of its shape precarious; however had it been otherwise than round, it might probably have been perceived; the phasis of its illumination at the time of observation being to a full disk as 1·6 to 2.

From this as well as from the high magnifying power, which a point so faint could not have borne with advantage, had it shone by reflected light, we may infer that it was visible by rays emitted from its own body.*

The Head of the Comet.

It has already been noticed that the brightest part of the comet seen by the naked eye, appeared to contain a small starlike nucleus. When this was viewed in a night glass, or finder, magnifying only 6 or 8 times, it might still have been mistaken for one; but when I applied a higher power, such as from 60 to 120, it

* On the subject of the nature of the light by which we see this comet, I may refer to what has been said in my paper of observations on that of the year 1807. Those who wish also to consult the opinion of an eminent philosopher, whose valuable works on meteorological subjects are well known, will find it expressed at large in a letter from Mr. DE LUC, addressed to Mr. BODE, so far back as the year 1799, and reprinted in Mr. NICHOLSON'S Journal, published the 1st of March 1809.

retained no longer this deceptive appearance; which evidently arises from an accumulation of light, condensed into the small compass of a few minutes; and which of course will vanish when diluted by magnifying.

Sept 2, I saw the comet at Glasgow, in a 14 feet Newtonian reflector; but being very low, the moon up, and the atmosphere hazy, it appeared only like a very brilliant nebula, gradually brighter in a large place about the middle.

The 9th and 10th of September at Alnwick, I viewed it with a fine achromatic telescope, and found that, when magnified about 65 times, the planetary disk-like appearance seen with the naked eye, was transformed into a bright cometic nebula, in which, with this power, no nucleus could be perceived.

The 18th of September the star-like object in my large 10 feet reflector, when magnified 110 times, had the appearance of a fine globular, luminous nebula; it seemed to be about 5 or 6 minutes in diameter, of which one or two minutes about the centre were nearly of equal brightness. The small 10 feet showed it in the same manner.

In all my instruments this bright appearance was equally transformed into a brilliant head of the comet, with this difference, that when high powers were applied, the central illumination which moderately magnified was pretty uniform, became diluted into a gradual decrease from the middle towards the outside; losing itself by imperceptible degrees, especially towards the sides and following parts, into a darkish space, which from observations that will be given hereafter, I take to be a cometic atmosphere.

The Colour and Eccentricity of the Light of the Head.

The colour of the head being very remarkable, I examined it with all my different telescopes; and in every one of them, its light appeared to be greenish, or bluish green. Its appearance was certainly very peculiar.

The disposition of the light of the head was likewise accompanied with some remarkable circumstances; for notwithstanding a general accumulation about the middle, there seemed to be a greater share of it towards the sun, than a portion in that situation of the circumference was entitled to, had it been uniformly arranged; and if we look upon the head as a coma to the planetary point, the eccentricity of its light will be still more evident; for this point was constantly more or less farther from the sun than the middle of the greatest brightness of the light surrounding it. The eccentricity of the head was indeed so considerable, that considering the difficulty of seeing the point, it might easily have escaped the notice of one who looked for it in the centre of the head.

The apparent and real Magnitudes of the Head.

With an intention to ascertain the dimensions of the various parts of the comet, I viewed the head in the 7, 10, and 20 feet telescopes, and estimated its size by the proportion it bore to the known fields of the eye-glasses that were used. I shall only mention two estimations: September 29, the 10 feet gave its apparent diameter 3′ 0″. With the 20 feet Oct. 6, it was 3′ 45″.

From a calculation of the 20 feet measure, which I prefer, it appears that the real diameter of the head at this time was about 127 thousand miles.

A transparent and elastic Atmosphere about the Head.

In every instrument through which I have examined the comet, I perceived a comparatively very faint or rather darkish interval surrounding the head, wherein the gradually diminishing light of the central brightness was lost. This can only be accounted for by admitting a transparent elastic atmosphere to envelope the head of the comet.

Its transparency I had an opportunity of ascertaining the 18th of September, when I saw three very small stars of different magnitudes within the compass of it; and its elasticity may be inferred from the circular form under which it was always seen. For being surrounded by a certain bright equi-distant envelope, we can only account for the equality of the distance by admitting the interval between the envelope and the head of the comet to be filled with an elastic atmospherical fluid.

The Extent of the cometic Atmosphere.

When I examined the comet in the 20 feet telescope the 6th of October, the circular darkish space, which surrounded the brightness, just filled the field of the eye-glass; which gives its apparent diameter 15 minutes. This atmosphere was therefore more than 507 thousand miles in diameter; but its real extent of which, as will be seen, we can have no observation, must far exceed the above calculated dimensions.

The bright Envelope of the cometic Atmosphere.

When I observed the comet at Alnwick in an achromatic refractor with a magnifying power of 65, I perceived that the head of it was partly surrounded by a train of light, which was kept at some considerable distance by an interval of comparative darkness; and from its concentric figure I call this light an envelope.

The Figure, Colour, and Magnitude of the Envelope.

On viewing this envelope in telescopes that magnify no more than about 16 times, or in finders and night glasses with still lower powers, I found that its shape, as far as it extended, was apparently circular; but that in its course it did not reach quite half way round the head of the comet. A little before it came so far it divided itself into two streams, one passing by each side of the head.

The colour of the envelope in my 7, 10, and 20 feet telescopes had a strong yellowish cast, and formed a striking contrast with the greenish tint of the head.

The distance of the outside of the envelope from the centre of the head, in the direction of a line drawn from it to the sun, was about 9′ 30″; and supposing it to have extended sideways, without increase of distance as far as a semi-circle, this would give its diameter about 19 minutes. By computation therefore its real diameter must have exceeded 643 thousand miles.

The Tail of the Comet.

The most brilliant phenomenon that accompanies a comet is the stream of light which we call the tail. Its length is well known to be variable, but the measures or estimations of its extent cannot be expected to be very consistent from several causes foreign to its actual change.

The 2d of September, the moon being up, the comet very low, and the atmosphere hazy, I could perceive no tail.

The 9th, it had a very conspicuous one, about 9 or 10 degrees in length.

On the 18th, the length was 11 or 12 degrees.

The 6th of October it was 25 degrees.

The 12th I estimated it to be only 17 degrees long.

The 14th it appeared to extend to 17½ degrees.

The 15th, by very careful attention, and in a very clear atmosphere, I found the tail to cover a space of 23½ degrees in length.

The greatest real Length of the Tail.

Of the two observations which were made of the greatest length of the tail of the comet, I prefer that of the 15th of October, on account of the clearness of the night.

The apparent length being 23½ degrees, its real extent, taking into the calculation the oblique position in which we saw it, must have been upwards of 100 millions of miles.

The Breadth of the Tail.

The variations in the breadth of the tail will hardly admit of any description; the scattered light of the sides being generally lost by its faintness in such a manner as to render its termination very doubtful.

The 12th of October its breadth in the broadest part was 6¾ degrees, and about 5 or 6 degrees from the head it began to be a little contracted.

The 15th, it was nearly of the same breadth about the middle of its length.

By calculating from the observation of the 12th, we find that the real breadth of the tail on that day was nearly 15 millions of miles.

The Curvature of the Tail.

The shape of the tail with respect to its curvature is generally considered only as it relates to the direction of the motion of the comet; it is nevertheless subject to variations arising from causes that will be noticed in the next article, but which are not taken into the account of the following observations.

The 9th and 10th of September the curvature of the tail was very considerable.

The 18th, I remarked, that towards the end of the tail its curvature had the appearance as if, with respect to the motion of the comet, that part of the tail were left a little behind the head.

The 17th of October the tail appeared to be more curved than it had been at any time before.

Dec. 2, the flexure of the curvature of the tail, contrary to its former direction, was convex on the following side.

The general Appearance of the Tail.

On account of the great length and breadth of the tail of the comet, a night glass with a large field of view is the most proper instrument for examining its appearance. Mine takes in 4° 41′.

By viewing the comet with this glass I found the tail to be inclosed at the sides by the two streams which I have described as the continuation of the bright arch, or envelope surrounding the head.

Sept. 18, I observed that the two streams or branches arising from the sides of the head, scattered a considerable portion of their light as they proceeded towards the end of the tail, and were at last so much diluted that the whole of the farthest part of the tail, contained only scattered light.

Oct. 12, I remarked that the two streams remained sufficiently condensed in their diverging course to be distinguished for a length of about six degrees, after which their scattered light began to be pretty equally spread over the tail.

Oct. 15. The preceding branch of the tail was 7° 1′ in length. The following was only 4° 41′; which caused the appearance of an irregular curvature.

Nov. 3. The two branches were nearly of an equal length.

Nov. 5. The length of the preceding stream was 5° 16′; that of the following about 4° 41′.

Nov. 9. The two branches might still be seen to extend full 4 degrees, but their light was much scattered.

Nov. 10. The preceding branch was 5° 16′ long; the following one only 3° 31′; the preceding one was also fuller and broader.

In the course of these observations I attended also to the appearance of the nebulosity of the tail.

Sept. 18. The appearance of the nebulosity, examined with a 10 feet reflector, perfectly resembled the milky nebulosity of the nebula in the constellation of Orion, in places where the brightness of the one was equal to that of the other.

Nov. 9. The tail of the comet being very near the milky-way, the appearance of the one compared to that of the other, in places where no stars can be seen in the milky-way, was perfectly alike.

The Return of the Comet to the nebulous Appearance.

From the observations of the decreasing length of the tail, the diminution of brightness and increased scattering of the streams, and from the gradually fainter appearance of the transparent atmosphere, brought on by the contraction and more scattered condition of the envelope, I had reason to suppose that all the still visible cometic phenomena of planetary body, head, atmosphere, envelope, and tail, would soon be reduced to the semblance of a common globular nebula; not from the increase of the distance of the comet, which could only occasion an alteration in the apparent magnitude of the several parts, but by the actual physical changes which I observed in the construction of the comet.

The gradual vanishing of the planetary Body.

Nov. 4. 10 feet reflector. I saw the planetary disk with 289. It was rather more eccentric than usual.

Nov. 9. I saw it imperfectly with 169. It was more visible with 240; but the nebulosity of the envelope overpowered its light already so much that no good observations could be made of it.

Nov. 10. Large 10 feet. I had a glimpse of the disk and its eccentricity.

Nov. 13. I tried all magnifiers, but could no longer perceive the planetary body.

The Disappearance of the transparent Part of the Atmosphere under the Cover of the scattered Light of the contracted Envelope.

Nov. 4. In the night-glass, that part of the atmosphere which used to separate the envelope from the head, could no longer be distinguished. In the 10 feet reflector, with a large double eye-glass, I found the envelope drawn nearer to the head, its central distance at the vertex being less than 7′ 10″; and the atmosphere was almost involved in the scattered haziness of the streams.

Nov. 5. The envelope was still disengaged from the head, but much scattered light had nearly effaced the cometic atmosphere on the side towards the sun.

Nov. 9. The atmosphere was nearly covered by the approximation, or scattering light, of the envelope. Its vertical distance was 5′ 45″.

Nov. 10. The envelope could only be distinguished from the head by a small remaining darkish space, in which the atmosphere might still be seen. The vertical distance of the envelope was 4′ 46″.

Nov. 13. The atmosphere was almost effaced by scattered light towards the sun, but on the opposite side it was darker, or rather more transparent.

Nov. 14, 15, and 16. The atmosphere was gradually more covered in.

Nov. 19. I found in the 10 feet telescope, the envelope so broad and scattered as to leave no room for seeing the atmosphere; and the comet seemed to be fast returning to the mere appearance of a nebula.

Nov. 24. The envelope was turned into haziness; and on the side towards the sun, the comet had already the appearance of a globular nebula, with a faint hazy border.

Dec. 2. The haziness of the border was of a different colour from the light of the head, which preserved its former greenish appearance.

Dec. 9. The envelope, which had been turned into a hazy border of light, in which state I saw it again the 5th, was very unexpectedly renewed. It was however very narrow and much fainter than it used to be. By four measures I found its distance from the centre of the head to be about $4\frac{3}{4}$ minutes.

Dec. 14. The narrow faint envelope of the 9th existed no longer. If the scattered light near the head should not be raised again, all observations of the atmosphere must be at an end; for the space beyond this light being equally clear, we have nothing left to point out any extent that might be supposed to contain a transparent elastic fluid, notwithstanding it should remain in its former situation.

Uncommon Appearances in the Dissolution of the Envelope.

Nov. 4. 10 feet. The envelope was double towards the sun, and divided itself at each side into three streams; the outside ones being very faint, and of no great length.

Nov. 5. On the preceding side the envelope was very faintly accompanied by an outer one, but not on the following side.

Nov. 13. On the following side the envelope diverged into three streams, the two outside ones being very faint and narrow; but on the preceding side there was but one additional streamlet, which was at the distance of the outermost one of the opposite side.

Nov. 14. On the preceding side there was a very faint outward stream, and on the following side there was a still fainter and shorter stream, also on the outside.

Dec. 14. There was only one short and faint outside stream at the preceding side.

Uncommon Variations in the Length of the Streams.

It has already been mentioned, that the streams or branches were subject to a considerable difference in their respective lengths; in order if possible to discover the cause of the observed changes, I continued my observations of them.

Oct. 15 and Nov. 5 and 10, the preceding branch was the longest.

The 3d and 9th of November the branches were of equal length.

The 13th, the following was 4° 6′ long, the preceding only 3° 31′.

The 14th. They were both of the equal length of about 3° 31′.

The 15th. The preceding branch was 3° 31′ long, the following 4° 6′.

The 16th. The preceding was 3° 13′ long, the following 3° 48′.

The 19th. The branches were equal, and about 4° 23′ long.

Dec. 2. The branches were nearly equal and about 3° 12′ long; they joined more to the sides than the vertex, and had lost their former vivid appearance; their colour being changed into that of scattered light.

The 9th and 14th. The branches were already so much scattered that observations of them could no longer be made with any accuracy.

Alterations in the Angle of the Direction of the Envelope.

Nov. 4. 10 feet reflector. Large double eye-glass. The streams departed from their source in a greater angle of divergence. This probably arose from a contraction of the envelope towards the sun, but not about the root of the streams, where it remained extended as before.

Nov. 13. 10 feet. The angle of the bending of the envelope at its vertex was considerably enlarged. In the night-glass the divergence of the streams themselves was certainly not increased.

Nov. 24. 10 feet. The divergence of the light, which may still be called the envelope, although no longer to be distinguished from the head, was from 60 to 65 degrees; but in the night-glass, the branches which were hardly to be seen were closer together than formerly.

The additional faint duplicates of the envelope Nov. 4, 5, 13, and 14 always departed from the vertex in an angle considerably greater than the permanent interior streams.

The Shortening of the Tail.

The 5th of November, the air being very clear, I found, when attending to the tail of the comet, that its length was much reduced; its utmost extent not exceeding 12½ degrees.

The 9th. It was 10 degrees long.

The 15th. In the night glass the tail was much shortened.

The 16th. With the naked eye the tail was nearly 7½ degrees long.

The 19th. Its length was about 6° 10′.

Dec. 2. The tail was hardly 5 degrees long and of a very feeble light.

The 9th. The length of the tail was not materially altered.

The 14th. It still remained as before, but the end of it was much fainter.

Increasing Darkness between the Streams that inclose the Tail.

The 4th of November the darkness near the head on the side from the sun was grown more conspicuous, and much less filled up with scattered light.

The 5th. The darkness of the atmosphere on the side opposite the sun was stronger than on the sun side.

The 10th. A considerable darkness prevailed between the two branches of the tail.

The 14th. In the tail, close to the head, there was a large space almost free from scattered light; where the small stars of the milky-way are as bright as if nothing had intercepted their light.

The 16th. The space between the streams was of a considerable darkness.

The 19th. 10 feet reflector. The darkness between the streams was increased.

Dec. 9. The space close to the head on the side from the sun was quite dark, or rather transparent.

The 14th. Many small stars of the milky-way were in the dark interval of the tail close to the head of the comet.

Of the real Construction of the Comet, and its various Parts.

Hitherto I have only related the appearances of the several parts of the comet, in order to determine their linear extent; but the observations which are now before us, contain facts that will allow me also to ascertain the construction of the comet and its various parts in their solid dimensions.

From the laws of gravitation we might be allowed to conclude that the planetary body containing the solid matter of the comet must be spherical; but actual observation will furnish a more substantial argument; for in no part of the long, geocentric path described by the comet, did I see its little disk otherwise than round; whereas it would not have preserved this appearance, if its construction were not spherical.

If what has been said in my last paper, when treating of round nebulæ, be remembered, the head of the present comet, which by observation appeared round like a nebula, cannot be supposed to be of any other than a spherical construction. With my collection of round nebulæ the arguments, however, which proved their globular form, rested only, though very soundly, upon the doctrine of chances, and the known effects of gravitation; but here, on the contrary, while nebulæ remain in their places, the geocentric position of the head of the comet has undergone a change amounting to a whole quadrant; in all which time I have observed it to retain its roundness without any visible alteration; from which it necessarily follows that its form is globular.

With regard to its transparent cometic atmosphere, we have not only the constant observations of its roundness, during the above-mentioned long period of the comet's motion, to prove it to be spherical; but in addition to this, I have already shown that it is of an elastic nature, for which reason alone, had we no other, its globular figure could not be doubted.

A most singular circumstance, which however must certainly be admitted, is, that the constant appearance of the bright envelope, with its two opposite diverging branches, can arise from no other figure than that of an inverted hollow cone, terminating at its vertex in an equally hollow cap, of nearly a hemispherical construction; nor can the sides or caps of this hollow cone be of any considerable thickness.

The proof of this assigned construction is, that the bright envelope has constantly been seen in my observation as being every where nearly equidistant from the transparent atmosphere; now if that part of it which in a semi-circular form surrounds the comet, on the side exposed to the sun, were not hemispherical, but had the shape of a certain portion of a ring, like that which we see about the planet Saturn, it must have been gradually transformed from the appearance of a semicircle into that of a straight line, during the time that we have seen it in all the various aspects presented to us by a geocentric motion of the comet, amounting to 90 degrees.

That this hemispherical cap is comparatively thin, is proved from the darkness and transparency of that part of the atmosphere which it covers; for had the curtain of light, which was drawn over it, been of any great thickness, the scattered rays of its lustre would have taken away the appearance of this darkness; nor would the atmosphere have remained sufficiently transparent for us to see extremely small stars through it.

It remains now only to account for the semi-circular appearance of the bright envelope; but this, it will be seen, is the immediate consequence of the great depth of light near the circumference, contrasted with its comparative thinness towards the centre. The 6th of October, for instance, the radius of the envelope was 9′ 30″ on the outside, and 7′ 30″ on the inside; and as the greatest brightness was rather nearer to the outside, we may suppose its radius to have been about 8′$\frac{3}{4}$. Then if we compute the depth of the luminous matter at this distance from the centre, we find that it could not be less than 248 thousand miles; whereas in the place where the atmosphere was darkest, its thickness would be only about 50 thousand; so that a superior intensity of light in the ratio of about 5 to 1, could not fail to produce the remarkable appearance of a bright semi-circle, enveloping the head of the comet at the distance at which it was observed.*

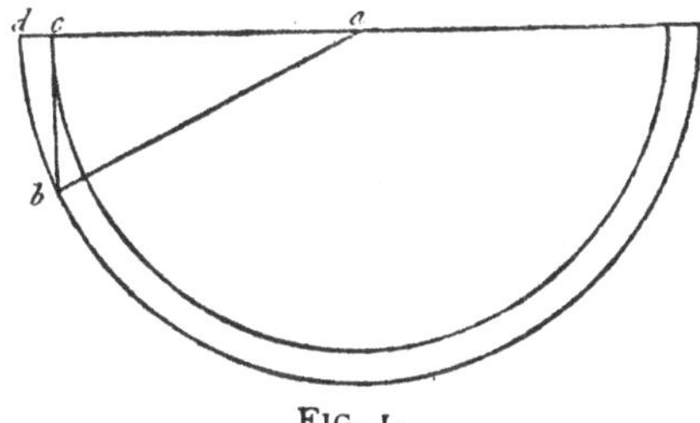

Fig. 1.

I have entered so fully into the formation of the envelope, as the argument, by which its construction has been analysed, will completely explain the appearance of the streams of light inclosing the tail of the comet, and indeed its whole construction.

* From the measure of the envelope, whose diameter the 6th of October was 643032 miles, we have the radius *ab*, fig. 1, 321516. Then if *cd* be 25000, we find the angle *bac*, of which *ac* is cosine 22° 44′ 37″; and the sine *bc*, which is the depth, will be to the versed sine *cd*, which is the thickness, as 4·972 to 1. And if *ad* is 9′ 30″, the greatest brightness which is at *c* will give the distance *ac* equal to 8′ 45″·7. This calculation being made for that part which is convex towards us, the addition of the concave opposite side will double the dimensions of the depth and thickness.

The luminous matter as it arises from the envelope, of which it is a continuation, is thrown a little outwards, and assumes the appearance of two diverging bright streams or branches; but if the source from which they rise be the circular rim of an hemispherical hollow shell, the luminous matter in its diverging progress upwards can only form a hollow cone; and the appearance of the two bright streams inclosing the tail, after what has been said of the envelope, will want no farther explanation.

Add to this that, having actually seen these brilliant streams remain at the borders of the tail in the same diverging situation during a motion of the comet through more than 130 degrees, the hollow conical form of the comet's tail is in fact established by observation.

The feebler light of the tail between its branches is sufficiently accounted for by the thinness of the luminous matter of the hollow cone through which we look towards the middle of the tail, compared with its great depth about the sides; and indeed the comparative darkness of the inside of the cone and transparency of the atmosphere seen through the envelope, bear witness to their hollow construction; for, were these parts solid, both the cone and the hemispherical termination of it must have been much brighter in the middle than towards the circumference, which is contrary to observation.

Of the solar Agency in the Production of Cometic Phenomena.

As we are now in a great measure acquainted with the physical construction of the different parts of the present comet, and have seen many successive alterations that have happened in their arrangement, it may possibly be within our reach to assign the probable manner in which the action of such agents as we are acquainted with has produced the phenomena we have observed.

In its approach to a perihelion, a comet becomes exposed to the action of the solar rays, which, we know, are capable of producing light, heat, and chemical effects. That their influence on the present comet has caused an expansion, and decomposition of the cometic matter, we have experienced in the growing condition of the tail and shining quality of its light, which seems to be of a phosphoric nature. The way by which these effects have been produced may be supposed to be as follows.

The matter contained in the head of the comet would be dilated by the action of the sun, but chiefly in that hemisphere of it which is immediately exposed to the solar influence; and being more increased in this direction than on the opposite side, it would become eccentric, when referred to the situation of the body of the comet; but as the head is what draws our greatest attention, on account of its brightness, the little planetary body would appear to be in the eccentric situation in which we have seen it.

Now, as from observed phenomena, we have good reason to believe the comet to be surrounded by a very extensive, transparent, elastic atmosphere ; the nebulous matter, which probably, when the comet is at a distance from the perihelion, is gathered about the head in a spherical form, would on its approach to the sun be greatly rarefied, and rise in the cometic atmosphere till it came to a certain level, where it could remain suspended, for some time, exposed to the continued action of the sun.

In this situation we have had an opportunity of seeing the transparent atmosphere, which, but for the suspension of the nebulous matter, we might never have discovered ; and indeed, how far it may extend beyond the region which contained the shining substance, we can have no observation to ascertain, on account of its transparency. In consequence of the darkish interval, occasioned by the atmospheric space, the suspended light appeared to us in the shape of a very bright envelope.

The brilliancy of the envelope, and its yellowish colour, so different from that of the head, and probably acquired by its mixture with the atmospheric fluid, are proofs of the continued action of the sun upon the luminous matter, already in so high a state of rarefaction ; and if we suppose the attenuation and decomposition of this matter to be carried on till its particles are sufficiently minute to receive a slow motion from the impulse of the solar beams, then will they gradually recede from the hemisphere exposed to the sun, and ascend in a very moderately diverging direction towards the regions of the fixed stars.

That some such operation must have been carried on, is pretty evident from our having seen the gradual rise, and increased expansion of the tail of the comet ; and if we saw the shining matter, while suspended in the cometic atmosphere, in the shape of an envelope, it follows that, in its rising condition, it would assume the appearance of those two luminous branches which we have so long observed to inclose the tail of the comet.

The seemingly circular form, and the stream-like appearance of the luminous matter having been already explained, we may now see the reason why it can rise in no other form than the conical ; for a whole hemisphere of it being exposed to the action of the sun, it must of course ascend equally every where all around it.

That the luminous matter ascending in the hollow cone, received no addition to its quantity from any other source than the exposed hemisphere, we may conclude from its appearance ; which notwithstanding the great circumference of the cone it filled, at the altitude of 6 degrees from the head, was never seen with increased lustre ; although the diameter of an annular section of it, in that place, must have been nearly 15 millions of miles, and was but little more than half a million at its rising from the envelope.

This consideration points out the extreme degree of rarefaction of the luminous matter about the end of the tail ; for its expansion, while still much confined in the streams, at the altitude which has been mentioned, must have exceeded the density it had at rising about 524 times ; but when afterwards it extended itself so as to produce nearly an evenly scattered light over the whole compass of the end of the tail, we may easily conceive to what an extreme degree of rareness its expansion must have been carried.

The vacancy occasioned by the escape of the nebulous matter, which after rarefaction passed from the hemisphere exposed to the sun into the regions of the tail, was probably filled up, either by a succession of it from the opposite hemisphere, or by a rotation of the comet about an axis ; and the gradual decomposition of this matter would therefore be carried on as long as any remained to replace the deficiency.

That such a kind of process took place, seems to be supported by the observations which were made during the regression of the comet from its perihelion. For the space between the branches of the tail, very near the head of the comet, became gradually of a darker appearance than before ; which indicated the absence of the nebulous matter that had formerly been lodged there.

A rotatory motion of the comet, which has been suggested, would also explain the frequent variations in the length of the opposite branches which inclosed the tail ; for if any portion of the cometary matter should be more susceptible of being thrown into a luminous decomposition than some others, a rotatory motion would bring such more susceptible matter into different situations, and cause a more or less copious emission of it in different places.

The additional short and faint double streams of nebulous light which issued from the vertex or side of the enfeebled envelope, in the gradual regress of the comet, tend likewise to add probability to the conception of a rotatory motion ; for the changeable appearance of the situation of these streamlets might arise from a periodical exposition of some remaining small portions of less rarefied matter, when nearly the whole of it had been exhausted.

Of the Result of a Comet's Perihelion Passage.

After having given a detail of phenomena, and entered into a research of the most likely manner in which they were produced, I shall only mention what appears to me to be the most probable consequence of the perihelion passage of a comet.

The quality of giving out light, although it may always reside in a comet, as it does in the immensity of the nebulous matter, which I have shown to exist in the heavens, is exceedingly increased by its approach to the sun. Of this we should not be so sensible, if it were not accompanied with an almost inconceivable expansion and rarefaction of the luminous substance of the comet about the time of its perihelion passage.

It is admitted, on all hands, that the act of shining denotes a decomposition in which at least light is given out ; but that many other elastic volatile substances may escape at the same time, especially in so high a degree of rarefaction, is far from improbable.

Then, since light certainly, and very likely other subtile fluids also escape in great abundance during a considerable time before and after a comet's nearest approach to the sun, I look upon a perihelion passage in some degree as an act of consolidation.

If this idea should be admitted, we may draw some interesting conclusions from it. Let us, for example, compare the phenomena that accompanied the comet of 1807 with those of the present one. The first of these in its approach to the sun came within 61 millions of miles of it ; and its tail, when longest, covered an extent of 9 millions. The present one in its perihelion did not come so near the sun by nearly 36 millions of miles, and nevertheless acquired a tail 91 millions longer than that of the former. The difference in their distances from the earth when these measures were taken was but about 2 millions.

Then may we not conclude, that the consolidation of the comet of 1807, when it came to the perihelion, had already been carried to a much higher degree than that of the present one, by some former approach to our sun, or to other similarly constructed celestial bodies, such as we have reason to believe the fixed stars to be ?

And that comets may pass round other suns than ours, is rendered probable from our knowing as yet, with certainty, the return of only one comet among the great number that have been observed.

Since then, from what has been said, it is proved that the influence of the sun upon our present comet has been beyond all comparison greater than it was upon that of 1807 ; and since we cannot suppose our sun to have altered so much in its radiance as to be the cause of the difference ; have we not reason to suppose that the matter of the present comet has either very seldom, or never before passed through some perihelion by which it could have been so much condensed as the preceding comet ? Hence may we not surmise that the comet of 1807 was more advanced in maturity than the present one ; that is to say, that it was comparatively a much older comet.

Should the idea of age be rejected, we may indeed have recourse to another supposition, namely, that the present comet, since the time of some former perihelion passage, may have acquired an additional quantity (if I may so call it) of *unperihelioned* matter, by moving in a parabolical direction through the immensity of space, and passing through extensive strata of nebulosity ; and that a small comet, having already some solidity in its nucleus, should carry off a portion of such matter, cannot be improbable. Nay, from the complete resemblance of many comets to a number of nebulæ I have seen, I think it not unlikely that the matter they contain is originally nebulous. It may therefore possibly happen that some of the nebulæ, in which this matter is already in a high state of condensation, may be drawn towards the nearest celestial body of the nature of our sun ; and after their first perihelion passage round it proceed, in a parabolic direction, towards some other similar body; and passing successively from one to another, may come into the regions of our sun, where at last we perceive them transformed into comets.

The brilliant appearance of our small comet may therefore be ascribed either to its having but lately emerged from a nebulous condition, or to having carried off some of the nebulous matter, situated in the far extended branch of its parabolic motion. The first of these cases will lead us to conceive how planetary bodies may begin to have an existence; and the second, how they may increase and, as it were, grow up to maturity. For if the accession of fresh nebulous matter can be admitted to happen once, what hinders us from believing a repetition of it probable? and in the case of parabolic motions, the passage of a comet through immense regions of such matter is unavoidable.

WM. HERSCHEL.

Slough, near Windsor,
Dec. 16, 1811.

LXV.

Observations of a second Comet, with Remarks on its Construction.

[*Phil. Trans.*, 1812, pp. 229–237.]

Read March 12, 1812.

As we have lately had two comets to observe at the same time, I have called that of which the following observations are given, the second.* Its appearance has been so totally different from that of the first, that every particular relating to its construction becomes valuable; and notwithstanding the unfavourable state of the weather at this time of the year, I have been sufficiently successful to obtain a few good views of the phenomena which this comet has afforded.

A short detail of the observations, in the order of their relation to the different cometic appearances, is as follows:

The Body of the Comet.

January 1, 1812. I viewed the second comet with several of my telescopes, and found it to have a considerable nucleus surrounded with very faint chevelure.

Jan. 2. The comet had a large round nucleus within its faint nebulosity. Not seeing it very well defined, and of so large a diameter, I doubted whether it could be the body of the comet; but although it might be called very large when supposed to be of a planetary construction, it was much too small for the condensed light of a head; its diameter, by estimation not exceeding 5 or 6 seconds.

By way of comparing the two comets together I viewed them alternately. The first, within a nebulosity which in the form of a brilliant head was of great extent, had nothing resembling a nucleus: the light of this head was very gradually much brighter up to the very middle; its small planetary body being invisible. The second comet, on the contrary, although surrounded by a faint chevelure, seemed to be all nucleus; for the abrupt transition from the central light to that of the chevelure would not admit of the idea of a gradual condensation of nebulosity, such as I saw in the head of the first comet; but plainly pointed out that the nucleus and its chevelure were two distinct objects.

Jan. 8. The comet had a pretty well defined nucleus with very faint chevelure. When magnified 170 times the nucleus, though less bright, was rather better defined.

* [Comet 1811 II., discovered by Pons, Nov. 16.—Ed.]

Jan. 18. Within a very faint chevelure I saw the nucleus as before.

Jan. 20. The air being uncommonly clear, I saw the body of the comet well defined; and as the moon was already so far advanced in its orbit as to render future opportunities of viewing the comet very improbable, I ascertained the magnitude of its body, with a very distinct 10 feet reflector, by the following three observations.

First with a low power, which gave a bright image of the nucleus, I kept my attention fixed upon its apparent size; then looking away from the telescope, I mentally reviewed the impression its appearance had made on the imagination, in order to see whether it was a faithful picture of the object; and by looking again into the telescope I was satisfied of the similitude.

In the next place I used a deeper magnifier, and alternately viewed and remembered the appearance of the nucleus. It was fainter with this power.

The third observation was made in the same manner with a magnifier of 170. This showed the nucleus of a larger diameter, but much less bright, and not so well defined.

The next morning, having recourse to my usual experiment with a set of globules, by viewing them at a given distance with the same telescope and eye-glasses, I found that one of them, on which I fixed, gave me, as nearly as could be estimated, the same magnitude with the first eye-glass, and was proportionally magnified by the second and third, with only this difference, that the highest power showed the globule with more distinctness than it did the nucleus; and by trigonometry the angle under which I saw the globule was found to be 5″·2744.*

It will be necessary to mention that in the calculations belonging to this comet, I have used the elements of Mr. GAUSS, with a small correction of the longitude of the perihelion, which I found would answer the end of giving the observed place with sufficient accuracy from the 1st of January to the 20th. These calculations may however be repeated, if hereafter we should obtain elements improved by additional observations, made with fixed instruments; but the result, I may venture to say, will not be materially different.

The distance of the comet from the earth, the 20th of January when its apparent diameter was determined, was 1·0867, the mean distance of the earth from the sun being 1; whence we deduce a very remarkable consequence, which is, that the real diameter of its nucleus cannot be less than 2637 miles.

The Chevelure of the Comet.

Instead of that bright appearance, which in the first comet has been considered as the head, there was about the nucleus of the second a faint whitish scattered light, which may be called its chevelure.

* I prefer this method of ascertaining the small diameter of a faint object to measuring it with a micrometer, which requires light to show the wires, and a high magnifying power to give an image sufficiently large for mensuration; neither of which conditions the present comet would admit.

Jan. 1. Examining the chevelure of the comet with a 10 feet reflector, I found that it surrounded the nucleus, not in the form of a head consisting of gradually much condensed nebulosity, but had the appearance of a faint haziness, which although of some extent, was not much brighter near the nucleus than at a distance from it.

Jan. 2. I viewed the two comets alternately. The first could only be distinguished from a bright globular nebula by the scattered light of its tail, which was still 2° 20′ long. The second comet, on the contrary, had nothing in its appearance resembling such a nebula: it consisted merely of a nucleus, surrounded by a very faint chevelure; and had it not been for an extremely faint light in a direction opposite to the sun, it would hardly have been intitled to the name of a comet; having rather the appearance of a planet seen through an atmosphere full of haziness.

Jan. 8. The chevelure consisted of so faint a light that, when magnified only 170 times, it was nearly lost.

Jan. 18. The chevelure was extremely faint and of very little extent.

Jan. 20. The light of the moon, which was up, would not admit of further accurate observations on the chevelure.

The Tail of the Comet.

Jan. 1. With a low magnifying power, I saw in the 10 feet reflector an extremely faint scattered light, in opposition to the sun, forming the tail of the comet. It reached from the centre of the double eye-glass half way toward the circumference.

Jan. 8. The narrow, very faint scattered light beyond the chevelure remains extended in the direction opposite the sun.

Jan. 18. I estimated the length of the tail by the proportion it bore to the diameter of the field of the eye-glass, which takes in 38′ 39″, and found that it filled about one quarter of it, which gives 9′ 40″.

Jan. 20. On account of moonlight the tail was no longer visible.

From the angle which it subtended in the last observation, it will be found that its length must have been about 659 thousand miles.

Remarks on the Construction of the Comet.

The method I have taken in my last paper of comparing together the phenomena of different comets appears to me most likely to throw some light upon a subject which still remains involved in great obscurity. When the comet of which the observations have been given in this paper is compared with the preceding one, it will be found to be extremely different. Its physical construction appears indeed to approach nearly to a planetary condition. In its magnitude it bears a considerable proportion to the size of the planets; the diameter of its nucleus being very nearly one-third of that of the earth.

The light by which we see it is probably also planetary; that is to say reflected from the sun. For were it of a phosphoric, self-luminous nature, we could hardly account for its little density: for instance, the very small body of the first comet, at the distance of 114 millions of miles from the earth bore a magnifying power of 600, and was even seen better with this than with a lower one; * whereas the second, notwithstanding its large size, and being only at the distance of 103 millions, had not light enough to bear conveniently to be magnified 107 times; but if we admit this nucleus to be opaque, like the bodies of the planets, and of a nature not to reflect much light, then its distance from the sun, which the 20th of January was above 174 millions of miles, will explain the cause of its feeble illumination.

That the nucleus of this comet was surrounded by an atmosphere appears from its chevelure, which, though faint, was of considerable extent; and the elasticity of this atmosphere may be inferred from the spherical figure of the chevelure, proved by its roundness and equal decrease of light at equal distances from the centre.

The transparency of the atmosphere is partly ascertained from our seeing the nucleus through it, but may also be inferred by analogy from an observation of the first comet. It will be remembered that an atmosphere of great transparency, which had been seen for a long time, was lost when the comet receded from the sun, by the subsidence of some nebulous matter not sufficiently rarified to enter the regions of the tail.* Now as the existence of this atmosphere, when it was no longer visible, might have been doubted, the luminous matter suspended in it, which had already 20 days obstructed our view of it, happened fortunately to be once more elevated the 9th of December, and thereby enabled us, from its transparency and capacity of sustaining luminous vapours, to ascertain the continuance of its existence. By analogy, therefore, we may surmise that the faint chevelure of the second comet consists also of the condensation of some remaining phosphoric matter, suspended in the lower regions, of an elastic, transparent fluid, extending probably far beyond the chevelure without our being able to perceive it.

We might ascribe the little extent and extreme faintness of the tail to the great perihelion distance of the comet, if it had not already been proved, by the comparative view which in my last paper has been taken of the two comets of 1807 and 1811, that the effect of the solar agency depends entirely upon the state of the nebulous matter, which the comet in its approach exposes to the action of the sun. Our last comet therefore had probably but little *unperihelioned* matter in its atmosphere.

The high consolidation of the matter contained in the second comet is also much supported by the different appearance of the two comets in the observation of the 2d of January. In order to judge of them properly, we must consider their

* See Observations of the First Comet.

situation with regard to the sun and the earth; the first comet was 192 millions of miles from the sun; the second only 164: the first was at the same time 262 millions from the earth: the second only 83; but notwithstanding the great disadvantage of being 28 millions of miles farther from the sun, and about 179 millions farther from the earth, the first comet had the luminous appearance of a brilliant head accompanied by a tail 45 millions of miles in length; whereas the second comet, so advantageously situated, had only a very faint chevelure about its large but faint nucleus, with a still fainter tail, whose length has been shown not much to exceed half a million.

If then the effect of the action of the sun on the comets at the time of their perihelion passage is more or less conspicuous, according to the quantity of unperihelioned nebulous matter they contain, we may by observation of cometic phenomena arrange these celestial bodies into a certain order of consolidation, from which, in the end, a considerable insight into their nature and destination may be obtained. The three last observed comets, for instance, will give us already the following results.

The comet of which this paper contains observations, is of such a construction that it was but little more affected by a perihelion passage than a planet would have been. This may be ascribed to its very advanced state of consolidation, and to its having but a small share of phosphoric or nebulous matter in its construction.

That of the year 1807 was more affected, and although considerably condensed, showed clearly that it conveyed a great quantity of nebulosity to the perihelion passage.

The comet of last year contained with little solidity a most abundant portion of nebulous matter, on which, in its approach to the perihelion, the action of the sun produced those beautiful phenomena, which have so favourably afforded an opportunity for critical observations.

LXVI.

Astronomical Observations relating to the sidereal part of the Heavens, and its Connection with the nebulous part: arranged for the purpose of a critical Examination.

[*Phil. Trans.*, 1814, pp. 248–284.]

Read February 24, 1814.

IN my paper of observations of the nebulous part of the heavens, I have endeavoured to shew the probability of a very gradual conversion of the nebulous matter into the sidereal appearance. The observations contained in this paper are intended to display the sidereal part of the heavens, and also to shew the intimate connection between the two opposite extremes, one of which is the immensity of the widely diffused and seemingly chaotic nebulous matter; and the other, the highly complicated and most artificially constructed globular clusters of compressed stars.

The proof of an intimate connection between these extremes will greatly support the probability of the conversion of the one into the other; and in order to make this connection gradually visible, I have arranged my observations into a series of collections, such as I suppose will best answer the end of a critical examination.

I. *Of Stars in remarkable situations with regard to Nebulæ.**

Among the great number of stars, with nebulosity dispersed between them are some in situations that deserve to be remarked.

IV. 5 † is "A pretty bright star situated exactly north of the center of an extended milky ray, which is about 15 or 20 minutes in length." By a second observation, two years after the first, it appeared that the star was then included in part of the nebulosity.

V. 46 ‡ is "A pretty bright star in the middle of a very bright nebula, about 10 minutes in length and 2′ broad." See fig. 1.

III. 616 § is "A star of the 6th magnitude, about 5′ north of a very faint nebula, of an irregular figure." By an observation of the same star, two years

* See five stars in remarkable situations. II. 246. III. 201, 616. IV. 5, 46. The places of these objects will be found in three catalogues of nebulæ and clusters of stars, published in the *Philosophical Transactions* for 1786, 1789, and 1802. The reference IV. 5 for instance, points out number 5, in the IVth class.

† [N.G.C. 4517.]

‡ [N.G.C. 3556.]

§ [N.G.C. 3930. The star is G. 1830, the large P.M. of which is towards *sf*.—ED.]

before, the two objects were then so near each other, as, at first sight, to cause a suspicion that some damp had settled upon the eye-glass and affected the star.

The singularity that five stars should be similarly situated with regard to nebulæ is not very striking; but the difference in the additional observation is worthy of notice, and may suggest a surmise that nebulæ may have considerable proper motions, by which they are occasionally carried towards neighbouring stars: the difference in the clearness of the atmosphere at different times, however, ought to make us cautious about assigning the cause of the difference in the observations.

2. *Of two Stars with nebulosity between them.**

A more remarkable situation than the former is that of two stars with nebulosity between them, or both included in the same nebulosity.

III. 67 † is "An extremely faint nebulosity extended from one star to a smaller one, at the distance of about 2 minutes south of the former." See fig. 2.

II. 706.‡ "Two considerable stars are involved in a very faint nebulosity of 3 or 4 minutes in extent." See fig. 3.

Here I have referred to 19 instances, where two stars have an extended nebulosity between them, or at least are both contained within it. Now, if we were to enter into a calculation of chances to investigate the probability that in every one of these 19 objects, the stars and the nebulosity should be unconnected, we should have to consider that in order to produce this appearance by three objects at a distance from each other, it would be required that every one of them should be precisely in a given line of sight, and that the nebulosity should not only be in the middle of them, but that it also should be extended from the situation of one star to that of the other; and that all this should happen in the confined space of a few minutes of a degree; which cannot be probable. Then, if on the other hand we recollect that in the 8th, 9th, and 10th articles of my paper on the nebulous part of the heavens, I have given 139 double nebulæ joined by nebulosity between them, and that we have now before us 19 similar objects, with no other difference than that instead of nebulæ we have stars with nebulosity remaining between them, should we not surmise that possibly these stars had formerly been highly condensed nebulæ, like those that have been mentioned, and were now by gradually increasing condensation turned into small stars; and may not the nebulosity still remaining shew their nebulous origin?

When to this is added that we also have an account of 700 double stars entirely free from nebulosity,§ many of which are probably at no great real distance from

* See nineteen double stars joined by intermediate nebulosity II. 16, 706, 732. III. 19, 32, 67, 68, 113, 126, 182, 200, 312, 376, 540, 637, 757, 785, 820, 854.

† [N.G.C. 3473.] ‡ [N.G.C. 7538.]

§ See *Phil. Trans.* for 1782, page 112; and for 1785, page 40. [Vol. I. pp. 58 and 167.]

each other, it seems as if we had these double objects in three different successive conditions: first as nebulæ; next as stars with remaining nebulosity; and lastly as stars completely free from nebulous appearance.

3. *Of Stars with nebulosities of various shapes attached to them.*

When a nebula seems to be joined to a star, or closely pointing to it, the manner of its appearance deserves our attention. Here follow three different sorts of such conjunctions.*

First sort; I. 143.† "On the north preceding side of a pretty bright star is a considerable, bright nebulosity. It is joined to the star so as to appear like a brush to it." See fig. 4.

Second sort; IV. 4.‡ "A very small star has an extremely faint, and very small nebula attached to it in the shape of a puff." See fig. 5.

Third sort; IV. 35.§ "A small star has a small, faint, fan-shaped nebulosity joining to it on the north preceding side." See fig. 6.

Here we have a list of fourteen objects, in which the probability of a union between the nebulosities and the stars will gradually become more apparent. With regard to the first nine, the particularity of their construction is already very pointed: the conditions are that the nebulosity must be extended; the direction of its extension must be exactly towards the star, and it must also be apparently just near enough to touch it; but that all this should happen cannot be probable; whereas a real contact of the objects, held together by mutual gravitation, will readily account for the whole appearance.

In the two next objects there is already some indication of a union between the nebula and the star, for the roundness of the nebulosity appeared to be a little drawn out of its figure towards the star.

But the last three instances, in which the whole mass of nebulous matter is pointedly directed to the stars, and in contact with them, can hardly leave any room for doubting a union between them.

Now if we admit a contact, or union between these nebulæ and the stars, it deserves to be remarked that stars, in the situation of these fourteen, cannot have been formed from their adjoining nebulosities; for a gradual condensation of the nebulous matter would have been central; whereas the stars are at the extremity of the nebulæ. It is therefore reasonable to suppose that their conjunction must be owing to some motion either of the stars or of the nebulous matter: a mutual attraction might draw them together. In either of these cases it would follow,

* See fourteen stars connected with nebulæ.
Nine with a brush I. 143. II. 214, 683. III. 643. IV. 10, 17, 29, 40, 77.
Two with a puff IV. 3, 4.
Three with fan-shaped nebulosity IV. 2, 35, 66.

† [N.G.C. 4900.] ‡ [N.G.C. 3662.] § [N.G.C. 2610.]

that if the nebulosity should subside into the star, as seems to be indicated by the assumed form of the fan-shaped nebulæ, the star would receive an increase of matter proportional to the magnitude and density of the nebulosity in contact with it. This would give us the idea of what might be called the *growth* of stars.

4. *Of Stars with nebulous branches.*

That an intimate connection between the nebulous matter and a star is not incompatible with their nature will clearly appear by the following instances, in which a union is manifested that cannot be mistaken for a deceptive appearance.*

IV. 42 † is " A star of about the 8th or 9th magnitude with very faint nebulous branches extended in the direction of the meridian: each branch is about one minute in length. Other stars of the same size, and at the same time in view, are free from these branches." See fig. 7.

The three objects to which I have referred shew sufficiently that stars and nebulæ may be connected; for a little swelling and increase of light of the branches, at their junction with the star, which generally takes place, seems evidently to be an effect arising from the gravitation of the nebulous matter towards a center, in which the star is situated.

Here again the visible effect of gravitation supports the idea of the *growth* of stars by the gradual access of nebulous matter; for in the present case I may refer to the observations already published in the *Phil. Trans.* for 1811, where, page 301,‡ we have an account of twenty-four extended nebulæ, gradually a little brighter in the middle; page 303 there are fifty extended nebulæ, with an increased brightness towards the middle; § page 304, we have fifty-four extended nebulæ, with a much greater accumulation of brightness; || page 307 there are seven extended nebulæ, in which the central increase of brightness approaches towards the formation of a nucleus.¶ Page 309, we have twenty-seven extended nebulæ, in which the central nucleus is already formed; ** and finally, page 311 contains the account of twenty-three extended nebulæ, where the nebulosity seems to have so far subsided into the nucleus, as to leave only two opposite faint branches.†† Who then that has followed up the gradual condensation of an extended nebula till it appeared in the shape of a bright nucleus with faint branches, and finds now in the center of two such opposite faint branches, instead of a condensed nucleus, a star—who, I may ask, would not rather admit that the nucleus had gradually cleared up in brightness, and assumed the lustre of a star, than have recourse to the most improbable of all hypotheses, that a fortuitous central meeting of a star and a nebula should be the cause of such a singular appearance?

* See three stars with nebulous branches IV. 42, 43, 48.

† [N.G.C. 676.] ‡ [Above, p. 477.] § [Above, p. 478.] || [Above, p. 479.]

¶ [Above, p. 481.] ** [Above, p. 482.] †† [Above, p. 483.]

5. *Of nebulous Stars.**

The conjunction of the nebulous and sidereal condition is still more clearly manifested in nebulous stars. Having already described many of them in a paper read before the Royal Society in 1791, I shall here only mention two of them.

IV. 45.† "A star of about the 9th magnitude has a pretty strong milky nebulosity equally dispersed all around it."

IV. 69 ‡ is "A star of about the 8th magnitude with a faint luminous atmosphere of a circular form of about 3 minutes in diameter; the star is perfectly in the center, and the atmosphere is so diluted, faint, and equal throughout, that there can be no surmise of its consisting of stars, nor can there be a doubt of the evident connection between the atmosphere and the star." See fig. 8.

Among the thirteen objects referred to, there are many so variously constructed as to prove not only that nebulous stars are intimately connected with a nebulosity, which from its great regularity might be taken for an atmosphere, but also with the luminous appearances, which have been described as belonging to the nebulous matter that is so widely expanded over various regions of the starry heavens. For instance, in IV. 45, 58, 65 and 69 the stars are perfectly central, which proves that the chevelure is connected with them. In IV. 36, 71 and 74 the nebulosity is likewise attached to the stars, but their nebulosity is more irregularly and extensively expanded, so as to resemble the general mass of nebulous matter.

What has been said of the gradual condensation of the nebulous matter in the case of extended nebulæ, is supported by a much greater number of nebulosities of a spherical form. The different gradations of their condensation are pointed out in the same paper from page 301 to 308; § and contain 322 cases in which the fact of the gradual condensation is rendered so evident as not to admit of a doubt. Then, if instead of the last 13 globular nebulæ, page 309,‖ each of which has a nucleus in the middle, we now look at our 13 stars, each of which is situated in the very center of a globular nebulosity, it will evidently point out the high probability, or rather the certainty, that nebulous stars only differ from round nebulæ containing a nucleus, in the order of condensation, which in the case of the nebulous stars, has been carried to a somewhat higher degree than in the nebulous nuclei.

6. *Of Stars connected with extensive windings of nebulosity.*¶

The nebulosity which has been shewn to be connected with stars may be fully proved to be of the same nature as the general mass of nebulous matter by the following instances.

IV. 33.** "A star situated upon a ground of extremely faint milky nebulosity

* See thirteen nebulous stars IV. 19, 25, 36, 38, 44, 45, 52, 57, 58, 65, 69, 71, 74.

† [N.G.C. 2392.] ‡ [N.G.C. 1514.] § [Above, pp. 477-481.] ‖ [Above, p. 482.]

¶ See three stars connected with diffusions of nebulosity IV. 24, 33. V. 27.

** [N.G.C. 1999.]

diffused over this part of the heavens, has a milky chevelure surrounding it, which is brighter than the nebulosity of the ground ; but which loses itself imperceptibly in the extreme faintness of the general diffusion of the nebulous matter." See fig. 9.

The formation of these objects is extremely instructive, as it manifests the affinity between the matter of which stars are composed, and that of the most unshapen chaotic mass of nebulosity. For the vanishing chevelure of a star being equally connected, on the one hand, with the generally diffused nebulous matter, and on the other with the star itself, round which it is in a state of gradual condensation ; this double union denotes the mutual gravitation of the whole mass of nebulosity and the star towards each other ; and unless this proof can be invalidated, we must admit the fact of the growing condition of stars, that are in the situation which has been pointed out.

This argument also adds greatly to the probability of stars being originally formed by a condensation of the nebulous matter ; for, as it now appears that stars must receive an addition to their solid contents, when they are in contact with nebulosity, there is an evident possibility of their being originally formed of it. Moreover, the affinity between the nebulous and sidereal condition being established by these observations, we may be permitted to conceive both the generation and growth of stars, as the legitimate effects of the law of gravitation, to which the nebulous matter, as proved by observation, is subject.

7. *Of small patches consisting of Stars mixed with nebulosity.*

When a small patch of stars is mixed with nebulosity, there is a possibility of its being a deception arising from their being accidentally in the same line of sight ; but it has already been shewn that in such appearances the probability is much in favour of a real union ; especially when the objects are numerous ; * and in that case there are but two ways of accounting for it.

First, admitting from what has been said, that stars may be formed of nebulous matter, it may happen that the nebulosity still mixed with them is some remaining unsubsided part of that from which they were formed ; and in the next place, a union of stars and nebulosity, originally at a distance from each other, may have been effected by the motion of either the stars or the nebulosity.

That such motions may happen has been shewn in the third article, which contains instances of the conjunction of stars with nebulosities of which they cannot have been formed, and which must, consequently, have been united by motion. We also know that nebulæ are subject to great changes in their appearance, which proves that some of the nebulous matter in their composition must be in motion ; instances of which have been given in the luminous nebulosity of the constellation

* See thirty-seven small patches, consisting of stars mixed with nebulosity I. 172, 192, 258. II. 21, 39, 103, 304, 489, 745, 878. III. 8, 43, 61, 64, 71, 143, 144, 146, 147, 165, 185, 204, 227, 256, 271, 349, 471, 538, 559, 560, 568, 583, 595, 697, 922. IV. 75. V. 49.

of Orion.* It may therefore be easily conceived that any moving patch of nebulous matter must be arrested on its meeting with stars; especially if several of them should happen to be pretty near each other; in which case there will be, as it were, a net spread out for intercepting every nebulosity that comes within the reach of their attraction.

II. 304.† "Three or four stars of various sizes are mixed with pretty strong nebulosity."

III. 165.‡ "Five or six stars forming a parallelogram, are mixed with very faint milky nebulosity."

III. 697.§ "Several small stars are contained in faint nebulosity about 3 or 4 minutes long and ¾ broad." See fig. 10.

IV. 75.|| "Three stars of about the 9th or 10th magnitude are involved in pretty strong milky nebulosity."

This collection of thirty-seven objects, consisting of 2, 3, 4, 5, 6, or more small stars that are mixed with nebulosity, contains a variety of instances in which the effect that has been mentioned of the interception of the nebulous matter may have taken place. It is very obvious that nothing positive can be said about the formation of so many starry-nebulous patches; for unless by long continued observation of the same patches we could be acquainted with every change that may happen in the nebulosity or in the magnitude of the stars which apparently compose them, their real union and construction must remain unknown. We can only hint, that every nebulosity which is carried into the region of a small patch of stars will probably be gradually arrested and absorbed by them, and that thus the *growth of stars* may be continued.

8. *Of objects of an ambiguous construction.*

From objects consisting decidedly of stars, but which either have nebulosity mixed with them, or are in such situations as to be seen in the same line with nebulosity, I proceed to give an account of some others, of which my observations have not ascertained into what order we ought to class them.

It has been remarked, on a former occasion, that clusters of stars, when they are at a great distance, may assume a nebulous appearance.¶ This may be experienced by observing a certain celestial object with a telescope of an inferior space-penetrating power, through which it will be seen as a nebula; whereas with an instrument which has a higher degree of this power, its appearance will be a mixture of nebulosity and stars; and if this power of the telescope is of a still higher order, the stars of the same object will then be distinctly perceived: the

* *Phil. Trans.* for 1811, p. 320. [Above, p. 488.]
† [N.G.C. 2316.]
‡ [N.G.C. 7186.]
§ [N.G.C. 4183.]
|| [N.G.C. 7129.]
¶ *Phil. Trans.* for 1811, page 270. [Above, p. 459.]

nebulosity will no longer be seen, and the object will be entitled to be placed into the rank of clusters of stars.

Other objects there are, where a greater space-penetrating power will only increase the brightness of the nebulosity, and at the same time make the tinge of it more uniformly united and of a milky appearance, which will decide it to be purely nebulous.

But when an object is of such a construction, or at such a distance from us, that the highest power of penetration, which hitherto has been applied to it, leaves it undetermined whether it belong to the class of nebulæ or of stars, it may then be called ambiguous. As there is, however, a considerable difference in the ambiguity of such objects, I have arranged 71 of them into the following four collections.*

The first contains seven objects that may be supposed to consist of stars, but where the observations hitherto made, of either their appearance or form, leave it undecided into which class they should be placed.

Connoiss. 31 † is "A large nucleus with very extensive nebulous branches, but the nucleus is very gradually joined to them. The stars which are scattered over it appear to be behind it, and seem to lose part of their lustre in the passage of their light through the nebulosity; there are not more of them scattered over the nebula than there are over the immediate neighbourhood. I examined it in the meridian with a mirror 24 inches in diameter, and saw it in high perfection; but its nature remains mysterious. Its light, instead of appearing resolvable with this aperture, seemed to be more milky."

The objects in this collection must at present remain ambiguous.

The next contains 26 nebulous objects, of which the figure has been ascertained to be round or nearly round.

II. 101 ‡ is "A pretty large, round, extremely faint, easily resolvable nebula. I can almost see the stars in it." See fig. 11.

Connoiss. 57 § is "An oval nebula with an eccentric oval dark space in the middle; there is a strong suspicion of its consisting of stars. The diameter, measured by the large 10 feet telescope, is 1′ 28″·3."

The globular form of the objects in this collection, which is deduced from their round figure, will so far ascertain the manner of their construction, that they must either be still in a condensed state purely nebulous, or else, if consisting of

* See seventy-one ambiguous objects, in four collections.

First collection II. 400. III. 379, 693, 745. V. 2. *Connoiss.* 1, 31.

Second collection I. 46, 50. II. 27, 78, 79, 180, 195, 199, 207, 554, 609, 771, 822, 850, 855. III. 3, 101, 239, 399, 455, 696, 725, 743. IV. 22. *Connoiss.* 57, 70.

Third collection I. 44, 47. II. 47, 48, 76, 105, 202, 279, 283, 469, 473, 500, 608, 808. III. 47, 53, 55, 134, 288, 580, 747, 910. V. 1. VI. 38. *Connoiss.* 81. 82.

Fourth collection I. 52, 103, 122, 249, 288. II. 4, 84, 584. V. 3. VI. 15, 20. *Connoiss.* 100

† [N.G.C. 224, the great nebula in Andromeda.]

‡ [N.G.C. 3489.]

§ [N.G.C. 6720, the nebula in Lyra.]

stars, that they must be already in a far advanced order of compression, and only appear nebulous on account of their very great distance from us. A middle state between the progressive condensation of a globular nebula and a cluster of stars can have no existence; because a globular nebulosity when condensed can only produce a single star. There is, however, a possibility that a mass of nebulous matter in motion may be intercepted by a globular cluster, in which case the nebulosity must soon assume the form of the cluster, and will finally be absorbed by it.

In the third collection I have placed 26 nebulæ, which not only are described as easily resolvable, but in most of which some stars have actually been seen.

II. 500 * is " A very large, easily resolvable, extended, nebulous object. I see a few of the largest stars in it." See fig. 12.

Here the uncertainty in which the descriptions leave us, is that the objects in this collection may be either clusters of stars mixed with nebulosity, or that in consequence of the great distance and compression of the small stars composing a cluster which contains no nebulosity, it may put on the nebulous appearance.

The fourth collection contains 12 nebulous objects, of which the description makes it probable that they belong to the order of clusters of stars.

I. 249 † is " A considerably bright extended nebula about 4′ long and 2′ broad; it is easily resolvable, and I suppose with a higher power and longer attention the stars would become visible. It is brighter about the middle."

Connoiss. 100 ‡ is " A nebula of about 10′ in diameter, but there is in the middle of it, a small, bright cluster of supposed stars."

9. *Of the sidereal part of the Heavens.*

The foregoing observations have proved the intimate connection between the nebulous and sidereal condition; and although in passing from one to the other we have met with a number of ambiguous objects, it has been seen that the apparent uncertainty of their construction is only the consequence of the want of an adequate power in our telescopes, to shew them of their real form. We have indeed no reason to expect that an increase of light and distinctness of our telescopes would free us from ambiguous objects; for by improving our power of penetrating into space, and resolving those which we have at present, we should probably reach so many new objects that others, of an equally obscure construction, would obtrude themselves, even in greater number, on account of the increased space of the more distant regions of their situation.

From stars mixed with nebulosity we are now to direct our attention to the purely sidereal part of the heavens; and as stars are the elementary parts of sidereal constructions, it will be proper to review what we know of their nature.

* [N.G.C. 4535.] † [N.G.C. 2742.] ‡ [N.G.C. 4321.]

Having already entered upon this subject in a former paper at some length,* I shall only give a few additional observations, with a summary outline of the former arguments.

The intensity of the light of a star of the first magnitude may be compared with solar light, by considering, that if the sun were removed to the distance at which we generally admit the brightest stars to be from us, its visible diameter could not exceed the 215th part of a second; and its appearance therefore would probably not differ much from the size and brightness of such stars. By reversing this argument we shall be authorised to conclude, from analogy, that stars, were they near enough, would assume the brightness, and some of them perhaps also the size, of the sun; and the consequences that have been drawn from the observations given in my paper on the nature and construction of the sun, may be legitimately applied to the stars; whence it follows that stars, although surrounded by a luminous atmosphere, may be looked upon as so many opaque, habitable, planetary globes; differing, from what we know of our own planets, only in their size, and by their intrinsically luminous appearance.

They also, like the planets, shine with differently coloured light. That of Arcturus and Aldebaran for instance, is as different from the light of Sirius and Capella, as that of Mars and Saturn is from the light of Venus and Jupiter. A still greater variety of coloured star-light has already been shewn to exist in many double stars, such as γ Andromedæ, β Cygni, and many more.† In my sweeps are also recorded the places of 9 deep garnet, 5 bright garnet, and 10 red coloured stars, of various small magnitudes from the 7th to the 12th.

By some experiments, on the light of a few of the stars of the first magnitude, made in 1798, by a prism applied to the eye-glasses of my reflectors, adjustable to any angle and to any direction, I had the following analyses.

The light of Sirius consists of red, orange, yellow, green, blue, purple, and violet.

α Orionis contains the same colours, but the red is more intense, and the orange and yellow are less copious in proportion than they are in Sirius.

Procyon contains all the colours, but proportionally more blue and purple than Sirius.

Arcturus contains more red and orange and less yellow in proportion than Sirius.

Aldebaran contains much orange, and very little yellow.

α Lyræ contains much yellow, green, blue, and purple.

The similarity of the general construction of the sun, the stars, and the planets, is also much supported by the periodical variations of the light of the stars observed

* *Phil. Trans.* for 1795, page 68 [Vol. I. p. 482].

† See Catalogue of double stars *Phil. Trans.* for 1782, III. 5. V. 5, &c.

in many of them; * for these variations can only be satisfactorily accounted for by admitting such stars to have a rotatory motion on their axes, like that which the sun and the planets are known to have.†

10. *Of the aggregation of Stars.*

That stars are not spread in equal portions over the celestial regions is evident to the eye of every one who directs his view to them in a clear night; but if this wanted any proof, the star-gages I have given in the *Phil. Trans.* for 1785, would abundantly shew that the greatest variety in their distribution takes place; for while in my sweeps many fields of view of the telescope were without a single star, others contained every assignable number, from one to more than six hundred.

In my examination of the heavens, I remarked that in many places there were patches of stars of such a particular appearance that I was induced to call them *forming* clusters. This expression was however only used to denote that some peculiar arrangement of stars in lines making different angles, directed to a certain aggregation of a few central stars, suggested the idea that they might be in a state of progressive approach to them. This tendency to clustering seemed chiefly to be visible in places that were extremely rich in stars. In order therefore to investigate the existence of a clustering power, we may expect its effects to be most visible in and near the milky way, and it is for this purpose I have distinguished the relative situation of the clusters to which I refer.‡

Connoiss. 6 § is "A cluster of stars of various sizes containing several lines that seem to be drawing to a centre like a forming cluster."

VIII. 35 || is "A large cluster of stars considerably compressed and rich; some of the stars are arranged in a long crooked line."

VIII. 44 ¶ is "A very coarsely scattered cluster of large stars; they form a cross and extend over a large space; not rich." See fig. 13. The stars about the cluster belong to the milky way.

VIII. 83 ** is "A cluster of scattered stars above 15′ in diameter; pretty rich and joining to the milky way, or a projecting part of it."

The 20 objects here referred to are not given as instances of the actual formation of clusters, which, being an effect that must undoubtedly require much time, cannot be visible; but merely to draw our attention to a seemingly aggregating arrangement of the stars, which must render it probable that in regions where stars are very numerous, but unequally scattered, a clustering of them may arise from the preponderating attractions residing in different places.

* See Mr. PIGOTT's Catalogue of variable stars *Phil. Trans.* for 1786, page 191.

† See Remarks on the rotatory motion of stars on their axes, *Phil. Trans.*, 1796, p. 456 [Vol. I. p. 561].

‡ See twenty clusters of stars; fifteen in the milky way VII. 40, 45. VIII. 16, 18, 35, 36, 42, 47, 50, 56, 60, 61, 64, 67. *Connoiss.* 6; and five near it VIII. 8, 40, 41, 44, 83.

§ [N.G.C. 6405.] || [N.G.C. 2374.] ¶ [N.G.C. 2394.] ** [N.G.C. 6895.]

11. *Of irregular Clusters.*

When clusters of stars are situated in very rich parts of the heavens, they are generally of an irregular form and very imperfectly collected; those which are in, and very near the milky way, may indeed be looked upon as so many portions of the great mass drawn together by the action of a clustering power, of which they tend to prove the existence.

I have divided the following 112 objects into two collections. The first of them contains 80 clusters of which the size or figure has not been particularized.*

VIII. 4 † is "A cluster of coarsely and irregularly scattered, pretty large stars, of nearly one size and colour." See fig. 14.

The stars of these clusters are in general very promiscuously scattered; they are however sufficiently drawn together to shew that they form separate groups; and in many places a defalcation of the number of stars surrounding the clusters is already so far advanced as to indicate a tendency to future insulation.

The second collection contains 32 irregular clusters that are from 2 to 30′ in diameter.‡

VII. 4 § is "A cluster of large stars about 20 or 25′ in diameter, considerably rich; it is of a coarsely circular figure."

The great number of clusters in these two collections is not only an indication that they owe their origin to a clustering power residing in the stars about their center; but the still remaining irregularity of their arrangement additionally proves that the action of the clustering power has not been exerted long enough to produce a more artificial construction. The length of time required for this purpose must, however, greatly depend upon the original situation of the stars exposed to the clustering power.

12. *Of Clusters variously extended and compressed.*

The outlines of clusters of stars in rich parts of the heavens, and even of those that are insulated, are seldom sufficiently defined to arrange such clusters by their figure; and as the following assortment contains some that are variously extended and differently compressed, it will be seen, from the descriptions of a few of them,

* See eighty irregular clusters of stars, of various unascertained sizes, fifty-three in the milky way; VII. 5, 35, 36, 42, 50, 62, 67. VIII. 4, 5, 6, 13, 15, 19, 20, 21, 22, 25, 27, 28, 30, 31, 33, 34, 37, 45, 46, 51, 52, 54, 55, 57, 58, 59, 63, 72, 76, 79, 82, 84, 85, 86, 87. *Connoiss.* 7, 8, 16, 18, 21, 24, 25, 26, 29, 36, 38. Eighteen near the milky way; VII. 6, 15, 46. VIII. 2, 11, 23, 43, 49, 62, 65, 68, 69, 73. *Connoiss.* 20, 34, 39, 41, 48: nine at a distance from it; VII. 3, 54. VIII. 7, 10, 29, 71. *Connoiss.* 44, 45, 73.

† [N.G.C. 1896.]

‡ See thirty-two irregular clusters from 2 to 30′ in diameter; twenty-two in the milky way; VI. 23. VII. 10, 12, 30, 52. VIII. 9, 12, 14, 17, 26, 32, 39, 48, 53, 70, 74, 77, 78, 80, 81. *Connoiss.* 23, 93. Ten near the milky way VI. 39. VII. 4, 11, 13, 14, 16, 32, 66. VIII. 66, 88.

§ [N.G.C. 1817.]

that the power which has drawn the stars together must have acted under different circumstances.*

VI. 3 † is " A cluster of very compressed, extremely small stars, containing a few large ones. It is of an extended figure, and, as it were, divided."

In this cluster the observed partial division points out the cause of its being extended, which may be ascribed to a double seat of preponderating attractions at some little distance from each other.

VI. 24 ‡ is " A very rich cluster of extremely small and very compressed stars ; it is about 6′ long and 4′ broad."

Here the stars of the cluster are not only much compressed, but the borders of it are moreover sufficiently determined to shew the limits of its extent ; from which we may infer that the cluster is advancing towards insulation, and that in the end a gradual concentration may bring it to a globular form.

VI. 36 § is " A very compressed cluster of small, and some large stars ; extended nearly in the meridian ; the most compressed part is about 8′ long and 2′ broad, with many stars scattered around it to a considerable distance." See fig. 15.

The construction of this cluster may have arisen from the situation of many stars in the same plane, drawn towards a centre by the clustering power, for any plane seen obliquely will have the appearance of an extended form.

VII. 64 || is " A large cluster of stars, of a middling size, irregularly extended, and considerably rich. The stars are chiefly in rows."

Here each row of stars may have a different preponderating attraction, but every row will attract all the other rows ; nay, from the laws of gravitation it is evident that there must be somewhere in all the rows together the seat of a preponderating clustering power, which will act upon all the stars in the neighbourhood.

13. *Of Clusters of Stars of a peculiar description.*

The great variety of ways in which the attractions of unequally scattered stars may produce a clustering power will be further exemplified in the following objects.¶

VII. 55 ** is " A pretty compressed cluster of very small stars ; it is of an irregular figure, and has a vacancy in the middle."

This appearance may be accounted for by supposing, for instance, three, four, or a greater number of preponderating attracting centres near each other, situated so as to inclose a certain space, the stars in which, then, cannot be accumulated,

* See fifteen extended clusters of stars ; twelve in the milky way II. 198. VI. 3, 14, 24, 36. VII. 18, 19, 27, 41, 44, 56. VIII. 3. And three near the milky way VII. 29, 64. VIII. 75.

† [N.G.C. 2269.] ‡ [N.G.C. 7044.] § [N.G.C. 2432.] || [N.G.C. 2567.]

¶ See six clusters of stars of a peculiar description ; one in the milky way VIII. 24. Three near the milky way VII. 26, 55. *Connoiss.* 4 ; and two at a distance from it VII. 1. *Connoiss.* 30.

** [N.G.C. 7762.]

while the clustering power arising from the combined attractions will be exerted on the surrounding stars.

Connoiss. 4 * is "A rich cluster of considerably compressed small stars surrounded by many straggling ones. It contains a ridge of stars running through the middle from south preceding to north following. The ridge contains 8 or 10 pretty bright stars. All the stars are red."

The curious construction of this cluster is sufficiently accounted for by the bright stars in what is called a ridge; the small stars accumulated about it having somewhat the appearance of the shelving sides of the ridge. The observed red colour was probably owing to the low situation of the object.

VII. 26 † is "A cluster of extremely small and pretty much compressed stars, with a few large ones in the shape of a hook."

From what has been remarked already it will not be necessary to enter into a consideration of the cause of the uncommon form of this cluster.

Connoiss. 30 ‡ is "A brilliant cluster, the stars of which are gradually more compressed in the middle. It is insulated, that is, none of the stars in the neighbourhood are likely to be connected with it. Its diameter is from 2′ 40″ to 3′ 30″. Its figure is irregularly round. The stars about the centre are so much compressed as to appear to run together. Towards the north, are two rows of bright stars 4 or 5 in a line."

In this accumulation of stars, we plainly see the exertion of a central clustering power, which may reside in a central mass, or, what is more probable, in the compound energy of the stars about the centre. The lines of the bright stars, although by a drawing made at the time of observation, one of them seems to pass through the cluster, are probably not connected with it.

14. *Of differently compressed Clusters of Stars.*

I have hitherto only considered the arrangement of stars in clusters with a view to point out that they are drawn together by a clustering power, in the same manner as the nebulous matter has, in my former paper, been proved to be condensed by the gravitating principle; but in the 41 clusters of the following two collections we shall see that it is one and the same power uniformly exerted which first condenses nebulous matter into stars, and afterwards draws them together into clusters, and which by a continuance of its action gradually increases the compression of the stars that form the clusters.

The first collection contains 33 clusters, the stars of which are considerably compressed.§

* [N.G.C. 6121.] † [N.G.C. 2225.] ‡ [N.G.C. 7099.]

§ See thirty clusters of considerably compressed stars; seventeen in the milky way VI. 16, 29, 33, 34. VII. 2, 7, 9, 22, 23, 33, 43, 65. VIII. 38. *Connoiss.* 11, 35, 50, 103. Fifteen near the milky way VI. 6, 22, 42. VII. 12, 17, 20, 21, 24, 34, 47, 57, 58, 59, 63. VIII. 1; and one at a distance from it *Connoiss.* 67.

VII. 12 * is "A beautiful cluster of pretty compressed stars, near half a degree in diameter. It is considerably rich, and most of the stars are of the same size."

The moderate compression of the stars in the clusters of this order renders them fine objects for good telescopes.

The second collection contains 8 clusters in which the compression of the stars is carried to a much higher degree.†

VI. 30 ‡ is "A very beautiful rich cluster of very compressed small stars."

The clusters in this collection are also fine objects; but, on account of their higher compression, require superior telescopes.

15. *Of the gradual concentration and insulation of Clusters of Stars.*

The existence of a clustering power is nowhere so visibly pointed out as in the 39 clusters which are given in the following collection.§ My remarks upon them will come with more clearness when applied to a particular description of some of them.

VI. 5 ‖ is "A beautiful cluster of very compressed small stars of several sizes. It is of an irregular round form, about 12 or 15′ in diameter, and the stars are gradually most compressed in the middle."

Here the gradually increasing compression of the stars points out the central situation of the clustering power; the form is also that of a solid, not much differing from a globular figure; and by the outline of the cluster we may consider it as already in an advanced state of insulation; from these circumstances we may therefore conclude that this cluster has been long under the influence of the clustering power. See fig. 16.

Connoiss. 68 ¶ is "A beautiful cluster of stars, extremely rich, and so compressed that most of the stars are blended together; it is near 3′ broad and about 4′ long, but chiefly round, and there are very few scattered stars about."

This oval cluster is also approaching to the globular form, and the central compression is carried to a high degree. The insulation is likewise so far advanced that it admits of an accurate description of the contour.

The clusters of this class are beautiful, but can hardly be seen to any advantage without a 20 feet telescope.

* [N.G.C. 2360.]

† See eight clusters of very compressed stars; five in the milky way VI. 27, 30. VII. 8. *Connoiss.* 22, 46. Two near the milky way VI. 10. VII. 48; and one at a distance from it VI. 4.

‡ [N.G.C. 7789.]

§ See thirty-nine clusters of gradual concentration; twenty-one in the milky way VI. 5, 13, 17, 18, 25, 26, 28, 32. VII. 25, 28, 31, 37, 38, 39, 51, 60, 61. *Connoiss.* 28, 37, 52, 71. Seven near the milky way VI. 2, 21, 31, 37, 40. VII. 49, 53; and eleven at a distance from it I. 41. IV. 63. VI. 1, 8, 9, 19. *Connoiss.* 33, 55, 68, 74, 77.

‖ [N.G.C. 2194.]

¶ [N.G.C. 4590.]

16. *Of globular Clusters of Stars.*

The objects of this collection are of a sufficient brightness to be seen with any good common telescope, in which they appear like telescopic comets, or bright nebulæ, and under this disguise, we owe their discovery to many eminent astronomers; but in order to ascertain their most beautiful and artificial construction, the application of high powers, not only of penetrating into space but also of magnifying are absolutely necessary; and as they are generally but little known and are undoubtedly the most interesting objects in the heavens, I shall describe several of them, by selecting from a series of observations of 34 years some that were made with each of my instruments, that it may be a direction for those who wish to view them to know what they may expect to see with such telescopes as happen to be in their possession.*

Oct. 30, 1810. 40 feet telescope. Space-penetrating power 191·68. Magnifying power 280. "Having been about 20 minutes at the telescope to prepare the eye properly for seeing minute objects, the 72d of the *Connoissance des Temps* † came into the field. It is a very bright object."

"It is a cluster of stars of a round figure, but the very faint stars on the outside of globular clusters are generally a little dispersed so as to deviate from a perfect circular form. The telescopes which have the greatest light shew this best."

"It is very gradually extremely condensed in the centre, but with much attention, even there, the stars may be distinguished."

"There are many stars in the field of view with it, but they are of several magnitudes totally different from the excessively small ones which compose the cluster."

"It is not possible to form an idea of the number of stars that may be in such a cluster; but I think we cannot estimate them by hundreds."

"The diameter of the cluster is about $\frac{1}{8}$ of the field, which gives 1′ 53″·6." See fig. 17.

Sept. 4, 1799. 40 feet telescope, power 240. "I examined the 2d of the *Connoiss.* ‡ It appeared very brilliant and luminous."

"The scattered stars were brought to a good, well determined focus, from which it appears that the central condensed light is owing to a multitude of stars that appeared at various distances behind and near each other. I could actually see and distinguish the stars even in the central mass. The Rev. Mr. VINCE, Plumian Professor of Astronomy at Cambridge, saw it in the same telescope as described."

May 27, 1791. 40 feet telescope, power 370. "The 5th of the *Connoiss.*§ is a

* See fourteen globular clusters of stars. One in the milky way *Connoiss.* 19.
Four near the milky way *Connoiss.* 10, 12, 56, 80; and nine at a distance from it *Connoiss.* 2, 3, 5, 13, 15, 53, 72, 79, 92.

† [N.G.C. 6981.] ‡ [N.G.C. 7089.] § [N.G.C. 5904.]

beautiful cluster of stars; I counted about 200 of them; but the middle of it is so compressed that it is impossible to distinguish the stars." *

January 5, 1807 20 feet telescope. Space-penetrating power 75·08. Magnifying power 157·3. "The 56th of the *Connoiss.*† is a globular cluster of very compressed and very small stars. They are gradually more compressed towards the centre."

May 26, 1786. 20 feet telescope. "The 80th of the *Connoiss.*‡ is a beautiful, round cluster of extremely minute and very compressed stars about 3 or 4′ in diameter; by the increasing compression of the stars the cluster is very gradually much brighter in the middle."

May 16, 1787. 20 feet telescope. "The 13th of the *Connoiss.*§ is a most beautiful cluster of stars. It is exceedingly compressed in the middle and very rich. The most compressed part of it is round and is about 2 or 2½′ in diameter, the scattered stars which belong to it extend to 8 or 9′ in diameter, but are irregular." ||

Sept. 24, 1810. Large 10 feet Newtonian telescope. Space-penetrating power 75·82. Magnifying powers 71, 108, 171, 220. "The 3d of the *Connoiss.*¶ is one of the globular clusters; very brilliant and beautiful. The compression of the stars begins to increase pretty suddenly from the outside at ¾ of the radius, and continues gradually up to the centre, its diameter taking in the outside is full half of the field of the glass magnifying 171 times, which gives 4′ 30″."

Nov. 23, 1805. Large 10 feet. "The 15th of the *Connoiss.*** is perfectly round, and insulated. The accumulation of the stars towards the centre is more sudden than the 13th of the *Connoiss.* and the scattered stars extend proportionally much farther. Its diameter is ⅙ of the field of the glass which magnifies 108 times, that is to say 4′ 0″. It passes the wire in 13″·0 of time which by calculation gives only 2′ 11″·3, but I rely more on the estimation by the known field of view which is 24′ 0″; because the limits of the cluster cannot be properly fixed upon for a transit."

* A 40 feet telescope should only be used for examining objects that other instruments will not reach. To look through one larger than required is loss of time, which, in a fine night, an astronomer has not to spare; but it ought to be known that the opportunities of using the 40 feet reflector are rendered very scarce by two material circumstances. The first is the changeable temperature of the atmosphere, by which the mirror is often covered with the condensation of vapour upon its surface, which renders it useless for many hours; and in cold weather by freezing upon it for the whole night, and even for weeks together; for the ice cannot be safely taken off till a general thaw removes it. The next is that, with all imaginable care, the polish of a mirror exposed like that in the 40 feet telescope, though well covered up, will only preserve its required lustre and delicacy about two years. The three observations I have given must consequently be looked upon as having been made by three different mirrors; but if we will have superior views of the heavens, we must submit to circumstances that cannot easily be altered.

† [N.G.C. 6779.] ‡ [N.G.C. 6093.] § [N.G.C. 6205.]

|| The 20 feet telescope, on account of the moderate weight of the mirror and the proportionally long wooden tube, has the great advantage that with proper precaution it may be used in any temperature. Sometimes, however, a sudden change from cold to heat towards morning has put a stop to the observations of the night. The mirror will also preserve an excellent polish for several years; and having a second one ready to supply the place of that which is in use the instrument may always be ready for observation.

¶ [N.G.C. 5272.] ** [N.G.C. 7078.]

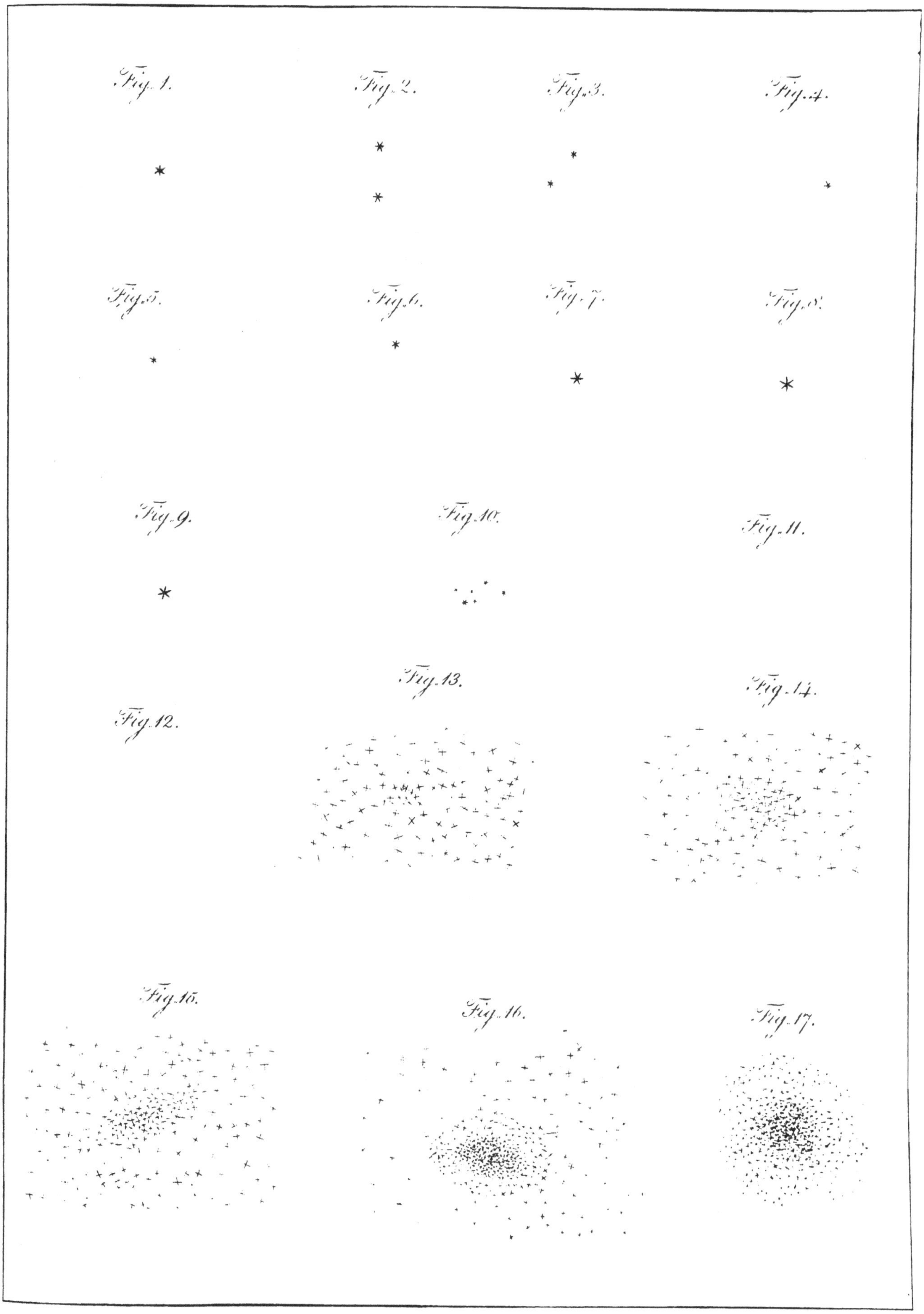

Fig. 1.

Fig. 2.

Fig. 3.

Fig. 4.

Fig. 5.

Fig. 6.

Fig. 7.

Fig. 8.

Fig. 9.

Fig. 10.

Fig. 11.

Fig. 12.

Fig. 13.

Fig. 14.

Fig. 15.

Fig. 16.

Fig. 17.

Jan. 13, 1806. Large 10 feet. "The 79th of the *Connoiss.** is a cluster of stars of a globular construction, and certainly extremely rich. Towards the centre the stars are extremely compressed, and even a good way from it. With 171 the diameter is a little less than $\frac{1}{3}$ of the field, and with 220 a little more; the field of one being 9′ 0″, and of the other 8′ 0″, a mean of both gives the diameter of the cluster 2′ 50″, but I suppose that the lowness of the situation prevents my seeing the thinly scattered stars, so that this cluster is probably larger than it appears." †

Common 10 feet telescope. Space-penetrating power 28·67. "When the 19th of the *Connoiss.*‡ is viewed with a magnifying power of 120, the stars are visible; the cluster is insulated; some of the small stars scattered in the neighbourhood are near it; but they are larger than those belonging to the cluster. With 240 it is better resolved, and is much condensed in the center. With 300 no nucleus or central body can be seen. The diameter with the 10 feet is 3′ 16″, and the stars in the centre are too accumulated to be separately seen."

7 feet telescope, space-penetrating power 20·25. "The 53d of the *Connoiss.*§ with 118 is easily resolvable, and some of the stars may be seen."

It will not be necessary to add that the two last mentioned globular clusters, viewed with more powerful instruments, are of equal beauty with the rest; and from what has been said it is obvious that here the exertion of a clustering power has brought the accumulation and artificial construction of these wonderful celestial objects to the highest degree of mysterious perfection.

17. *Of more distant globular Clusters of Stars.*

The objects contained in this assortment are so like those of the foregoing collection that in my observations I have called them miniatures of the former. Small instruments cannot reach them, I shall therefore describe them as they appear when proper powers are applied to them.||

VI. 35 ¶ is "A cluster of very faint exceedingly compressed stars, about one minute in diameter. It is the next step to an easily resolvable nebula."

VI. 11 ** is "A cluster of stars about $1\frac{1}{2}$ or 2 minutes in diameter. It is a good miniature of the 19th of the *Connoiss.* not only with respect to the size of the

* [N.G.C. 1904.]

† The large 10 feet telescope is in a considerable degree subject to the obstructions arising from change of temperature, and tarnish; but as it can be directed to any part of the heavens in a few minutes, and is easily prepared for observation, it becomes a very useful instrument when the clearness of the atmosphere is interrupted by flying clouds; or when the place of an object not visible in the finder, or night-glass, is to be ascertained.

‡ [N.G.C. 6273.]

§ [N.G.C. 5024.]

|| See eleven miniature globular clusters of stars, five in the milky way VI. 11, 12, 35. *Connoiss.* 9, 62. One near the milky way *Connoiss.* 14; and five at a distance from it I. 78. III. 709. VI. 7, 41. *Connoiss.* 75.

¶ [N.G.C. 136.]

** [N.G.C. 6284.]

cluster, but also with regard to the mutual distance and the reduced magnitude of the stars of which it consists."

Connoiss. 9 * is "A cluster of very compressed and extremely small stars. It is a miniature of the 53d."

Connoiss. 14 † is "Like an extremely bright, easily resolvable round nebula; but with a power of 300 I can see the stars of it. It resembles the 10th of the *Connoiss.* which probably would put on the same appearance as this, were it removed half its distance farther from us. The stars are much condensed in the middle."

Connoiss. 62 ‡ is "Extremely bright, round, very gradually brighter in the middle, easily resolvable, about 4′ in diameter. With 240 and strong attention I see the stars of it. It is a miniature of the 3d of the *Connoiss.*"

I. 78 § is "Very bright, suddenly much brighter in the middle, round, about 3′ in diameter. I take it to be a cluster of stars, as it seems to be a miniature of the 2d of the *Connoiss.*"

III. 709 ‖ is "Very faint, round, very gradually brighter in the middle; about 2½ minutes in diameter." A later observation says "I can perceive some of the stars."

Connoiss. 75 ¶ is "A globular cluster of stars, and is a miniature of the third."

I have supposed the clusters of this class to be at a greater distance from us than those of the preceding collection, because the stars of which they are composed are more minute than those of the clusters of which I have called them miniatures; their compression is also closer, and the size of the whole is much contracted, all which particulars are readily explained by admitting them to be more distant. This argument, however, does not extend so far as to exclude a real difference which there may be in different clusters, not only in the size but also in the number and arrangement of the stars.

18. *Of still more distant globular Clusters of Stars.*

It has already been shewn in the 8th article, that when our telescopes have extended vision as far as they can reach with distinctness, they will still shew objects at a greater distance if they are sufficiently bright to be seen, although we should not be able to ascertain exactly into what class we ought to place them; but as it frequently happened that I saw three objects in succession, the first of which was a brilliant globular cluster of stars, the second a miniature of the former of which the stars could but just be perceived, and the third in every respect a similar miniature of the second as the second was of the first, but in which the stars,

* [N.G.C. 6333.] † [N.G.C. 6402.] ‡ [N.G.C. 6266.]
§ [N.G.C. 2985.] ‖ [N.G.C. 2500.] ¶ [N.G.C. 6864.]

though suspected, were no longer to be distinguished, I called them second miniature globular clusters. The following collection contains five of them.*

I. 45 † is " A bright round nebula, much brighter in the middle, but the brightness decreasing very gradually. It is a perfect miniature of VI. 12,‡ which is itself a miniature cluster of the 19th of the *Connoiss.*"

I. 48 § is " A miniature of the 9th of the *Connoiss.*|| (which is itself a miniature of the 53d ¶). " I suppose if I had looked long enough, I might have perceived some of the stars which compose it."

I. 147 ** is " A miniature of the 62d †† of the *Connoiss.* which is a miniature of the 3d." ‡‡

I. 51 §§ and *Connoiss.* 69 |||| are second miniatures of the 53d.¶¶.

19. *Of a recurrence of the ambiguous limit of observation.*

In the 16th article I have given a description of the most magnificently constructed sidereal systems; and very little doubt can be entertained but that the objects of the 17th and 18th articles are of the same nature, and are only less beautiful in their appearance as they are gradually more remote. It has already been shewn in the 8th article, that in passing from faint nebulosity to the suspected sidereal condition, we cannot avoid meeting with ambiguous objects, to which I must now add, that the same critical situation will again occur, when from the distinctly sidereal appearance we endeavour to penetrate gradually farther into space. In consequence of this remark, it seems probable that among the numerous globular nebulæ which have been given in my last paper, many beautiful clusters of stars may lie concealed. To this we may add, that several of the great number of objects which have been given as stellar nebulæ, and are probably at a still greater distance from us, may be the last glimpses we can have of such clusters of stars as the 77th of the *Connoissance des Temps*, which will nearly put on the stellar appearance when it is viewed in a very good common telescope.

This ambiguity, however, being the necessary consequence of the faintness or distance of objects, when seen through telescopes that are not sufficiently powerful to shew them as they are, will not affect any of the arguments that have been used to establish the existence of a clustering power, the effects of which have gradually been traced from the first indication of clustering stars, through irregular as well as through more artificially arranged clusters, up to the beautiful globular form.

The extended views I have taken, in this and my former papers, of the various parts that enter into the construction of the heavens, have prepared the way for

* See five second miniature globular clusters of stars in the milky way I. 45, 48, 51, 147. *Connoiss.* 69.

† [N.G.C. 6316.] ‡ [N.G.C. 6293.] § [N.G.C. 6356.]
|| [N.G.C. 6333.] ¶ [N.G.C. 5024.] ** [N.G.C. 6304.]
†† [N.G.C. 6266.] ‡‡ [N.G.C. 5272.] §§ [N.G.C. 6638.]
|||| [N.G.C. 6637.] ¶¶ [N.G.C. 5024.]

a final investigation of the universal arrangement of all these celestial bodies in space; but as I am still engaged in a series of observations for ascertaining a scale whereby the extent of the universe, as far as it is possible for us to penetrate into space, may be fathomed, I shall conclude this paper by pointing out some inferences which the continuation of the action of the clustering power enables us to draw from the observations that have been given.

20. *Of the breaking up of the milky way.*

The milky way is generally represented in astronomical maps as an irregular zone of brightness encircling the heavens, and my star gages have proved its whitish tinge to arise from accumulated stars, too faint to be distinguished by the eye. The great difficulty of giving a true picture of it is a sufficient excuse for those who have traced it on a globe, or through the different constellations of an *Atlas Cœlestis*, as if it were a uniform succession of brightness. It is, however, evident that, if ever it consisted of equally scattered stars, it does so no longer; for, by looking at it in a fine night, we may see its course between the constellations of Sagittarius and Perseus affected by not less than eighteen different shades of glimmering light, resembling the telescopic appearances of large easily resolvable nebulæ; but in addition to these general divisions, the observations detailed in the preceding pages of this paper, authorise us to anticipate the breaking up of the milky way, in all its minute parts, as the unavoidable consequence of the clustering power arising out of those preponderating attractions which have been shewn to be everywhere existing in its compass.

One hundred and fifty-seven instances have been given of clusters situated within the extent of the milky way, and their places are referred to in nine preceding articles. They may also be found in BODE's *Atlas Cœlestis*, whose delineation of this bright zone I have taken for a standard. To these must be added 68 more, which are in the less rich parts, or what may be called the vanishing borders of the milky way: for this immense stratum of stars does not break off abruptly, as generally represented in maps, but gradually becomes invisible to the eye when the stars are no longer sufficiently numerous to cause the impression of milkiness.

Now, since the stars of the milky way are permanently exposed to the action of a power whereby they are irresistibly drawn into groups, we may be certain that from mere clustering stars they will be gradually compressed through successive stages of accumulation, more or less resembling the state of some of the 263 objects by which, in the tenth and six succeeding articles, the operation of the clustering power has been laid open to our view, till they come up to what may be called the ripening period of the globular form, and total insulation; from which it is evident that the milky way must be finally broken up, and cease to be a stratum of scattered stars.

We may also draw a very important additional conclusion from the gradual dissolution of the milky way; for the state into which the incessant action of the clustering power has brought it at present, is a kind of chronometer that may be used to measure the time of its past and future existence; and although we do not know the rate of going of this mysterious chronometer, it is nevertheless certain, that since the breaking up of the parts of the milky way affords a proof that it cannot last for ever, it equally bears witness that its past duration cannot be admitted to be infinite.

LXVII.

A series of observations of the satellites of the Georgian planet, including a passage through the node of their orbits; with an introductory account of the telescopic apparatus that has been used on this occasion; and a final exposition of some calculated particulars deduced from the observations.

[*Phil. Trans.*, 1815, pp. 293–362.]

Read June 8, 1815.

THE observations of the satellites of the Georgian planet, of which an account is given in this paper, are of such a nature that, in order to judge of them properly, and to make them useful to those who would continue them, it will be necessary to enter into some particulars relating to the telescopic powers required for critically viewing such difficult objects.

The great distance of the Georgian planet renders an attempt to investigate the movements of its satellites a very arduous undertaking; for their light, having to traverse a space of such vast extent before it can reach us, is so enfeebled, and their apparent diameter so diminished, that an instrument, to be prepared for viewing them, must be armed with the double power of magnifying and of penetrating into space.

With regard to the first of these requisites, I have already shown in a former Paper,* that the magnifying power of my ten feet telescope, when no uncommon degree of light is wanting, is fully equal to what may be required to view extremely small objects; but this branch of the properties of optical instruments seems not to be generally understood: the question how much a telescope magnifies, admits of various answers. To resolve it properly, we ought in all circumstances to consider how far the magnifying power of a telescope is supported by an adequate quantity of light; as without it, even the highest power and distinctness cannot be *efficient*. The question therefore ought to be limited to an inquiry into the extent of what may be called the *effective* magnifying power. It will however be found, that even then, the quantity of this power cannot be positively assigned. For if a card containing engraved letters of a certain size be put up at a given distance, the effective power of a telescope directed to it, will be that wherewith we can read these letters with the greatest facility; but if either the size of the letters, or their

* *Phil. Trans.* for 1805, page 31. [Above, p. 297.]

distance from the telescope, be changed, the quantity of this power will no longer remain the same.

An obvious consequence of this consideration is, that the effective power of telescopes has a considerable range of extent, and can only be assigned when the object to be viewed is given; and that in this determination two circumstances are concerned, which require a separate investigation; and this is abundantly confirmed when a ten feet reflector, such as has been mentioned, is directed to the Georgian planet; for with none of its highest powers can we possibly ascertain even the existence of the satellites.

Since, then, it is absolutely necessary that the power of magnifying should be accompanied with a sufficient quantity of light, to reach the satellites of this remote planet, it may be useful to cast an eye upon the action of a power which is become so essential. Its advantages and its inconveniences must equally be objects of consideration.

A very material inconvenience is that mirrors, which must be large in order to grasp much light, must also be of a great focal length; and that in consequence of this, we must submit to be incumbered with a large apparatus, which will require an assistant at the clock and writing desk, and also an additional person to work the necessary movements. The machinery of my twenty feet telescope is however so complete, that I have been able to take up the planet at an early hour in the evening, and to continue the observations of its own motion, together with that of its satellites, for seven, eight, or nine hours successively.

The forty feet telescope having more light than the twenty feet, it ought to be explained why I have not always used it in these observations. Of two reasons that may be assigned, the first relates to the apparatus and the nature of the instrument. The preparations for observing with it take up much time, which in fine astronomical nights is too precious to be wasted in mechanical arrangements. The temperature of the air for observations that must not be interrupted, is often too changeable to use an instrument that will not easily accommodate itself to the change: and since this telescope, besides the assistant at the clock and writing desk, requires moreover the attendance of two workmen to execute the necessary movements, it cannot be convenient to have every thing prepared for occasional lucid intervals between flying clouds that may chance to occur; whereas in less than ten minutes, the twenty feet telescope may be properly adjusted and directed so as to have the planet in the field of view.

In the next place I have to mention, that it has constantly been a rule with me, not to observe with a larger instrument, when a smaller would answer the intended purpose. To use a manageable apparatus saves not only time and trouble, but what is of greater consequence, a smaller instrument may comparatively be carried to a more perfect degree of action than a larger one; because a mirror of less weight and diameter may be composed of a metal which will reflect more light than that of a larger one; it will also accommodate itself sooner to a change of temperature; and when it contracts tarnish, it may with less trouble be repolished; to which may be added, that having two mirrors for the twenty feet always ready, my observations could never be interrupted by accidents which often happen to large mirrors, such as greatly injure, or even destroy their polish.

The quantity of light reflected by the mirror of a twenty feet telescope of my construction being known, and the satellites of the Georgian planet being the objects to be viewed, I may now examine the combined powers of this instrument, and assign the limits to which they may be stretched. It will however be proper first, to point out from experience some of the advantages that may be taken, if not to increase, at least not to obstruct, the penetrating power, by the full effect of which the magnifying power is to be supported.

The first precaution I ought to give is, that in these delicate observations, no double eye glass should be used, as it cannot be prudent to permit the waste of light at four surfaces, when two will collect the rays to their proper focus. The hole through which they pass in coming to the eye, should be much larger than the diameter of the optic pencils, and considerably nearer the glass than their focus; for the eye ought on no account to come into contact with the eye piece; and a little practice will soon enable the observer to keep his eye in the required situation. It is hardly necessary to add, that no hand should touch the eye piece.

With regard to the eye glasses, when merely the object of saving light is considered, I can say from experience, that concaves have greatly the advantage of convexes; and that they give also a much more distinct image than convex glasses.

This fact I established by repeated experiments about the year 1776, with a set of concave eye glasses I had prepared for the purpose, and which are still in my possession. The glasses, both double and plano-concaves, were alternately tried with convex lenses of an equal focus, and the result, for brightness and distinctness, was decidedly in favour of the concaves.

For the cause of the superior brightness and sharpness of the image which is given by these glasses, we must probably look to the circumstance of their not permitting the reflected rays to come to a focus.

Perhaps a certain mechanical effect, considerably injurious to clearness and distinctness, takes place at the focal crossing of the rays, which is admitted in convex lenses.*

* About the same time that the experiments on concave eye glasses were made, I tried also to investigate the cause of the inferiority of the convex ones; and it occurred to me, that an experiment might be made to ascertain whether the rays of light in crossing, jostled against each other, or were turned aside from their right lined course by inflections or deflections. With a view to this, I directed a 10 feet telescope to some finely engraved letters put up at a convenient distance. A convex eye glass was fixed to a skeleton apparatus, which left the focal point freely exposed. A concave mirror was placed so as to throw the focus of the sun's rays upon the focal image of the telescope, where, meeting with no intercepting body, they would freely pass through it at right angles. Then a screen being placed to keep off the solar rays, I fixed my attention upon the letters viewed in the telescope,

I have occasionally availed myself of the light of concave eye glasses, but a great objection against their constant use is, that none of the customary micrometers can be applied to them, since they do not permit the rays to form a focal image. Their very small field of view is also a considerable imperfection; in observations, however, that do not require a very extensive field, such as double stars or the satellites of Saturn and the Georgian planet, this inconvenience is not so material.*

As I have already shown that the *effective* power of a telescope arises from the combination of its magnifying and space penetrating powers; and have also proved that the effect of their union, when they are differently combined, must have a considerable range, it will now be easy to point out the extent of this range in the telescope by which the following observations have been made.

The magnifying power by which the satellites of the planet were discovered was only 157; but this power, which has been constantly used in my sweeps of the heavens, and was found to be very *effective* for the discovery of faint nebulæ and minute clusters of stars, is hardly sufficient to show the satellites steadily; for, unless every thing is favourable, their faint scintillation will only be perceived by interrupted glimpses.

The magnifiers 300, 460, 600 and 800, it will appear by the following observations, have gradually been found to be more effective on the objects on which they were used; according to the clearness of the air, the altitude of the planet, the absence of the moon, the high polish of the mirror, and other circumstances; on particular occasions, when doubtful points were to be resolved, even 1200 has been most effective. The higher magnifiers 2400, 3600 and 7200 have also been used to scrutinize the closest neighbourhood of the planet, in order to discover additional satellites; but, from the appearance of the known ones, which began to be nebulous, I concluded that these powers were not distinct enough to be used on this occasion.

As the following observations are given for the purpose of enabling astronomers to calculate the elements of the orbits and motions of the satellites with mathematical precision, I have endeavoured to save them some labour by giving a clear statement of the general outlines of them; and that some judgment may be formed of their accuracy, which I hope will be found considerable, a short detail of the method I have pursued will be necessary.

For ascertaining the position of the satellites from which their periodical revolutions were determined, three different methods have been used.

and the screen being alternately withdrawn and replaced, I could perceive no sensible alteration in the brightness or distinctness of the letters. Hence I surmised, that the rays of light did not sensibly jostle in an instantaneous right angled passage, but that possibly they might suffer inflections or deflections in their crossing at the focal point on account of their being longer in collateral proximity.

* See *Phil. Trans.* for 1794, p. 58 [Vol. I. p. 464].

Coarse estimations were made when they seemed to be sufficient to keep the satellites in view, by way of ascertaining their identity; for unless they were followed in their course and known to be satellites, it would have been endless to measure either the distance or position of every small star that might have the appearance of one; and as the opportunities for taking measures, which require a very clear and undisturbed atmosphere, were scarce, and often interrupted by cloudy or moonlight nights, the identity of the satellites would have been doubtful if their position had not been attended to, when seen in unfavourable circumstances. When no other stars interfered, it was often sufficient barely to mention the quadrant in which they were seen, by recording that such a satellite was np, nf, sf, or sp; or if necessary, some rather more determined account, such as 40 or 50 degrees np, sf, &c.

As a check upon the description of the situations, a figure was always added to represent the planet, its satellites and the neighbouring stars as they appeared in the telescope. Very often indeed the configuration itself was deemed to be sufficient to point out the situation of the satellites, which by way of distinction were marked by numbered points; 1 and 2 being used to distinguish the known satellites; 3, 4, 5, &c. those that might possibly be other suspected, but not ascertained ones. Stars instead of points, were marked by asterisks.

More careful estimations were made with a power not less than 300, and a wire in the focus of the eye glass, to ascertain the parallel; they are capable of considerable accuracy in situations that are only a few degrees north or south preceding or following, and also when the position of a satellite is nearly 90 degrees north or south of the planet.

Measures taken with the micrometer may always be supposed to be accurate, unless they are marked as being affected by some circumstances existing at the time they were taken: when these are favourable, they can hardly be liable to any great error.

The calculations which I have given with the observations, will show the appropriate confidence each of these three methods of obtaining the positions of the satellites may separately deserve.

A much greater difficulty attaches to taking measures of distances than to those of angular positions: when the latter are taken, we have the position of the satellite in view all the time the planet passes along the parallel; and, although the moment of ascertaining the angle is only that in which the planet is in the centre of the wires, yet a constant attention to the motion of the two bodies will sufficiently enable us to perceive any excess or defect in the parallelism between their situation and that of the adjustable wire; whereas in measures of distances, the telescope must be kept in motion to retain the two bodies in their contact with the two wires, which disturbance considerably affects the delicacy of vision, and moreover requires a divided attention, as the passage of each body over its respective wire must be viewed. The only exception is, when the satellite is at 90 degrees, in which case the distance of the two bodies may indeed be measured with great accuracy.

The lucid point micrometer which has been tried is subject to the same difficulties; * its application to my construction of the 20 feet telescope, with regard to situation, is very convenient. When the apparatus was preparing, I found that handles, 20 feet long, would be very cumbersome, and attempted to try the micrometer with the assistance of a person to arrange the points; but, when engaged in the first measure, I found that unless I had myself the command of the motions, a perfect adjustment could not be obtained; or would at least take up so much time as would bring on an alteration in the telescopic motions, not consistent with perfect vision. This micrometer has, however, the peculiar advantage that it may be used with a concave eye glass.

When a satellite is either directly preceding or following the planet, its distance may be measured by the difference of the time of their passing the meridional wire. This method, which has also been tried, is however not sufficiently delicate for very small intervals, and is moreover of little use, on account of the very limited situations.

The following observations on the satellites of the Georgian planet are given in the order of time they were made. They contain every thing that relates to the appearance and motion not only of the two principal large satellites, that are plainly within the reach of a 20 feet telescope of my construction, but also the more difficult researches that have been pursued for detecting additional satellites. That such there are I can have no doubt; but to determine their number and situation will probably require an increase of the illuminating power, such as I was in hopes, when I published my announce of their existence, would have been used by other astronomers, in pursuit of the subject pointed out to them; a 25 feet reflector which is mentioned in the observations, may probably be sufficient for the purpose.

To facilitate calculation, the observations are all given in mean time, and after each of them is added a theoretical exposition of the place of the satellites, which I have called an identification, and is denoted by the sign ‡; the great use of which will be to point out the validity of each observation, by comparing the observed places with the theoretical ones. The method of identification, which will be described hereafter, by giving not only the angle of position at which a satellite ought to have been seen, but also its proportional distance in 600dth part of the radius of its orbit, is of great consequence when the orbits of the satellites are much contracted. These distances indeed become at last the only criterion by which we may know the satellites, for the angle of position, when the planet is

* For a description of this micrometer, see *Phil. Trans.* for 1782, page 163 [Vol. I. p. 91].

near the node of the orbits, admits of so little change that it ceases to be a direction for identifying them.

The same distance will also give us the total value of the measure of any distances taken by the micrometer, so far at least as to show which of them may be the most proper to be chosen for a more rigorous investigation.

An identification of supposed satellites cannot be made by calculation ; but the observations of following and also of preceding nights, accompanied by accurate configurations, may ascertain whether the object in question be of a sidereal or planetary nature. For if by the removal of the planet a supposed satellite be left in its former place, it is decidedly a star ; whereas a well ascertained absence from the observed place will make its planetary nature highly probable. Then also, if a configuration and description of every small star, that is situated in, and very near the path of the planet, has been previously made, and additional stars are afterwards found to be near the planet, which cannot be accounted for, it becomes again probable that such questionable objects are of a planetary nature. And this being a kind of identification, I have added it after the calculated one, to every observation of doubtful objects, except where a supposed satellite is pointed out which there is reason to believe may be a real one ; for in that case, the observations relating to the object in question are given in their regular order.

It will not be necessary to give the configurations that were made at the time of observation ; they generally contained the planet, its satellites, and some of the neighbouring stars, especially those that were in the path of the planet's motion ; nor will it be necessary to mention lines and descriptions of situations of stars pointed out by letters affixed to them, as the observations are generally so redundant, that I found it highly necessary to compress them.

Observations of the satellites of the Georgian planet, accompanied by a theoretical determination of their situation, whereby their identity may be ascertained.

1787, January 11^{d} 12^{h} 13^{m}. There is a supposed first satellite about 42 or 43 degrees south following the planet ; and a second about 45 degrees north preceding. A third supposed satellite is south following the planet.

‡ By the identifying method, it appears that a real satellite, called the first, was visible at the time mentioned about 45½ degrees south following ; which agrees with the estimation of the angle of its situation, and also with a configuration of the stars and planet, drawn at the time of observation. By the same method it appears that a real satellite called the second, was visible about 65 degrees north preceding the planet ; which situation agrees with the configuration and also sufficiently well with a coarse estimation, which, as there was no wire for the parallel in the focus of the eye glass, could not be accurate. The supposed third satellite, by subsequent observations, was found to be a star.

1787, January 12. The first and second satellites are not to be seen in the place where I saw them last night. The supposed third is left where it was. I can see no small star near the planet, but the evening is not sufficiently clear.

1787, January 14, 12^{h} 3′. There are again three supposed satellites ; I have marked them 1st, 2d, and 3d, without any particular reason for that order.

‡ The first satellite was 88° nf; which agrees with the situation of that which in the configuration was marked 1st. The second satellite was 22¼ sp, and agrees with that which in the configuration was marked 3d. In the configuration the numbers are placed according to their distances from the planet, and that which is marked 2d was found to be a star remaining in its place.

1787, January 17, 11^h 51′. There are now again three supposed satellites. The first is south preceding the planet, and makes a right angle with the 2d and 3d. The second is at the angular point and is south of the planet but a little preceding. The 3d is north following the planet. I have also added a 4th and 5th. The night is very fine and my telescope bears a high power.

‡ The first was 34° sp, which agrees with the configuration. The second was south of the planet but a little following, namely 80¾° sf; which agrees sufficiently well with an estimation made without a direction for the parallel. The supposed 3d, 4th, and 5th, by next night's observation, remained in their places as small stars.

1787, January 18, 11^h 45′. There are two supposed satellites; the first is directly south of the planet; the second is about 45 degrees south following, and a little farther from the planet than the first. With 480 the first is about 4 diameters of the planet distant from it; the second is about 4½ or 5 diameters from the planet; the first is from the second about 2½ diameters of the planet. There is no small star in the path of the planet that might be taken for a satellite to morrow.

‡ The first was 76⅔ sf. The second was 59⅓ sf; both these positions agree sufficiently well with the delineated configuration.

1787, January 24, 11^h 23′. The first and second satellites of January 18, are no longer in the place in which they were that night. There are two satellites; the first is about 45° np the planet; the second is about 80° np; it is brighter than the first. I had a glimpse of a 3d and 4th.

‡ The first satellite was 49° np; the second was 75¼ np, which agrees well with the estimations and with the configuration. The observations of the third and fourth were lost, the planet not being seen again till eight days after, when it would have taken up too much time to look for them.

1787, February 4, 6^h 21′. The first satellite is about 80° sp; the second is about 30° nf. There is too much day-light to see the satellites well. A third supposed satellite is south preceding the first; it is extremely small. There is but one single small star in the path of the planet which to morrow night may be taken for a satellite.

‡ The first satellite was 50° sp; the second was 40° nf. This differs considerably from the estimations, probably owing to the remaining day-light; the satellites however could not be mistaken, as there were no other stars near the planet.

1787, February 5, 9^h 3′. Both satellites are certainly absent from the place where I saw them last night. The first is about 85° sf; the second satellite (miscalled a small star) is by the configuration at a great angle nf. The small star in the path of the planet observed last night remains in its place.

‡ The first satellite was 89° sf; the second was 69° nf.

1787, February 6. I compared the configurations of January 11, 14, 17, 18, 24, February 4 and 5 together, and found that, admitting one of the satellites to make a revolution round the planet in about 8¾ days, and supposing its orbit to be very open to the visual ray, there was always one that would answer to a projection made on that scale.

1787, February 7, 6^h 54′. A satellite (miscalled the third) is a few degrees south following. 6^h 30′, another (miscalled the first) is about 65° np. A small star (miscalled the second satellite) is about 60° sp.

‡ The first satellite was 11¼ sf. The second was 68⅔ np. Two days after, the miscalled second, was seen remaining in its observed place. In the course of about nine hours of observation, I saw the planet accompanied by its two satellites, very evidently moving together in the path of the planet.

1787, February 9, 10^h 39′. Both satellites are gone from the place where I saw them the 7th of February. The first satellite (miscalled the second) is directly north of the planet; the second (miscalled the first) is a few degrees np.

‡ The first was 81⅔° nf; the second 11° np.

1787, February 10, 8^h 57′. The first satellite is about 53° np. 8^h 33′. The second satellite is about 20° sp; a supposed third is about 45° sf. In a little more than four hours, I saw the satellites go on with the planet, and also in their orbits.

‡ The first satellite was 67¾° np; the second was 20° sp. The supposed third was lost, no subsequent observation having been made of it. Before I began observing, I had delineated their places

on paper, on a supposition that one of them moved at the rate of $8\frac{3}{4}$, the other at that of $13\frac{1}{2}$ days the revolution.

1787, February 11, 13^h 28′. Between flying clouds I saw the second satellite.

‡ The satellite was $57\frac{1}{2}°$ sp, which agrees with the configuration.

1787, February 13, 10^h 0′. The first satellite, with 300, is about 75 or 80° sp; its distance from the planet is about $\frac{3}{4}$ of a minute. The second is about 85° sf; its distance is one full minute; the estimations are by the field of view of the sweeping piece. Third, fourth, and fifth supposed satellites were marked.

‡ The first satellite was $68\frac{3}{4}°$ sp; its distance was 553, the radius of its orbit being 600. The second was $80\frac{1}{2}°$ sf; its distance was 599, the radius of its orbit being also 600. The third, fourth, and fifth supposed satellites proved to be stars. No great accuracy can be expected from the estimated distances given in the observation, the field of the eye piece, which took in 15 minutes, being much too large for the purpose.

1787, February 16, 9^h 38′. The two satellites are in the places where I had drawn them on paper. With a power of 300, and a wire for the parallel in the focus of the eye glass, the first satellite is, by very accurate estimation, about 5 degrees north following; at the same time, and with the same power and accuracy, the second is about 3 degrees south following. A third supposed satellite is pointed out.

‡ The position being so near the parallel, and by calculation also near the conjugate axis of the elliptical projection of the orbits, and therefore less liable to an error arising from the application of correction, have been fixed upon as standards for the calculation of the periodical revolutions of the satellites. The supposed third was next evening observed to remain in its place.

1787, February 17, 7^h 58′. I tried to measure the distance of the second satellite from the planet by a lamp micrometer. The lucid points were 246·4 inches from the eye, and when they were 14·4 inches from each other, I found that the adjustment of the distance and angle of position could not be made to my satisfaction by an assistant, and gave up the measure. The magnifying power being 157, the opening of the points gives the angular distance 1′ 17″; but the measure when given up was still much too large.

‡ The satellite was $28\frac{1}{2}°$ nf, and the distance 505.

1787, February 19, 7^h 55′. Having delineated the situation of the satellites on paper, I found them in the expected situation. A third and fourth were added in the configuration.

‡ The first satellite was 58° np; the second was 82° nf. The supposed third and fourth proved to be stars.

1787, February 22, 7^h 14′. By the configuration the first satellite is at a considerable angle sp; the second is at a moderate angle np. Third, fourth, and fifth satellites were noticed.

‡ The first was 76° sp; the second was 31° np. A long interval happening to prevent subsequent observations, the supposed satellites were lost.

1787, March 5, 7^h 14′. The first satellite is about 6° sf. 7^h 17′, the second is about 87° nf; a third is about 40° nf.

‡ The first satellite was $20\frac{1}{4}°$ sf; the second was 87° np; the third proved to be a star. The planet was only observed about 3 or 4′, and it does not appear that great accuracy in the estimations was attempted.

1787, March 7, 7^h 12′. The first satellite is 82° np. 7^h 13′ the second is about 30° np. Very coarsely estimated. A third is about 6° nf; it seems to have a fourth close to it. Having some doubts about the fourth, I viewed it with 600 and 800; I saw it also well with 1200, and had a glimpse of it with 2400. These high magnifiers require a fine apparatus for adjusting the focus.

‡ The first satellite was $78\frac{3}{4}°$ nf; the second was 41° np; the third and fourth proved to be stars.

1787, March 8, 8^h 52′. Both satellites were seen for a few minutes.

‡ The first was 70° np; the second was $9\frac{1}{2}°$ np.

1787, March 11. I found that some friends who came to view the satellites saw them best with 480, when the planet was drawn to the margin just out of the field.

1787, March 15, 8^h 7′. The first satellite is about 48° nf; the second is 5° sf. The second satellite being so nearly following the planet, I tried to measure its distance by sidereal time. Of eight transits, four gave 3″; three gave 3″·5; and one gave 4″; a mean of them is 3″·31; and the declination of the planet being 21° 57′ north, we have the apparent distance 45″·99; but I do not trust much to measures by time, in the manner these were taken without a system of wires in the focus of the eye glass, and with the clock and assistant at a considerable distance.

‡ The first satellite was 46° nf; the second was 2¼° sf, and its distance from the planet was 481; this would give the greatest elongation 57″·36 which is probably much too large.

1787, March 18, 8^h 3′. The satellites are in the place where I expected them. The first is 5 or 6° np; the second is about 75° nf; it seems to be farther from the planet than when it was near the parallel. I attempted to measure its distance by the parallel wire micrometer; eclipsing the satellite with one wire, and bisecting the planet with the other. The measure gave the distance 46″·46.

‡ The first was 21° np; the second was 83½° nf; and its distance was 588; which gives the greatest elongation 47″·41.

1787, March 19. Both the satellites are in their expected situation, which for the first is 36° sp; for the second 79° np. At 7^h 48′ I took a good measure of the distance of the second satellite; it gave 44″·24. I attempted a second measure, but was interrupted before I had quite finished it to my liking; it gave 45″·98.

‡ The first satellite was 29¾° sp; the second was 74⅓° np, and its distance 596. The expected situations, though calculated from imperfect tables, were sufficient to show that the satellites were not mistaken.

1787, March 20, 7^h 44′. I took three measures of the distance of the second satellite; the first gave 40″·23; the second, with the remark, pretty full measure, gave 41″·89; the third with the addition, not too large, gave 40″·20.

‡ The satellite was 52⅝° np, and the distance 564.

1787, April 9, 10^h 22′. I took two very accurate measures of the distance of the second satellite from the planet; the first gave 44″·54, the second 44″·35. By temporary tables its expected place was 57° sf.

‡ The satellite was 54° sf; and its distance 563.

1787, April 11, 9^h 18′. By temporary tables the expected situation of the second satellite was 4° sf. I took three good measures of its distance from the planet; the first gave 34″·47; the second 35″·32; the third 35″·74. A mean of them is 34″·99.

‡ The satellite was 1½° nf; and its distance 477.

1787, April 17, 8^h 53′. The two satellites are on opposite sides of the planet.

‡ The first was 40½° sf; distance 531. 9^h 6′, the second was 21⅝° np; distance 503.

1787, September 19, 15^h 55′. The first satellite (miscalled second) is 85° sp; the second (miscalled first) is about 30° sf.

‡ The first was 87¾° sp; the second was 10¾° sf. This being the first time of seeing the planet after its conjunction, accounts for the mistakes of the names.

1787, October 11, 16^h 49′. The first satellite (miscalled second) is 78° np. Two good measures of its distance from the planet were taken; the first gave 35″·18, the second 35″·96; a mean is 35″·57. 16^h 51′. The second satellite (miscalled first) is 40° sp.

‡ The first was 84° np, and its distance 599; the second was 52⅝° sp; and its distance was 492. The long interruption in the observations was again the cause of a mistake of the names, which the calculation sets right.

1787, October 14, 15^h 59′. The angle of position of the first satellite by the micrometer is 48° 22′ sp; that of the second at 16^h 29′ is 66° 2′ sf.

‡ The first satellite was 49½° sp; the second was 65½° sf.

1787, October 20, 15^h 36′. Position of the first satellite by the micrometer 72° 0′ np. Position of the second at 16^h 8′, 80° 12′ np.

‡ The first was 76⅝° np; the second was 80½° np.

1787, November 9, 15^h 56′. The second satellite is about 87° sf. The distance by four good measures 46″·15; 43″·92; 42″·94; 46″·57; mean 44″·89.

‡ The satellite was 84° sf; distance 594.

1788, January 14, 12^h 3′. The two satellites are almost in opposition; but the first precedes a line continued from the second through the planet.

‡ The first satellite was 77¼° nf; the second was 66° sp.

1789, February 22, 9^h 48′. The first satellite is about 80° sp; the second is about 85° sp; too much wind for measuring.

‡ The first satellite was 69½° sp; the second was 85⅓° sp.

1789, February 24, 9^h 13′. The first satellite is a few degrees more advanced in its orbit than the second.

‡ The first satellite was 48⅓° sf; the second was 55° sf.

1789, March 13, 9^h 1′. The first satellite is 60° sf. 7^h 47′, the second is about 45° nf, third and fourth satellites were marked.

‡ The first satellite was 63⅓° sf; the second was 57⅔° nf. The third and fourth were found to be stars.

1789, March 14, 9^h 22′. The first satellite is 8° sf; the second is 70° nf.

‡ The first satellite was 19¼° sf; the second was 81¼° nf.

1789, March 16, 7^h 33′. The first satellite is 83° nf; the second is about 60° np. A third, about 2° sf; a fourth, about 8 or 10° np.

‡ The first satellite was 73⅓° nf; the second was 61° np. The third and fourth were stars.

1789, March 20, 7^h 50′. The two satellites were coarsely estimated to be at considerable angles sp.

‡ The first satellite was 63° sp; the second was 65⅓° sp.

1789, March 26, 10^h 44′. A star was mistaken for the first satellite; the second satellite (miscalled the first) is 45° nf.

‡ The first satellite was 63¼° np; the second was 50° nf.

1789, December 15, 10^h 54′. The first satellite is about 71° sp. 10^h 49′, the second is about 75° sp; a third is about 75° sf.

‡ The first satellite was 72° sp; the second was 81⅓° sp; the third was a star.

1789, December 16, 10^h 12′. The first satellite is about 83 or 84° sf; the second is 85° sf. By the configuration they are very nearly in conjunction.

‡ The first was 83⅔° sf; the second was 83° sf.

1790. January 18, 9^h 32′. The first and second satellites are in the places I had calculated. There is a supposed third satellite about two diameters of the planet following, extremely faint and only seen by glimpses; 1^h 6′ after I could not perceive it; a fourth is about 70° np.

‡ The first was 38¾° sp; the second 85¼° nf.

1790, January 19, 9^h 34′. There is a very small star left in the place where the supposed fourth satellite was last night. 10^h 47′, I can see no fourth satellite near the second where it would be now if it had been a real satellite. With the assistance of a field bar to hide the planet, and a power of 300, I can see the first and second satellites very steadily, even the very first moment I look into the telescope.

‡ The first satellite was 76⅔° sp; the second was 77¼° np. It is very strange that the third supposed satellite should not have been attended to when two observations are given to prove that the supposed fourth was not a satellite.

1790, January 20, 12^h 5′. The first and second satellites are in the places I had calculated; a third satellite is 45° np, and in a line with the planet and the second satellite.

‡ The first satellite was 77¼° sf; the second was 54½° np. The third was not accounted for.

1790, February 6, 9^h 28′. I viewed the planet and satellites with three concave eye glasses, power about 240, 320, and 460. I see very clearly with these glasses. Cloudy.

‡ The first was 89¾° sf; the second was 64° sp.

1790, February 9, 9^h 19′. By a configuration the first satellite is at a considerable angle nf; the second at a great angle sf. A third is in a line with the planet and the second satellite; its distance from the planet by the configuration is about twice that of the second satellite.

‡ The first was 48¾° nf; the second was 61½° sf; the third was 61½° sf; two succeeding observations are decisive that the supposed third satellite was not a star remaining in its place.

1790, February 11, 8^h 30′. The satellites are in the places I had calculated. 8^h 56′, the small star of the 9th of February I believe is wanting; at least I cannot see it though the weather is very clear, but windy. An additional third and fourth are pointed out.

‡ The first satellite was 74° np; the second was 7° nf. The third and fourth of this night were found to be stars.

1790, February 12, 11^h 27′. The first and second satellites are in the places I had calculated. The third and fourth of last night are small fixed stars remaining in their places. The supposed third satellite of the ninth is not in the place where I saw it that night.

‡ The first satellite was 27° np; the second was 48⅓° nf.

1790, February 16, $8^h\ 2'$. The first and second satellites are in the places I had calculated; the situation of a supposed third is described.

‡ The second was $56\frac{1}{3}°$ np; the supposed third proved to be a star.

1790, February 17. A configuration of stars situated in the planet's path is delineated.

1790, March 3, $7^h\ 58'$. The first satellite is 40° sp. $8^h\ 42'$, the second is 3 or 4° np.

‡ The first satellite was 56° sp; the second was $0\frac{1}{3}°$ np.

1790, March 5, $10^h\ 38'$. The first and second satellites are in the places I had calculated; a 3d, with 600 is 56° sf; a fourth is delineated.

‡ The first satellite was 63° sf; the second was 66° sp; the third and fourth proved to be stars.

1790, March 8, $10^h\ 43'$. Forty feet telescope. I saw the satellites with great ease. The speculum being extremely tarnished, I did not expect to have seen so well as I did.* Twenty feet telescope. The first satellite is 85° 7′ nf. $8^h\ 39'$, the second is 67° 36′ sf. My wire is too fine and the power 460 too high for great accuracy.

‡ The first satellite was $79\frac{1}{3}°$ nf; the second was $59\frac{1}{2}°$ sf.

1790, April 3, $9^h\ 39'$. The first satellite is on the opposite side of the second; the position of the second is 77° 53′ sf.

‡ The first satellite was 75° nf; the second was $76\frac{2}{3}°$ sf.

1791, January 31, $11^h\ 5'$. The second satellite is 74 or 75° np. A supposed satellite in opposition to the second, and at double its distance from the planet, is marked in the configuration.

‡ The first satellite was $0\frac{3}{4}°$ nf; its distance was 336, and not being noticed it was probably invisible; the second was $78\frac{3}{4}°$ np; the supposed exterior satellite was $78\frac{3}{4}$ np.

1791, February 2, $8^h\ 23'$. The first satellite is about 70° nf. $8^h\ 10'$, the second is gone with the planet from the stars of the configuration of the 31st of January.

‡ The first satellite was 81° nf; the second was 30° np. The lettered stars of the configuration were all named as being left in their places, but the supposed exterior satellite of that day is not mentioned among them.

1791, February 4, $8^h\ 13'$. The second satellite is 40° 48′ sp, but the measure is imperfect and may be out 5 or 6 degrees; a supposed satellite was marked, and a small star pointed out in the path of the planet.

‡ The second satellite was 52° sp. The supposed satellite was found to be a star.

1791, February 5, $11^h\ 5'$. The first satellite is 20° sp. $10^h\ 45'$, the second is 65° sp. With 600, third, fourth, and fifth satellites are marked: but as they are also visible with 300, they are probably stars.

‡ The first satellite was 41° sp; the second was $74\frac{1}{2}°$ sp. The third, fourth and fifth, were lost for want of subsequent observations.

1791, February 22, $8^h\ 23'$. I cannot perceive the first satellite, probably owing to its nearness to the planet; I am pretty sure the orbits are contracted, so that the planet is approaching towards their node. $7^h\ 30'$, a measure of the position of the second satellite is 36° 18′ sf.

‡ The first satellite was $8\frac{3}{4}°$ sp; distance 333, which may account for its not having been seen. The second was 39° sf.

1791, February 23, $7^h\ 59'$. Position of the first satellite 56° 33′ sp.

‡ The first was $60\frac{1}{2}°$ sp; the second was $7\frac{1}{2}°$ nf; the distance was 331, at which the satellite is sometimes invisible.

1791, March 1, $11^h\ 47'$. The two satellites are in the places I had calculated.

‡ The first was $73\frac{1}{3}°$ np; the second was $20\frac{3}{4}°$ np.

1791, March 2, $9^h\ 18'$. The first satellite is hardly to be seen; it seems to be in about the most contracted part of its orbit; the second is only about two diameters of the planet from the edge of the disk, but the estimation cannot be very accurate, as I am obliged to hide the planet to see the satellite.

‡ The first was $37\frac{1}{3}°$ np; its distance 377. The second was $22\frac{1}{2}°$ sp; its distance 358.

1791, March 5, $7^h\ 56'$. The first satellite is about 75° sp; the second about 85° sp; third, fourth, and fifth satellites were pointed out.

‡ The first satellite was 87° sf; the second was $89\frac{1}{2}°$ sf; the third, fourth and fifth proved to be large stars, the nearness of the planet having diminished their lustre when observed as satellites.

* See *Phil. Trans.* for 1814, p. 275. A note relating to the polish of the 40 feet mirror. [Above, p. 536.]

1791, March 6, 12^h 2′. The first satellite is much nearer the planet than it was last night; the second is also nearer, but not much.

‡ The first satellite was 50½° sf; the second was 70½ sf.

1791, March 9, 9^h 52′. The first satellite, with a wire for the parallel and 300, is about 86° nf. The position being so near the perpendicular cannot be much out. By the micrometer it is 86° 25′ nf. 9^h 36′, the second is on the following side; it is nearer the planet than the first, and on that account appears smaller.

‡ The first satellite was 86¾° nf; distance 599; the second was 32½° nf; distance 379.

1791, April 4, 8^h 43′. The first satellite is 84° 56′ nf; the second was not observed.

‡ The first was 82¾° nf; the second was 9¾° sf; its distance being 343 it might not be visible.

1791, December 19, 11^h 45′. I do not perceive the first satellite; the second is about 75° nf.

‡ The first was 7⅓° sf, and its distance being 263, it was therefore invisible. The second was 82½° nf.

1792, January 27, 11^h 58′. The first satellite was not observed; the second is about 40° nf, the estimation may be out 6 or 8 degrees. Cloudy.

‡ The first satellite was 20⅓° np; its distance being 283 it could not be seen. The second was 62¼° nf.

1792, February 12, 8^h 28′. The first and second satellites are in the same line, and I measured their position together, it is 88° 19′ np; a supposed third is 84° 23′ sp.

‡ The first satellite was 86⅓° np, and its distance 576; the second satellite was 83¾° np; the third proved to be a star.

1792, February 13, 8^h 42′. Forty feet reflector, with 360, I saw the disk of the planet very well defined. Twenty feet. The satellites are advanced in their orbits; the first is drawn much nearer to the planet than it was yesterday; a very small star is 41° 22′ nf.

‡ The first satellite was 56½° np; its distance 389; the second was 66⅓° np. The very small star was left in its place.

1792, February 20, 12^h 57′. The first satellite is 89° 58′ nf; 13^h 8′, the second is 53° 21′ sf; a supposed third is 66° 17′ np.

‡ The first satellite was 83⅓° nf; the second was 53⅓° sf. By an increase of 25 or 30° in the angle of the third it was the same evening proved to be a star.

1792, February 21, 9^h 10′. Position of the first satellite 73° 52′ np.—9^h 30′, I suspected the second satellite to be in its calculated place, but even 600 would not verify it.

‡ The first satellite was 78⅔° np; the second was 17⅓° sf; distance 292.

1792, February 26, 11^h 30′. The position of the first satellite is 42° 49′ sf.—8^h 2′, that of the second is 73° 49′ np. A very small star between the planet and the second satellite is pointed out, and another towards the south at double the distance of the first is marked in the configuration.

‡ The first satellite was 52¾° sf; the second was 75⅓° np. The small star was left in its place; but the distant one is not accounted for.

1792, February 28, 10^h 52′. The position of the first satellite is 69° 43′ nf. The second was not seen.

‡ The first satellite was 65½° nf; the second was 3½° sp; distance 293, and therefore invisible.

1792, March 15, 10^h 3′. I cannot see the first satellite with 300, 480, nor with 600. The second satellite is 73° 22′ sp.

‡ The first satellite was 17° sf; its distance was 302 and therefore invisible; the second was 74½ sp.

1792, March 18, 8^h 19′. The first satellite is 82° 35′ np. 8^h 37′, the second is 60° 16′ sf.

‡ The first satellite was 81⅔° np; the second was 56° sf.

1792, March 19, 8^h 20′. The first satellite is 38° 4′ np, I see it very well notwithstanding it is near the planet. 8^h 42′, I cannot see the second with 300. With 480 I see it very well; I see it also with 800 and 1200; I tried 2400 and 4800, but a whitish haziness in the air prevents my seeing it with these powers.

‡ The first was 46° np and its distance 364; the second was 15° sf; its distance 299 which accounts for the difficulty of seeing it.

1792, March 23, 8^h 21′. I see the first satellite through flying clouds; the second is 89° 21′ np.

‡ The first satellite was 61⅔° sf; distance 430. The second was 88⅔° np.

1792, March 27, 11^h 6′. The first satellite (miscalled a very small star) is about 80° np; the second is by the configuration about 45° sp; a third (miscalled the first) was pointed out.

‡ The first satellite was 70° np; the second was 46½° sp; the third was not accounted for.

1792, March 30, 11^h 18′. The first satellite by the configuration is at a great angle sp; the second is at a great angle sf.

‡ The first satellite was 78½° sp; the second was 83° sf. The first satellite (miscalled a star the 27th) is gone from the place where it was.

1793, February 5, 9° 18′. Neither the first nor second satellites are visible. A very small star is 19° 3′ sp.

‡ The first satellite was 30° sp; distance 282; the second was 23° sf; distance 245. There is no subsequent observation of the small star.

1793, February 7, 9^h 38′. The first satellite is 79° 39′ sp. 9^h 20′, the second is 59° 51′ nf. The wind being very troublesome, the measures cannot be very accurate: The difficulty was in finding the parallel.* I viewed the planet with 240, 320, 480, 600, 800 and 1200, but saw no satellites nearer than the two known ones. The north following satellite being farther from the planet than the south following one, I take it to be the second; the difference of their distance appeared plainest with 1200. I viewed the planet also with 2400, 3600, 7200.

‡ The first satellite was 86¾° sp. Distance 590. The second was 68¼° nf; distance 513; and supposing the radius of the orbit of the first to be to that of the second as 3 to 4, we have the apparent distance of the first to that of the second as 177 to 205.

1793, March 8, 11^h 21′. The first satellite is about 65° nf; the second is 90° nf; a third is about 75° nf.

‡ The first was 57¾° nf; the second was 89⅔° nf; the third was a star.

1793, March 9, 10^h 35′. The first satellite is 85° nf; the second is about 82° np; a third is about 65° sp.

‡ The first was 80° nf; the second 77½° np; there is no subsequent observation of the third.

1793, March 14, 9^h 37′. The first and second satellites are seen in their places; the situation of a third and fourth is pointed out. The first satellite is brighter than the second.

‡ The first was 89⅔° sf, and 16° 26′ from its greatest elongation; distance 574. The second was 80° sp, and 7° 56′ from its greatest elongation; distance 588. The superior brightness of the first therefore could not arise from its greater distance. The third and fourth supposed satellites had no subsequent observations.

1793, April 3, 10^h 53′. The first satellite is 50° nf; the second is 80° nf.

‡ The first was 52⅝° nf; the second was 79° nf.

1794, February 21, 8^h 24′. The first satellite is about 88° nf; the second is about 86° nf.

‡ The first was 89° nf; the second 85° nf.

1794, February 25, 8^h 24′. By a configuration the first satellite is at a great angle sp; and its distance from the planet is greater than that of the second, which is at a much smaller angle sp. Several small stars are pointed out.

‡ The first satellite was 84° sp; distance 593. The second was 47¾° sp; distance 323. The stars were left in their places.

February 26, 8^h 28′. The first satellite is 70° 53′ sf. 8^h 7′, the second is 66° 56′ sp; many small stars are pointed out.

‡ The first satellite was 78° sf; the second was 67⅓° sp; the stars remained in their places.

1794, February 28, 9^h 43′. The first satellite is 62° 55′ nf. 9^h 26′, the second is 86° 44′ sp. 8^h 15′, there is a very small star which I did not see the 26th; it is brighter than a lettered star not far from it. Its position is pointed out by the stars of the configuration.

‡ The first satellite was 61½ nf; the second was 87½° sp. The stars remained in their places. The position of the small star of the 26th, by identification was about 24° nf.

1794, March 2, 8° 25′. The first satellite seems to be at its greatest elongation; the second satellite was not seen.

‡ The first was 86° np; distance 507; the second was 55¾° sf; distance 275, therefore not visible.

* Telescopic vision in windy weather is generally very perfect, and except in cases which require an uninterrupted steadiness of the instrument, will admit of the highest magnifying powers.

1794, March 4, 11^h 22′. I can see neither the first nor the second satellite. A third satellite is 61° 32′ nf. Many small stars are pointed out.

‡ The second satellite was $57\frac{1}{2}$° nf ; its distance 383 ; it was therefore visible, and its position agrees with the measure taken of a satellite miscalled the third : the inaccuracy of my tables in 1794, occasioned the mistake. The small stars remained.

1794, March 5, 11^h 10′. The first satellite is 75° 50′ sp. 10^h 57′, the second is 72° 27′ np. There is no star in the place where the supposed third was last night. Many small stars are again pointed out.

‡ The first was 76° sp ; the second was $72\frac{1}{4}$° nf. The absence of the miscalled third confirms the mistake, and is a proof of the great attention that was paid to ascertain the nature of supposed satellites. The small stars remained.

1794, March 7, 11^h 18′. I cannot perceive the first satellite. 10^h 57′, the second is nearer the planet than it was the last time I saw it. Small stars are pointed out.

‡ The first was 63° sf ; distance 320, invisible. The second was $87\frac{1}{3}$° np. There is no subsequent observation of the small stars.

1794, March 17, 7^h 38′. I can see neither of the two satellites.

‡ The first was 41° nf ; distance 292, invisible. The second was $38\frac{2}{3}$° nf ; distance 285, invisible.

1794, March 21, 11^h 53′. I cannot see the first satellite. I looked at several different hours for it. 10^h 53′, the second satellite is 88° 8′ np. 9^h 19′, the place of a small suspected star is pointed out, but it cannot be verified with 460 and 600.

‡ The first satellite was 27° sp ; distance 243, invisible. The second was $81\frac{1}{2}$° np. The suspected star was seen in its place the following night.

1794, March 22, 8^h 47′. There is no mention of the first satellite ; the second is 61° 46′ np.

‡ The first was $66\frac{1}{4}$° sp, and its distance being 480, it must have been seen and taken for a star. The second satellite was $60\frac{3}{4}$° np ; distance 311.

1794, March 23, 8^h 32′. The first satellite is one of two small stars that are south of the planet ; it is the preceding and largest of the two. 8^h 42 , the second satellite is not visible.

‡ The first was 82° sp, 1° 26′ past its greatest elongation. The second was 1° np, 1° 49′ past its shortest elongation, distance 207 ; invisible.

1794, March 26, 9^h 2′. Position of the first satellite 61° 53′ nf, as accurate as the faintness of the satellite will permit. 8^h 48′, the second satellite is 77° 0′ sp, very accurate. 9^h 17′, I suspect a third satellite directly north a little farther from the planet than the first, and the power 480 almost verifies the suspicion. 9^h 26′, with 600 I still suspect the same, but cannot satisfy myself of the reality. 11^h 24′, I see the supposed third satellite perfectly well now ; it is much smaller than the first, and in a line with the planet and the first. An extensive configuration is delineated.

‡ The first satellite was $56\frac{2}{3}$° nf ; the second was $79\frac{1}{2}$° sp ; the third satellite being in the position of the first at 11^h 14′, must have been $59\frac{1}{2}$° nf.

1794, March 27, 10^h 25′. The first satellite is 75° 59′ nf. 10^h 12′, the second is 88° 35′ sf. 8^h 15′, the small star observed last night at 11^h 14′ is gone from the place where I saw it. From its light last night compared to a star marked *r* in the configuration which to night is very near the planet, and scarcely visible, I am certain that it must be bright enough to be perceived immediately, if it were in the place pointed out by the configuration. 11^h 19′, I have many glimpses of small stars, one of them is in a place a little north following the first satellite, agreeing with what would probably be the situation of the third satellite of last night if it had moved with the planet. A supposed third of this evening is preceding the first satellite, but nearer the planet. A supposed fourth is sf ; its distance is almost double that of the second satellite. Some other small stars or supposed satellites are seen to the south at a good distance.

‡ The first satellite was 79° nf ; the second was $81\frac{1}{4}$° sf. The fourth proved to be a star.

1794, March 28, 9^h 12′. The first satellite is 88° 31′ np. 9^h 1′, the second is 82° 7′ sf.

‡ The first satellite was 87° np ; the second was 78° sf.

1794, April 1, 9^h 14′. The first satellite is 83° 2′ sp. 9^h 23′, the second is 70° 26′ nf.

‡ The first satellite was $87\frac{1}{4}$° sp ; the second was 72° nf.

1796, March 4, 11^h 10′. The Georgian planet is about 13′ of space preceding and 5 or 6′ north of a nebula. An extensive configuration was made, but no satellites were noticed.

‡ The nebula was No. 272 of the first class of my third catalogue. (See *Phil. Trans.* for 1802, page 505.) The first satellite was $42\frac{1}{3}$° sp ; distance 158, invisible. The second was $87\frac{1}{2}$° np, distance

364, and after an interval of near two years might possibly be overlooked. The stars pointed out remained in their places.

1796, March 5, 12^h 25′. The first satellite is 72° 20′ sp. 10^h 53′, I suspect a very small star between two of the lettered stars of last night's configuration which at the time it was made was not there. I had a pretty certain glimpse of it.

‡ The first satellite was 75° sp.—10^h 25′, the second was 72° np; distance 123, invisible. By the configuration the suspected star was at a considerable distance, about 72° np.

1796, March 9, 11^h 35′. The first satellite is 70° 36′ nf. 11^h 22′, the second is 83° 35′ sp. As the probability is that other supposed satellites move in the same plane with the first and second, I chiefly look for them in the direction of the position of their orbits which is now nearly a straight line; a star that may possibly be a distant satellite is pointed out.

‡ The first satellite was 70⅔° nf. The second was 81° sp; the star remained in its place.

1796, March 10, 11^h 43′. The first and second satellites by a configuration, are not far from being in opposition, the first not being come to a line drawn from the second through the planet. With 600 there is no star between the satellites and the planet that may be supposed to be an inner satellite; with this power the satellites are very large and visible, I see them better than with a lower.

‡ The first satellite was 79¾° nf; the second was 86½° sp.

1796, March 27, 10^h 6′. The first satellite is in the place I had calculated.

‡ It was 75½° nf.

1796, March 28, 10^h 7′. The first and second satellites are in the places I had calculated; the apparent contraction of their orbits is such as to approach to a straight line.

‡ The first satellite was 84° nf; the second was 74° nf.

1796, April 4, 11^h 16′. The first satellite is not visible; the second is near a small sp. star. There is no star in the transverse of the apparent elliptical orbit that could be taken for a satellite, unless that near the second should be one going towards its greatest elongation, or coming from it.

‡ The first satellite was 68° nf; its distance 415; how it happened not to be visible I cannot account for; the configuration has no star near the place; the second satellite was 76⅓° sp. The star near the second remained in its former situation.

1796, April 5, 10^h 48′. The first and second satellites are apparently in opposition, the same wire covers them both and the planet. There is no star in the line of the transverse that can be taken for a satellite: the night being beautiful, I examined that line with 300 at a distance, and with 600 within the orbit of the two satellites.

‡ The first satellite was 74⅔° nf; the second was 81⅔° sp.

1797, March 15, 9^h 44′. The first and second satellites are not far from their opposition; by the configuration they are short of it.

‡ The first was 78¾° sp; the second was 81¼° nf.

1797, March 17, 9^h 51′. The first and second satellites are both invisible; the night is very beautiful and I have a field bar to hide the planet; but notwithstanding this, I cannot see either of the satellites; many stars are pointed out.

‡ The first was 69¾° sf; distance 110, invisible. The second was 76° np; distance 126, invisible. The stars had no subsequent observation.

1797, March 21, 10^h 9′. The first satellite is not visible; the second is nearly at its greatest elongation, it is about 70° south preceding; many stars are pointed out.

‡ The first satellite was 88½° np; distance 240, invisible; the second was 79½° sp; distance 588. The stars remained.

1797, March 23, 10^h 28′. With 320, I see neither of the satellites. 10^h 32′, having just been told where the second should be, I perceived it in its place; with 600 I see it very well; many stars are pointed out.

‡ The first satellite was 76° sp; distance 518; it does not appear why it could not be seen. The second was 88° sp; distance 298. The stars remained.

1797, March 25, 10^h 40′. With 320 I cannot perceive either of the satellites; with 600 I can see neither of them; many stars are pointed out.

‡ The first satellite was 87° sp; distance 320, invisible. The second was 66⅔° nf; distance 251, invisible. The stars remained.

1797, March 28, 10^h 36′. The two satellites seem to be nearly in a line drawn from the second to the centre of the planet; the second is 80° 26′ nf. There is an exceeding small star about four times the distance of the second satellite in the line of the greatest elongation, I do not remember to have seen it among the lettered stars which are pointed out, the 25th.

‡ The first satellite was 78½° nf; distance 597. The second was 80⅓° nf; distance 592. There is no subsequent observation of the small star.

1797, March 29, 9ʰ 39′. I see one of the satellites. Cloudy. I suppose it to have been the second.

‡ The first was 83° nf; the second was 83⅓° nf.

1798, February 6, 11ʰ 29′. The second satellite (miscalled a star) is at a great angle sp; three stars are pointed out.

‡ The first satellite was 74⅓° nf; distance 251; invisible. The second was 79° sp; distance 552. The three stars remained.*

1798, February 7, 11ʰ 7′. I cannot see the first satellite. The second is at a great angle sp near one of the stars marked down last night, which is now so small that I cannot distinguish it from the satellite. An extremely small star preceding a line that joins two lettered stars may be an exterior satellite. Position of the satellite (called the 6th) 64° 41′ np. 12ʰ 36′, there is no star in the path of the planet which to morrow or next day can be taken for this 6th satellite.

‡ The first satellite was 85° sp; distance 159, invisible. The second 78⅓° sp; distance 592. The supposed satellite, called the 6th by way of readily referring to it, and also partly to express its distance, was found to remain in its observed place.

1798, February 11, 9ʰ 17′. The path of the planet is marked by a configuration of lettered stars, taking in those from which it comes and those towards which it will go.

11ʰ 35′, the situation of a supposed exterior satellite (called the 5th) with regard to the lettered stars is pointed out. It is excessively faint, but the night is very beautiful.

11ʰ 46′, the position of the 5th satellite is 89° 19′·5 nf; but the satellite is so faint that the measure cannot be very accurate.

12ʰ 16′, I cannot see either of the two old satellites. There is an extremely small north preceding star *x*, which may be a more distant satellite; it is much smaller than the 5th, and will therefore become invisible when the planet comes near it.

‡ The first satellite was 89¾° nf; its distance was 41, and it was therefore invisible. The second was 79½° nf; its distance was 241; and it was therefore also invisible.

1798, February 13, 10ʰ 17′. The old satellites are in the place I had calculated. The 5th satellite and the small star *x*, observed February 11, are not visible; but the weather is very indifferent.

11ʰ 49′, I do not see the 5th satellite where it was February 11.

12ʰ 0′, the position of the two old satellites is 76° 48′ nf.

12ʰ 15′, I see the extremely small star *x* remaining in its former place.

12ʰ 44′, the first and second satellites are exactly in a line pointing to the centre of the planet. A second measure of their position is 75° 45′. I cannot see the 5th satellite in the place where it was February 11.

‡ The first satellite was 78¼° nf; distance 594. The second was 78½° nf; distance 576.

1798, February 15, 11ʰ 21′. There is a very small star in the line of the greatest northern elongation, it may possibly be an interior satellite, the first and second being invisible.

I see the extremely small star *x* of the 11th perfectly well, but the 5th satellite of the same night is gone from the place where it was that evening. It was considerably brighter than *x*, so that if it were in its place, I must certainly see it.

11ʰ 41′, the star which at 11ʰ 21′ I supposed might be an interior satellite is too far from the planet; it may possibly be the 5th satellite of the 11th on its return from the northern elongation towards the planet.

I believe there is another satellite or star between this last mentioned one and the planet; I do not suppose the second satellite to be visible, otherwise it would agree well enough with the situation of the star between the 5th and the planet. By the configuration the intermediate star is at about half the distance of the farthest of the two.

12ʰ 13′, position of the supposed 5th satellite 84° 49′ nf.

‡ The first satellite was 73° nf; distance 124, it was therefore invisible and could not be the supposed 5th satellite. The second was 77° nf; distance 455; it was therefore visible, and agrees

* The planet being past the node, the angular distance of the satellites from zero, and their apparent motions are inverted.

very well with the satellite miscalled the 5th; the star between the second satellite and the planet must have been an interior satellite at its greatest northern elongation. At the time of observation, my defective tables made me suppose the nearest of the two to be the second satellite.

1798, February 16, 9^h 25′. The supposed fifth satellite observed last night at 11^h 41′, I believe is gone from the place where I saw it at that time. The night is very beautiful; the planet however is still low, and I shall look for it again when it is higher.

10^h 57′, the supposed fifth satellite is gone from its former place. It was so visible last night when near the planet, that I should certainly see it without difficulty if it remained in the same place, as the planet is now removed from it.

11^h 5′, the first and second satellites are invisible.

11^h 12′, there is a very faint satellite in the southern elongation; probably the sixth; and if it be the sixth satellite it is probably a little before or after its greatest elongation. It is excessively faint.

12^h 27′, the weather is not so clear now, though still fine, but the sixth satellite cannot be seen; it is plain, therefore, that the least haziness will render it invisible.

‡ The first satellite was 80⅓° sp; distance 289, invisible; the second was 75½° nf; distance 244, invisible. The interior satellite observed the fifteenth, being taken for the second, was not looked after; but as the supposed fifth was scrupulously ascertained to be removed, the interior satellite, had it been a star remaining in its former situation, must unavoidably have been seen; for by the configuration they could not be much more than a diameter of the planet asunder.

1798, February 18, 9^h 19′. I see the sixth satellite observed February 16, at 11^h 12′, it has left the place where it was at that time. It is nearer the planet than it was that evening, I suppose it therefore to be on its return from its southern elongation.

There is a seventh satellite near the sixth, rather a little fainter than the sixth; a supposed eighth satellite is pointed out.

11^h 25′, the position of the sixth satellite is 80° 53′ sp.

11^h 31′, with 480 I see the satellite near the sixth perfectly well; the distance between the two is about ¾ or one diameter of the planet.

11^h 44′, I see the satellite much better with 600; that which is farthest from the planet is the largest.

‡ The first satellite was 78° sp; distance 563. This therefore might be the satellite which was seen near the sixth. The second satellite was 80½ sp; distance 290, and was invisible. The supposed eighth, proved to be a star.

1798, February 19, 11^h 12′. The first satellite is invisible; the second is near the greatest elongation.

‡ The first was 81° sp; distance 269; invisible. The second was 79½° sp; distance 490.

1798, February 22, 9^h 9′. The first satellite by its distance is not far from its greatest northern elongation; it is very large. There is a satellite to the south exactly opposite to the first; it is very small but may be the second. The moon is too bright to see very faint satellites.

‡ The first satellite was 78° nf; distance 594. The second was 78° sp; distance 431. The small appearance of the second satellite is not easily to be accounted for; its distance from the planet was not much less than that of the first; for if the greatest elongations of the satellites be as 3 to 4, the above distance will be 1782 to 1724.

1798, February 26, 9^h 52′. The first and second satellites are in opposition; the first being sp, the second nf. The moon is so bright that their light is very feeble. Position 79° 53′ from sp to nf. The first satellite is small, the second is large.

‡ The first satellite was 78½° sp; distance 584. The second was 79° nf; distance 527. The different proportional light of the satellites in different situations, will lead us to suppose that they have a rotation on their axes. The twenty-second, when the second satellite was 78° sp, it was fainter than the first, and this evening when it was nf, it was brighter.

1798, March 11, 8^h 13′. With 300 the first and second satellites are close together, like a very faint double star of the first class. The second satellite is the most north of the two.

9^h 45′, the position of the two satellites is 78° 15′·3 nf. There is hardly a division between them.

‡ The first satellite was 78¼° nf; distance 580. The second was 78¼° nf; distance 445; and supposing the diameter of the orbits of the two satellites to be as 3 to 4, their distances at the time of observation would be as 174 to 178.*

* The angular distance of the satellites from zero and their apparent motions are inverted.

1798, March 12, 9^h 11′. The first satellite is nearly at the same distance from the planet as it was last night; the second is farther from the planet. The two satellites and the centre of the planet are exactly in a line. Their position is 78° 12′·6. With 480 I had a glimpse of a south preceding satellite; but could not verify it with 600.

12^h 6′. Distance of the second satellite 50″·02. I contrived to throw a little light upon the wires, as the satellite was bright enough this evening to bear it.

‡ The first satellite was 78¼° nf; distance 528. The second was 78¼° nf; distance 577.

1798, March 13, 11^h 46′. The first satellite is invisible. The second is much nearer the planet than it was last night. The weather is not clear, owing to easterly winds.

‡ The first satellite was 78¼° nf; distance 170; invisible. The second was 78¼° nf; distance 577.

1798, March 14, 11^h 55′. The first satellite is invisible; 8^h 31′, the second is still at a considerable distance south preceding.

11^h 47′, twenty-five feet reflector, power 200. The Georgian planet is better defined in this instrument than I have ever seen it before. With 300, its disk is as sharp and well defined as that of Jupiter. The second satellite is brought to a sharp point. A little while ago I had a glimpse of a south preceding satellite, and just now I have seen it again. 12^h 0′, I cannot verify the satellite, but can hardly believe it a deception.

Twenty feet reflector, power 300. I tried to measure the distance of the second satellite, but its present faintness will not afford light enough to see the wires of the micrometer.

‡ The first satellite was 78¼° sp; distance 260, invisible; but the 25 feet telescope with a mirror of 24 inches in diameter, it appears, had light enough to show it. The second satellite was 78¼° nf; distance 473.

1798, March 16, 8^h 37′. I see a south preceding satellite at a good distance; it may be the first at its greatest elongation, but it is certainly smaller than it should be; unless the state of the air should be worse for seeing than it appears to be.

9^h 31′, the distance of the first satellite is 36″·05. The satellite is so faint that it is impossible to be very accurate; it will not bear any light to the wires.

11^h 34′, twenty-five feet reflector. With 300 I see the satellite very distinctly, but the evening is not fine.

‡ The first satellite was 78¼° sp; distance 576. The second at 8^h 28′ was 78¼° sp; distance 13; invisible.

1798, March 18, 8^h 26′. The first satellite is invisible; the second is south preceding at a considerable distance; it is farther off than the greatest elongation of the first.

‡ The first satellite was 78¼° nf; distance 51; invisible. The second was 78¼° sp; distance 484.

1798, March 19, 9^h 51′. I see a north following satellite which I suppose to be the first. The second is near its south preceding greatest elongation. Distance 49″·90, I can only apply a very distant lantern, which will hardly give light enough to show the wires. The satellite is not so bright in its southern elongation as it was March 12th in its northern one, though the weather is now very beautiful. In the south preceding elongation is a distant star that may be a satellite; many other stars are pointed out.

‡ The first satellite was 78¼° nf; distance 445. The second was 78¼° sp; distance 588. The stars remained in their places.

1798, March 21, 10^h 25′. The first satellite is north following; it is faint, and at nearly the same distance from the planet as it was on March 19. The second satellite is near one of the stars pointed out the 19th, both being at the same distance from the planet. 10^h 40′, there is such a multitude of small stars in the neighbourhood of the planet, that it would be endless to look for the additional satellites among them.

‡ The first satellite was 78¼° nf; distance 438. The second was 78¼° sp; distance 409.

1798, March 22, 10^h 35′. The first and second satellites are both invisible.

‡ The first was 78¼° nf; distance 58; invisible. The second was 78¼° sp; distance 173; invisible.

1798, April 6, 8^h 31′. The first satellite is north following; I suspect the second to be between the first and the planet, but cannot verify the suspicion. There is a supposed south preceding satellite, but it is too near the planet to be seen steadily.

‡ The first satellite was 78° nf; distance 564. The second was 75⅓° nf: distance 230; it was

therefore the satellite suspected between the first and the planet. The supposed south preceding satellite was lost among the numerous small stars.

1798, April 7, 9^h 26′. There are two satellites north following; they are very near together. The distance between them is less than half the diameter of the planet. The centre of the planet and the two satellites are exactly in a line. There are so many small stars that it is next to impossible to look for the additional satellites.

‡ The first satellite was 79° nf; distance 546. The second at 8^h 45′ was 77⅓° nf; distance 453.

1798, April 8, 10^h 19′. There is no satellite visible between the second and the planet; the second satellite is north following, at a greater distance from the planet than last night. There is a very small star at a little more than twice the distance of the second satellite north following.

‡ The first satellite was 80½° nf; distance 239; invisible. The second was 78° nf; distance 576. The small star remained in its place.

1798, April 9, 9^h 34′. I cannot see the first satellite. The second is at a distance north following, rather farther from the planet than last night. With 480 and 600, there is no satellite between the second and the planet.

‡ The first was 74½° sp; distance 171; invisible. The second was 79° nf; distance 578.

1798, April 11, 9^h 8′. The first satellite is south preceding at a considerable distance; but not at its greatest elongation. I cannot see the second satellite. I suspect a very small star in the line of the north following greatest elongation, a little farther from the planet than the first satellite. With 480 I cannot verify the suspicion.

‡ The first satellite was 78½° sp; distance 587. The second was 80¼° nf; distance 247; invisible. This satellite could hardly be the suspected star, as it was but at little more than half the distance of the first satellite from the planet.

1798, April 12, 9^h 54′. I cannot see the first satellite, nor the second. With 480 there is no satellite either new or old visible. The night seems to be very clear, but the wind is in the north-east.

‡ The first satellite was 80⅔° sp; its distance was 382, and it ought to have been seen. The second was 61° sp; distance 46; invisible; if the north-east wind, which is always unfavourable for astronomical observations, prevented my seeing the first satellite, it could not be expected that the suspected star of the eleventh would be visible.

1798, April 13, 9^h 5′. The first satellite was not seen. There is a south preceding satellite at a little greater distance than half the greatest elongation of the first. It took some time to verify its existence. 400 shows it very well; it is the second satellite. The very small stars in this neighbourhood are so numerous, that it is impossible to look for the new satellites among them. A great many of the stars are pointed out.

‡ The first was 1° sf; distance 16; invisible. The distance being so small, the method here used is not sufficient to give an accurate position. The second satellite was 76° sp; distance 301.

1798, May 3, 10^h 3′. The first satellite is north following. 10^h 7′, I suspect another north following a little nearer the planet than the first; 460 almost verifies it.

‡ The first satellite was 79⅓° nf; distance 556. The second, which was the suspected one, was 74½° nf; distance 259; it was therefore seen at a distance considerably less than half the greatest elongation.

1799, April 3, 9^h 51′. I viewed the Georgian planet with 157 and 300; one of the satellites is about 10° sp.

‡ The first satellite was 77° sp; distance 581; the second was 66° sp; distance 172; this being invisible and the first at so great an angle, a star must have been taken for a satellite, which might well happen after an interval of eleven months. The observation was chiefly made to try two newly polished mirrors.*

1799, April 8, 9^h 51′. Both the old satellites are in a line near the greatest north following elongation. The nearest is very faint.

‡ The first satellite was 74⅓° nf; distance 452; it was the faint satellite and the nearest of the two. The second was 75¾° nf; distance 529.

1800, March 26, 11^h 0′. With a new mirror and power 300 I saw the planet beautifully well defined, and one of its satellites south preceding.

‡ The first was 74¾° sp; distance 566, it was therefore the observed satellite. The second was 31⅔° nf; distance 134; invisible.

* The planet being now again past the node, the angular distance of the satellites from zero, and their apparent motions are again inverted, and will remain so.

1800, April 26, 9h 30′. I see two satellites almost in opposition; the south preceding one is the largest and at the greatest distance. I see also several extremely small stars; but without a regular succession of observations, it is impossible to determine whether any of them may be satellites.

‡ The first satellite was 71° nf; distance 469. The second was 76⅓° sp; distance 592. They were the observed satellites, and the apparently inverted direction of their motion is evident.

1801, March 8, 12h 0′. The first satellite is at a great angle south preceding; the second at a great angle north following; but a line from the second drawn through the planet, leaves the first satellite on the following side.

‡ The first was 85¾° sp; distance 533. The second was 78¼° nf; distance 599.

1801, April 17, 10h 30′. The first and second satellites are in view at great angles north following the planet. There is a third satellite at a great angle south preceding; in the configuration it is marked exactly in opposition to the second, and at half the distance of the first. Six stars are pointed out.

‡ The first satellite was 77¼° nf; distance 598. The second was 81° nf; distance 586; and the third by the configuration was 81° sp.

1801, April 18, 10h 26′. The first and second satellites are in the configuration at great angles north following. The six stars of last night are in their places, but I do not see any star where the third satellite was marked.

‡ The first satellite was 65° nf; distance 438. The second was 74° nf; distance 578. The third was probably the interior satellite at its greatest southern elongation, which cannot be visible two days together. The configuration of this evening, compared with that of the night before, shows by the situation of the satellites, that their apparent motion is in an inverted direction.

1801, April 19, 10h 24′. The first satellite was not observed. I saw the second satellite advanced in its orbit in the inverted order. The moon is too bright to make observations on additional satellites.

‡ The first was 7½° nf; distance 144; invisible. The second was 66° nf; distance 438.

1808, May 27, 10h 0′. The planet is too low to admit the use of very high magnifying powers. With 300, however, I have a glimpse of what I suppose to be the two large satellites. A haziness coming on will not permit the angle of position to be taken.

‡ The first was 47° sf; distance 478; the second was 56° np; distance 509. They were therefore both visible.

1809, May 12, 11h 0′. I viewed the planet with 300, but could not perceive the satellites. The planet is too low, and there is a strong twilight.

‡ The first satellite was 72½° sp; distance 578. The second was 53° np; distance 529.

1810, May 25, 10h 40′. I viewed the Georgian planet with the 40 feet telescope, power 400. The disk of it is very bright. Several small stars are near it, but without a series of observations, it cannot be possible to ascertain which of them are satellites. What I suppose to be the second is 65° or 70° nf.

‡ The first was 87¼° np; distance 594. The second was 76° np; distance 588. Both satellites were therefore visible, but being among surrounding stars, could not be distinguished from them.

Investigation of several particulars deduced from the foregoing observations, with an exposition of the method by which they have been obtained.

The first use to be made of the numerous angles of position that have been taken, must be an investigation of the place of the node, and the inclination of the orbits of the satellites. When these two particulars are obtained, the times of the periodical revolutions of the satellites about the planet, may be settled. It will then be necessary to calculate the places of the satellites for the times in which they were observed, in order to identify them; for, notwithstanding all possible care was taken to keep them in view, yet after the long unavoidable annual interruptions, and the periodical interference of the moon, it will be seen, that several mistakes have been made in naming the satellites, which by that means may be easily corrected.

The place of the ascending node, the inclination of the orbits, and the retrograde motions of the satellites determined.

When the observations of the satellites in the year 1797, and the beginning of 1798, are examined, it will be found that the first satellite could seldom be seen, and that its positions, when observed, were always at a great angle from the parallel; the second was also frequently invisible; and its observed positions were likewise at great angles. From these appearances it may be concluded that the planet was approaching to the node of the satellites' orbits. At the latter end of February, and the beginning of March 1798, the position of both the satellites approached to a settled angle, which at last, for two successive days, namely the 11th and 12th of March, remained stationary; I have therefore supposed that the planet was then in its passage through the node, and have in my calculations admitted its place to be five signs, 15 degrees, 30 minutes.

Then a mean of the angles of position 78° 15′·3 and 78° 12′·9 taken the 11th and 12th of March, being 78° 14′ north following, it will appear from the method of calculation which will be explained, that the inclination of the orbits of the satellites to the ecliptic is 78° 58′.

From these data it also follows, that the motion of the satellites in their revolutions round the planet, by which they are carried from their ascending node to their greatest elongation, is retrograde.

Consideration of the principles by which the periodical revolution of the satellites may be obtained from the observed angles of position.

It will be necessary to premise, that, in order to simplify the investigation of the periodical revolution of the satellites, I have supposed the orbit of the Georgian planet, which differs only about $\frac{3}{4}$ of a degree from the ecliptic, to be coincident with it.

When the rate of the motion of a satellite in its orbit is to be determined from two observed situations, it is required to reduce its first apparent place on the plane of projection, to its real situation in its orbit; I have therefore taken the ascending node for a fixed point, from which we may begin to number the degrees of the satellite's situation; then, as in the second observation, the satellite must also be brought from its apparent place to its real one, we have to allow for three material alterations, that will more or less affect the calculation, according to the length of the interval of time between the two observations.

The first of these alterations is that which takes place in the situation of the parallel, from which the angles are measured; the second is a change in the inclination of the plane of the orbit of the satellite to the plane of projection; and the third is an alteration in the distance between the ascending node and the extreme point of the transverse axis of the ellipsis, into which the orbit is projected.

In figure 1, suppose the circle PSFN to represent the plane of the orthographic projection, in which the angles of position are counted from the parallel P and F towards S and N, and expressed by the number of degrees 10, 20, 30 to 90 in each quadrant. Then let there be a moveable circle within the former, of which the degrees should be marked in succession, 10, 20, 30 continued to 360.

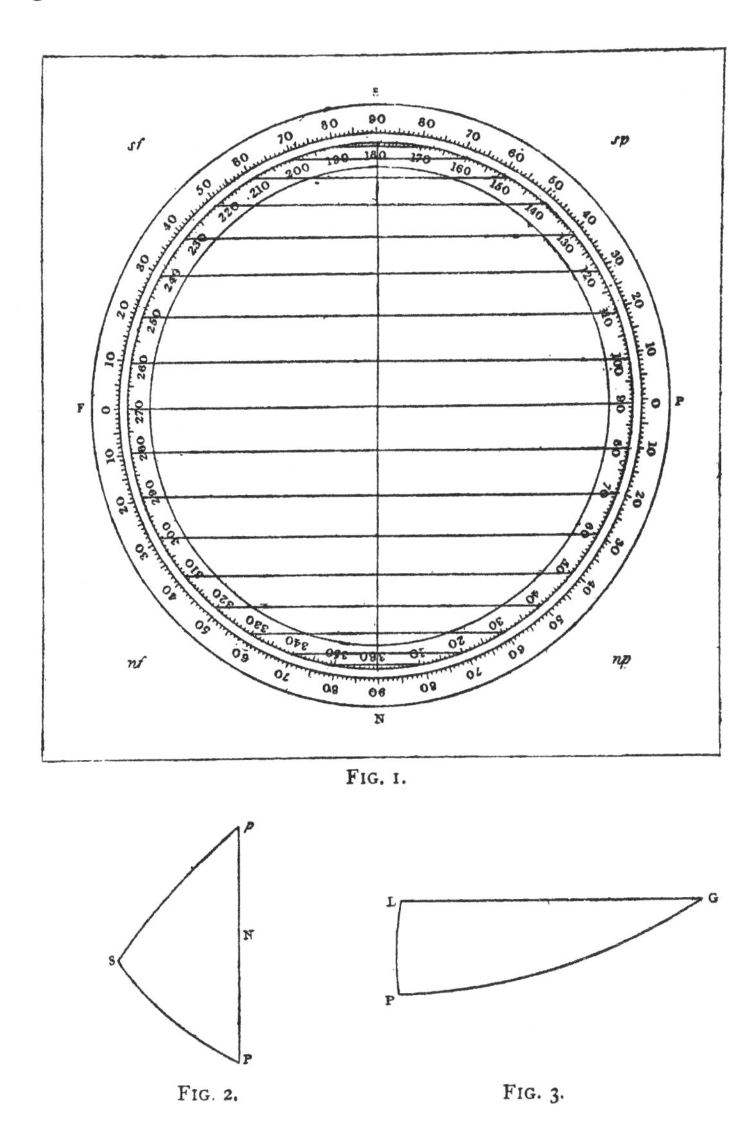

FIG. 1.

FIG. 2.

FIG. 3.

The inner circle being moveable, the line from 180 to 360 will express, by its different situations, the position of the transverse axis of the ellipsis into which the orbits of the satellites at any given time are projected. The conjugate extending from 90 to 270, and the lines parallel to it, will point out the direction in which the orbits will be more or less contracted according to the different inclinations of their planes to the plane of the projection.

In a triangle PS*p*, figure 2, let P be the pole of the ecliptic. S, a point in the orbit of the satellite, 90 degrees distant from the ascending node. *p*, the point of the greatest northern elongation of the satellites, which on the moveable circle is marked 360. It is the zero of the ellipsis into which the orbits of the satellites are projected, and its calculated situation will regulate the adjustment of the moveable circle. N, a point in a meridional direction, 90 degrees distant from the geocentric place of the planet. The arch PS then being the complement of the inclination of the satellite's orbit to the ecliptic, is therefore given; and the angle at P is equal to the distance of the longitude of the planet from the node. The angle at S is a right one; and the arch PN, being the measure of the angle of position of that point of the ecliptic where the planet is situated, may be had by Table LIII., published in Dr. MASKELYNE'S first volume of *Observations*.

To find what alteration has taken place in the situation of the parallel with regard to the point p, we calculate the position of this point for any given time by the following analogy. (1) cos P : rad : : tan PS : tan Pp; and the difference between PN and Pp will give Np; the complement of which is the position of p, with regard to the parallel. When Pp is greater than PN, the position of P being north following, that of p will be north preceding; but when Pp is less than PN, it will be north following.

To find the inclination of the plane of the satellite's orbit to the plane of projection, we have only to calculate the distance of the poles of these planes; and since the place of the planet is the pole of the projecting plane, and since also the situation of the pole of the orbits is known, which is in longitude 15 degrees 30 minutes of Gemini, and latitude south 11° 2′, therefore in the right angled triangle figure 3, of which GL is the distance of the planet from the longitude of the pole, and PL the given latitude of the pole, we have the analogy (2) rad : cos GL : : cos PL : cos GP; which is the required inclination of the two planes to each other.

To find the distance of the point p, or zero of the projected ellipsis from the ascending node, by figure 2 we have the analogy (3) rad : sin PS : : tan P : tan Sp; and 90° + Sp, before the node, and 90° − Sp, after it, will give the distance of p, or zero point of the inner circle from the ascending node.

The periodical revolutions of the satellites determined.

In the natural order of investigating the motions of the satellites, the first consideration ought to be to identify the observations, lest a star should have been mistaken for a satellite, or one satellite for another; but as the calculations required for this purpose cannot be made without proper tables of their periodical revolutions, I have proceeded in the following manner.

The earliest angles of the positions of the satellites which appeared to be sufficiently accurate for the purpose of settling their motions were taken 1787, February 16, 9^h 38′ mean time. With a wire for the parallel in the focus of the eye-glass, and a magnifying power of 300, the position of the first satellite was five degrees north following; and that of the second was three degrees south following:

the motion of the satellites being so near the parallel, there can be no material error in the estimation of the angles; and to prevent the influence of a diversity of errors, I have fixed upon the above-mentioned time as a general epoch to which every calculation of the motion of the satellites has been referred, not only in the determination of the periods, but also in every identification.

With the assistance of the analogies that have been given, seven single periods of the revolution of the first satellite were calculated, from the union of which a general compound period has been deduced. The single periods were calculated from a combination of the observation of the 16th of February 1787, with one of the same year, and with six more of the years 1790, 91, 92, 93, 94 and 96.

In the same manner, nine single periods of the second satellite have been calculated by combining the observation of the 16th of February 1787, with one or more of the succeeding, as far as 1797: the observations of 1798 being too close to the node to give a result that might be depended upon.

It will be proper to give the particulars of one of the single periods, to show what degree of accuracy has been used in the calculation.

By observation, the position of the second satellite, 1787 February 16, 9^h 38′ was 3° sf. The situation of the parallel of declination, by analogy (1) was such that the point F, upon the outer circle of Fig. 1, was opposite to 261° 14′ of the inner circle; 3° sf therefore, was at 258° 14′ of the inner circle. By analogy (2), we then find the inclination of the plane of the orbit to the plane of projection, which enables us, by the argument of the satellites being 78° 14′ from the greatest southern elongation, to reduce the apparent place in the circle to the real one, in the orbit elliptically projected. The correction for this reduction will be +2° 27′; which being applied, gives 260° 41′. But this being the situation which is numbered from the moveable zero, marked 360, it must be brought to its fixed distance from the ascending node by analogy (3); which gives the distance of the zero from that node for this day 104° 25′; and this being added to the former quantity, gives the real place of the satellite in its orbit from the ascending node 5° 6′.

To combine this with the observation of 1797 March 28, 10^h 36′ when the same satellite was 80° 44′ nf; we find that F in the parallel should now point to 281° 4′ of the inner circle; and that consequently 80° 44′ on the outer circle, will be opposite to 1° 48′ of the inner circle; and that to reduce it by the inclination of the orbit, a correction of +15° 41′ must be applied, which gives its situation in the apparent elliptical orbit 17° 41′ from zero. And when the distance of this point from the ascending node, which now is 91° 6′, is added, we have the satellite's real place in its orbit 108° 47′. Then as in 1787, it was at 5° 6′, and is now at 108° 47′, it must have moved over an arch of 103° 41′ of its orbit, to which, if we add 274 revolutions, we find that the sum of its motion amounts to 98743·68 degrees. The interval of time, in which it has moved over this number of degrees, will be found to be 3693·040277 . . days; from this we obtain the required periodical time, which is $13^d\ 11^h\ 8'\ 19''$. This single period differs only $40''$ from the compound mean period of the revolution of the second satellite.

The seven detached periods of the first satellite, and the nine of the second have all been calculated in the same manner; and in order to obtain a mean value of them, I judged it proper to allow to the duration of every interval of the time, for which they were calculated, its due weight in the scale, by compounding them together. This was done by adding together the single intervals of time in each period, and also adding together the number of degrees passed over in each single period, and computing then the compound period by these collected sums of times and motions, the result of which is, that the first satellite makes a synodical revolution about the planet in $8^d\ 16^h\ 56'\ 5''\cdot2$, and the second in $13^d\ 11^h\ 8'\ 59''$.

Explanation of the identifying method.

It is evident, that we cannot be satisfied with a conclusion that is drawn from the apparent situation of an observed satellite, if a doubt should remain whether it actually was the satellite which it is said to be; and where such numerous observations are to be examined, a method of identifying the satellites becomes absolutely necessary.

When the periodical revolution is known, the place where a satellite at any given time should be seen, may be strictly calculated; but a method somewhat less rigorous, and much more expedient, will be sufficient for the purpose; but even this will be found to require tedious computations; for in the first place, the motion of the satellite to be identified for any day, must be cast up by the table of its motion in days, hours, and minutes; and for this purpose, the interval of time for which it is calculated, must first be ascertained; this has been done for every day the satellites have been observed. The amount of the motion in orbit being obtained, it must be added to the number of degrees from which the motion proceeded; this at the already mentioned general epoch of 1787, Feb. 16, $9^h\ 38'$, was for the first satellite $11°\ 27'$ from the node, and for the second $5°\ 6'$. The sum then will be the real place of the satellite in its orbit.

Now, to obtain the apparent place of a satellite from a given real one, a table must be made, the first column of which must contain the degree of the geocentric longitude of the planet, for which the rest of the columns are calculated. The three analogies that have been given, are to be used for obtaining the contents of the second, third, and fourth columns; the fifth, contains the natural cosine of the inclination given in the fourth column multiplied by 6.

Such a Table has also been calculated for every degree, from three signs 20° to seven signs 12°, which takes in the whole compass of the observations that have been given. I insert the three first lines of the Table as a sample of its construction.

Geoc. longitude of the planet.	Position of 360 or zero.	Distance of zero from ditto.	Inclin. of the plane of the orbits to the plane of projection.	Natural cosine of the Inclin. × 6.
3ˢ 20°	79° 27′ np	105° 34′	36° 1′	4·85
21	80 17	105 1	36 57	4·79
22	81 5	104 30	37 54	4·73

In order now to use these preparatory calculations, I made an apparatus consisting of a square piece of pasteboard, upon which a circle was drawn and graduated as in figure 1. To the centre of this, I joined a moveable circle also drawn upon pasteboard, and graduated as in the figure. The radius of the inner circle was exactly six inches, and its circumference was nearly in contact with the inside of the outer circle. From what has already been said of the construction of this figure, its use in the identification of the satellites, will easily be understood by a few examples of it.

With the geocentric longitude of the planet taken from the *Nautical Almanack*, I take out the required quantities from the different columns of the Table. In this operation it might be sufficient to take only the nearest degree for entering the Table; but as the difference between any two degrees may be had by inspection, I have always used the nearest half degree. For instance, the quantities for the half degree between three signs 21° and 22°, in the second, third, and fifth columns, will be 80° 41′; 104° 45′; and 4·76. And the same quantities will do for any day from March 7, 1787, till April 23, for which day the quantities must be taken from three signs 22°, &c. &c.

Now, suppose it be required to ascertain whether a satellite called the second, which the 15th of March 1787, at 8^h 7′ was observed to be five degrees south following the planet, was indeed the second satellite? Then I see in the general list of calculated motions, that from February 16, 19^h 38′ to March 15, 8^h 7′, is an interval of 26^d 22^h 29′; in which the second satellite has moved 0° 12′ from its place; and as it was then at 5° 6′, it is therefore now 5° 18′ from the ascending node of its orbit.

In using the identifying apparatus, the first thing to be done is, the adjustment of the inner circle to the position it ought to have for the day of observation, which is pointed out by the geocentric longitude of the planet three signs 21½ degrees; the zero must consequently be adjusted to 80° 41′ north preceding; I therefore turn the inner circle upon its centre, till the point 360 is opposite to 80⅔° np; for in the adjustment of the circles to each other, and in reading off the angles pointed out by them, a critical estimation of minutes has not been attempted; whenever, therefore, minutes are given, they must be understood to relate to calculations, or to measures taken with a micrometer.

In the next place, the point of the inner circle, answering to the calculated situation of the satellite in its orbit, is to be found by the tabular quantity 104° 45′, which is the distance of the ascending node from the zero of that circle; and as the satellite is 5° 18′ from the same node, the quantity given by the Table must be deducted from the same; that is from 365° 18′, and the remainder 260° 33′ will be the place of the satellite on the moveable circle.

Finally, to get the angle of position at which the satellite will be seen, the two ends of a proportional compass must be adjusted to each other, so that when one end of it is opened to six inches, the other may give the quantity in the last column of the Table, which in the present case is 4·76 inches. The distance from 260½° to the transverse must then be measured by the long end of the compass, in a direction parallel to the lines in the figure; and the opening of the short end must be set back again in the same direction, from the transverse towards 260½°; a fine point must be marked with the end of it upon the pasteboard. A black silk thread fixed to the centre of the circle, may then be stretched over the impressed point to intersect the degrees of the outer circle, upon which the positions are reckoned in the order they are marked; and this being done, it will be seen that the intersection of the thread falls upon 2¼ degrees of the south following quadrant; which sufficiently identifies the satellite.

In observations that are made when we are in, or very near the node of the orbits of the satellites, their angular positions undergo hardly any change, and can therefore be of no use for identifying them; but they may then be distinguished by their proportional distances from the planet; and these may be very conveniently had in six hundredth parts of the respective radii of the satellites, and the impression of the fine point whereby the angle of position is obtained will be of eminent use; for by putting one leg of the compass upon this point, and extending the other to the centre of the circle, we shall in the present case have 4·81 inches for the measure of the required distance, which, as the radius of the circle is six inches, will be $\frac{481}{600}$ parts of it; and in such parts all the distances which are given in the foregoing observations have been expressed.

I have called this manner of obtaining the angles of position and proportional distances, the identifying method, that it may remain distinguished from strict computation; there is, however, so much real calculation mixed with it, that I may confidently draw the following interesting conclusions from it.

I. With the light of my 20 feet telescope, the first satellite generally becomes invisible at the distance of a little more than half its greatest elongation; I suppose it to be when the identified measure of it is from 302 to about 310.

II. The second satellite becomes invisible at very nearly half the distance of its greatest elongation; I suppose it to be when its identified distance is from 295 to about 305.

III. An interior satellite as large as the first, must be more than half the greatest elongation of the first satellite from the planet; and if it be smaller, it must be at so much greater a distance from the planet, to be seen at its greatest elongation. Nor can there be any chance for seeing it two nights together, when the orbits are contracted by projection.

IV. Exterior satellites that are very faint when at their greatest elongation, can hardly ever be seen at any other time when the orbits are contracted.

V. The first satellite is probably larger than the second; for though the latter is generally the brightest, it seems to be only in consequence of its being farthest from the planet. On comparing the limit of its disappearance with the number 302, expressing that at which the first satellite generally ceases to be visible, we find that the second satellite, upon its own scale, should not be lost in the light of the planet till it came within the limit of 224, instead of 295.

VI. Both the satellites are subject to great variations of light, not owing to the changeable clearness of the air at different times; for by comparing the brightness of one satellite with that of the other when they are seen together, the state of the air will be of equal clearness to both, and yet their comparative brightness has been observed to be very different: for instance, March 14, 1793, the first satellite was brighter than the second, when the distance of the former was to that of the latter as 172 to 235; and February 26, 1798, the first was small, and the second larger when the distance of the former was to that of the latter as 175 to 210.

VII. The variable brightness of the satellites may be owing to a rotation upon their axes, whereby they alternately present different parts of their surfaces to our view. These variations may also arise from their having atmospheres that occasionally hide or expose the dark surface of their bodies, as is the case with the sun, Jupiter and Saturn.

VIII. The real angular distances of the satellites from the planet may be determined from the measures that are given in the observations, but to enter critically into this subject would extend far beyond the compass of this Paper. The disagreement of the measures is very considerable; this will, however, not appear so remarkable, when the faintness of the satellites is considered, which will not admit of an illumination of the wires of the micrometer. The two measures of the distance of the first satellite that were taken Oct. 11, 1787, and March 12, 1798, should both be considered; but a selection of about six of the most consistent measures of the second satellite, will probably be necessary to give the truest result. For instance, those of 1787, March 18, 20, April 9, and November 9, with those of 1798, March 12 and 19, might be taken. If these measures are brought to their greatest elongation by the identified distances that are given with them, some kind of judgment may be formed of the probable result, when calculation is applied to them.

In my observations I have supposed the distances of the first and second satellites to be 36 and 48 seconds, and by this proportion I have occasionally reduced the identified distances of the two satellites to an equal scale.

IX. The existence of additional satellites has already been considered in a former paper.* Many remarks on them were given under the four heads of interior, intermediate, exterior, and more distant satellites; and, as many additions are contained in the foregoing observations, I shall review the former remarks, with the assistance of the light which the identifying method has thrown upon them, and afterwards, in the same order consider, in each class, what evidence of the existence of such satellites may be derived from the additional observations, especially from those that were made in the year 1798, when the orbits of the satellites were contracted into a line, which might be examined with greater facility than a more expanded space; and where even the very situation of a star in this given direction, rather than in the numberless others, in which it might be placed, must be a presumption of its being a satellite, provided its distance at the same time should not exceed a certain probable limit.

An interior satellite.

The supposed interior satellite observed January 18, 1790, could not be a star, whose existence was doubtful, as it had light enough for an estimation of its distance in diameters of the planet; its absence, however, not having been noticed the 19th of January, although great attention has been shown to ascertain the sidereal nature of another supposed satellite, observed at the same time with the former, leaves the observation of the eighteenth unsupported.

The observation of the 4th of March 1794, which has been supposed to relate to an interior satellite, is by the identifying method proved to belong to the second satellite; and its observed absence on the 5th from the place where it was the 4th, being thus verified and accounted for, shows that great confidence may be placed on such observations.

The observation of an interior satellite of the 27th of March, 1794, is without a subsequent observation; but then it has already been noticed in remark III., that an interior satellite cannot be seen two successive days, when its orbit is already contracted, as it was on the day of observation.

Addition.

The 15th of February, 1798, an interior satellite was seen about its greatest northern elongation; as it was between the planet and the second satellite (miscalled the fifth), its position must have been 84° 49′ nf. On account of its faintness

* *Phil. Trans.* for 1798, page 59 [above, p. 9].

it was not seen immediately, but as soon as it was perceived, it was surmised to be the second satellite ; but the identified distance of the second, which was 455, is much too far for the observed distance of the faint satellite, and proves that in reality the supposed fifth was the second satellite. This is moreover confirmed by its brightness, and by the angle of position which was taken. The first satellite was invisible, its distance being only 124 ; it even remained invisible the next day, when its distance was 289 ; the observed faint one, therefore, must have been an interior satellite in a distant part of its northern elongation. The 16th of February, the place where the satellite had been the day before was scrupulously examined in looking for the supposed fifth, and as there was no star remaining in that place, the removal of the interior satellite from its former situation was thereby also ascertained. It has already been noticed that in the contracted position of the orbits an interior satellite observed the 15th could not possibly be seen the 16th, which accounts for its not being noticed the last day of observation. This is one of the cases where the singular situation of a star alone, is almost sufficient to prove it to be a satellite.

The 17th of April, 1801, the interior satellite, which had been seen in its greatest northern elongation, was now seen about its greatest southern elongation. Its situation, by identification, was 81° sp.* The 18th of April, when the stars that were pointed out were looked after, and were all found remaining in their places, no star could be seen where the interior satellite had been situated the 17th ; nor could it be expected to be visible, as, by its motion towards the planet, it must already have been involved in the splendour of its light.

An intermediate satellite.

The 26th of March, 1794, an intermediate satellite was seen ; by the configuration its distance was greater than that of the first, but less than that of the second. By identification its situation must have been 59½° nf. The 27th of March, the satellite was no longer in the place where it had been seen the 26th, and moreover, a very small star was seen in a place that agreed with what would be the situation of an intermediate satellite, had it accompanied the planet.

An exterior satellite.

The 9th of February, 1790, an exterior satellite was observed. It was by the configuration at double the distance of the second satellite, and by identification, its position was 61½° sf. The observations of two succeeding days proved, that it remained no longer in the place where it had been seen on the 9th.

The 27th of March, 1794, some distant stars south of the planet were observed as being supposed satellites ; but they are not sufficiently supported by succeeding observations.

* [This was very probably the satellite Ariel; HOLDEN, *Monthly Notices R.A.S.*, xxxv. p. 20.—ED.]

The 5th of March 1796, a star was seen, which the night before had not been in the place where it was at the time of the observation. By the configuration its identified place must have been 72° np; and its distance exceeded that of the second satellite.

Addition.

The 31st of January, 1791, a satellite in opposition to the second, and at about double the distance from the planet was observed; its identified position was 78½° np. The 2d of February all the stars of the configuration that had been pointed out, were seen remaining in their places, but the exterior satellite was not among them.

The 26th of February, 1792, a star at double the distance of the first satellite was pointed out, but it has not been accounted for in succeeding observations. By remark IV., however, faint exterior satellites can hardly be expected to be seen at any other time, than when they are about their greatest elongation.

The 11th of February, 1798, an exterior satellite, called the fifth, was observed; its situation was measured 89° 19′·5 nf. The 13th of February, this satellite was no longer in the place where it had been seen the 11th. The 15th of February it was again ascertained that the satellite had left its former situation. The orbits of the satellites, at the time of these observations, were already contracted into a line, and a very faint satellite like this could not remain visible two, and four days successively; its motion, according to remark IV., would in a short time immerge it again into the effusive light of the planet, and render it invisible.

More distant satellites.

The 28th of February, 1794, a small star was seen in a place where the 26th there was none. By the configuration of that day, and the identifying method, it was at a considerable distance, about 24 degrees north following the planet, and not far from a lettered star which was smaller than the new star. It cannot be supposed that a larger star should have been omitted to have been marked in the situation pointed out by smaller lettered stars, where it must have been seen the 26th, if it had been there.

The 27th of March, 1794, south of the planet, at a considerable distance, were small stars, that had the appearance of satellites; but there are no subsequent observations of them.

The 28th of March, 1797, a distant star is mentioned that was not seen the 25th, although the situation of the lettered stars of that day was carefully examined.

Addition.

The 16th of February, 1798, at 11^h $12'$ a very faint satellite, called the sixth, was observed, and from its distance, supposed to be a little before or after its greatest southern elongation. It was so faint, that a small alteration in the clearness of the air, rendered it invisible. On the 18th the sixth satellite was seen again, and, being nearer the planet than it was on the 16th at 11^h $12'$, it was supposed to be on its return from the greatest southern elongation. It was also ascertained on the 18th, that it had left the place where it was seen on the 16th. The angle of its position, by a measure taken of it, was $82°$ $55'$ south preceding.

LXVIII.

Astronomical observations and experiments tending to investigate the local arrangement of the celestial bodies in space, and to determine the extent and condition of the Milky Way. By Sir WILLIAM HERSCHEL, *Knt. Guelp, LL.D. F.R.S.*

[*Phil. Trans.*, 1817, pp. 302–331.]

Read June 19, 1817.

THE construction of the heavens, in which the real place of every celestial object in space is to be determined, can only be delineated with precision, when we have the situation of each heavenly body assigned in three dimensions, which in the case of the visible universe may be called length, breadth, and depth ; or longitude, latitude, and Profundity.

The angular positions of the stars and other celestial objects, as they are given in astronomical catalogues, and represented upon globes, or laid down in maps, enable us, in a clear night, to find them by the eye or to view them in a telescope ; for, in order to direct an instrument to them, a superficial place consisting of only two dimensions is sufficient ; but although the line in which they are to be seen is thus pointed out to us, their distance from the eye in that line remains unknown ; and unless a proper method for obtaining the profundity of objects can be found, their longitude and latitude will not enable us to assign their local arrangement in space.

With regard to objects comparatively very near to us, astronomers have completely succeeded by the method of parallaxes. The distance of the sun ; the dimensions of the orbits of the planets and of their satellites ; the diameters of the sun, the moon, and the rest of the bodies belonging to the solar system, as well as the distances of comets, have all been successfully ascertained. The parallax of the fixed stars has also been an object of attention ; and although we have hitherto had no satisfactory result from the investigation, the attempt has at least so far succeeded as to give us a most magnificent idea of the vast expansion of the sidereal heavens, by showing that probably the whole diameter of the earth's orbit, at the distance of a star of the first magnitude, does not subtend an angle of more than a single second of a degree, if indeed it should amount to so much ; with regard to more remote objects, however, such as the stars of smaller size, highly compressed clusters of stars and nebulæ, the parallactic method can give us no assistance.

I. *Of the local situation of the stars of the heavens.*

The superficial situation of the stars having already been carefully assigned in the catalogues of astronomers, it will be proper to examine how far the arrangement of the stars into a certain order of magnitudes can assist us to determine their local situation.

When we look at the heavens in a clear night, and observe the different lustre of the stars, we are impressed with a certain idea of their different magnitudes; and when our estimation is confined to their appearance only, we shall be justified in saying, for instance, that Arcturus is larger than Aldebaran; the principle on which the stars are classed is, therefore, entirely founded on their apparent magnitude, or brightness. Now, as it was thought convenient to arrange all the stars which in fine weather may be seen by the eye into seven classes, the brightest were called of the first, and the rest according to their gradually diminishing lustre, of the 2d, 3d, 4th, 5th, 6th, and 7th magnitudes. Then, since it is evident that we cannot mean to affirm that the stars of the 5th, 6th, and 7th magnitudes are really smaller than those of the 1st, 2d, or 3d, we must ascribe the cause of the difference in the apparent magnitudes of the stars to a difference in their relative distances from us; and on account of the great number of stars contained in each class, we must also allow that the stars of each succeeding magnitude, beginning from the first, are one with another farther from us than those of the magnitude immediately preceding. It may therefore be said, that since in our catalogues the magnitudes are added to the two dimensions which give the superficial place of the stars, we have also at least a presumptive value of the third dimension; but admitting that the naked eye can see stars as far from us as those of the seventh magnitude, this presumptive value, which can only point out their relative situation, will afford us no information about the real distance at which they are placed.

II. *Of a standard by which the relative arrangement of the stars may be examined.*

It is evident, that when we propose to examine how the stars of the heavens are arranged, we ought to have a certain standard of reference; and this I believe may be had by comparing their distribution to a certain properly modified equality of scattering. Now, the equality I shall here propose, does not require that the stars should be at equal distances from each other; nor is it necessary that all those of the same nominal magnitude should be equally distant from us. It consists in allotting a certain equal portion of space to every star, in consequence of which we may calculate how many stars any given extent of space should contain. This definition of equal scattering agrees so far with observation, that it admits, for instance, Sirius, Arcturus, and Aldebaran to be put into the same class, notwithstanding their very different lustre will not allow us to suppose them to be at equal distances from us ; but its chief advantage will be, that instead of the order of magnitudes into which our catalogues have arranged the stars, it will give us an order of distances, which may be used for ascertaining the local distribution of the heavenly bodies in space.

To explain this arrangement, let a circle be drawn with any given radius about the point S fig. 1, and with 3, 5, 7, 9, &c. times the same radius draw circles, or circular arcs, about the same centre. Then if a portion of space equal to the solid contents of a sphere, represented by the circle S, be allotted to each star, the circles, or circular arcs drawn about it will denote spheres containing the stars of their own order, and of all the orders belonging to the included spheres, and on the supposition of an equality of scattering, the number of stars of any given order may be had by inspection of the figure, which contains all the numbers that are required for the purpose; for those in front of the diagram express the diameters of spherical figures. The first row of numbers enclosed between the successive arcs, are the cubes of the diameters; the next column expresses the order of the central distances; and the last gives the difference between the cube numbers of any order and the cube of the next enclosed order.

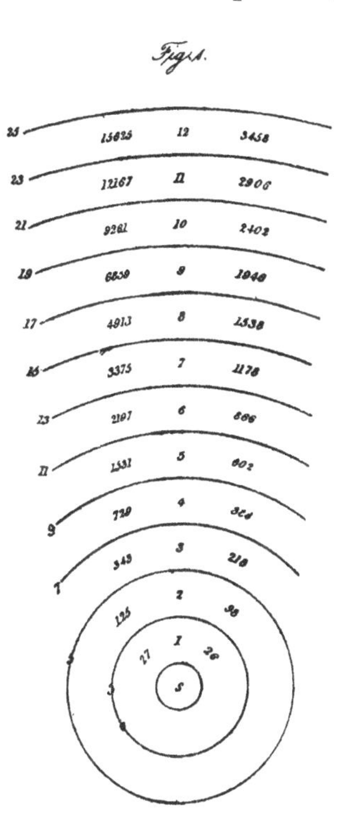

The use to be made of these columns of numbers is by inspection to determine how many stars of any particular order there ought to be if the stars were equally scattered. For instance, let it be required how many stars there should be of the 4th order. Then No. 4, in the column of the orders points out a sphere of nine times the diameter of the central one, and shows that it would contain 729 stars; but as this sphere includes all the stars of the 3d, 2d, and 1st order as well as the sun, their number will be the sum of all the stars contained in the next inferior sphere amounting to 343; which being taken from 729 leaves 386 for the space allotted to those of the 4th order of distances.

III. *Comparison of the order of magnitudes with the order of distances.*

With a view to throw some light upon the question, in what manner the stars are scattered in space, we may now compare their magnitudes, as we find them assigned in Mr. BODE's extensive catalogue of stars, with the order of their distances which has been explained.

The catalogue I have mentioned contains 17 stars of the 1st magnitude; but in my figure of the order of the distances we find their number to be 26.

The same catalogue has 57 stars of the 2d magnitude; but the order of distances admits 98.

Of the third magnitude the catalogue has 206, and the order of distances will admit 218.

The number of the stars of the 4th magnitude is by the catalogue 454, and by the order of distances 386.

Before I proceed, it may be proper to remark, that, by these four classifications of the stars into magnitudes, it appears already, that, on account of the great difference in the lustre of the brightest stars, many of them have been put back into the second class; and that the same visible excess of light has also occasioned many of the stars of the next degree of brightness to be put into the third class; but the principle of the visibility of the difference in brightness would have less influence with the gradually diminishing lustre of the stars, so that the number of those of the third magnitude would come nearly up to those of the third distance. And as the difference in the light of small stars is less visible than in the large ones, we find that the catalogue has admitted a greater number of stars of the 4th magnitude than the 4th order of distances points out; this may, however, be owing to taking in the stars that were thrown back from the preceding orders; and a remarkable coincidence of numbers seems to confirm this account of the arrangement of the stars into magnitudes. For the total number of the catalogued stars of the 1st, 2d, 3d, and 4th magnitudes, with the addition of the sun, is 735; and the number contained in the whole sphere of the 4th distance is 729.

Now the distinguishable difference of brightness becoming gradually less as the stars are smaller, the effect of the principle of classification will be, as indeed we find it in the 5th, 6th, and 7th classes, that fainter stars must be admitted into them than the order of distances points out.

The catalogue contains 1161 stars of the 5th magnitude, whereas the 5th order of distances has only room for 602.

Of the 6th magnitude the catalogue contains not less than 6103 stars, but the 6th order of distances will admit only 866.

And lastly, the same catalogue points out 6146 stars of the 7th magnitude, while the number of stars that can be taken into the 7th order of distances is only 1178.

The result of this comparison therefore is, that if the order of magnitudes could indicate the distance of the stars, it would denote at first a gradual, and afterwards a very abrupt condensation of them; but that, considering the principle on which the stars are classed, their arrangement into magnitudes can only apply to certain relative distances, and show that taking the stars of each class one with another, those of the succeeding magnitudes are farther from us than the stars of the preceding order.

IV. *Of a criterion for ascertaining the Profundity, or local situation of celestial objects in space*

It has been shown that the presumptive distances of the stars pointed out by their magnitudes can give us no information of their real situation in space. The statement, however, that one with another the faintest stars are at the greatest distance from us, seems to me so forcible, that I believe it may serve for the foundation of an experimental investigation.

It will be admitted, that the light of a star is inversely as the square of its distance; if therefore we can find a method by which the degree of light of any given star may be ascertained, its distance will become a subject of calculation. But in order to draw valid consequences from experiments made upon the brightness of different stars, we shall be obliged to admit, that one with another the stars are of a certain physical generic size and brightness, still allowing that all such deviations may exist, as generally take place among the individuals belonging to the same species.

There may be some difference in the intrinsic brightness of starlight: that of highly coloured stars may differ from the light of the bluish white ones; but in remarkable cases allowances may be made.

With regard to size, or diameter, we are perhaps more liable to error; but the extensive catalogue which has already been consulted, contains not less than 14,144 stars of the seven magnitudes that have been adverted to; it may therefore be presumed that any star promiscuously chosen for an experiment, out of such a number, is not likely to differ much from a certain mean size of them all.

At all events it will be certain that those stars the light of which we can experimentally prove to be $\frac{1}{4}$, $\frac{1}{9}$, $\frac{1}{16}$, $\frac{1}{25}$, $\frac{1}{36}$, and $\frac{1}{49}$ of the light of any certain star of the 1st magnitude, must be 2, 3, 4, 5, 6, and 7 times as far from us as the standard star, provided the condition of the stars should come up to the supposed mean state of diameter and lustre of the standard star, and of this, when many equalisations are made, there is at least a great probability in favour.

V. *Of the equalisation of starlight.*

In my sweeps of the heavens, the idea of ascertaining the Profundity of space to which our telescopes might reach, gave rise to an investigation of their space penetrating power; and finding that this might be calculated with reference to the extent of the same power of which the unassisted eye is capable, there always remained a desideratum of some sure method by which this might be ascertained.

Of various experiments I have long ago tried, the equalisation of starlight, which about four years ago I began to put into execution, appeared to be the most practicable. A description of the apparatus and the method of making use of it is as follows.

Of ten highly finished mirrors I selected two of an equal diameter and focal length, and placed them in two similarly fitted up seven feet telescopes. When they were completely adjusted, I directed them both, with a magnifying power of 118, to the same star, for instance, Arcturus: and upon trial I found the light not only of this, but of every other star to which they were directed, perfectly equal in both telescopes.

The two instruments, when I viewed the stars, were placed one a little before the other, and so near together that it would require little more than one second of time to look from one into the other. This convenient situation of the instruments is of great importance. The impression of the light made by the view of one star should be succeeded as soon as possible by the view of the other; and these alternate inspections should also be many times repeated, in order to take away some little advantage which the last view of a bright object has over that immediately preceding.

In comparing the light of one star with that of another, I laid it down as a principle, that no estimation but that of perfect equality should be admitted; and as the equal action of the instruments was now ascertained, I calculated the diameters of several apertures to be given to one of the telescopes as a standard, so that the other, called the equalising telescope, might be employed, with all its aperture unconfined, to examine a variety of stars, till one of them was found whose light was equal to that of the star to which the standard telescope was directed.*

In order to be sufficiently accurate in the calculation of the diameter of the limiting apertures, I thought it necessary to take into consideration not only the obstruction of incident light occasioned by the interposition of the small mirror, but also of the arm to which it is fastened, and proceeded as follows:

If A be the diameter of the large mirror; b that of the small one; $\frac{A-b}{2}$ the length of the arm; t its thickness; π the circumference, diameter being unity; x an assumed quantity for finding the correction; A′ the aperture corrected for the interposition of the arm; L the light of the equalising telescope; p the proportion of the light required for the standard telescope; D the diameter of an aperture to give that light; D′ the diameter corrected for the interposition of the arm.

Then will the diameters of the limiting apertures be had by the following equations. $\frac{A-b}{2} \times t = \pi A x$; $\frac{A-b}{\pi A} \times t = 2x$; $A - 2x = A'$; $A'^2 - b^2 = L$; $pL = D^2 - b^2$; $\sqrt{pL + b^2} = D$; $\frac{D-b}{\pi D} \times t = 2y$; $D + 2y = D'$ the required diameter.

* I preferred the limitation of the light by circular apertures to the method of obtaining it by the approach or recess of two opposite rectangular plates, in order to avoid the inflections which take place in the angles.

In the calculation of a set of apertures for the intended purpose, I admitted none that gave less than $\frac{1}{4}$ of the light; for by a greater contraction of the aperture of the mirror, an increase of the spurious diameters would render a judgment of equality liable to deception; * when therefore a star of the third order of distances was to be found, I rejected the direct way of reducing a star of the first order to $\frac{1}{9}$ of its light, but selected a star previously ascertained to be of the second order; and by taking $\frac{4}{9}$ of its light, the equalising telescope, with all its light, was used to examine all such stars as appeared likely to give the required equality, till one of them was found; nor was it necessary to have a great number of limiting apertures, as it soon appeared that with eight or ten of them I could have many different gradations of light, which would ascertain even fractional degrees, and reach as far as the stars of any order of distances I could expect to be visible to the unassisted eye.

This method of equalising the light of the stars, easy as it may appear, is nevertheless subject to great difficulties; for as the brightness of a star is affected by its situation, with regard to the ambient light of the heavens, the stars to be equalised should, if possible, be in nearly the same region. When the sun is deep under the horizon, this is however not of so much consequence as the altitude of the star to be equalised, which ought to be as nearly as possible equal to that of the standard star. At great elevations some difference in the altitudes of the stars to be equalised may be admitted; but, if they are far from each other, the circumstance of the equal illumination of the heavens, and the equal clearness of the air, must still be attended to.

VI. *Of the extent of natural vision.*

The method of equalising star light may be rigorously applied to ascertain the extent of natural vision; for in this case it will not be required that the star on which the experiment is tried, should be of the same size or diameter with the standard star; nor is it necessary that the intrinsic brightness of the light of the two stars should be the same in both. It will be sufficient, that the star we choose for an equalisation is one of the smallest that are still visible to the natural eye. It is also to be understood that, till we can have a well ascertained value of the parallax of any one star of the first magnitude, the extent of natural vision can only be given in a measure of which the distance of the standard star is the unit.

The following equalisations were made in August and December 1813, and February 1814, and are given as a specimen of the method I have pursued.

* This was fully proved by the following experiment. July 27, 1813, I viewed Arcturus in a 10 feet reflector; first with all its light; next with circular diaphragms, which confined its aperture to $\frac{1}{4}$, $\frac{1}{9}$, $\frac{1}{16}$, $\frac{1}{25}$, $\frac{1}{36}$, and $\frac{1}{49}$ of it; but I found that the different spurious diameters, arising from the smallness of the apertures, made estimations of what is generally called the magnitude of the stars impossible.

See also experiments on the spurious diameters of the celestial objects, *Phil. Trans.* for 1805, page 40 [above, p. 302].

Taking Arcturus for the standard of an experiment, I directed the telescope, with one quarter of its light, upon it; while the equalising telescope, with all its light, was successively set upon such stars as I supposed might be at double the distance of the standard star; which, as Arcturus is a star of the first magnitude, I expected to find among those of the second.

The first I tried was β (FL. 53) Pegasi, but I found it not quite bright enough

The light of α Andromedæ, which next I tried, was nearly equalised to that of Arcturus; and the observation being repeated on a different night gave it equal.

Now as in these experiments the standard star is supposed to be one of the first order of distances, it follows that, if Arcturus were put at twice its distance from us, it would then appear like α Andromedæ, as a star of the 2d magnitude, and would also at the same time be really a star of the 2d order of distances.

In order to obtain some other stars whose light might be equalised by one quarter the light of Arcturus, I tried many different ones, and found among them α Polaris, γ Ursæ, and δ Cassiopeæ. These stars therefore may also be put into the class of those whose light is equal to the stars of the second order of the distance of Arcturus.

For the purpose of ascertaining the extent of natural vision, it will not be necessary here to give the equalisation of stars of the 3d, 5th, 6th, or 7th order of distances; but taking now the light of one of the stars of the 2d order of distances for a standard, I tried many that might be expected to have the required light, and found that when α Andromedæ, with its light reduced to one quarter, was in the standard telescope, the equalising one gave μ (FL. 48) Pegasi for a star of the 4th order of distances. That is to say, the equalisation proved that, if Arcturus were placed at four times its distance from us, we should see it as a star of the 4th magnitude, and also as one of the 4th order of distances.

Proceeding in the same manner with μ Pegasi taken as a standard, I found that its light reduced to $\frac{1}{4}$ was equal to that of *q* (FL. 70) Pegasi, when seen in the equalising telescope; and that consequently Arcturus, removed to 8 times its present distance from us, would put on the appearance of a star which in our catalogues is called of the 5th or 6th magnitude, but which would in fact be of the 8th order of distances.

As the foregoing experiments can only show that a star of the light of Arcturus might be removed to 8 times its distance, and still remain visible to the naked eye as a star of between the 5th and 6th magnitude; it will be proper to take also other stars of the first magnitude for the original standards.

For instance, if we begin from Capella as the standard star, we may with $\frac{1}{4}$ of its light equalise β Aurigæ and β Tauri, which stars will therefore be of the 2d order of distances. With $\frac{1}{4}$ of the light of β Tauri we equalise ζ Tauri and ι Aurigæ; they will then be of the 4th order. With $\frac{1}{4}$ of the light of ι Aurigæ we can equalise *e* Persei and H Geminorum which will be of the 8th order. And with $\frac{16}{25}$ of the light of H Geminorum we equalise *d* Geminorum, which makes it a star of the 10th order. That is to say, if Capella were successively removed to 2, 4, 8 and 10 times the distance at which it is from us, it would then have the appearance of the stars which have been named.

A similar deduction may be made from α Lyræ, as $\frac{1}{4}$ of its light equalises it with β Tauri; for it will be α Lyræ 1, β Tauri 2, ι Aurigæ 4, H Geminorum 8, and *d* Geminorum 10: the numbers annexed to the stars expressing their orders of distances in terms of the distance of α Lyræ from us.

To find stars of the intermediate orders of distances, the following Table gives the proportional light that should be used with the star which is made the standard; for instance, a star of the 2d order of distances, with $\frac{4}{9}$ of its light, will equalise a star of the 3d order; $\frac{9}{25}$ of the light of a star of the 3d order of distances will give one of the 5th order, and so on.

A star of the order of distances.	With the proportion of its light.	Gives one of the order of distances.
1	$\frac{1}{4}$	2
2	$\frac{4}{9}$	3
	$\frac{1}{4}$	4
3	$\frac{9}{25}$	5
	$\frac{1}{4}$	6
4	$\frac{16}{49}$	7
	$\frac{1}{4}$	8
5	$\frac{25}{81}$	9
	$\frac{1}{4}$	10
6	$\frac{36}{121}$	11
	$\frac{1}{4}$	12

Some other proportions of light useful for fractional distances are $\frac{9}{16}$, $\frac{16}{25}$, $\frac{36}{49}$, $\frac{64}{81}$, and $\frac{100}{121}$.

The results of equalisations that are made with different standard stars, may be connected together by an equalisation of the standards; by which means many different sets may be brought to support each other. For instance, Capella with $\frac{36}{49}$ of its light is of an equal lustre with Procyon, which therefore is of the $1\frac{1}{6}$ order of Capella, and Sirius with $\frac{16}{49}$ of its light is also of an equal lustre with Procyon, which consequently, with regard to Sirius, is of the $1\frac{3}{4}$ order; then, by compounding, it follows that Capella to Sirius is a star of the $1\frac{1}{2}$ order, and from this we obtain the following series. Sirius 1, Capella $1\frac{1}{2}$, Procyon $1\frac{3}{4}$, β Tauri 3, ι Aurigæ 6, H Geminorum 12, and *d* Geminorum 15. By this connection we shall be able to obtain an equalisation of the same ultimate star with all the standards; for if Sirius must be removed to the 15th order, to appear as faint as *d* Geminorum; and if Capella, and also α Lyræ must be removed to the 10th order of distances to appear as faint as the same star, then any star of the size and brightness of Sirius, Capella, and α Lyræ must generally appear as faint as *d* Geminorum, when it is removed to nearly 12 times its distance; and the more stars of the first order are admitted in these general equalisations reduced to the same faint star, the more will the probability of the result be extended. Now as *d* Geminorum is a star of the 6th magnitude, we may expect that a still fainter visible star will give a somewhat greater extent to the reach of the natural eye; if however I take its vision, including other stars of the 1st magnitude, to extend to the 12th order of distances, there will probably be no material error, at least none but what a diligent astronomer, who is provided with the necessary apparatus, may correct by observation.

But the extent of natural vision is not limited to the light of solitary stars only; the united lustre of a number of them will become visible when the stars themselves cannot be seen. For instance, the milky way; the bright spot in the sword handle of Perseus; the cluster north of η and H Geminorum; the cluster south of FL. 6 and 9 Aquilæ; the cluster south of η Herculis, and the cluster north preceding ε Pegasi. But their distances cannot be ascertained by the method of equalising starlight: their probable situation in space may however be deduced from telescopic observations.

To these very faintly visible objects may be added two of a different nature, namely, the nebulosity in the sword of Orion, and that in the girdle of Andromeda.

VII. *Of the extent of telescopic vision.*

The powers of telescopes to penetrate into the Profundity of space is the result of the quantity of light they collect and send to the eye in a state fit for vision. The method of calculating the quantity of this power has been fully explained in a Paper read before the Royal Society, November 21, 1799; and the formulæ which have been given in that Paper have already been applied to show to what extent this power has been carried in the telescopes I used for astronomical observations. The calculated results, however, give this power only in reference to that of natural vision, and the uncertainty in which we were left with regard to its extent, was equally thrown over that of telescopic vision.

The equalisation of starlight, when carried to a proper degree of accuracy, will do away with the cause of the error to which the telescopic extent of vision has been unavoidably subject; we may therefore safely apply this vision to measure the Profundity of sidereal objects that are far beyond the reach of the natural eye; but for this purpose the powers of penetrating into space of the telescopes that are to be used must be reduced to what may be called gaging powers; and as the formula $\frac{\sqrt{x \, . \, \overline{A^2 - b^2}}}{a}$* gives the whole quantity of the space penetrating power, a

* See *Phil. Trans.* for 1800, page 66 [above, p. 41].

reduction to any inferior power p, may be made by the expression $\sqrt{\frac{p^2 a^2}{x} + b^2} = A$; when the aperture is then limited to the calculated value of A, the telescopes will have the required gaging power. Or we may prepare a regular set of apertures to serve for trials, and find the gaging powers they give to the telescope by the original formula.

In the formula by which the required apertures for the gaging powers were calculated, a has been put equal to two tenths of an inch, and to show that this assumption is founded upon observation, I give the following extract from my astronomical Journal.

Dec. 27, 1801. I looked at α Lyræ with one eye shut and the other guarded by a slip of brass with holes of various sizes in it. Through the hole which was 0·28 inch in diameter, I saw the star just as well as without the limiting diaphragm, which shows that the opening of the pupil of the eye does not exceed 0·28 inch.

I tried the same star through 0·24 and still saw it equally well. I tried next 0·21 and still saw it as well.

The slip of brass was held as close to the eye as possible. The next I tried was 0·17 in diameter, and through this I could perceive a small deficiency of light, so that the opening of the pupil exceeds 0·17 inch. The night is hardly dark enough yet for great accuracy.

Having been out long in the dark, and trying the same experiment upon many different large and small stars, they all concur to show that 0·21 does not sensibly stop any light; but that less does certainly render the object rather less luminous; so that the opening may be put at two-tenths of an inch in my eye.

VIII. *Application of the extent of natural and telescopic vision to the probable arrangement of the celestial bodies in space.*

When the extent of natural and telescopic vision is to be applied to investigate the distance of celestial objects, the result can only have a high degree of probability; for it will then be necessary to admit a certain physical generic size and brightness of the stars. But when two hypotheses are proposed to explain a certain phenomenon, that which will most naturally account for it ought to be preferred as being the most probable. Now as the different magnitudes of the stars may be ascribed to a physical difference in their size and lustre, and may also be owing to the greater distance of the fainter ones, we cannot think it probable that all those of the 5th, 6th and 7th magnitude, should be gradually of a smaller physical construction than those of the 1st, 2d, and 3d; but shall, on the contrary, be fairly justified in concluding that, in conformity with all the phenomena of vision, the greater faintness of those stars is owing to their greater distance from us. The average size and brightness of several stars of the first magnitude being also taken as a standard, in the manner that has been shown, the conclusion drawn from different series of equalisations will support one another; so that we shall be able to say, a distant celestial object is so far from us, provided the stars of which it is composed are of a size and lustre equal to the size and lustre of such stars as Sirius, Arcturus, Capella, Lyra, Rigel, and Procyon, &c.

I proceed now to consider some conclusions that may be drawn from a known extent of natural vision, a very obvious one of which is, that all the visible stars are probably contained within a sphere of the 12th order of distances. Now as on the principle of equal scattering, we should see about 15625 of them, it may be remarked that the stars of the catalogue, including all those of the 7th magnitude, amount to 14144, which agrees sufficiently well with the calculated number; but the next inference is, that if they were equally scattered, there would be 2402 of the 10th, 2906 of the 11th, and 3458 of the 12th order of distances, which added together amount only to 8766, whereas the number of stars of the 6th and 7th magnitudes that must come into these three orders, is not less than 12249, which would indicate that the stars in the higher order of distances are more compressed than they are in the neighbourhood of the sun; but from astronomical observations, we also know that the stars of the 6th and 7th magnitude are very sparingly scattered over many of the constellations, and that consequently the stars which belong to the 10th, 11th, and 12th order of distances, are not only more compressed than those in the neighbourhood of the sun, but that moreover their compression in different parts of the heavens must be very unequal.

IX. *Of the construction and extent of the milky way.*

Of all the celestial objects consisting of stars not visible to the eye, the milky way is the most striking; its general appearance, without applying a telescope to it, is that of a zone surrounding our situation in the solar system, in the shape of a succession of differently condensed patches of brightness, intermixed with others of a fainter tinge.

To enumerate a partial series of them, we have a very bright patch under the arrow of Sagittarius; another in the Scutum Sobiescii; between these two there are three unequally bright places; north preceding α, β and γ Aquilæ is a bright patch; between Aquila and the Scutum are two very faint places; a long faint place follows the shoulder of Ophiuchus; near β Cygni is a bright place; near γ is another, and a third near α. A smaller brightish place follows in the succession of the milky way, and a large one towards Cassiopea. A faint place is on one side; a second towards Cassiopea, and a third is within that constellation; a very bright place is in the sword handle of Perseus; and α and γ Cassiopeæ inclose a dark spot.

The breadth of the milky way appears to be very unequal. In a few places it does not exceed five degrees; but in several constellations it is extended from ten to sixteen. In its course it runs nearly 120 degrees in a divided clustering stream, of which the two branches between Serpentarius and Antinous are expanded over more than 22 degrees.

That the sun is within its plane may be seen by an observer in the latitude of about 60 degrees ; for when at 100 degrees of right ascension the milky way is in the east, it will at the same time be in the west at 280 ; while in its meridional situation it will pass through Cassiopea in the Zenith, and through the constellation of the cross in the Nadir.

From this survey of the milky way by the eye I shall now proceed to show what appears to be its construction by applying to it the extent of telescopic vision ; but as I had prepared a gradually increasing series of reductions of the space-penetrating powers of my instruments for the purpose of measuring the Profundity of sidereal objects not visible to the eye, which I have called gaging powers, it will be necessary to give the following account of it.

From the formula which has been given, I calculated a set of apertures, which by limiting the light of the finder of my seven feet reflector would reduce its space-penetrating power to the low gaging powers 2, 3, and 4. I then limited in the same manner the space-penetrating power of my night glass, by using calculated apertures such as would give the gaging powers 5, 6, 7 and 8. From the space-penetrating power of the 7 feet reflector, I obtained by limitation the successive gaging powers 9, 10 and upwards to 17. And lastly, by limiting the space-penetrating power of my 10 feet reflector, I carried the gaging powers from 17 to 28.

For the purpose of trying these powers, I selected the bright spot in the sword handle of Perseus, as being probably a protuberant part of the milky way, in which it is situated. Its altitude at the time of observation was about 30 degrees, and no star in it was visible.

In the finder with the gaging power 2, I saw many stars ; and admitting the eye to reach to stars at the distance of the 12th order, we may conclude that the small stars which were visible with this low power, are such as contribute to the brightness of the spot, and that their situation is probably from between the 12th to the 24th order of distances ; at least we are certain, that if stars of the size and lustre of Sirius, Arcturus, Capella, &c. were removed into the Profundity of space which I have mentioned, they would then appear like the stars I saw with the gaging power of the finder. I then changed this power from 2 to 3, and saw more stars than before ; and changing it again from 3 to 4, a still greater number of them became visible. The situation of these additional stars was consequently between the 24th, 36th, and 48th order of distances.

With the gaging power 5 of the night glass I saw a great number of stars ; with 6, more stars and whitishness became visible ; with 7, more stars with resolvable whitishness were seen ; and with 8, still more. The stars that gradually made their appearance, therefore, were probably scattered over the space between the 48th and 96th order of distances.

In the 7 feet reflector, with the gaging powers 9 and 10, I saw a great number of stars; with 11 and 12, a greater number of stars and resolvable whitishness were seen; with 13 and 14, the number of visible stars was increased, and was so again with 15; with 16 and 17 in addition to the visible stars, there were many too faint to be distinctly perceived. These gages therefore extend the space over which the additional stars were scattered from the 96th to the 204th order of distances.

With a 10 feet reflector, reduced to a gaging power of 18, I saw a great number of stars: they were of very different magnitudes, and many whitish appearances were so faint that their consisting of stars remained doubtful. The power 19, which next I used, verified the reality of several suspected stars, and increased the lustre of the former ones. With 20, 22, and 25, the same progressive verifications of suspected stars took place, and those which had been verified by the preceding powers, received subsequent additional illumination. With the whole space-penetrating power of the instrument, which is 28·67, the extremely faint stars in the field of view acquired more light, and many still fainter suspected whitish points could not be verified for want of a still higher gaging power. The stars which filled the field of view were of every various order of telescopic magnitudes, and, as appears by these observations, were probably scattered over a space extending from the 204th to the 344th order of distances.

As the power of the 10 feet reflector could not reach farther into space, I shall have recourse to some of my numerous observations made with the 20 feet telescope. In addition to 863 gages already published,* above 400 more have been taken in various parts of the heavens,† but with regard to these gages, which on a supposition of an equality of scattering were looked upon as gages of distances, I have now to remark that, although a greater number of stars in the field of view is generally an indication of their greater distance from us, these gages, in fact, relate more immediately to the scattering of the stars, of which they give us a valuable information, such as will prove the different richness of the various regions of the heavens.

July 30, 1785. Right ascension 19^h 4′. Polar distance 87° 5′. The milky way is extremely rich in stars that are too small for the gage.

Dec. 7, 1785. Right ascension 5^h 33′. Polar distance 66° 6′. There are about 66 stars in the field of view, and many more so extremely small as not to admit of being gaged.

Sept. 20, 1786. Right ascension 20^h 40′. Polar distance 54° 36′. There are about 80 stars in a quadrant, or 320 in the field of view, besides many more too small to be distinctly seen.

Oct. 14, 1787. Right ascension 21^h 57′. Polar distance from 35° 18′ to 38° 50′. In this part of the heavens the large stars seem to be of the 9th and 10th magnitude. The small ones are gradually less till they escape the eye, so that appearances here favour the idea of a succeeding, more distant clustering part of the milky way.

Sept. 18, 1784. Right ascension 20^h 8′. Polar distance from 70° 9′ to 72° 49′. The end of the stratum of the stars of the milky way cannot be seen.

By these observations it appears that the utmost stretch of the space-penetrating power of the 20 feet telescope could not fathom the Profundity of the milky

* See *Phil. Trans.* for 1785, p. 221 [Vol. I. p. 228]. † [Printed below, in the Appendix.—Ed.]

way, and that the stars which were beyond its reach must have been farther from us than the 900dth order of distances.

I am far from limiting the milky way to the extent deduced from these observations; but as even the distance which has been stated may appear doubtful, I must repeat the argument which has been used with stars visible to the eye, but which now is greatly supported by telescopic vision. If the stars of the 5th, 6th, and 7th magnitudes cannot be supposed to be gradually of a smaller physical size and brightness than those of the 1st, 2d, and 3d, how much less can a supposition be admitted that would require that the stars, which by a long series of gaging powers have been proved to make their gradual telescopic appearance from the 12th to the 900dth order of distances, should also be gradually of a different construction, with regard to physical size and brightness, from those which we see with the naked eye.

From the great diameter of the mirror of the 40 feet telescope we have reason to believe, that a review of the milky way with this instrument would carry the extent of this brilliant arrangement of stars as far into space as its penetrating power can reach, which would be to the 2300dth order of distances, and that it would then probably leave us again in the same uncertainty as the 20 feet telescope. When I made some sweeps of the heavens with the 40 feet telescope with a magnifying power of 370, I found it necessary to reduce the intended breadth of the sweep from one degree to 30 minutes, and the great length of time this would have taken up to examine only the ecliptic, to which I had directed the telescope, soon proved that by continuing to use this instrument for sweeping, I should have been obliged to neglect the necessary observations of the 20 feet telescope.

The following observations are extracted from my sweeps to support a few general remarks relating to the construction of the milky way.

Dec. 7, 1785. Right ascension 4^h 39′. Polar distance from 64° 0′ to 66° 12′. The straggling stars of the milky way seem now to come on gradually; most of them are small. Right ascension 4^h 43′. They begin now to be intermixed with some large ones.

This observation proves that the telescopic breadth of the milky way, considerably exceeds the extent which in our maps is assigned to it. In this situation it began to appear at 6 or 7 degrees from where it might have been expected to enter the telescope.

Aug. 21, 1791. Right ascension 18^h 59′. Polar distance from 84° 15′ to 86° 17′. The milky way comes on very suddenly, and is amazingly crowded with very small stars intermixed with many of several sizes.

By our maps this place is already within the limits of the milky way.

Jan. 1, 1786. Right ascension 5^h 17′. Polar distance from 89° 28′ to 91° 47′. Most of the stars are larger than usual, but the whole breadth of the sweep contains a great mixture of all sizes.

From the brightness of the stars we may conclude that the constellation of Orion to which the observation belongs, is one of those that are nearest to our own situation.

Dec. 27, 1786. Right ascension 6^h 49′. Polar distance from 87° 37′ to 89° 55′. From the appearance of the heavens in this neighbourhood, there is reason to suppose that there is a break or vacancy among the stars between our situation and the more remote parts of the milky way.

The place to which this observation refers is in the breast of Monoceros.

Oct. 14, 1787. Right ascension 22^h 14′. Polar distance from 35° 18′ to 38° 50′. It is very evident that in this part of the heavens there is some distance between us and the milky way, which is not equally scattered over with stars.

The situation of the place pointed out by the observation is near the crown of Cepheus.

Sept. 15, 1792. Right ascension 19^h 46′. Polar distance 52° 29′. There are 153 stars in a quadrant, or 612 in the field of view, and the whole breadth of the sweep, which is 2° 2′, is equally rich.

The gage was taken in the preceding branch of the milky way, in the neck of the Swan.

Aug. 22, 1792. Right ascension 19^h 35′. Polar distance 75° 5′. The field of view is extremely crowded, but the stars are too small and too numerous to be counted; there cannot be less than 100, or probably 150 in a quadrant. From some careful trials I suppose there were 150; this would give 600 stars in the field of view. The stars continue to be equally crowded throughout the whole breadth of the sweep, which was 2° 35′ till right ascension 19^h 54′, when there still were 440 in the field.

This gage was taken in the following branch of the milky way, in the wing of Aquila.

Sept. 13, 1784. Right ascension 20^h 43′. Polar distance from 54° 15′ to 57° 1′. This branch of the milky way is less rich than the preceding one.

The same sweep passed through both the branches, and the observation relates to a place in the following wing of the Swan.

Oct. 19, 1788. Right ascension 21^h 13′. Polar distance from 43° 35′ to 46° 13′. The milky way is very rich, but the stars are very unequally scattered.

This observation belongs to the tail of the Swan.

Nov. 26, 1788. Right ascension 23^h 40′. Polar distance 29° 13′. The milky way is very rich, but the stars are very unequally scattered. The stars are clustering. Right ascension 0^h 14′. Polar distance 29° 51′. Clustering small stars considerably rich. Right ascension 0^h 27′. Polar distance 30° 5′. Clustering small stars.

These observations belong to a place preceding the back of Cassiopea's chair.

Dec. 27, 1786. Right ascension 6^h 36′. Polar distance 88° 5′. Clustering stars of the milky way, almost close enough, and so far separated from the rest of the stars as to be called a cluster, but still evidently joining to them.

This observation of the clustering of the stars of the milky way relates to a place that precedes the breast of Monoceros.

X. *Concluding Remarks.*

What has been said of the extent and condition of the milky way in several of my papers on the construction of the heavens, with the addition of the observations contained in this attempt to give a more correct idea of its profundity in

Fig. 2.

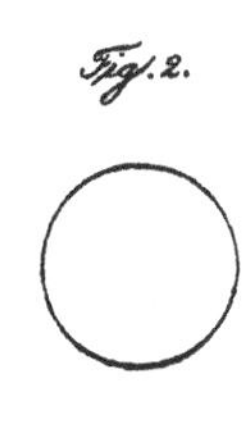

space, will nearly contain all the general knowledge we can ever have of this magnificent collection of stars. To enter upon the subject of the contents of the heavens in the two comparatively vacant spaces on each side adjoining the milky way, the situation of globular clusters of planetary nebulæ, and of far extended nebulosities, would greatly exceed the compass of this Paper; I shall therefore only add one remarkable conclusion that may be drawn from the experiments which have been made with the gaging powers.

In fig. 2, let a circle, drawn with the radius of the 12th order of distances, represent a sphere containing every star that can be seen by the naked eye; then, if the breadth of the milky way were only 5 degrees, and if its profundity did not exceed the 900dth order of distances, the two parallel lines in the figure, representing the breadth of the milky way, will, on each side of the centre of the inclosed circle, extend to more than the 39th order of distances.

From this it follows, that not only our sun, but all the stars we can see with the eye, are deeply immersed in the milky way, and form a component part of it.

WILLIAM HERSCHEL.

Slough, near Windsor,
May 10, 1817.

LXIX.

Astronomical observations and experiments, selected for the purpose of ascertaining the relative distances of clusters of stars, and of investigating how far the power of our telescopes may be expected to reach into space, when directed to ambiguous celestial objects. By Sir WILLIAM HERSCHEL, *Knt. Guelp. LL.D. F.R.S*

[*Phil. Trans.*, 1818, pp. 429–470.]

Read June 11, 1818.

IN my last paper on the local arrangement of the celestial bodies in space, I have shown how, by an equalization of the light of stars of different brightness, we may ascertain their relative distances from the observer, in the direction of the line in which they are seen; and from this equalization, a method of turning the space-penetrating power of a telescope into a gradually increasing series of gaging powers has been deduced, by which means the profundity in space, of every object consisting of stars, can be ascertained, as far as the light of the instrument which is used upon this occasion will reach. This method has already been applied to fathom the milky way, and may with equal propriety be used to ascertain the profundity of globular and other clusters of stars in space; I shall therefore make use of some of the numerous observations, contained in my journals and sweeps of the heavens, to show how the distances of these objects may be obtained; and shall also attempt to represent their situation in space by a figure, in which their distances are made proportional to the diameter of a globular space, sufficiently large to contain all the stars that in the clearest nights are visible to the eye of an observer.

I. *Of the distance of globular and other clusters of stars.*

In observations which are made for ascertaining the distance of a cluster of stars, it is necessary that the gaging power should be marked, which will just make some of the stars belonging to it visible in the telescope that is used for this purpose. If the cluster is of a globular form, but is not insulated, the stars that belong to it may be easily distinguished from those which may happen to be scattered about, or upon it. In clusters of a different construction, the compression, or the apparent size of the stars, must direct the observer.

It is to be remarked, that neither the brightness, nor the diameter of the clusters of which the distance is to be ascertained, are to be considered : some of them are bright enough to be perceived by the eye ; others are visible in the finder of the telescope, and many of them can only be seen in the telescope itself. These are circumstances that have no influence on the exactness of the result of the gaging power ; but as they regard our knowledge of the construction of these magnificent sidereal systems, an abridged account of them is given, with the observations by which their profundity in space is ascertained ; and in the arrangement of these observations, I have followed the order of the space-penetrating power of the instruments by which they were made.

In recording the examination of celestial objects, I have often applied to them the expressions resolvable, or easily resolvable, when, from their appearance, I could not decide whether they belonged to the class of nebulæ, properly so called, or whether they might not consist of an aggregation of stars, at too great a distance from us to be distinctly perceived ; but it is evident that the distance of a cluster of stars cannot be ascertained, as long as it remains doubtful whether the object consists of stars ; and that, consequently, their first perceptibility must be the gaging power by which its profundity in space is to be ascertained.

II. *A series of observations of clusters of stars, from which the order of their profundity in space is determined.*

Observation of the 7th cluster of stars in the VIth *class of my catalogues of celestial objects.** [N.G.C. 5053.]

" 1784, 20 feet telescope, power 157. An excessively faint cluster of stars, intermixed with resolvable nebulosity, 8 or 10 minutes in diameter. The stars are so small that they cannot be seen without the greatest attention ; 240 verified it beyond all doubt. I have suspected many such in this neighbourhood."

At the time of this observation, the 20 feet telescope was of the NEWTONIAN construction, and its power to penetrate into space was 61·18 times that of the eye, which it has been shown can see stars of the 12th order : † and since it appears from the foregoing observation, that with this power the telescope could but just reach the stars of the cluster, we may conclude that its profundity in space cannot be less than of the 734th order.

The 9th cluster in the VIth *class.* [N.G.C. 5466.]

" 1784, 20 feet telescope. A large cluster of exceedingly small, and compressed stars, about 6 or 7 minutes in diameter ; a great many of the stars are visible, the rest so small as to appear nebulous ; those that are visible are of one size, and are scattered all over equally. The cluster is of an irregular round form."

The profundity of the cluster by this observation is of the 734th order.

The 10th cluster in the VIth *class.* [N.G.C. 6144.]

" 1784, 20 feet telescope. A very compressed, considerably large cluster of the smallest stars imaginable ; all the stars are of a dusky red colour. This cluster is the next step to an easily resolvable nebula."

The ruddy colour of the stars is probably owing to its low situation ; the profundity of the cluster is of the 734th order.

* For these catalogues, see *Phil. Trans.* for 1786, page 471 ; 1789, page 226, and 1802, page 503. [Vol. I. pp. 268 and 337, Vol. II. p. 215.]
† *Phil. Trans.* for 1817, page 317 [above, p. 584].

The 11*th cluster in the* VI*th class.* [N.G.C. 6284.]

" 1784, 20 feet telescope. A cluster of stars, which, in respect of the size of the whole as well as the distance and magnitude of the stars, is a good miniature of the 19th of the *Connoissance* observed a few minutes ago. The stars, like those of the foregoing cluster, preserve a faint red tint. It may be called the next step to an easily resolvable nebula. It is about 1½ or 2 minutes in diameter."

The profundity of this cluster cannot be much less than of the 734th order. It is in the preceding branch of the milky way.

The 12*th cluster in the* VI*th class.* [N.G.C. 6293.]

" 1784, 20 feet telescope. This cluster of stars is another miniature of the 19th of the *Connoissance,* but rather coarser than my 11th cluster."

The profundity of the 19th of the *Connoissance* being of the 344th order, this cluster, as rather a coarse miniature of it, may be of the 466th order ; it is in the preceding branch of the milky way.

The 17*th cluster in the* VI*th class.* [N.G.C. 2158.]

" 1784, 1785, 20 feet telescope. A very rich cluster of very compressed and extremely small stars ; 4 or 5 minutes in diameter."

This cluster is probably of a profundity of about the 600th order. It is in the preceding branch of the milky way.

The 20*th cluster in the* VI*th class.* [N.G.C. 288.]

" 1785, 1786, 20 feet telescope. Considerably bright, irregularly round, 8 or 9 minutes in diameter ; a great many of the stars are visible, so that there can remain no doubt of its being a cluster of the most minute stars imaginable."

The profundity of this cluster cannot be less than of the 734th order. It is near the south pole of the milky way.

The 26*th cluster in the* VI*th class.* [N.G.C. 1605.]

" 1786, 20 feet telescope. A very faint cluster of very compressed extremely small stars ; near 4 minutes in diameter."

The 20 feet telescope being of the construction of the front view, and having a gaging power of 75·08 gives the profundity of this cluster of the 900th order. It is in the milky way.

The 35*th cluster in the* VI*th class.* [N.G.C. 136.]

" 1788, 20 feet telescope. A small cluster of very faint, exceedingly compressed stars, about 1 minute in diameter ; the next step to an easily resolvable nebula."

The profundity of this cluster is of the 900th order ; it is in the milky way.

The 38*th cluster in the* VI*th class.* [N.G.C. 6804.]

" 1791, 20 feet telescope. Considerably bright, small, of an irregular figure ; easily resolvable : some of the stars are visible."

The profundity of this cluster is of the 900th order. It is in the milky way.

The 41*st cluster in the* VI*th class.* [N.G.C. 6412.]

" 1797, 20 feet telescope. Round, resolvable, about 3 minutes in diameter ; very gradually brighter in the middle. I suppose it to be a cluster of extremely compressed stars ; 320 confirms the supposition, and shows a few of the stars ; it must be immensely rich."

The profundity of this cluster is of the 900th order.

The 63*rd cluster in the* IV*th class.* [N.G.C. 5204.]

" 1789, 20 feet telescope. Considerably bright, considerably large, irregularly round, very gradually much brighter in the middle ; about 4 minutes in diameter."

The profundity of this cluster must be at least of the 900th order,

The 1st of the Connoissance des Temps. [N.G.C. 1952.]*

" 1783, 1794, 7 feet telescope. With 287, light without stars."

" 1805, 1809, 10 feet telescope. It is resolvable. There does not seem any milky nebulosity mixed with what I take to be small lucid points."

" 1783, 1784, 1809, 20 feet telescope. Very bright, of an irregular figure; full 5 minutes in the longest direction. I suspect it to consist of stars."

" 1805, large 10 feet telescope. With 220 the diameter is 4′ 0″; with this power and light it is what must be called resolvable."

" 1809. With a fine new 10 feet. It is resolvable, and with 170 some of the faint stars may be seen; it is however possible that faint milky nebulosity may be mixed with a few very small stars."

As all the observations of the large telescopes agree to call this object resolvable, it is probably a cluster of stars at no very great distance beyond their gaging powers; its profundity may therefore be of about the 980th order. It is near the milky way.

The 2nd of the Connoissance. [N.G.C. 7089.]

" 1799, 7 feet finder of the telescope. It is visible as a star. 1810, it may just be perceived to have rather a larger diameter than a star."

" 1783, 2 feet sweeper. It is like a telescopic comet."

" 1794, 7 feet telescope. With 287 I can see that it is a cluster of stars, many of them being visible."

" 1810, 10 feet telescope. A beautiful bright object."

" 1784, 1785, 1802, 20 feet telescope. A cluster of very compressed exceedingly small stars."

" 1805, 1810, large 10 feet telescope. Its diameter with 108 is 4′ 59″; with 171 and 220, it is 6′ 0″."

" 1799, 40 feet telescope. A globular cluster of stars." †

By the observation of the 7 feet telescope, which has a power of seeing stars that exceeds the power of the eye to see them 20·25 times, the profundity of this cluster is of the 243rd order.

The 3rd of the Connoissance. [N.G.C. 5272.]

" 1813, 7 feet finder. It is at a small distance from a star of equal brightness; the star is clear, the object is hazy, and somewhat larger than the star."

" 1783, 7 feet telescope. With 460 the light is so feeble that the object can hardly be seen; I suspect some stars in it. 1813, with 80, many stars are visible in it."

" 1799, 10 feet telescope, power 120; with an aperture of 4 inches it is resolvable; with 5 easily resolvable; with 6 it is resolved; with 7 and all open the stars may be easily perceived."

" 1784, 1785, 20 feet telescope. A beautiful globular cluster of stars, about 5 or 6 minutes in diameter."

" 1810, Large 10 feet telescope. With 171 the diameter is full 4′ 30″." ‡

By the observation of the 7 feet telescope this cluster must be of the 243rd order.

The 4th of the Connoissance. [N.G.C. 6121.]

" 1783, 10 feet telescope. All resolved into stars. I can count a great number of them, while others escape the eye by their minuteness."

" 1783, small 20 feet telescope. All resolved into stars."

" 1784, 20 feet telescope. The cluster contains a ridge of stars in the middle, running from south preceding to north following."

The 10 feet telescope having a power to show stars exceeding that of the eye 28·67 times, gives the profundity of this cluster of the 344th order.

* [The "Crab Nebula" in Taurus. There are no observations with the 40 feet telescope.—ED.]

† For the particulars of this observation see *Phil. Trans.* for 1814, page 274 [above, p. 535].

‡ [Compare above, p. 536.]

The 5th of the Connoissance. [N.G.C. 5904.]

" 1813, 7 feet finder. It is near a star of equal brightness ; the star is clear but the object is hazy."

" 1783, 7 feet telescope. It consists of stars ; they are however so small that I can but just perceive some, and suspect others. 1810, the globular figure is visible."

" 1783, 10 feet telescope. With 600, all resolved into stars."

" 1785, 1786, 20 feet telescope. A very compressed cluster of stars, 7 or 8 minutes in diameter ; the greatest compression about 2 or 2½ minutes.

" 1791, 40 feet telescope. With 370 I counted about 200 stars ; the middle of it is so compressed, that it is impossible to distinguish the stars."

The profundity of this cluster, by the observation of the 7 feet telescope, is of the 243rd order.

The 9th of the Connoissance. [N.G.C. 6333.]

" 1783, 10 feet telescope, power 250. I see several stars in it ; and have no doubt a higher power and more light will resolve it all into stars."

" 1784, 1786, 20 feet telescope. A cluster of extremely compressed stars ; it is a miniature of the 53d."

By the observations of the 10 feet the profundity is at least of the 344th order. It is in the preceding branch of the milky way.

The 10th of the Connoissance. [N.G.C. 6254.]

" 1783, 7 feet telescope. With 227 I suspected it to consist of stars ; with 460 I can see several of them, but they are too small to be counted."

" 1784, 1791, 20 feet telescope. A beautiful cluster of extremely compressed stars ; it resembles the 53d ; and the most compressed part is about 3 or 4 minutes in diameter."

The profundity of this cluster, by the observation of the 7 feet telescope, is of the 243d order.

The 11th of the Connoissance. [N.G.C. 6705.]

" 1799, 10 feet finder. The cluster is visible ; and, directed by neighbouring stars, it may be seen by the eye."

" 1783, 1799, 10 feet telescope. Power 300. With 3 inches of aperture, the small stars are not to be distinguished ; with 4 inches I can see then."

" 1803, 1810, large 10 feet telescope. The cluster is of an irregular form, from 9 to 12 minutes in diameter."

The 10 feet telescope with an aperture of 4 inches, had a gaging power of 12·02 ; the profundity of this cluster is therefore of the 144th order. It is in the milky way.

The 12th of the Connoissance. [N.G.C. 6218.]

" 1799, 10 feet finder. The object is visible in it."

" 1783, 1799, 10 feet telescope. With 120, and an aperture of 4 inches, easily resolvable ; with 5 inches, stars become visible ; with 6 inches, pretty distinctly visible ; and with all open, the lowest power shows the stars."

" 1785, 1786, 20 feet telescope. A brilliant cluster, 7 or 8 minutes in diameter ; the most compressed parts about 2 minutes."

With an aperture of 5 inches the 10 feet telescope had a gaging power of 15·53 ; and this cluster is consequently of a profundity of the 186th order.

The 13th of the Connoissance. [N.G.C. 6205.]*

" 1799, 1805. It is very plainly to be seen by the eye."

" 1799, 7 feet finder. Very visible."

" 1783, 7 feet telescope. With 227 plainly resolved into stars."

" 1799, 10 feet telescope. With an aperture of 4 inches the stars cannot be distinguished; with 9 inches, very beautiful."

" 1787, 1799, 20 feet telescope. The stars belonging to the cluster extend to 8 or 9 minutes in diameter; the most compressed part about 2 or $2\frac{1}{2}$; the latter is round, the former irregular."

" 1805, large 10 feet telescope. A brilliant cluster all resolved into stars. The space penetrating power of this NEWTONIAN reflector is $\frac{\sqrt{\cdot 41 \times (240^2 - 39^2)}}{\cdot 2} = 75\cdot 82$."

By the observation of the 7 feet telescope, the profundity of this cluster is nearly of the 243d order.

The 14th of the Connoissance. [N.G.C. 6402.]

" 1783, 7 feet telescope. With 227, there is a strong suspicion of its consisting of stars.

" 1783, 1784, 1791, 1799, 20 feet telescope. Extremely bright, round, easily resolvable; with 300 I can see the stars. The heavens are pretty rich in stars of a certain size, but they are larger than those in the cluster, and easily to be distinguished from them. The cluster is considerably behind the scattered stars, as some of them are projected upon it."

From the observations of the 20 feet telescope, which in 1791 and 1799 had the power of discerning stars 75·08 times as far as the eye, the profundity of this cluster must be of the 900th order.

The 15th of the Connoissance. [N.G.C. 7078.]

" 1799. It is visible to the eye."

" 1783, 1794, 7 feet telescope. With 278 the stars of the cluster may be seen."

" 1799, 10 feet telescope. With an aperture of 4 inches, no trace of stars is visible. 1817, with an aperture of 4·56 inches, which gives a gaging power of 14, it appears like a nebulous patch, gradually brighter in the middle; with a gaging power of 16, the hazy border of it is larger; with 18, the whole of it much larger and brighter; with 20, resolvable; and with 22, the stars are visible."

" 1784, 1787, 1807, 20 feet telescope. A globular cluster of stars, about 6 minutes in diameter."

" 1810, large 10 feet telescope. The diameter, with 171, is full 4′ 30″, and taking in the stars that probably belong to it, it is 6′ 45″."

By the observation of the 7 feet telescope, the profundity of this cluster is of the 243d order.

The 19th of the Connoissance. [N.G.C. 6273.]

" 1783, 10 feet telescope. With 250, I can see 5 or 6 stars, and all the rest appears mottled like other objects of this kind, when not sufficiently magnified or illuminated." †

" 1784, 20 feet telescope. A cluster of very compressed stars, much accumulated in the middle; 4 or 5 minutes in diameter."

By the observation of the 10 feet telescope, the profundity of this cluster is of the 344th order. It is in the preceding branch of the milky way.

The 22d of the Connoissance. [N.G.C. 6656.]

" 1783, 7 feet telescope. 460 has not light enough to show it; with 227, I see it very imperfectly."

" 1801, 10 feet telescope. With 600 it is a cluster of stars."

" 1783, small 20 feet telescope. With 350, all resolved into stars."

" 1784, 20 feet telescope. An extensive cluster of stars."

" 1810, large 10 feet telescope. The stars are condensed in the middle. The diameter is 8′ 0″; the greatest condensation is about 4′ 0″."

By the observation of the 10 feet telescope, the profundity of this cluster must be nearly of the 344th order. It is near the following branch of the milky way.

* [The great cluster in Hercules. There are no observations with the 40 feet telescope.—ED.]

† [Compare above, p. 537.]

The 30th of the Connoissance. [N.G.C. 7099.]

" 1794, 7 feet finder. It is but just visible."

" 1794, 7 feet telescope. It seems to be resolvable, but is too faint to bear a high power."

" 1810, 10 feet telescope. With 71, it appears like a pretty large cometic nebula, very gradually much brighter in the middle. 1783, with 250 it is resolved into very small stars."

" 1783, small 20 feet NEWTONIAN, 12 inch diameter. Power 200 ; it consists of very small stars ; with two rows of stars, 4 or 5 in a line."

" 1783, large 20 feet NEWTONIAN. Power 120 ; by a drawing of the cluster, the rows of stars probably do not belong to the cluster."

" 1784, 1785, 1786, 20 feet telescope, power 157. A brilliant cluster."

" 1810, large 10 feet telescope. With 171 and 220 the diameter is 3′ 5″ ; it is not round."

By the observation of the 10 feet telescope, the profundity of this cluster is of the 344th order.

The 33d of the Connoissance. [N.G.C. 598.]*

" 1799, 10 feet finder. It is visible as a faint nebula."

" 1783, 1794, 7 feet telescope. With 75, it has a nebulous appearance ; it will not bear 278 and 460, but with 120 it seems to be composed of stars."

" 1799, 1810, 10 feet telescope. The brightest part is resolvable ; some of the stars are visible."

" 1805, 1810, Large 10 feet telescope. The condensation of the stars is very gradual towards the middle ; but with the four powers 71, 108, 171, and 220, some nebulosity remains. The stars of the cluster are the smallest points imaginable. The diameter is nearly 18 minutes."

The profundity of this cluster, by the observation of the 10 feet telescope, must be of the 344th order.

The 34th of the Connoissance. [N.G.C. 1039.]

" 1799, 7 feet finder. It is visible."

" 1783, 1794, 7 feet telescope. A cluster of stars ; with 120, I think it is accompanied with mottled light, like stars at a distance."

" 1784, 1786, 20 feet telescope. A coarse cluster of large stars of different sizes."

By the observation of the 7 feet telescope, the profundity of this cluster does probably not exceed the 144th order.

The 35th of the Connoissance. [N.G.C. 2168.]

" 1794, It is visible to the naked eye as a very small cloudiness."

" 1783, 1794, 1801, 1813, 7 feet telescope. It is a rich cluster of stars of various sizes."

" 1806, 10 feet telescope. There is no central contraction to denote a globular form."

" 1783, 1785, 20 feet telescope. A cluster of pretty compressed large stars."

The profundity of this cluster does probably not exceed the 144th order. It is in the milky way.

The 53d of the Connoissance. [N.G.C. 5024.]

" 1813, 7 feet finder. It appears like a very small haziness."

" 1783, 7 feet telescope. With 460 the object is extremely faint. 1813, with 118 it is easily resolvable, and some of the stars may be seen."

" 1783, 10 feet telescope. With 250, I perceive 4 or 5 places that seem to consist of very small stars."

" 1784, 1786, 20 feet telescope. A globular cluster of very compressed stars."

From the observation of the 7 feet telescope, it appears that the profundity of this cluster is of the 243d order.

* [Large spiral nebula in Triangulum. There are no observations with the 40 feet telescope.—ED.]

The 55th of the Connoissance. [N.G.C. 6809.]

" 1783, small 20 feet telescope. With 250 fairly resolved into stars ; I can count a great many of them, while others are too close to be distinguished separately."

" 1784, 1785, 20 feet telescope. A rich cluster of very compressed stars, irregularly round, about 8 minutes long."

By the observation of the small 20 feet telescope, which could reach stars 38·99 times as far as the eye, the profundity of this cluster cannot be much less than of the 467th order: I have taken it to be of the 400th.

The 56th of the Connoissance. [N.G.C. 6779.]

" 1783, 7 feet telescope. A strong suspicion of its being stars."

" 1783, 1799, 10 feet telescope. 120 will not resolve it; 240 wants light: 350 however shows the stars, but they are so exceedingly close and small that they cannot be counted."

" 1784, 1807, 20 feet telescope. A globular cluster of very compressed small stars about 4 or 5 minutes in diameter."

" 1805, 1807, large 10 feet telescope. With 171 it is 3′ 36″ in diameter."

The profundity of this cluster, by the observation of the 10 feet telescope, must be of the 344th order. It is near the preceding branch of the milky way.

The 57th of the Connoissance. [N.G.C. 6720.]*

" 1782, 7 feet telescope. I suspect it to consist of very small stars ; in the middle it seems to be dark."

" 1783, 1805, 1806, 10 feet telescope. With 130 it seems to be a rim of stars, but with 350 there remains a doubt. It is a little oval ; the dark place in the middle is also oval ; one side of the bright margin is a little narrower than the other."

" 1784, 1799, 20 feet telescope. It is an oval with a dark place within ; the light is resolvable. 240 showed several small stars near, but none that seem to belong to it. It is near 2 minutes in diameter."

" 1805, large 10 feet telescope. By a meridian passage of 7 seconds of sidereal time, the diameter is 1′ 28″·4."

By the observation of the 20 feet telescope, the profundity of the stars of which it probably consists must be of a higher than the 900th order ; perhaps 950.

The 62d of the Connoissance. [N.G.C. 6266.]

" 1783, 10 feet telescope. With 250, a strong suspicion, amounting almost to a certainty, of its consisting of stars."

" 1785, 1786, 20 feet telescope. Extremely bright, round, very gradually brighter in the middle, about 4 or 5 minutes in diameter ; 240 with strong attention showed the stars of it. The cluster is a miniature of the 3d of the *Connoissance.*"

By the 20 feet telescope, which at the time of these observations was of the NEWTONIAN construction, the profundity of this cluster is of the 734th order. It is in the preceding branch of the milky way.

The 67th of the Connoissance. [N.G.C. 2682.]

" 1783, 7 feet telescope. A cluster of stars."

" 1809, 10 feet telescope. A cluster of vS. stars, there seems to be F. milky nebulosity among them."

" 1784, 20 feet telescope. A most beautiful cluster of stars ; not less than 200 in view."

By estimation, the profundity of this cluster may be of the 144th order.

* [The nebula in Lyra. There are no observations with the 40 feet telescope.—ED.]

The 68th of the Connoissance. [N.G.C. 4590.]

" 1786, 1789, 1790, 20 feet telescope. A cluster of very compressed small stars, about 3 minutes broad and 4 minutes long. The stars are so compressed, that most of them are blended together."

Probably the stars of this cluster might be perceived by a 10 feet telescope, so that its profundity may be of the 344th order.

The 69th of the Connoissance. [N.G.C. 6637.]

" 1784, 20 feet telescope. Very bright, pretty large, easily resolvable, or rather an already resolved cluster of minute stars. It is a miniature of the 53d of the *Connoissance.*"

By this observation, the profundity of the cluster must be of the 734th order.

The 71st of the Connoissance. [N.G.C. 6838.]

" 1794, 7 feet telescope. With 120 and 160 the stars of it become just visible."

" 1783, 1799, 1810, 10 feet telescope. A cluster of stars of an irregular figure."

" 1784, 1799, 1807, 20 feet telescope. It is situated in the milky way, and the stars are probably in the extent of it; it is however considerably condensed; about 3 minutes in diameter."

" 1805, large 10 feet telescope. An irregular cluster of very small stars, 2′ 35″ in diameter."

By the observation of the 7 feet telescope, the profundity of this cluster is of the 243d order. It is in the following branch of the milky way.

The 72d of the Connoissance. [N.G.C. 6981.]

" 1805, 7 feet telescope. With a power of 80 the stars may just be perceived."

" 1783, 1810, 10 feet telescope. With 150 fairly resolved."

" 1784, 1788, 20 feet telescope. A cluster of very small stars."

" 1810, large 10 feet telescope. A globular cluster; its diameter is 2′ 40″."

" 1810, Oct. 30, 40 feet telescope. A beautiful cluster of stars." [For further particulars see above, p. 535.]

By the observation of the 7 feet telescope, the profundity of this cluster must be of the 243d order.

The 74th of the Connoissance. [N.G.C. 628.]

" 1783, 1784, 7 feet telescope. With 100 and 120 it is a collection of very small stars; I see many of them."

" 1799, 1801, 10 feet telescope. Several of the stars are visible; it is a very faint object."

" 1784, 20 feet telescope. Some stars are visible in it; the edges are not resolvable."

" 1805, 1810, large 10 feet telescope. With 108 it consists of extremely small stars, of an irregular figure; a very faint object of nearly 12 minutes in diameter."

" 1799, Dec. 28, 40 feet telescope. Very bright in the middle, but the brightness is confined to a very small part, and is not round; about the bright middle is a very faint nebulosity to a considerable extent. The bright part seems to be of the resolvable kind, but my mirror has been injured by condensed vapours."

By the observation of the 7 feet telescope, the profundity of the nearest part of this cluster must be of the 243d order, but most probably a succession of more distant stars was seen in the larger telescopes.

The 75th of the Connoissance. [N.G.C. 6864.]

" 1799, 7 feet finder. It is but just visible."

" 1799, 7 feet telescope. There is not the least appearance of its consisting of stars, but it resembles other clusters of this kind, when they are seen with low space-penetrating and magnifying powers."

" 1810, 10 feet telescope. With 71 it is small and cometic."

" 1784, 1785, 20 feet NEWTONIAN. Easily resolvable; some of the stars are visible."

" 1810, 20 feet, front view. It is a globular cluster."

" 1799, 1810, large 10 feet. Its diameter with 171 is 1′ 48″; with 220 it is 2′ 0″."

By the observation of the 20 feet NEWTONIAN telescope, the profundity of this cluster must be of the 734th order.

The 77th of the Connoissance. [N.G.C. 1068.]

" 1783, 7 feet telescope. An ill defined star, surrounded by nebulosity."

" 1801, 1805, 1809, 1810, 10 feet telescope. It has almost the appearance of a large stellar nebula."

" 1783, 1785, 1786, 20 feet telescope. Very bright; an irregular extended nucleus with milky chevelure, 3 or 4 minutes long, near 3 minutes broad."

" 1801, 1805, 1807, large 10 feet telescope. A kind of much magnified stellar cluster; it contains some bright stars in the centre. With 171 its diameter is 1′ 17″; with 220 it is 1′ 36″."

From the observations of the large 10 feet telescope, which has a gaging power of 75·82, we may conclude that the profundity of the nearest part of this object is at least of the 910th order.

The 79th of the Connoissance. [N.G.C. 1904.]

" 1783, 7 feet telescope. With 57 nebulous; with 86 a strong suspicion of its being stars."

" 1799, 10 feet telescope. 300 shows the stars of it with difficulty."

" 1784, 20 feet telescope. A beautiful cluster of stars, nearly 3 minutes in diameter."

" 1806, large 10 feet telescope. A globular cluster, the stars of which are extremely compressed in the middle; with 171 and 220 the diameter is 2′ 50″, but the lowness of the situation probably prevents my seeing the whole of its extent."

By the observation of the 10 feet telescope the profundity of the cluster is of the 344th order.

The 80th of the Connoissance. [N.G.C. 6093.]

" 1784, 1786, 20 feet telescope. A globular cluster of extremely minute and very compressed stars of about 3 or 4 minutes in diameter; very gradually much brighter in the middle; towards the circumference the stars are distinctly to be seen, and are the smallest imaginable."

The profundity of this cluster is probably not much less than of the 734th order.

The 92d of the Connoissance. [N.G.C. 6341.]

" 1799, 7 feet finder. It may just be distinguished; it is but very little larger than a star."

" 1783, 2 feet sweeper. With 15 it appears like a clouded star."

" 1783, 7 feet telescope. With 227 resolved into very small stars; with 460 I can count many of them."

" 1799, 10 feet telescope. With 240 the stars are much condensed in the centre."

" 1783, 1787, 1799, 20 feet telescope. A brilliant cluster; about 6 or 7 minutes in diameter."

" 1805, large 10 feet telescope. The most condensed part is 3′ 16″ in diameter."

The profundity of this cluster, by the observation of the 7 feet telescope, is of the 243d order.

The 97th of the Connoissance. [N.G.C. 3587.]

" 1799, 7 feet finder. The object is not visible in it."

" 1789, 20 feet telescope; considerably bright, globular, of equal light throughout, with a diminishing border of no great extent. About 3 minutes in diameter."

" 1805, large 10 feet telescope. The constellation being too low it had the appearance of a faint nebula."

From the observation of the 20 feet telescope, it appears that the profundity of this object is beyond the gaging power of that instrument; and as it must be sufficiently distant to be ambiguous, it cannot well be less than of the 980th order.

III. *Of a method to represent the profundity of celestial objects in space by a diagram.*

In order to represent the profundity of celestial objects in space, I shall have recourse to the construction of an astronomical globe, on the surface of which the situations of the heavenly bodies are pointed out to us in the given two dimensions of right ascension and polar distance; but as their distance from an eye placed in the centre of the globe cannot be expressed by their situation on the surface, I shall endeavour to show that this deficiency may be artificially supplied in a figure representing such a globe, by the addition of lines that are of a length which is proportional to the diameter of it.

It has been shown in my last paper, that all the stars which may be seen in the clearest nights, are probably contained within a globular space, of which the radius does not exceed the 12th order of distances; I shall therefore suppose the circle *c* in the centre of fig. 1 to represent a celestial globe, containing all the stars that are generally marked on its surface; their arrangement within this globular space, however, must be supposed to be according to their order of distances, the stars of the first order being placed nearest the centre, and those of the 2d, 3d, and 4th, &c. gradually farther off; but they must all be placed in their well ascertained directions, so that a line from the centre drawn through any one of them may come to the surface at the place where its situation is marked.

According to this assumption it follows, that all those celestial objects which are farther than the 12th order of distances from the centre, must be represented as being at the outside of the globular space; but as our celestial globes represent not only the situation of the stars of the heavens, but give us also many additional objects, such as clusters of stars, nebulæ, and the milky way, it is evident that the point, where the line of sight from the centre to any one of these distant objects leaves the surface of the globular space, is ascertained; and since any celestial object not inserted on our globes, of which the right ascension and polar distance are given, may be easily added, the position of the visual ray directed to such an object will thereby also be determined.

In my last paper I have drawn the attention of astronomers to the condition of the milky way, as being the most brilliant, and beyond all comparison the most extensive sidereal system; and have also shown that the globular space containing all our visible stars is situated within its compass; I shall therefore now make the plane of it the principal dimension of my figure; then if the line *ab* represent this plane, a perpendicular drawn from the centre *c* of the figure to *d* and to *e*, will be directed towards the north and south poles of it, and the situation of the globular space in the figure will be like that of a celestial globe adjusted to the latitude of 30 degrees, having the milky way in the horizon, the 190th degree of right ascension in the meridian, and the 60th degree of north polar distance in the zenith.

From this description of the arrangement of the stars within the globular space, and its situation in the plane of the milky way, it is evident that, having already an expression for the position of a celestial object in two dimensions, the addition of the third, which is its profundity or central distance, may be represented by a line of a length that is proportional to the diameter of the globular space; and if this line be a continuation of the direction in which the object is seen from the centre, its termination will show the real place of the object, and point out its situation with respect to the great sidereal stratum of the milky way.

An observer who looks at a celestial globe, and wishes to see the angle of the direction of the line in which an object is seen from the centre, will for this purpose turn the globe horizontally till the plane of the azimuth circle is at right angles to the line in which he looks at it; or, if more convenient, he will change his

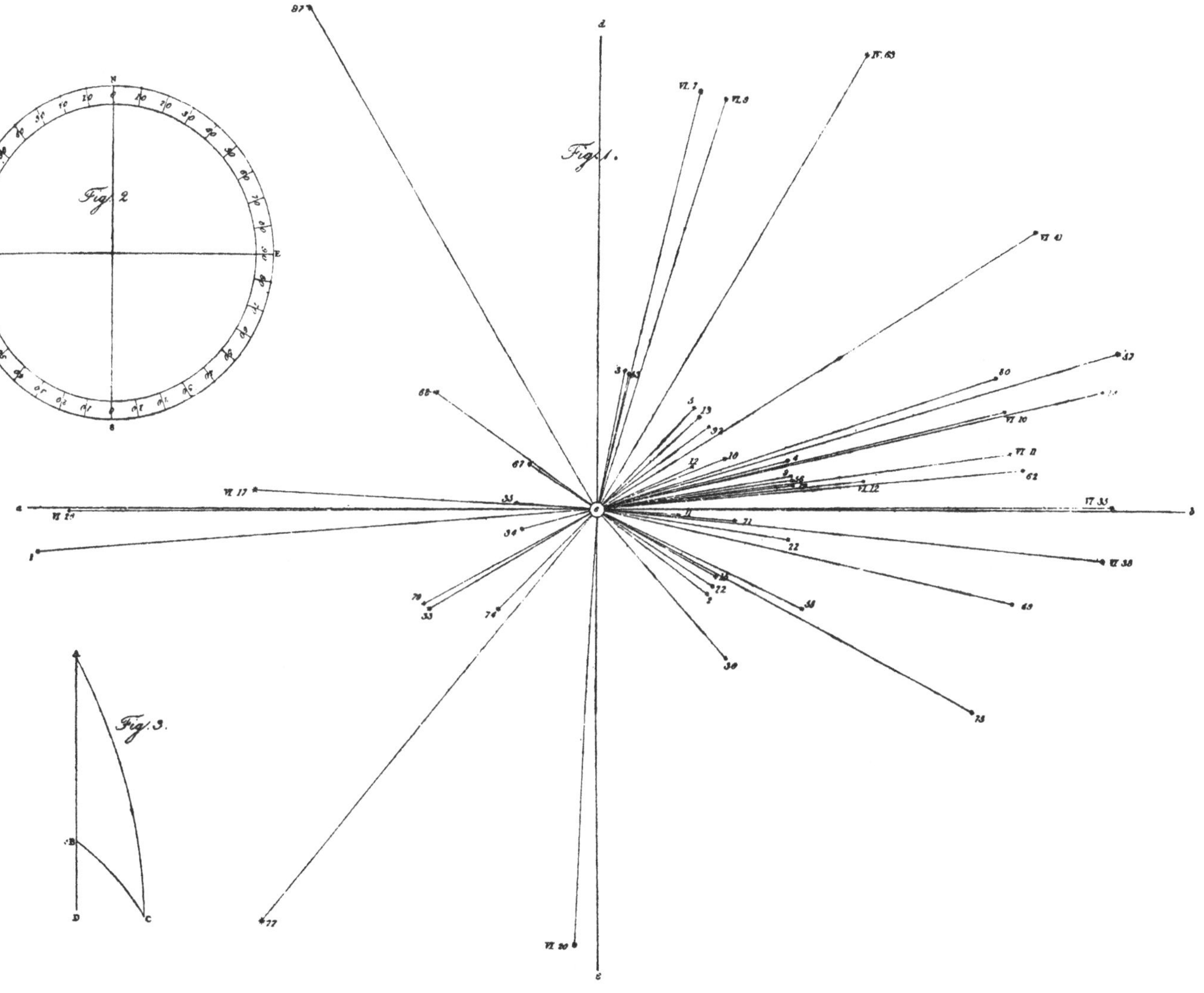

position by going round the globe till he comes to the situation in which this angle will appear of its true magnitude.

In illustration of this, let NESW, figure 2, be the circle on which the azimuths of celestial objects are to be reckoned, and let the meridional line NS pass through the 190th degree of right ascension at S; then will the numbers at the circumference of the circle point out the degrees, and the quadrant of the azimuth of the situation in which any object is to be seen when referred to the milky way. The particular use of this azimuth circle will appear, when the construction of the figure which expresses the profundity of the clusters of stars, of which I have given the observations, has been explained.

Having fixed upon the plane of the milky way as the region of the heavens to which the situation of the clusters of stars is to be referred, their right ascension and polar distance, which are required for this purpose, must be reduced to this plane; and will appear under the denomination of elevation and azimuth. The elevation from the plane of the milky way will be either north or south, and the azimuths in either the northern or southern hemisphere of it, will be in the north-east, south-east, north-west, and south-west quadrants. In order to make this reduction, we have the construction of the triangle ABC, figure 3, in which A is the pole of the heavens; B the north pole of the milky way, and C the situation of the cluster of stars; and there is given the side AB, which is the distance of the two poles; the side AC, which is the polar distance of the cluster, and the angle A, which is the difference between the right ascension of the pole of the milky way and that of the cluster of stars. From these data we find the side BC, the complement of which is the angle of the elevation of the cluster; and the angle ABC, or its supplement CBD, which gives the degree and the quadrant of the azimuth of the cluster. When to these two particulars the profundity of a cluster is added, we have its local situation, with regard to the plane of the milky way, in the required three dimensions of space.

The following table is the result of a set of calculations made for the purpose of obtaining the above mentioned particulars.

Clusters of stars taken from my catalogues.

Class	No.	Profundity.	Elevation.	Azimuth.	Point of sight.
VI.	7	734	76° 58′ N	31° 43′ SE	58° 17′ SW
	9	734	73 25 N	87 2 SE	2 58 SW
	10	734	14 11 N	48 42 SE	41 18 SW
	11	734	8 35 N	55 10 SE	34 50 SW
	12	466	6 26 N	54 30 SE	35 30 SW
	17	600	2 52 N	63 49 NW	26 11 SW
	20	734	87 39 S	10 57 NW	79 3 SW
	26	900	0 5 S	35 38 NW	54 22 SW
	35	900	0 27 N	1 55 NE	88 5 SE
	38	900	4 31 S	77 5 NE	12 55 SE
	41	900	32 55 N	16 9 NE	73 51 SE
IV.	63	900	59 47 N	10 23 NE	79 37 SE

Clusters of stars taken from the *Connoissance*.				
Class No.	Profundity.	Elevation.	Azimuth.	Point of sight.
1	980	4° 42′ S	61° 24′ NW	28° 36′ SW
2	243	35 29 S	68 17 NE	21 43 SE
3	243	78 29 N	89 38 NE	0 22 SE
4	344	14 31 N	47 41 SE	42 19 SW
5	243	45 36 N	59 27 SE	30 33 SW
9	344	9 35 N	62 19 SE	27 41 SW
10	243	22 11 N	71 28 SE	18 32 SW
11	144	3 10 S	84 21 SE	5 39 SW
12	186	25 26 N	71 57 SE	18 3 SW
13	243	41 19 N	65 31 NE	24 29 SE
14	900	14 6 N	77 48 SE	12 12 SW
15	243	26 38 S	57 3 NE	32 57 SE
19	344	7 56 N	53 51 SE	36 9 SW
22	344	8 45 S	67 2 SE	22 58 SW
30	344	47 26 S	86 5 SE	3 55 SW
33	344	29 25 S	10 37 NW	79 23 SW
34	144	13 48 S	20 33 NW	69 27 SW
35	144	3 13 N	63 58 NW	26 2 SW
53	243	77 58 N	28 6 SE	61 54 SW
55	400	24 19 S	66 30 SE	23 30 SW
56	344	8 59 N	60 43 NE	29 17 SE
57	950	16 51 N	61 28 NE	28 32 SE
62	734	5 54 N	50 29 SE	39 31 SW
67	144	31 44 N	83 4 SW	6 56 SE
68	344	34 19 N	3 1 SW	86 59 SE
69	734	11 35 S	59 6 SE	30 54 SW
71	243	4 10 S	66 6 NE	23 54 SE
72	243	32 58 S	86 40 NE	3 20 SE
74	243	43 53 S	15 28 NW	74 32 SW
75	734	26 29 S	78 9 SE	11 51 SW
77	910	50 32 S	47 36 NW	42 24 SW
79	344	29 25 S	76 47 SW	13 13 SE
80	734	18 41 N	48 39 SE	41 21 SW
92	243	35 33 N	55 50 NE	34 10 SE
97	980	58 52 N	26 5 NW	63 55 SW

In order to explain the construction of the table, and the use that is to be made of it when the situation of any one of the clusters delineated in figure 1 is to be examined, I shall take the first cluster it contains for an example.

The first column points out the class and number, where the clusters taken from my catalogues are to be found, and only the number of those that are taken from the *Connoissance des Temps* for 1784. In the figure, the place of the cluster whose situation is to be examined is distinguished by the same mark as in the table, namely VI. 7.

The second column contains the distance of the same cluster from an eye placed in the centre of the globular space, the profundity of which is 734, as determined by the observations that have been given. In the figure it is expressed by the length of the line *c* VI. 7 drawn from the centre of it to the cluster, whose length is 734, the radius of the circle representing the globular space being 12.

The third column gives the angle of elevation of the cluster, which in the present instance is 76° 58′ above the northern plane of the milky way. In the figure it is expressed by the central meeting of the lines *bc* and *c* VI. 7 : one of which denotes this plane, and the other the profundity of the cluster.

To find the quantity of this angle, it is necessary to have the right ascension and polar distance of the cluster ; and here it will be proper to notice that I have deduced these requisites from my own observations of the clusters, brought to the beginning of the year 1800. Then to find the elevation of the present cluster by the method which has been explained, we have in figure 3, the side AB = 60° ; the side AC = 71° 17′, being the polar distance of the cluster ; and the angle A = 16° 56′ 45″, being the right ascension of the cluster 196° 56′ 45″ minus 180°. By these quantities we find BC = 13° 2′ 28″, and its complement 76° 57′ 32″, which is the required elevation of the cluster VI. 7.

The fourth column assigns the azimuth of the cluster ; and as the degrees of the quadrants of the azimuth circle in figure 2, are numbered one from the south, the other from the north, the letter S is prefixed to E, to show that the degrees of it are to be looked for in the south-east quadrant ; the quantity of the angle, in consequence of the foregoing calculation, is easily obtained ; for as we now already have the side BC, the opposite angle A, and the side AC, we find the supplemental angle CBD, which gives the azimuth 31° 43′ 9″. By this result the situation of the direction, in which an observer in the centre of the globular space must look to see the cluster, is determined.

The fifth column contains the point of sight, or situation in which the eye of an observer should be placed, when, by the assistance of a celestial globe, the profundity of any cluster marked in the figure is to be examined. This point, for the cluster VI. 7, is 58° 17′ south-west, which denotes that the globe must be turned horizontally till the 58th degree of the south-west quadrant directly faces the observer, or that, by changing his situation, he must place himself so as to face the globe in the assigned position.

I have called the construction of the figure which gives the profundity of the clusters, an artificial one ; because, as soon as the celestial globe is brought into the situation where it can be seen from the tabular point of sight, the figure will always be found already prepared to show by inspection the azimuth, the elevation and the profundity of the cluster under examination ; for as the globe, which in its adjusted situation has the azimuth of the cluster VI. 7, at right angles to the line of sight, so the globular space in the centre of the figure being supposed similarly arranged, has the tabular azimuth 31° 43′ SE also at right angles to the line drawn to the figure, when seen from the point of sight 58° 17′ SW.

The direction from the centre of the globe to the place on its surface where the cluster is inserted, is also preserved in the continuation of it beyond the surface of the globular space, by the angle of its elevation 76° 58′ above the northern plane of the milky way.

The profundity of the cluster, as has already been noticed, is expressed by the continuation of the line of elevation to 734 such parts as the radius of the globular space contains 12 ; and it may not be amiss, by way of assisting our conception of the vast distance of the situation at which this cluster is placed, to state, that if a line directed to it were added to an 18 inch globe, supposed to contain all the visible stars of the heavens, its length to express this distance would be above 45 feet.

This figure which, from its construction, represents all the different aspects in which a celestial globe should be seen, when its horizontal position for any cluster is adjusted by the foregoing table, has the imperfection that, on account of the different azimuths of their situation, they cannot all be collected into one perspective view ; but as it affords the means of examining them separately, which may even be done without the assistance of the globe, this inconvenience is compensated by the advantage it has of showing all the angles of elevation, and the comparative lengths of the lines expressing the profundity of the clusters in their true magnitude, which an orthographic projection of their situation could not have done.

IV. *Of ambiguous celestial objects.*

When the nature or construction of a celestial object is called ambiguous, this expression may be looked upon as referring either to the eye of the observer, or to the telescope by which it has been examined. In the foregoing observations we find that the 11th, 13th, 15th, and 35th of the *Connoissance*, when they are at a sufficient altitude for the purpose, may be seen by the eye ; but as, without artificial vision, they appear only under the semblance of very small, faint cloudy spots, we should not be able to decide whether they were of a nebulous or sidereal condition, if we were not informed by the telescope that they are brilliant clusters of stars; the eye therefore sees them as ambiguous objects.

If these objects are ambiguous when only viewed by the unassisted eye, there are others that will appear to be so, when they are seen through such small telescopes as are generally attached to large ones, and are called finders, because they point out objects that are not visible to the eye. With regard to these finders, I have occasionally used them of different sizes and constructions; but from experience I can say, that a small one of a most simple composition, with a power of penetrating into space of about four times that of the eye, has generally been sufficient for all the purposes of a 7 or 10 feet telescope; because these instruments may easily be made to act as finders to themselves, by using a double eye glass with a large field of view and a small magnifying power. It is indeed very obvious that when a small telescope, acting as a finder to a larger one, has not sufficient light to show the objects we look for, a more powerful one must be used. In this manner I have often been obliged to have recourse to a 10 feet reflector as a finder, to point out the situation of an object to be viewed in the 20 feet telescope.

It may have appeared singular, that among the observations which have been given, there are many that were made by the 7 and 10 feet telescope finders, but the important use of these observations will appear in the consequences that may be drawn from them; for the clusters of stars, No. 2, 3, 5, 12, 30, 33, 34, 53, 75, and 92 of the *Connoissance* were all to be seen in these finders; they were, however, not seen as clusters of stars, but as ambiguous objects. No. 12, 30, 34, and 75 were but just to be perceived; No. 2 and 92 appeared like stars with rather a large diameter; No. 3 and 5 like hazy stars; No. 33 and 53 like small hazinesses or nebulosities; and yet they were all proved by the telescopes in which they were critically examined to be clusters of stars. If then a cluster of stars in a very small telescope will appear like a star with rather a larger diameter than stars of the same size generally have, we shall certainly be authorised to conclude, that an object seen in a larger and more perfect telescope as a star with rather a larger diameter, is also an ambiguous object, and might possibly be proved to be a cluster of stars, had we a superior instrument by which we could examine its nature and construction.

This seems to throw some light upon a species of objects called stellar nebulæ, one hundred and forty of which have been inserted in my catalogues. For as it has just been mentioned that a 10 feet telescope may become a finder to a 20 feet one, the 20 feet telescope itself will be but a finder to objects that are so far out of its reach as not to appear otherwise than ambiguous; nay, the 40 feet telescope, when it is but just powerful enough to show the existence of an object which decidedly differs from the appearance of a star, may then truly be called a finder.

V. *The milky way, at the profundity beyond which the gaging powers of our instruments cannot reach, is not an ambiguous object.*

Celestial objects can only be said to remain ambiguous, when the telescopes that have been directed to them leave it undetermined whether they are composed of stars or of nebulous matter. Six observations of different parts of the milky way, relating to this subject, have already been given in my last paper,* to which the following four may be added.

Dec. 27, 1786. Right ascension 6^h 42′. Polar distance 88° 33′ There are 116 stars in the field of view, besides many too small for the gage.

Sept. 21, 1788. Right ascension 21^h 29′. Polar distance 41° 1′. There are about 360 stars in the field of view, but most of them are so small that it requires the utmost attention to see them.

Sept. 27, 1788. Right ascension 21^h 17′. Polar distance 52° 50′. With 157 there are small stars with suspected nebulosity; 300 shows a great many smaller stars intermixed with the former.

Sept. 11, 1790. Right ascension 19^h 50′. Polar distance 47° 0′. About 240 stars in the field of view, with many too small to be counted.

In these ten observations the gages applied to the milky way were found to be arrested in their progress by the extreme smallness and faintness of the stars; this can however leave no doubt of the progressive extent of the starry regions; for when in one of the observations a faint nebulosity was suspected, the application of a higher magnifying power evinced, that the doubtful appearance was owing to an intermixture of many stars that were too minute to be distinctly perceived with the lower power; hence we may conclude, that when our gages will no longer resolve the milky way into stars, it is not because its nature is ambiguous, but because it is fathomless.

VI. *Of the assumed semblance of clusters of stars, when seen through telescopes that have not light and power sufficient to show their nature and construction.*

The variety of telescopes used in the long series of observations that have been given, will afford us many instances to ascertain the various deceptive appearances that clusters of stars may put on when they are observed with an inadequate apparatus.

An examination of some particulars relating to this subject may assist us to ascertain in what class we ought probably to place the numerous observations of ambiguous objects that in my sweeps of the heavens were seen by the 20 feet telescope; and having already compared the different forms under which clusters of stars appeared in the finders of the instruments, I shall now also notice how they were seen in the gradually larger telescopes.

* See *Phil. Trans.* for 1817, pages 325, 326, and 329 [above, pp. 588 and 590].

In the 2 feet NEWTONIAN sweeper,

No. 92 of the *Connoissance* appeared like a clouded star, with a magnifying power of 15. No. 2, with a power of 24, appeared like a telescopic comet.

In the 7 feet telescope,

No. 77 was like an ill defined star, surrounded by nebulosity. No. 79, with a power of 57, appeared nebulous. With 460 No. 3 could hardly be seen, for want of light. No. 10, with 227, could not be resolved into stars, for want of power. With 460 No. 22 wanted light, and with 227 it wanted power. With a magnifier of 171 No. 33 had a nebulous appearance. No. 1 was seen as light without stars.

In the 10 feet telescope,

The light of No. 19 appeared mottled. With a power of 71 No. 30 appeared like a pretty large cometic nebula, very gradually much brighter in the middle. With the same power No. 75 was small and cometic. No. 77 had nearly the appearance of a large stellar nebula.

In the large 10 feet telescope,

No. 97, being too low for examination, had the appearance of a faint nebula.

The numerous ambiguous objects that have been seen in the 20 feet telescope do not properly come under this head; for as none of them have been critically examined by superior telescopes, they must still remain ambiguous; and it is for the purpose of being able to form some probable conjecture about the nature of these doubtful objects, that the foregoing results of the appearance of such as have been ascertained to be clusters of stars, have been pointed out.

It would be far too extensive to enter into particular applications, I shall therefore confine myself to a few general remarks. In the depth of the celestial regions, we have hitherto only been acquainted with two different principles, the nebulous and the sidereal. The light of the nebulous matter is comparatively very faint, and, except in a few instances, invisible to the eye. It is also in general widely diffused over a great expanse of space, in which, by an increase of faintness, it generally escapes the sight: the light of stars on the contrary, is comparatively very brilliant, and confined to a small point, except when many of them are collected together in clusters, when their united lustre sometimes takes up a considerable number of minutes of space; but in this case the stars of them may be seen in our telescopes, and by the observations that have been given, it appears that when they are viewed with instruments gradually inferior to those which prove them to be clusters of stars, their diameters, seen with less light and a smaller magnifying power, are generally contracted; a globular cluster is reduced to a cometic appearance; to an ill defined star surrounded by nebulosity, and to a mere small star with rather a larger diameter than stars of the same size generally have. In consequence of these considerations, it seems to be highly probable that some of the cometic, many of the planetary, and a considerable number of the stellar nebulæ, are clusters of stars in disguise, on account of their being so deeply immersed in space, that none of the gaging powers of our telescopes have hitherto been able to reach them. The distance of objects of the same appearances, but which are of a nebulous origin, on the contrary, must be so much less than that of the former, that their profundity in space may probably not exceed the 900th order.

VII. *Of the extent of the power of our telescopes to reach into space, when they are directed to ambiguous celestial objects.*

The method of equalising the light of stars on which the gaging power of telescopes has been established, may also be applied to give us an estimate of the extent of their power to reach ambiguous celestial objects.

When the united light of a cluster of stars is visible to the eye, there will then be a certain maximum of distance to which the same cluster might be removed so as still to remain visible in a telescope of a given space-penetrating power ; and if the distance of this cluster can be ascertained by the gaging power of any instrument that will just show the stars of it, the order of the profundity at which the cluster could still be seen as an ambiguous object, may be ascertained by the space-penetrating power of the telescope through which it is observed. But as the aggregate brightness of the stars depends entirely on their number and arrangement, this method can only be used with clusters of stars that have been actually observed.

The 35th of the *Connoissance*, for instance, being visible to the eye as a small cloudiness, its profundity in space was, by an observation of the 7 feet telescope, shown probably not to exceed the 144th order ; then, as the stars that enter into the composition of this cluster are of such an arrangement that their united lustre may be seen by the eye at the distance of the 144th order, the 10 feet telescope, by which this cluster was viewed, having a power of penetrating into space 28·67 times that of the eye, would be able to show this cluster as a small cloudiness, if it were removed to the distance of the 4128th order. The 20 feet NEWTONIAN telescope, in which it was also observed, having a space-penetrating power 61·16 times that of the eye, would still be sufficient to discover it as an ambiguous object, if it were removed to the distance of the 8809th order.

To investigate how far the 15th cluster, which is also visible to the eye, might be removed, so as still to be seen in the front view of the 20 feet telescope, we find, by inspecting the table in which the profundities are given, that the eye can reach it at the distance of the 243d order ; therefore this telescope, with a power 75·08 times that of the eye, would still be able to show it at the distance of the 18244th order, and being a globular cluster, its appearance would be that of a small star with rather a large diameter.

As there are but few clusters of stars that can be seen by the eye, the observations of their visibility in the finders of telescopes, and their appearance in them, are of eminent use in ascertaining the distance at which we can expect to see celestial objects in large telescopes; when, therefore, a cluster of stars cannot be seen by the eye, its visibility in the finder must first of all be reduced to the standard of the eye. I have already noticed that the power of my finders to show stars, has generally been about four times that of the eye; then, as they would show a star at the distance of the 48th order, a celestial object, situated at this distance, would require to be brought to one quarter of that distance to become visible to the eye.

The 2d cluster of the *Connoissance*, for instance, was seen in a finder with the above mentioned power, and its profundity having been ascertained to be 243, we may conclude that it would be visible to the eye, if it were only of the 60·75th order; this being admitted, it will follow that the 20 feet telescope would still show this cluster of stars as an ambiguous object, if it were removed to the 4561st order; and with a space-penetrating power of 191·69, the 40 feet telescope, by which it was also observed, would have shown this cluster under the semblance of a star that might be distinguished from others by having rather a larger diameter, if it had been at the distance of the 11645th order.

In the foregoing instances, I have assigned the extent of the power to reach celestial objects, as it is in the same instruments whereby they were observed, but this is not a necessary condition; for when the visibility, and the particular manner of its appearance of any cluster of stars in a finder or in a small telescope of any known gaging power is ascertained; and when also by any superior instrument its profundity in space has been assigned, so that it may be reduced to the station at which it would be visible to the eye, it may then be viewed with any telescope of which the space-penetrating power is known; and if we put e for the power of the eye, f for that of the telescope which acts the part of the finder, p for the ascertained profundity of the cluster, and S for the space-penetrating power of a superior telescope, then will the extent E of this telescope to reach the same cluster, as an ambiguous object of any required appearance, be had by the formula $E = \frac{epS}{f}$.

It will not be necessary to calculate, by this formula, the order of distances at which in large telescopes some of the clusters of stars would be seen like telescopic comets, others as large stellar nebulæ, and others again as ill defined stars surrounded by nebulosity, as all these appearances must fall within the compass of the full stretch of their power; I shall therefore only add a calculation of the ambiguous visibility of one of the very distant clusters of stars.

The 75th of the *Connoissance* is not visible to the eye, but may be seen in the finder; and the telescopic observations of it have ascertained its profundity to be of the 734th order; the station to which it should be brought, that it might be visible to the eye, is therefore of the 183·5th order. From this it follows, that with any telescope which has the space-penetrating power of the front view of my 20 feet reflector, this cluster might still be perceived if it were removed to the distance of the 13707th order; and that the 40 feet telescope, which in this case would really act the part of a finder, would still show this cluster of stars as an ambiguous object at a profundity in space amounting to the 35175th order.

W. HERSCHEL.

Slough, near Windsor.

LXX.

On the places of 145 new Double Stars. By Sir WILLIAM HERSCHEL, *President of the [Astronomical] Society.*

[*Memoirs Astron. Soc.*, Vol. I. pp. 166–181.]

Read June 8, 1821.

AN account of the places of 145 new double stars, which were intended to be arranged like those of my two catalogues printed in the *Philosophical Transactions*, as soon as the four particulars of their comparative magnitudes and colour, mutual distance and angle of position could be ascertained. Some of the places of these double stars are taken from a first and second review of the stars with my seven-feet NEWTONIAN telescope; the former being made with a magnifying power of 227, the latter with 460. Some of these double stars are also collected from a review of the ecliptic with a very high magnifier. The rest are taken from my sweeps of the heavens, with the twenty-feet telescope, and a power of 157.

The places of the double stars will be found sufficiently accurate for finding them, as these objects, by their singular appearance, will be easily discovered in a field of view of a large diameter. It should be mentioned, that 15 of these objects were communicated to the late Rev. Mr. FRANCIS WOLLASTON, and inserted by him in his catalogue, distinguished by (M. S.).

It will be seen, by my observations of these double stars, that few of them contain more than one or two of the required particulars, and that the distance and position of the two stars when given, are only in terms of general estimations; so that any lover of astronomy, furnished with a proper telescope and micrometers, who wishes to undertake the work of completing these observations, will find sufficient employment in this interesting pursuit. If this should be the case, it will be an apology for my laying the following observations, in their imperfect state, before this Astronomical Society.

WILLIAM HERSCHEL.

Slough, near Windsor,
February 1, 1821.

Observations of the new Double Stars.

(1.) 125 Sweep, Jan. 24, 1784. A very pretty treble star, making an equilateral triangle, all equal and w. 3d class far, or 4th near. 93 (τ) *Virginis* f. . . . n. $0^\circ\ 5'$, R.A. $14^h\ 4'\pm''$, P.D. $87^\circ\ 24'$.

(2.) 162 Sw. March 11, 1784. A double star preceding the head of *Monoceros*, not in Fl., a very considerable star. 15 *Monocerotis* p. $10'\ 30''$, n. $1^\circ\ 12'$, R.A. $6^h\ 18'\ 43''$, P.D. $78^\circ\ 43'$.

682 Sw. Jan. 11, 1787. Double. 75 (l) *Orionis* f. $14'\ 3''$, n. $1^\circ\ 24'$, R.A. $6^h\ 19'\ 22''$, P.D. $78^\circ\ 35'$.

(3.) 183 Sw. March 21, 1784. A very pretty treble star, making an isosceles triangle, the vertex preceding, and the base in the same meridian. All equal stars w. of the 4th class, near I suppose. 48 *Serpentis* f. $20'\ 15''$ n. $0^\circ\ 19'$, R.A. $16^h\ 22'\ 0''$, P.D. $72^\circ\ 28'$.

(4.) 210 Sw. May 9, 1784. About $18'$ south of 51 (ξ) *Libræ*, double 3d class far.

(5.) 218 Sw. May 16, 1784. Suspected an extended nebulosity between two stars, but 240 showed two double stars, making a parallelogram without nebulosity. 58 (ϵ) *Herculis* f. $22'\ 42''$, n. $1^\circ\ 20'$, R.A. $17^h\ 15'\ 16''$, P.D. $57^\circ\ 27'$.

(6.) 236 Sw. July 12, 1784. Between three nebulæ (10, 11, 12, V. class) is a double star of the 2d or 3d class. 5 (i) *Sagittarii* f. $2'\ 42''$, n. $0^\circ\ 49''$: :, R.A. $17^h\ 49'\ 39''$, P.D. $113^\circ\ 27'$: : .

(7.) 236 Sw. July 12, 1784. A very close treble star, making a triangle, whose vertex is following. 16 (ψ) *Capricorni* p. $11'\ 48''$, s. $0^\circ\ 25'$, R.A. $20^h\ 21'\ 28''$, P.D. $116^\circ\ 26'$.

(8.) 240 Sw. July 18, 1784. A double star, 33 : : *Vulpeculæ* f. $11'\ 18''$, n. $0^\circ\ 8'$, R.A. $20^h\ 59'\ 53''$, P.D. $68^\circ\ 22'$.

301 Sw. Oct. 20, 1784. Double, equal, 4th class near. 33 *Vulpeculæ* f. $12'\ 18''$, n. $0^\circ\ 7'$, R.A. $21^h\ 0'\ 52''$, P.D. $68^\circ\ 23'$.

963 Sw. Oct. 3, 1790. Double 4th class, equal, both considerably large. 33 *Vulpeculæ* f. $12'\ 21''$, n. $0^\circ\ 3'$, R.A. $21^h\ 1'\ 12''$, P.D. $68^\circ\ 26'$.

(9.) 243 Sw. July 22, 1784. 5 *Aquilæ*, treble, the 3d excessively small. Position following the other two, the line bending a little towards the south. Distance almost the same from the 2d, as the 2d from the 1st.*

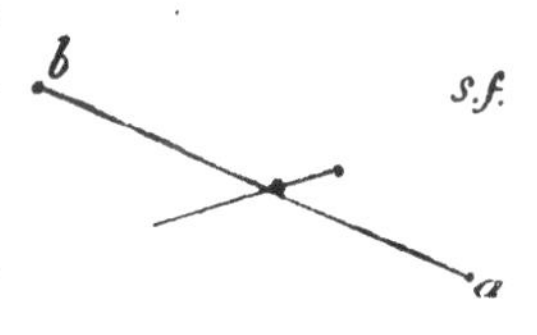

Review, August 5, 1796. 5 *Aquilæ*, treble. Distance of the largest and next to it 0 rev. 24·5 parts $+4\frac{1}{2}$ for zero $= 11''\cdot9$. Position 2 rev. $-61\cdot3$ parts $+1\cdot1$ for zero $=$ $31^\circ\ 27'\cdot3$, considerably unequal. The 2d and 3d very unequal. The 1st and 3d extremely unequal. sf. The 3d is more sf. still, and requires some attention to be seen. Lw. S. dr. 3d very obscure; 460 shows it better than a lower power.

* [=III. 33.—ED.]

(10.) 247 Sw. August 10, 1784. Double, 3d class. 19 *Capricorni* p. 6′ 30″, s. 0° 14′, R.A. 20h 36′ 7″, P.D. 108° 58′.

40 feet Journal, 1st Sw. eclip. Sept. 21, 1791. Double 19 *Capricorni* p. 6′ 30″, s. 0° 15′, R.A. 20h 36′ 31″, P.D. 108° 57″.

Rev. of eclip. Oct. 18, 1792. Double, a little unequal, 3d or 4th class. Position p. It is the preceding of two pretty L. stars; they are near 2 degrees following 15 (υ) in a line parallel to 12 (o) and 43 (κ) *Capricorni*.

Rev. Oct. 12, 1801. Double, the preceding of 2 p. L. stars, about the middle between 12 (o) and θ *Capricorni*. 3d class, a little unequal. The preceding is the smallest.

(11.) 258 Sw. Sept. 6, 1784. Double 3d class near, 7 m., both taken together in time and number. 64 *Pegasi* p. 14′ 18″, n. 1° 9′, R.A. 22h 57′ 2″, P.D. 58° 13′.

(12.) 264 Sw. Sept. 10, 1784. Double. 20 *Arietis* f. 14′ 36″, s. 0° 30′, R.A. 2h 17′ 56″, P.D. 65° 44′.

(13.) 269 Sw. Sept. 13, 1784. Double. 21 (η) *Cygni* p. 10′ 12″, n. 1° 7′, R.A. 19h 37′ 55″, P.D. 54° 22′.

(14.) 275 Sw. Sept. 16, 1784. Two large stars, the time and number taken between them; the second is double. 5 *Pegasi* f. 16′ 6″, n. 0° 19′, R.A. 21h 43′ 41″, P.D. 71° 19′.

(15.) 279 Sw. Sept. 20, 1784. Double. 63 (χ) *Aquarii* f. 19′ 18″, n. 0° 57′, R.A. 22h 45′ 52″, P.D. 94° 24′.

(16.) 313 Sw. Nov. 12, 1784. 57 (m) *Pegasi* double. Position about 20 or 30° sp. L. r. S. b., considerably unequal, 4th class.

(17.) 316 Sw. Nov. 16, 1784. A double star of the 2d class. 65 (1st κ) *Tauri* p. 16′ 35″, n. 0° 45′, R.A. 3h 55′ 52″, P.D. 67° 28′.

(18.) 312 Sw. Nov. 17, 1784. Double 2d class, near : sp. perhaps 1°; a large star, followed by two more. 11 *Eridani* f. 15′ 0″ n. . ., R.A. 3h 8′ 41″ : : P.D. 113° 38′ : :.

(19.) 326 Sw. Nov. 20, 1784. Double 2d class, both L. 11 (e) *Navis* p. 22′ 23″, s. 0° 43′, R.A. 7h 25′ 12″, P.D. 113° 1′. Is n *Argus* in *Puppi*, L. C. 656.

(20.) 326 Sw. Nov. 20, 1784. 6 m. double, 6th class. Position directly preceding, considerably unequal. 15 *Navis* f. 1h 32′ 3″, n. 1° 2′, R.A. 9h 30′ 22″, P.D. 112° 38′.

(21.) 340 Sw. Dec. 13, 1784. 17 (1st ρ) *Orionis*. Double, 2d class. Position n. f. unequal.

Rev. Jan. 17, 1809. Distance between 3 and 4 diameters of L. A pretty object.

(22.) 353 Sw. Jan. 6, 1785. Double, 3d class, equal; position nearly in the meridian, 8. 8 m. 87 (c) *Leonis* f. 47′ 38″, s. 0° 59′, R.A. 12h 6′ 57″, P.D. 92° 48′.

674 Sw. Dec. 29, 1786. Double, of the 4th class, near, equal, nearly in the meridian ; but the most south is the preceding. Position about 82°, 8. 8m. 29 (ν) *Virginis* p. 23′ 41″, s. 2° 29′, R.A. 12^h 7′ 13″, P.D. 92° 46′.

Rev. of eclip. March 14, 1793. Double, 4th class, 3½ degrees south of 15 (η) *Virginis* ; a large star.

(23.) 360 Sw. Jan. 29, 1785. Double, equal 8. 8m., nearly in the meridian ; 3d class near, or 9. 9m. 42 (ψ) *Tauri* p. 26′ 12″, s. 0° 20′, R.A. 3^h 27′ 25″, P.D. 61° 56′.

(24.) 362 Sw. Jan. 31, 1785. 39 (A) *Eridani* has a very small star to the south. 2d class very near. Position in the meridian.

Rev. Jan. 17, 1809. Extremely unequal. I see it best with the double eye-piece. Very unequal will be more proper. With 240 it will not bear light enough to see the wires : it is, however, about 85° sp. With 160 the distance is about between 2 and 3 diameters of L.

b ——:—— *a*

(25.) 368 Sw. Feb. 7. 1785. Double, nearly equal, both w. about 9 m. 3d class far. Position nearly in the parallel. 6 (3d b) *Hydræ et Crat.* p. 49′ 7″, n. 1° 53′, R.A. 9^h 53′ 51″, P.D. 107° 5′. (Is 40 *Felis* of BODE's Cat.)

(26.) 379 Sw. March 5, 1785. 7·6 m. has a star about 8·9 m. following 6th class. 20 *Sextantis* f. 65′ 29″, n. 0° 18′, R.A. 11^h 8′ 23″, P.D. 96° 1′.*

(27.) 380 Sw. March 5, 1785. 72 (1st L) *Virginis*, double, extremely unequal. Position about 30° n. f. 4th class near. L. w. S. r.

913 Sw. March 20, 1789. 72 (1st L) *Virginis*, double.

(28.) 383 Sw. March 10, 1785. Double, very unequal ; 3d class. 24 *Libræ* p. 15′ 2″, s. 1° 28′, R.A. 14^h 45′ 0″, P.D. 110° 26′.

1008 Sw. May 25, 1791. Double, considerably unequal. Position sp. but near the parallel ; 3d class, 7·8 m. MAYER's 575 z. f. 22′ 10″, s. 0° 58′, R.A. 14^h 45′ 17″, P.D. 110° 28′.

(29.) 385 Sw. March 12, 1785. 7 m. has a very small star just preceding, about 4th or 5th class. 72 (τ) *Cancri* f. 25′ 36″, s. 1° 5′, R.A. 9^h 20′ 39″, P.D. 60° 34′. (Is 29 *Leonis* of BODE's Cat.)

(30.) 386 Sw. March 13, 1785. A very small and close double star (with 240) ; the sweeping power made me suspect it to be nebulous. 72 (τ) *Cancri* f. 1′ 6″, n. 1° 15′, R.A. 8^h 57′ 3″, P.D. 58° 18′.

(31.) 393 Sw. April 6, 1785. Double, equal, 3d class, nearly in the same parallel. 12 (e) *Comæ Ber.* p. 1′ 54″, n. 1° 12′, R.A. 12^h 9′ 47″, P.D. 61° 45′.

(32.) 405 Sw. May 1, 1785. Double, 4th or 5th class. 7 (ζ) *Coronæ* f. 7′ 6″, s. 0° 13′, R.A. 15^h 38′ 36″, P.D. 58° 52′.

* [Not 20 Sextantis, but P. X. 6 ; the double star is Crateris 36, R.A. 11^h 5^m 55^s, P.D. 95° 57′.—ED.]

(33.) 411 Sw. May 28, 1785. Double, 4th class near, equal. 50 *Libræ* p. 22′ 8″, s. 0° 17′, R.A. 15^h 27′ 9″, P.D. 98° 5′.

Rev. May 22, 1797. My double star of 411 sweep is about 1° 40′ nf. 37 *Libræ*. (Fl. star, observed page 45, north of 37, is in its place.)

(34.) 430 Sw. Sept. 1, 1785. Double. 19 *Piscis Austr.* p. 11′ 37″, n. 1° 11′, R.A. 22^h 18′ 44″, P.D. 119° 17′.

(35.) 478 Sw. Nov. 27, 1785. 6 m. double, very unequal. Position . . . following. 74 *Aquarii* f. 44′ 16″, s. 1° 30′, R.A. 23^h 26′ 25″, P.D. 104° 16′.

(36.) 521 Sw. Feb. 2, 1786. 35 *Sextantis*. Double, 3d class near. Position south preceding.

675 Sw. Dec. 30, 1786. 35 *Sextantis*, double, 3d class far, a little unequal. Position south preceding.

(37.) 521 Sw. Feb. 2, 1786. Double, both 8 m.; 3d class near. 64 *Virg.* f. 1^h 42′ 7″, n. 0° 3′, R.A. 14^h 53′ 27″, P.D. 83° 41′.

557 Sw. April 29, 1786. Double, equal, 3d class, 8·8 m. 3 *Serpentis* p. 11′ 2″, n. 0° 37′, R.A. 14^h 53′ 31″, P.D. 83° 40′.

(38.) 548 Sw. March 27, 1786. Double, equal $1\frac{1}{2}$ diameter, 7·7 m. 24 (ι) *Crateris* f. 1^h 2′ 24″, n. 0° 11′, R.A. 12^h 30′ 11″, P.D. 101° 50′. (Is 58 *Corvi* in BODE'S Cat., a star of Hev.)

(39.) 559 Sw. April 30, 1786. Double, 3d class near, a little unequal. Position almost in the meridian. 23 (τ) *Scorpii* p. 11′ 22″, s. 1° 24′, R.A. 16^h 11′ 11″, P.D. 119° 10′. (It is MAYER'S 644 z. L. c. 1366.)

(40.) 566 Sw. May 26, 1786. A double star within neb. IV. 41. 14 *Sagittarii* p. 11′ 58″, s. 1° 15′, R.A. 17^h 49′ 30″, P.D. 113° 1′.*

(41.) 595 Sw. Sept. 20, 1786. 53 *Aquarii*, double (cloudy).

1050 Sw. Sept. 6, 1783. 53 *Aquarii*, double, equal; 2d class, or 3d class near. Position about 15° from np. to sf.

(42.) 613 Sw. Oct. 17, 1786. 13 *Lacertæ* has an extremely small star following, 3d class.

(43.) 616 Sw. Oct. 18, 1786. 10 (κ) *Pegasi*, double, extremely unequal, the small star almost n. but a little preceding; 3d class near I suppose.

(44.) 619 Sw. Oct. 18, 1786. Double, equal, 3d class. 4 (ω) *Aurigæ* p. 28′ 0″, n. 2° 1′, R.A. 4^h 16′ 33″, P.D. 50° 26′.

Rev. Oct. 16, 1795. 1° 40′ sp. 58 (e) *Persei* in a line parallel to β and ι *Aurigæ*, double; 3d class, equal.

(45.) 621 Sw. Oct. 24, 1786. Double. 41 (δ) *Andromedæ* p. 7′ 55″, n. 0° 46′, R.A. 0^h 47′ 51″, P.D. 46° 26′.

Jour. Sept. 18, 1794. About 1° 45′ np. 41 *Andromedæ*, double, nearly equal, in a line parallel to 57 (γ) and 42 (ϕ) nearly; a considerable star,

* [=N. 6.—ED.]

2d or 3d class. The southmost is the smallest. Position not far from the meridian; 7 feet. 41 *Andromedæ* p. 7′ 55″, n. 0° 46′.

Rev. August 5, 1796. The double star 7′ 55″ p. 41 *Andromedæ*. Position 3 rev. + 31·5 parts − 1·1 for zero = 74° 20′·4 sp. Considerably unequal. Distance 0 rev. 13·9 parts + 2½ for zero = 7″·2 L. w. S. w. rather pretty unequal.

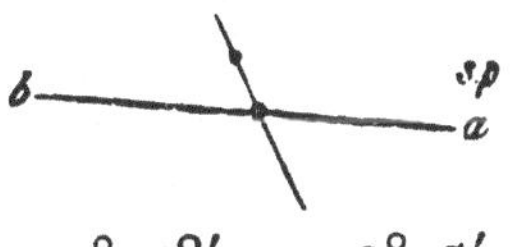

(46.) 654 Sw. Dec. 19, 1786. Double. 69 (λ) *Eridani* p. 0° 48′, n. 0° 5′, R.A. 4^h 58′ 47″, P.D. 98° 56′.

Rev. Jan. 17, 1809. 69 (λ) *Eridani* ½° preceding the nearest of two. Considerably unequal. L. w. S. r. Position with 240, 0 rev., 27·2 parts + 2·5 for zero = 6°·683 or 6° 41′·2. It is IV. 43 of my first catalogues; λ 69 is a single star.

(47.) 692 Sw. Jan. 17, 1787. 7 m. double L. r. S. b., extremely unequal. 14 *Trianguli* f. 3′ 43″, n. 1° 11′, R.A. 2^h 23′ 16″, P.D. 53° 38′.

(48.) 700 Sw. Feb. 13, 1787. Double. 1 *Lupi* p. 21′ : : 43″, n. 0° 44′, R.A. 14^h 39′ : : 45″, P.D. 119° 58′.

(49.) 704 Sw. Feb. 22, 1787. 8 *Sextantis* 5 m. Fl. 6 m. Double, 4th or 5th class, extremely unequal. Position n. p.

(50.) 710 Sw. March 15, 1787. δ *Ant. Pneum.*, L. C. 933. WOLL. Cat. zone 119° 10^h . . . . Double, very unequal; 2d class. Position about 40° sp.

(51.) 711 Sw. March 15, 1787. 4 (h) *Centauri*. Double, very unequal. Position 80° sp.; 3d or 4th class.

(52.) 714 Sw. March 17, 1787. Double, very unequal: 2d class, about 80° sp. L. r. S. b. 7 m. 6 *Canum Ven.* p. 5′ 37″, s. 1° 7′, R.A. 12^h 9′ 45″, P.D. 50° 54′.

(53.) 722 Sw. March 20, 1787. 54 (ν) *Ursæ*, double, very unequal, 60 or 70° sf.; 2d class.

(54.) 748 Sw. July 10, 1787. 7 m. Double, 3d class. Position preceding, a very little south L. w. S. d. 37 (k) *Aquilæ* p. 1^h 3′ 46″, n. 0° 7′, R.A. 18^h 19′ 27″, P.D. 100° 55′.

Rev. Aug. 6, 1796. Of 748 Sw. South of 1 *Aquilæ*, double, L. w. S. d. very unequal. Distance 0 rev. 27·0 parts + 2·5 for zero = 12″·9. It is difficult to measure on account of the position. Position 0 rev. + 65·4 parts − 1·1 for zero = 14° 28′·1. It is the preceding of two large stars near 3° south of 1 *Aquilæ*.

(55.) 752 Sw. August 19, 1787. Double, 2d class, equal, nearly in the meridian. 17 (θ) *Sagittæ* f. 4′ 16″, n. 1° 19′, R.A. 20^h 4′ 46″, P.D. 68° 25′.

(56.) 754 Sw. Sept. 11, 1787. 41 *Aquarii*, double, 2d class near, very unequal. Position sf.

(57.) 765 Sw. Oct. 14, 1787. Double 7·7 m. 3 *Lacertæ* p. 30′ 30″, n. 3° 38′, R.A. 21^h 44′ 42″, P.D. 35° 11′.

768 Sw. Oct. 16, 1787. Double. 14 *Cephei* p. 10′ 15″, s. 2° 11′, R.A. 21^h 44′ 22″, P.D. 35° 11′.

(58.) 794 Sw. Dec. 13, 1787. Double 7 m., 2d class, near, equal. 25 (σ) *Andromedæ* p. 17′ 42″, s. 3° 7′, R.A. 23^h 49′ 19″, P.D. 57° 31′.

981 Sw. Nov. 26, 1790. Double, 2d class, nearly equal, not far from the meridian. 73 *Pegasi* f. 24′ 30″, n. 0° 12′, R.A. 23^h 48′ 44″, P.D. 57° 28′.

Journal, Sept. 18, 1794. Sp. 25 (σ) *Andromedæ*, a pretty considerable star; the largest of two. A pretty double star, 1st or 2d class, very nearly equal. Position not much from the meridian. 25 *Andromedæ* p. 17′ 42″, s. 3° 7′.

(59.) 806 Sw. Feb. 3, 1788. Double, unequal. 49 (π) *Hydræ* p. 10′ 16″, n. 1° 7′, R.A. 13^h 43′ 29″, P.D. 114° 33′.

(60.) 813 Sw. March 4, 1788. Double, 4th class, equal, from sp. to nf. 50 *Aurigæ* p. 2′ 21″, s. 1° 23′, R.A. 6^h 21′ 49″, P.D. 48° 42′.

(61.) 815 Sw. March 9, 1788. 20 *Lyncis*, double, equal, sp. to nf. 8·8 m. 4th class.

(62.) 834 Sw. April 27, 1788. Double, equal, 4th or 5th class, nearly in the meridian. 37 (ξ) *Bootis* f. 21′ 37″, n. 0° 4′, R.A. 15^h 3′ 10″, P.D. 69° 57′.

835 Sw. April 28, 1788. Double, that to the north is a very little smaller, and a little following the meridian of the other. 37 (ξ) *Bootis* f. 21′ 44″, n. 0° 6′, R.A. 15^h 3′ 17″, P.D. 69° 55′.

1006 Sw. May 24, 1791. Double. 37 (ξ) *Bootis* f. 21′ 41″, n. 0° 5′, R.A. 15^h 3′ 23″, P.D. 69° 57′.

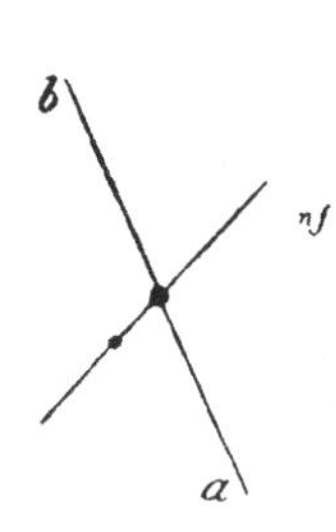

Rev. July 25, 1796. The most north of three that form an arch; double. Position 3 rev. +40 parts +8·4 for zero = 78° 23′·4. It is the double star following (ξ) *Bootis* of the 834 Sweep. A little unequal.

Rev. August 6, 1796. The most north of three, double. Distance 0 rev. 54·6 parts + 2·5 for zero = 25″·0. A little unequal. L. r. S. dr.

(63.) 842 Sw. May 5, 1788. Double, 5th or 6th class, equal, 7·7 m. Nebula observed in this sweep at 15^h 1′ 36″ [II. 757], p. 10′ 38″, s. 3° 8′, R.A. 14^h 52′ 51″, P.D. 35° 23′.

927 Sw. April 24, 1789. Double, 7·7 m.

(64.) 872 Sw. Oct. 30, 1788. Double, of the 2d class, nearly in the parallel. 2 *Lacertæ* f. 2′ 16″, s. 1° 13′, R.A. 22^h 14′ 30″, P.D. 45° 44′.

(65.) 880 Sw. Nov. 4, 1788. 35 *Cassiopeæ Hevelii*, double, 2d class, very unequal. (In zone 23° . . . 2^h 11′, &c. WOLL. Cat.)

(66.) 894 Sw. Dec. 18, 1788. 4 *Camelopardalis Hevelii*, double, very unequal, 3d or 4th class. (WOLL. Cat., zone 35° . . 3^h . .)

(67.) 894 Sw. Dec. 18, 1788. 1 *Camelopardalis*, double, 3d class, a little unequal. Position np.

Rev. Mar. 22, 1795. 1st *Camelopardalis*, double, considerably unequal. Position with 164 np. 2 Rev. $-26{\cdot}7$ parts $+3{\cdot}5$ for zero $=39^\circ\ 46'{\cdot}8$: 2d measure 2 rev. $-26{\cdot}1$ parts $+3{\cdot}5$ for zero $=39^\circ\ 54'{\cdot}9$.

(68.) 919 Sw. April 12, 1789. Double, pretty unequal. 64 (γ) *Ursæ* p. 38′ 46″, s. 0° 56′, R.A. 11^h 3′ 59″, P.D. 36° 4′.

(69.) 940 Sw. March 10, 1790. f. *Hydræ* 1153 L. C. double, a little unequal, 3d class, pretty near. Position about 75° sp. (WOLL. Cat. zone 115° . . . 13^h . . .)

(70.) 953 Sw. March 19, 1790. Double, 3d class. Position a little nf. a little unequal. 76 *Ursæ* p. 1^h 13′ 7″, s. 3° 5′, R.A. 11^h 18′ 12″, P.D. 29° 13′.

1039 Sw. April 9, 1793. Double, equal, 3d class, near. Position nearly in the parallel. 42 *Ursæ* f. 41′ 4″, n. 0° 26′, R.A. 11^h 19′ 29″, P.D. 29° 9′.

(71.) 956 Sw. May 3, 1790. Double, unequal. 9 *Draconis* f. 1^h 24′ 8″, n. 1° 4′, R.A. 14^h 15′ 53″, P.D. 21° 12′.

(72.) 958 Sw. Sept. 7, 1790. Double, very unequal, with a third at no great distance preceding. 20 *Cygni* f. 23′ 8″, n. 0° 1′, R.A. 20^h 8′ 29″, P.D. 37° 31′.

(73.) 959 Sw. Sept. 11, 1790. 50 (α) *Cygni*, 2 m. It has a very small star directly following, about 1′ distance.

Rev. Jan. 17, 1809. The small star is extremely small, and in the 10 feet with 240 will bear no illumination for seeing the wires. Its position is a few degrees from the parallel, on the following side.

(74.) 970 Sw. Oct. 9, 1790. Double, very unequal, 3 or 4° nf., 3d or 4th class, 6 m. 5 *Pegasi* f. 13′ 48″, n. 0° 27′, R.A. 21^h 41′ 40″, P.D. 71° 10′.

(75.) 980 Sw. Nov. 13, 1790. Double, equal, both 7 m. 26 *Aurigæ* f. 2′ 47″, s. 1° 0′, R.A. 5^h 27′ 59″, P.D. 60° 39′.

(76.) 983 Sw. Dec. 2, 1790. 7 m., double, extremely unequal. Position about 80° sf., 2d class, very near. 65 (i) *Piscium* p. 16′ 53″, n. 0° 34′, R.A. 0^h 21′ 41″, P.D. 62° 52′.

(77.) 989 Sw. Dec. 28, 1790. Double, equal, 3d class. 41 *Persei Hevelii* f. 35′ 50″, s. 0° 37′, R.A. 4^h 38′ 25″, P.D. 40° 51′.

(78.) 999 Sw. March 24, 1791. Treble, the two largest equal, 3d class. The first star very small, north preceding the other two; a little further from the preceding of the two, than they are from each other. (R.A. By the sweep 11^h 37′ : :, P.D. 121° : : , no star in the sweep to settle its place.)*

(79.) 1000 Sw. April 2, 1791. Double, 4th class, near, equal. 14 (τ) *Ursæ* f. 2′ 6″, s. 1° 52′, R.A. 8^h 55′ 38″, P.D. 27° 30′.

* [Cannot be identified. In this short sweep, the only one that night, there are only two objects, a ∗ 7·8 and the triple star 5^m 6^s f., 1° 27′ s. of it.—ED.]

(80.) 1008 Sw. May 25, 1791. Double, equal, 7·7 m. Distance about 1′, or a little more. MAYER'S 575 z. p. 9′ 16″, n. 0° 30′, R.A. 14^h 13′ 51″, P.D. 109° 0′.

(81.) 1014 Sw. May 28, 1791. Double. Position sp., extremely unequal; not in WOLL. 21 *Coronæ* p. 0′ 8″, n. 0° 8′, R.A. 16^h 14′ 26″, P.D. 55° 34′.

Rev. March 20, 1795. Fl. 20 (ν) *Coronæ*, consists of two equal stars 6 m. 6 m. The most north and preceding of them has a very small star on the preceding side.

Rev. March 22, 1795. The preceding and the most north of the two stars 6 m. 6 m. has its little star about 50° sp., which is also nearer to the star than the small one of the former double star is to its larger one. (See VI. 18.)

(82.) 1021 Sw. April 20, 1792. Double, equal, 4th or 5th class. 8 (η) *Bootis* f. 7′ 24″, n. 0° 58′, R.A. 13^h 52′ 10″, P.D. 69° 34′.

Rev. August 6, 1796. I cannot find the double star of the 1021 sweep 7′ 24″ following η *Bootis*.

(83.) 1024 Sw. August 22, 1792. Double, extremely unequal. Position directly preceding 7 m. 4 (ϵ) *Sagittæ* p. 2′ 57″, s. 0° 11′, R.A. 19^h 24′ 52″, P.D. 74° 11′.

Rev. Oct. 17, 1795. I cannot see the small star of the double star in the 1024 sweep observed at 19^h 25′ 33″.

Rev. August 7, 1796. I cannot see the small star of the double star in the 1024 sweep observed at 19^h 25′ 33″, with the 7-feet telescope. I see a very small star following 6th class, but the star I look for should be preceding.*

(84.) 1024 Sw. August 22, 1792. Double, considerably unequal. Position about 25° np. 5 α *Sagittæ* p. 0° 46′, s. 1° 24′, R.A. 19^h 30′ 0″, P.D. 73° 51′.

Rev. Oct. 17, 1795. The double star observed in 1024 sweep at 19^h 30′ 37″. Of the 5th or 6th class, very unequal. L. deep red: S. blueish or dusky. Position np.

Rev. August 6, 1796. 1° south of 6 *Sagittæ*. Double, very unequal. Position np. 2 Rev. $-59·8$ parts $+1·1$ for zero $=31°$ 47′·6 L. r. S. b.

Rev. August 7, 1796. 1° south of 6 *Sagittæ*, in a line parallel to 5 and 4 *Sagittæ*; a pretty small star. Distance 0 rev. 59·6 parts + 2·5 for zero = 27″·2.

(85.) 1025 Sw. August 23, 1792. Double, very unequal L. r. S. b. Position nf. 5 *Vulpeculæ* f. 0′ 9″, s. 0° 12′, R.A. 19^h 17′ 16″, P.D. 70° 30′.

(86.) 1027 Sw. Sept. 15, 1792. 8 m. Double. Neb. IV. 72 joins to it. 34 *Cygni* p. 5′ 10″, n. 0° 23′, R.A. 20^h 4′ 53″, P.D. 52° 13′. (Time not accurate.)

(87.) 1027 Sw. Sept. 15, 1792. Double, of the 2d class, unequal. 34 *Cygni* f. 21′ 5″, n. 0° 28′, R.A. 20^h 31′ 8″, P.D. 52° 8′.

(88.) Rev. of eclip. Sept. 15, 1792. Double (between 87 and ϕ 90 *Aquarii*).

(89.) 1028 Sw. Sept. 16, 1792. Double, a little unequal, 4th or 5th class. 3 *Cephei Hevelii* f. 13′ 10″, n. 0° 3′, R.A. 20^h 21′ 44″, P.D. 33° 58′.

* [The star 7 m has not any companion p.—ED.]

(90.) Rev. of eclip. Oct. 16, 1792. Double: *c* is the double star, and *a b c d e* are about the 80 *Aquarii*.

(91.) Rev. of eclip. Oct. 16, 1792. Double, 1st class, very near, a little unequal. It is a very small star, about 2 degrees south of 18 (λ) *Piscium*. With 900 I saw them very well. The line goes to 18 and 17 (λ and ι) *Piscium*.

(92.) Rev. of eclip. Oct. 21, 1792. Double, a pretty object, a little unequal, less than a diameter asunder. Position nf., a third star following at some distance. It is the preceding of two in a line between 98 (μ) and 110 (ο) *Piscium*, and about half way between them. The line from μ, in which the two stars are (of which it is the preceding), passes a little north of 110 (ο).

a b c d e

* double. (91.)

Rev. Oct. 5, 1801. Double, 1st class. A beautiful minute object with 400. It is a star sp. 110 (ο) towards μ, the largest of two.

Rev. Dec. 9, 1801. Double, 1st class, extremely close, equal. It is a star 1° 40′ nf. μ, the first of two in that line. It is a very beautiful object. A third large star in view. They are less than half a diameter asunder. Position about 80° nf. The northern star is rather the smallest.

Rev. Dec. 10, 1801. The double star nf. μ *Piscium*; as described last night.

(93.) Rev. of eclip. Jan. 4, 1793. Double, 2d class, a little unequal. np. 37 (A) *Tauri* 1½° in a line parallel to 54 (ν) and *Pleïades*.

(94.) Rev. Jan. 8, 1793. 55 (δ) *Geminorum*, 6 m. One double towards 43 (ζ).

Rev. March 25, 1795. Sp. δ *Geminorum*, near 2° in a line parallel to 60 and 27 (ε) *Geminorum*. Double, with a third star near. About the 4th class.

(95.) Rev. of eclip. Jan. 8, 1793. Double, 3d class, sf. 55 (δ) *Gemin.*

Rev. Dec. 14, 1795. Sf. δ *Geminorum* towards r, and about 25′ from r. Double, 3d class, a little unequal.

(96.) 1033 Sw. March 4, 1793. ε *Pixidis Naut.* L. C. 831 6 m. Double, very unequal, 5th or 6th class. Position about 50 or 60 degrees sf. L. r. S. dr. (WOLL. Cat. Zone 119° . . . 9h . . .)

(97.) 1038 Sw. April 8, 1793. Double, considerably unequal. Position nf. 39 *Ursæ* f. 16′ 14″, n. 1° 43′, R.A. 10h 46′ 51″, P.D. 30° 0′.

1039 Sw. April 9, 1793. Double, as described last night. 42 *Ursæ* f. 8′ 42″, s. 0° 26′, R.A. 10h 47′ 7″, P.D. 30° 1′.

(98.) 1042 Sw. May 12, 1793. Double, 1st class, equal. Position directly in the meridian, 1½ diameter asunder. *Bootis* 19 *Hevelii* p. 10′ 50″, s. 0° 22′, R.A. 14h 3′ 5″, P.D. 83° 35′.

(99.) 1042 Sw. May 12, 1793. Double, 1st class, very unequal. Position directly preceding: 1 diameter of L. asunder. *Bootis* 19 *Hevelii* p. 7′ 52″, n. 0° 16′, R.A. 14h 6′ 3″, P.D. 82° 57′.

(100.) 1047 Sw. August 25, 1793. Double, equal. Position from np. to sf. 2d class. 3 *Vulpeculæ* f. 3′ 47″, s. 0° 46′, R.A. 19ʰ 18′ 6″, P.D. 64° 54′.

(101.) 1053 Sw. Sept. 27, 1793. Double, 3d class. Position from np. to sf., equal. 1st *Piscis Austr.* p. 18′ 59″, n. 1′ 46″, R.A. 20ʰ 29′ 30″, P.D. 121° 17′.

(102.) 1054 Sw. Sept. 28, 1793. Double, equal, 3d class. Position from np. to sf. but nearer the parallel. 31 (*o*) *Aquarii* f. 12′ 41″, s. 0° 47′, R.A. 22ʰ 5′ 17″, P.D. 93° 56′.

(103.) Journal, Feb. 25, 1794. ½ degree south of the 15th *Monocerotis*; double, a pretty considerable star, very unequal, 3d class far.

(104.) 1058 Sw. April 19, 1794. Double, 3d class, a little unequal, a few degrees np. 12 (*δ*) *Hydræ Crat.* p. 3′ 37″, s. 1° 33′, R.A. 11ʰ 5′ 27″, P.D. 105° 14′.

(105.) Rev. of eclip. Jan. 13, 1795. Double, the middle one of an arch, almost in the meridian: 2d class, unequal; the southern one is the smallest. It is near 2 degrees south of 19 *Arietis*.

Journal, Jan. 15, 1795. Double: it is the most south but one of four small stars in a crooked row, which is nearly in the meridional direction, and it is about 1° 50′ south of the 19th *Arietis*. 1st class, unequal.

(106.) Rev. of eclip. Jan. 13, 1795. Double, 3d class, the middle one of three in the meridian nearly, the most south of which I suppose to be 29 *Arietis*; or ¾° north of 29 *Arietis*.

Journal, Jan. 15, 1795. Double; it is the middle one of an arch of three stars, that are nearly in a meridional direction, the most south of which is the 27 *Arietis*. Or it is about ¾ degree north of, and a little following, the 27 *Arietis*. 2d class, unequal.

(107.) Rev. of eclip. Jan 13, 1795. Double, very unequal, 3d class, 1° 25′ sp. 37 (*o*) *Arietis*.

Journal, Jan. 15, 1795. Double, 1° 25′ sp. 37 (*o*) *Arietis*: 2d class, very unequal.

(108.) Rev. March 20, 1795. 2° 40′ sf. 54 (λ) *Geminorum* towards *β Cancri*, double, 1st class, pretty unequal.

(109.) Rev. Oct. 16, 1795. About 10′ south of 17 (χ) *Cygni* in a line parallel to 58 and 21, is a very small star, which is double, 1st class, nearly equal; the preceding however is the largest: 1 diameter of S.

(110.) Rev. Oct. 16, 1795. About ¾° or 50′ south of 17 (χ) *Cygni* in a line parallel to 6 and 10. A considerable star, double, 5th class, very unequal.

a
b

(111.) Rev. Oct. 16, 1795. About 25 or 30′ nf. 18 (ν) *Geminorum*. A very small star, double, 5th class. L. r. S. d., very unequal, or rather extremely unequal. Position 3 rev. +20 −7 for zero = 77° 12′ sf.

(112.) Rev. Oct. 30, 1795. 1° 40′ north following 93 *Aquarii*. A considerable star, double, pretty unequal. The preceding is the smallest. It is in a line parallel to γ and ω *Piscium*. 3d class I believe.

(113.) Journal, Nov. 9, 1795. A small telescopic star nf. 15 *Cygni*, double, 2d class, very unequal. It is about 5′ or 6′ from 15 *Cygni*, and its position with 15 is 4 rev. − 37 parts − 23·2 for zero = 71° 55′.

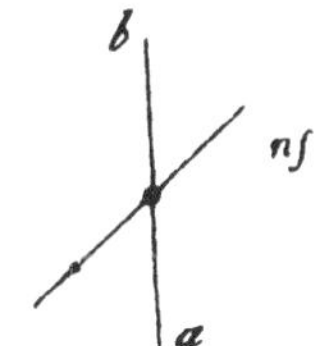

Journal, Dec. 30, 1795. The small double star north following 15 *Cygni* follows it 17″·5 in time: 7-feet reflector, power 115.

(114.) Journal, April 5, 1796. 7-feet reflector, power 460. (ζ) *Bootis*, double, 1st class. Very nearly in contact; I can however see a small division. A little unequal, the preceding is the smallest.

Rev. August 6, 1796. ζ *Bootis*, double. Position 2 rev. − 14·5 parts + 1·1 for zero = 41′ 59″·1 np. With 460 a division is but barely visible $\frac{1}{4}$ of S. Both w. A little, or pretty unequal.

Rev. July 12, 1807. ζ Bootis. They are fine, equal, whitish stars: the interval between their apparent disks with 460 is $\frac{1}{3}$ of the diameter of either.*

(115.) Rev. July 26, 1796. 2° 50′ np. *Arcturus*. Double, 2d class I believe. Considerably unequal. In a line parallel to ϵ and ρ: it is a very small telescopic star.

Rev. August 6, 1796. 2° 40′ np. *Arcturus*. Double. Distance 4 diameters of L. In a line parallel to 32 and α *Bootis*. Position nf. 2 rev. + 78·2 parts − 1·1 for zero = 62° 20′·9.

(116.) Rev. August 10, 1796. About 4 degrees nf. 23 (θ) *Bootis*, the second star in the line from θ to this, double, 3d or 4th class, considerably unequal. Almost directly following I believe, instead of nf.; but the evening is bad.

(117.) Rev. Oct. 25, 1797. The star most south of my double star VI. 119, is double of the 1st class. Considerably unequal. Position nf. 1st class. The angle is such, that a line continued and met by one from the other star, so as to make an isosceles triangle, would meet the line of position at a little more than twice the distance of the two large stars. I verified it with 460, after having looked a considerable time with 920, by way of getting the eye in order. A division can hardly be perceived. But the situation is so low, that certainly the greatest difficulty of seeing the stars arises from that cause. Both together might conveniently stand between the two stars of ζ *Aquarii*, and leave a considerable interval on each side.

(118.) 1066 Sw. Dec. 10, 1797. Double, close of the 2d class, considerably unequal. Distance 1 diameter of L. L. r. S. dr. (See 1068 Sweep.)

1068 Sw. Dec. 12, 1797. The double star of 1066 sweep. 5 *Draconis Hevelii* f. 13′ 54″, s. 0° 23′, R.A. 12h 24′ 12″, P.D. 14° 4′.

* [Identical with VI. 104.—Ed.]

(119.) 1076 Sw. Sept. 5, 1798. 7 m. Double, extremely unequal. Position sf. 52 (h^2) *Sagittarii* p. 7′ 2″, s. 2° 1′, R.A. 19^h 17′ 34″, P.D. 117° 20′, 1st class. (It is q. L. C. 1600. And MAYER'S 786 z.)

(120.) 1078 Sw. Sept. 13, 1798. Double, considerably unequal. The small star is blue, 3d or 4th class. 66 *Draconis* p. 12′ 16″, n. 2° 10′, R.A. 19^h 50′ 15″, P.D. 26° 25′.

(121.) 1081 Sw. Oct. 7, 1798. Double, considerably unequal, 3d or 4th class. Position preceding, or a few degrees sp. 16 *Cephei* f. 8′ 30″, s. 3° 0′, R.A. 22^h 4′ 31″, P.D. 20° 46′.

(122.) 1082 Sw. Oct. 7, 1798. 21 *Cassiopeæ*, double, 6th class, very unequal. Position sf.

(123.) 1089 Sw. Jan. 30, 1799. 8 m. Double, a very small star. Position directly north. 2d class, extremely unequal. 15 (π 1st) *Canis* f. 2′ 1″, n. 0° 5′, R.A. 6^h 46′ 47″, P.D. 109° 55′.

(124.) 1093 Sweep. January 21, 1800. Double, 9·9 m., 2d or 3d class. 15 *Orionis* f. 22′ 24″, n. 1° 33′, R.A. 5^h 20′ 38″, P.D. 73° 9′.

(125.) Rev. Sept. 4, 1801. The 2d of two nf. 22 (λ) *Sagittarii*, probably double; or has a larger diameter. It is about 25′ from λ towards the stars 23, 24, 25. I am pretty sure it is double.

Rev. Sept. 12, 1801. 20′ nf. 22 (λ) *Sagittarii*, double, 1st class, both very small. The smallest of 2 stars.

(126.) Rev. Sept. 4, 1801. About 10′ np. 39 (o) *Sagittarii*, double, very close. (It is N° 191 in Cat. of omitted stars.)

(127.) Rev. Sept. 4, 1801. The middle one of 3 nf. α *Capricorni* is double, 2d or 3d class.

(128.) Rev. Sept. 7, 1801. 67 *Aquarii*, double, 1st class.

(129.) Rev. Sept. 12, 1801. 1 degree south of 39 (o) *Sagittarii*, double, 2d class near, considerably unequal.

(130.) Rev. Sept. 12, 1801. Double, 1st class. It is a small star, equally distant from d and μ *Capricorni*, but a little more south than either. It is a little nearer μ than d.

(131.) Rev. Sept. 12, 1801. Within the triangle δ μ *Capricorni* and ι *Aquarii*, 18 more. 1 double, 3d class, very unequal.

(132.) Rev. Sept. 12, 1801. Double, 2d class, near. It is between α and ϵ *Tauri*, rather nearer α, and it is a little following the line that joins α and ϵ; a considerable star.

(133.) Rev. Sept. 15, 1801. Double, 1st class, 2 degrees sp. 73 (λ) *Aquarii* towards σ.

Rev. Sept. 16, 1801. The double star 2 degrees sp. λ *Aquarii* is very unequal. Position np. Distance 1 diameter of L. It is not towards σ, but rather in a line between 57 (σ) and 43 (θ). The third star in view is north of the double star, or a little nf. The distance of the D. star, after long looking at it, is nearly 2 diameters of L.

(134.) Rev. Oct. 2, 1801. Double, 2 degrees np. 14 *Capricorni*, in a line parallel to α and β. It is the middle one of three small telescopic stars in that line; 2d or 3d class, considerably unequal. There is a star very near it in WOLLASTON'S Catalogue.

(135.) Rev. Oct. 4, 1801. Double, 1st class, both very small. One-third from 74 *Aquarii* towards 93 (ψ). In the finder it appears to be double, owing to a star very near it.

(136.) Rev. Oct. 6, 1801. Double, 2d class, equal. It is a star 35′ nf. 17 (ι) *Piscium*, in a line from κ through ι.

(137.) Rev. Oct. 6, 1801. Double, 2d class, equal. It is south, and a little following θ *Piscium*; about 1° 10′ from it, in a line towards 16.

(138.) Rev. Oct. 12, 1801. Double, 1st class, very near. Very small stars. It is the angular star of a triangle of very small stars: $1\frac{1}{2}$° np. 11 (ρ) *Capricorni* towards 63 *Sagittarii*. Considerably unequal. The preceding is the smallest.

(139.) Rev. Oct. 12, 1801. Double, 1st class, very minute stars. It is a very small star south of 2 that appear coarsely double in the finder. It follows 29 *Capricorni* $\frac{3}{4}$° towards δ; and forms a triangle with 29 and the above-mentioned very coarse double star of the finder.

(140.) Rev. Nov. 27, 1801. Double, 2d class, unequal. The south-preceding star is the smallest. It is 1° 40′ sf. κ *Aquarii* towards ψ.

(141.) Rev. Dec. 7, 1801. Double, 2d class. It is 1° 20′ nf. 18 (ν) *Geminorum*, in a line parallel to γ and ϵ. Equal; or the preceding perhaps the smallest.

(142.) Rev. Dec. 7, 1801. Double, 1st class, very near. $1\frac{1}{2}$° sf. 70 (θ) *Leonis*, in a line from b through θ continued.

(143.) Rev. Jan. 29, 1802. $1\frac{3}{4}$ or $1\frac{1}{4}$ degree north of 32 and 31 *Virginis*, double, 1st class, extremely near, less than half a diameter of either; nearly equal. Position sp. The most south is the smallest.

(144.) 1112 Sw. Sept. 30, 1802. Double, 2d class. 32 *Ursæ* of BODE'S Catalogue p. 2′ 3″, n. 0° 8′, R.A. 8ʰ 36′ 34″, P.D. 18° 29′.

(145.) 1112 Sw. Sept. 30, 1802. Double 7 m. 8 m.: the 8 m. about $\frac{3}{4}$ of a minute sf. the 7 m. 133 *Ursæ* of BODE'S Catalogue p. 3′ 53″, n. 2° 17′, R.A. 10ʰ 2′ 47″, P.D. 17° 59′.

LXXI.

Fifth and Sixth Catalogues of the Comparative Brightness of the Stars—in Continuation of those printed in the "Philosophical Transactions of the Royal Society" for 1796–99.

Prepared for Press from the Original MS. Records by Colonel J. HERSCHEL, *R.E., F.R.S.*

[*Phil. Trans.*, Series A, vol. ccv., pp. 399–447.]

Received July 24,—Read December 7, 1905.

IN the *Philosophical Transactions of the Royal Society* for 1796, 1797, and 1799 there appeared a series of four papers by Sir WILLIAM (then Dr.) HERSCHEL containing the description and results of observations made by him of the "Comparative Lustres of Stars" visible to the naked eye in northern latitudes. They were arranged in six "Catalogues," of which four were actually published, as above. Apparently two more were to have followed, containing the remaining constellations. The annexed Tables show the distribution of the constellations among the six Catalogues.*

It is not known what prevented the completion of the design at the time. Drafts of the intended Fifth and Sixth Catalogues exist among Sir WILLIAM'S papers, prepared, as the previous four had been, by Miss CAROLINE HERSCHEL, by abstraction from the body of his observations of various kinds, entitled "Abstract of Sweeps and Reviews."

Circumstances which it is unnecessary to detail have now led to the revision (and correction where called for) of these drafts and to their publication in the following pages, in the same form as those in the earlier volumes.

DISTRIBUTION of Constellations in Catalogues.

Constellation.		Catalogue Number.	Number of stars in constellation.	Constellation.		Catalogue Number.	Number of stars in constellation.
An	Andromeda	III.	66	Ca	Canis major	II.	31
Aq	Aquarius	I.	108	Ci	Canis minor	II.	14
Al	Aquila	I.	71	Cp	Capricornus	I.	51
Ar	Aries	II.	66	Cs	Cassiopeia	II.	55
Au	Auriga	IV.	66	Cn	Centaurus	III.	5
Bo	Bootes	III.	54	Ce	Cepheus	III.	35
Cm	Camelopardalus	V.	58	Ct	Cetus	II.	97
Cc	Cancer	III.	83	Co	Coma Berenices	VI.	43
Cv	Canes venatici	VI.	25	Cb	Corona borealis	III.	21

* [A table of the observations of variable stars will be found at the end of this paper.—ED.]

DISTRIBUTION of Constellations—*continued*.

Constellation.		Catalogue Number.	Number of stars in constellation.	Constellation.		Catalogue Number.	Number of stars in constellation.
Cr	Corvus	II.	9	Na	Navis	III.	22
Cy	Cygnus	I.	81	Or	Orion	III.	78
Dl	Delphinus	I.	18	Pg	Pegasus	I.	89
Dr	Draco	IV.	80	Pr	Perseus	IV.	59
Eq	Equuleus	I.	10	Ps	Pisces	V.	113
Er	Eridanus	II.	69	Pa	Piscis austrinus	VI.	24
Gm	Gemini	II.	85	Sa	Sagitta	I.	18
Hr	Hercules	I.	113	Sr	Sagittarius	V.	65
Hy	Hydra	V.	60	Sc	Scorpio	VI.	35
HC	Hydra et Crater	V.	31	Ss	Serpens	VI	64
Lc	Lacerta	III.	16	St	Serpentarius	VI.	74
La	Leo major	II.	95	Sx	Sextans	IV.	41
Li	Leo minor	V.	53	Ta	Taurus	IV.	141
Lp	Lepus	III.	19	Tr	Triangulum	IV.	16
Lb	Libra	VI.	51	Ua	Ursa major	VI.	87
Lu	Lupus	VI.	5	Ui	Ursa minor	V.	24
Lx	Lynx	IV.	45	Vr	Virgo	VI.	110
Ly	Lyra	IV.	21	Vl	Vulpecula	V.	35
Mn	Monoceros	IV.	31				

NUMBER of Stars Catalogued.

	Number of stars.		Number of stars.		Number of stars.
CATALOGUE I.		CATALOGUE III.		CATALOGUE V.	
Aquarius	108	Andromeda	66	Camelopardalus	58
Aquila	71	Bootes	54	Hydra	60
Capricornus	51	Cancer	83	Hydra et Crater	31
Cygnus	81	Centaurus	5	Leo minor	53
Delphinus	18	Cepheus	35	Pisces	113
Equuleus	10	Corona borealis	21	Sagittarius	65
Hercules	113	Lacerta	16	Ursa minor	24
Pegasus	89	Lepus	19	Vulpecula	35
Sagitta	18	Navis	22		
		Orion	78		
CATALOGUE II.		CATALOGUE IV.		CATALOGUE VI.	
Aries	66	Auriga	66	Canes venatici	25
Canis major	31	Draco	80	Coma Berenices	43
Canis minor	14	Lynx	45	Libra	51
Cassiopeia	55	Lyra	21	Lupus	5
Cetus	97	Monoceros	31	Piscis austrinus	24
Corvus	9	Perseus	59	Scorpio	35
Eridanus	69	Sextans	41	Serpens	64
Gemini	85	Taurus	141	Serpentarius	74
Leo	95	Triangulum	16	Ursa major	87
				Virgo	110

CATALOGUE V.

A Fifth Catalogue of the Comparative Brightness of the Stars.

Lustre of the Stars in Camelopardalus.

1		6	3 . 1
2		5	2 . 3 7 - 2
3		6	2 . 3 . 1
4		6	7 -, 4 , 5
5		6	4 , 5 - 8
6		6	8 , 6
7		5	7 -; 9 Aur 7 - 2 7 -, 8 7 -, 4
8		7	7 -, 8 5 - 8 , 6
9		4 . 5	10 , 9
10		5 . 4	33 Aur ; 10 10 , 9
11		5	11 , 9 Aur
12		6	9 Aur -, 12
13		4 . 5	Does not exist
14		5	17 , 14 , 19
15		6	30 Aur (32) , 15
16		6	16 . 30 Aur (32)
17		6	31 , 17 , 30 17 , 14
18		6	24 ; 18
19		6	14 , 19
20		7	22 . 20
21		6 . 7	30 , 21 . 23
22		7 . 8	24 , 22 . 20 28 . 22
23		6	21 . 23
24		6	26 . 24 , 22 20 ; 18
25		7 . 8	25 . 34
26		5 . 6	26 . 24
27		5 . 6	Does not exist
28		6 . 7	29 , 28 . 22
29		5 . 6	29 , 28
30		6	31 - 30 17 , 30 , 31
31		5	30 Aur (32) - 31 - 30 31 , 17 37 , 31 -, 38
32		5	32 - - 33 16 . 30 Aur (32) , 15 30 Aur (32) - 31 42 . 30 Aur (32)
33		7	32 - - 33 . 34
34		6	33 . 34 25 . 34 34 ; 35
35		5 . 6	34 ; 35
36		6	42 , 36
37		5 . 6	37 , 31 37 , 40
38		7	31 -, 38
39		6 . 7	40 - 39
40		6 . 7	37 , 40 - 39
41		7	8 Lyn , 41 - 10 Lyn
42		4 . 5	43 ; 42 . 30 Aur (32) 43 , 42 , 36
43		4 . 5	43 ; 42 43 , 42
44		6	46 ; 44 -; 45
45		7	44 -; 45
46		7	47 , 46 ; 44
47		6	18 Lyn -, 47 , 46
48		6	56 - 48
49		5	51 , 49
50		6	27 Lyn - 50
51		5	55 -, 51 , 49
52		5	58 - 52 - 54
53		6	53 , 56 57 . 53
54		6	52 - 54
55		5	55 -, 51
56		6	29 Lyn - 56 58 , 56 53 , 56 56 - 48
57		5	58 . 57 57 . 53
58		5	29 Lyn , 58 , 56 58 . 57 58 - 52

Lustre of the Stars in Hydra.

1		4	1 , 2
2		4	1 , 2 - 10
3		6	15 ; 3 - 17
4	δ	4	22 , 4 . 7 4 . 12 35 ; 4 , 31
5	σ	5	7 , 5 13 . 5 , 18
6		6	9 . 6
7	η	4	4 . 7 , 5 7 , 13
8		6	
9		6	22 . 9 . 6
10		5	2 - 10
11	ε	4	16 -, 11 16 -, 11 - 22 4 Crat . 11 11 , 4 Crat
12		6	4 . 12
13	ρ	5	7 , 13 . 5
14		5 . 6	18 . 14
15		6	15 ; 3
16	ζ	4	16 -, 11 17 Leo -, 16 -, 11
17		6	3 - 17

Lustre of the Stars in Hydra—*continued.*

18	ω	6	5 , 18 . 14	41	λ	4	41 -, 39 4 Crat -, 41
19		6	19 - 20 27 - 19 - 20 23 ; 19 , 21	42	μ	4	2 Crat - - 42 . 43 42 , 7 Crat
20		6	19 - 20 . 24 19 - 20	43	ϕ^1	5	42 . 43 . 1 Crat 1 Crat -, 43
21		6	19 , 21				1 Crat - - 43
22	θ	4	11 - 22 , 4 22 . 9	44		6	44 , 3 Crat
23		6	23 ; 19	45	ψ	6	8 Corvi - 45
24		6	20 . 24 - 29 24 - 25	46	γ	3	7 Corvi =, 46 46 -; 49
25		6	24 - 25	47		6	47 - 48
26		6	27 - 26	48		6	47 - 48
27		6	27 - 19 27 - 26	49	π	4	46 -; 49 49 - 20 Lib.
28	A	6	28 . 33	50		6	52 , 50 50 -, 1 Lib
29		6	24 - 29	51		5	51 , 52
30	α	2	46 Or - 30 - 53 Or	52		5	51 , 52 , 50
31	τ^1	5	4 , 31 ; 32	53		6	58 . 53 , 56 4 Lib . 56
32	τ^2	5	31 ; 32 15 Sext - 32				54 - 4 Lib
			32 - 30 Sext	54		5 . 6	54 , 58 6 Lib - 54 - 4 Lib
33		6	28 . 33	55		6	57 . 55 . 59 57 -; 55 -, 3 Lib
34		6	27 . 34 - 36				12 Lib , 55 -, 3 Lib
35	ι	4	35 ; 4 35 , 15 Sext - 32	56		6	53 , 56 . 57 4 Lib . 56 . 57
36		6	34 -, 36	57		7	56 . 57 . 55 56 . 57 -; 55
37		6	37 . 34	58		5	54 , 58 . 53 6 Lib - 54
38	κ	4 . 5	40 -; 38	59		6	55 . 59 - 60
39	υ^1	5	41 -, 39 -, 40	60		6 . 7	59 - 60
40	υ^2	5	39 -, 40 -; 38				

Lustre of the Stars in Hydra et Crater.

1	ϕ^2	6	43 Hy . 1 3 . 1 -, 43 Hy	15	γ	4	12 - 15 12 -, 15 , 7
			1 - - 43 Hy 2 -, 1 3 , 1	16	κ	5	24 - 16
2	ϕ^3	5	2 - - 42 Hy 2 -, 3 2 -, 1	17		6	19 -, 17 , 18 31 . 17 . 29
			2 - 3	18		6	17 , 18 - 26 28 , 18
3	b^1	6	2 -, 3 . 1 3 , 13 44 Hy , 3	19	ξ	4	19 -, 17
			2 - 3 , 1 3 , 6	20		6	26 -; 20 , 23 25 . 20
4	ν	4	4 . 11 Hy 4 - 12 4 ; 9 Corvi	21	θ	4	21 - 14
			4 -, 12 4 -, 41 Hy 11 Hy , 4	22		7	23 . 22
5	b^2	6	6 , 5	23		6	20 , 23 23 . 22
6	b^3	6	3 , 6 , 5 13 - 6	24	ι	5	14 - 24 - 16
7	α	4	42 Hy , 7 15 , 7 . 11	25	ο	5	25 . 20
8	ι	6	10 - 8	26		6	18 - 26 -; 20
9	χ	5	9 , 10	27	ζ	4	27 , 13
10		6	9 , 10 - 8	28	β	4	28 , 18
11	β^1	3 . 4	7 . 11	29		6	17 . 29
12	δ	4	4 - 12 - 15 4 -, 12 12 -, 15	30	η	4	13 , 30 . 31
13	λ	5 . 6	3 , 13 13 - 6 27 , 13 , 30	31		5 . 6	30 . 31 . 17
14	ε	4	21 - 14 - 24				

Lustre of the Stars in Leo minor.

1		7	1 , 4 5 , 1	29		6	26 . 29
2		6	3 . 2	30		5 . 4	30 – 28
3		6	4 – 3 . 2 3 . 6	31		5	31 ; 21
4		7	1 , 4 – 3	32		6	38 , 32
5		7	5 , 1	33		4 . 5	42 , 33
6		6	3 . 6	34		4 . 5	34 –, 36 34 – 35
7		6	8 , 7 19 Ursæ maj , 7	35		5 . 6	34 – 35 , 36
8		5	8 , 7 11 . 8 8 – 19 Ursæ maj	36		6	34 –, 36 35 , 36
9		6	9 Leonis maj , 9 . 13 Leonis maj	37		3	37 – 42
10		4 . 5	39 Lyncis –, 10 –, 11	38		6	38 , 32
11		6	10 –, 11 . 8 11 – 13	39		6	40 – 39
12		5	13 . 12	40		6	41 –, 40 – 39 40 – 44
13		6	11 – 13 . 12	41		5	41 –, 40 41 – 53 41 – 52 Leonis maj
14		6	42 Lyncis – – 14				
15		6	15 . 42 Lyncis	42		4 . 5	37 – 42 , 33 42 , 44
16		6	17 , 16	43		6	44 , 43 – – 45 44 ; 43 – 45
17		6	19 – – 17 , 16	44		6	40 – 44 , 43 42 , 44 ; 43
18		6	20 – – 18	45		6	43 – – 45 43 – 45
19		5 . 6	19 – – 17	46		4 . 5	36 Leonis maj – 46 , 24 Leonis maj
20		6	21 – – 20 – – 18				
21		5	31 ; 21 – – 20	47		6	46 Leonis maj – 47 47 , 25 46 Ursæ maj –, 47
22		6 . 7	24 – 22				
23		5 . 6	23 –, 24	48		6	48 , 50
24		6	23 –, 24 – 22	49		6	51 Leonis maj – – 49
25		6	47 , 25	50		6	48 , 50 50 , 52
26		6	27 – 26 . 29	51		6	52 , 51 52 , 51
27		6	28 . 27 – 26	52		5 . 6	53 – – 52 , 51 50 , 52 , 51
28		6	30 – 28 . 27	53		5 . 6	41 – 53 – – 52

Lustre of the Stars in Pisces.

1		7	2 . 1 – 3	17	ι	6	28 , 17 17 . 18
2		6	5 , 2 . 1	18	λ	5	17 , 18 . 10
3		6	1 – 3	19		5	19 . 7
4	β	5	4 , 5	20		5 . 6	27 , 20 , 24
5	A	6	4 , 5 , 2 7 . 5	21		6	21 . 22
6	γ	4	6 –, 28	22		6	21 . 22 – 25
7	b	5 . 6	10 , 7 . 5 7 – 16 19 . 7 7 – 32 7 ; 34	23		6	23 – 83 Pegasi
				24		6	20 , 24
8	κ^1	5	9 – – 8	25		6	22 – 25
9	κ^2	7 . 6	9 – – 8	26		6	28 – – 26
10	θ	5	10 , 7 18 . 10	27		5	29 . 27 , 20
11		6	14 , 11 , 12	28	ω	5	6 –, 28 – – 26 28 , 17
12		6	11 , 12 , 13	29		5	30 – 29 . 27
13		6	12 , 13	30		5	33 . 30 – 29
14		6	14 , 11	31	ζ^1	6	32 , 31
15		6	16 – 15	32	ζ^2	5 . 6	7 – 32 , 31
16		6	7 – 16 – 15	33		4	33 . 30

Lustre of the Stars in Pisces—continued.

34		6	7 ; 34	77		6	80 - 77 , 73
35		6	41 , 35 , 36 35 , 51	78		6	82 - 78 . 76
36		6	35 , 36 , 38	79	ψ^2	6	84 - 79 ; 81
37		6	39 , 37 42 , 37 43 -; 37	80	e	5	80 - 77
38		7	36 , 38 - 45	81	ψ^3	6	79 ; 81 , 72
39		6	40 ; 39 40 ; 39 , 37	82	g	6	69 . 82 - 78
40		6	40 ; 39 40 ; 39	83	τ	5	83 , 69 83 ; 90
41	d	6	41 , 35	84	χ	5	74 , 84 - 79
42		6	43 . 42 , 37 42 ; 43	85	ϕ	5	90 . 85
43		6	43 . 42 44 , 43 42 ; 43 -; 37	86	ζ	4	71 -, 86 86 , 89
44		6	44 , 43 44 - 10 Ceti	87		7	72 , 87
45		6	38 - 45	88		6 . 7	73 , 88
46		6	52 -, 46	89	f	6	86 , 89
47		6	47 . 52 47 - 48	90	υ	5	83 ; 90 - 91 90 , 95 90 . 85
48		6	47 - 48 . 49 48 -, 49	91	l	6	90 - 91 95 , 91
49		6	48 . 49 , 53 48 -, 49	92		7	97 , 92
50		6	See note at foot as to this number and 55	93	ρ	5	93 . 94
				94		5	93 . 94 - 97 94 , 107
51		6	35 , 51 [footnote	95		7	90 , 95 , 91 96 , 95
52		6	47 . 52 -, 46 56 , 52 , 54 See	96		6 . 7	96 , 95
53		7	49 , 53	97		6 . 7	94 - 97 , 92
54		6	56 -; 54 52 , 54 -; 61 54 , 59	98	μ	5	51 Ceti (106) , 98 106 - 98
55		6	See note at foot	99	η	4	99 , 5 Arietis 2 Trianguli -
56		6	56 -; 54 56 , 52 See footnote				99 - 5 Arietis
57		6	58 ; 57 58 ; 57	100		6	102 -, 100 101 , 100 101 -
58		7	58 ; 57 64 - 58 ; 57				100 , 104
59		6	54 , 59 , 61 66 . 59 66 - 59 - 61	101		6	101 -, 104 105 , 101 , 103 102
60		6	62 . 60				- 101 , 100 102 - 101 - 100
61		7	54 -; 61 59 , 61 59 - 61				101 . 105
62		6	63 -, 62 . 60	102	π	5	102 -, 100 102 - 101
63	δ	4	63 -, 62				107 -; 102 - 109 102 - 101
64		6	64 - 58 64 - 66	103		8 . 7	101 , 103 105 , 103 , 104
65	i	6	65 . 68	104		6 . 7	101 -, 104 100 , 104 103 , 104
66		6	64 - 66 . 59 66 - 59	105		6 . 7	105 , 101 101 . 105 , 103
67	k	6	68 , 67	106	ν	5	110 - 106 , 98 110 - 106 - 98
68	h	6	65 . 68 , 67				111 , 106 -, 112
69	σ^1	5	83 , 69 . 82	107		6 . 7	107 -; 102 94 , 107 - - 109
70		6	Does not exist 71 - - - 70	108		6	Does not exist
71	ϵ	4	71 -, 86 71 - - - 70	109		8	102 - 109 107 - - 109
72		6	81 , 72 - 75 72 , 87	110	o	5	110 - 51 Ceti (106) 5 Arietis -,
73		6	77 , 73 , 88				110 110 - 106
74	ψ	5	74 , 84	111	ξ	6	111 , 106
75		6	72 - 75	112		6 . 7	106 -, 112
76	σ^2	5	78 . 76	113	a	3	113 , 5 Arietis

[NOTE to 50, 52, 55, 56.—The following entries occur: January 1, 1796, "Either 50 or 52 is wanting. By 46 it is 52 that is wanting" . . . . "56 is wanting." On the same date are comparisons involving 50 and 55, to which asterisks are affixed, referring to a footnote, *in W. H.'s hand* and obviously of later date, "* As it appears by Index that 50 and 55 have no observation, put 52 and 56 for them." In drawing up Catalogue V, C. H. has evidently done this, adding, "does not exist" opposite 50 and 55. With this exception, the same substitutions have been made in this Abstract—J. H.] [50 and 56 are non-existent, caused by errors of reduction.—ED.]

[108 is shown to be (by an error of FLAMSTEED'S, transferred to the Atlas) the same as 109, but 3° out of place.]

Lustre of the Stars in Sagittarius.

1		6	33 Scorpii, 1
2		6	2, 52 Ophiuchi
3	*p*	6	51 Ophiuchi - 3
4	*b*	6 . 7	7 . 4 ⁏ 9
5	*i*	7	5, 7 5 . 12
6		7	54 Ophiuchi - 6 . 8
7	*a*	6	5, 7 . 4 12, 7
8		7	6 . 8 8 does not exist
9		7	4 ⁏ 9
10	γ	3	19 = ⁏ 10
11		7	Does not exist
12		7	5 . 12, 7
13	μ¹	4	27, 13, 40 39 - 13 = ⁏ 15 13 -, 21
14		7	15, 14 . 16
15	μ²	6	13 = ⁏ 15, 14 21 ⁏ 15
16		7	14 . 16 - 17
17		7	16 - 17
18		7	
19	δ	3	38 . 19, 27 22 ⁏ 19 ; 20 19 = ⁏ 10
20	ε	3	19 ; 20
21		6	21 ⁏ 15 13 -, 21
22	λ	4	41 . 22, 38 22 ⁏ 19
23		7	25 - 23
24		7	24 -, 26
25		7	26, 25 25 - 23
26		6	24 -, 26, 25
27	φ	5	19, 27, 40 27, 36 27, 13
28		7	28 - 31
29		6	36, 29, 33
30		6	33, 30 . 31 31, 30
31		6	30 . 31 28 - 31, 30
32	ν¹	5	32 ; 35
33		6	35 - 33, 30 29, 33

34	σ	4 . 3	34——41 34 - - 41 50 Aquilæ, 34 . 33 Capricorni
35	ν²		32 ; 35 35 - 33
36	ξ¹		27, 36, 39 37——36 36, 29 36, 39
37	ξ²	6	37——36
38	ζ	3	22, 38 . 19
39	ο	4	36, 39 36, 39 39 . 44 39 - 13
40	τ	4	27, 40 13, 40
41	π	4	34 —— 41 . 22 34 - - 41
42	ψ	5	42, 49
43	*d*	6	46 - 43 ⁏ 45
44	ρ¹	5	39 . 44 -, 46
45	ρ²	6	43 ⁏ 45 50 ; 45
46	υ	6	44 -, 46 - 43
47	χ¹	5	47 —— 48 47 ; 49
48	χ²	5	47 —— 48
49	χ³	6	47 ; 49 42, 49
50		6	50 ; 45
51	*h*¹	6	52 - 51 51, 53 . 53
52	*h*²	6	52 - 51
53		6	51, 53 . 53
54	*e*¹	6	55 ; 54 . 61
55	*e*²	6	55 ; 54
56	*f*	6	56 -, 57
57		6	56 -, 57
58	ω	5	62, 58 . 60
59	*b*	5	60, 59
60	*a*	5	58 . 60, 59
61	*g*	6	54 . 61
62	*c*	6	62, 58
63		6	63 - 64
64		6	63 - 64, 65
65		6	64, 65

Lustre of the Stars in Ursa minor.

1	α	3	7 ; 1 - 14 Draconis 1, 7 α (1) - β (7) Polaris (1) ⁝ 7 1, 7 1 - 7 1 - 7 1 - 7 α (50) Ursæ maj ⁏ 1 ⁏ 7 1, 7	7	β	3	7 ; 1 1, 7, γ (33) Draconis 1 - 7 50 Ursæ maj, 7 50 Ursæ maj ⁝ 7 1 ⁝ 7 7, 50 Ursæ maj 1, 7 1 - 7 1 - 7 50 Ursæ maj ⁝ 7 1 - 7 1 ⁏ 7 79 Ursæ maj ⁝ 7 7 - 64 Ursæ maj 7 - 33 Draconis 1, 7
2		6	Is wanting				
3		6	4 -, 3				
4	*b*	5	5 - 4 4 -, 3				
5	*a*	4	22 - 5 - 4				
6		7	11 - - 6 9 - 6	8		6	

Lustre of the Stars in Ursa minor—continued.

9		7	9 - 6 9 , 10	17		7	19 - 17 , 20
10		7	9 , 10 , 14 14 . 10	18		6	15 - - 18
11		5	13 - - 11 - - 12 11 - - 6	19		5	21 , 19 . 20 21 , 19 - 17
12		7	11 - - 12 12 . 8 . 8	20		6	21 - 20 19 . 20 17 , 20
13	γ	3	13 - - 11	21	η	5	21 - 20 21 , 19 21 , 19
14		7	10 , 14 14 . 10	22	ε	4	22 - 5
15	θ	5	16 - 15 16 -, 15 15 - - 18	23	δ	3	23 -, 24
16	ζ	4	16 - 15 16 -, 15	24		6 . 7	23 -, 24

Lustre of the Stars in Vulpecula.

1		5	1 - 1 Sagittæ 1 - 2 1 -, 2	16		5	16 , 13
2		6	1 - 2 , 1 Sagittæ 1 -, 2 , 1	17		4 . 5	13 . 17 17 . 22
			Sagittæ	18		6 . 5	19 . 18 , 20
3		6	6 - 3 , 3 Cygni 3 - - 3 Cygni	19		6	19 . 18
			3 - - 3 Cygni	20		5 . 6	18 , 20
4		6	9 , 4 . 5	21		5 . 6	23 , 21 -, 24
5		6	4 . 5 , 7	22		5	17 . 22
6		4	6 - - 8 6 - 3	23		4 . 5	15 , 23 , 21
7		5	9 - 7 5 , 7	24		5	21 -, 24 24 . 25
8		6	6 - - 8 8 . 3 Cygni 8 . 3 Cygni	25		6	24 . 25
9		6	5 Sagittæ - 9 , 8 Sagittæ 9 - 7	26		6	27 , 26
			9 , 4 9 - 10 14 - 9	27		5	27 , 26
10		6	9 - 10 , 13 10 - - 11 13 - - 10	28		6	29 . 28 . 32
			10 , 14	29		5	31 , 29 . 28
11			10 - - 11	30		6	32 . 30
12		5	13 - 12 -, 14	31	r	6	31 , 29
13		6	10 , 13 13 - 12 13 - - 10	32	q	5	28 . 32 . 30 35 . 32
			16 , 13 . 17	33		6	33 - 34
14		5	14 - 9 12 -, 14 10 , 14	34		6	33 - 34
15		4 . 5	15 , 23	35		6	35 . 32

NOTES.

[N.B.—A long dash between two notes or remarks under the same number indicates that they are disconnected, and occur at an interval of time—of days or months even—in the course of the "reviews." The only connecting link is the number of the star to which they refer.—J. H.]

Notes to Camelopardalus.

8 Is not in the place where it is marked in Atlas: the R.A. should be + to make it agree with a star that is thereabout, or – to make it agree with another. Either of them will be 7 –, 8.——The star following 7 and 8, observed by FLAMSTEED, p. 286, is in its place, but is much less than 6m. I should call it 8m.

9 Has no time in FLAMSTEED's observation. It seems to be placed in Atlas considerably too late, so as perhaps to require a correction –10′ in time.

13 Does not exist.——13 does not exist. My double star VI. 35, is 9 Aurigæ.

17 The time in FLAMSTEED's observation is marked "*circiter*," but I find that my viewing instrument cannot, for want of other near stars, determine whether it is properly placed in the Atlas and catalogue.

27 Does not exist. FLAMSTEED never observed it.——27 28 There is an observation by FLAMSTEED, p. 286, on a star S. of 28, but it does not exist, nor 27.——27 is wanting. A star observed by FLAMSTEED, p. 286, is not in the place where it should be. 27 was never observed by FLAMSTEED.

32 Is the same with 30 Aurigæ.——The stars 32 33 34, as I have called them, Oct. 30, are small stars nearly in a line, but I doubt whether my 32 is FLAMSTEED's star. The Atlas does not give it as it is in the heavens.——The star taken for 32 Cam. is a small star between 33 and 30 Aurigæ, not given in any catalogue.

35 Has no time, but seems to be very properly placed in Atlas and catalogue.

39 My instrument will not determine its place. It is without time in FLAMSTEED's observation.

42 A star observed by FLAMSTEED, p. 288, who calls it 4m, preceding 42 and 43 is in its place.

45 and 46 By FLAMSTEED's observations should have their P.D. reversed, but in the heavens they seem to stand as they are placed in Atlas and catalogue.

49 I cannot determine the time of 49, which FLAMSTEED's observations have : :

52 54 58 – 52 – 54 but I am not quite sure of 52 and 54. There are so many small stars, that it is not possible without fixed instruments to ascertain them positively.——54 in FLAMSTEED's observations by *strias* (screws) requires P.D. –2°, but it is not possible to ascertain its place positively.

Notes to Hydra.

8 There is but one star, which if it be 31 Monocerotis, then 8 is not there. FLAMSTEED never observed it.

36 Is not in the place where it is marked in Atlas. The time in FLAMSTEED's observations is marked : :

43 Is hardly visible in my small telescope.——1 Crateris – – 43 . Dec. 15, 1795, it is 43 Hydræ . 1 Crateris, but now (Jan. 26, 1797) it is 1 Crateris – – 43 Hydræ. I suppose 43 to be changeable.

Notes to Hydra et Crater.

1 See note to 43 Hydræ, above.

22 I cannot see 22 in the place where FLAMSTEED has given it, but 1° above is a star which I suppose is it; calling that, therefore, 22, it is 23 . 22.

Notes to Leo minor.

12 Near 12 is a star observed by FLAMSTEED, p. 438. 12 wants a correction + in R.A.

17 Requires −10′ in P.D.

22 Is not to be seen. 23 −, 24 − 22. There is a star pointed out by 23 and 24 which may be 22, but then its situation is faulty about 30′, being too far from 28.*

32 The star north of 32 observed by FLAMSTEED, p. 220, is in the place.

41 54 54, β Leonis, and the star in Leo minor's tail-end, 41 Leonis minoris, are in succession of magnitude.

49 Is a very small star, and a much larger between 49 and 60 Leonis majoris is not down in catalogue and Atlas.

Notes to Pisces.

1 Which has the time "*circiter*" in FLAMSTEED seems to be placed in Atlas and in the catalogue a little later than it should be; perhaps 5′ or 6′ of space.

40; 39 A larger star than either is 1° 4′ towards α Androm. If this was mistaken for 39 perhaps it might give rise to the supposition of the loss of 40.——40 is not lost.

48 Has no time. In the heavens it seems to be nearly in the place where the catalogue gives it.—— 48 −, 49. The observation 48 . 49, Jan. 1, 1796, is probably owing to a mistake of the star, as there is one nearly equal to 48 near it which is not in FLAMSTEED's catalogue nor Atlas.

50 52 56 Either 50 or 52 is wanting. By 46 it is 52 that is wanting; 56 is wanting. [Note by W. H.: "As it appears by Index that 50 and 55 have no observations, put 52 and 56 for them."] †

59 FLAMSTEED has no observation of 59, but there is a star in the place where the catalogue gives it.

70 Does not exist.——70 is a very small star. FLAMSTEED observed it, p. 406.

71 Is so small that it may, perhaps, not be FLAMSTEED's star, but there is no other.

72 A star between 72 and 78, observed by FLAMSTEED, pp. 149, 180, is in its place. It is = 72 nearly.

104 Is 8° lower than 1 Arietis (which does not exist) is marked; perhaps it was by mistake placed 8° more north and called 1 Arietis.‡

108 Does not exist, or is invisible.——There is a large star 1¼° from 6 Arietis and 2¾ from 107, not in Atlas.——108 does not exist. 109 is just 3° south of it and is, perhaps, the same. On p. 332 of FLAMSTEED's observations the number is cast up 3° wrong, which has produced 108 Pisc. The observation belongs to 109.

Notes to Sagittarius.

1 FLAMSTEED has no observation of 1, but there is a star exactly in the place where 1 is marked in the Atlas.

8 Does not exist. There is a small star at rectangles to 17 15 13 towards the place where 8 is marked in the Atlas, but it is much too near 13 to be 8.

11 Does not exist.§

12 The R.A. of 12 requires a correction of about 1° minus, for in the place where 12 is marked in Atlas is no star, but 1° before there is one which answers to it.||

14 The star observed by FLAMSTEED, p. 171, is in its place; it is 1½° S. of 14.

18 I see many small stars north of 19, but cannot see 18 south of it.——18 is not in the place assigned by FLAMSTEED's catalogue, but about 1° more in R.A. is a star which is probably the one intended. It was observed by FLAMSTEED, p. 115.

23 24 The star between 25 and 26 north of them observed by FLAMSTEED, p. 374, is in its place. 23 does not exist. There is a star that answers pretty well to 23. It is a little farther from 25 than it is laid down in Atlas. 24 should be nearer to 25 than in Atlas. The observation of FLAMSTEED, p. 532, gives it right.¶

53 Is double, and I cannot say which is FLAMSTEED's star.

* [Flamsteed's position of 22 Leonis min. is correct.—ED.]

† [50 and 56 Piscium are non-existent, caused by errors of computation.—ED.]

‡ [The R.A. of 1 Arietis was 10^m too small.—ED.]

§ [It is P. XVII. 366.—ED.]

|| [It is the P.D. which is 52′ too great.—ED.]

¶ [The P.D. of 23 was 40′ too small.—ED.]

Notes to Ursa minor.

1 α appears uncommonly bright.——The pole star seems to be decreased, or β is increased. The place of the moon may possibly influence appearances.—— α, β The night is not favourable.——Very clear. α – β.

2 Is not as in Atlas, or rather it exists not.——FLAMSTEED observed a star, pp. 213, 214, 215, which has been misplaced and called 2 Ursæ minoris. It should be 2° further from 1, and it is in the place where it was observed.

4 By FLAMSTEED'S observation the R.A. of 4 should be −3° 50′ in time; but without a fixed instrument I cannot perceive that 4 is misplaced, being so near the pole.

8 Either exists not, or is at least not in the place marked in Atlas.——8 is misplaced in Atlas: there are two small stars about 1° from 7 towards 15: one of them is probably 8. They are equal, 12 . 8 . 8.

10 14 There is a larger star than either 10 or 14, between but following these two, which is not in FLAMSTEED.

12 Appears too small for 7m. It is 8 or 9m. FLAMSTEED has no observation of 12.

14 Has no time in FLAMSTEED'S observation, but it seems to be placed very justly in the Atlas.

15 Requires P.D. −10′.

16 19 There is a large star between 16 and 19 not in FLAMSTEED. The mistake of Sept. 14, 1795, is owing to the large above-mentioned star.

18 There are seven stars about the place of 18.——FLAMSTEED has no observation of 18.

19. 20 Sept. 14, 1795, I suppose this to be a mistake of the star.

24 Requires + 10 minutes or 2½° in R.A.

Notes to Vulpecula.

Vulpecula in Atlas is laid down so confusedly and erroneously that it is impossible to ascertain the stars without a fixed instrument.

2 Is misplaced. It requires a correction of ¾° minus in R.A. and 30′ + in P.D.

3 The observation of Sept. 17, 1795, 6 – 3 , 3 Cygni does not agree with this 3 – – 3 Cygni [*i.e.*, of this date, Nov. 3, 1795].——Nov. 15, 1795, 3 – – 3 Cygni.

7 Requires a correction, - near ¾° in R.A.

11 10 – – 11, but 11 is very small and FLAMSTEED has no observation of it. I suppose therefore that this is not the star which is given in Atlas and catalogue.——11 is forgot in Atlas.

13 9 – 10 , 13, Sept. 17, 1795, but 13 is further from 10 and nearer to 14 than it is marked in Atlas.

12 Is placed too far north in Atlas—at least 15 by 12 Sagittæ.——Large star in the breast near 14 – 9.

13 The expression 10 , 13, Sept. 17, 1795, cannot be right; it is 13 – – 10.——Dec. 4, 1796, I have my doubts about the expression 10 , 13 used Sept. 17, 1795. I could hardly mistake the star 10 as [? and] there is none in the neighbourhood that exceeds 13.

14 A large star in Atlas preceding 14 is not in the heavens, nor do I know how it comes into the Atlas, as FLAMSTEED has it nowhere. This constellation must be reviewed again, when it is higher.

16 A considerable star near 16.

24 25 A star larger than either, north of 24, observed by FLAMSTEED, p. 64, is in its place.

31 32 Are contrary in magnitude to what they are in Atlas.

CATALOGUE VI.

A Sixth Catalogue of the Comparative Brightness of the Stars.

Lustre of the Stars in Canes venatici.

1		6	5 - - 1 , 7	14		5	5 , 14
2		5	10 . 2	15		6 . 5	15 . 17
3		6	3 , 7	16		6	17 - 16
4		6	9 . 4	17		6	15 . 17 - 16
5		6	5 - - 1 8 -, 5 , 14	18		6	19 - 18
6		5	8 -, 6 , 10	19		7	23 . 19 - 18 Note
7		7	1 , 7 3 , 7 -, 11	20		6	20 - 23
8		4 . 5	25 - 8 8 -, 5 8 -, 6	21		6	24 - 21
9		6 . 7	10 , 9 . 4	22		6	Does not exist
10		6	10 , 9 6 , 10 . 2	23		7	20 - 23 . 19
11		6	7 -, 11 Note	24		5 . 6	24 - 21
12		2 . 3		25		5	25 - 8 Note
13		4 . 5	41 Com Ber , 13 (=37 Com Ber)				

Lustre of the Stars in Coma Berenices.

1		7	2 . 1 , 3	24		5	24 . 7 24 =, 27
2		6	5 , 2 . 1	25		6	25 , 8
3		6	1 , 3	26		5	18 . 26
4		6	13 , 4	27		5	24 =, 27 , 29 27 , 36
5		6	5 , 2 Note	28		6	29 - 28
6		5	6 , 11	29		5	27 , 29 - 28 Is the same with 36 Virginis
7	*h*	4 . 5	14 . 7 -, 8 7 - 20 24 . 7				
8		7	7 -, 8 20 , 8 25 , 8	30		6	31 , 30
9		6	9 . 10	31		4 . 5	41 . 31 , 30
10		6	9 . 10	32		7	38 , 32 . 33
11		4 . 5	6 , 11	33		7	32 . 33
12	*e*	5	15 , 12 . 16	34		5	Does not exist
13	*f*	4 . 5	17 . 13 , 4	35		4 . 5	35 - - 39 Note
14	*b*	4 . 5	16 . 14 . 17 14 . 7	36		5	27 , 36 - 38
15	*c*	4 . 5	15 , 12	37		5 . 6	41 , 37 or 13 Can venat
16	*a*	4 . 5	12 . 16 . 14	38		6	36 - 38 , 32
17	∂	4 . 5	14 . 17 . 13	39		5	35 - - 39 39 , 40
18		5	21 , 18 - - 22 18 . 26	40		6	39 . 40
19		6	Does not exist	41		5 . 4	43 -, 41 . 31 42 -, 41 41 , 37 Note
20		6	7 - 20 , 8				
21	*g*	5	23 - 21 , 18.	42		4 . 5	5 Boot , 42 , 4 Boot 43 , 42 -, 41 Note
22		7	18 - - 22				
23	*k*	4	23 - 21	43		5 . 4	43 -, 41 43 , 42

Lustre of the Stars in Libra.

1		5 . 6	Does not exist See note	28		6	25 , 28
2		7	2 – 96 Virginis Note	29	o^1	7	32 , 29 . 34
3		6	55 Hyd –, 3 55 Hyd –; 3 . 14	30	o^2	6	33 , 30
4		6	54 Hyd – 4 4 . 56 Hyd 4 is 53 Hyd	31	ϵ	4	37 , 31 . 35 37 , 31 , 19 37 . 31
5		6	5 . 18 5 . 10	32	ζ^1	6	32 , 34 32 , 29
6		5	45 . 6 . 7 6 – 54 Hyd 6 is 58 Hyd	33	ζ^2	7	35 – 33 , 30
				34	ζ^3	6	32 , 34 . 35 29 . 34
7	μ	5	6 . 7 . 21 7 , 19 7 – – 15	35	ζ^4	4	31 . 35 . 44 34 . 35 – 33
8		6	24 – 8 – 25	36		6	40 –, 36
9	α	2	27 , 9 – 20 27 . 9 – 20 27 –, 9	37		6	24 , 37 , 31 37 , 31 37 . 31 Note
10		6	5 – 10				
11		6	105 Virg – 11	38	γ	3 . 4	39 . 38 . 51 51 , 38 , 46 20 – 38 51 . 38 38 –, 46
12		6	12 , 55 Hyd				
13	ξ^1	6	15 , 13 ; 18	39		4	40 , 39 . 38 46 . 39 , 40 39 ; 40
14		6	3 . 14 23 , 14	40		4	20 , 40 , 39 39 , 40 39 ; 40 –; 36
15	ξ^2	6	7 – – 15 , 13	41		6	21 . 41 47 , 41
16		5 . 6	16 , 105 Virginis	42		6	1 Scorp – 42 – 4 Scorp
17		7	18 . 17	43	κ	4	19 . 43 . 45 43 , 45
18		6	13 ; 18 . 17 5 . 18 19 . 18	44	η	4	35 . 44 . 19 48 , 44 48 , 44 44 , 49
19	δ	4 . 5	44 . 19 . 43 31 , 19 7 , 19 19 . 18	45	λ	4	43 . 45 . 6 45 – 47 43 , 45 –; 47
20	γ	3	20 , 40 20 , 51 49 Hyd – 20 – 38	46	θ	4	51 . 46 , 48 88 , 46 . 39 46 , 48 46 – 48 38 –, 46 –, 48
21	ν^1	5	7 . 21 . 41 21 –, 22 21 – 26	47		6	45 – 47 45 –; 47 , 41
22	ν^2	6	21 –, 22 26 , 22	48	ψ	4	46 , 48 , 24 48 , 15 Scorp 46 , 48 , 44 46 – 48 , 44 46 –, 48 – – 49
23		7	23 , 14 Note				
24	ι^1	4 . 3	48 , 24 , 37 24 – 8				
25	ι^2	6	8 – 25 25 , 28	49		6	48 – – 49 44 , 49
26		6	21 – 26 , 22	50		6	42 Serpii – 50 50 – 43 Serpentis
27	β	2	27 , 9 27 . 9 27 –, 9 27 , 24 Serpentis Note	51	ξ	4 . 5	38 . 51 . 46 20 , 51 , 38 51 . 38

Lustre of the Stars in Lupus.

1		5	5 , 1	4		5 . 6	3 , 4
2	δ	5 . 6	2 , 5	5	λ	5	5 –, 3 2 , 5 , 1
3	γ	5 . 6	5 –, 3 , 4				

Lustre of the Stars in Piscis austrinus.

1		5	See Note	6		6	
2		6		7		6	
3		6		8		4 . 5	41 Cap –, 8
4		4 . 5		9	ι	4	10 – 9
5		6		10	θ	4	10 – 9

Lustre of the Stars in Piscis austrinus—*continued.*

11		6	13 , 11	19		5	23 –, 19 , 21 20 – 19
12	η	5	12 – 14 12 – 16	20		6	20 – 19
13		6	14 –, 13 , 11	21		6	19 , 21
14	μ	4	14 . 15 14 – 15 12 – 14 –, 13	22	γ	5	22 . 23 17 – 22
15		5 . 6	14 . 15 14 – 15	23	δ	5	22 . 23 –, 19
16	λ	4 . 5	12 – 16	24	α	1	8 Peg , 24 , 44 Peg 44 Peg is 19 Aquar Note
17	β	3	17 – 22				
18	ε	3 . 4	88 Aquar – 18 . 86 Aquar Note				

Lustre of the Stars in Scorpius.

1	b	6	2 , 1 – 3 1 – 42 Libræ	19		6	19 – 11 22 , 19 24 , 19
2	A^1	5	5 – 2 –, 3 2 – 4 2 , 1	20	σ	5	6 , 20
3	A^2	7	2 –, 3 4 , 3 1 – 3	21	α	1	21 , 50 Cyg α Cyg – – 21 –, α Ophiuchi Note
4		6	2 – 4 , 3 42 Lib – 4				
5	ρ	4	5 – 2	22		5 . 6	22 , 19 22 = –, 25
6	π	3	8 –, 6 23 , 6 , 20	23	τ	4	23 , 6 42 Oph , 23
7	δ	3	21 – 7 , 8 7 – 8 7 ; 8 8 . 7	24		6	24 , 19
8	β	2	7 , 8 – 20 7 – 8 7 ; 8 8 . 7 7 ; 8 –, 6	25		6	22 = –, 25 Note
				26	ε	3	14 , 26 – – – 27 26 , 9 Oph Note
9	ω^1	5	9 – 10 14 , 9 –, 10 14 . 9 , 10				
10	ω^2	5	9 – 10 9 –, 10 9 , 10	27		6	26 – – – 27 9 Oph – –, 27
11		6	19 – 11 17 , 11	28		6	33 , 28
12	o^1	6	13 – 12	29		6	30 . 29 , 31 29 , 38 Oph (=31)
13	o^2	6	13 – 12	30		6	30 . 29
14	ν	4	14 , 9 14 , 26 14 . 9	31		6 . 7	29 , 31 29 , 38 Oph (=31)
15	χ	5	15 , 16	32		6	33 . 32 32 – 50 Oph
16		6	15 , 16 . 18 Note	33		7	33 , 1 Sagitt 33 . 32 33 , 28
17		6	17 , 11	34	υ	4	35 –, 34
18		4	16 . 18	35		3	35 –, 34

Lustre of the Stars in Serpens.

1		7	4 , . 1 , 2	14	A^1	6	11 – 14
2		7	1 , 2	15		6	22 – 15
3		6 . 7	3 , 5	16		7	6 . 16
4		6	6 , 4 . 1 4 –, 8 4 . 11	17		6 . 7	19 –, 17 12 , 17 Note
5		6	5 , 10 3 , 5	18	τ^2	6	41 , 18
6		6	10 – 6 , 4 6 . 16	19	τ^3	6	19 –, 17 26 – 19 – 29
7		7	9 – 7	20	χ	6	20 , 9
8		7	4 –, 8	21	ι	5	35 , 21 – – 22 21 –, 44
9		6	20 , 9 – 7	22		6	21 – – 22 – 15
10		6	10 – 34 5 , 10 – 6	23	ψ	6	34 , 23
11		6	4 . 11 – 14 25 – 11	24	α	2	27 Lib , 24 , 27 Herc
12	τ^1	7	12 , 17	25	A^2	6	25 – 36 25 – 11
13	δ	3	13 –, 27 13 . 37	26		6	26 – 19 26 , 31

Lustre of the Stars in Serpens—*continued.*

27	λ	4	13 -, 27	46		6	46 - 40 46 -̦ 47
28	β	3	28 , 37 28 , 41	47		6	46 -̦ 47
29		5 . 6	19 - 29	48		6	8 Herc ; 48 48 -, 49
30		6	36 - 30 50 , 30	49		6	48 -, 49
31	ν	6	26 , 31 ; 39	50	σ	5	36 , 50 , 30
32	μ	4	32 - 37	51		6	51 , 25 Oph
33		6	Does not exist	52		6	
34	ω	6	34 , 23 10 - 34	53	ν	4	
35	κ	4	35 , 21	54		6	47 Oph - - 54
36	b	6	25 - 36 - 30 36 , 50	55	ξ	4	
37	ε	3	37 . 10 Oph 32 - 37 28 . 37 - 41 13 . 37	56	ο	5	56 -, 57 Oph
				57	ζ	3	57 - 69 Oph 57 -, 69 Oph
38	ρ	4 . 3	44 , 38	58	η	3	58 - 64 Oph
39		6	31 ; 39	59	d	6	59 -, 61
40		7	46 - 40 . 45	60	o	6	61 . 60 60 -, 47 Oph
41	γ	3	10 Oph - 41 37 - 41 28 , 41 , 18	61	e	6	59 -, 61 . 60
42		6	Does not exist Note	62		6	64 - 62
43		6	50 Lib - 43	63	θ	3	
44	π	4	21 -, 44 , 38	64		6	64 - 62 Note
45		6	40 . 45				

Lustre of the Stars in Serpentarius (or Ophiuchus).

1	δ	3	35 . 1 , 13 35 , 1 . 13 60 - 1 1 - 13	22		7	22 . 28 22 . 18
				23		6	
2	ε	3 . 4	13 , 2 13 - 2 - 10	24		7	24 - 26
3	ν	5	3 , : : 18 Lib Note	25	ι	4	51 Serpentis , 25
4	ψ	5	4 - 5 8 -̦ 4 -̦ 7	26		6	26 -, 28 24 - 26 Note
5	g	5	4 - 5 - 9	27	κ	4	[A number of comparisons of 27 with α (64) Herculis have been printed in the 2nd of these papers on the " Lustre of the Stars "—*see Phil. Trans.*, 1796, p. 453 [Vol. I. p. 560]—and it is needless to repeat them here. There are others, of 27 with δ (65) Herc. and with 60 (β) Serpentarii. The former may be represented by 27 ; δ Herc and δ Herc ; 27. For the latter, see below, number 60.—J. H.]
6		6	Does not exist				
7	χ	6	4 -̦ 7				
8	φ	4	8 -̦ 4				
9	ω	5	26 Scorpii , 9 - -, 27 Scorpii 5 - 9				
10	λ	4	37 Serpentis . 10 - 41 Serpentis 2 - 10				
11		6	21 , 11				
12		6	19 , 12				
13		3	1 , 13 1 , 13 13 , 2 13 - 2 1 - 13 Note				
14		6	21 , 14 , 19				
15		6					
16		6	19 . 16				
17		6	Is 43 Herculis	28		6	26 -, 28 . 31 22 . 28
18		6 · 7	22 . 18	29		6	
19		6	14 , 19 . 16 19 , 12	30		6	Note
20		5 . 6		31		6	28 . 31
21		6	21 , 14 21 , 11	32		6	32 , 33

Lustre of the Stars in Serpentarius (or Ophiuchus)—*continued.*

33		6	32, 33, 34
34		6	33, 34
35	η	3	35 . 1 35, 1
36	A	6 . 5	44, 36, 51
37		6	66 Herc, 37 66 Herc - 37
38		6 . 7	29 Scorp, 38 (or 31 Scorpii)
39		6	39 ; 51
40	ρ	4	
41		6	
42	θ	4 . 3	42 - 50 Lib 42, 23 Scorp
43		4 . 5	
44	B	5 . 4	44 -, 51 44, 36
45		6	
46		6	Does not exist Note
47		6	60 Serpentis -, 47 - - 54 Serpentis
48		6	Does not exist
49	σ	5	67 - 49
50		7	32 Scorp - 50
51	e	6	51 - 3 Sagitt 44 -, 51 39 ; 51 36, 51
52		6	2 Sagitt, 52 58 - 52 ; 2 Sagitt
53		6	
54		6	54, 56
55	α	2	55, α Coronæ 55 - - 60 55, 5 Coronæ 55 - 33 Drac α Cygni - - - 55 - α Coronæ α Scorp -, 55 -̣ α Coronæ

56		6	54, 56
57	μ	4	56 Serpentis -, 57 Note
58	D	6	58 - 52
59		6	Does not exist
60	β	3	60 - α Herc (3 times) 60 -̦ α Herc (3 times) 60 27 (β) Herc 60, β Herc (twice) 55 - - 60 60, 17 Aquilæ 60 - 1 60 -̦ 27 60 . 27 27, 60 60 - - 62 Note
61		6	66, 61
62	γ	3	60 - - 62, 67 72 -̦ 62 72 -̦ 62 62 -̦ 72 72 . 62 72 - 62 -, 71 64 - 62
63		5	
64	ν	4	58 Serpentis - 64 - 62
65		6	65 - 6 Sagittarii
66	n	4 . 5	68, 66, 61 66, 73
67	o	4	62, 67, 70 67 - 49 72 : 67
68	k	4	70, 68, 66
69	τ	5	57 Serpentis - 69 57 Serpentis -, 69
70	p	4	67, 70, 68
71	S	6	72 - 71 72 -, 71 62 -, 71
72	S	6	72 - 71 72 -, 71 72 -̦ 62 72 -̦ 62 62 -̦ 72 ; 67 72, 62 Note
73	q	6	74, 73 66, 73
74	r	6	74, 73

Lustre of the Stars in Ursa major.

1	o	4 . 5	1 -, 23 1, 69
2	A	5	3, 2, 4 2, 5
3	π¹	5	3, 2 14, 3
4	π²	6	2, 4 . 6 5, 4
5		5	2, 5, 4
6		5	4 . 6
7	b	6	7 is lost
8	ρ	5	13 . 8, 11
9	ι	4	9 - 25
10	n	4	39 Lyncis, 10
11	σ¹	5	8, 11
12	κ	4	41 Lyncis - 12, 39 Lyn 33 - 12
13	σ²	5	13 . 8
14	τ	5	14, 3 14, 16 24 - 14
15	f	5	15, 18 15 - 24 30 . 15 -, 18
16	o	5	14, 16 16 - -, 20
17		5	18 - 17

18	e	5	15, 18 - 17 26 -, 18 15 -, 18 -, 31
19		6	8 Leo min - 19, 7 Leo min
20		7	16 - -, 20 Does not exist Note
21		6	Does not exist Note
22		7	27 - 22
23	h	4	1 -, 23 . 29
24	d	4 . 5	24 - 14 15 - 24
25	θ	3 . 4	25 . 41 Lync 9 - 25 - 69
26		5 . 6	30 -, 26 -, 18
27		6	27 - 22
28		5	Does not exist
29	ν	4	23 . 29 29 - - 45 Lyncis
30	φ	5	30 -, 26 30 . 15
31		6	18 -, 31
32		5	32 . 38
33	λ	3 . 4	34 -, 33 - 12 52 -, 33 - 63

Lustre of the Stars in Ursa major—*continued.*

34	μ	3	34 -, 33 34 - 52 Note
35		6	
36		5	36 - 37 45 - 36
37		5	36 - 37 -, 39 37 , 44
38		5	
39		6	37 -, 39 , 43 39 , 42
40		6	41 -, 40
41		6 . 7	43 - 41 -, 40
42		5 . 6	39 , 42
43		6	39 , 43 - 41
44		6	37 , 44 . 45
45	ω	4 . 5	44 . 45 - 36 45 - 55
46		6	46 -, 47 Leo min
47		6	47 . 49
48	β	2	50 - - 48 79 - 48 . 64 79 -, 48 . 64 48 ⁊ 64 (twice) 64 ; 48 48 - 64 (3 times)
49		6	47 . 49 - 51
50	α	1 . 2	50 - 77 (5 times) 50 - - 48 50 ⁊ 77 (twice) 50 ; 77 50 ⁝ 77 50 ⁝ 77 77 , 50 85 ⁝ 50 ⁝ 77 50 , β Urs min 50 ⁝ β Urs min 50 ⁝ 7 (β) Urs min Note
51		7	49 - 51
52	ψ	3 . 4	34 - 52 -, 33
53	ξ	4	63 - 53
54	ν	4	54 - 63
55		5	45 - 55 55 - 67
56		6	56 -, 59 57 . 56
57		6	57 . 56 67 - 57
58		6	59 , 58 58 . 60
59		6	56 -, 59 , 58 61 . 59 , 62
60		6	65 , 60 58 . 60
61		6	61 . 59
62		6	59 , 62
63	χ	4	33 - 63 54 - 63 - 53
64	γ	2	48 . 64 -, 69 48 , 64 = = - δ or 69 48 ⁊ 64 7 Urs min - 64 64 ; 48 48 -, 64 48 - 64 (3 times)
65		7	65 , 60
66		6	71 . 66 70 . 66
67		6	55 - 67 - 57
68		7	70 - - 68 73 - 68 . 72
69	δ	2 . 3	69 -, 70 69 - 74 1 , 69 64 -, 69 64 = = - 69 25 - 69
70		6	69 -, 70 - - 68 75 . 70 . 71 74 , 70 . 75 70 ⁊ 71 70 . 66
71		7	70 . 71 . 73 71 . 66 70 ⁊ 71 . 73
72		7	73 -, 72 73 - 72 68 . 72
73		6	71 . 73 -, 72 71 . 73 - 72 73 - 68
74		6	69 - 74 , 75 74 , 70 76 - 74 76 . 74
75		6	74 , 75 . 70 70 . 75 Note
76		6	76 . 76 - 74 Note
77	ϵ	3	77 , 85 50 - 77 (3 times) 50 ⁊ 77 ; 85 50 ; 77 (3 times) 50 -, 77 50 ⁝ 77 - 79 50 ⁝ 77 77 , 50 77 ⁊ 85 77 - 79 1 Urs min . 77
78		6	78 ; 80 Note
79	ζ	3	85 , 79 - 48 79 ⁝ 7 Urs min 77 - 79 -, 48
80	g	5	83 , 80 , 81 80 , 83 78 ; 80
81		5 . 6	80 , 81 84 , 81 . 86 83 , 81 . 84
82		6	86 . 82 Note
83		6	87 - 83 , 80 83 , 84 80 , 83 , 81
84		6	83 , 84 , 81 81 . 84 . 86
85	η	3	77 , 85 , 79 77 ; 85 , 79 85 ⁝ 50 77 ; 85
86		6	81 . 86 84 . 86 . 82 Note
87		5	87 - 83 87 - 8 Draconis

Lustre of the Stars in Virgo.

1	ω	6	4 - 1 2 , 1 2 - 1 , 4
2	ξ	5	8 - 2 - 4 4 - 2 , 1 2 - 11 2 - 1 8 , 2
3	ν	5	9 - 3 - 8 9 - 3 - 8 9 , 3 - 8
4	A[1]	6	2 - 4 - 1 4 - 2 1 , 4 - 6
5	β	3	43 . 5 - 15 43 , 5 - 15
6	A[2]	6	43 - 6 . 109 4 - 6 7 , 6 12 ; 6
7	b	5 . 6	8 - 7 7 . 13 7 , 6 7 -, 10 7 - 11 7 , 13
8	π	5	3 - 8 - 2 3 - 8 - 7 9 - 8 , 16 3 - 8 , 2 8 - 16 51 ; 8 - 78
9	o	5	9 - 3 9 - 3 9 - 8 9 , 3
10	r	6	12 - 10 7 -, 10 11 , 10 10 - 17
11	s	6	2 - 11 - 12 7 - 11 , 10

Lustre of the Stars in Virgo—*continued.*

12		6 . 7	11 – 12 – 10 12 . 17 12 ; 6
13		6	7 . 13 7 , 13 – 14
14		6	13 – 14
15	η	3	5 – 15 15 – 51 15 – 93 109 . 15 , 107 5 – 15
16	c	4 . 3	8 , 16 8 – 16
17		6	12 . 17 10 – 17
18		6	Does not exist Note
19		6	Does not exist Note
20		6	27 . 20 27 . 20
21	q	6	26 – 21 , 25
22		6	27 . 22 31 . 22 Does not exist Note
23		6	Does not exist Note
24		6	Does not exist Note
25	f	6	21 , 25 –, 28 Note
26	χ	5	26 – 21
27		6	33 , 27 27 . 22 27 . 20 30 – – 27 . 20 33 , 27 – – 42 (see note) 41 , 27
28		6	25 –, 28
29	γ	3	47 – 29 – 79 67 – 29 . 47 29 , 47
30	ρ	5	30 – 32 30 – – 27 30 , 32
31	d^1	6	32 – 31 – 33 32 , 31 32 . 31 32 , 31
32	d^2	6	30 – 32 – 31 32 , 31 30 . 32 . 31 32 , 31 See note
33		6 . 7	31 – 33 33 , 27 33 – 34 33 , 27
34		6	33 – 34 36 , 34 , 41
35		6	37 , 35
36		6	36 , 34
37		6	37 , 35
38		6	48 . 38
39		6	40 – 39 40 – – 39
40	ψ	5	40 – 39 40 – – 39
41		6	34 , 41 , 27
42		6	27 – – 42 Note
43	δ	3	79 – 43 79 , 43 , 5 79 . 43 – 6 47 –, 43 , 5 79 – 43
44	k	6	46 . 44 , 48
45		6	Does not exist Note
46		6	46 . 44
47	ε	3	67 – 47 – 29 29 . 47 – 79 29 , 47 –, 43
48		6	44 , 48 . 38
49	g	5	49 –, 50 49 –, 50
50		6	49 –, 50 , 52 50 , 52 49 –, 50 – – 56
51	θ	4	15 – 51 – 74 51 ; 8
52		6	50 , 52 . 62 Does not exist Note
53		4 . 5	53 , 61 61 – 53 , 55
54		6	61 , 54 61 –, 54 73 . 54 57 – 54 57 – 54
55		6	55 . 57 61 – 55 . 57 55 . 57 53 , 55 . 57
56		6	58 . 56 56 , 58 50 – – 56
57		6	55 . 57 – 61 55 . 57 55 . 57 – 54 55 . 57 – 54
58		6	62 . 58 . 56 56 , 58 Note
59	e	6 . 7	60 , 59 , 64 70 – 59 , 71
60	σ	5	84 – 60 – 78 60 – 64 60 , 59
61		4 . 5	61 , 69 57 – 61 , 54 53 , 61 –, 54 61 – 55 61 – 53 Note
62		6	62 . 58 52 . 62
63		6	69 –̦ 63
64		6	60 – 64 59 , 64
65		6	74 – 65 . 66
66		6 . 7	65 . 66 , 72 66 , 80
67	α	1	67 – 47 β Gem , α Virg . α Leon 67 – 29 67 –̦ –, 32 Leon Note
68	i	4	69 –, 68 , 75
69		5 . 6	69 –, 68 61 –̦ 69 –̦ 63
70		6	70 – 59
71		6	59 , 71 Note
72	l^1	6	80 – 72 66 , 72 . 80 76 –, 72 – 77 80 , 72 82 – 72 – –, 77
73		6	73 . 54
74	l^2	6	51 – 74 – 80 74 – 65 74 –, 82
75		6	68 , 75
76	h	6	82 . 76 –, 72 76 – 80
77		7	72 – 77 , 81 72 – –, 77 . 81
78		6	60 – 78 8 – 78 –, 84
79	ζ	6	29 – 79 – 43 79 , 43 47 – 79 . 43 79 – 43
80		6	74 – 80 – 72 72 . 80 76 – 80 66 , 80 , 72
81		6	77 , 81 77 . 81 . 88 Note
82	m	6	74 –, 82 . 76 82 – 72
83		6	89 , 83 , 87 Note
84	o	6	93 – 84 – 60 78 –, 84
85		6	87 , 85 86 , 85
86		6	87 . 86 , 85
87		6	83 , 87 , 85 87 . 86
88		6	81 . 88 Note
89		5 . 6	89 , 83
90	p	6	93 – – 90 , 92
91		6	Does not exist

Lustre of the Stars in Virgo—*continued.*

92		6	93 - 92 90 , 92	101		6	20 Bootis -, 101 Note
93	τ	5	15 - 93 - 84 93 - 92 107 . 93 . 99 93 - - 90	102	υ1	5	105 - 102 -, 103 102 - 104
				103	υ2	5	102 -, 103 106 , 103
94		6	95 , 94 - 97 94 , 96	104		6	102 - 104 . 106 104 , 108
95		6	98 -, 95 , 94	105	φ	4	105 - 102 16 Lib , 105 - 11 Lib
96		5	94 , 96 , 97 2 Lib - 96	106		6	104 . 106 , 103
97		6	94 - 97 96 , 97	107	μ	4	15 , 107 . 93 107 -, 99 109 - 107
98	κ	4	99 - 98 . 100 98 -, 95 98 , 100 98 ; 110	108		6	104 , 108
99	ι	4	93 . 99 - 98 107 -, 99 Note	109		4	6 , 109 . 15 109 - 107
100	λ	4	98 . 100 98 , 100 110 ; 100	110		6	98 ; 110 ; 100

Notes to Canes venatici.

July 22, 1797. 11 There are two stars about the place of 11 nearly alike in brightness.

13 Is 37 Comæ Berenices.

19 A considerable star sp 19 is omitted: much larger than 18.

22 Does not exist. It was never observed by FLAMSTEED.

25 Is misplaced: the P.D. should be +10°. It is not in the place where the catalogue has it, but is 10° more south. 25 - 8 A star observed by FLAMSTEED, p. 228, is in its place about $\frac{3}{4}$° or 1° north of this 25, and a little preceding it is * , 14.

A star observed by FLAMSTEED, p. 225, from 64 Ursæ towards 54 Ursæ is in its place. It is 1 -, *

Notes to Coma Berenices.

5 December 27, 1786. I looked for 5 Comæ, but could not find it.

19 April 19, 1797. 19 does not exist. FLAMSTEED never observed it.*

29 Is the same with 36 Virginis.

34 Does not exist, nor did FLAMSTEED observe it.

35 39 A star between 35 and 39 observed by FLAMSTEED, p. 165, is in its place. It is 39 =, *

41 A star near 41 observed by FLAMSTEED, p. 165, is in its place. A star south following 41 observed by FLAMSTEED, p. 165, is in its place. It is 41 - - *

42 A star south of 42 observed by FLAMSTEED, p. 164, is in its place. Calling it in general * it will be 38 , *

Notes to Libra.

1 Does not exist: there is a star of a considerable magnitude near 50 Hydræ, but the place does not agree with 1——1 is not in the place where it is marked in Atlas, but there is a star which FLAMSTEED observed, p. 166, which is probably 1. It is R.A. −30′ and P.D. +2° and is in its place. I shall call it 1 and it is 50 Hydræ -, 1.†

2 There are two about the place of 2, but I suppose the largest, and nearest to 98 Virginis, to be FLAMSTEED's star. It agrees best with the place.

23 Is not in the place where Atlas gives it, nor did FLAMSTEED observe it there. He has a star, p. 531, which is 1° 26′ more in R.A. This is probably 23, and it is 23 , 14 and is in its place.

27 Does not seem larger than 9, at least not very decidedly, and so as to be denoted 27 , 9, but 9 has a small star near it, not visible to the naked eye, which increases its lustre; but in my glass it is evident that 27 is a little brighter than 9.

37 North of 37 is a star nearly as large as 37, but 37 is a very little larger in the finder.——FLAMSTEED's star observed, p. 45, north of 37 is in its place 37 - *

* [It is = 18 Comæ, see Baily's note to No. 1729.—ED.]

† [1 Libræ is 50 Hydræ with an error of 1° in R.A.—ED.]

Note to Piscis austrinus.

September 22, 1795. This constellation, on account of its low situation, can be of no use for comparative magnitudes. The opportunities of observing it must be so scarce that no discoveries of changes can be made in it. I can see no other star with the naked eye but those I have equated [viz., 24 and 18. The observations of other stars of this constellation were made two years later.—J. H.].

Notes to Scorpius.

16 Should be about 3 or 4 minutes nearer to 15. FLAMSTEED'S observation, p. 197, leaves the Z.D. doubtful.

21 Is of a very brilliant ruddy light.——Is of a pale garnet colour : it seems to be the most coloured of all the large stars. Its low situation probably contributes to it.

25 Either does not exist or is misplaced. There is a star about 4° from 23 and $2\frac{3}{4}$° from 22, which may be the star if misplaced. In that case the R.A. of 25 should be -1° and it will be 22 = ; 25. Several stars of Serpentarius are so small that 25 may exist.

26 Being low it may be larger than 14, for I make no allowance in my observations.

Notes to Serpens.

17 There are two of 17 but little different in brightness. I have taken the brightest of them.

33 Does not exist. FLAMSTEED never observed it.

42 Does not exist. The place where it should be, according to the catalogue, cannot be mistaken. FLAMSTEED never observed it. 50 Libræ not far from it is in its place

θ (63) $=l$ Aquilæ and less than λ Aquilæ.

64 Is the largest of two.

Notes to Serpentarius (Ophiuchus).

3 (ν) is misplaced in Atlas 1°. It should be about +1° in R.A. A star $\frac{3}{4}$° north of it, observed by FLAMSTEED, pp. 442, 443, is in its place.

6 Does not exist. FLAMSTEED never observed it.

13 3° np 13 is a star not marked in FLAMSTEED =20.

26 28 26 has another near it larger than 28.

30 Seems not to be rightly placed.

38 31 Scorpii is 38.

46 A larger star than 46 is just by, but not marked in Atlas.——46 does not exist.

48 Does not exist. FLAMSTEED never observed either of them.

57 A large star np 57 observed by FLAMSTEED, p. 442, is in its place.* It is 57.*

59 Does not exist. FLAMSTEED never observed it.

60 I suspect 27 Herculis to be changeable, for it is now 60 . 27 Herc, or even 27 Herc , 60. There is great difference in the weather.

72 Is much too large for 6m.

Notes to Ursa major.

20 There is a very small star about the place of 20, which I can hardly take for one of FLAMSTEED'S. It is 16 - -, 20.——20 does not exist in the place where it is marked in the Atlas. There is no star but of the 9th mag. within a degree of the place.

21 I think does not exist. There is a star not far from the place where the Atlas has it, but it is much too small.——21 does not exist. I cannot mistake the place.†

34 The star south of 34 observed by FLAMSTEED, p. 439, is in the place.

35 Is not as laid down in Atlas.

50 α (Oct. 25, 1795) Appears unusually large 8^h 20^m. When I saw it at 6^h I thought so immediately. I suspect it to be changeable, or rather am pretty sure it is so. It is as large as β Ursæ minoris, but that is so much higher that no fair comparison can be made between them.

Oct. 26, 1795. 50 is not so bright as last night.

[= P. XVII. 99.—ED.]

† [Mag. is 7·5 and not 6 as in the *British Catalogue*.—ED.]

Oct. 28, 1795. It is much less than it was Oct. 26. The place of the moon may possibly influence appearances.

Nov. 28, 1795. It would not be proper to compare α Urs. maj. with ϵ and η, as they are much lower, but α seems to be remarkably bright.

75 Has no time in FLAMSTEED's observations and is misplaced in Atlas. It is but very little following 74, being almost in the same R.A. with it.

76 There are two of 76, at a distance of nearly $\frac{3}{4}$° from each other.

77 June 25, 1796, 77 is very bright. July 21, 1796, 77 is decreased.

78 Is missing; at least is not as marked in Atlas.——78 has no time. In the observation of FLAMSTEED in the Atlas, it is placed about 20′ of a degree too far East.

82 Is missing.*

86 The place of 86 is not right in Atlas by many minutes, perhaps 15′.

Notes to Virgo.

18 Is lost.——18 does not exist, or is reduced to 9m at least.——18 does not exist.†

19 Is lost; or, as there are 4 or 5 stars about its place, if it is among them, it is at least reduced to the 10th mag.——19 exists not, or is less than 9m. There are 3 or 4 stars near the place, but extremely small.——19 does not exist, or is at least 9m or 10m.——19 exists not, but there is a star sp 20 about the same distance as 19 is marked np.†

22 Is in its place and 7m.——22 23 are both either 7 or 7.8 mag.——22 does not exist. The observation 31 , 22, April 9, 1796, can not be right. I mistook very probably a star sf 32 and 31 instead of np, as there is such a one.——22 and 23 do not exist. FLAMSTEED has no observation of them. There is a pretty considerable star near the place of 22.——23 is not to be seen. There is no star that can be taken for it.——23 does not exist. There is no star that can be taken for it.

24 is lost. There is no small star to represent it.——24 does not exist. There is no star that can be taken for it.——24 does not exist. FLAMSTEED has no observation of it.

25 By FLAMSTEED's observations requires −19′ in R.A. and by the heavens it does the same.

42 Does not exist. There is no star nearer than 1° of any size to the place of 42 given in Atlas. FLAMSTEED never observed this star. The star estimated April 9, 1796, 27 - - 42 is one of these small stars nearest the place, which is rather larger than 2 or 3 others thereabout.

45 I cannot see 45. There is no star so large as 10 or 11m near the place of 45.——45 does not exist. FLAMSTEED never observed it.

52 Does not exist. There is a very small star not far from the place. FLAMSTEED has no observation of 52.

58 The P.D. of 58 should be +11′.——58 by FLAMSTEED's observations requires +11′ in P.D. and by the heavens it does the same.

56 58 They are very small stars. 58 is double in my finder. There are two other stars situated like 56 and 58 in Atlas, which were probably taken for them, May 2, 1796, when they were estimated 58 . 56. Not knowing then that 58 wants a correction of P.D. +11′, occasioned the mistake.

61 There seems to be a change in the brightness of 61 since last night.

67 Is of a sparkling bluish white colour: a beautiful star.

71 A star following 71 observed by FLAMSTEED, pp. 194, 478 is in its place * . 71 59 , * . 71.

77 . 81 . 88 The three last are very small stars. About the place of 88 there are two nearly equal. I cannot determine which is FLAMSTEED's star.

83 The R.A. of 83 should be +22′ by FLAMSTEED's observations, and it requires the same by the heavens.

91 Does not exist.‡

99 The star nf 99 observed by FLAMSTEED, p. 41, is in its place. It is 108 - - *

101 Is misplaced in the *British Catalogue*: it should be +1° in P.D. Then it is 20 Bootis -, 101.

* [Flamsteed's R.A. is 6′ too small.—ED.]
† [Non-existent, error of computation.—ED.]
‡ [Identical with 92.—ED.]

[In *The Observatory*, vol. vii. p. 256, Professor E. C. Pickering gave the following Table of W. Herschel's observations of Variable Stars, from data supplied by Colonel J. Herschel.

Cat.	Star.	Fl.	Date.	Comparison.	Mag.
I	η Aquilæ	55	1795 July 19	65 , 30 , 55 . 60	3·8
			,, ,, ,,	60 . 55	4·1
			,, ,, ,,	55 – 41 – 38 . 44	3·9
			1795 July 25	41 , 55	4·5
I	*g* Herculis	30	1795 May 23	1 , 30	4·7
			1795 May 25	30 – 25	5·1
			,, ,, ,,	11 . 35 – 6 , 30	4·9
			,, ,, ,,	6 , 52 , 30	5·2
			,, ,, ,,	52 , 42 , 30 , 34	5·6
			1795 Sept. 20	30 . 1	4·4
			,, ,, ,,	30 –, 25	4·9
I	*u* Herculis	68	1795 May 22	71 . 68 , 72	5·2
			,, ,, ,,	68 , 90 , 72	5·1
			1795 Aug. 18	68 , 59 , 61	5·1
			,, ,, ,,	69 – 68	5·3
II	*o* Ceti	68	1779 Nov. 30	$\beta > o > \alpha$	2·4
			1798 Feb. 18	62 . 82	4·0
II	η Geminorum	7	1795 Mar. 29	$\mu - \eta - \xi''$	3·3
			,, ,, ,,	$\mu - \epsilon - \eta$	3·6
			1795 Nov. 7	27 ; 13 , 7 – 31	3·2
			1796 Feb. 1	13 ; 7	3·3
			,, ,, ,,	13 . 7 –, 31	3·1
			1796 Nov. 30	13 –, 123 Tauri , 7	3·2
II	ζ Geminorum	43	1795 Nov. 7	55 . 77 , 34 – 43	4·1
III	δ Cephei	27	1796 Nov. 5	32 , 27 [than β.	3·8
IV	β Lyræ	10	1782 May 12	To the n. eye, γ much larger	—
			1795 May 5	β much less than γ, $10^h 45'$ M.T.	—
			1795 Sept. 15	14 –, 10	3·8
IV	R Lyræ	13	1796 Aug. 28	12 . 13	4·5
			,, ,, ,,	21 . 20 , 13	4·7
IV	ρ Persei	25	1795 Aug. 21	23 , 25 – 41 , 46	3·4
			1796 Sept. 7	45 – – 39 – 25	3·6
IV	λ Tauri	35	1796 Jan. 1	1 – 2 , 35 . 38	4·0
			1796 Nov. 30	123 –, 35	3·6
V	X Sagittarii	3	1795 Sept. 15	2 , 3	—
VI	δ Libræ	19	1795 May 11	37 , 31 . 35 , 44 . 19 . 43 . 45 . 6	5·2
			1795 May 18	37 , 31 , 19	5·4
			,, ,, ,,	7 , 19	5·6
			1797 May 22	19 . 18	5·9

All W. Herschel's observations of the brightness of stars were reduced and discussed by Professor E. C. Pickering in the *Annals of the Observatory of Harvard College*, vol. xiv. Part 2, and vol. xxiii. Part 2.—ED.]

Caroline Herschel

From an oil painting by Tielemann 1829.
In the possession of Sir W. J. Herschel, Bart.

APPENDIX

I.—*Unpublished Observations of Messier's Nebulæ and Clusters*

With References to the Observations published in the *Philosophical Transactions* 1800, 1814, and 1818

[THE objects catalogued by Messier in the *Connaissance des Temps* for 1783 and 1784 were not included by Herschel in the eight classes into which he divided the nebulæ and clusters found by himself. Whenever he came across one of them in the course of his sweeps with the 20-feet telescope, he determined its place and gave a brief description of it. In later years, after the conclusion of his sweeps, he often looked up some of these objects with one of his telescopes and gave more lengthy descriptions of them. Many of these are quoted in his two last papers on the Construction of the Universe; of the remaining ones, all which possess any interest are given in the following, including at least one observation of every object and *all* the observations of nebulæ made with the 40-feet telescope (very few in number) which have not already been printed. Wherever no instrument is mentioned, the 20-feet telescope was used.]

M. 1 [N.G.C. 1952]. See above, p. 595.
M. 2 [N.G.C. 7089]. See above, pp. 44, 535, and 595.
M. 3 [N.G.C. 5272]. See above, pp. 536 and 595.
M. 4 [N.G.C. 6121]. See above, pp. 533 and 595.
M. 5 [N.G.C. 5904]. See above, pp. 44, 46, 535, and 596.

M. 6 [N.G.C. 6405].

1783, July 30. 20 feet, lowest power. I counted above 50 stars; it contains the greatest variety of magnitudes of any nebula I recollect. The compound eye-piece shows more of them variously scattered and intermixt.

1786, Apr. 30 (Sw. 559). Contains several lines that seem to be drawing to a center like a forming cluster.

M. 7 [N.G.C. 6475].

1783, July 30. About 20 small stars. [Only seen once.]

M. 8 [N.G.C. 6523].

1784, May 22 (Sw. 223). L. E. pB. broad. The nebulosity of the milky kind, there are some pB. stars in it, but they seem to have no connection with it, being of very different sizes and colours and resembling the other stars that are everywhere scattered about in this neighbourhood. This is probably the star surrounded with nebulosity mentioned by Messier. There is indeed one of the stars which are in the nebula that is somewhat larger than the rest and may be the only one he saw. The nebula follows 51 Ophiuchi 32'·4 in time and is 35' more south.

M. 9 [N.G.C. 6333]. See above, pp. 45, 538, and 596.
M. 10 [N.G.C. 6254]. See above, p. 46, footnote, and p. 596.
M. 11 [N.G.C. 6705]. See above, p. 596.
M. 12 [N.G.C. 6218]. See above, p. 596.
M. 13 [N.G.C. 6205]. See above, pp. 536 and 597.
M. 14 [N.G.C. 6402]. See above, pp. 45, 538, and 597.
M. 15 [N.G.C. 7078]. See above, pp. 536 and 597.

M. 16 [N.G.C. 6611].

1783, July 30. Large stars with small ones among them; within a small compass I counted more than 50, and there must be at least 100 without taking in a number of straggling ones, everywhere dispersed in the neighbourhood.

M. 17 [N.G.C. 6618].

1783, July 31. A very singular nebula; it seems to be the link to join the nebula in Orion to others, for this is not without a possibility of being stars. I think a great deal more of light and a much higher power would be of service.

1784, June 22 (Sw. 231). A wonderful nebula. Very much extended, with a hook on the preceding side; the nebulosity of the milky kind; several stars visible in it, but they seem to have no connection with the nebula, which is far more distant. I saw it only through short intervals of flying clouds and haziness; but the extent of the light including the hook is above 10'. I suspect besides, that on the following side it goes on much farther and diffuses itself towards the north and south. It is not of equal brightness throughout and has one or more places where the milky nebulosity seems to degenerate into the resolvable kind; such a one is that just following the hook towards the north. Should this be confirmed on a very fine night, it would bring on the step between these two nebulosities which is at present wanting, and would lead us to surmise that this nebula is a stupendous stratum of immensely distant fixed stars, some of whose branches come near enough to us to be visible as resolvable nebulosity, while the rest runs on to so great a distance as only to appear under the milky form.

M. 18 [N.G.C. 6613].

1783, July 31. About 20 L. and sev. S. stars irregularly scattered.

1784, June 22 (Sw. 231). A cl. of coarsely scattered L. stars, not rich.

M. 19 [N.G.C. 6273]. See above, p. 537* and p. 597.

M. 20 [N.G.C. 6514].

1783, May 3. Two nebulæ close together, both resolvable into stars; the preceding however leaves some doubt, though I suppose a higher power and more light would confirm the conjecture. 10 feet, power 350: the instrument will not bear a higher power in this low altitude.

M. 21 [N.G.C. 6531].

1786, May 26 (Sw. 556). A rich cluster of large stars.

M. 22 [N.G.C. 6656]. See above, pp. 45 and 597.

M. 23 [N.G.C. 6494].

1784, June 18 (Sw. 230). A cluster of beautifully scattered, large stars, nearly of equal magnitudes (visible in my finder), it extends much farther than the field of the telescope will take in, and in the finder seems to be a nebula of a lengthened form extending to about half a degree.

M. 24 [N.G.C. 6611].

1783, Aug. 2. Considerable stars in great number.

M. 25 [Not in N.G.C., $18^h\ 23^m$, 109° 2' for 1860].

1783, July 30. Very large stars and some small ones; I counted 70, and there are many more within no considerable extent.

M. 26 [N.G.C. 6694].

1784, June 16 (Sw. 228). A cluster of scattered stars, not rich.

* [The observation quoted in that place cannot be found.—ED.]

M. 27 [N.G.C. 6853, The "Dumb-bell"].

1782, Sept. 30. My sister discovered this nebula this evening in sweeping for comets; on comparing its place with MESSIER's nebulæ we find it is his 27. It is very curious with a compound piece; the shape of it though oval as M. calls it, is rather divided in two; it is situated among a number of small stars, but with this compound piece no star is visible within it. I can only make it bear 278. It vanishes with higher powers on account of its feeble light. With 278 the division between the two patches is stronger, because the intermediate faint light vanishes more.

1783, Aug. 2. A distant suspicion of its being all stars; I want light.

1784, July 19 (Sw. 241). This nebula I suppose to be a double stratum of stars of a very great extent. The ends next to us are not only resolvable nebulosity, but I really do see very many of the stars mixt with the resolvable nebulosity; farther on the nebulosity is but barely resolvable, and ends at last in milky whitishness of the same appearance at that in Orion. The idea I form of the shape of the strata is this:—

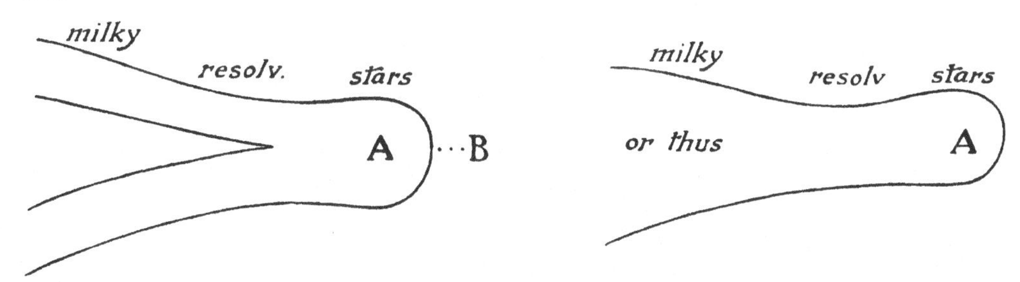

These two being laid on each other, A on A and viewed from B, so as to have the small round end A foremost, may produce the appearance of this curious nebula.

1794, October 27. With 287, 7 feet reflector, I see only two patches of light joined together, like two nebulæ without stars, running into one another. There are a few very small stars visible in it, but no more than what in the rest of the heavens are here scattered about. They therefore are not connected with the nebula or nebulæ.

M. 28 [N.G.C. 6626].

1799, August 1. It may be called insulated though situated in a part of the heavens that is very rich in stars. It may have a nucleus, for it is much compressed towards the centre, and the situation is too low for seeing it well. The stars of the cluster are pretty numerous.

M. 29 [N.G.C. 6913].

1794, Oct. 27. Is not sufficiently marked in the heavens to deserve notice, as 7 or 8 small stars together are so frequent about this part of the heavens that one might find them by hundreds.

M. 30 [N.G.C. 7099]. See above, pp. 533 and 598.

M. 31 [N.G.C. 224, nebula in Andromeda]. See above, p. 527.

M. 32 [N.G.C. 221].

1813, Dec. 26. A vB. R. nebula, vgbM. up to a nucleus.

M. 33 [N.G.C. 598].

See above, pp. 48 and 598. Two observations are recorded as being of V. 17, the outlying parts of the great nebula; they were made September 11 and 12, 1784, and are described in vol. i. pp. 255–256.

M. 34 [N.G.C. 1039]. See above, p. 598.

M. 35 [N.G.C. 2168]. See above, p. 598.

M. 36 [N.G.C. 1960].

1794, Oct. 28. 7 feet reflector. With 120, a pretty rich cluster of small stars, seems to have many more than are visible, very small.

M. 37 [N.G.C. 2099].

1782, Nov. 4. Is an astonishing number of small stars with 227; they are almost all of the 2nd or 3rd classes. I see no kind of nebulosity in the spot. With 460 the whole is resolvable into stars without nebulosity.

1783, August 24. A useful, coarse step; it will serve to learn to see nebulæ, because it contains many small stars mixed with others of various magnitudes, many of which are not to be seen without great and long attention.

M. 38 [N.G.C. 1912].

1805, Nov. 23, Review. Large 10 feet reflector. A cluster of scattered, pretty large stars of various magnitudes, of an irregular figure. It is in the Milky Way.

M. 39 [N.G.C. 7092].

1788, Sept. 27 (Sw. 866). Consists of such large and straggling stars that I could not tell where it began nor where it ended. It cannot be called a cluster.

M. 40.

1799, Aug. 5. Not visible in the finder.*

M. 41 [N.G.C. 2287].

1784, Oct. 20 (Sw. 304). A large cluster of very coarsely scattered large stars.

M. 42 and 43 [N.G.C. 1976 and 1982, Great Nebula in Orion].†

1774, March 1. Observed the Lucid Spot in Orion's sword belt; but the air not being very clear, it appeared not distinct.

1774, March 4. Saw the lucid spot in Orion's sword thro' a 5½ foot reflector; its shape was not as Dr Smith has delineated in his *Optics*; tho' something resembling it, being nearly as follows [Plate III. fig. 37]. From this we may infer that there are undoubtedly changes among the fixt stars, and perhaps from a careful observation of this spot something might be concluded concerning the nature of it.

1776, Nov. 11. The lucid spot in Orion, about 10 o'clock just rising at 5° high [see diagram, 1779 Oct. 7]. The greatest glare immediately about the 3 small stars 2, 5, 6 in the corner, the middle whereof is one or two magnitudes larger than the other two. The three succeeding stars 1, 3, 4 were almost on the upper side of this figure free from any glare, and there was a total darkness in the corner by the 3 before mentioned close in the corner. There was a very faint glimmering of a seventh star which I have marked but which must be several magnitudes less than the other 6. The two Nos. 1 2 were of one size, the two 3 4 next, the two 5 6 considerably less and 7 very much less again, almost invisible. The whole was exceedingly distinct. Instrument 10 ft. reflector, power 120. There was also an 8th star visible near the 6th with the same small size with the 7th or rather less.

1778, Jan. 25. Die Nebula im Orion from 10 to 12 o'clock. In the east the lucid ray seems to make an equilateral ‡ triangle with the stars 1 & 3 where one is the vertex and seems from the base to go on in the direction of 1 3 4 rather approaching to 4 and afterwards bends round 4 in an angle of about 110 or 120 degrees to the east. From 2 to 7 the lucid part is concave, the concave part turned towards 3, and goes to the northward about ¾ of the distance from 2 to 7; it turns from thence to west in angle of about 75 or 70.

1778, Jan. 26. From 10 to 12, observation on the Nebula.

6 . 2 . 1 make a straight line.
6 . 8 . 7 make a straight line.

The lines 2 . 5 and 1 . 3 . 4 diverge; 5 a little larger than 6; 4 , 5 , 8 make a straight line.

1778, Feb. 7. 11^h 30′. I observed the Lucid Spot again and saw all the 4 stars very well, and their place agreed with the observation of Jan. 25.

1778, Feb. 25. I observed the Lucid Spot in Orion when on and near the meridian and found everything in regard to the situation of the 4 little stars to agree with the observation of Jan. 25.

* [Thus described by Messier: "Deux étoiles très près l'une de l'autre et très petites, placées à la naissance de la queue de la grande ourse; on a de la peine à les distinguer avec une lunette ordinaire de 6 pieds." Only once looked for by Herschel; Messier only inserted it because he found it when looking for a nebula alleged to be in that neighbourhood.—ED.]

† [The following observations have already appeared in print in Holden's "Monograph of the Central Parts of the Nebula of Orion," *Washington Observations* for 1878, Appendix I., but only copied from Caroline Herschel's transcript and not from the original Journal. They are nere given in full on account of the special interest attaching to this object. The rough pen-and-ink sketch made on March 4, 1774, was copied on the plate belonging to the paper of 1811, see Pl. III. fig. 37; compare p. 465.—ED.]

‡ [Over this word is written in pencil "isosceles."—ED.]

1778, Dec. 15. Lucid Spot in Orion's sword handle.

6 . 8 . 7 make a straight line.
6 . 2 . 1 make a straight line.
4 . 5 . 8 make a straight line.

The lines 2 . 5 and 1 . 3 . 4 diverge. This agrees with the observation of Jan. 26. But there is a visible alteration in the figure of the lucid part.

1779, Oct. 7. θ Orionis. The line 6 . 2 . 1 is a little convex towards 5, when that line is taken into the middle of the field; this I mention, as it is possible there might be a little curvature arising from the spherical figure of the eyeglass, tho' I believe there is not. If a line be drawn from 6 to 7, the star 8 stands outwards I suppose no less than 15°, so that 6 . 8 . 7 is concave towards the side 1 . 3 . 4. The line 4 . 5 . 8 I cannot very well compare, being rather too far distant by the power I now use, but I believe it is not far from a right line. I see a 9th star which is marked in the annexed figure, where however not the least exactness is intended. Altitude about 26 degrees. 14^h 10′. The figure of the lucid part is much altered.

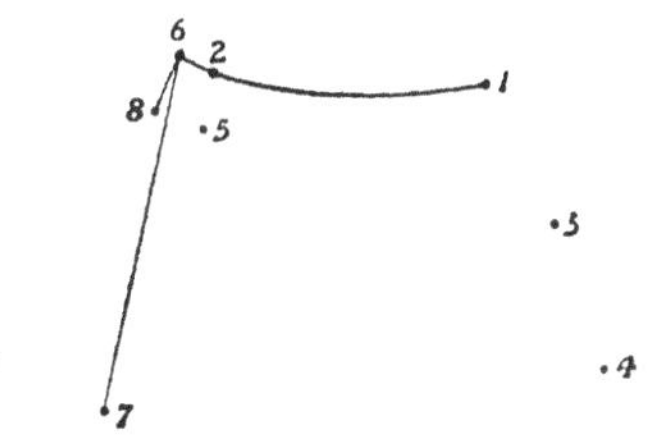

1779, Dec. 5. 6 . 2 . 1 concave, the concave part turned to the south. 8 . 6 . 7 still make an angle at 6, tho' very small. I see the 9th star mentioned [on Oct. 7].

1780, Jan. 22. 10^h 30′. The stars 6 . 2 . 1 instead of seeming concave towards the north appear convex. This may however be a deception as the star 2 is the largest, and since there is a pretty strong aberration on account of a fog, its diameter is more increased than that of 6 and consequently may give the balance towards the north.

1780, Feb. 19. Exactly as described [Oct. 7, 1779].

1780, Feb. 28. The two stars 6 . 2 measure 12″·812; the measure is pretty narrow, but I believe true enough. The two stars 2 . 5 measure 14″·271, this is also a pretty narrow but just measure. The two stars 6 . 8 measure about 9″·062, but this is doubtful on account of the obscurity of the star 8, which is hardly perceptible when the field of view is illuminated properly so as to make the parallel hair very distinct. The two stars 5 . 8 measure 20″·521. This is also doubtful on the same account.

1780, Oct. 10. θ. The upper stars concave by the hair, the spot extremely fine. The 4 stars all full and well defined.

1780, Nov. 24. 12^h. The Nebula in Orion is very fine indeed. I perceive not the least alteration.

1783, Jan. 31. θ Orionis. The Nebula quite different from what it was last year. The 9th star very strong.

1783, Sept. 20. The Neb. in Orion has evidently changed its shape since I saw it last. The star under the nebula is nebulous. 20 ft., 200.

1783, Sept. 28. Surprising changes in the Nebula Orionis.

1783, Nov. 3. The Nebula in Orion is beautiful, and I see several circumstances which I never observed with other instruments, viz. just close to the four stars it is totally black for a very short space, a few seconds. Below, in the open black part is a small distinct nebula of an extended shape.* The eastern branch of the great nebula extends very far; it passes between two very small stars and runs on so far as to meet a pretty bright star. The nebulous star below the nebula is not equally surrounded, but most towards the south; on the north of this lesser nebula it is joined by one still fainter, which makes a rectangular corner by its meeting the small nebula.

1784, Oct. 16 (Sweep 296). The beginning of the neb., 5 Monocerotis p. 41′ 6″ n. 0° 43′. My small neb. is just under the south following corner of the great one. The 43rd is not a nebulous star, the star not being at all in the center of it; my little one makes a part of it. It is altogether the most wonderful object in the heavens.

1784, Dec. 20. The nebula in Orion as before described, but the moon being bright it seems to extend hardly ¾ of a degree.

1785, Feb. 13. With a new 10 feet reflector I examined the neb. in Orion, and with long attentiom could just perceive my small faint nebula in the dark part of the great nebula.

* [This is III. 1, an appendage to Messier 43.—ED.]

1785, Oct. 5 (Sweep 458). The nebula in Orion. A wonderful phænomenon. One of the clearest nights I have ever had.

1786, Nov. 28 (Sw. 640). The nebula in Orion which I saw by the front-view was so glaring and beautiful that I could not think of taking any place of its extent.

1799, Dec. 30, Review.* No. 43 of the *Connois.* is not visible in the finder but is the star north of No. 42. 10 feet, 300. The nebulosity about the star is not central and belongs probably to the rest of the nebulosity, the star being one of the scattered ones.

1801, Jan. 14, Review. Large X feet telescope, power 120. As before described. The nebulosity of the 43rd does not seem to belong to the star.

1806, Feb. 11, Review. Large 10 feet. The 4 stars are completely in the nebulosity. The 3 stars are entirely out of it with 270. With the double eyeglass, appearances are very different.

1810, Feb. 4, Review. 10 feet. The nebulosity is entirely of the milky kind and extends a great way. The 43rd is not a nebulous star, but a star which happens to be situated in a place where some of the milky nebulosity of the great nebula happens to be. The star is not central, nor is there any condensation towards the star.

1810, Dec. 31, Review. 10 feet, double eye piece. The 4 stars are within the nebulosity. The star No. 7 (see the figure of Oct. 7, 1779) is upon the borders of the dark vacancy. I see No. 9 very well. The little star between 3 and 4 is still within very faint nebulosity. The nebulosity reaches beyond 4, as far as from 1 to 4 nearly. It touches a very small star and from that star goes on to two very bright ones, in the direction from the small star to the preceding one of the two. The black space near the four stars is much contracted. The nebulosity from 1 to 4 is concave, the concavity being to the following side. The parallel is nearly in the line of 1 . 3 . 4. I can see eight different condensations notwithstanding the moon is very bright. The nebulous star is pretty equally involved, it has the appearance of a star shining through a very faint mist. The star is a little larger than 4. The concavity from 2 to 7 goes beyond 7.

1811, Jan. 19, Review. 10 feet. Two of the four stars are within the nebulosity. No. 7 is very near the borders of the black. The little star between and following 3 and 4 is still within very faint nebulosity. The nebulosity reaches beyond 4 rather farther than from 1 to 4.

X feet. I perceive 7 or 8 different condensations. The place near the 4 stars is much contracted. The nebulous star is exactly what we might expect to see if a star were to shine through whitish nebulosity.

40 feet. 5^h 16′ B. affected.†

17 B. much affected.

22 The 4 stars are intirely involved in nebulosity. The 7th and 9th stars are very bright. In the brightest part are four places brighter than the rest. I see the small detached nebula, it is extremely faint. It is between the corner and a small star. The star called nebulous is within a nebulosity nearly detached; but the small stars marked nebulous in the figure of the 4th March 1774 † are free from nebulosity. There is a very small, nearly detached, nebulosity north of the nebulous star. The nebulous star has some resemblance to a star shining through a very thin mist.

1811, March 13, Review.† 7 feet, double eye piece. The following or rather the southern branch (for I find the parallel is nearly in the line 1 . 3 . 4) goes towards the preceding star *e* of the two large stars *d e*, or rather a little preceding it, but it partly includes the star *e* and makes it appear a little nebulous. The light about the nebulous *b* is a little denser nearer the star than at a distance. A line from 5 through 7 goes to *b* or rather a little south of it; and 7 is about a ¼ of the distance towards *b*. The star south of 3 and 4 makes an equilateral triangle with them. The two large stars *d* and *e* are parallel to 1 . 3 . 4 nearly. A line from the four stars parallel to 1 . 3 . 4 passes a little south of the small formerly nebulous star *c*. There are many other stars connected with the nebula which I do not notice.

1811, March 15, Review. 7 feet, double eye piece. The northern branch is parallel to the stars *a b*. The nebulosity reaches nearly up to the stars *g h*. A very faint nebulosity still joins the star *b* to the northern branch, but *b* is more nebulous than the intermediate nebulosity. The southern nebulosity goes towards the star *e*, and some part of the very faint nebulosity incloses the star.

* [This and the following observations were not taken in the course of sweeps.—ED.]
† [Compare Plate III. fig. 37 and p. 489.—ED.]

1811, March 16, Review. 10 feet reflector, power 100. The stars 1 . 3 are in the parallel; 4 is a very little south of their parallel. The nebulosity about *b* is brightest about the star.

M. 46 [N.G.C. 2437].

1786, Mar. 19 (Sw. 540). A beautiful, very rich, compressed cluster of stars of various magnitudes.

M. 48 [Not in N.G.C., 8^h 7^m, 91° 32′ for 1860].

1790, March 5 (Sweep 935). Looked for the 48 of the *Connoiss. des Temps* but found that it does not exist in the place mentioned by Wollaston.

Mem.* I looked with the Sweeper for the 48th of the *Conn.*, &c., and found a parcel of coarsely scattered stars, not deserving the name of a cluster; which on account of their being too far from each other could not be seen clustering in my 20 ft. Teles. They are scattered over a place near 2° in extent.

M. 49 [N.G.C. 4472].

1785, Dec. 28 (Sw. 498). Very bright, considerably large, gradually much brighter in the middle, extended with faint branches.†

M. 50 [N.G.C. 2323].

1785, March 4 (Sw. 377). A very brilliant cluster of large stars, considerably compressed and rich, above 20′ in diameter, the stars of various sizes, visible in the finder.

M. 51 [N.G.C. 5194].

1783, Sept. 17. 7 feet, 57. Two nebulæ joined together; both suspected of being stars. Of the most north [I. 186] I have hardly any doubt. 7 ft., about 150. A strong suspicion next to a certainty of their being stars. I make no doubt the 20 ft. will resolve them clearly, as they want light and prevent my using a higher power with this instrument.

1783, Sept. 20. 20 ft., 200. Most difficult to resolve, yet I do no longer doubt it. In the southern nebula I saw several stars by various glimpses, in the northern also three or four in the thickest part of it, but never very distinctly. Evening very bad.

1787, May 12 (Sw. 734). Bright, a very uncommon object, nebulosity in the center with a nucleus surrounded by detached nebulosity in the form of a circle, of unequal brightness in three or four places, forming altogether a most curious object. I. 186 B. R. S. vgbM. just north of the former.

1788, April 29 (Sw. 836). vB. L., surrounded with a beautiful glory of milky nebulosity with here and there small interruptions that seem to throw the glory at a distance. I. 186 cB. pL., a little E., about 3′ p. Mess. 51 and about 2′ more north.‡

M. 52 [N.G.C. 7654].

1783, Aug. 29. All resolved into innumerable small stars without any suspicion of nebulosity. 7 ft., 57. The sweeper, 30, shews nebulosity, the stars being too obscure to be distinguished with its light tho' considerable.

1805, December 23, Review. Large 10 feet. This is a cluster of pretty condensed stars of different sizes. It is situated in a very rich part of the heavens and can hardly be called insulated, it may only be a very condensed part of the Milky Way which is here much divided and scattered. It is however so far drawn together with some accumulation that it may be called a cluster of the third order.

M. 53 [N.G.C. 5024]. See Vol. I. p. 159, and above, pp. 537 and 598.

M. 54 [N.G.C. 6715].

1784, June 24 (Sw. 232). A round, resolvable nebula. Very bright in the middle and the brightness diminishing gradually, about 2½′ or 3′ in diameter. 240 shews two pretty large stars in the faint part of the nebulosity, but I rather suppose them to have no connection with the nebula. I believe it to be no other than a miniature cluster of very compressed stars resembling that near the 42nd Comæ.§ It is like that under δ Sagittarii, ‖ but rather larger and brighter tho' not much.

* At the end of the same sweep.—Ed.]

† [This is the only observation recorded as being of M. 49. But see Vol. I. pp. 294–295, under I. 7 and II. 19.—Ed.]

‡ [These are the only observations recorded of the great spiral.—Ed.]

§ [M. 53.]

‖ [I. 50=N.G.C. 6624.]

1784, July 13 (Sw. 237). A cL. vB. R. nebula, mbM. and breaking off suddenly, the rest being much fainter.

M. 55 [N.G.C. 6809]. See above, p. 599.
M. 56 [N.G.C. 6779]. See above, pp. 536 and 599.
M. 57 [N.G.C. 6720, annular in Lyra]. See Vol. I. p. 257, and above, p. 599.

M. 58 [N.G.C. 4579].

1784, March 15 (Sw. 174). pB. pL.
1784, April 17 (Sw. 199). L. F.

M. 59 [N.G.C. 4621].

1784, April 17 (Sw. 199). pB. R., not S., mbM.

M. 60 [N.G.C. 4649].

1784, March 15 (Sw. 174). Two nebulæ, one of them very bright.

1784, April 17 (Sw. 199). Two nebulæ, the p. vF. S. (III. 44), the following which is the 60th of the *Conn. des temps*, B. cL.

M. 62 [N.G.C. 6266]. See above, pp. 538 and 599.

M. 63 [N.G.C. 5055].

1787, March 18 (Sw. 717). E. npsf., 5 or 6′ long and near 4′ broad, bright nucleus, very brilliant.

1787, April 9 (Sw. 725). vB., 9 or 10′ long, considerably broad, the brightness confined to a small place.

M. 64 [N.G.C. 4826].

1785, April 27 (Sw. 403). Extends not less than 25′ and the southern branch loses itself more imperceptibly and is much broader and more diffused than the northern one; the night being very fine I viewed it to the greatest advantage. I suspected the ground of the heavens about the northern branch to be tinged with a very faint, milky nebulosity for a considerable way.

1787, Feb. 13 (Sw. 699). A very remarkable object, mE., about 12′ long, 4 or 5′ broad, contains one lucid spot like a star with a small black arch under it, so that it gives one the idea of what is called a black eye, arising from fighting.

M. 65 [N.G.C. 3623]. See above, p. 483.

M. 66 [N.G.C. 3627].

1784, Apr. 12 (Sw. 188). A vB. mE. nebula of an irregular figure; the extension is chiefly in the direction of the meridian and the greatest brightness near the middle.

M. 67 [N.G.C. 2682]. See above, p. 599.
M. 68 [N.G.C. 4590]. See above, pp. 534 and 600.
M. 69 [N.G.C. 6637]. See above, pp. 539 and 600.

M. 70 [N.G.C. 6681].

1784, July 13 (Sw. 237). er. cB. pL. iR. A very faint red perceivable.

M. 71 [N.G.C. 6838]. See above, p. 600.

M. 72 [N.G.C. 6981].

See above, pp. 535 and 600.

1810, Oct. 30, Review. 40 feet. Having been about 20 minutes at the telescope to prepare the eye properly for seeing critical objects, the 72nd of the *Connois.* came into the field. It is a very bright object. It is a cluster of stars of a round figure, but the very faint stars on the outside of these sorts of clusters are generally a little dispersed so as to deviate from a very perfect circular form; the telescopes which have the greatest light shew this best. It is very gradually extremely condensed in the center, but with much attention even there the stars may be distinguished. Power 280. There are many stars in the field of view with it, but they are of many magnitudes and totally different from the excessively small ones which compose the cluster. It is not possible to form an idea of the number of stars that may be in such a cluster, but I think we cannot estimate them by hundreds. The diameter is about $\frac{1}{5}$ of the field = 1′ 53″·6. 10 feet telescope. I viewed the same object. The contrast is very striking, it appears eF.

M. 73 [N.G.C. 6994].

1783, Sept. 28. Consists of a few stars arranged in a triangular form. No nebulosity among them. 10 ft., 150.

M. 74 [N.G.C. 628]. See above, pp. 43 and 600.

M. 75 [N.G.C. 6864]. See above, pp. 538 and 600.

M. 76 [N.G.C. 650].

1787, Nov. 12 (Sw. 780). Two, close together, their nebulosities run into each other; distance of their centers 1½ or 2′ [M. 76 and I. 193].

M. 77 [N.G.C. 1068]. See above, p. 601.

M. 78 [N.G.C. 2068].

1783, Dec. 19 (Sw. 59). Two large stars, well defined, within a nebulous glare of light resembling that in Orion's sword. There are also three very small stars just visible in the nebulous part which seem to be component particles thereof. I think there is a faint ray near ½° long towards the east and another towards the south east less extended, but I am not quite so well assured of the reality of these latter phenomena as I could wish, and would rather ascribe them to some deception. At least I shall suspend my judgement till I have seen it again in very fine weather, tho' the night is far from bad.

1786, Jan. 1 (Sw. 506). Very large milky nebulosity, terminating suddenly on the north side; contains 2 pL. stars, they are on the north side.

1786, Dec. 22 (Sw. 661). Milky nebulosity containing three stars, iF., 5 or 6′ long.

M. 79 [N.G.C. 1904]. See above, pp. 537 and 601.

M. 80 [N.G.C. 6093]. See above, pp. 536 and 601.

M. 81 [N.G.C. 3031].

1801, Nov. 8 (Sw. 1100). eB., the bright part confined to a very small place; the nebulosity is of the milky kind, vm. E. npsf.

1802, Sept. 30 (Sw. 1112). vB. eL.; it very nearly fills all the field, it loses itself imperceptibly, m. E. npsf.; I can trace it nearly ½° in extent beyond the bright part.

1810, Nov. 26, Review. I viewed this nebula with the large 10 feet. It has a bright, resolvable nucleus, certainly consisting either of 3 or 4 stars or something resembling them. It is about 15 or 16′ long. The object was already too low to be seen to an advantage.

M. 82 [N.G.C. 3034].

1801, Nov. 8 (Sw. 1100). eB. m. E. spnf., about 10′ long.

1802, Sept. 30 (Sw. 1112). A vB., beautiful ray of light, brightest in the middle of all the length, about 8′ long, 2 or 3′ broad.*

1810, Nov. 26, Review. Viewed with the large 10 feet. It is mottled in its length as containing 5 or 6 vS. stars affected with nebulosity. With No. 1 about ⅛ of the field or less, about 6 or 7′ in length; the breadth is about 1½ or 1¾, the object is too low.

M. 83 [N.G.C. 5236].

1787, March 15 (Sw. 711). vB., a B. resolvable nucleus in the middle with F. branches about 5′ or 6′ long, E. spnf.

1793 May 5 (Sw. 1041). vB., a SBN. with very extensive and vF. nebulosity; it more than fills the field, it seems to be rather stronger from sp. to nf. It may be ranked among the nebulous stars.

M. 84 [N.G.C. 4374].

1784, Apr. 17 (Sw. 199). A bright nebula. [The only observation.]

M. 85 [N.G.C. 4382].

1784, Mar. 14 (Sw. 170). Two resolvable nebulæ; the preceding is the largest, and with 157 seems to have another small nebula joining to it, but with 240 it appears to be a star. The following neb. is II. 55.

M. 86 [N.G.C. 4406].

1784, Apr. 8 (Sw. 187). Two resolvable nebulæ at 4 or 5′ distance.†

1784, April 17 (Sw. 199). Two B. cL. nebulæ. One is the 86th of the *Connoiss.*, the other is I. 28.

* [Entered in the *Cape Observations*, p. 128, as a new nebula, IV. 79.—ED.]

† [In the sweep they are $3^{m}\cdot0$ p. and 1° 16′ s. of M. 88. They are therefore = II. 121 and 122. H. took them for M. 86 and another which he called I 28. The comparison of I. 28 with 34 Virginis was not made in this sweep but in Sw. 199.—ED.]

M. 87 [N.G.C. 4486].

1784, Apr. 12 (Sw. 189). Two S. F. and one L. B. nebula, the two first like the p. two [*i.e.* II. 121, 122], the B. one R. mbM.

1784, Apr. 17 (Sw. 199). Three nebulæ, the two first vF. S., the third L. B. mbM., but diminishing very gradually in brightness. The two first are II. 123, 124.

M. 88 [N.G.C. 4501].

1784, Apr. 8 (Sw. 187). B. pL. r. with a S. one after it; ☾ light so strong that I had nearly overlooked the latter.*

1787, Jan. 14 (Sw. 691). vB. vL. E.

M. 89 [N.G.C. 4552].

1784, Apr. 17 (Sw. 199) B. pS.

M. 90 [N.G.C. 4569].

1784, Apr. 8 (Sw. 187). pL., with a nucleus, perhaps cometic, but moonlight permits not to give a proper description.

1784, Apr. 17 (Sw. 199). pL.

M. 92 [N.G.C. 6341]. See above, p. 601.

M. 93 [N.G.C. 2447].

1784, Nov. 20 (Sw. 326). A cluster of scattered stars, pretty close and nearly of a size, the densest part of it about 15′ diam., but the rest very extensive.

M. 94 [N.G.C. 4736].

1787, March 18 (Sw. 717). Very brilliant, a large, luminous nucleus of more than 20″ diameter with faint chevelure and branches extending 6 or 8′.

1787, Apr. 9 (Sw. 725). Very brilliant, with much F. nebulosity on the sp. and more on the f. side.

M. 95 [N.G.C. 3351].

1784, March 11 (Sw. 164). A fine, bright nebula, much brighter in the middle than at the extremes, of a pretty considerable extent, perhaps 3 or 4′ or more. The middle seems to be of the magnitude of 3 or 4 stars joined together, but not exactly round; from the brightest part of it there is a sudden transition to the nebulous part, so that I should call it cometic.

M. 96 [N.G.C. 3368].

1784, Mar. 11 (Sw. 164). A fine, bright nebula, much like the former, but the brightest part in the middle is more joined to the nebulosity than in the former, and the bright part is rather longer, tho' not quite so vivid as in the former. It may still be called cometic, tho' it begins to depart a little from that kind.

M. 97 [N.G.C. 3587]. See above, pp. 485 and 601.

M. 98 [N.G.C. 4192].

1787, Jan. 14 (Sw. 691). vB. mE., above 15′ long, a BN. in the middle.

M. 99 [N.G.C. 4254].

1787, Jan. 14 (Sw. 691). vB. vL. vgmbM., and the brightness taking up a great space.

M. 100 [N.G.C. 4321]. See above, p. 528 (one obs.).

M. 101 [N.G.C. 5457].

1783, Sept. 20. 20 feet, 200. In the northern part is a large star pretty distinctly seen, and in the southern I saw 5 or 6 small ones glitter through the greatest nebulosity which appears to consist of stars. Evening bad. This and the 51st are both so far removed from the appearance of stars that it is the next step to not being able to resolve them. My new 20 feet will probably render it easy.

1789, April 14 (Sw. 921). vB. SN. with extensive nebulosity, pretty well determined on the preceding side, but very diffused to the north following. Includes the two following nebulæ [III. 788 and 789], and seems to extend 20′, perhaps 30′ or more.

M. 103 [N.G.C. 581].

1783, Aug. 8. 14 or 16 pL. stars with a great many eS. ones. Two of the large ones are double, one of the 1st the other of the 2nd class.† The compound eye glass shews a few more that may be taken into the cluster so as to make about 20. I exclude a good many straggling ones, otherwise there would be no knowing where to stop.

* [See Vol. I. p. 297, note to II. 118.—Ed.]

† [Neither was catalogued by H. One is Σ 131.—Ed.]

II.—*Synopsis of all Sir William Herschel's Measures of Double Stars*

By Sir JOHN F. W. HERSCHEL, Bart.

[THE following Synopsis was published in 1867 in Vol. XXXV. of the *Memoirs of the Royal Astronomical Society*. It is here given in a slightly condensed form, omitting the approximate Right Ascensions, the dates of discovery, and many of the references to later observers, but with the addition of the numbers of the objects in Burnham's *General Catalogue*. The errors pointed out by Mr. Burnham and Mr. Sadler * have been corrected, in many cases after referring to the original observations. Many of the corrections given by Sir John Herschel in the last column, on the authority of Caroline Herschel,† had already been announced among errata in subsequent volumes of the *Phil. Trans.*; having in this edition been taken into account in their proper places, these notes have been omitted from this Synopsis.]

CLASS I.

Burnham Gen. Cat.	Class I. No.	Star.		Date of Measures. A.D.	Angle of Position. ° dec.	No. of Meas.		Date of Measures. A.D.	Distance. "		Synonyms and Remarks.
6993	1	36 Bootis ε	A B	1779·816	297	est.		1779·747	1·67		Σ. 1877.
				1780·320	302·32	1		1780·309	4·06:		
				1781·194	300·35	1		1780·320	2·97		
				353	303·02	1					
				887	310·05	1					
				1782·129	308·43	1					
				1783·153	306·40	1					
				1796·633	315·54	1					
				1802·074	319·30	1					
				663	316·78	1					
				1803·222	313·72	3					
				230	313·30	3					
5734	2	53 Ursæ ξ	A B	1781·961	143·78	1					Σ. 1523.
				1802·093	97·52	1					
				1804·077	92·63	1					
7563	3	17 Coronæ B σ	A B	1781·786	347·53	1					Σ. 2032.
				1802·679	11·40	1	A C	1780·600	15 ... 20	est.	
			A C	1781·786	65·	est.	A C	1781·786	24	est.	
7703	4	17 Draconis	A B	1781·879	114·00	1					Σ. 2078. A 3d * precedes.
				1802·827	117·68	1					
12666	5	8 Cassiop. σ	A B	1781·964	330·47	1					Σ. 3049; Sh. 359.
				1804·433	319·23	1					

* *Monthly Notices R.A.S.*, vol. xxxiii. p. 567, xxxiv. p. 98, and xlv. p. 316.

† *I.e.* of the "Register Sheets" of all the observations of double stars, now belonging to the Roy. Astron. Society. These have been quoted as "MS." in the footnotes to the two catalogues of double stars in Vol. I. of this edition.

CLASS I. (*continued*).

Burnham Gen. Cat.	Class I. No.	Star.		Date of Measures. A.D.	Angle of Position. ° dec.	No. of Meas.		Date of Measures. A.D.	Distance. ″		Synonyms and Remarks.
3559	6	12 Lyncis	A B	1782·367	181·38	1					B.A.C. 2187; Σ. 948; Sh. 74.
			A C								Triple. For A C, see III. 22.
8574	7	39 Draconis *b*	A B	1780·775	12·68	1					Σ. 2323.
				1802·827	6·32	1					
			A C	1780·775	26·09	1					
9713	8	63 Draconis ϵ	A B	1781·800	333·23	1					Σ. 2603.
				1804·389	354·48	1					
			A C	1781·800	1·73	1	A C	1781·800	3′ or 4′	est.	
5014	9	38 Lyncis	A B	1782·405	244·15	1					Σ. 1334.
3402	10	11 Monoc.	B C	1781·800	101·53	1					Σ. 919.
				1802·170	101·50	1					Is also II. 17, which see for A C.
4421	11	Cancri 17	A B	1782·285	355·17	1					Σ. 1177.
8562	12	59 Serpen. *d*	A B	1781·786	314·55	1					Σ. 2316.
				1802·337	312·42	1					
9464	13	Aquilæ 136	A B	1782·762	304·80	1::	A C	1781·561	7″ or 8″		Triple, P. XIX. 185; Σ. 2545. The 2d measure of 1782 "more exact" than the 1st.
					309·70	1					
				1802·750	314·75	1					
9189	14	23 Aquilæ	A B	1781·575	18°±	1::					Σ. 2492.
				1802·747	20·45	1					
7120	15	44 Bootis *i*	A B	1781·624	60·10	1					Σ. 1909.
				1802·246	62·98	1					
7251	16	2 Coronæ η	A B	1781·687	30·68	1					Σ. 1937.
				1794·660	*nf.*	...					* 180° subtracted from the angle of 1802 whose reading (*np*) should be *sf*.
				1802·679	179·67	2*					
7259	17	51 Bootis μ_2	B C	1782·671	357·23	1					B.A.C. 5084; Σ. 1938; Sh. 203.
				1802·660	346·23	1					μ is VI. 17, which see for A C.
7341	18	Coronæ	A B	1781·805	291·00	1*					Σ. 1963.
				1802·679	292·23	1					
7834	19	20 Draconis h^2	A B	1781·758	240..245	est.					Σ. 2118.
				1783·263	251·50	1					
2976	20	52 Orionis	A B	1781·747	200·32	1					Σ. 795.
				1802·058	200·05	1					
1278	21	Trianguli	A B	1781·805	325 ±	est.					P. ii. 89; Σ. 269.
2783	22	33 Orionis n^1	A B	1781·805	28·62	1					Σ. 729.
				1802·030	32·58	1					
				1802·058	31·52	1					
4193	23	Can. Min.	A B	1781·906	117·35	1					Σ. 1126.
			A C		P		A C	1781·906	119·65	1	Central measure.

CLASS I. (*continued*).

Burnham Gen. Cat.	Class I. No.	Star.		Date of Measures. A.D.	Angle of Position. ° dec.	No. of Meas.		Date of Measures. A.D.	Distance. "		Synonyms and Remarks.
4477	24	16 Cancri ζ	A B	1781·906	3·47	1					Σ. 1196. Is also III. 19, which see for A C.
2780	25	32 Orionis A	A B	1782·052	217·83	1					Σ. 728.
				1802·030	204·37	1::					
				1802·058	216·57	2					
5103	26	2 Leonis ω	A B	1782·107	95°..100°	est.					Σ. 1356.
				1782·838	270 +	? ?					* The estimation of 1782·838 is of no authority and must be rejected. To the measure of 1782·846 is appended a note in the original MS. "This was probably a deception."
				1782·846	307·20	1*					
				1782·865	110·90	1					
				1802·093	131·47	1					
				1804·096	130·28	1					
5833	27	90 Leonis	A B	1782·285	208·85	1					Σ. 1552.
				1802·101	214·33	1					
				1802·173	207·00	1					
				1802·260	209·47	1					
			A C	1782·285	234·92	1					
				1783·293	234·80	1	A C	1783·293	53·72	1	
5388	28	41 Leonis γ	A B	1782·126	82·38	1				1	Σ. 1424.
				1783·293	84·60	1					
				1800·063	93·27	1					
				1800·134	94·47	2					
				1800·230	93·78	1					
				1802·058	96·07	1					
				1803·110	93·55	1					* 1 Revolution subtracted from micrometer reading (=24°), conforming with a surmise to that effect in MS.
				1803·219	96·54	2					
				1803·230	96·35	1					
			A C	1782·991	295·20	1*::	A C	1782·991	111·38	1	
				1783·293	301·00	1†					† Very accurate measure with 20 ft. of 18¾ in. ap.
5426	29	Leonis	A B	1782·129	63·47	1					Σ. 1431.
4839	30	57 Cancri σ^2	A B	1782·285	338·20	1					Σ. 1291.
5011	31	Lyncis	A B	1782·865	38·65	1					Σ. 1333.
5171	32	Leo. Min. 30	A B	1783·058	261·55	1					Σ. 1374.
7487	33	51 Scorpii ξ	A B	1782·359	7·97	1					Σ. 1998. Is also II. 20, which see for A C.
1262	34	Cassiop. ι	A B	1782·756	290·50	1					Σ. 262. Is also III. 4 (which see for A C) and H.N. 65.
				1804·438	274·42	1					

CLASS I. (*continued*).

Burnham Gen. Cat.	Class I. No.	Star.		Date of Measures. A.D.	Angle of Position. ° dec.	No of Meas.		Date of Measures. A.D.	Distance. ″		Synonyms and Remarks.
7923	35	38 Ophiuchi	A B	1783·178	330·80	1					B.A.C. 5822 ; S. 685.
7717	36	40 Herculis ζ	A B	1782·550	69·30	1					Σ. 2084.
				1795·219	*n s d*						
				1795·813	*n f*						
				1802·717	*n s d*						
				·742	219·17	1::					Wedge-shaped with 600; *n s d* with 400
				·791	*n s d*						*n s d* with 901.
				·805	5·67	1::					Measure taken with 527.
				1803·230	*n s d*						Power 1243, a little disturbed.
				·263	*n s d*						Perfectly round with 460.
				·274	*n s d*						A little distorted with 2140.
				·285	116·82	1::					Very doubtful if really seen elongated.
				·402	*n s d*						Completely round and well defined.
7534	37	Herculis	A B	1782·553	135·±	est.					Σ. 2015.
				·761	149·80	1					
				1802·717	155·00	1					
1459	38	Persei 85	A B	1782·633	278·40	1					Σ. 314.
				1804·181	290·57	2					
12755	39	Cassiopeæ	A B	1782·646	319·40	1					Σ. 3062.
				1783·047	319·15	2					
422	40	Cassiop. 78	A B	1783·337	140·50	1					Σ. 59.
8166	41	Draconis	A B	1782·660	350·	est.					LL. 32725.
				1783·263	354·35	1					
7318	42	13 Serpen. δ	A B	1782·671	227·20	1					Σ. 1954.
				·988	227·20	1					
				1802·101	208·55	1					
8966	43	Draconis	A B	1782·676	355·	est.					Σ. 2438.
				1783·263	358·40	1					
10559	44	4 Aquarii	A B	1783·556	351·50	1					Σ. 2729.
				1802·654	29·78	1					Is also VI. 58, which
				·657	28·05	1					see for A C.
2543	45	Aurigæ	A B	1782·676	222·45	1					Σ. 644.
				1783·058	222·90	1					
10834	46	Aquarii	A B	1783·578	332·45	1					Σ. 2776.
				1802·657	337·45	1					" Too far asunder for
			C A	1783·578	54·15	1	C A	1783·578	82·70	1	one of the first class."
10801	47	Capricorni	A B	1782·745	315..320*	est.					* 180° added to the
				1783·556	354·80	1					estimated angle, the stars being very nearly equal.
				1802·665	336·83	1					h. 5252.

Class I. (*continued*).

Burnham Gen. Cat.	Class I. No.	Star.		Date of Measures. A.D.	Angle of Position. ° dec.	No. of Meas.		Date of Measures. A.D.	Distance. ″		Synonyms and Remarks.
10863	48	Cephei	A B	1782·731	*s p*						A. CLARK, 1859.
				1783·181	255·85	1					Binary.
11578	49	Cephei	A B	1783·060	4·20	1					Probably Σ. 2880.
11914	50	Aquarii	A B	1782·745	311·20	1					Σ. 2935.
				1802·665	312·43	1					Is also H.N. 128 and H.N. 133.
11997	51	Cephei	A B	1782·745	86·40	1					Σ. 2947.
				1788·901	*n f*	est.					
2718	52	Orionis	A B	1782·750	320·±	est.					P. v. 84 : Σ. 708.
				1783·033	326·70	1					
				·049	322·80	1					
				1802·066	320·02	1					
2731	53	Orionis	A B	1782·750	50·±	est.					Σ. 712.
				·769	40·35	1					
				1783·047	46·60	1					
				1802·058	45·90	1					
2606	54	Orionis	A B	1783·058	305·70	1					Σ. 667.
				1802·068	311·45	1					
1738	55	Tauri	A B	1783·049	172·80	1					Σ. 403.
981	56	Ceti	A B	1782·778	2·35	1					Σ. 178.
3171	57	Orionis	A B	1782·824	*s f*	est.					Σ. 848.
				1783·181	109·80	1					"Multiple."
				1802·066	106·42	1					
8950	58	Lyræ	A B	1783·181	283·00	1					Σ. 2429.
9012	59	Lyræ	A B	1782·816	185..190	est.					Σ. 2448.
				1783·263	195·00	1					
9000	60	Lyræ	A B	1783·101	286·80	1					Certainly = Σ. 2441.
10605	61	Equulei	A B	1782·846	288·40	1					Σ. 2735.
				1802·791	287·85	1					
10707	62	Equulei	A B	1783·402	234·85	1					β 269.
				1802·791	237·10	1					
10794	63	Equulei	A B	1783·233	264·05	1					Σ. 2765.
				1802·786	264·33	1					STRUVE'S position differs by 180°. The stars are almost exactly equal.
1448	64	42 Arietis π	A B	1782·824	109·15	1					Σ. 311.
				1802·791	124·18	1					
				1804·099	121·24	2					
			A C	1782·824	109·15	1	A C	1782·824	25″..26″	est.	
9538	65	Sagittæ	A B	1783·203	284·00	1					Σ. 2563.
							A C				Triple. A 3d star, cl. V. or VI.
8057	66	Draconis	A B	1782·841	267·60	1					P. xvii. 147 ; Σ. 2180.
2984	67	Aurigæ	A B	1782·846	66·05	1					P. v. 225 ; Σ. 796.
2443	68	Orionis	A B	1782·865	0±	est.					P. iv. 258 ; Σ. 622.
				1783·058	5·10	1					

CLASS I. (*continued*).

Burnham Gen. Cat.	Class I. No.	Star.		Date of Measures. A.D.	Angle of Position. ° dec.	No. of Meas.		Date of Measures. A.D.	Distance. ″		Synonyms and Remarks.
3793	69	Lyncis	A B	1782·865	167·40	1					Σ. 1009.
2835	70	Tauri 380	A B	1782·865	233·60	1					Σ. 742.
5422	71	Ursæ	A B	1783·060	87·90	1					Σ. 1428.
5962	72	65 Ursæ	A B	1782·884	36·25	1					Σ. 1579.
			A C	1782·884	112·35	1	A C	1782·884	60·1	1	
963	73	Piscium 304	A B	1782·975	167·40	1					P. i. 179 ; Σ. 174.
1405	74	Arietis	A B	1782·972	290·60	1					P. ii. 160 ; Σ. 300.
2696	75	Orionis	A B	1783·022	359·60	1 :					Σ. 700.
				1802·058	10·80	1					
3875	76	Lyncis	A B	1783·060	270·00	1					Σ. 1033.
			A C	1783·060	266·30	1	A C	1783·060	67·77	1	
5650	77	Crateris	A B	1783·082	5°±	est.					L.L. 21178.
				·183	7·60	1					
7019	78	Libræ	A B	1783·181	148·40	1					Σ. 1885.
7749	79	46 Herculis	A B	1783·096	156·60	1					Σ. 2095.
				1802·742	166·30	1					
6558	80	81 Virginis	A B	1783·101	48·80	1					Σ. 1763.
				1802·304	47·17	1					
7461	81	Serpentis 112	A B	1783·178	220·20	1					Σ. 1990.
				1802·660	210·91	1 :					* Measured from the two stars A, B, as one.
			A C	1783·293	238·20	1	A C*	1783·293	56·47	1	
7551	82	49 Serpentis	A B	1783·178	291·55	1					Σ. 2021.
				1802·381	302·87	1					
				1804·252	305·17	1					
7649	83	10 Ophiuchi λ	A B	1783·183	75·50 ?	1					Σ. 2055.
				1802·381	69·32	1					
3490	84	Aurigæ	A B	1783·208	76·00	1					Σ. 941.
4835	85	Lyncis	A B	1783·224	1·05	1					Σ. 1289.
8501	86	Herculis	A B	1783·318	349·40	1					Σ. 2309.
				1802·742	292·45	1*					* Must refer to some other star.
8380	87	73 Ophiuchi *q*	A B	1783·318	267·20	1					Σ. 2281.
				1802·381	264·73	1					
8303	88	69 Ophiuchi τ	A B	1783·337	331·60	1					Σ. 2262.
				1802·737	360 −	est.					
				1804·441	360 ±	est.					
988	89	Androm. 241	A B	1783·570	165·50	1					Σ. 179.
				1802·715	157·07	1					
10849	90	Aquarii	A B	1783·578	167·60	1					Σ. 2781.
				1802·717	166·83	1					
9574	91	Aquilæ	A B	1783·597	278·30	1					P. xix. 257 ; Σ. 2570.
				1802·717	282·38	1					
9634	92	52 Aquilæ π	A B	1783·652	124·40	1					Σ. 2583.
				1802·717	127·53	1					
9813	93	Aquilæ	A B	1783·695	289·15	1					
				1802·750	283·35	1					

CLASS I. (*continued*).

Burnham Gen. Cat.	Class I. No.	Star.		Date of Measures. A.D.	Angle of Position. ° dec.	No. of Meas.		Date of Measures. A.D.	Distance. "		Synonyms and Remarks.
9605	94	18 Cygni δ	A B	1783·723	71·65	1					Σ. 2579. 1803·402 near ½ diameter of smaller.
10135	95	Cygni	A B	1783·723	342·25	1					Σ. 2671.
9868	96	Cygni	A B	1783·726	179·30	1					Σ. 2624.
			A C	1783·726	326·05	1					
10659	97	Cygni 280	A B	1783·728	43·60	1					Σ. 2741.

CLASS II.

Burnham Gen. Cat.	Class II. No.	Star.		Date of Measures. A.D.	Angle of Position. ° dec.	No. of Meas.		Date of Measures. A.D.	Distance. "		Synonyms and Remarks.
4122	1	66 Gemin. α	A B	1779·843	302·78	1	A B	1779·843	5·31	1	Σ 1110
				1783·635	293·05	1		·923	5·78 –	1	
				1791·145	292·95	1		1780·058	4·69	1	
				1792·153	297·27	1		·260	5·00	1	
				1795·953	283·88	1		1781·140	5·43	1	
				1800·230	288·13	1		·159	5·53	1	
				·307	280·51	1					
				1801·997	277·97	1					
				1802·025	280·88	1					
				·060	280·47	1					
				·151	291·62	1::					
				·159	283·00	1					
				1803·107	275·49	2					
				·112	277·88	1					
				·222	284·59	3					
				·233	280·88	2					
7914	2	64 Herculis α	A B	1779·753	117 ±	est.	A B	1779·753	3·91	1	Σ. 2140.
				1781·381	111·47	1		1780·375	5·62	1	
				·786	120·58	1		·537	5·94	2	
				1783·022	120·00	1		·657	4·06	2	
				·252	115·47	2		1781·381	5·20	1:	
				1795·813	111·60	1		·676	4·57	1	
				1802·140	119·08	1:		1782·301	6·00	1::	
				·170	124·20	1					
				1803·274	124·68	2					
				1804·403	118·50	2					
				·408	123·15	1					
				·422	122·83	1					

CLASS II. (*continued*).

Burnham Gen. Cat.	Class II. No.	Star.		Date of Measures. A.D.	Angle of Position. ° dec.	No. of Meas.		Date of Measures. A.D.	Distance. "		Synonyms and Remarks.
8003	3	75 Herculis ρ	A B	1779·764	308 ±	est.	A B	1779·764	2·89	2	Σ. 2161.
				1781·786	300·35	1		1780·657	3·12	1	
				1802·170	301·20	1					
8340	4	70 Ophiuchi *p*	A B	1779·764	90·00	1	A B	1779·764	3·59	1	Σ. 2272.
				1781·728	99·23	1*		1780·359	5·47	1 ::	* In the review in May 1804, it is remarked, "how to reconcile the observations of the years 1781 and 1802 together I cannot comprehend. I have the greatest reason to suppose them both equally authentic."
				1802·255	336·13	1		·465	4·37		
				1804·408	318·02	1		·657	4·53		
				·422	319·58	1					
8783	5	4 Lyræ ε¹	A B	1779·764	30·	est.	A B	1779·925	3·44 −	1	Σ. 2382.
				·835	33·92	1			..		
				1802·717	34·90	1			..		
				1804·400	27·32	1			..		
				·408	30·07	1			..		
8785	6	5 Lyræ ε²	A B	1779·764	165 ±	est.					Σ. 2383.
				·835	173·47	1					
				1802·717	162·05	1					
				1804·400	167·13	2					
				·408	169·08	1					
				·444	169·90	1					
11743	7	55 Aquarii ζ	A B	1779·901	18·92	1	A B	1779·728	4·22	1	Σ. 2909.
				1781·728	18·35	1		·942	5·31 −	1	
				1782·463	17·88	1		1780·485	5·62	1	
				1802·006	11·95	1		·602	4·38	1	
7352	8	7 Coronæ ζ	A B	1779·747	*n p*	est.	A B	1779·747	4" or 5"	est.	Σ. 1965.
				·767	312·5±	est.		·767	4·69	1	
				1781·690	295·85	1		1780·485	6·25	1	
				1802·246	295·50	1					
2821	9	39 Orionis λ	A B	1779·884	44·77	1	A B	1779·884	4·53	1	Σ. 738. Triple.
				1802·101	222·75	1*		1780·058	6·09	1	* The position is stated at 47° 15′ *s p*, were it *n f* it would give 42°·75, agreeing well with the former observations and later results.
								·161	6·87	1 ::	
2883	10	48 Orionis σ²	A B	1781·882	84·92	1	A B	1779·791	20 ±	est.	Σ. 762.
			A C	1781·882	60·92	1		1780·161	13·44	1	
							A C	1781·679	43·20	1	

Class II. (*continued*).

Burnham Gen. Cat.	Class II. No.	Star.		Date of Measures. A.D.	Angle of Position. ° dec.	No. of Meas.		Date of Measures. A.D.	Distance. ″		Synonyms and Remarks.
2881	11	Orionis σ^1	A B	1781·882	87° or 88°	est.	A B	1779·791	8″ or 10″	est.	Σ. 761.
			A C		203·42	1					
1061	12	113 Piscium α	A B	1779·797	345 ±	est.	A B	1779·797	4·375	2	Σ. 202.
				1781·797	337·38	1		1779·925	5·	1	
				1802·074	331·95	1		1780·058	6·09	1	
				·093	333·52	2					
7878	13	21 Dracon. μ	A B	1781·728	232·37	1	A B	1779·797	4″ or 5″	est.	Σ. 2130.
				1802·170	219·47	1		·925	4·69	1	
				1804·096	221·00	1		1780·537	4·38		
				·099	219·93	1					
2435	14	4 Aurigæ ω	A B	1781·800	352·62	1	A B	1780·257	6″ or 10″	est.	Σ. 616.
				1802·827	349·43	1					
9765	15	24 Cygni ψ	A B	1779·884	180·53	1					Σ. 2605.
				1781·800	194·08	1					
				1802·006	183·10	1					
11483	16	17 Cephei ξ	A B	1781·964	290·30	1	A B	1780·372	5	1	Σ. 2863.
				1787·843	*n p*	est.					
				1803·225	293·77	1					
3402	17	11 Monocer. (Triple.)	A B	1781·800	121·63	1					Σ. 919.
				1802·093	117·57	1					
				·170	117·55	2					
7034	18	37 Bootis ξ	A B	1782·285	24·12	1	A B	1780·662	1½ D.	*	Σ. 1888.
				1791·392	*n f*			1781·266	3·23	1	* "Interval 1½ diam. of the largest *."
				1792·298	355·72	1		1804·252	>3·23	†	† "The small * is farther off than formerly."
				1795·219	354·93	1					
				1802·246	352·95	1					
				1804·252	353·90	1					
7613	19	5 Ophiuchi ρ	A B	1782·293	7·85	1					Sh. 228.
				1804·441	7·87						
7487	20	51 Libræ ξ	A C	1782·359	88·62	1	A C	1782·359	6·38	1	Σ. 1998. Is also I. 33.
				1784·350	90·	est.		1784·350	Cl. III.		
7488	21	Libræ					A B	1780·392	5 or 6 D.		Σ. 1999.
1950	22	45 Persei ϵ	A B	1782·449	8·53	1					Σ. 471.
				1802·827	7·25	1					
7655	23	Ophiuchi	A B	1782·375	316·40	1					Σ. 2056.
				1802·381	336·93	1*					* Probably an error of 20° in this.
12543	24	107 Aquarii									Σ.C.P. 786 ; Sh. 356.
10506	25	52 Cygni *k*	A B	1781·654	58·95	1					Σ. 2726.
2842	26	Orionis	A B	1781·800	63·92	1::					Σ. 750.
3970	27	55 Geminor. δ	A B	1781·879	184·15	1					Σ. 1066.
				1787·027	188 ±	?					
				1802·074	193·63	1					
				·093	196·92	1					
				1804·099	200·13	1					
9682	28	Aquilæ	A B	1781·805	306·47	1					P. xix. 307 ; Σ. 2590.
				1802·789	309·30	1					

CLASS II. (*continued*).

Burnham Gen. Cat.	Class II. No.	Stars.		Date of Measures. A.D.	Angle of Position. ° dec.	No. of Meas.		Date of Measures. A.D.	Distance. ″		Synonyms and Remarks.
9926	29	Aquilæ 227	A B	1781·805	345·80	1					Σ. 2628.
				1802·758	349·02	1					
9643	30	8 Sagittæ ζ	A B	1781·641	315 ±	est.	A B	1781·893	8·83	1 ::	Σ. 2585.
				·882	304·17	1					
				1802·435	310·68	1					
8997	31	Draconis 233	A B				A B	1781·679	5·12	1	Σ. 2452.
								1782·673	10″or12″	est.	
9498	32	Sagittæ					A B	1781·641	27·50	2	Σ.C.G. 2332. Is also V. 51 = N. 84.
2605	33	19 Orionis β	A B	1781·747	201·80	1	A B	1781·805	6·62	1	Σ. 668.
				1782·676	203·53	1		1782·676	12·67	1	
				·887	196·34	2		·977	10·98	1*	* Central measure.
				·977	198·30	1		1783·041	9·10	2	
				1783·033	203·95	1		·153	8·92	1	
				·041	204·35	1					
				·153	203·35	1					
				·717	193·02	3†					† 3 measures of Pos. "all bad."
				·772	203·15	1					
				1784·173	199·83	3					
				1791·145	216·27	1					
				1796·022	195·81	2					
				1798·126	197·82	1					
				1799·903	195·75	1					
				1800·055	201·68	1					
				·063	199·43	2					
				·134	204·08	2					
				·211	209·57	1					
				·227	212·50	1					
				·230	211·40	1					
				1801·997	204·13	2					
				1802·003	200·43	2					
				·025	200·63	1					
				·058	199·93	1					
				·074	204·48	1					
				·189	202·72	1					
				1803·107	200·47	1					
				·110	202·65	1					
				·131	199·83	2::					
1137	34	6 Trianguli ι	A B	1781·767	85·62	1					Σ. 227.
1159	35	Trianguli 28									Σ. 232.
1939	36	32 Eridani	A B	1781·805	343·38	1	A B	1781·805	4·32	1	Σ. 470.
				1804·099	347·32	1		1783·082	5·78	1	
3442	37	Monocerotis									Σ. 927 rej., one of STRUVE'S "plures duplices in acervo."
6842	38	Bootis	A B	1781·977	186·92	1	A B	1781·977	5·17	1	Σ. 1835.

CLASS II. (*continued*).

Burnham Gen. Cat.	Class II. No.	Star.		Date of Measures. A.D.	Angle of Position. ° dec.	No. of Meas.		Date of Measures. A.D.	Distance. "		Synonyms and Remarks.
4227	39	Canis Min.	A B	1782·088	144·47	1					Σ. 1134.
4601	40	23 Cancri ϕ^2	A B	1783·060	33·30	1					Σ. 1223.
4602	41	24 Cancri υ^1	A B	1783·060	57·85	1					Σ. 1224. There is doubtless a mistake of 1 Rev. = 24° in the micrometer reading. This would make the Position 33·85, agreeing with the subsequent history of this star.
6311	42	Virginis	A B	1783·183	142·40	1	...				Σ. 1690.
				1802·307	144·43	1					
5397	43	Leonis 145	A C*	1782·129	4·97	1	A C	1782·129	5 ±	est.	Σ. 1426. * A is a very close double star.
6599	44	84 Virginis *o*	A B	1782·129	240·92	1					Σ. 1777.
				1802·304	239·83	1					
6422	45	54 Virginis	A B	1783·178	33·00	1	A B	1782·252	6 ±	est.	Σ.C.P. 433; Sh. 161.
				1802·307	35·43	1*					* Called *s p* instead of *n f*; but the stars are very nearly equal.
				·329	*n f*						
6438	46	Comæ	A B	1782·285	96·70	1					Σ. 1733. Probably an error of 1 rev. = 24°, as the P.A. should be 125° (fixed).
			A C	1782·285	*p*		A C	1782·285	60 +	est.	
6018	47	2 Comæ	A B	1782·293	242·30	1					Σ. 1596.
2610	48	Aurigæ	A B	1783·183	74·20	1					Σ. 666.
907	49	Piscium	A B	1783·635	329·10	1					Σ. 155.
				1802·663	330·32	1					
116	50	38 Piscium	A B	1783·493	244·95	1					Σ. 22.
				1802·663	235·28	1					
10228	51	11 Capricorni ρ	A B	1783·504	174·0	1					Sh. 323; same as VI. 29.
				1802·657	183·08	1*					* Should probably be 176°·92.
1799	52	Persei	A B	1783·058	278·40	1					Σ. 425.
2465	53	Camelopardi	A B	1783·334	108·55	1:					Σ. 625.
2200	54	Tauri	A B	1782·997	190 ±	*					Σ. 546.
				1783·646	201·30	1					* "By wires"—a mere rude estimate.
55	55	Ceti 27	A B	1782·969	295 ±	est.	A B	1782·687	6...7	est.	Σ. 8.
				1783·635	291·70	1					
				1802·750	292·77	1					

CLASS II. (*continued*).

Burnham Gen. Cat.	Class II. No.	Star.		Date of Measures. A.D.	Angle of Position. ° dec.	No. of Meas.		Date of Measures. A.D.	Distance. "		Synonyms and Remarks.
974	56	Arietis	A B	1782·690	*n p*						Σ. 175.
				1783·578	293·20	1					
11968	57	Aquarii 231	A B	1782·682	244·15	1					Σ. 2944.
				1802·750	242·12	1					
			A C	1782·682	140..145	est.					
1047	58	Ceti 292	A B	1782·682	*n p*	AB		1782·682	5 ±		Σ.C.P. 51 ; Sh. 24.
				1783·055	295·20	1					
11767	59	Aquarii	A B	1782·748	340 ±	est.					Σ. 2913.
				1783·578	331·20	1					
				1802·758	331·98	1					
3510	60	Canis	A B	1783·003	325..330	est.					L.L. 12755, h. 3876.
				·033	337·60	1					
2940	61	Orionis	A B	1783·022	92 or 93	est.					Σ. 788.
				·723	85·1	1					
			A C	1783·022	140 ±	est.					
11100	62	Pegasi	A B	1783·334	358·40	1					OΣ. 443.
				1802·663	349·63	1					
4169	63	Argus	A B	1783·134	300·20	1					Σ. 1121.
4261	64	Geminor. 201	A B	1782·780	270 −	est.					Σ. 1140.
				1783·203	274·15	1					
4262	65	Geminorum	A B	1783·162	0·80	1					Σ. 1144.
10504	66	Delphini	A B	1783·293	348·70	1					Σ. 2725.
8797	67	Lyræ	A B	1783·203	158·10	1					Σ. 2390.
9114	68	Lyræ	B C	1783·334	81·60	1					Triple. Σ. 2481.
			A C	1783·334	244·05	1					
9344	69	Cygni	A B	1783·183	60·80	1					P. xix. 149 ; Σ. 2534.
9949	70	Sagittæ	A B	1783·334	17·05	1					Σ. 2634.
3607	71	Aurigæ	A B	1783·290	45·40	1	C D	1783·208	17·68	1	Σ. 961 rej.
3587	72	Lyncis	A B	1782·865	259·00	1					Σ. 958.
5059	73	21 Ursæ	A B	1782·876	306·75	1					Σ. 1346.
				1802·381	317·62	1					
5563	74	Crateris	A B	1783·033	18·45	1					Σ. 1474.
			A C	1783·033	201..202	est.	A C		Cl. VI.		
2751	75	118 Tauri	A B	1782·931	200·±	est.	A B	1782·931	7·±	est.	Σ. 716.
				1783·739	192·75	1		1783·739	4·85	2	
				1801·928	*s p*						
1664	76	Arietis	A B	1782·977	254·60	1	A B	1782·977	5·77	1*	Σ. 376.
											* Central measure.
4859	77	17 Hydræ	A B	1782·988	353·±	est.					Σ. 1295.
				1783·033	0·00	1					
				1802·101	0·65	1					
5671	78	Leonis	A B	1783·323	165·35	1					P. x. 239 ; Σ 1507.
				1802·329	161·70	1					
7031	79	39 Bootis	A B	1783·019	51·65	1					Σ. 1890.
				1802·660	48·20	1					
2109	80	40 Eridani	B C	1783·134	326·70	1					Σ. 518.
			A B	1783·134	107·55	1	A B	1783·134	81·78	1	

CLASS II. (*continued*).

Burnham Gen. Cat.	Class II. No.	Star.		Date of Measures. A.D.	Angle of Position. ° dec.	No. of Meas.		Date of Measures. A.D.	Distance. "		Synonyms and Remarks.
2312	81	Eridani	A B	1783·082	321·60	1					Σ. 583.
6987	82	Bootis	A B	1783·090	91·0	1 :					Σ. 1873.
			A C	1783·090	290..300	est.					
37	83	Androm. 51	A B	1783·153	84·20	1					Σ. 3.
				1802·665	83·27	1					
439	84	65 Piscium	A B	1783·156	300·95	1					Σ. 61.
				1790·917	285..290	est.					
				1802·663	297·37	1					
7433	85	Serpentis	A B	1783·170	*n p*						Σ. 1985.
				·323	316·15	1					
7540	86	Serpentis	A B	1783·178	143·15	1					Σ. 2016.
4506?	87	Monoc.	A B	1783·181	176·20	1					Probably = h. 2435.
7605	88	Serpentis	A B	1783·181	*n p*						Σ. 2040.
				·293	314·75	1					
3396	89	Monoc.	A B	1784·173	39·15	1					Σ. 915.
8301	90	Herculis	A B	1783·326	165·15	1					Σ. 2263.
				1802·742	167·90	1					
9863	91	Sagittæ	A B	1783·293	195·10	1					Σ. 2622.
			A C	1783·293	310..320	est.					
2333	92	Camelop.	A B	1783·326	112·70	1 ::					Σ. 584.
8956	93	Aquilæ	A B	1783·550	286·00	1					P. xviii. 263 ; Σ. 2428.
				1802·758	286·23	1					
12420	94	Andromedæ	A B	1783·630	304·40	1					Σ. 3024.
				1802·674	305·93	1 :					
9679	95	Aquilæ	A B	1783·695	299·05	1					Σ. 2589.
				1802·758	300·03	1					
9982	96	Aquilæ	A B	1783·695	213·80	1					P. xx. 26 ; Σ. 2644.
				1793·737	217·13	1					
				1802·758	212·08	1					
10773	97	Cygni	A B	1783·704	315·25	1					P. xxi. 1 ; Σ. 2762.
			A C	1783·704	220·±	est.					
10437	98	49 Cygni	A B	1783·704	58·20	1					Σ. 2716.
9394	99	Cygni	A B	1783·704	2·20	1					P. xix. 169 ; Σ. 2539.
10560	100	Cygni	A B	1783·728	74·15	1					Σ. 2732.
4215	101	Camelop.	A B	1783·734	337·25	1					Probably Σ. 1127.
2705	102	Orionis 88	A B	1783·758	142·40	1					Σ. 701.

Class III.

Burnham Gen. Cat.	Class III. No.	Star.		Date of Measures. A.D.	Angle of Position. ° dec.	No. of Meas.		Date of Measures. A.D.	Distance. "		Synonyms and Remarks.
2837	1	41 Orionis θ^1					A B	1780·134	12·81	1	Σ. 748.
							A C	·134	14·27	1	
							B D	·134	9·06	1	
							C D	·134	20·52	1	
6482	2	79 Ursæ ζ	A B	1781·879	146·77	1	A B	1779·728	12·11	2	Σ. 1744.
				1802·753	141·23	1		·747	12·34	1*	* Diameters included.
								·816	12·50	1	
								1780·162	15·62	2	
								·192	13·98	2	
								·334	13·75	2	
								·372	14·58 +	1	
								·465	13·44	1	
								·578	14·37	1	
								·663	13·75	1	
								1781·131	13·12	1	
								·487	15·08	1	
426	3	24 Cassiop. η	A B	1779·816	70 ±	est.	A B	1779·816	11·09	1	Σ. 60.
				1782·449	62·07	1		1780·523	11·46	1	
				1803·112	70·77	1					
1262	4	Cassiop. ι	A B	1782·449	100·62	1	A B	1780·665	7·50	1	Σ. 262.
				1804·433	108·95	1					Is also I. 34, and H.N. 65.
1070	5	57 Androm. γ	A B	1781·786	70·38	1	A B	1780·523	9·69	1	Σ. 205.
				1802·090	63·43	1		·663	9·38	1	
				1803·112	63·92	1		1781·550	9·04	3	
				1804·096	62·35	1					
11046	6	8 Cephei β	A B	1781·966	254·53	1	A B	1780·372	13·12	1	Σ. 2806.
				1803·224	252·70	1					
7493	7	8 Scorpii β	A B	1782·285	25·15	1	A B	1780·359	14·37	1	Sh. 217.
				1783·030	28·05	1		1781·392	15·08	1	
				1802·304	24·95	1		1783·030	13·12	1	
6954	8	29 Bootis π	A B	1781·824	96·47	1	A B	1779·737	5·47	1	Σ. 1864.
				1802·090	96·18	1		1780·482	6·87	1	
				1804·252	99·07	1					
993	9	5 Arietis γ	A B	1779·835	174·00	1	A B	1779·737	10·16	1	Σ. 180.
				1780·772	176·08	1		1780·058	9·90	1	
				1802·101	179·17	1		·101	11·25 −	1	
								·134	10·94	1	
								·161	10·31	1	
								·485	10·00	1 ::	
								·643	9·77	1	
								·663	8·75	1	
								·775	10·31	1	
10509	10	12 Delphini γ	A B	1779·835	272·77	1	A B	1779·737	10·31	1	Σ. 2727.
				·843	276·47	1		·835	11·87 +	1	
				·884	274·15	1		·843	10·78	1	

CLASS III. (*continued*).

Burnham Gen. Cat.	Class III. No.	Star.		Date of Measures. A.D.	Angle of Position. ° dec.	No. of Meas.		Date of Measures. A.D.	Distance. "		Synonyms and Remarks.
10509	10	12 Delphini γ	A B	1783·033	274·85	1		1779·701	12·03	1	
cont.				1804·438	273·33	1		·884	11·51	3	
								·925	12·58	2	
								·942	12·50	3	
								1780·375	11·87	1	
								·465	11·41	1	
								·586	12·50	1	
								·633	11·41	1	
								·775	12·34	1	
								1781·580	15·95 –	1	
6778	11	17 Bootis κ	A B	1779·753	240 ±	est.	A B	1779·753	11·09	1	Σ. 1821.
				1782·293	242·53	1		1780·559	12·08	1	
				1802·663	240·68	1		1781·697	14·33 –	1	
2843	12	44 Orionis ι	A B	1781·800	133·85	1	A B	1780·161	12·50	1	Σ. 752.
			A C	1781·800	101·32	1	A C	1781·800	48·52	1	
2833	13	Orionis 133									Σ. 747. No measures given.
2830	14	Orionis					A B	1780·16	36·25	1	Σ. 745.
11214	15	78 Cygni μ	A B	1779·884	110·08	1	A B	1779·884	6·41	1	Σ. 2822.
				1781·800	108·47	1		1780·375	7·50	1	
								1780·521	6·87	1	
10281	16	Delphini	A B	1781·821	260·30	1	A B	1780·657	12·50	1	Σ. 2690.
11641	17	Lacertæ	A B	1781·797	193·73	1	A B	1781·797	13·72	1 ::	P. xxii. 65 ; Σ. 2894.
6243	18	29 Virginis γ (Well known before as a double star.)	A B	1781·887	130·73	1	A B	1780·055	6·98 –	1	Σ. 1670.
				1802·077	118·37	1		·216	8·75*	1	* "Occulted by the moon. As fast as I could, lest I should be too late."
				1803·285	120·45	3		·263	6·25†	1	† "Very carefully taken, both diameters included, but too large a measure."
				·403	122·03	1					
				·430	120·20	3					
4477	19	16 Cancri ζ	A C	1781·887	181·73	1	A C	1780·260	7·97	1	Σ. 1196.
				1802·156	171·78	1		1781·312	8·12	1	Is also I. 24, which see for A B.
6852	20	Bootis	A B	1782·298	329·53	1	A B	1781·364	7·60	1	Σ. 1838.
10643	21	1 Equulei ε	A B	1781·805	84·35	1	A B	1780·654	8·75	1	Σ. 2737.
								·663	10·00	1	
3559	22	12 Lyncis	A C	1782·367	302·55	1	A C	1782·367	9·38	1	Σ. 948. Is also I. 6, which see for A B.
691	23	34 Cassiop. φ	A C	1783·657	268·15	1	A B	1780·602	12...15	est.	Triple, Classes III. and VI.
9955	24	17 Sagittæ θ					A B	1781·641	11·07	1	Σ. 2637.
							A C	1781·641	57·82	1	
7928	25	39 Ophiuchi	A B	1782·455	357·23	1	A B	1782·455	10·37	1	Sh. 245. Is also VI. 54.

Class III. (*continued*).

Burn-ham Gen. Cat.	Class III. No.	Star.		Date of Measures. A.D.	Angle of Position. ° dec.	No. of Meas.		Date of Measures. A.D.	Distance. ″		Synonyms and Remarks.
8302	26	95 Herculis	A B	1781·805	265·85	1	A B	1781·805	6·10	1	Σ. 2264.
				1802·307	262·65	1					
4197	27	Puppis k¹					A B	1781·123	15±	est.	β 1061.
4320	28	Argûs					A B	1781·123	8±	est.	L.L. 15389.
3349	29	8 Monocerot.					A B	1781·123	12	est.	Σ. 900. Is identical with III. 44.
5603	30	54 Leonis	A B	1782·129	99·23	1	A B	1781·140	6·41	1	Σ. 1487
				1802·101	100·65	1		1782·266	7·80	3	
8105	31	Herculis					A B	1781·381	10 ±	est.	Unidentifiable.
8940	32	11 Aquilæ	A B	1802·758	238·42	1	A B	1781·561	7 ±	est.	Σ. 2424.
8779	33	5 Aquilæ					A B	1781·575	11·58	1	Σ. 2379.
								1783·383	12·03	1	Is also H.N. 9.
12292	34	94 Aquarii	A B	1802·674	342·75	1	A B	1781·633	13·75	1	Σ. 2998.
8065	35	54 Ophiuchi	A B	1783·318	82·30	1	A B	1781·635	5...6	est.	Σ. 2184.
								·786	8 ±	est.	
								1783·318	16·93	1 ::	
1510	36	Persei					A B	1781·701	11·88	1	P. ii. 220; Σ. 331. Same as III. 59, which see.
1832	37	Persei									Σ. 437.
1836	38	Persei									Σ. 439.
1818	39	40 Persei *o*					A B	1781·728	15·20	1	Σ. 431.
8192	40	Herculis	A B	1781·805	137·32	1	A B	1781·772	10·33	1	Σ. 2232.
				1783·230	131·10	1					
				·233	138·75	1					
				1802·170	140·65	1					
8377	41	100 Herculis *i*	A B	1781·805	1·62	1	A B	1781·772	11·72	1	Σ. 2280.
				1802·786	1·37	1					
1195	42	Trianguli						1781·77	6 or 7″		Σ. 246.
3388	43	Monocerotis	A B	1781·800	293·65	1					Σ. 914.
3349	44	8 Monocer.	A B	1781·800	29·77	1	A B	1781·800	12·50	1	Σ. 900.
1787	45	Tauri	A B	1781·805	234·47	est.*					Σ. 422. * Believed to be 6° or 8° too great.
3547	46	Monocerotis	A B								Probably = Σ. 952.
3692	47	38 Geminor. *e*	A B	1782·750	179·90	1	A B	1782·750	7·95	3	Σ. 982.
				1802·260	183·90	1					
4038	48	Geminorum	A B	1782·887	60 ±	est.	A B	1783·003	6·25	1	Σ. 1083. Identical with H.N. 95.
				1783·003	46·10	1					
4677	49	Hydræ 18	A B	1783·337	27·20	1	A B	1783·101	12·50	1	Σ. 1245.
				1802·173	24·75	1					
6405	50	51 Virginis θ	A B	1782·991	339·30	1	A B	1782·099	7 ±	est.	Σ. 1724.
				1802·307	341·17	1		·991	7·13*		* Central measure. Is also vi. 43, which see for A C.

Class III. (*continued*).

Burnham Gen. Cat.	Class III. No.	Star.		Date of Measures. A.D.	Angle of Position. ° dec.	No. of Meas.		Date of Measures. A.D.	Distance. "		Synonyms and Remarks.
5812	51	88 Leonis	A B	1782·298	317·55	1	A B	1782·107	9 ... 10	est.	Σ. 1547.
				1801·931	*n p*			1783·003	14·63		
2481	52	Orionis	A B	1783·726	52·95	1	A B	1782·129	9 ±	est.	Σ. 630.
								·843	15 ±	est.	
								·915	15 ±	est.*	* " By wires."
								1783·726	13·67	1	
6261	53	Virginis	A B	1783·323	349·00	1	A B	1783·323	12·97	1	Σ. 1677.
				1802·307	345·47	1					
4923	54	13 Ursæ σ^2	A B	1782·416	*s p*	est.	A B	1783·257	7·93	1	Σ. 1306.
				1783·337	270 ±	est.					
				·679	283·00	1					
7570	55	18 Coronæ υ	A B	1783·181	216·20	1					See V. 37.
			C A	·181	25·60	1	C A	1783·181	78·13	1	
8348	56	Ophiuchi	A B	1782·455	*s p**	est.	A B	1783·225	7·62	1	P. xvii. 362 ; Σ. 2276.
				1783·225	260·22	1					* " A few degrees *s p* (270 −)."
9315	57	Anseris	A B	1783·203	148·60	1	A B	1783·203	7·02	1	Σ. 2523.
1393	58	13 Persei θ	A B	1783·657	290·00	1	A B	1782·633	17 ±	est.	Σ. 296.
				1787·912	*n p*			1783·657	13·52	1	
1510	59	Persei	A B	1783·058	90·00	1	A B	1782·633	13 ±	est.	P. ii. 220 ; Σ. 331.
								1783·058	12·03	1	" 19 Persei, or what I have called so." See III. 36.
1471	60	20 Persei	A B	1783·230	239·50	1	A B	1783·578	14·03	1	Σ. 318.
				·271	240 ±	est.					
7004	61	Draconis 60	A B	1783·096	357·70	1	A B	1783·096	12·50	1	Σ. 1882.
87	62	35 Piscium	A B	1783·493	148·90	1	A B	1783·493	12·5	1	Σ. 12.
9891	63	Capricorni	A B	1783·578	16·20	1	A B	1782·676	15·02	1	P. xix. 396 ; Σ. 2625.
								1783·578	14·33	1	
2857	64	26 Aurigæ	A B	1783·182	272·60	1	A B	1783·183	13·42	1	Σ. 753.
				1802·268	273·92 −	1*					* Too large ; nearly preceding.
2260	65	Persei	A B	1782·203	41·10	1:	A B	1782·203	11·28	1	Σ. 563.
								1783·791	11·37	1	
1886	66	30 Tauri *e*	A B	1782·997	45...50	est.	A B	1783·126	11·27	1	Σ. 452.
				1783·723	72·75	1					
2581	67	3 Leporis ι	A B	1783·060	359·35	1	A B	1783·060	12·33	1	Σ. 655.
				1785·082	0·00	est.					
1120	68	Arietis	A B	1783·134	145·70	1	A B	1783·134	8·08	1	Σ. 221.
11940	69	Aquarii	A B	1783·635	69·95	1*	A B	1783·635	12·77 −	1	Σ. 2939.
				1802·789	58·60	1					* Error of 10° (fixed).
10085	70	1 Cephei κ	A B	1782·969	120 ±	est.	A B	1782·969	10 ±	est.	Σ. 2675.
				1783·189	122·50	1		1783·189	5·78	1	
				1802·170	128·88	1					
				1804·099	122·43	2					
11160	71	Cephei	A B	1783·203	125·40	1	A B	1783·203	11·58	1	Σ. 2816.
			A C	·203	343·95	1	A C	·203	18·62	1	

CLASS III. (*continued*).

Burnham Gen. Cat.	Class III. No.	Star.		Date of Measures. A.D.	Angle of Position. ° dec.	No. of Meas.		Date of Measures. A.D.	Distance. "		Synonyms and Remarks.
11182	72	Cephei	A B	1783·225	58·00	1	A B	1783·225	13·12	1	P. xxi. 256 ; Σ. 2819.
560	73	Ceti	A B	1783·082	180·80	1	A B	1782·750	10...11	est.	Σ. 86.
								1783·635	14·83	1	
11387	74	Pegasi	A B	1782·945	*n f*	est.	A B	1782·945	10...12	est.	Σ. 2848.
				1783·550	58·47	1		1783·550	14·48	1	
3434	75	Monoc.									Σ. 926.
2829	76	Orionis	A B	1783·726	283·10	1	A B	1783·726	9·20	1	Σ. 741.
1716	77	Arietis	A B	1783·047	163·30	1	A B	1783·047	8·53	1	Σ. 394.
1766	78	Tauri	A B	1783·047	357·95	1	A B	1783·047	7·17	1	Σ. 414.
1352	79	Ceti	A B	1783·649	224·80	1	A B	1783·649	10·80	1	Σ. 288.
1268	80	Ceti 378	A B	1783·649	292·40	1	A B	1783·649	11·27	1	S. 412.
8792	81	Lyræ	A B	1783·203	23·70	1	A B	1782·797	8...9	est.	Σ. 2393.
								1783·203	9·45	1	
3181	82	41 Aurigæ	A B	1783·183	350·00	1	A B	1783·183	8·53	1	Σ. 845.
3973	83	19 Lyncis	A B	1783·060	316·90	1	A B	1782·865	13·70	1	Σ. 1062.
								1783·060	14·18	1	
5034	84	Lyncis	A B	1783·225	318·20	1	A B	1783·225	7·18	2	Σ. 1342.
6102	85	2 Canum	A B	1783·337	259·00	1	A B	1783·260	12·20	1	Σ. 1622.
5793	86	57 Ursæ	A B	1783·096	14·40	1					Σ. 1543.
5858	87	Ursæ 290	A B	1783·096	270·00	1	A B	1783·096	12·50	1	Σ. 1561.
			A C	·096	86·00	1	A C	·096	32·35	1	
1826	88	Tauri	A B	1783·739	0·15	1	A B	1783·739	13·62	1	Σ. 435.
7858	89	Herculis 210	A B	1783·203	42·20	1	A B	1783·203	11·88	1	Σ. 2120.
2544	90	Aurigæ 47	A B	1783·739	26·00	1	A B	1782·904	13 ±	est.	Σ. 645.
								1783·739	13·10	1	
1679	91	Arietis	A B	1782·948	92...93	est.	A B	1782·948	11·28	1:*	The stars being equal, the angles may be 272 . . . 273 and 282·40.
				1783·657	102·40	1					* Central measure.
4929	92	Cancri 194	A B	1783·134	204·80.	1	A B	1782·934	7...8	est.	Σ. 1311.
				1784·876	220..225	est.		1783·134	8·83	1	
2789	93	Tauri	A B	1783·739	142·45	1	A B	1782·996	12·20	1*	Σ. 730. Is also H.N. 124. * Central measure.
2662	94	Leporis	A B	1783·041	94·00	1	A B	1783·041	11·73	2	Σ. 688.
2277	95	Eridani	A B	1783·728	260·70	1	A B	1783·728	15·35	1 ::	Σ. 571.
5820	96	17 Crateris	A B	1783·025	205·55	1	A B	1783·025	9·77	1	
				1791·178	210..220	est.					
6989	97	54 Hydræ	A B	1783·025	128·25	1	A B	1783·025	11·28	1	Sh. 184.
3421	98	Monocer.	A B	1783·025	208·05	1					Σ. 920.
2330	99	55 Eridani	A B	1783·082	314·15	1	A B	1783·082	9·15	1	Σ. 590.
2269	100	Eridani	A B	1783·134	253·60	1	A B	1783·134	11·88	1	Σ. 570.
...	101	3 Centauri *k*	A B	1783·082	112·00	1	A B	1783·082	11·58	1	B.A.C. 4263.
7644	102	Herculis	A B	1783·635	22·80	1	A B	1783·255	14·03	1	Σ. 2051.
7441	103	Serpentis	A B	1783·627	320·20	1	A B	1783·627	12·47	2	P. xv. 220 ; Σ. 1987.
8114	104	Herculis	A B	1783·635	6·20	1	A B	1783·255	14·33	1	P. xvii. 200 ; Σ. 2194.

CLASS III. (*continued*).

Burnham Gen. Cat.	Class III. No.	Star		Date of Measures. A.D.	Angle of Position. ° dec.	No. of Meas.		Date of Measures. A.D.	Distance. "		Synonyms and Remarks.
9701	105	Sagittæ	A B	1783·649	219·60	1	A B	1783·263	14·48	1 ::	Σ. 2595.
7213	106	5 Serpentis	A B	1783·383	50...60	est.					Σ. 1930.
8282	107	Sagittarii	A B	1783·646	215·20	1	A B	1783·646	15·17	1:	North of Cluster M. 20 ; h. 4355.
9021	108	Aquilæ	A B	1783·512	148·45	1	A B	1783·512	12·97	1	Σ. 2446.
9028	109	Aquilæ	A B	1783·512	292·10	1	A B	1783·512	10·22	1	Σ. 2449.
11145	110	Cygni	A B	1783·808	157·60	1	A B	1783·808	13·90	1	OΣ. 447.
			A C	·808	49·45	1	A C	1783·808	25·97	1	
2873	111	Orionis									Σ. 758.
9660	112	Cygni	A B	1783·726	161·0	1	A B	1783·726	10·13	1	Σ. 2588.
9916	113	Cygni	A B	1783·747	296·0	1	A B	1783·747	11·27	1	Σ. 2630 rej.
			A C	·747	32·80	1	A C	·747	29·45	1	
			D E	·747	242·40	1	D E	·747	19·33	1	
3534 or 3548	114	Monocer.									Σ. 951 or 953.

CLASS IV.

Burnham Gen. Cat.	Class IV. No.	Star.		Date of Measures. A.D.	Angle of Position. ° dec.	No. of Meas.		Date of Measures. A.D.	Distance. "		Synonyms and Remarks.
713	1	1 Polaris α	A B	1781·964	203·30	1	A B	1779·884	19·69 −	1 :	Σ. 93.
				1782·457	202·62	1		1781·666	17·25	1	
				1802·170	208·28	1					
9144	2	20 Lyræ η	A B	1782·298	238·15	1*	A B	1781·731	25·70	1	Σ. 2487. * Pos. Angle possibly belongs to V. 42.
9861	3	64 Sagittarii	A B	1782·542	10 ±	est.	A B	1780·649	25 ±	est.	P. xix. 382.
1440	4	15 Persei η	A B	1781·964	290·08	1	A B	1780·586	26·	1 ::	Σ. 307.
1364	5	33 Arietis	A B	1781·786	2·77	1	A B	1779·742	12 ±	est.	Σ. 289.
								1781·786	25·53	1 ::	
8914	6	63 Serpen. θ					A B	1779·791	19·06	1	Σ. 2417.
								1780·537	19·38	1	
								1781·539	18·32	1	
8182	7	31 Draconis ψ^1					A B	1779·798	30 ±	est.:	Σ. 2241.
								1780·756	30 ±	est.	
								1781·682	28·23	1	
648	8	86 Piscium ζ	A B	1781·876	67·38	1	A B	1779·797	20 ±	est.	Σ. 100.
								1780·646	22·19	1	
570	9	74 Piscium ψ^1	A B	1779·827	170 ±	est.	A B	1779·827	27·50	1	Σ. 88. Identical with IV. 116.
2147	10	59 Tauri χ					A B	1779·827	15 +	est.	Σ. 528.
								1780·731	18·75	1 ::	

CLASS IV. (*continued*).

Burnham Gen. Cat.	Class IV. No.	Star.		Date of Measures. A.D.	Angle of Position. ° dec.	No. of Meas.	•	Date of Measures. A.D.	Distance. ″		Synonyms and Remarks.
9617	11	17 Cygni χ	A B	1783·723	*n f*		A B	1779·884	15 ±	est.	Σ. 2580.
								1781·679	24·87	1	
								·758	22 ±	est.	
12257	12	91 Aquarii ψ	A B	1782·654	320 ±	est.	A B	1779·862	60 ±	est.	S. 827.
				·739	314·70	1		1780·649	25..30	est.	
								1781·635	23·08	1	
5779	13	83 Leonis	A B	1782·088	144·93	1	A B	1780·263	20 ±	est.	Σ. 1540.
								1781·392	25..30	est.	
								1782·285	29·50	2	
9707	14	57 Aquilæ	A B	1781·824	188·08	1	A B	1781·824	29·47	1	Σ. 2594.
6303	15	Camelop. 212					A B	1780·586	20 ±	1 ::	Σ. 1694.
								1781·665	20·08	1	
601	16	31 Cassiop. A					A B	1780·568	30 ±	est.	P.O. 293.
								·602	20 ±	est.	
6313	17	12 Canum α	A B	1782·293	228·22	1	A B	1780·600	20 ±	1 ::	Σ. 1692.
								1783·041	18·32	1	
10732	18	61 Cygni	A B	1781·821	53·53	1	A B	1780·720	16·42	1	Σ. 2758.
				·906	54·10	1		1781·747	16·12	1	
								1783·022	16·47	1	
2591	19	14 Aurigæ	A B	1781·827	232·37	1	A B	1780·731	15 ±	est.	Σ. 653.
								1781·827	16·13	1 ::	
8906	20	47 Draconis *o*	A B	1780·756	360	est.	A B	1780·756	20 ±	est.	Σ. 2420.
								1781·679	26·65	1	
2902	21	50 Orionis ξ	A B	1781·906	6·58	1 ::	A B	1780·775	60 ±	est.	Σ. 774.
								1781·747	25 ±	est.	
10670	22	59 Cygni *f*²					A B	1780·821	15 ±	est.	Σ. 2743.
								1781·786	18·18	1	
10309	23	Cygni	A B	1782·455	277·38	1	A B	1780·821	30 −	est.	P. xx. 199, S. 755.
10315	24	Cygni ω³	A B	1782·455	314·32	1	A B	1780·821	30 −	est.	Triple. h. 1534.
			A C		*p*				30 −	est.	
1149	25	66 Ceti	A B	1783·003	235·240	est.	A B	1780·975	16·87	1 ::	Σ. 231.
4480	26	19 Argûs						1782·797	VI.		P. viii. 11.
6212	27	24 Comæ	A B	1782·293	273·47	1	A B	1781·159	16·20	1	Σ. 1657.
								·392	20·60	1	
3435	28	20 Geminorum	A B	1782·285	213·00	1	A B	1781·194	30 ±	est.:	Σ. 924, =IV. 46.
								1782·285	19·67	1	
5104	29	23 Ursæ	A B	1782·301	273·23	1	A B	1781·312	20±	est.	Σ. 1351.
				·416	270 ±	est.		1782·301	19·43	1	
			A C	1792·249	*s p*		A C	1792·249	39 ±	est.	
5145	30	Lyncis					A B	1781·397	24·88	1 ::	Σ. 1369.
11778	31	Cephei					A B	1781·400	20 ±	est.	Probably OΣ. 473.
8136	32	61 Ophiuchi	A B	1782·655	90	est.	A B	1781·534	19·07	1 :	Σ. 2202.
9342	33	Aquilæ					A B	1781·545	21·98	1 ::	
9915	34	Aquilæ					A B	1781·561	30 ±	es ::	Near 64 Aquilæ. D.M. −1° 3896.
10363	35	6 Delphini β	A B	1781·580	350 ±	est.	A B	1781·580	25·90 +	1	Σ. 2704.
				1784·674	360 −	est.					

CLASS IV. (*continued*).

Burnham Gen. Cat.	Class IV. No.	Star.		Date of Measures. A.D.	Angle of Position. ° dec.	No. of Meas.		Date of Measures. A.D.	Distance. "		Synonyms and Remarks.
7386	36	28 Serpentis β	A B	1781·728	270 +	est.	A B	1781·613	20 ±	est.	Σ. 1970.
				·805	266 +	est.		·728	20 −	est.	
				1782·671	270	est.		·805	24 ±	est.	
10829	37	7 Equulei δ	A B	1781·805	78·35	1	A B	1781·613	25 ±	est.	Σ. 2777; S. 782.
				1785·594	*n f*	*		1781·805	19·53	1	* "A few degrees *n f*" (80 to 90).
11110	38	Aquarii					A B	1781·616	25 ±	est.	Σ. 2811 rej.
10902	39	Cygni	A B	1781·821	120·47	1	A B	1781·747	18..19	est.	h 614, OΣ. 434.
1183	40	Trianguli					A B	1781·767	17·15	1	P. ii. 38 +39; Σ. 239.
8162	41	86 Herculis μ	A B	1781·772	240 ±	est.	A B	1781·772	18 ±	1	Σ. 2220.
				1784·578	*s p*						
8279	42	Herculis	A B	1781·805	274·97	1	A B	1781·772	18·32	1	Σ. 2259.
2546	43	Eridani	A B	1809·044	83·32	1					Σ. 649; H.N. 46.
1750	44	Tauri									Σ. 409 rej.; D.M. +11° 487.
2771	45	Orionis	A B	1783·726	152·40	1	A B	1783·726	20·05	1	Σ. 721.
3435	46	20 Geminor.					A B	1782·079	24 ±	est.	Σ. 924.
								·887	25 ±	est.	=IV. 28.
5105	47	3 Leonis	A B	1783·000	105 ±	est.	A B	1782·107	18 ±	est.	
								·846	24 ±	est.	
3093	48	Geminorum	A B	1783·810	262·55	1	A B	1782·099	15 ±	est.	D.M. +23·1148.
								1783·000	20·45	1	
5944	49	Virginis	A B	1783·019	225 ±	est.	A B	1783·019	27·47	1	Σ. 1575.
				·340	213·50	1					
6147	50	17 Virginis	A B	1782·099	330	est.	A B	1783·019	20·15	1	Σ. 1636.
				1783·340	328·35	1					
6337	51	44 Virginis *k*	A B	1782·099	*n f*	est.	A B	1782·099	14 ±	est.	Σ. 1704.
				1783·340	57·50	1		1783·019	22 ±	est.	
								·340	22·28	1:	
4763	52	48 Cancri ι	A B	1783·145	309·90	1	A B	1782·104	30 +	est.	Σ. 1268.
								·988	29·90	1*	* Central measure.
4249	53	80 Gemin. π	A B	1782·865	210 ±	est.	A B	1782·112	18 ±	est.	Σ. 1135.
				1788·096	*s p*	est.		·865	40 −	est.	
			A C	1782·865	*n p*			·988	21·50	1	
							A C	·865	Cl. VI.		
4709	54	Hydræ	A B	1783·101	30·60	1	A B	1782·112	14 ±	est.	Σ. 1255.
				·337	30·60	1		1783·101	28·08	1	
								·337	23·35 +	1	
5090	55	Lyncis	A B	1783·173	320·80	1:	A B	1782·173	12 ±	est.	S. 598.
								1783·173	15·87	1	
7077	56	18 Libræ	A B	1783·030	45·75	1	A B	1783·030	17·98	1	Σ. 1894.
				1784·329	*n f*	est.					
6459	57	Comæ	A B	1782·285	223·48	1	A B	1783·082	17·08	2	Σ. 1737.
6289	58	Comæ	A B	1783·153	202·05	1	A B	1783·153	15·87	1	Σ. 1685.
8688	59	Lyræ	A B	1783·652	303·95	1	A B	1783·808	22·33	1 ::	Σ. 2354 rej.
			C A	1783·162	210·80	1	C A		150 ±	est.	C. is α Lyræ itself.
4706	60	Ursæ					A B	1783·233	30 ±	est.	Not identifiable.

CLASS IV. (*continued*).

Burnham Gen. Cat.	Class IV. No.	Star.		Date of Measures. A.D.	Angle of Position. ° dec.	No. of Meas.		Date of Measures. A.D.	Distance. "		Synonyms and Remarks.
7344	61	Coronæ	A B	1783·225	85·05	1	A B	1783·225	16·77	1	Σ. 1964.
7672	62	Herculis	A B	1783·074	182 ±	est.	A B	1782·608	30 ±	est.	Σ. 2063.
				·271	197·75	1		1783·074	24 ±	est.	
								·271	16·85	1	
7714	63	42 Herculis	A B	1783·646	93·70	1	A B	1782·608	30 ±	est.	Σ. 2082.
				1788·279	90 +	est.		1783·233	21·52	1	
1383	64	Persei	A B	1783·181	212·05	1	A B	1783·181	21·98	1	Σ. 292.
12411	65	Cassiop.	A B	1783·657	228·80	1	A B	1783·183	20·77	1	Σ. 3022.
576	66	Cassiop. 106	A B	1783·047	76·80	1	A B	1782·654	24·03	1	L.L. 1901.
8441	67	40 Draconis	A B	1782·778	235·55	1	A B	1782·778	20·85	1	Σ. 2308. Observed
								·991	20·45	1	by FLAMSTEED.
			A C	1782·778	120 ±	est.	A C	1782·778	196·55	1	
574	68	77 Piscium	A B	1783·145	85·20	1	A B	1783·145	29·60	1	Σ. 90.
8	69	Andromedæ	A B	1783·643	340·60	1	A B	1782·674	15 ±	est.	Probably Σ. 3064 rej.
								1783·643	21·97 −	1	
274	70	51 Piscium	A B	1783·630	89·40	1	A B	1782·674	20·57	1	Σ. 36.
								1783·630	22·48	1	
10246	71	12 Capric. *o*	A B	1783·504	230 ±	est.	A B	1783·504	23·50	1	Sh. 324.
				·621	239·25	1					
2160	72	Persei	A B	1783·657	62·60	1	A B	1782·682	17 ±	est.	Σ. 533.
								1783·657	16·85	1	
2338	73	Camelop.	A B	1783·189	185·00	1	A B	1783·189	19·53	1	Σ. 587.
2201	74	Tauri	A B	1782·997	70..75	est.	A B	1782·682	16·52	1	Σ. 545.
				1783·668	64·25	1					
2209	75	Tauri	A B	1783·123	151·60	1	A B	1783·123	22·60	1	Σ. 549.
298	76	Ceti	A B	1783·649	49·60	1	A B	1782·687	17..18	est.	Σ. 39.
								1783·649	18·58	1	
653	77	Ceti	A B	1782·737	*n p*		A B	1782·966	20 +	est.	Σ. 101.
				·969	*n p*			·969	19..20	est.	
				1783·649	333·40	1		1783·649	19·60	1	
10412	78	Cephei	A B	1783·224	49·40	1	A B	1783·224	19·53	1	Σ. 2712 rej.
11323	79	Cephei 147	A B	1783·622	192·20	1	A B	1782·969	21·22	1	Σ. 2840.
3360	80	Canis	A B	1782·745	90 ±	est.	A B	1783·058	17·98	1	S. 518.
				1783·058	87·60	1					L.L. 12304.
3503	81	6 Canis ν^1	A B	1782·745	270 ±	est.	A B	1782·997	18·32	1	Sh. 73.
				·997	270 ±	est.					
11617	82	Cephei	A B	1783·203	349·30	1	A B	1783·203	28·08	1	Σ. 2893.
553	83	26 Ceti	A B	1782·750	255·40	1	A B	1782·750	17·08	1	Σ. 84.
2692	84	23 Orionis *m*	A B	1783·208	30·45	1	A B	1782·988	32·80	1*	Σ. 696.
											* Central measure,
								1783·728	26·42	3	a little inaccurate.
12068	85	16 Lacertæ	A B	1782·756	10 ±	est.	A B	1783·687	20·45	1	Σ. 2960.
				1783·687	349·55	1	A C	1783·266	55·78	2	
			A C	1783·687	45·60	1					
			A D	1782·756	15 ±	est.					
11839	86	8 Lacertæ	A B	1783·627	185·50	1	A B	1783·263	17·23	1	Σ. 2922.
2673	87	Orionis 82	A B	1783·022	1..2	est.	A B	1782·756	32 ±	est.	Σ. 692.
				·737	7·70	1		1783·022	29·30	1	

CLASS IV. (*continued*).

Burnham Gen. Cat.	Class IV. No.	Star.		Date of Measures. A.D.	Angle of Position. dec.	No. of Meas.		Date of Measures. A.D.	Distance. "		Synonyms and Remarks.
1761	88	7 Tauri	A B	1783·134	66·75	1		1783·134	19·83	1	Σ. 412.
1730	89	Arietis	A B	1783·732	152·00	1	A B	1783·732	20·05	1	Σ. 399 rej.
7362	90	18 Urs. Min. π^1	A B	1783·504	86·80	1	A B	1782·778	30 −	1	Σ. 1972.
								·909	28	1	
								1783·504	26·40	1	
4250	91	2 Argûs	A B	1783·134	339·20	1	A B	1783·134	17·38	1	Σ. 1138.
10361	92	Delphini	A B	1783·646	288·45	1	A B	1782·909	26 ±	est.	Σ. 2703.
								1783·550	30·67 −	1 ::	Triple.
								1783·646	21·55 +	1	
8963	93	Lyræ	A B	1783·627	246·00	1	A B	1783·233	19·83	1	Σ. 2431.
8748	94	Lyræ	A B	1782·797	90 ±	est.	A B	1782·797	20 ±	est.	Σ. 2372.
				1783·627	84·60	1		1783·233	22·88	1	
4034	95	Monocerotis					A B	1783·153	20·45	1 ::	Probably Σ. 1084.
4219	96	Monocerotis	A B	1783·033	246·00	1	A B	1782·797	19..20	est.	Σ. 1132.
								1783·033	18·32	1	
4456	97	29 Monoc.	A B	1783·115	105·20	1	A B	1782·797	30 ±	est.	Σ. 1190.
								1783·115	29·90	1	
3043	98	Orionis					A B	1783·022	17·98	1	Σ. 817.
9655	99	Sagittæ	B C	1783·649	90·00	1	B C	1783·255	21·37 −	1	D.M. + 17°, 4110.
			A B	1783·649	259·40	1	A B	1783·649	Cl. V.		
9800	100	13 Sagittæ χ	A B	1783·649	259·80	1	A B	1783·257	23·03	1 ::	Σ. 2608 rej.
9797			A C	1783·649	290..285	est.	A C	1783·266	60 +	est.	C. is χ Sagittæ.
2699	101	Aurigæ	A B	1783·731	346·00	1	A B	1783·257	25·48	1	Σ. 698.
3653	102	59 Aurigæ	A B	1783·224	219·95	1	A B	1782·846	24..25	est.	Σ. 974.
								1783·224	23·50	1	
10951	103	Draconis	A B	1783·096	44·20	1	A B	1782·865	25 ±	est.	Σ. 2796.
								1783·096	22·58	1	
1025	104	Andromedæ	A B	1783·622	67·45	1	A B	1782·865	18..19	est.	Σ. 190 rej.; IV. 128.
								1783·622	18·95	1	
6183	105	7 Corvi δ	A B	1783·033	216·00	1	A B	1783·033	23·50	1	Sh. 145.
				1802·235	215·70	1					
5686	106	Ursæ	A B	1783·337	130 ±	est.	A B	1783·337	18·92	1 ::	Σ. 1513 rej.
				·808	134·55	1					
12552	107	Pegasi	A B	1783·621	39·65	1	A B	1782·972	28..29	est.	Σ. 3039.
								1783·621	26·20	1 :	
6040	108	Ursæ	A B	1783·337	79·80	1	A B	1783·260	14·78	1	Σ. 1603.
								·337	19·25	1 ::	
2162	109	62 Tauri	A B	1782·997	289 ±	1	A B	1782·997	28·08	1*	Σ. 534.
				1783·739	291·20	1					* Central measure.
2758	110	Tauri	A B	1782·977	355 ±	est.	A B	1782·977	16·08	1*	Σ. 719.
				1783·739	344·90	1					* Central measure.
				1790·865	345 ±	est.					
4800	111	Cancri	A B	1783·134	119·00	1	A B	1783·134	17·23	1	Σ. 1283.
5796	112	Crateris	A B	1783·340	148·70	1	A B	1783·000	26·25	1	S. 627.
10699	113	Cygni	A B	1783·758	298·40	1	A B	1783·750	17·50	1	P. xx. 452; Σ. 2748 rej.
6084	114	Virginis 75	A B	1782·019	295..298	est.	A B	1782·019	23·34	2	Σ. 1616.
				1783·340	285·90	1					

CLASS IV. (*continued*).

Burnham Gen. Cat.	Class IV. No.	Star.		Date of Measures. A.D.	Angle of Position. ° dec.	No. of Meas.		Date of Measures. A.D.	Distance. "		Synonyms and Remarks.
7553	115	Herculis	A B	1783·646	46·20	1	A B	1783·260	20·9	1	Σ. 2024.
570	116	74 Piscium ψ^1	A B	1783·033	158·35	1	A B	1783·033	28·98	1	IV. 9 ; Σ. 88.
2173	117	Eridani	A B	1783·134	238·20	1	A B	1783·134	19·53	1	Σ. 539 rej. ; h. 342.
4741	118	Cancri	A B	1783·096	65 ±	est.	A B	1783·096	24·10	1	Σ. 1266.
6464	119	Virginis	A B	1783·178	306·90	1	A B	1783·178	21·82	1	Σ. 1736 rej.
641	120	Piscium	A B	1783·156	249·00	1	A B	1783·156	18·32	1	Σ. 98.
7581	121	20 Scorpii σ	A B	1783·162	270 +	est.	A B	1783·162	21·67	1	Sh. 224.
7836	122	Herculis 192	A B	1783·630	244·95	1	A B	1783·233	21·05	1	Σ. 2115.
7758	123	19 Ophiuchi	A B	1783·627	93·15	1	A B	1783·255	20·45	1	Σ. 2096.
7579	124	Ophiuchi	A B	1783·225	27·10	1	A B	1783·225	15·40	1	P. xvi. 48 ; Sh. 226.
2959	125	29 Camelop.	A B	1783·742	137·60	1	A B	1783·268	22·43	1 ::	h. 2278.
11542	126	Cephei	A B	1783·621	315·65	1	A B	1783·298	18·85	1	Sh. 338.
9026	127	Aquilæ 39	A B	1783·600	339·90	1	A B	1783·600	16·55	2	Communicated by Mr. PIGOTT. Σ. 2447.
1025	128	Andromedæ	A B	1783·570	65·80	1	A B	1783·570	15·70	1	Σ. 190 ; same as IV. 104.
1125	129	59 Androm.	A B	1783·570	34·85	1	A B	1783·570	15·25	1	Σ. 222.
			A C		210..212	est.					
794	130	Piscium	A B	1783·635	27·75	1	A B	1783·635	15·82	1	Σ. 132.
813	131	100 Piscium	A B	1783·583	85·00	1	A B	1783·583	15·87	1	Σ. 136.
9556	132	Aquilæ	A B	1783·594	312·40	1	A B	1783·594	22·73	1	P. xix. 250 ; Σ. 2567.

CLASS V.

Burnham Gen. Cat.	Class V. No.	Star.		Date of Measures. A.D.	Angle of Position. ° dec.	No. of Meas.		Date of Measures. A.D.	Distance. "		Synonyms and Remarks.
7922	1	65 Herculis δ	A B	1781·805	162·47	1	A B	1779·764	34·69	1	Σ. 3127.
								1780·534	33·75	1	
8788	2	6 & 7 Lyræ ζ	A B	1782·298	152·30	1	A B	1781·728	41·97	1	Sh. 279.
8868	3	10 Lyræ β	A B	1782·359	150·47	1	A B	1781·712	30..35	est.	Sh. 281.
								·893	43·95	1	Quadruple.
11772	4	27 Cephei δ	A B				A B	1781·487	30 ±	est.	Sh. 347.
								·679	38·30	1	
9374	5	Cygni β	A B	1781·821	53·53	1	A B	1781·679	39·70	1	Sh. 297.
				1783·096	56·20	1		·761	29	est.	
								1783·096	34·84	2	
7533	6	14 Scorpii ν	A B	1782·293	334·85	1	A B	1780·334	40..60	est.	Sh. 220.
								1781·392	38·33	1	
8413	7	13 Sagittar. μ					A B	1780·646	30 ±	est.	h. 2822.
7514	8	7 Herculis κ	A B	1781·824	7·62	1	A C	1780·580	20 ±	est.	Σ. 2010.
								1781·728	30 −	est.	
								1782·463	39·98	1	

CLASS V. (*continued*).

Burnham Gen. Cat.	Class V. No.	Star.		Date of Measures. A.D.	Angle of Position. ° dec.	No. of Meas.		Date of Measures. A.D.	Distance. ″		Synonyms and Remarks.
6802	9	21 Bootis ι	A B	1782·293	37·15	1	A B	1779·737	16..20	es ::	Sh. 175.
								·767	36·56	1	
								1781·685	33·42	1	
2796	10	34 Orionis δ	A B	1781·906	358·17	1	A B	1779·816	20 ±	es ::	Sh. 60.
								1780·775	52·97	1	
8076	11	24, 25 Drac. ν	A B	1781·824	314·32	1	A B	1779·797	75 ±	es ::	Sh. 250.
								1781·665	54·80	1	
1028	12	9 Arietis λ	A B	1781·827	48·00	1	A B	1781·624	60 : ::	est.	Sh. 23.
								·827	36·62	1 :	
2130	13	52 Tauri ϕ					A B	1780·731	55·63	1 ::	Sh. 40.
3150	14	Monocerotis					A B	1779·925	60 −	est.	Unidentifiable. Multiple.
4962	15	16 Ursæ c	A B	1782·301	190·15	1	A B	1780·334	15..20	es ::	
								1781·312	15 ±	es ::	
								1782·301	48·99	1	
573	16	76 Piscium σ^2	A B	1781·87±	285·47	1	A B	1780·660	48·13 +	1	S. 393.
								1781·679	48 ±	est.	
329	17	29 Androm. π					A B	1780·646	25..30	est.	Sh. 4.
								1781·550	34·20	1	
361	18	18 Cassiop. α	A B	1781·964	275·43	1	A B	1780·665	52·81	1	Sh. 5, h. 1993.
								1781·964	56·17	1	
7596	19	20 Herculis γ	A B	1782·545	250·50	1	A B	1780·676	30 ±	est.	Sh. 227.
				1783·090	247·50	1		1781·731	25 ±	est.	
				1791·392	*s p*			1782·545	41·82	1 :	
								1783·090	37·08	1	
10932	20	1 Pegasi ϵ	A B	1781·821	308·32	1	A B	1780·685	30 ±	est.	S. 787.
								1781·821	40·75	1	
2968	21	29 Aurigæ τ					A B	1780·737	30 ±	est.	
								·747	30 ±	est.	
2627	22	15 Aurigæ λ					A B	1780·747	20..30	est.	4 or 5 near 2 of which 20 – 30″ fr. each other.
3266	23	Orionis	A B	1793·126	225 ±	est.	A B	1780·775	40 ±	est.	D.M. +15° 1139.
655	24	37 Ceti	A B	1783·649	332·60	1	A B	1780·780	42·81	1	Sh. 17.
								1782·737	44·09	2	
2639	25	20 Orionis τ	A B	1781·986	90 −	est.	A B	1780·810	30 ±	est.	
5131	26	6 Leonis h	A B	1782·293	77·08	1	A B	1781·140	34·06	1	Sh. 107.
								1782·088	30 ±	est.	
								·266	36·15	1	
7268	27	Libræ	A B	1782·359	130·28	1	A B	1782·359	44·42	1	Sh. 202.
11074?	28	Cephei					A B	1781·372	30 ±	est.	"Near β Cephei." No other indication to be found.
7957	29	53 Serpentis ν					A B	1781·537	30..35	est.	Sh. 247.
8067	30	53 Ophiuchi	A B	1782·375	192·80	1	A B	1781·545	32·35 +	1	Sh. 249.
								·635	30 ±	est.	
9241	31	Aquilæ					A B	1781·545	30 ±	est.	Several pairs near.

CLASS V. (*continued*).

Burnham Gen. Cat.	Class V. No.	Star.		Date of Measures. A.D.	Angle of Position. ° dec.	No. of Meas.		Date of Measures. A.D.	Distance. ″		Synonyms and Remarks.
19	32	21 Androm. α	A B	1781·964	259·38	1	A B	1781·578	60 ±	est.	
								·964	55·70	1	
9005	33	15 Aquilæ h					A B	1781·561	33·88	1 :	Sh. 286.
9207	34	28 Aquilæ A					A B	1781·561	35 ±	est.	S. 717.
10170	35	Aquilæ					A B	1781·561	40 ±	est.	P. xx. 116 ; Σ. 2677.
8725	36	2 Aquilæ					A B	1781·575	42·73	1 :	
7570	37	18 Coronæ υ					A B	1781·712	48 ±	est.	Sh. 223.
							A C	1781·712	90 ±	est.	
7612	38	23 Herculis	A B	1783·025	21·35	1	A B	1781·72	36·45	1 :	S. 678.
8692	39	3 Lyræ α	A B	1782·359	116·77	1	A B	1781·728	37·73	1	Sh. 272.
				1792·312	116·23	1		1792·312	42·99	1	
8862	40	8 Lyræ ν^1	A B	1782·463	118·62	1	A B	1781·728	35..40	est.	S. 706.
			A C	1781·731	*n p*			·731	35..40	est.	
								1782·463	56·78	1	
1933	41	43 Persei A					A B	1781·728	40..60	est.	S. 440.
9141	42	Lyræ	A B	1782·463	63·70	1	A B	1781·731	30 ±	est.	Sh. 289.
								1782·463	38·13	1	
11184	43	76 Cygni					A B	1781·747	48 ±	est.	S. 796.
10983	44	69 Cygni	A B	1781·747	*p* *						S. 790.
				1783·709	*f* *						* *Sic* in MSS. Obs. of 1781, Oct. 1, is not in the Journal (only V. 45). 1783, Sept. 17, only one companion mentioned, following.
			A C	1781·747	*p* *						
10867	45	Cygni	A B	1781·747			A B	1781·747	43..45	est.	
9560	46	16 Cygni c^1					A B	1781·758	30 ±	est.	Sh. 299.
9854	47	26 Cygni	A B	1793·679	135 ±	est.	A B	1781·767	39 ±	est.	
12387	48	Piscium					A B	1781·767	45 ±	est.	D.M. +5° 5175, 5174.
1332	49	30 Arietis					A B	1781·786	31·73	1	Sh. 32.
								1782·972	34·20	1	
2948	50	13 Leporis γ	B C	1782·805	300..310	est.	B C	1781·805	40 ±	est.	S. 498. Is also VI. 40, which see for A B.
9498	51	Sagittæ					A B	1781·682	27·50	2	=II. 32 =H.N. 84.
								·893	32·80	1	
3383	52	15 Geminor.					A B	1782·887	35·	1 :::	H.V. 56 ; Sh. 70.
4059	53	63 Geminor.					A B	1782·167	40 ±	est.	Sh. 368.
								·887	33 ±	est.	
								1783·000	44·25	1 :	
4984	54	22 Hydræ θ	A B	1782·988	165 ±	est.	A B	1782·112	60 −	est.	h. 2489 ; H. VI. 108.
3288	55	Geminorum					A B	1782·079	43 ±	est.	"Near 12 Geminorum." Not identifiable.
								83·077	60 −	est.	
							A C	83·077	60 −	est.	
3383	56	15 Geminor.	A B	1783·000	210 ±	est.	A B	1782·079	28 ±	est.	Sh. 70.
								1783·000	32·65	1	Identical with V. 52.

CLASS V. (*continued*).

Burnham Gen. Cat.	Class V. No.	Star.		Date of Measures. A.D.	Angle of Position. ° dec.	No. of Meas.		Date of Measures. A.D.	Distance. ″		Synonyms and Remarks.
2448	57	Orionis 26	A B	1783·728	303·60	1	A B	1782·093	30 −	est.	P. iv. 257. Sh. 49.
			A C	1783·728	f ±	est.		·843	34 ±	est.	Same as V. 113.
								1783·728	36·43	1	
5154	58	7 Leonis	A B	1783·096	81·40	1	A B	1782·093	16 ±	e. ::	Sh. 108.
								·846	35..36	est.	
								1783·096	42·42	1	
4655	59	31 Cancri θ	A B	1782·099	n f	est.	A B	1783·055	44·87	1 ::	h. 2452.
5959	60	Leonis	A B	1783·096	19·20	1	A B	1782·107	20 ±	e. ::	Sh. 132.
								·129	40 ±	est.	
								1783·096	37·25	1	
5775	61	81 Leonis					A B	1782·107	38 ±	est.	h. 4433.
								1783·096	57·38	1	
5590	62	Leonis					A B	1782·112	20 ±	est.	Probably S. 617.
								1783·096	33·27	1	
5269	63	Leonis 91	A B	1795·950	330..340	est.	A B	1783·096	52·17	2	
5412	64	Leonis 155					A B	1783·096	59·67	1	Sh. 115.
3713	65	17 Canis	A B	1783·000	140..150	est.	A B	1783·115	44·87	1	S. 540.
				·115	154·20	1					Quadruple.
4056	66	Geminorum	A B	1783·000	271..272	est.	A B	1782·167	35 ±	est.	S. 548.
								1783·000	34·65	1	
4258?	67	Geminorum					A B	1783·162	47·62	1	S. 560 (no other seen by β, but distance 89″).
5684	68	Leonis					A B	1783·162	54·62	1	Triple. D.M. +3° 2463, 2462.
5113	69	7 Leon. Min.					A B	1783·101	58·30	1	h. 1166.
6512	70	Bootis	A B	1783·178	263·00	1	A B	1783·178	56·93	1	h. 2657; OΣ. 268.
3496	71	Geminorum									VI. 91. North of γ Gemin.
7711	72	36 & 37 Hercul.	A B	1783·090	233·05	1	A B	1783·090	67·77	1	Σ. 2074 rej. Sh. 234.
								·271	59·98	1	
4930	73	14 Ursæ τ	A B	1782·441	n f		A B	1783·263	54·77	1	
				1783·263	45 ±	est.		·670			
				·679	45 ±	est.					
8369	74	Ophiuchi	A B	1783·627	129·25	1 ::	A B	1782·455	30 −	est.	
								1783·233	40·90	1	
7479	75	Coronæ	A B	1783·225	106·00	1	A B	1783·225	41·20	1	D.M. +26° 2767.
11026	76	22 Aquarii β	A B	1783·600	325·80	1	A B	1783·600	33·27	1 ::	h. 936.
							A C	1785·695	60 ±	est.	
9168	77	Sagittarii 215	A B	1783·621	168·75	1	A B	1783·621	36·05	1 :::	P. xix. 43.
8965	78	38 Sagittar. ζ	A B	1783·575	297·80	1 :::	A B	1782·589	Cl. V.		B.A.C. 6489.
				·646	298·10	1 :					
12727	79	9 Cassiop.	A B	1783·657	320·60	1	A B	1783·263	52·65	1	See Vol. I. p. 211.
11967	80	69 Aquarii τ	A B	1783·000	100..105	est.	A B	1783·000	34·50	1	Σ. 2943.
				·600	109·90	1		·600	36·78	1	
697	81	35 Cassiop.	A B	1783·263	0·00	est.	A B	1783·263	42·58	1	S. 397.
				·657	4·80	1					

CLASS V. (*continued*).

Burnham Gen. Cat.	Class V. No.	Star.		Date of Measures. A.D.	Angle of Position. ° dec.	No. of Meas.		Date of Measures. A.D.	Distance. "		Synonyms and Remarks.
417	82	Cassiop.	A B	1783·047	85 ±	est.	A B	1783·047	43·43	1	Sh. 7; D.M. +50° 141.
				·657	82·20	1					
732	83	36 Cassiop. ψ	A B	1782·654	s f		A B	1783·263	33·42	1	Σ. 117.
				1783·622	100·20	1					
1094	84	Cassiop.	A B	1783·657	273·55	1	A B	1783·602	50·97	1	S. 405, OΣ. 22.
144	85	Andromedæ	A B	1783·041	10·60	1	A B	1782·657	30 ±	est.	L.L. 335; S. 384.
								·838	25..28	est.	
								1783·630	30·95	1	
7241	86	12 Urs. Min.	A B	1783·260	90 +	est.	A B	1782·255	60 −	est.	B.A.C. 5078.
			A C	·260	90 + +	est.	A C	·255	60 −	est.	
10070	87	7 Capricor. σ	A B	1783·600	175·20	1	A B	1782·676	50·12	1	Sh. 380.
2630	88	Aurigæ	A B	1783·676	215·90	1	A B	1783·183	33·27	1 :::	D.M. +39·1250.
								·260	35·25 −	1 ::	
3074	89	37 Aurigæ θ	A B	1783·731	286·00	1	A B	1782·676	35·30 +	1	Sh. 68. Same as VI. 34.
2996	90	32 Aurigæ ν	A B	1783·183	208·20	1	A B	1783·183	53·72	1	
3073	91	Aurigæ 161	A B	1783·723	315·10	1	A B	1783·257	30·05	1 :	
946	92	Arietis	A B	1783·041	322·75	1	A B	1783·041	51·27	1	=AC of β 510.
8431	93	Herculis	A B	1783·646	135·70	1	A B	1783·260	47·77	1	D.M. +28° 2955, 2956.
11862	94	Cephei	A B	1783·200	135·25	1	A B	1783·200	41·67	1	OΣ. (App.) 236.
11691	95	51 Aquarii	A B	1782·805	n p	est.					AC of β 172.
			A C	1782·805	180..190	est.					
			A D	1782·805	120 ±	est.					
11841	96	Aquarii	A B	1783·635	250..255	est.	A B	1738·635	V. near		Not identifiable.
11877	97	10 Lacertæ	A B	1783·627	51·25	1	A B	1783·260	57·83	1	S. 813.
								·263	58·30	1	
								·630	52·57	1	
11103	98	3 Pegasi	A B	1782·972	360 ±	est.	A B	1782·756	33..34	est.	Sh. 331.
				1783·334	352·80	1		1783·334	34·72	1	The D. * II. 62 is near, to *n*.
11690	99	33 Pegasi	A B	1783·622	0·80	1	A B	1782·756	45 ±	est.	Σ. 2900.
								1783·622	45·05	1	
3079	100	59 Orionis	A B	1783·022	205 ±	est.	A B	1783·022	37·25	1	
2774	101	Orionis	A B	1783·022	105 ±	est.	A B	1783·022	44·25	1	S.D. −7°·1092.
1083	102	61 Ceti	A B	1783·170	190 ±	est.	A B	1783·170	34·50	est*	* "Too low to call it a measure."
				·649	193·65	1	A B	·649	37·88	1	
9042	103	Lyræ	A B	1783·627	60·80	1	A B	1783·627	45·53	1	L.L. 35845.
9444	104	Sagittæ	A B	1783·649	106·30	1					D.M. +15° 3877.
9875	105	Sagittæ	A B	1783·726	164·25	1	A B	1783·257	38·60	1	OΣ. 397 rej.
9705	106	Vulpeculæ	A B	1783·630	150·70	1	A B	1783·263	38·90	2	P. xix. 320; Sh. 310.
3585	107	56 Aurigæ	A B	1783·734	17·40	1	A B	1783·257	52·95	1	Sh. 75.
	108	Canis *h*	A B	1782·846	50..60	est.	A B	1783·033	42·88	1	B.A.C. 2251; Δ 36.
				1783·033	66·70	1					
4600	109	Cancri 64	A B	1783·145	325·00	1	A B	1783·145	35·40	1	h. 785.
2703	110	111 Tauri	A B	1783·740	273·80	1	A B	1782·997	46·70	1*	S. 478. * "Central measure."

Class V. (*continued*).

Burnham Gen. Cat.	Class V. No.	Star.		Date of Measures. A.D.	Angle of Position. ° dec.	No. of Meas.		Date of Measures. A.D.	Distance. "		Synonyms and Remarks.
5627	111	Ursæ	A B	1783·657	38·55	1	A B	1783·263	30·67	1	Σ. 1495 ; H.N. 97.
3455	112	Geminor.									S. 524.
											D.M. +22° 1386, 1384.
2448	113	Orionis 26	A B	1783·764	303·90	1	A B	1783·115	37·85	1	P. iv. 257 ; Sh. 49.
			A C	1783·764	90 –	est.					Also V 57.
2531	114	103 Tauri	A B	1782·931	200 ±	est.	A B	1782·931	45..50	est.	
				1783·791	197·60	1		1783·810	30·05	2	
2734	115	114 Tauri *o*	A B	1782·931	205 ±	est.	A B	1782·997	51·57	1 ::*	h. 365.
				·997	195..200	est.					*"Central measure."
				1783·739	192·10	1					
1450	116	41 Arietis	A B	1783·687	189·20	1	A B	1783·731	39·33	1	Sh. 36.
											A closer companion to VI. 5.
1590	117	Arietis	A B	1782·975	300..310	est.	A B	1782·975	34·80	1	Σ. 359 rej.
				1783·654	317·55	1					
2818	118	Orionis	A B	1782·988	270 ±	est.					D.M. –1° 949.
				1783·230	256·90	1					
2808	119	Orionis	A B	1782·988	*s p*		A B	1783·115	30·20	1	Σ. 734.
				1783·764	248·45	1					
			A C	1783·764	92..93	est.					
4828	120	15 Hydræ	A B	1783·033	340 ±	est.	A B	1783·033	43·03	1	
6148	121	12 Comæ	A B	1783·082	167 ±	est.	A B	1783·082	58·92	1	Sh. 143.
7103	122	Bootis 346	A B	1783·312	157·1	1	A B	1783·257	34·35	1	Sh. 189; OΣ. 291 rej.
415	123	Androm. 142	A B	1783·033	237·60	1	A B	1783·033	45·02	1	P.O. 175 –76 ; Sh. 6.
6657	124	Centauri					A B	1783·082	54·02	1*	Lacaille 4726. Principal star close D.
											*"Very inaccurate."
7165	125	Bootis	A B	1783·646	234·45	1	A B	1783·257	33·88	1	Sh. 196 ; D.M. +28° 2412, 2411.
7466	126	Herculis	A B	1783·635	217·90	1	A B	1783·260	37·88	1	Σ. 1993.
7731	127	Herculis	A B	1783·652	289·75	1	A B	1783·260	48·67	1	Sh. 235; D.M. +6° 3282, 3281.
6516	128	Virginis					A B	1783·260	41·96	1	Sh. 165.
6230	129	Virginis	A B	1783·162	96..97	est.	A B	1783·162	46·70	1	P. xii. 143 ; S. 639.
6296	130	35 Comæ	A B	1783·153	126·85	1	A B	1783·153	31·28	1	Σ. 1687.
				1785·279	*s f*						
7164	131	Libræ 97					A B	1783·257	47·77	1	Sh. 195; P. xv. 14.
7223	132	Libræ					A B	1783·255	39·98	1 :::	L.L. 27966.
7855	133	60 Herculis	A B	1783·178	*n p*		A B	1783·260	48·67	1	
				·635	307·00	1					
7545	134	Scorpii					A B	1783·257	45·78	1	P. xvi. 45 ; Sh. 225.
4203	135	Camelopardi	A B	1783·734	185·00	1	A B	1783·268	14·78	1*	P. vii. 159 + 160 ; Σ. 1122.
								1783·734	38·30	1	* " Perhaps a little inaccurate."
9961	136	Aquilæ	A B	1783·695	204·20	1	A B	1783·695	47·08	1	P.xx. 11 + 12; S. 735.
9609	137	Cygni	A B	1783·808	32·95	1	A B	1783·808	35·02	1	S. 725.

Class VI.

Burnham Gen. Cat.	Class VI. No.	Star.		Date of Measures. A.D.	Angle of Position. ° dec.	No. of Meas.		Date of Measures. A.D.	Distance. ″		Synonyms and Remarks.
1209	1	68 Ceti o	A B	1782·646	92·53	1	A B	1779·797	107·83	2	Sh. 362.
								·925	111·72	2	
								1780·011	104·69 +	1	
								·687	110·47	2	
								·717	110·	1	
								1781·624	107·90	1	
								·821	112·62	1	
								1782·646	114·60	1	
8284	2	67 Ophiuch. o					A B	1779·764	60 ±	est.	Sh. 255.
								1780·633	90 ±	est.	
								1781·641	50·30	1	
8907	3	11 Lyræ δ					A B	1781·893	240 ±	est.	
	4	Capric.	A B	1782·449	$s\ p$		A B	1781·624	75 ±	est.	F and unequal pair
								·682	75 ±	est.	p α^1 and α^2 Capr.
1450	5	41 Arietis	A B	1782·972	260 ±	est.	A B	1779·723	150 ±	est.	Sh. 36.
								1708·797	150 ±	est.	This is A.C. of V. 116.
								1781·786	125·58	1	
11077	6	39 Capric. ϵ					A B	1780·652	75 ±	est.	h. 3040.
								1781·624	75 ±	est.	
2313	7	94 Tauri τ					A B	1779·761	60 +	est.	S. 455.
								1780·731	61·42	1	
2177	8	65 & 67 Tauri κ									
3797	9	43 Geminor ζ	A B	1781·827	351·23	1	A B	1781·827	91·87	1	Sh. 77.
			A C	1782·088	f		A C	1782·088	92 ±	est.	
10036	10	31 Cygni o^2	A B	1781·827	177·23	1	A B	1781·800	110	est.	S. 742.
								1781·827	99·95	1	
5331	11	32 Leonis α	A B	1781·838	305.08	2	A B	1779·868	180 −	est.	Sh. 111.
								1781·838	168·33	1	
5790	12	84 Leonis τ	A B	1782·282	165·35	1	A B	1780·263	150 ±	est.	Sh. 125.
								1781·392	120 ±	est.	
								1782·282	82·70	1	
5967	13	95 Leonis o	A B	1782·449	10	est.	A B	1780·263	90 ±	est.	
8498	14	58 Serpen. η	A B	1780·465	f		A B	1780·465	60 ±	est.	
				1781·805	99·12	1		1781·805	81·03	1	
6647	15	Bootis					A B	1780·482	60 ±	es ::	=VI. 89.
								1781·685	120 +	est.	
								·624	60 +	est.	
7194	16	49 Bootis δ	A B	1782·457	84·23	1	A B	1780·559	135 ±	est.	Sh. 199.
7258	17	51 Bootis μ	A C	1781·808	170·42	1	A C	1780·578	127 ±	est.	Sh. 204.
								1781·786	128 ±	est.	Is also I. 17, which see for B.C.
7608	18	21 Coronæ ν	A B	1782·449	10 ±	est.					
				1789·326	10 ±	est.					
				1795·219	10·91	1					

Class VI. (*continued*).

Burn-ham Gen. Cat.	Class VI. No.	Star.		Date of Measures. A.D.	Angle of Position. ° dec.	No. of Meas.		Date of Measures. A.D.	Distance. "		Synonyms and Remarks.
1175	19	7 Persei χ									S. 409. Multiple. (In fact a cluster, see h. Neb. 512, VI. 33, of which it is a distant outlier.)
2073	20	51 Persei μ					A B	1780·586	120 ±	est.	Sh. 364.
								1781·728	75 ±	est.	
11924	21	44 Pegasi η					A B	1780·589	135 ±	est.	S. 816.
6510	22	Ursæ 426					A B	1780·600	210 ±	est.	P. xiii. 113; S. 649.
3152	23	Lyncis					A B	1780·600	120 ±	est.	Unidentifiable. "In the nose of Lynx."
12354	24	4 Cassiop. *d*.					A B	1780·613	120 ±	est.	
							A C	1780·613	105 ±	est.	
12349	25	3 Cassiop.					A B	1782·646	Cl. V.	est.	
							A C	1780·630	135 ±	est.	
9458	26	4 Sagittæ ε	A B	1782·301	81·47	1	A B	1781·641	78·85	1	S. 721.
								·893	91·87	1	
9960	27	Aquilæ					A B	1780·646	60 ±	est.	North of θ Aquilæ.
10112	28	9 Capric. β	A B	1782·449	*p*		A B	1780·597	180 ±	est.	S. 745.
10228	29	11 Capric. ρ					A C	1780·652	165 ±	est.	Sh. 323; II. 51, which see for A B.
2597	30	13 Aurigæ α	A B	1782·301	151·38	1	A B	1782·301	169·10	1	Sh. 51 (A C).
2267	31	88 Tauri *d*					A B	1780·731	70·63	1	Sh. 45.
10533	32	54 Cygni λ	A B	1781·906	102·70	1	A B	1780·720	120 ±	e. ::	S. 765.
								1781·720	60 ±	e. ::	
10060	33	32 Cygni					A B	1780·720	150 ±	e. ::	S. 743.
								1781·871	120 ±	e. ::	
3074	34	37 Aurigæ θ					A B	1780·737	150 ±	e. ::	Sh. 68. Same as V. [89.
2495	35	9 Aurigæ	A B	1783·334	62·20	1	A B	1780·737	120 ±	e. ::	
								1783·257	79·50	1	
2455	36	10 Camelop. β					A B	1780·747	90 ±	e. ::	S. 459.
8781	37	46 Draconis *c*					A B	1780·756	210 ±	e. ::	
9892	38	65 Draconis *e*					A B	1780·756	120 −	e. ::	
3048	39	58 Orionis α	A B	1781·882	152·30	1	A B	1781·882	161·77	2	
				1802·101	150·83	1					
2948	40	13 Leporis γ					A C	1782·682	85·90	1	S. 498. Is also V. 50, which see for A B.
4891	41	67 Cancri ρ	A B	1782·282	320·55	1	A B	1782·285	95·98	1	Sh. 101.
4233	42	78 Gemin. β	A B	1781·9±	65·53	1	A B	1781·9±	116·75	1	S. 559.
			A C	1781·9±	74·07	1	A C	1781·9±	160·70	1	
6405	43	51 Virginis θ	A C	1782·301	294·92	1	A C	1781·364	63·88	1	Σ. 1724; Sh. 160. Is also III. 50, which see for A B.

CLASS VI. (*continued*).

Burn-ham Gen. Cat.	Class VI. No.	Star.		Date of Measures. A.D.	Angle of Position. ° dec.	No. of Meas.		Date of Measures. A.D.	Distance. ″		Synonyms and Remarks.
7150	44	24 Libræ ι	A B	1782·359	112·50	1	A B	1781·392	59·07	1 ::	Sh. 376.
12470	45	Andromedæ					A B	1781·550	90 ±	e. ::	Not identifiable.
9657	46	53 Aquilæ α	A B	1781·821	334·73	1	A B	1781·556	160 −	e. ::	Sh. 308.
								·821	143·30	1	
9275	47	Aquilæ									D.M. +1° 3986–87.
9297	48	Aquilæ									D.M. +1° 3995–96.
8810	49	Aquilæ									One of the two following stars of a trapezium near *l* Aquilæ. Σ. 2391.
8827	50	Aquilæ									P. xviii. 197.
7068	51	1 Serpentis									
6867	52	Bootis					A B	1781·624	60 +	e. ::	Unidentifiable.
7121	53	Bootis					A B	1781·624	60 +	e. ::	Unidentifiable. ? D.M. +48° 2257.
..	54	Ophiuchi					A B	1781·633	60 +	e. ::	Unidentifiable; it is not III. 25.
12202	55	2 Cassiop.					A B	1781·679	120 +	e. ::	S. 823.
9186	56	21 Lyræ θ	A B	1782·449	*n f*		A B	1781·731	90 ±	e. ::	Sh. 292.
11208	57	79 Cygni					A B	1781·747	100 ±	e. ::	S. 799.
10559	58	4 Aquarii					A B	1781·758	60 +	est.	=I. 44.
9959	59	Cygni					A B	1781·758	73 ±	est.	
9854?	60	Cygni					A B	1781·767	88 ±	est.	? =V. 47.
12307	61	Piscium					A B	1781·767	60 ±	est.	Not identified.
							A C	1782·687	60 ±	est.	
							B C	1782·687	60 ±	est.	
12369	62	8 Piscium κ	A B	1783·041	0 ±	est.	A B	1781·767	120 −	est.	S. 830.
	63	Sagittæ	A B	1782·301	265·85	1	A B	1781·778	91·42	2*	* Diameters not included.
								·893	90·93	1	North following ε Sagittæ.
2248	64	Eridani	A B	1783·041	110 ±	est.	A B	1781·805	105 ±	est.	L.L. 8588.
								1783·041	112·00	1	
3542	65	15 Monocer.									Multiple.
2266	66	87 Tauri α	A B	1781·964	37·03	1	A B	1781·964	87·75	1	S. 452.
				1802·101	35·87	1					
2712	67	28 Orionis η	A B	1782·843	60 ±	est.	A B	1783·737	110·95	1	
				1783·737	54·80	1					
2719	68	Orionis	A B	1783·764	277·90	1	A B	1783·758	120·18	1	L.L. 10165.
1116	69	14 Arietis	A B	1783·657	281·20	1	A B	1781·986	66 ±	est.	S. 406.
								1782·887	105 ±	est.	
								1783·657	89·47	1	
4164	70	70 Gemin.					A B	1781·986	60 −	est.	
								1782·887	60 +	est.	
							A C	1782·887	60 +	est.	
5110	71	31 Hydræ τ	A B	1782·003	*s p*	est. :	A B	1782·988	61·67	1	Sh. 106.
				·988	357··358	est.					
				1783·340	358·60	1					

CLASS VI. (*continued*).

Burnham Gen. Cat.	Class VI. No.	Star.		Date of Measures. A.D.	Angle of Position. dec.	No. of Meas.		Date of Measures. A.D.	Distance. "		Synonyms and Remarks.
3206	72	68 Orionis	A B	1783·033	235 ±	est.	A B	1782·079	60	est.	
				·791	229·00	1		1783·033	61 +	est.	
								·791	72·83	1	
3568	73	27 Geminor. ϵ					A B	1782·142	66 ±	est.	S. 533.
								1783·049	110·5 −	1	
3893	74	51 Geminor.	A B	1782·096	40..50	est.	A B	1782·096	90 ±	est.	
			A C	1782·096	40..50	est.	A C	1782·096	120 ±	est.	
4383	75	4 Cancri ω^2	A B	1782·096	300 ±	est.	A B	1782·096	75 ±	est.	
							A C	1782·096	120 +	est.	
5175	76	14 Leonis o	A B	1783·077	40·40	1	A B	1783·077	63·48	1	Sh. 109.
6701	77	93 Virginis τ	A B	1793·361	p		A B	1782·969	68·37	1*	Sh. 171.
											* Central measure.
4499	78	Cancri 37					A B	1783·096	63·78	1	
5729	79	74 Leonis ϕ	A B	1783·000	280..282	est.	A B	1782·107	90 ±	est.	Sh. 121.
								1783·000	98·58	1	
5921	80	93 Leonis					A B	1782·293	82 ±	est.	Sh. 129.
								1783·096	70·22	1	
6242	81	27 Virginis					A B	1783·101	88·80	1	
4746	82	31 Monocer.	A B	1782·969	290 ±	est.	A B	1782·969	90 ±	est.	S. 579.
				1783·115	310·00	1		1783·115	70·22	1	
2379	83	Orionis	A B	1783·022	350 ±	est.	A B	1783·022	60 +	est.	D.M. +6° 765.
				·791	1·75	1		·791	80·97	1	
4361	84	14 Can. Min.	A B	1783·033	63·60	1	A B	1783·033	65·47	1	Sh. 87.
			A C	1782·107	170 ±						
5039	85	27 Hydræ	A B	1783·033	210 ±	est.					Sh. 105.
4822	86	51 Cancri	A B	1782·988	$n\ f$						S. 583.
4874	87	64 Cancri σ	A B	1783·134	295·20	1	A B	1783·134	85·75	1	Sh. 100.
3064	88	34 Aurigæ β	A B	1782·909	15 ±	es ::	A B	1782·173	120 ±	es ::	
				1783·791	35·80	1		1783·791	169·10	1	
			A C	1782·909	40..50	est.	A C	1782·909	350 ±	es ::	
6647	89	Bootis	A B	1782·909	$n\ f$ or $s\ p$		A B	1783·173	79·65	1	P. xiii. 219–220; S. 656; =VI. 15.
				1783·173	211·90	1					
6447	90	61 Virginis	A B	1783·000	345 ±	est. :	A B	1782·252	60 ±	est.:	
								1783·000	73·25	1	
3496	91	Geminor.									=V. 71.
9962	92	Capricorni	A B	1783·622	267·95	1	A B	1782·737	62·27	1	S.D. −12° 5663, 5662.
7480	93	15 Coronæ ρ	A B	1782·542	f		A B	1782·715	90 ±	est.	S. 676.
				1783·633	144·45	1		1783·271	87·73	1	
7442	94	12 Coronæ λ	A B	1783·646	56·80	1	A B	1782·715	120 ±	es ::	
								1783·646	95·23	1	
6670	95	8 Bootis η	A B	1782·586	$s\ f$		A B	1782·586	90 ±	es ::	Sh. 169.
				·909	120..130	est.					
				1783·550	115..120	est.					
1921	96	44 Persei ζ	A B	1783·058	203·40	1	A B	1783·058	71·43	1	S. 441.
			A C	1783·058	195..200	est.	A C	1783·058	100 ±	est.	
11985	97	71 Aquarii τ^2	A B	1782·654	280..285	est.	A B	1782·654	90 ±	es ::	S. 818.
				1783·622	288·50	1		1783·600	123·61	2	

Class VI. (*continued*).

Burnham Gen. Cat.	Class VI. No.	Star.		Date of Measures. A.D.	Angle of Position. ° dec.	No. of Meas.		Date of Measures. A.D.	Distance. "		Synonyms and Remarks.
2106	98	Tauri	A B	1783·126	313·80	1	A B	1783·126	62·57	1	P. iv. 24–25.
2239	99	57 Persei *m*	A B	1783·657	198·15	1	A B	1783·268	96·45	1	Sh. 44.
12038	100	Cephei	A B	1783·622	278·15	1	A B	1783·257	61·90	1	OΣ (App.) 238.
2183	101	68 Tauri δ	A B	1782·997	230 ±	est.	A B	1783·739	63·62	2	
				1783·739	234·60	1	A C	1782·997	Cl. VI.		
			A C	1782·997	320 ±	est.					
3338	102	5 Lyncis	A B	1783·687	272·00	1	A B	1783·271	88·33	1	S. 514.
11205	103	8 Pegasi ε	A B	1782·972	*n p*		A B	1782·972	90·93	1*	S. 798.
				1783·622	322·75						* Central measure.
6955	104	30 Bootis ζ	A B	1782·909	270±	est.	A B	1782·909	90 ±	es	Σ. 1865; Sh. 182. Is also Class I. = H.N. 114.
2528	105	105 Tauri	A B	1782·931	*p*		A B	1783·739	101·48	1	S. 466.
				1783·739	252·00	1					
2432	106	62 Eridani *b*	A B	1783·041	74·85	1	A B	1783·041	60·43	1	Sh. 48.
4737	107	Monocer. 231	A B	1782·969	140..150	est.	A B	1782·997	90 ±	est.	
				·988	140..150	est.					
				1783·115	140..150	est.					
4984	108	22 Hydræ θ	A B	1782·988	*p*						H.V. 54.
				1783·033	271..272	est.					
4597	109	22 Cancri ϕ^1									S. 566.
1158	110	Ceti	A B	1783·003	123·70	1	A B	1783·003	80·87	1	S.D. −3° 340, 341.
				·016	125·40	1					
5101	111	30 Hydræ α	A B	1783·019	*s f*		A C	1783·019	120 −	est.	
			A C	1783·019	*s f*		A C	1783·019	120 −	est.	
				·033	155 ±	est.	A D	1783·019	210 ±	est.	
			A D	1783·019	*f*						
6736	112	13 Bootis	A B	1783·633	277·40	1	A B	1783·019	60 ±	est.	
								·257	77·97	1	
5919	113	4 Virginis					A B	1783·318	145·73	1 ::	Sh. 131.
3192	114	Orionis	A B	1783·764	112·10	1	A B	1783·115	90·63	1	D.M. +15° 1087.
5924	115	Crateris	A B	1783·025	77·80	1					L.L. 22302.
7747	116	43 Herculis *i*	A B	1783·025	*p*		A B	1783·260	74·62	1	Sh. 239.
				·652	231·20	1					
7009	117	Libræ	A B	1783·025	230 ±	est.					S. 663.
4612?	118	30 Monocer. ??					A B	1783·115	210·90	1	See Vol. I. p. 222, note ‡.
11863	119	Piscis Austr.	A B	1783·570	157·75	1	A B	1783·570	86·97	1	h. 5356.—H.N. 117 (I. Class).
9216	120	Sagittarii 226	A B	1783·622	319·00	1	A B	1783·622	84·15	1	P. xix. 67.
11910	121	12 Lacertæ	A B	1783·627	17·00	1	A B	1783·627	60·17	1	S. 815.

145 *New Double Stars.*

Burnham Gen. Cat.	Novæ No.	Star.		Date of Measures. A.D.	Angle of Position. ° dec.	No. of Meas.		Date of Measures. A.D.	Distance. ″		Synonyms and Remarks.
6786	1	Virginis							III..IV.		Triple, forming an equilateral triangle. [$14^h\ 10^m{\cdot}4$, 87° 41′ for 1880].
3427	2	Monocerotis									Σ. 921 ; S. 523.
7660	3	Serpentis	A B	1784·219	0 ±	est.			IV.		An isosceles triangle of equal stars whose vertex precedes.
			A C	1784·219	*s p*	est.					
			B C	1784·219	*n p*	est.					
7489?	4	Libræ							III.		Σ. 2003 ?
8006	5	Herculis	A B	1784·372	*s f*	*					Double-double.
			C D	1784·372	*s f*	*					* Situations as per diagram.
			A C	1784·372	*n f*	*					D.M. +32° 2909.
8292	6	Sagittarii							II...III.		h. neb. 4355, =N. 40, Sh. 379.
10303	7	Capricorni							I...II.		I. II.? "Very close triple *."
10795	8	Vulpeculæ							III..IV.		Σ. 2769 ; S. 776.
8779	9	5 Aquilæ	A B	1796·594	121·47	1	A B	1796·594	11·15	1	Σ. 2379. Same as III. 33.
			A C	1796·594	122 +	est.					
10514	10	Capricorni	A B	1792·794	*p*				III.		S. 763, h. 2996.
12173	11	Pegasi							II...III.		P. xxii. 306 ; Σ. 2978.
1289	12	Arietis									Σ. 271.
9607	13	Cygni									P. xix. 276 + 277 ; Σ. 2578.
11335	14	Pegasi									Σ. 2841.
12069	15	Aquarii									Σ. 2959.
12188	16	57 Pegasi *m*	A B	1784·865	240..250				IV.		Σ. 2982.
2051	17	Tauri							II.		Σ. 494.
1648	18	Eridani							II.		h. 3563.
4147	19	Argûs *n*							II.		S. 552.
5180	20	Argûs	A B	1784·887	270·00	est.			VI.		L.L. 19034.
2584	21	17 Orionis ρ^1	A B	1784·950	*n f*				II.		Σ. 654.
6113	22	Virginis	A B	1785·014	0 ±	est.			III.		Σ. 1627.
				1786·991	8 ±	est.			IV.		
				1793·170					IV.		
1810	23	Tauri 34	A B	1785·077	180 ±	est.			III. −		Σ. 427.
2102	24	39 Eridani A	A B	1785·082	0·00	est.			I...II.		Σ. 516.
				1809·044	185·00	1 ::					
5303	25	Felis 40	A B	1785·101	90 ±	est.			III. +		Sh. 110.
5732	26	Crateris 36	A B	1785·173	*f*				VI.		Sh. 120.
6509	27	72 Virginis l^1	A B	1785·173	60 ±	est.			IV. −		Σ. 1750.
7060	28	Libræ	A B	1791·394	270 −	est.			IV.		P. xiv. 212 ; Sh. 190.
5134	29	Leonis 29	A B	1785·192	270·00	est.			IV....V.		

145 *New Double Stars* (continued).

Burnham Gen. Cat.	Novæ. No.	Star.		Date of Measures. A.D.	Angle of Position. ° dec.	No. of Meas.		Date of Measures. A.D.	Distance. "		Synonyms and Remarks.
4925	30	Cancri							I.		Probably D.M. +31° 1933.
6134	31	Comæ 55	A B	1785·260	90 ±	est.			III.		Σ. 1633.
7391	32	Coronæ							IV....V.		Σ. 1973.
7334	33	Libræ 178							IV. –		Σ. 1962.
11744	34	Pisc. Austr.									h. 3118 (no other found by β).
12465	35	Aquarii 355	A B	1785·904	*f*						L.L. 46271; h. 316.
5539	36	35 Sextantis	A B	1786·088	*s p*				III. –		Σ. 1466.
				·994	*s p*				III. +		
7111	37	Virginis							III.		Σ. 1904.
6239	38	Corvi 58							I.		Σ. 1669.
7599	39	Scorpii	A B	1786·326	0 ±	est.			III.		P. xvi. 60; h. 4850.
8292	40	Sagittarii									H.N. 6; Sh. 379.
11715	41	53 Aquarii	A B	1793·679	105 ±	est.			II...III.		Sh. 345.
11938	42	13 Lacertæ	A B	1786·791	*f*				III.		h. 1803; OΣ. 479.
11222	43	10 Pegasi κ	A B	1786·794	360 –	est.			III. –		Σ. 2824.
2229	44	Persei							III.		Σ. 552.
520	45	Androm. 164	A B	1794·712	180 ±	est.	A B	1796·602	7·2	1	Σ. 79.
				1796·602	195·33	1					
2546	46	Eridani	A B	1809·044	83·32	1					H. IV. 43; Σ. 649.
1320	47	Trianguli									Σ. 279.
7020	48	Lupi									h. 2748.
5235	49	8 Sextantis	A B	1787·142	*n p*				IV...V.		h. 4256.
5453	50	Antliæ δ	A B	1787·200	230 ±	est.			II.		h. 4321.
..	51	4 Centauri *h*	A B	1787·200	190 ±	est.			IV.		P. xiii. 221.
6130	52	Canum 20	A B	1787·205	190 ±	est.			II.		Σ. 1632.
5735	53	54 Ursæ ν	A B	1787·214	150..160	est.			II.		Σ. 1524.
8605	54	Scuti 29	A B	1787·520	270 –	est.	A B	1796·597	12·15	1	Σ. 2325.
				1796·597	255·53	1					
10022	55	Sagittæ	A B	1787·630	0 ±	est.					Σ. 2655.
11576	56	41 Aquarii	A B	1787·693	*s f*				II. –		Sh. 339.
11323	57	Cephei 147									Σ. 2840; H. IV. 79;
12675	58	Andromed. 37	A B	1790·901	0 ±	est.			II.		Σ. 3050.
				1794·712	0 ±	est.			I...II.		
6673	59	Hydræ									Arg. Oe. S., 13248.
3470	60	Aurigæ	A B	1788·170	*n f = s p*	est.			IV.		Σ. 933.
3974	61	20 Lyncis	A B	1788·183	*n f = s p*	est.			IV.		Sh. 79; Σ 1065.
7162	62	Bootis	A B	1788·326	0 ±	est.	A B	1788·326	IV.		Σ. 1919.
				1796·564	11·63	1	A B	1796·597	25·00	1	
7099	63	Draconis							V...VI.		Sh. 191.
11702	64	Lacertæ	A B	1788·830	90	est.			II.		Σ. 2902.
1262	65	Cassiopeiæ									Σ. 262; I. 34; III. 4 (which see).
1711	66	Camelopardi							III..IV.		Σ. 390.
2220	67	1 Camelop.	A B	1788·961	*n p*						Σ. 550.
				1795·219	309·85	2					

145 *New Double Stars* (continued).

Burnham Gen. Cat.	Novæ. No.	Star.		Date of Measures. A.D.	Angle of Position. ° dec.	No. of Meas.		Date of Measures. A.D.	Distance. ″		Synonyms and Remarks.
5722	68	Ursæ 234									Σ 1520.
6546	69	Hydræ 369	A B	1790·186	195 ±	est.					S. 651.
5806	70	Ursæ	A B	1790·211	90 −	est.			III.		Σ. 1544.
				1793·268	90 ±	est.					
6844	71	Draconis									Σ. 1840.
10044	72	Cygni	A C	1790·682	*p*						Σ. 2658.
10453	73	50 Cygni α	A B	1790·693	90	est.	A B	1790·693	60 ±	est.	
				1809·044	90 ±	est.					
11301	74	Pegasi	A B	1790·769	86..87	est.			III..IV.		h. 947; P. xxi. 312.
2887	75	Aurigæ									Σ. 764.
279	76	Piscium	A B	1790·917	170 ±	est.			I...II.		P.O. 103. OΣ. 14.
2385	77	Persei							III.		Σ. 603.
..	78	Hydræ									Triple; not identifiable.
4949	79	Ursæ 53							IV. −		Σ. 1315.
6857	80	Libræ					A B	1791·394	60 +	est.	L.L. 26320; Sh. 179.
7608	81	ν^1 Coronæ	A B	1791·403	*s p*				VI.		
				1795·219	220 ±	est.					
6705	82	Bootis							IV....V.		Σ. 1797.
..	83	Sagittæ	A B	1792·638	270	est.					D.M. +15° 3866 (single).
9498	84	Sagittæ	A B	1792·635	295 ±	est.	A B	1796·597	27·20	1	=II. 32, =V. 51.
				1795·791	*n p*						
				1796·594	301·80	1					
9308	85	Vulpeculæ	A B	1792·638	*n f*						P. xix. 128; Σ. 2521.
10008	86	Cygni									OΣ. 401.
10402	87	Cygni							II.		Σ. 2708.
12208	88	Aquarii									Unidentifiable. "Between 87 and 90 Aquarii."
10256	89	Cephei 37							IV....V.		Σ. 2687.
12097	90	Aquarii							I.		Σ. 2964.
12500	91	Piscium							I.		Σ. 3030.
830	92	Piscium	A B	1801·936	10 ±	est.			I.		Σ. 138.
1985	93	Tauri							II.		Σ. 479.
3878	94	Geminorum							IV.		Triple. OΣ. 168 rej.
4038	95	Geminorum							III.		Σ. 1083; III. 48
4963	96	Pixidis ε	A B	1793·170	140..150	est.			V....VI.		h. 4183.
5627	97	Ursæ	A B	1793·266	*n f*						Σ. 1495; H. V. 111.
6762	98	Bootis	A B	1793·359	0·00	est.			I.		Σ. 1813.
6785	99	Bootis	A B	1793·359	270·00	est.			I.		Σ. 1824.
9318	100	Vulpeculæ	A B	1793·646	*n p..s f*	est.			II.		Σ. 2524.
10417	101	Capricorni							III.		Arg. Oe. S. 20747.
11606	102	Aquarii	A B	1793·739	300 ±	est.			III.		h. 5322; S.D. −3° 5414.
3546	103	Monocerotis							III.		Σ. 954.
5724	104	Crateris	A B	1794·298	270 +	est.			III.		Sh. 372.
1129	105	Arietis							I.		Σ. 224.

145 *New Double Stars* (continued).

Burn-ham Gen. Cat.	Novæ. No.	Star.		Date of Measures. A.D.	Angle of Position. ° dec.	No. of Meas.		Date of Measures. A.D.	Distance. "		Synonyms and Remarks.
1303	106	Arietis							II...III.		Σ. 273.
1353	107	Arietis							II...III.		Σ. 287.
4058	108	Geminorum							I.		Σ. 1094.
9602	109	Cygni							I.		Σ. 2576.
9619	110	Cygni							V.		S. 726.
3412	111	Geminorum	A B	1795·789	167·20	1			V.		D.M. +20°, 1454.
12340	112	Aquarii	A C	1795·827	*p*				III.		Σ. 3008.
9591	113	Cygni							II.		β makes it = Σ. 2578, 1^m *f*, 1° 27′ south.
6955	114	30 Bootis ζ	A B	1796·597	311·98	1			I.		Σ. 1865. Same with VI. 104.
6729	115	Bootis 76	A B	1796·597	27·65	1			II.		Σ. 1804.
6981	116	Bootis							III..IV.		Unidentifiable.
11863	117	Piscis Austr.	A B	1797·813	*n f*				I.		See VI. 119.
6204	118	Draconis							I...II.		Σ. 1654.
9330	119	Sagittarii	A B	1798·676	*s f*				I.		P. xix. 126.
9747	120	Draconis							III..IV.		Σ. 2604.
11582	121	Cephei 189	A B	1798·764	270 −	est.			III..IV.		Σ. 2883.
391	122	21 Cassiopeiæ	A B	1798·764	*s f*				VI.		
3721	123	19 Canis Maj.	A B	1799·079	0·00	est.			II.		
2789	124	Tauri							III.		Σ. 730; H. III. 93.
8565	125	Sagittarii							I.		L.L. 3408.
8993	126	Sagittarii							I.		B.A.C. 6504.
10052	127	Capricorni							II...III.		
11914	128	Aquarii 213							I.		Is I. 50, which see.
8991	129	Sagittarii							II. −		L.L. 35530.
11253	130	Capricorni							I.		Some error in the place.
11377	131	Capricorni							III.		L.L. 42770; P. xxi. 338; h. 3071.
2250	132	Tauri							II.		Σ. 559.
11914	133	Aquarii							I.		Is I. 50, which see.
10352	134	Capricorni							II...III.		Σ. 2699.
12117	135	Aquarii							I.		Probably Σ. 2970.
12506	136	Piscium							II.		Σ. 3031.
12406	137	Piscium							II.		Σ. 3019.
10131	138	Capricorni							I.		S.D. −17° 5954.
10919	139	Capricorni							I.		L.L. 41483.
11907	140	Aquarii							II.		S.D. −5° 5843.
3400	141	Geminorum							II.		
5739	142	Leonis 339							I.		Σ. 1527.
6253	143	Virginis	A B	1802·077	*s p*	est.			I.		Σ. 1674.
4815	144	Ursæ							II.		Σ. 1280.
5356	145	Ursæ							IV.		Σ. 1415.

III.—*Star-Gages from the 358th to the 1111th Sweep*

Brought together by CAROLINE HERSCHEL

[THE following table has already appeared in vol. ii. of the *Publications of the Washburn Observatory*, where the results are also given in order of R.A. and reduced to 1860. The table is here printed from Caroline Herschel's MS., with the addition of the dates and a few notes and a list of "vacant places" extracted from the sweep-books.]

The following Gages begin with the 358 Sweep.

As far as 357, they are printed in the paper on the Construction of the Heavens,* and their places have been given in Flamsteed's time and polar distance. But these gages are calculated for the time when the observations were made; though as far as the 438 Sweep the places are down in the Journals in Flamsteed's Time & P.D. But every gage is calculated twice, and after having been brought to the time of observation carried into this book.

Sweep.	R.A.	P.D.	Stars.	Fields.	Memorandums.
358	2 51 43	95 59	7·4	10	
Jan. 27	3 4 43	95 59	9·8	10	
1785					
359	3 50 57	96 41	8·0	10	
Jan. 28					
360	2 54 17	61 27	16·1	10	Most of them excess S. and
Jan. 29	3 23 17	61 27	22	5	some pL. but very few be-
	3 30 23	62 11	33	1	tween.
	3 34 13	62 33	32	1	
	4 0 43	62 27	0	1	
	4 2 13	61 29	12·4	10	Unequally scattered.
	4 4 13	61 29	11·3	4	
	4 5 37	62 25	1	1	
	4 22 13	61 29	18·2	5	
	4 25 13	61 29	16·8	5	
	4 30 12	61 29	15·6	5	
	4 37 36	61 5	39	1	
	5 11 32	61 15	80	½	
	5 15 12	60 56	72	½	
	5 41 44	61 23	120	½	
	5 45 44	62 4	112	½	See Journal.*
	5 49 14	61 54	118	½	
	5 50 14	62 17	116	½	
	5 52 19	61 35	108	½	
	5 56 19	62 42	62	1	Faint ☾ light.

* [Vol. I. p. 228.]

† [The number of stars counted was 56, but in the sweep-book there is written in pencil: "I suppose was ½ F," *i.e.* half a field.—ED.]

Sweep.	R.A.			P.D.		Stars.	Fields.	Memorandums.
361	6	47	37	62	58	36	1	L & innumerable too S. etc.
Jan. 30	6	49	37	63	23	92	½	With S. S.
1785	6	52	37	62	54	73	1	Many more suspected etc.
	7	7	37	62	52	66	½	
362	3	37	8	101	34	7·1	10	
Jan. 31	3	50	6	101	35	8·8	10	
	4	12	6	101	35	7·3	10	
	4	25	6	101	35	7·7	10	
	4	34	4	101	38	8·4	10	
	4	47	4	101	38	10·7	10	
	4	59	10	101	37	10·6	10	
	5	7	11	101	36	13·5	10	
	5	33	53	101	37	10·3	10	See Journal.*
	5	36	47	100	33	4	1	
	5	37	41	101	9	2	1	
	5	38	11	102	43	20	1	
	5	46	53	101	37	12·2	10	
	6	9	53	101	37	21·1	10	
	6	20	53	101	37	41·8	5	
363	7	8	27	101	32	37	1	
Jan. 31	7	22	27	102	10	120	½	Changeable focus.
	7	25	57	101	58	96	½	
	7	27	27	102	14	100	½	
	7	33	27	102	33	46	1	
	7	35	27	101	17	52	1	
	7	36	57	102	2	58	1	
	7	45	27	102	45	47	1	
	7	45	57	102	40	56	1	
	7	56	27	102	3	54	1	
	8	12	27	101	33	38	1	
364	3	56	16	103	38	8·2	10	
Feb. 1	4	59	15	103	38	12·3	10	
365	3	26	4	105	53	7·9	10	
Feb. 4	3	58	4	105	48	8·9	10	
	4	25	4	105	48	7·8	10	
	4	52	17	105	51	12·3	10	
	5	18	17	105	51	15·8	10	
	5	29	6	105	50	20·5	10	
	5	46	6	105	50	25·8	5	
	6	11	15	106	54	50	1	
	6	12	15	106	17	54	½	
	6	13	45	106	3	40	1	
	6	18	15	105	42	46	1	
	6	23	15	105	25	75	1	
	6	34	15	105	11	56	1	

* [In sweep-book nine minutes after the last: "Too unequally scattered to be cast up together, probably some small vacancy." At $5^h\ 20^m$ "I perceive a flying cloud which may have influenced the gages these 2 or 3 minutes," and at $5^h\ 33^m$ "Perfectly clear."—ED.]

Sweep.	R.A.	P.D.	Stars.	Fields.	Memorandums.
366 Feb. 4 1785	7 14 15	105 5	90	1	
	7 30 15	106 10	184	½	Many such F.
	7 32 15	105 6	212	¼	
	7 45 15	104 43	120	½	But a little hazy.
	7 49 15	106 39	216	½	Most S.
	8 43 15	105 0	42	1	
	9 44 15	105 49	16·0	10	
367 Feb. 6	4 49 49	108 9	13·5	10	
	5 13 30	108 9	16·2	10	
	5 47 33	108 9	23·8	5	
	6 2 33	108 9	31·2	5	
	6 10 58	109 26	52	1	
	6 15 58	106 52	34	1	
	6 16 58	108 45	50	1	
	6 23 30	109 28	32	1	
	6 24 0	109 4	30	1	
	6 24 30	108 42	29	1	
	6 25 0	108 18	32	1	
	6 25 30	107 54	43	1	
	6 26 0	107 31	25	1	
	6 26 30	107 8	40	1	
	6 42 29	109 27	40	1	
	6 42 59	109 3	32	1	
	6 43 20	108 40	41	1	
	6 44 17	108 17	37	1	
368 Feb. 7	9 32 40	107 58	18·0	10	
	10 6 40	107 58	17·0	10	
	11 25 41	107 59	8·6	10	
	11 36 41	107 59	5·9	10	
	12 36 41	107 59	9·6	10	
370 Feb. 8	6 42 6	103 7	100	½	
	6 44 6	103 29	116	½	
	6 58 6	104 18	136	½	
371 Feb. 8	9 13 59	103 52	19·0	5	
	9 35 16	103 51	15·8	4	
	10 42 38	104 48	8	1	
	10 59 40	103 53	8·9	10	
372 Feb. 8	11 58 12	103 52	10·3	10	
	13 5 34	103 57	8·4	10	
	13 26 9	103 53	6·9	10	
	13 46 13	103 53	8·9	10	
	13 52 13	103 53	10·2	10	
	14 3 13	103 53	11·6	10	
	14 12 12	103 51	10·3	10	
	14 30 12	103 51	11·6	10	
	14 46 15	103 53	12·7	10	

Sweep.	R.A.	P.D.	Stars.	Fields.	Memorandums.
373 Feb. 17, '85	15 18 58	64 51	9,9	10	
374 Feb. 28	5 50 38	64 29	152	¼	Of all sizes.
	6 11 36	64 45	104	½	
	6 22 6	65 36	140	¼	
	6 39 6	66 3	94	½	
	6 42 22	64 21	96	½	
	7 20 15	66 7	48	½	
	7 20 45	65 39	54	½	
376 Mar. 1	5 55 36	63 30	160	½	
377 Mar. 4	6 31 59	97 26	60	1	
	6 38 59	98 35	136	¼	
	6 39 29	97 46	136	½	
	6 44 59	98 14	120	¼	
	7 0 59	97 31	128	¼	
	7 1 59	97 5	184	¼	
	7 7 59	96 49	132	½	
	7 16 29	98 5	160	¼	
378 Mar. 5	6 33 13	94 56	216	¼	
379 Mar. 5	9 33 27	95 5	14·3	10	
	10 21 28	95 9	9·6	10	
	11 3 58	95 9	9·6	10	
	11 50 42	95 7	9·6	10	
	12 23 42	95 7	7·3	10	
380 Mar. 5	13 31 29	95 38	7·6	10	
	14 45 14	95 36	12·6	10	
381 Mar. 6	6 34 14	113 36	40	1	
	6 37 32	114 27	33	1	
	6 40 14	114 5	46	1	
	6 47 44	113 40	42	1	
	6 50 9	113 47	59	1	
	7 1 45	114 12	46	1	
	7 8 45	115 28	73	1	
	7 12 45	115 30	150	⅓	Of all sizes.
	7 23 13	114 40	78	1	
	7 31 43	115 3	184	¼	
	7 39 13	114 52	212	¼	
	7 42 43	114 35	232	¼	
	7 57 41	115 48	200	¼	
	8 3 41	114 15	108	¼	Most S.
	8 6 11	115 21	86	½	
	8 14 41	115 16	58	1	

Sweep.	R.A.	P.D.	Stars.	Fields.	Memorandums.
382 Mar. 10 1785	10 56 7 11 12 52 12 19 16 12 48 5 13 45 5 14 8 5	110 22 110 22 110 22 110 23 110 23 110 23	9·9 10·5 7·9 8·3 10·3 9·4	10 10 10 11 10 10	
384 Mar. 11	6 50 36 7 4 29 7 54 33 8 23 35	64 59 62 16 63 5 63 8	44 36 17·8 14·7	1 1 10 10	Faint Twilight etc.*
385 Mar. 12	7 22 29 7 22 59 7 23 39 7 39 39 7 42 39 8 53 11 9 38 41	61 55 61 21 60 49 61 44 60 48 60 44 60 43	32 30 22 39 23·4 13·8 9·2	1 1 1 1 5 10 10	
386 Mar. 13	7 45 20 9 36 44 9 47 39 10 11 39 10 32 9	58 58 58 41 58 38 58 38 58 38	25 8·7 8·0 6·6 5·5	1 10 10 10 10	F. ☾ light.
387 Mar. 13	13 20 54	58 39	7·7	10	
388 Mar. 13	15 25 44 15 38 44 15 54 20	56 43 56 43 56 39	9·2 12·5 12·9	10 10 10	
389 Mar. 16	14 7 49 15 33 56 16 0 26	16 19 16 18 16 18	11·6 10·1 15·5	5 10 10	
390 Apr. 3	10 11 19 10 47 19 11 23 19	16 4 16 4 16 4	11·2 10·0 11·2	10 10 10	
391 Apr. 3	13 32 28	16 6	9·3	10	
392 Apr. 3	15 46 4	14 12	13·9	10	
393 Apr. 6	10 18 28 11 16 16 11 53 12	62 59 63 0 62 58	8·3 6·6 4·3	10 10 10	

* [In sweep-book : "But I have not been out long enough, and some faint twilight."—ED.]

Sweep.	R.A.	P.D.	Stars.	Fields.	Memorandums.
393	12 5 12	62 58	5·3	10	
Apr. 6	13 9 19	62 59	5·3	10	
1785	13 47 19	62 59	5·4	10	
396	13 16 43	60 47	5·6	10	
Apr. 11	13 17 43	60 47	6·6	10	
397 Apr. 11	14 55 39	60 46	7·0	10	
399	13 29 22	67 22	5·4	10	
Apr. 13	13 50 23	67 25	5·0	10	Most of them S.
	14 5 30	67 22	5·8	10	
	14 45 20	67 13	6·5	10	
400	15 26 47	91 32	7·1	10	Chiefly S.
Apr. 14	15 55 40	91 34	11·9	10	Chiefly S.
	16 0 41	91 33	16·2	10	Of various sizes.
403	13 5 16	69 0	6·4	10	
Apr. 27	13 9 16	69 0	4·6	10	
405	14 44 28	53 11	9·0	10	
May 1	14 58 58	53 11	7·2	10	
	15 47 6	53 12	9·3	10	
407 May 3	12 48 52	55 15	6·4	10	Most S. and many vS. suspected.
408 May 3	15 26 24	55 12	7·0	10	
409 May 5	16 15 5	87 19	17·9	9	
... *	15 28 41	87 24	15·0	10	Mr. AUBERT.
May 6	15 35 41	87 24	14·2	10	W. H.
	15 39 41	87 24	14·5	10	Mr. AUBERT.
	16 22 40	87 23	11·8	10	Mr. AUBERT.
415 July 17	20 6 24	67 29	93	1	
416	18 54 56	87 29	90	1	
July 30	19 18 21	87 4	150	1	
417	20 12 5	88 4	46	1	
July 31	20 13 35	87 4	41·4	5	

* [These four gauges were taken on the 6th May 1785, in the presence of M. von Zach and Count Brühl. Sw. 410 was on the 11th May, but only M. 62 was observed: "This sweep must not be depended upon, as I saw not very distinctly, being still affected with the remains of the Ague."—ED.]

Sweep.	R.A.	P.D.	Stars.	Fields.	Memorandums.
418 Aug. 1 1785	19 2 23	118 21	26·2	5	
	19 55 24	118 24	10·5	10	
	20 27 21	118 21	10·9	10	
	20 42 21	118 21	6·9	10	
	21 2 21	118 20	6·5	10	
419 Aug. 6	21 12 13	79 54	28	1	
421 Aug. 7	19 59 57	84 31	92	½	
	20 16 27	85 2	45	1	
	20 20 57	85 40	67	1	
	20 21 27	85 29	50	1	
	20 22 38	85 5	48	1	
	20 23 40	84 41	52	1	
	20 24 48	84 16	40	1	
	20 25 46	83 52	43	1	
425 Aug. 12	22 20 3	90 42	9·2	10	
	22 28 33	90 42	9·2	10	
427 Aug. 30	19 15 17	86 59	220	¼	
	19 25 32	86 27	172	¼	
	19 38 32	87 22	144	¼	
	21 1 37	86 41	18·2	10	
	22 12 32	86 42	10·7	10	
	23 17 35	86 41	5·0	10	
430 Sept. 1	20 19 25	120 29	7·2	10	Very clear and with great attention.
	20 53 25	120 29	4·3	10	
	21 30 25	120 29	4·5	10	
	22 34 34	120 31	2·9	10	
	23 16 28	120 30	2·3	10	
	23 37 28	120 30	2·4	10	
	0 30 28	120 30	1·5	10	
431 Sept. 3	20 42 15	94 26	18·4	5	
438 Sept. 12	23 34 40	96 21	3·4	10	The small spec. affected by damp.
445 Sept. 28	23 22 7	110 18	5·6	10	
	23 49 7	110 18	3·9	10	
447 Oct. 1	22 31 14	92 20	7·8	10	
	22 45 15	92 20	7·2	10	
451 Oct. 3	0 2 19	105 5	6·3	10	
	0 13 19	105 5	4·7	10	
	2 5 53	105 4	4·5	10	

Sweep.	R.A.	P.D.	Stars.	Fields.	Memorandums.
459 Oct. 6, '85	2 21 5	108 17	4·8	10	
460 Oct. 6	5 41 13	119 0	8·1	10	
461 Oct. 8	22 54 20	88 35	7·1	10	
	0 11 17	88 41	5·6	10	
462 Oct. 8	2 19 21	88 38	7·5	10	
465 Oct. 26	21 52 52	115 0	5·8	10	
	22 3 52	115 0	5·8	10	
	22 35 52	115 2	4·9	10	
	22 56 52	115 2	2·8	10	
	23 9 52	115 2	3·5	10	
	23 49 52	115 2	4·0	10	
	0 8 52	115 2	2·3	10	
	1 15 52	115 0	3·6	10	
	1 34 52	115 0	3·6	10	
466 Oct. 26	3 17 52	117 6	3·6	10	
467 Oct. 27	21 48 47	117 0	6·2	10	
	22 39 43	117 2	3·3	10	
	23 32 43	117 3	4·8	10	
	1 6 43	117 2	2·5	10	
469 Nov. 7	23 34 3	90 39	7·2	10	
	0 4 13	90 39	5·5	10	
473 Nov. 22	23 24 43	92 30	6·3	10	
474 Nov. 22	1 46 6	96 12	3·5	2	
477 Nov. 23	0 28 24	80 8	7·1	10	
478 Nov. 27	23 2 29	103 17	6·1	10	
	23 13 32	103 17	5·3	10	
	2 31 16	103 16	6·6	10	
	3 13 41	103 16	6·0	10	Most S.
479 Nov. 28	22 56 21	101 21	6·3	10	
	23 47 15	101 21	6·0	10	
	1 4 22	101 24	6·1	10	

Sweep.	R.A.	P.D.	Stars.	Fields.	Memorandums.
481 Nov. 29, '85	1 9 50 2 12 2	71 31 71 29	11·2 6·9	10 10	
482 Nov. 29	4 31 27 4 44 43	71 29 71 29	14·3 22·8	10 5	 All sizes.
485 Dec. 7	5 32 21	66 6	66	1	Many S. st. too S. for the gage.
486 Dec. 7	6 21 31	63 47	88	½	
496 Dec. 28	5 48 44 6 2 52 6 12 22	81 42 81 38 81 47	120 136 200	½ ½ ¼	
497 Dec. 28	9 2 54	81 17	12·3	10	
498 Dec. 28	11 17 55	81 11	6·3	10	
499 Dec. 30	1 21 40	106 47	6·4	10	
500 Dec. 30	4 41 30 5 23 30	116 46 116 49	5·5 11·6	10 10	
502 Dec. 30	9 22 29 10 17 54	117 47 117 48	14·8 9·9	10 10	
506 Jan. 1 1786	6 51 47 6 52 37	90 59 91 44	180 192	⅓ ¼	
508 Jan. 2	11 26 5	54 22	5·6	10	
511 Jan. 27	4 35 37	84 32	22·6	5	
523 Feb. 15	16 47 18 17 1 2	22 58 23 55	15·8 18·6	10 5	
526 Feb. 22	7 6 42	88 48	160	¼	
529 Feb. 24	7 28 40 7 39 30 7 50 55	97 54 98 8 97 3	104 92 45	½ ½ 1	

Sweep.	R.A.	P.D.	Stars.	Fields.	Memorandums.
558 Apr. 30, '86	13 10 28 13 54 0	85 53 85 53	6·8 8·3	10 10	
559 Apr. 30	16 40 35	119 1	27·6	10	
561 May 1	16 41 32	83 59	19·1	10	
568 May 28	15 22 36	89 32	11·4	10	
569 May 28	17 27 13 17 27 33 17 27 53	 109 47 110 40	 0 34	 1 1	The stars very uneq. scattered.
584 Sept. 14	19 54 45	79 53	252	¼	
594 Sept. 20	20 40 44	54 33	320	¼	Besides many too S. to be distinctly seen.
598 Sept. 21	21 12 46	52 32	272	¼	Most v. S.
600 Sept. 22	20 55 42	52 16	220	¼	
616 Oct. 18	21 55 12	66 34	46	1	
626 Oct. 26	1 2 42 2 17 33	63 25 63 26	14·8 13·2	10 10	
641 Nov. 28	6 54 35	95 2	232	¼	
660 Dec. 21	12 18 51	111 53	5·0	10	
667 Dec. 26	7 25 19 7 26 49 7 28 19	86 11 87 33 86 22	11·2 52 48	¼ ½ ½	
668 Dec. 27 1786	6 4 23 6 41 2	89 22 88 31	104 116	¼ ½	But it is one of the richest places at present. See Memor. 8′ after the gage.*

* [" From the appearance of the heavens in this neighbourhood there is no reason to suppose that there is any break or vacancy among the stars between our situation and the remotest part of the milky way."—ED.]

Sweep.	R.A.	P.D.	Stars.	Fields.	Memorandums.
681 Jan. 11 1787	3 18 48 3 32 18 3 47 18 3 57 48	75 18 75 18 75 18 75 18	9·2 7·5 8·7 6·1	10 10 10 10	
711 Mar. 15	12 21 18	119 33	9·4	10	
712 Mar. 15	16 7 51	121 35	13·6	10	
715 Mar. 17	15 54 37	50 50	16·7	10	
716 Mar. 18	9 43 0 10 32 47	48 34 48 33	10·4 7·0	10 10	
720 Mar. 19	.. .. .. time forgot	74 18	11·2	10	Betw. 14^h 31′ and 15^h 15′.*
722 Mar. 20	14 12 43	56 44	10·0	10	
736 May 15	13 45 38	39 19	8·4	10	
762 Oct. 10	22 12 9	12° 1′	20·3	10	Most extr. S.
765 Oct. 14 1787	21 48 20 21 50 50	38 4 38 9	280 336	¼ ¼	See mem. on appearance of the Heavens.†
815 Mar. 9 1788	8 54 40	39 49	12·6	10	
816 Mar. 9	12 36 29	39 48	8·0	10	
820 Mar. 11	14 0 44	107 52	10·8	10	
858 Sept. 9	2 17 12 2 18 12 2 19 32	44 38 44 59 44 57	60 72 68	½ ½ ½	

* [Between transits of ξ Bootis and ρ Serpentis.—Ed.]

† [" In this part of the heavens most of the large stars of the milky way seem to be of the 9 and 10 magnitude, the small ones gradually less till they escape the eye, so that appearances here favour the idea of an adjoining cluster." And ten minutes later: " It is very evident in this part of the heavens, that there is some distance between us and the milky way, not equally scattered over with stars."]

Sweep.	R.A.	P.D.	Stars.	Fields.	Memorandums.
860 Sept. 21, '88	21 28 49	41 4	360	¼	But most of them too S. &c.
862 Sept. 26	20 52 55	39 49	15	1	These gages show, etc.*
	20 53 15	38 54	41	1	
	20 54 55	40 20	17	1	
866 Sept. 27 1788	20 59 32	43 49	360) btw. 400)	¼	
	21 2 32	43 41	280	¼	
934 Mar. 4 1790	6 57 29	107 32	100	¼	
	6 58 54	107 27	128	¼	Of all sizes.
	8 34 54	105 44	38	1	
	8 36 22	107 10	43	1	The stars have been gradually reduced to this number, and are still decreasing.
	8 41 54	107 43	17	1	
	8 43 24	107 24	26	1	
	8 43 54	108 0	15	1	
	8 44 39	107 13	16	1	
	8 47 54	107 42	14·2	10	
	8 55 54	107 42	11·6	10	
959 Sept. 11	19 49 56	47 3	240	¼	With many too S. to be counted.
	19 52 56	48 4	264	¼	
963 Oct. 3	20 52 52	66 45	184	¼	
969 Oct. 9	20 30 58	44 55	172	¼	
	20 31 58	43 52	232	¼	
1014 May 28 91	17 26 58	55 35	41·2	5	
1022 Aug. 14 '92	21 13 49	82 10 ::	40	1	
1023 Aug. 19	19 34 5	72 11	360	¼	
1024 Aug. 22	19 35 23	75 5	600 ::	¼	†
	19 53 20	74 11	440	¼	
	20 5 20	72 54	360	¼	
	20 16 20	73 5	260	½	

* "These gages shew that close to the crowded milky way there is a considerable space which contain but few stars, and looking with the naked eye at a place north following α Cygni it plainly appears much darker than the rest."

† [Note at $19^h\ 35^m$: "The field extremely crowded, I tried to count but found the stars too small and too crowded to count them; however there can be no less than 100 in a quadrant, but probably near 150; from some careful trials I suppose there were 150." At $19^h\ 51^m$: "By calculation it appears that not less than 133084 stars have passed thro' the telescope in 16′ of time." At $19^h\ 57^m$: "36598 stars have been seen in 6′." At $20^h\ 5^m$: "74859 stars have been seen in 15′"; and at $20^h\ 16^m$: "14417 stars have been seen in 4′."—Compare Vol. I. p. 483.—ED.]

Sweep.	R.A.	P.D.	Stars.	Fields.	Memorandums.
1027 Sept. 15 1792	19 41 54 19 45 54	52 26 52 29	480 612	¼ ¼	The whole breadth eq. rich.
1036 Apr. 6 1793	10 36 53	23 40	18·2	10	
1043 May 12	17 37 2	103 11	60	1	But is one of the richest fields now.
1056 Oct. 5	20 16 18	98 16	80	½	
1067 Dec. 10 '97	15 36 0	13 16	14·8	10	
1070 Dec. 12	14 38 0	13 24	17·2	10	
1077 Sept. 9 '98	19 46 15	30 23	112	½	
1080 Oct. 2	20 40 41 21 20 41	20 34 21 47	48 32	1 1	Probably affected by haziness.
1082 Oct. 7	0 17 58	15 32	96	1	
1093 Jan. 21 1800	5 33	73 18	120	1	
1095 Mar. 15 1801	7 41	108 28	176	¼	
1111 Sept. 26 1802	8 58 11 9 50 11	9 6 9 8	60 or 70 from 20 to 30	1 1	

VACANT PLACES

[*Extracted from the Sweeps. Places for the Year of Observation.*]

Sweep.	R.A.			P.D.		Stars.	
383 Mar. 10 1785	16	5	22	109	25	0	
	16	6	22	109	20	0	
	16	6	32	109	31	0	
	16	7	22	109	49	0	
	16	7	42	109	12	0	
	16	11	52	110	17	0	
	16	12	22	109	11	0	
	16	12	40	110	25	0	
	16	13	0	111	29	2	
485 Dec. 7 1785	4	17	37	65	29		Upper border of a vacancy, but it is a very irregular one.
	4	18	30	65	27		Do.
	4	19	17	65	29		Do.
	4	21	35	64	31	0	
	4	22	26	64	22	0	
	4	23	53	64	4	0	and many such in the neighbourhood.
	4	25	17	..			There is a vacancy between the bright row of stars in the direction of Orion's belt and the Bull's head, Perseus' body and the Milky Way, and I am now in that vacancy.
	4	27	26	65	4	0 }	Intermixed with places that have many stars.
	4	28	6	64	10	0 }	
	4	28	42	65	11	0 }	
	4	29	24	65	15	0 }	
	4	30	54	65	16	0 }	
	4	37	51	65	16	0 }	
	4	39	16	..			The straggling stars of the Milky Way seem now to come on gradually, most small.
	4	43	20	..			They begin now to be intermixed with some larger ones.
516 Jan. 30 1786	5	32	16	98	30	0 }	Vacant spaces picked out, between stars sparingly and irregularly scattered.
	5	32	42	100	21	0 }	
	5	33	5	99	33	0 }	
	5	34	40	100	39	0 }	
566 May 26 1786	16	8	52	113	18 }		Vacant between these two places.*
	16	9	12	112	25 }		
	16	11	56	112	53		From this place to the bottom of sweep [113° 20′] vacant.
	16	16	8	112	36 }		From these places downwards vacant, the night very fine.
	16	17	6	112	37 }		
	16	18	25	112	27 }		
	16	20	32	112	54 }		

* [Compare Vol. I. p. 253.—Ed.]

Sweep.	R.A.	P.D.	Stars.	
627 Oct. 26 1786	4 6 3	63 30	0	And several more such vacant places.
	4 8 16	63 32	0	
	4 9 31	63 46	0	
	4 9 47	63 8	1	
	4 10 35	64 3	2	
	4 10 44	63 54	0	
	4 11 0	62 51	0	[? P.D. 62° 56′.]
	4 11 31	62 39	1	
	4 12 41	63 28 }		No stars in this place.
	4 12 58	63 8 }		
	4 13 58			Very few stars in the whole breadth of the sweep [62° 15′ to 64° 31′], this seems to be the continuation of the vacancy between Orion and the M. Way.
	4 16 0	63 41 }		Not a star in this place. Gaging the whole breadth of the sweep would not be right, as it is interspersed with places containing many straggling stars.
	4 16 15	63 20 }		
	4 18 44	63 16 }		Almost quite vacant.
	4 19 4	64 29 }		
	4 20 27	63 30 }		Not a star.
	4 20 45	64 30 }		
	4 23 47	..		The lower part of the sweep is very vacant.
	4 24 50	63 41		Up to this place, still very vacant.
	4 25 58	63 46		The same.
	4 27 0	64 3		From the bottom to this place not a star.
	4 30 30 ::	..		The whole breadth of the sweep is now again scattered over with the usual quantity of stars in this part of the heavens.
741 May 19 1787	16 11 24	108 2 }		All the enclosed place without stars. North of it considerably rich.
	16 11 46	109 21 }		
	16 12 9	108 48 }		
	16 12 56	108 51 }		
	16 13	109 31 }		
	16 16 12	109 15 }		The inclosed place vacant, I except however here and there a very few pretty large stars.
	16 16 26	108 58 }		
	16 17 17	109 3 }		
	16 17 32	109 12 }		
	16 18 13	108 3 }		
	16 18 29	108 56 }		
	16 18 41	109 30 }		
	16 19 7	108 24 }		
	16 19 43	107 56 }		
	16 20 9	109 2 }		
	16 21 8	107 56 }		Vacant. The night remarkably clear.
	16 21 40	109 5 }		
	16 21 51	108 25 }		
	16 24 0	108 44 }		Quite vacant.
	16 24 18	108 20 }		
	16 27			Left off [end of sweep].

INDEX

The references in Roman figures are to the Introduction to Vol. I., but I. and II. followed by an Arabic numeral refer to the rest of the two volumes.

END OF VOL II. AND OF THE WHOLE WORK

PRINTED BY
NEILL AND COMPANY, LIMITED,
EDINBURGH.

Printed in the United States